DNA Science

A FIRST COURSE

SECOND EDITION

DNA Science

A FIRST COURSE

SECOND EDITION

David A. Micklos
Dolan DNA Learning Center
Cold Spring Harbor Laboratory

Greg A. Freyer
Mailman School of Public Health
College of Physicians & Surgeons
Columbia University

with David A. Crotty
Cold Spring Harbor Laboratory

I.K. International Pvt. Ltd.
New Delhi • Mumbai • Bangalore

Published by
I.K. International Pvt. Ltd.
S-25, Green Park Extension
Uphaar Cinema Market
New Delhi 110 016 (India)
E-mail : ik_in@vsnl.net

Branch Offices:
A-6, Royal Industrial Estate, Naigaum Cross Road
Wadala, Mumbai 400 031 (India)
E-mail : ik_mumbai@vsnl.net

G-4, "Embassy Centre", 11 Crescent Road
Kumara Park East, Bangalore 560001 (India)
E-mail : ik_bang@vsnl.net

ISBN 81-88237-35-3

First Indian Reprint : 2005

This edition has been published in India by arrangement with Cold Spring Harbor Laboratory Press, Cold Spring Harbor, New York, U.S.A.

DNA Science
A First Course (Second Edition)

Publisher	**John Inglis**
Developmental Editors	**David Crotty (Text) and Emily Huang (Laboratories)**
Project Coordinators	**Mary Cozza and Maryliz Dickerson**
Permissions Coordinators	**Nora Rice**
Production Editor	**Dorothy Brown**
Desktop Editor	**Susan Schaefer**
Production Manager	**Denise Weiss**
Cover Designer	**Ed Atkeson**

Cover: Illustrated model of DNA courtesy of Slim Films (http://www.slimfilms.com; © Slim Films).

Printed by Rekha Printers Pvt. Ltd., A-102/1 Okhla Industrial Area, Phase II, New Delhi 110 020.

For Charles and Charlotte,
My connections to the past

For Richard, Marian, Carol, and Ann,
My connections to my childhood

For Dana, Alec, and Andrew,
My connections to the future

For Ellie Greenan,
Who will tell you that DNA is "the thing"

— DAVID A. MICKLOS

To my wife Joadie,
and my three sons Kurt, Eric, and Alec

— GREG A. FREYER

Contents

APPENDICES

LABORATORY SUPPLIES

The laboratory course in this book is supported by quality-assured products from the Carolina Biological Supply Company (U.S.: 800-334-5551; Canada: 800-387-2474; International: [336]-584-0381; Web: www.carolina.com).

Preface

IN 1984, THE LABORATORIES IN THIS BOOK were second nature to Greg, then a postdoctoral fellow with Rich Roberts at Cold Spring Harbor Laboratory. But to Dave they were the keys to the unseen world of DNA. At the time, there were virtually no simple experiments using these techniques that were aimed at advanced high school and beginning university students. So why not develop a lab sequence that would allow virtually any student the opportunity to make and analyze a recombinant DNA molecule?

Thus began this book, with Dave spending odd hours in the Roberts' lab trying and fleshing out a 14-page recipe book containing Greg's vision of a beginning laboratory course in recombinant DNA technology. In addition to Rich Roberts, who went on to win a Nobel prize, many prominent scientists were generous with their ideas and lab tips. Doug Hanahan shared with us his supremely simple method to transform *E. coli*. The hair dryers in Ed Harlow's lab provided a sensible, and stylish, means to dry a DNA pellet.

The result was a 124-page manuscript called "Recombinant DNA for Beginners," first tested with faculty and students on Long Island in summer 1985. By the following year, we had equipped the first of two "Vector Vans," crammed to the ceiling with centrifuges, pipettes, electrophoresis chambers, water baths, incubators, and reagents enough to clone a molecular genetics lab anywhere. For many years, summers became a blur of places and faces, as we traveled the United States training several thousand biology faculty to do the laboratories in this book. In the meantime, Greg left the Cold Spring Harbor campus ultimately taking a faculty position at Columbia University and Dave moved down the street to found the Dolan DNA Learning Center.

Jim Watson coined the term *DNA Science* over lunch in Cold Spring Harbor in 1988. For Jim, it was an everyday, throwaway comment—his world had revolved around DNA for 40 years by that time. But for us it captured the simplicity of the new DNA world we wanted to introduce to students: a science based on the molecule of life and a book to help bring that science to life for students.

In making this second edition, we preserved the successful formula of the first edition—one part well-tested laboratories and one part insightful, explanatory text. We maintained the core laboratory sequence, which first introduces the basic techniques of DNA restriction, transformation, isolation, and analysis, and then applies these techniques to the construction and analysis of a simple recombinant DNA molecule. We resisted the temptation to tinker very much with the laboratories. After all, they are the best-tested and most widely used teaching labs available on the basic techniques of gene manipulation. These labs, as well as numerous adaptations and analogs, provide biology students an introduction to molecular genetics at high schools and universities throughout the United States.

We have, however, included insights and refinements introduced in our teaching lab over the last several years. These include a method to spread *E. coli* with glass beads, shared by Steve Hughes of the National Cancer Institute, which puts an end to alcohol fires. We have included additional labs that focus on gene products to at least partially address the criticism that "there is more to biology than just DNA." One provides a simple colorimetric assay for the activity of β-lactamase (penicillinase), the enzyme produced by the ampicillin resistance gene. The other uses green fluorescent protein (GFP) to illustrate the principles of protein expression and purification (without the fuss of a column).

While maintaining the historical approach of the first edition, the text portion has been entirely reorganized and updated with almost 200 pages of new material—bringing it up to the minute with current research. More than a recitation of facts, the narrative takes students behind the scenes of modern research, introducing key people and their experiments. David Crotty, a molecular and developmental biologist, provided additional insight and material for the text portion.

The first three chapters cover essential principles of genetics, and DNA structure and function. Chapters 1 and 2 cover the historical foundations of DNA, by simply explaining "How We Learned That DNA Is the Genetic Material" and "How We Learned the Function of DNA." The third chapter, "How We Learned How Genes Are Regulated," moves from classic studies of the *lac* operon to the multiplicity of mechanisms that we now know to modulate gene expression.

The next three chapters introduce small- and large-scale methods for analyzing DNA. Chapter 4, "Basic Tools and Techniques of DNA Science," provides the theory behind the laboratories in this book. Chapter 5, "Methods for Finding and Expressing Important Genes," explains the arsenal of modern techniques for investigating individual genes. Chapter 6, "Modern Methods for Analyzing Whole Genomes," documents the race to sequence the human genome and new methods for working with hundreds of genes simultaneously.

The final chapters focus on human issues. Chapter 7, "The DNA Science of Cancer," describes the origin of the war on cancer and recent strides in fighting this most-dreaded disease. Chapter 8, "Applying DNA Science to Human Genetics and Evolution," explores the molecular basis of human variation, and its relation to human disease and our emergence as a species. This chapter also contains the first substantial treatment of American eugenics available in a general biology text.

Although much has changed in biology since the first edition, the ideas and techniques in this book are still the minimum requirements for any degree in DNA manipulation. At a time when molecular biology is increasingly accomplished on sequencers and microchips, agarose gel electrophoresis still provides a door through which anyone can enter the fraternity of DNA. Despite a growing emphasis on higher organisms, *E. coli* still provides insights to anyone with a warm incubator and a few hours to spare.

As with the first edition, we hope this second one correctly anticipates and explains trends that will occupy scientific thought for many years. We hope that *DNA Science* continues to provide a simple roadmap for beginning an exploration of the molecule of life—one that will take on added importance as more and more biology teachers around the world realize the value of giving students freedom to get their hands dirty with DNA.

November 2002 DAVE MICKLOS AND GREG FREYER

Acknowledgments

MANY PEOPLE HAVE BEEN GENEROUS WITH THEIR TIME and resources over the years. *DNA Science* would not have been possible without them:

To Rich Roberts, who provided us both a place to think and to tinker.

To Wendy Russell and Mary Jeanne and Henry Harris, who have supported Dave from the very beginning.

To Charles and Helen Dolan, who gave Dave a great place to work.

To Clarence Michalis and David Luke, of the Josiah Macy, Jr. Foundation, who twice took the chance to support new ideas in science education.

To the National Science Foundation, which has enabled us to help a lot of teachers move into the DNA world.

To Mark Bloom, who helped perfect these experiments and taught them to as many teachers as anyone.

To Sue Lauter, who conceived much of the art.

To Ellen Skaggs and Judy Cumella-Korabik, who minded the store when Dave was away.

To Shirley Chan and all who worked on "DNA From The Beginning" (http://www.dnaftb.org), which proved to be a valuable resource in assembling this book.

To Scott Bronson and Jennie Aizenman, who made and tested the new molecules and protocols in this edition.

To J. Jiji Miranda, who produced many of the photographs of gels found in the Laboratory section.

To the Carolina Biological Supply Company for their assistance throughout the development of this edition.

To Jamie Lee for his thorough and helpful review of the manuscript.

To Alex Gann, Laurie Goodman, and the many other scientists who contributed suggestions about the content and illustrations and figures.

To Emily Huang and Beth Nickerson for their editorial contributions, and to Siân Curtis and Tamara Howard for fact-checking portions of the material.

To Mary Cozza, Dorothy Brown, and Susan Schaefer for their hard work in organizing and assembling our material into book form.

To Jim Watson, the dean of DNA, who has made Cold Spring Harbor a place where any good idea can grow.

And to the teachers and students who have made this book a success.

Development and testing of the laboratory sequence was made possible through the support of the following:

Citicorp/Citibank
Brinkmann Instruments
Josiah Macy, Jr. Foundation
National Science Foundation
J.M. Foundation
Richard Lounsbery Foundation
The Banbury Fund
Esther A. and Joseph Klingenstein Fund

Amersham Corporation
Argonne National Laboratory
Bethany College
Biology Teachers' Organization, Winnipeg, Manitoba
Center for Biotechnology, State University of New York at Stony Brook
Cleveland Clinic Foundation
Cooperating School Districts of St. Louis Suburban Area, Inc.
Dorcas Cummings Memorial fund of the Long Island Biological Association
Samuel Freeman Charitable Trust
GIBCO/BRL Research Products, a division of Life Technologies, Inc.
Fotodyne Incorporated
Harris Trust
Eli Lilly and Company
New England Biolabs Foundation
North Carolina Biotechnology Center
Pioneer Hi-Bred International, Inc.
San Francisco State University
University of California at Davis

and THE COLD SPRING HARBOR CURRICULUM STUDY:
Cold Spring Harbor Central School District
Commack Union Free School District
East Williston Union Free School District
Great Neck Public Schools
Harborfields Central School District
Half Hollow Hills Central School District
Herricks Union Free School District
Huntington Union Free School District
Island Trees Union Free School District
Irvington Union Free School District
Jericho Union Free School District
Lawrence Public Schools
Lindenhurst Public Schools
Locust Valley Central School District
Manhasset Public Schools
Northport–East Northport Union Free School District
North Shore Central School District
Oyster Bay–East Norwich Central School District
Plainedge Public Schools
Plainview–Old Bethpage Central School District
Portledge School
Port Washington Union Free School District
Sachem Central School District at Holbrook
South Huntington Union Free School District
Syosset Central School District

CHAPTERS

Francis Crick and James Watson in Cambridge, England, 1953
(Courtesy of the James D. Watson Special Collection. Cold Spring Harbor Laboratory Archives.
From Watson J.D. 1968. *The Double Helix.* Atheneum Press, New York.)

CHAPTER 1

How We Learned That DNA Is the Genetic Material

DNA SCIENCE WAS BORN ON APRIL 25, 1953, when James Watson and Francis Crick announced in the British journal *Nature* that they had determined the "double helix" structure of the DNA molecule. The DNA era came of age in 2001 with the publication of a draft of the human genome, the entire DNA code that sets the parameters of human life. In between came a childhood of gene cloning, genetic engineering, gene splicing, and recombinant DNA. In the maturity of the future lie the answers to ancient questions: Where did we come from? Why must we die? How do we remember the past and anticipate the future?

To the general public, this new era is probably most frequently represented by the word "biotechnology." So let us begin our exploration of DNA with that word in mind. Literally translated, biotechnology means "life technology," applying knowledge about living things for the practical use of humankind. The ancient uses of yeasts in making bread and alcoholic beverages and of bacteria in making cheese are, in the broadest sense, biotechnology. However, the modern biotechnology revolution is based on a deep understanding of the technology of life, the mechanics of living machines.

Egyptian Hieroglyph of Wine Production
The fermenting of grapes to make wine is one of the earliest examples of biotechnology. (Courtesy of the Metropolitan Museum of Art.)

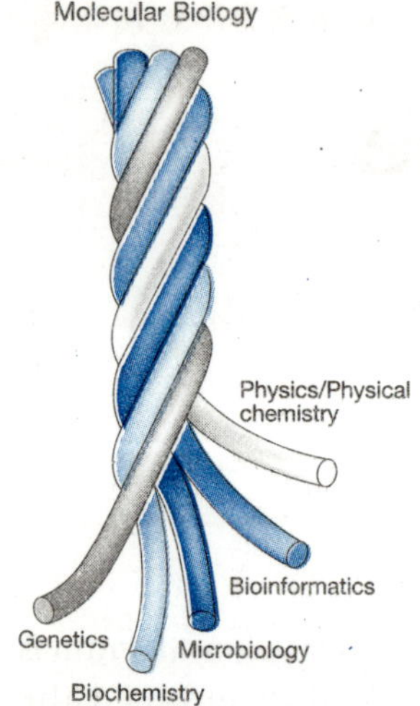

Molecular Biology Is a Synthesis of Several Disciplines
(Art concept developed by Lisa Shoemaker.)

The technical aspects of life involve the complex chemical interactions that take place among the several thousand different kinds of molecules found within any living cell. Of these, DNA (deoxyribonucleic acid) is the master molecule in whose structure is encoded all of the information needed to create and direct the chemical machinery of life. Analysis of the flow and regulation of this genetic information among DNA, RNA (ribonucleic acid), and protein is the subject of molecular genetics.

In a broad sense, the terms molecular genetics and molecular biology have become nearly synonymous. This change in our understanding of life has launched a dramatic and seemingly sudden biological revolution. However, it is prudent to remember that embedded in the word "revolution" are both "revolve" and "evolution." These remind us of the long-term, cyclical, and historical undercurrents of revolution, however sharp its break from the past appears. As we trace the development of the concepts that led to this biological revolution, it will serve us well to keep in mind some themes and trends that can help organize our understanding of molecular biology.

MOLECULAR BIOLOGY IS A HYBRID DISCIPLINE

The techniques of molecular genetics are now being applied to nearly every major field of biology—from neurophysiology to botany and from immunology to forensics. Molecular genetics has even blurred the lines between biology,

Max Delbrück (Top Right) with Students at Cal Tech, 1949
(Courtesy of the California Institute of Technology.)

physics, and chemistry, with many scientists moving freely among them. Despite this, too often students are led to think that the branches of the sciences are quite separate from one another. Physics is physics, chemistry is chemistry, biology is biology—and never the three shall meet. Unfortunately, this artificial division still persists even in the minds of some scientists.

The science of molecular biology is the antithesis of this notion of separateness. It arose from a confluence of disciplines from both the physical and natural sciences—notably, genetics, physical chemistry, X-ray crystallography, biochemistry, microbiology, bacteriology, and virology. In the beginning of the 20th century, physics and chemistry were united by quantum theory, which explained the fine structure of matter. Beginning in the fifth decade of the 20th century, biology, in turn, started to benefit from an influx of ideas from physical chemistry.

Two quantum physicists were especially influential in breaking down the thought barriers between the sciences: Max Delbrück and Erwin Schrödinger. Delbrück, who is rightly called the intellectual father of molecular biology, was trained under Niels Bohr, the great atomic physicist who deduced that electrons occupy discrete energy states (orbitals) surrounding the atomic nucleus. Schrödinger's wave equation defined the movement of electrons within the orbitals. Both men thought that the biological mystery of self-replication could be explained in quantum-mechanical terms. In mid-life, Delbrück changed over to biology and never looked back.

Schrödinger never made the switch to biology, but his brief book *What is Life? The Physical Aspects of the Living Cell* (1944) influenced a generation of physical scientists to take a closer look at biological systems. Although Schrödinger admitted to being a dilettante in the field of genetics, he speculated that "...from all we learnt about the structure of living matter, we must be prepared to find it working in a manner that cannot be reduced to the ordinary laws of physics." In so doing, Schrödinger tantalized a generation of scientists with the prospect that, in studying the nature of heredity, one might uncover entirely new principles of matter.

Although no novel laws were uncovered, Delbrück and other newcomers showed that the axioms and methods of the physical sciences apply equally well to biology. Molecular biology could not become a rigorous discipline until this notion was firmly established. It would have been impossible to study how molecules interact in even the simplest of organisms without a basic understanding of how they react on the chalkboard or in the test tube.

MOLECULAR BIOLOGY IS BASED ON PHYSICAL/CHEMICAL PRINCIPLES AND ABSTRACT MODEL SYSTEMS

Basic principles underlie all biological phenomena; living things abide by all the laws of physics and chemistry. The physical and chemical behavior of elementary particles ultimately defines the parameters of behavior of any complex biological system: a bacterium, a plant, a frog, or a human being. This "reductionist" explanation of life is a direct result of understanding that living systems are biology's cross-fertilization by chemistry and physics.

Historically, biology was based on direct observation of complex natural phenomena in the real world. Molecular genetics borrowed from the physical sciences the rigorous use of model systems—simplified abstractions of reality in which variables are limited and experimental situations can be controlled. The development of molecular biology was in large part driven by the quest to find increasingly purer and more powerful abstractions of essential biological processes. Experimental genetic systems thus progressed from complex, multicellular organisms (such as pea plants and fruit flies used in the early 1900s) to simple, one-cell organisms (bacteria and viruses, beginning in the 1940s) to purified cellular components (in vitro systems, beginning in the 1960s). The increasing sophistication of experimental systems culminated in the ability to routinely add new or altered genes into bacteria (1970s) and into mammals and plants (1980s) and to target gene alterations to specific cells and at specific times in development (1990s).

THE POSTGENOMIC ERA WILL REQUIRE A MORE SYNTHETIC APPROACH

The first 50 years of molecular biology took a primarily analytic approach; that is, scientists were mainly concerned with reducing biological problems to the level of individual genes. This approach was extremely useful in finding key genes

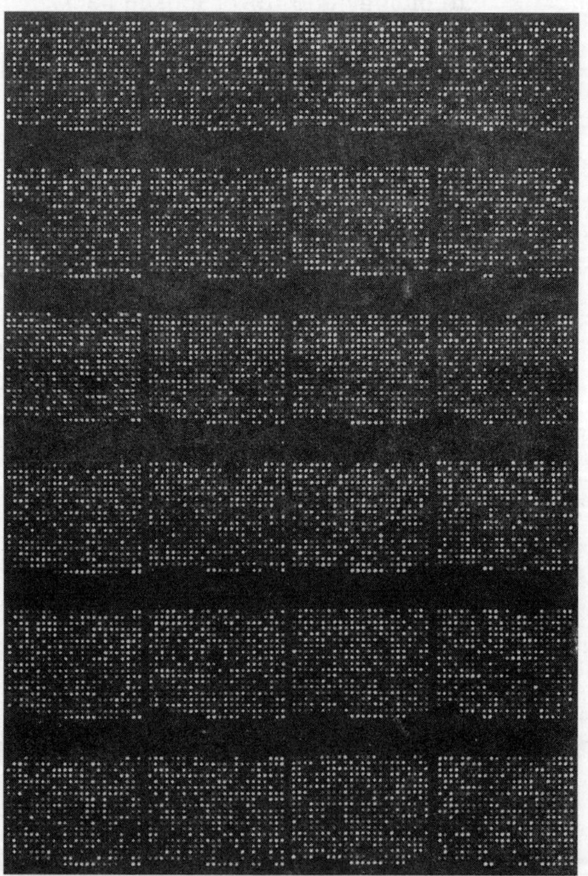

DNA Microarray

DNA microarrays allow scientists to analyze the expression of thousands of genes at once. (Reprinted, with permission, from Bowtell D. and Sambrook J. 2002. *DNA microarrays: A molecular cloning manual.* Cold Spring Harbor Laboratory Press, Cold Spring Harbor, New York.)

involved in replication and development, as well as identifying genes involved in a number of rare genetic disorders, such as cystic fibrosis, Huntington's disease, and neurofibromatosis. However, ahead lies the difficult task of teasing out the synchronized activity of numerous genes during key biological events, such as development and memory formation. Different approaches will be needed to identify the multiple genes presumably involved in common disorders such as asthma and diabetes. Adding to the complication are the facts that (1) several different proteins can be produced from a single gene by different arrangements of the RNA transcript and (2) additional forms are created by "posttranslational" modifications made after a protein is assembled at the ribosome. Even more difficult will be to examine the complex interactions between genes and the environment—the age-old question of nature versus nurture.

To build a more complete picture of gene, protein, and environmental contributions to biological processes, biologists increasingly will need to turn to synthetic approaches that allow them to examine the coordinated expression of multiple genes. DNA microarrays, or gene chips, have provided many of the initial clues, by showing that hundreds or even thousands of genes are coordinately expressed in response to various developmental and environmental cues. This sort of multitasking is beyond the ability of even the brightest human beings, so computer programs are required to detect and analyze the thousands of individual experiments contained on a single chip. This is one of many tasks for the new hybrid discipline of bioinformatics, which uses computer algorithms to manage and analyze large-scale experiments.

The functions of genes implicated in a specific pathway can then be explored by several means. Computer programs can be used to search gene databases for related genes that have been discovered previously in humans or other organisms. A mutated copy of a gene can be inserted into the mouse germ line, and the resulting transgenic animals can be analyzed for metabolic or behavioral changes. This sort of synthetic thinking demands that biologists integrate knowledge from experiments done in vitro ("in glass," or a test tube), in vivo (in a living cell or organism), and in silico (in a computer).

MOLECULAR BIOLOGY ARISES FROM THE STRUCTURE-FUNCTION TRADITION

William Harvey
(Courtesy of the Moody Medical Library.)

Natural scientists have always tried to find relationships between structure and function in living things. Molecular biology is the culmination of this tradition, which has led biologists to peel back successive layers of organization to ultimately reveal the molecular interactions that take place within the living cell. This pursuit began with the examination of obvious physical attributes.

Physicians from the time of the earliest civilizations tried to relate their knowledge of the human body to the treatment of illness. Thus, anatomy and physiology became the classic expression of structure-functionalism in the natural sciences. The 17th century anatomist William Harvey, for example, showed that a number of physical structures—organs including heart, lungs, veins, arteries, and valves—work together as a system to circulate blood throughout the body. The heart functions as a pump, and the blood vessels function as pipes.

Cell theory, advanced by Matthias Schleiden and Theodor Schwann in the late 1830s, was an important milestone; it moved structure-functionalism beyond systems directly observable with the naked eye. Schleiden and Schwann

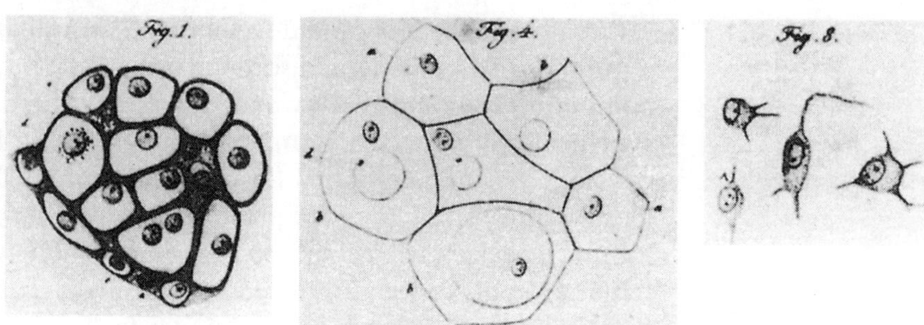

Theodor Schwann's Drawings of Vertebrate Cells Showing Nuclei
(From *Microscopial researches into the accordance in the structure and growth of animals and plants,* 1839. Sydenham Society, London.)

proposed that microscopic cells, defined essentially by the presence of individual nuclei, are the basic units of structure and function in both plants and animals. Organs were then seen to be composed of various tissues—groupings of cells with similar structures that perform a specific function. For example, epithelial cells that line the respiratory passages contain cilia that oscillate to help eject foreign particles such as smoke and dust. Cells, in turn, were found to be composed of substructures called organelles, each with their own specific functions: mitochondria for energy production, vacuoles for storage, chloroplasts for photosynthesis, and ribosomes for protein synthesis.

The stage was set for structure-functionalism to move to the level of biologically important molecules during the 1930s, when physical chemist Linus Pauling codified the physical laws that govern the arrangement of atoms within molecules. During this same period, J. Desmond Bernal showed that the structures of giant molecules, such as proteins, can be studied using X-ray crystallography.

MOLECULAR BIOLOGY ARISES FROM THE QUEST TO DEFINE THE NATURE OF HEREDITY

Reproduction, or autonomous replication, is perhaps the most distinctive attribute of life. To explain replication of cells and inheritance of traits over successive generations is, in large measure, to define life. During the development of molecular biology, scientists sought an increasingly explicit explanation of the nature of heredity. With these themes in mind, we now go back in time to trace the development of molecular biology through several successive explanations of the hereditary process:

- Diversity and changes in populations of organisms (Linnaeus, Darwin, Wallace).
- Traits inherited by individual organisms (Mendel).
- Specialized germ cells "set-aside" from the somatic cells (Weismann).
- Chromosome behavior within cells (Boveri, Sutton, Morgan, McClintock).
- Biochemical interactions within cells (Beadle and Tatum).
- DNA interactions within cells (Avery, Hershey).
- Molecular structure of DNA (Watson, Crick, Franklin, and Wilkins).

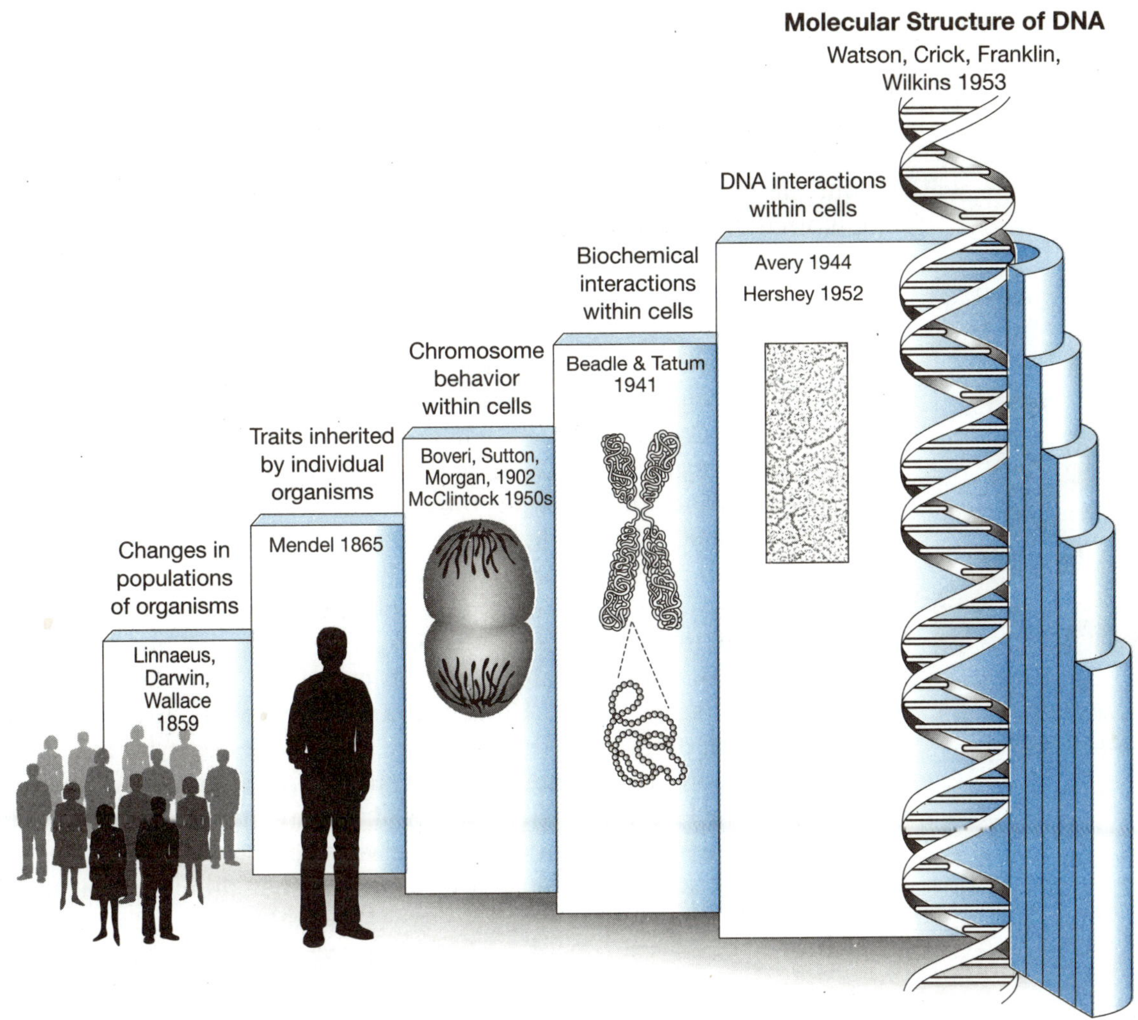

Seeking an Increasingly Explicit Definition of Heredity
(Art concept developed by Lisa Shoemaker.)

Beyond Watson and Crick's double helix lay the question that would fuel the biotechnology revolution: Can DNA, and therefore life, be manipulated? Before addressing this question—which is the subject of this book—we must understand the questions that came before the double helix. The answers to these "pre-DNA" questions provide a review of basic biology and a perspective on the DNA revolution.

Hopefully, many questions will come to mind as you read through this book, and, surprisingly, they may be the same sort of questions that led to important developments in molecular biology. Bear in mind that only 100 years ago, there was no explanation as to why some siblings have brown eyes and some have blue eyes. Seventy-five years ago, the physical structures of simple organic molecules were not known. Fifty years ago, we still did not know the correct number of human chromosomes. Twenty-five years ago, we did not know any of the genes behind cancer. We still do not know the precise number of genes in the human genome. There are still plenty of good questions left to ask.

HOW CAN WE ACCOUNT FOR THE DIVERSITY (AND SIMILARITY) OF SPECIES?

The Swedish biologist Carolus Linnaeus set up the first systematic hierarchy for classifying all living things—based on the unit "species." The cataloging of a multitude of organisms by Linnaeus and his disciples during the 17th century pointed out the incredible diversity of life forms. At the same time, his taxonomy emphasized the strong to relatively subtle similarities that exist among members of the same genus, family, order, class, phylum, and kingdom.

Charles Darwin, ca. 1859
(Courtesy of the American Museum of Natural History.)

Even so, the question of how to account for diversity and similarity among living things was not one worth asking in the 18th and 19th centuries. It had already been answered, by inference, through interpretation of the biblical story of creation, which stated that each of the diverse life forms was present in perfect form at the Creation. Nowhere in the Bible is the age of the earth, or human beings, explicitly given. However, a 17th century Irish clergyman, Archbishop James Ussher, concluded through study of biblical genealogies that the earth was created in 4004 BC. That left little time for any major alterations to living organisms. Each major type of organism must have been "perfectly created" as a fixed form; similarities between species were explained as minor changes that had occurred within the fixed forms. So it was presumptuous, if not downright blasphemous, to ask questions about the variety of life without mentioning the Creation in the same breath. However, questions did stir in the 19th century minds of English naturalists Alfred Wallace and Charles Darwin. Independently, they arrived at the same impudent conclusion: All living things had evolved from preexisting forms through a process of incremental change over millennia.

The publication in 1859 of Charles Darwin's epic book *On the Origin of Species* marked the first step in the biological revolution that culminated nearly a cen-

Flightless Cormorant

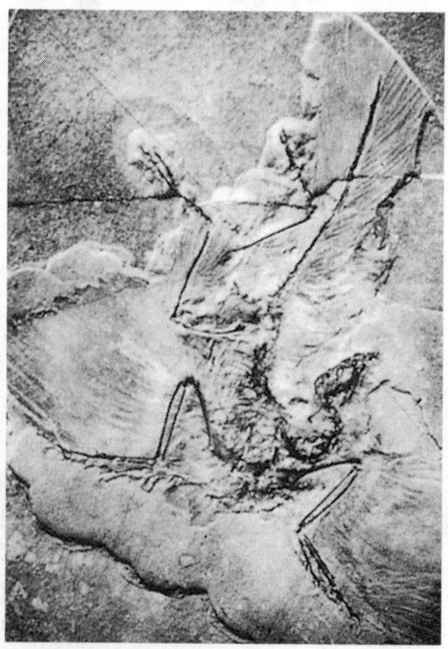

Archaeopteryx

Darwin observed several mainland species, such as the flightless cormorant, that had adapted in specialized ways to life in the Galápagos Islands. Fossils, such as the birdlike reptile *Archaeopteryx*, provided evidence of the gradual evolution of life forms over long periods of geological time. (Courtesy of Taurus Photos; the American Museum of Natural History.)

tury later in the Watson-Crick structure of DNA. The theory of evolution described how heredity operates in large populations of living things:

- There is a natural selection, over great periods of evolutionary time, for the "fittest" forms of life.
- Natural selection arises from competition for limited food and other resources among members of the same species as well as between species.
- Only the fittest members of a population survive to reproduce.

On rare occasions, a random physical change increases an individual's ability to adapt to environmental conditions or exploit new food resources. This "adaptive" change increases the individual's chances to survive and to reproduce. Adaptive changes are passed on to offspring as part of their hereditary endowment. These individuals are, in turn, fitter than their peers and survive to pass on their physical characteristics to succeeding generations. Through the process of "adaptive radiation," populations of organisms evolve to exploit specialized food resources, thus limiting competition and increasing chances for survival.

Darwin was profoundly influenced by the British geologist Sir Charles Lyell and his doctrine of uniformitarianism. This stated that the earth's physical features can be accounted for by climatic and geological processes that have occurred continuously and *uniformly* over very long periods of time. Ordinary processes that can be observed today, such as volcanism, sedimentation, coral deposition, and erosion, are the *same* processes that built the Himalayas and other seemingly "fixed" physical features. This was in direct opposition to the biblical interpretation that all geological phenomena are the products of several unique, catastrophic events: a single creation and a single flood. Lyell argued that sedimentary rock layers (strata) are a geological time line extending back millions of years into the earth's history. He demonstrated that one could date the emergence and extinction of species by observing their fossilized remains in rock strata.

Darwin avidly studied Lyell's *Principles of Geology* (1830) on his world-circling voyage aboard the H.M.S. Beagle, during which he made many of the observations on which he based his theory of evolution. By tracing the distinctive anatomical changes in related fossil forms, Darwin showed the operation of evolution over long periods of prehistory. His studies of the comparative anatomy of finches that colonized the volcanic Galápagos Islands, for example, vividly illustrated adaptive radiation during relatively recent history. Darwin ultimately came up with the biological equivalent of geological uniformitarianism: The same selective processes that had recently shaped the new species of the Galápagos had also differentiated new life forms in the dim past.

HOW ARE TRAITS PASSED FROM ONE GENERATION TO THE NEXT?

Darwin did not know the source of individual variation upon which his evolutionary processes acted or how it was passed on to successive generations. In his later theory of pangenesis, he proposed that "gemmules" are shed by cells of the body, collected in the sex organs, and transmitted to the next generation. This idea won little support from other scientists, and his own cousin, the multifaceted scientist and mathematician Francis Galton, published an experiment that refuted it.

Gregor Mendel, ca. 1860
(Courtesy of the Austrian Press and Information Service.)

It was the Augustinian monk Gregor Mendel who brought the hereditary process down to the individual organism and provided a mechanism to drive evolution. Mendel's paper "Experiments in Plant-Hybridization," published in 1865, provided a basis for the mathematical analysis of inheritance. From the results of controlled crosses of garden peas, he showed that traits are inherited in a predictable manner as "factors," which we now call genes.

Common sense tells us that, on the whole, offspring are a mixture of parental traits. However, Mendel showed that the genes governing individual traits do not blend. Instead, genes are maintained as discrete bits of hereditary information, unchanged through generations. Mendel proposed that genes behave like atoms that compose a pure substance—they can combine in various ways, but they always maintain their distinct identities.

Centuries of breeding of domestic plants and animals had shown that useful traits—speed in horses, strength in oxen, and larger fruits in crops—can be accentuated by controlled mating. However, there was no scientific way to predict the outcome of a cross between two particular parents. Mendel's rigorous approach transformed agricultural breeding from an art to a science. He started with parents of known genetic background—to provide a baseline against which to compare patterns of inheritance in the resulting offspring. Then he carefully counted the traits found in successive generations of offspring. In addition, rather than looking at the pea plant as a whole, he focused on seven individual physical (or visible) traits that he could readily distinguish. Each trait had two

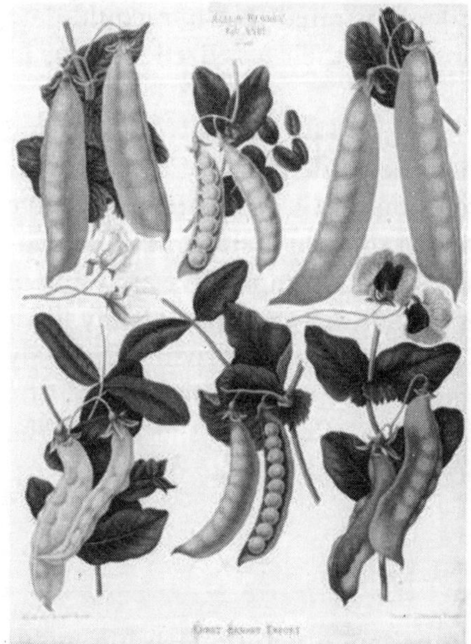

Some Pea Traits Examined by Mendel and Album Bernay
Album Bernay (1876–1893) shows some of the pea traits that Mendel used. (John Innes Foundation Historical Collections. Courtesy of the John Innes Foundation.)

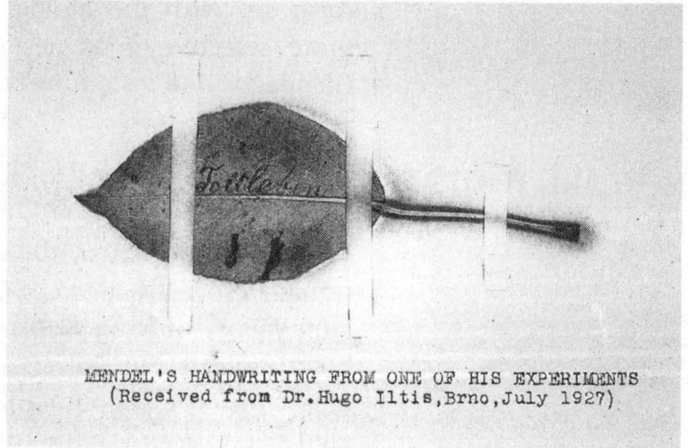

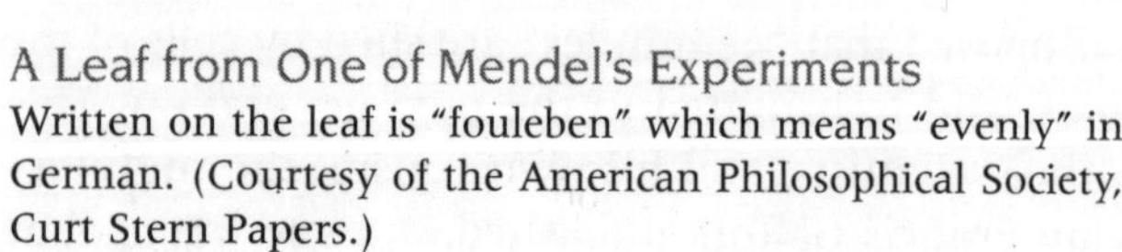

A Leaf from One of Mendel's Experiments
Written on the leaf is "fouleben" which means "evenly" in German. (Courtesy of the American Philosophical Society, Curt Stern Papers.)

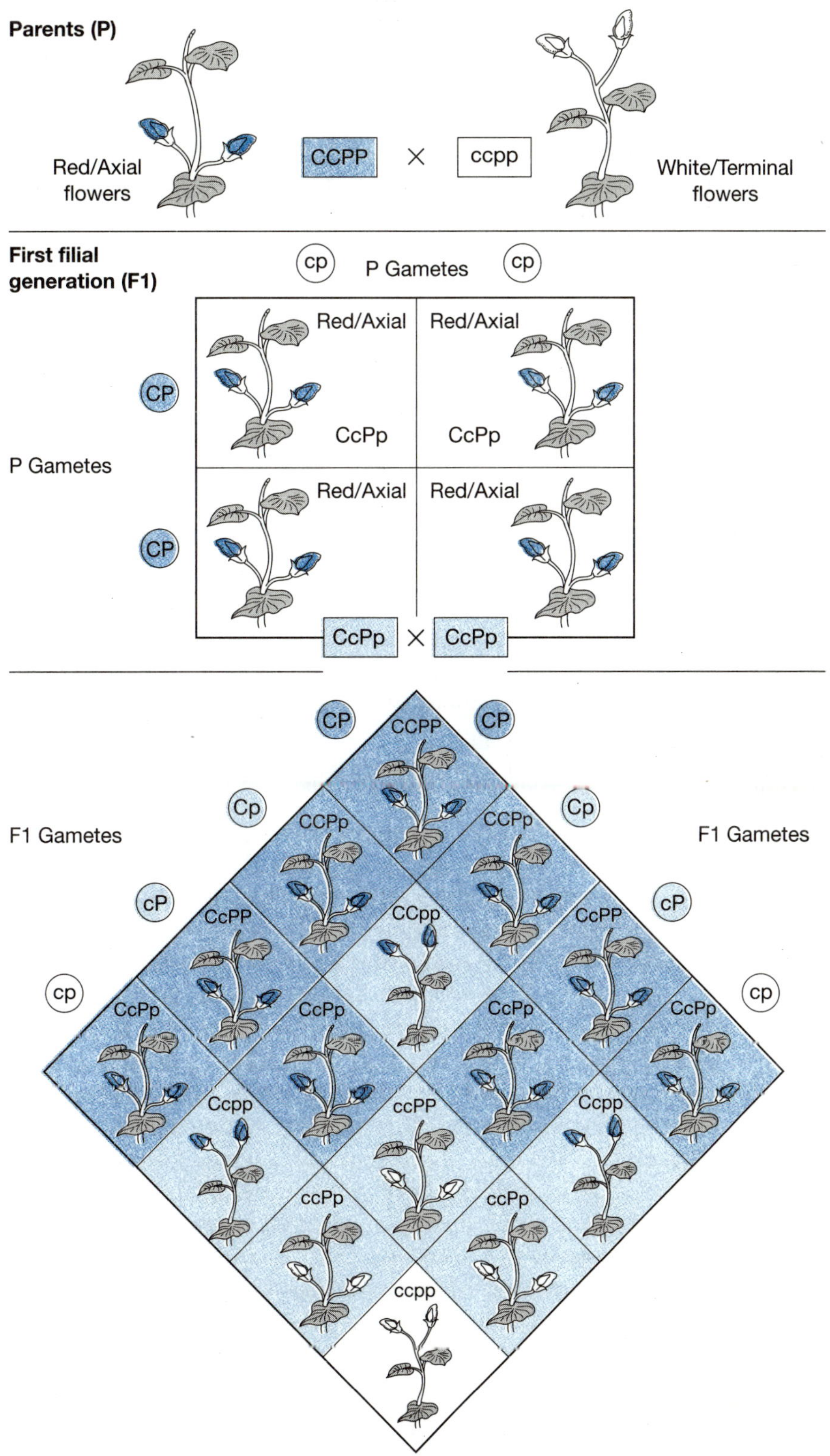

Dihybrid Cross

A dihybrid cross of garden peas showing the inheritance of two traits: flower color (C) and flower position (P). (*Dark blue*) The presence of at least one dominant allele for each trait yields a dominant phenotype for both. (*Light blue*) The presence of at least one dominant allele for one trait and two recessive alleles for the other yields a mixed phenotype. (*White*) The presence of two recessive alleles for both traits yields a recessive phenotype for both.

alternate forms. For example, pea flowers are either red or white in color. Mendel reasoned that each parent plant must contribute one gene alternate to each offspring, resulting in a pair of genes for each trait.

To follow the inheritance of genes from parent to offspring, Mendel first needed to be sure which genes each parent carried. So he developed "pure-bred" stocks by repeatedly breeding a pea plant to itself. This resulted in pea stocks that show only one trait alternate and that consistently pass on only one gene alternate to offspring. Crossing two pure-bred plants with red flowers produces only offspring with red flowers. Crossing two pure-bred plants with white flowers produces only offspring with white flowers.

One might expect that a cross between a pure-bred plant with red flowers and a pure-bred plant with white flowers would produce offspring with pink flowers or a mixture of red and white flowers. Rather, Mendel found that this cross produces only red-flowered offspring. Although these hybrid offspring must have received a different gene from each parent, there is no blending of color. If, as Mendel reasoned, the hybrid had inherited one copy of a different color gene from each parent, then the gene for red flowers (C, for color) must be "dominant" over the gene for white flowers (c). When he crossed two hybrid plants, white-flowered plants reappeared in the following generation. From this, he reasoned that the "recessive" white trait is shown only when two copies of the recessive gene are inherited (cc)—one from each hybrid parent. Thus, he related the outward appearance of each plant (phenotype) to its inner genetic constitution (genotype).

Carl Correns
(Courtesy of the American Philosophical Society, Curt Stern Papers.)

Mendel found that there was a consistent mathematical relationship between genotype and phenotype. For example, a cross between two hybrid plants with red flowers produces three times as many red-flowered plants as white-flowered ones. Reproducing this experiment with hybrids of each of the seven distinctive trait pairs resulted in Mendel's famous 3-to-1 phenotypic ratio of dominant-to-recessive traits.

When Mendel proposed that traits are determined by a pair of genes, it presented a potential problem. If parents pass on both copies of a gene pair, then their offspring would end up with four genes for each trait. This doubling of genetic material would continue in ensuing generations. Mendel deduced that parents contribute only half of their genes to their offspring. He hypothesized that the gene number is reduced during gametogenesis, so that each gamete (sex cell) receives one copy of the gene pair. During fertilization, the male and female gametes fuse to yield a pair of genes in the offspring.

Hugo de Vries, ca. 1920
(Courtesy of Cold Spring Harbor Laboratory Archives.)

In a hybrid cross, mixing of parental genes makes three genotypic combinations possible in the offspring: CC, Cc, and cc. The dominant phenotype (red flowers) is expressed when two copies of the gene are present, CC. Red flowers also result from the mixed genotype Cc, where the C gene dominates the recessive c gene. The recessive phenotype (white flowers) is expressed only by the genotype cc, where two copies of the recessive gene are present.

Now imagine simultaneously following a second trait, flower position, where axial flowers are dominant (P, for position) and terminal flowers are recessive (p). Given two hybrid parents with genotype CcPp, Mendel showed that genes for each trait "segregate" so that each sex cell contains only one sort. Thus, each contrasting member of a gene pair is equally likely to occur in gametes: C or c and P or p. Genes for each different trait "assort" into gametes "independently" of one another, making every conceivable combination possible: CP, Cp, cP, and cp. This "dihybrid" cross yields a phenotypic ratio of 9 red/axial flowers : 3 red/terminal flowers : 3 white/axial flowers : 1 white/terminal flowers (9:3:3:1).

WHERE ARE GENES LOCATED?

Jean-Baptiste de Lamarck, ca. 1821 (Courtesy of the Wellcome Institute Library, London.)

Although Mendel was a contemporary of Darwin, his work lay fallow, unrecognized until the beginning of the 20th century. In 1900, the Dutch scientist Hugo de Vries, the German scientist Carl Correns, and the Austrian scientist Erich von Tschermak-Seysenegg rediscovered Mendel's paper and published research data that confirmed his earlier work. de Vries realized that Mendel's "factors" were the same entities that he called "pangenes," and in 1909, Wilhelm Johansson shortened the term to "gene."

Historians still ponder the reasons for this 35-year lapse in acceptance. Although Mendel published his research in a relatively obscure journal, at least 140 copies were circulated, predominately in Europe. However, his abstract notion of genes was not appreciated by naturalists of his time—they had been trained primarily to observe and categorize living things. Clearly, Mendel was ahead of his time. He envisioned a hereditary process in advance of a broader theoretical context in which it could be understood by other bright scientists. This might well be the working definition of genius.

There is little doubt that the work of Jean-Baptiste de Lamarck, the self-trained French scientist, seriously muddied thought about inheritance during most of the 19th century. In his book *Systeme des Animaux sans Vertebres* (1800), he claimed that organisms acquire physical features in response to environmental changes and that these characteristics are then inherited by successive generations of offspring. On the face of it, Lamarck's theory of inheritance of acquired characteristics was consistent with evolution and even Darwin favored it. However, to social reformers, Lamarckism offered a brighter alternative to the grim application of Darwinism in the burgeoning factories and sweat shops of Europe. Social Darwinism justified harsh working and living conditions as a means of natural selection of the sturdiest individuals. Lamarckism countered that good environments ultimately produce good heredity, providing a rationale for charitable treatment of the poor, the sick, and the mentally ill.

August Weismann (Courtesy of the National Library of Medicine.)

By the 1880s, key findings in cell biology and chromosomal behavior provided a physical context in which to understand Mendel's abstract genetic work. Schleiden and Schwann's cell theory had been extended to explain the process of fertilization—where separate sperm and egg cells fuse to give rise to a zygote. Aniline dyes, a by-product of coal, revealed thread-like chromosomes in the nucleus. Different organisms proved to have different numbers of chromosomes, suggesting that the chromosomes might carry information specific for each life form. In 1882, Walther Flemming described the process of mitosis, the duplication and movement of chromosomes that occur when a cell divides to produce two daughter cells.

During the years 1883 to 1885, the German doctor August Weismann delivered an extraordinary series of lectures at the University of Frieburg that countered Lamarckism and provided a mechanism for the continuity of inheritance

Theodor Boveri
(Courtesy of the American Philosophical Society, Curt Stern Papers.)

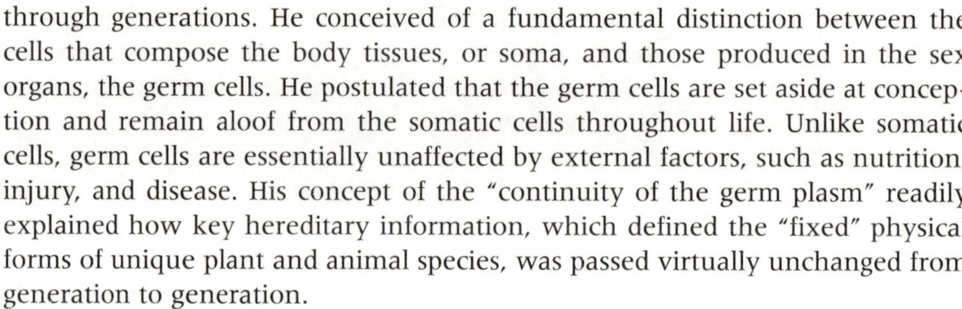

through generations. He conceived of a fundamental distinction between the cells that compose the body tissues, or soma, and those produced in the sex organs, the germ cells. He postulated that the germ cells are set aside at conception and remain aloof from the somatic cells throughout life. Unlike somatic cells, germ cells are essentially unaffected by external factors, such as nutrition, injury, and disease. His concept of the "continuity of the germ plasm" readily explained how key hereditary information, which defined the "fixed" physical forms of unique plant and animal species, was passed virtually unchanged from generation to generation.

With the scientific world thus prepared, genetics seemingly burst onto the scene fully formed in 1900. In 1902, the German scientist Theodor Boveri and Walter Sutton, a student at Columbia University, were the first to directly relate heredity to chromosome behavior, or cytology. Boveri found that sea urchin zygotes died or developed abnormally when a single egg was fertilized by two sperm. Fertilization intitially results in three daughter cells, which then divide asymmetrically, producing cells with incomplete sets of chromosomes. Sutton's analysis of chromosome movements during meiosis in the grasshopper *Brachystola* formed the basis of the chromosomal theory of heredity. Sutton showed that the grasshopper genetic material consists of 11 pairs of homologous chromosomes and that gametes formed during meiosis receive only one chromosome from each homologous pair. This behavior paralleled exactly the segregation of Mendel's hereditary factors and suggested that genes are physically located on the chromosomes.

Edmund Wilson, ca. 1925
(Courtesy of the American Society of Zoologists.)

The behavior of sex chromosomes, described independently in 1905 by Nettie Stevens and Edmund Wilson (Sutton's mentor at Columbia), provided the first direct evidence to support the chromosomal theory of heredity. They showed that sex is determined by separate X and Y chromosomes. Femaleness is characterized by two copies of the X chromosome (XX), and maleness is determined by a single copy of each type of chromosome (XY). The movements of X and Y chromosomes during formation of sperm and egg cells are exactly as pre-

Nettie Stevens, ca. 1904
(Courtesy of the Carnegie Institute of Washington.)

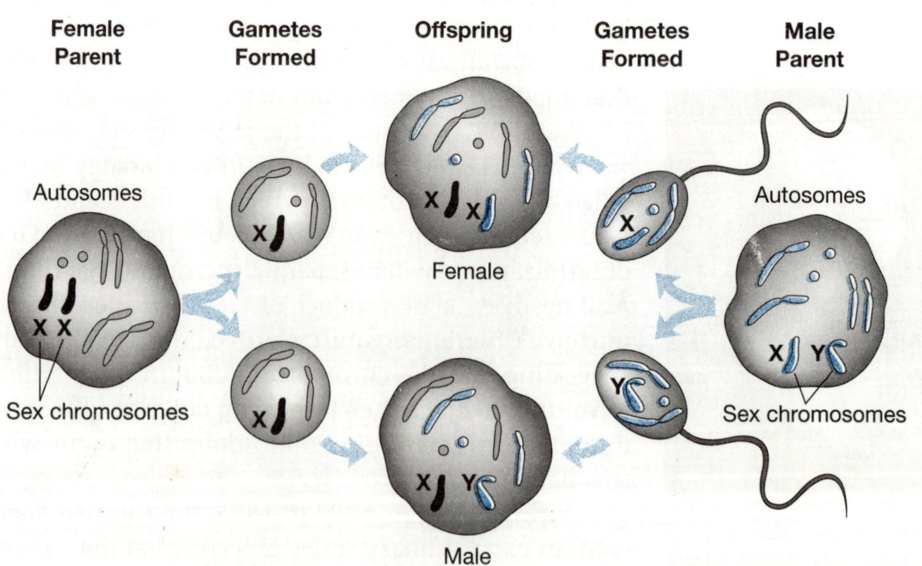

Segregation of X and Y Chromosomes in *Drosophila*

dicted by Mendelian genetics: Each egg receives a single copy of an X chromosome; each sperm receives either an X or a Y chromosome.

Evidence that traits other than sex determination are located on the chromosomes came during the second decade of the 20th century. During this period, Thomas Hunt Morgan and his astoundingly bright cadre of students—Alfred Sturtevant, Calvin Bridges, and Hermann Muller—established the intellectual basis of modern genetics. They also established the humble fruit fly, *Drosophila melanogaster*, as the organism of choice for genetic research. Morgan's modestly sized laboratory in the zoology department at Columbia University became known simply as the Fly Room.

WHAT IS THE LINK BETWEEN EVOLUTION AND GENETICS?

Charles Davenport, ca. 1919 (Courtesy of the National Library of Medicine.)

Cytology was one clear path to genetics at the turn of the 20th century; evolution was another. Although a connection between evolutionary theory and genetic theory makes sense to the student of biology, the direct transference of ideas occurred during a brief and little known moment in the history of modern biology. This moment—when the nature of the hereditary process was first focused from the level of populations to individual organisms—is best captured as the forerunner of the current Cold Spring Harbor Laboratory.

Two years before the rediscovery of Mendel's work, the Harvard-trained biologist Charles Davenport assumed the directorship of The Biological Laboratory at Cold Spring Harbor, a progressive, if somewhat sleepy, "summer camp" for the study of evolution. Founded in 1890, The Biological Laboratory followed in the footsteps of the Marine Biological Laboratory founded at Woods Hole, Massachusetts in 1888. Both had been established in the tradition of seaside biological stations that were founded along the European coastlines following the publication of Darwin's *Origin of Species*. The intersection of land and water was thought to be the ideal place to study how organisms had evolved and adapted to fill a multitude of aquatic, semi-aquatic, and terrestrial niches. The biologists who occupied these field stations were mainly content to observe the results of evolution in the natural world. Many became especially interested in the dynamic interactions between organisms and their environments, thus becoming the first generation of ecologists.

Although Davenport had been trained in the classical observation methods of zoology and comparative morphology, he became interested in the new movement that sought to directly recreate evolution in the laboratory. He and a growing number of biologists believed that controlled breeding experiments with plants and animals would yield new insight into evolutionary processes. Davenport gained the ear of the Board of the Carnegie Institution of Washington, one of several philanthropies formed under Andrew Carnegie's will. From them, Davenport secured funding to establish at Cold Spring Harbor a Station for Experimental Evolution.

The dedication of the Station's first building, in 1904, offered a prediction about the future of experimental evolution—the keynote speech was delivered by Hugo de Vries, one of three researchers who had recently rediscovered Mendel's seminal work. Davenport populated the Station with like-minded researchers, who made agricultural plants and animals their research subjects.

Chicken Coops at the Station for Experimental Evolution, ca. 1910
(Courtesy of Cold Spring Harbor Laboratory Archives.)

The Station took on the aspect of a farm with chicken coops, goat sheds, Manx cats, canaries, and fields of corn and Jimsonweed. Many of the researchers at the Station for Experimental Evolution embraced Mendel's laws, and the shorthand popularized by Reginald Punnet, as a means to follow traits through their experimental crosses. For many, the intricacies of genetics eventually subsumed the experimental study of evolution, and in 1920, the Station for Experimental Evolution was quietly renamed the Carnegie Department of Genetics. Thus, the experimental evolutionists had become the first generation of geneticists.

ARE GENES PHYSICAL ENTITIES?

Thomas Hunt Morgan would have counted himself in the camp of experimental evolutionists. He, like many others, was interested in testing whether the variations that result in new species happen gradually or occur in abrupt fits and starts—the still contentious problem of gradual versus punctuated evolution. Working in New York City, Morgan lacked the space to cultivate the domestic plants and animals favored by many experimental evolutionists. He chose the fruit fly *Drosophila melanogaster* as his experimental model because it had a short generation time, produced numerous offspring, and was easy to culture. In addition, he lacked sufficient funds and facilities to maintain higher organisms, such as mice.

The appearance of a single white-eyed fly in Morgan's laboratory in the spring of 1910 was clearly a fluke—wild fruit flies have red eyes. Luckily, Calvin Bridges, then an undergraduate student, recognized its potential importance and saved that unique male fly. The white-eyed variant might have been viewed only as a tiny, nonadaptive evolutionary step. Instead, it became the cornerstone

The Fly Room at Columbia University, ca. 1920
Note the bananas used as fruit fly food. The room no longer exists at Columbia. (Courtesy of the American Philosophical Society, Curt Stern Papers.)

Thomas Hunt Morgan, ca. 1917
(Courtesy of the American Society of Zoologists.)

upon which Mendelian, chromosomal, and sexual inheritance were built into a cohesive whole. The white-eyed male also established forever the importance of an observable mutation, or variation from the norm, as the starting point for genetic analysis.

The white-eyed male fly was mated with its red-eyed sisters, and the appearance of the trait was followed through several successive generations. Mendelian analysis revealed that, in most types of crosses, white eyes appear in males only. Morgan's group showed that white eyes (like some types of baldness and hemophilia in humans) is a sex-linked recessive trait. In so doing, the gene for eye color was localized to the X chromosome.

During the following decade, Morgan and his disciples identified numerous other mutations in *Drosophila* and used these to expand their physical and chromosomal description of heredity. Early on, they showed that certain genes are

Calvin Bridges, ca. 1926
(Courtesy of the American Society of Zoologists.)

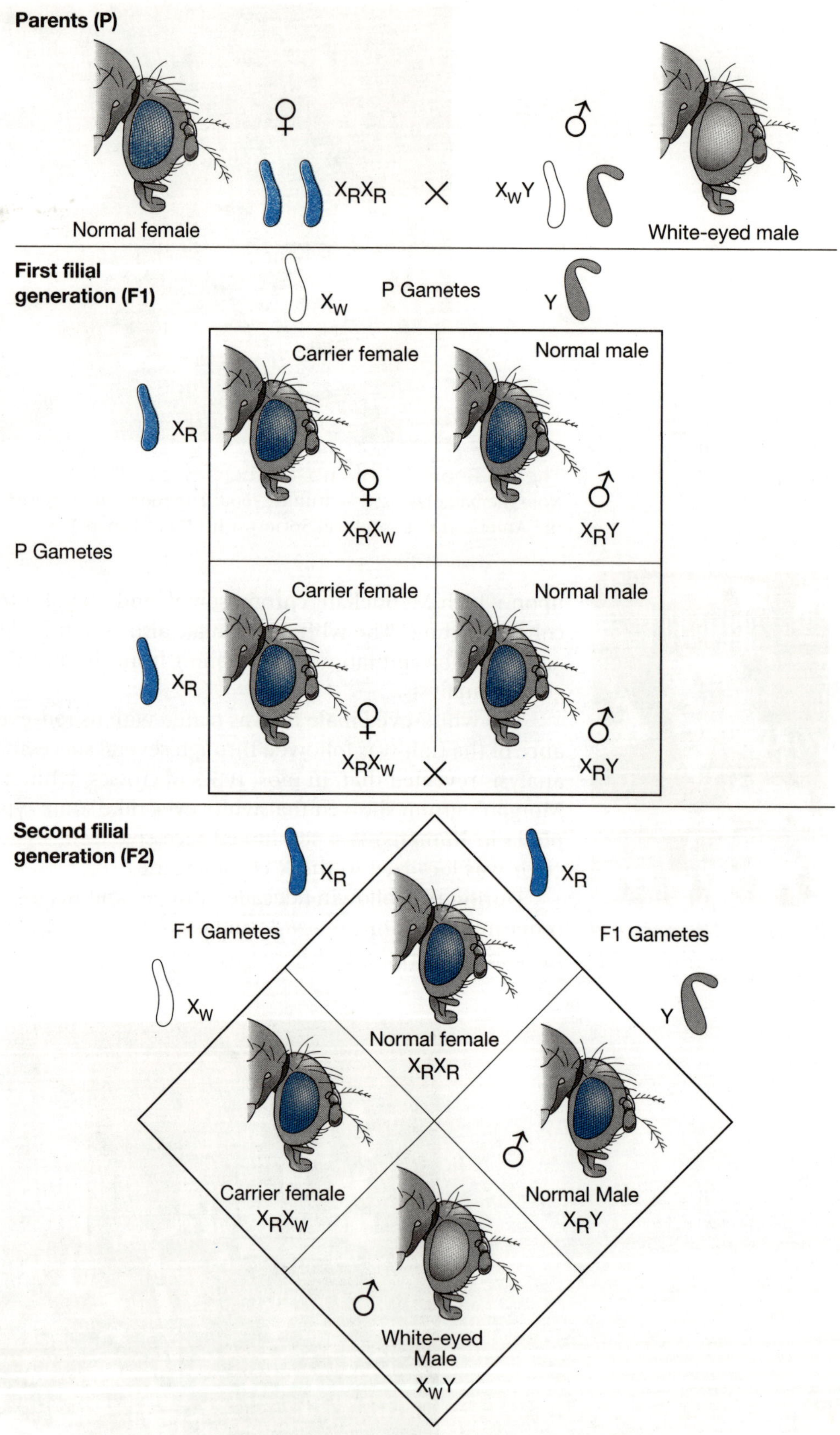

Sex-linked Inheritance of White Eyes in *Drosophila*

Alfred Sturtevant
(Courtesy of the American Philosophical Society, Curt Stern papers.)

"linked," or inherited together as if they are a single physical unit. All genes sorted into *four* linkage groups, which corresponded to the number of *Drosophila* chromosomes seen under a microscope.

The Belgian cytologist F.A. Janssens found that, early in meiosis, homologous chromosomes intertwine and exchange pieces. This process later became known as "crossing over." Morgan realized that crossing over could provide a measure of the relative distance between two genes. Whereas closely linked genes will rarely be separated by crossing over, genes that are far apart will be frequently separated. Therefore, the lower the crossover frequency between two

Barbara McClintock (Left) and Harriet Creighton (Right) at a Meeting at Cold Spring Harbor
(Courtesy of Cold Spring Harbor Laboratory Archives.)

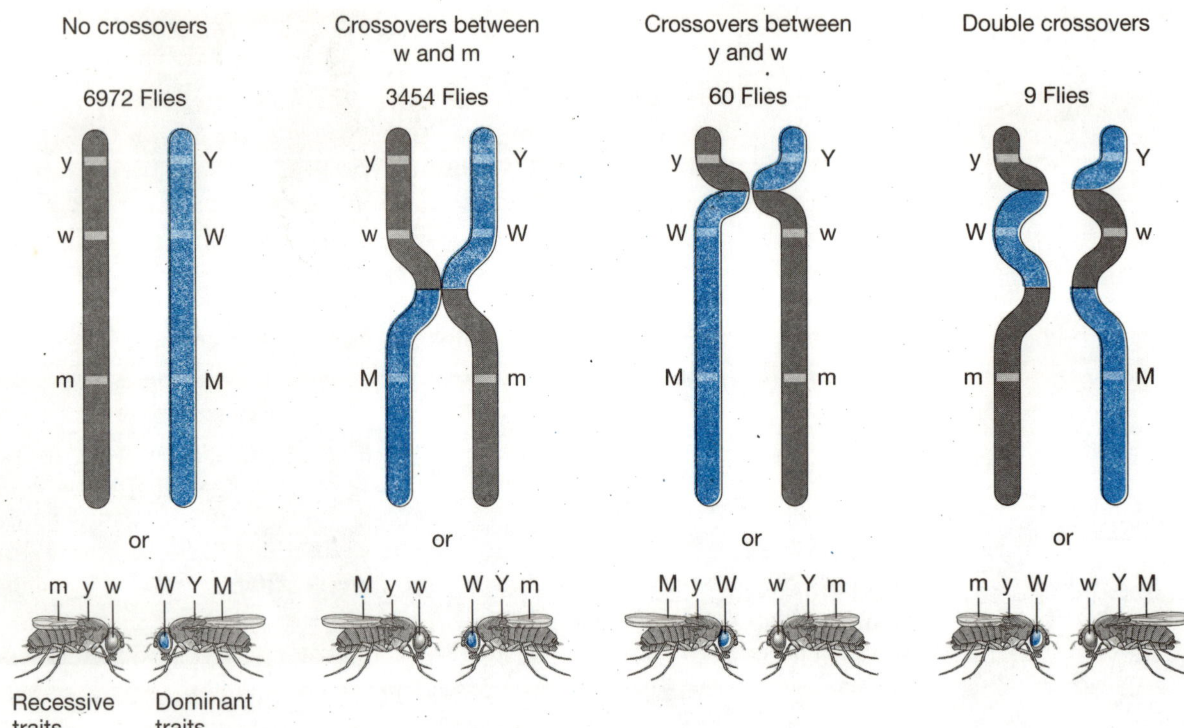

Sturtevant's Linkage Experiment in *Drosophila*, 1913

Sturtevant examined the X-linked inheritance of three recessive traits: yellow body (*y*), white eyes (*w*), and miniature wings (*m*). He crossed recessive males (*y,w,m*) with heterozygous females having recessive genes on one X chromosome (*y,w,m*) and dominant genes on the other (*Y,X,M*). Because a male parent *only* can pass on recessive genes on its single X chromosome, the phenotypes of both male and female offspring are due entirely to the inheritance of the maternal X chromosomes. Mendelian analysis predicts that all of the 10,495 offspring in Sturtevant's experiment would show either a purely dominant phenotype, normal body/eye color/wings, or a purely recessive phenotype, yellow body/red eyes/miniature wings. However, offspring inherited various mixtures of dominant and recessive traits. Sturtevant deduced that the mixed phenotypes were caused by genetic exchange between a female's two X chromosomes during gamete formation. The frequency of exchange is a measure of the distance between two genes located on the same chromosome.

genes, the closer together they are on the chromosome. Alfred Sturtevant provided proof of this concept in his doctoral research, in which he used linkage data to construct the first gene map of a chromosome.

It was not until 1931, however, that Barbara McClintock and Harriet Creighton obtained cytological proof that the inheritance of novel gene combinations during crossing over is due to the physical exchange of specific chromosome segments. Key to their experiments was a corn strain whose 9th chromosome contained a distinctive knob, composed of centromere-like heterochromatin, which made it easy to distinguish under the light microscope. This strain was constructed to carry the dominant alleles of several visible traits and then bred into a stock with unmarked chromosomes bearing only recessive alleles. Crossovers between the marked and unmarked chromosomes yielded offspring with mixed dominant and recessive phenotypes. Microscopic examination showed that the knob marker always accompanied the dominant phenotypes. In the same year, Curt Stern, at the University of Berlin, reported the use of a similar approach to study the X chromosome of *Drosophila*.

WHAT IS THE JOB OF THE GENE?

Archibald Garrod, ca. 1910 (From *Genetics: A periodical record of investigations bearing on heredity and variation 56:* frontispiece, 1967.)

Throughout the first 40 years of genetics, the gene was merely a concept whose true identity was shrouded in mystery. Genes were known only by their outward manifestation as visible traits. Nothing was known about their function in the biochemical life of an organism.

The British physician Sir Archibald Garrod proposed, in 1908, that some human diseases are "inborn errors of metabolism" that result from the lack of a specific enzyme needed to perform a biochemical reaction. He speculated that lack of enzyme function resulted from a defective gene inherited at birth. It would take more than 30 years to prove this prophetic hypothesis, for it could not be seriously tested by the geneticists of the day. The experimental systems they used—primarily fruit flies, mice, and domestic plants and animals—were all highly evolved, multicellular organisms. They were far too complex, as models, to shed light on the connection between genes and cellular biochemistry.

It was George Beadle and Edward Tatum, at Stanford University, who finally introduced a fungus as the first genetic model in which it was possible to study metabolism. The red bread mold *Neurospora* can thrive on minimal medium containing only sucrose, inorganic salts, and the vitamin biotin. Beadle and Tatum reasoned that it must therefore possess a number of specific enzymes that metabolically convert these simple nutrients (plus water and oxygen) into amino acids, vitamins, and all the other complex molecules necessary for life.

Edward Tatum, 1965 (Courtesy of the Rockefeller Archive Center.)

They hypothesized that if they could induce a mutation in a gene that produces one of these enzymes, it would be lethal to the mold. They employed the powerful genetic tool that had been introduced in 1927, when Hermann Muller (then at the University of Texas) showed that mutations can be generated at will by irradiating organisms with X-rays. Beadle and Tatum's ingenious experimental design allowed them to select for X-ray-induced mutations that would, under normal circumstances, be lethal.

First, they irradiated *Neurospora* with X-rays and crossed them to produce spores; then they examined the spores' growth on minimal medium. Spores that failed to grow on minimal medium were, in turn, tried on several types of media, each containing a single nutritional supplement. The 299th spore they examined would grow only when supplemented with vitamin B_6. Thus, irradiation had mutated a gene that produces an enzyme necessary for its synthesis. Numerous metabolic mutants for other essential vitamins and amino acids were found. Moreover, multiple mutant strains were isolated that failed to produce the amino acid arginine. Beadle and Tatum found that each mutant strain lacked a different enzyme at different points along the arginine synthesis pathway.

George W. Beadle, ca. 1940 (Courtesy of the Stanford University News Service.)

Beadle and Tatum's work, published in 1941, introduced the concept that genes mediate cellular chemistry through the production of specific enzymes. Garrod's essential hypothesis, now directly stated as "one gene/one enzyme," was confirmed. (It is most useful to broaden that adage to read "one gene/one protein.") Their experiments also pointed to the advantages of microorganisms as model systems. Using their method of plating onto selective media, it became possible to identify extremely rare genetic events, such as one-in-a-million biochemical mutants.

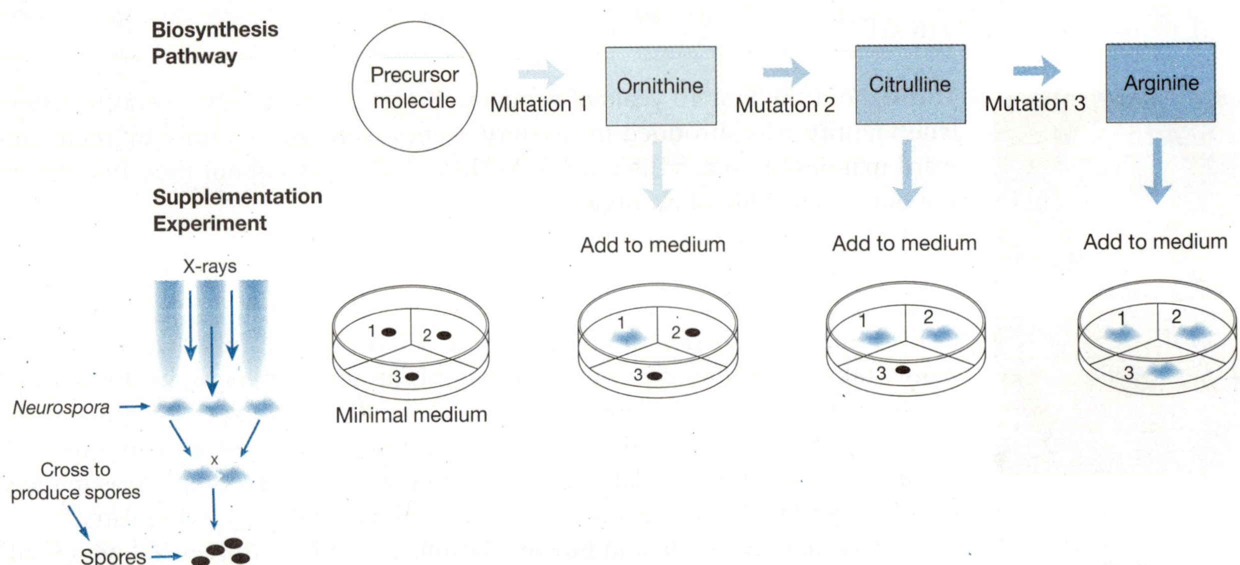

Beadle and Tatum's "One Gene/One Enzyme" Experiment, 1941
Following irradiation, *Neurospora* spores that failed to germinate on minimal medium were sequentially screened for growth on media supplemented with three nutrients in the arginine biosynthesis pathway. The pattern of supplementation for spores 1, 2, and 3 indicates mutations in genes for enzymes at corresponding points in the pathway.

WHAT MOLECULE IS THE GENETIC MATERIAL?

In a series of experiments with *Diplococcus pneumoniae*, the ball-shaped bacterium that causes pneumonia, English microbiologist Fred Griffith discovered the model system that provided the key to answer the next question in the development of molecular biology. Two naturally occurring strains of the pneumococcus bacterium have markedly different properties. The virulent smooth (S) strain possesses a smooth polysaccharide capsule that is essential for infection. The nonvirulent, rough (R) strain lacks this outer capsule, giving its surface a rough appearance.

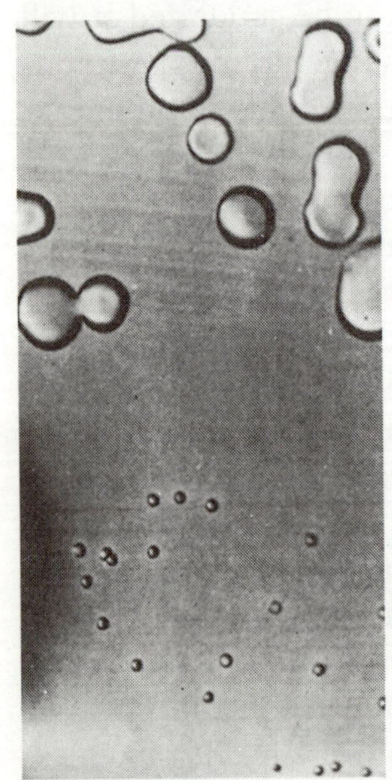

Colonies of Rough (Bottom) and Smooth Pneumococci
(Reprinted, with permission, from Avery et al. 1944. Studies on the chemical nature of the substance inducing transformation of pneumococcal types. *J. Exp. Med. 79:* 137–158; © The Rockefeller University Press.)

Oswald Avery (center foreground) and Associates, 1932
(Courtesy of the Rockefeller Archive Center.)

Following injection with the S strain, mice succumb in several days to pneumonia. Although neither living R strain nor heat-killed S strain causes illness when injected alone, Griffith was surprised to find that coinjection of the two produces a lethal infection. Furthermore, he was able to retrieve virulent S strain bacteria from mice infected with this mixture of bacteria. Through some unknown process, the innocuous R strain had been transformed into the infective S strain. Griffith presented his hypothesis in 1928: Some "principle" transferred from the killed S strain converted the R strain to virulence by enabling it to synthesize a new polysaccharide coat.

During the next decade and a half, the "transforming principle" was earnestly pursued by a research team headed by Oswald T. Avery at the Rockefeller Institute. Between 1930 and 1933, they achieved transformation outside the body of a living mouse. They observed microscopically the formation of polysaccharide coats in R pneumococci when the cultured bacteria were treated with extracts purified from the heat-killed S strain. A newcomer to Avery's laborato-

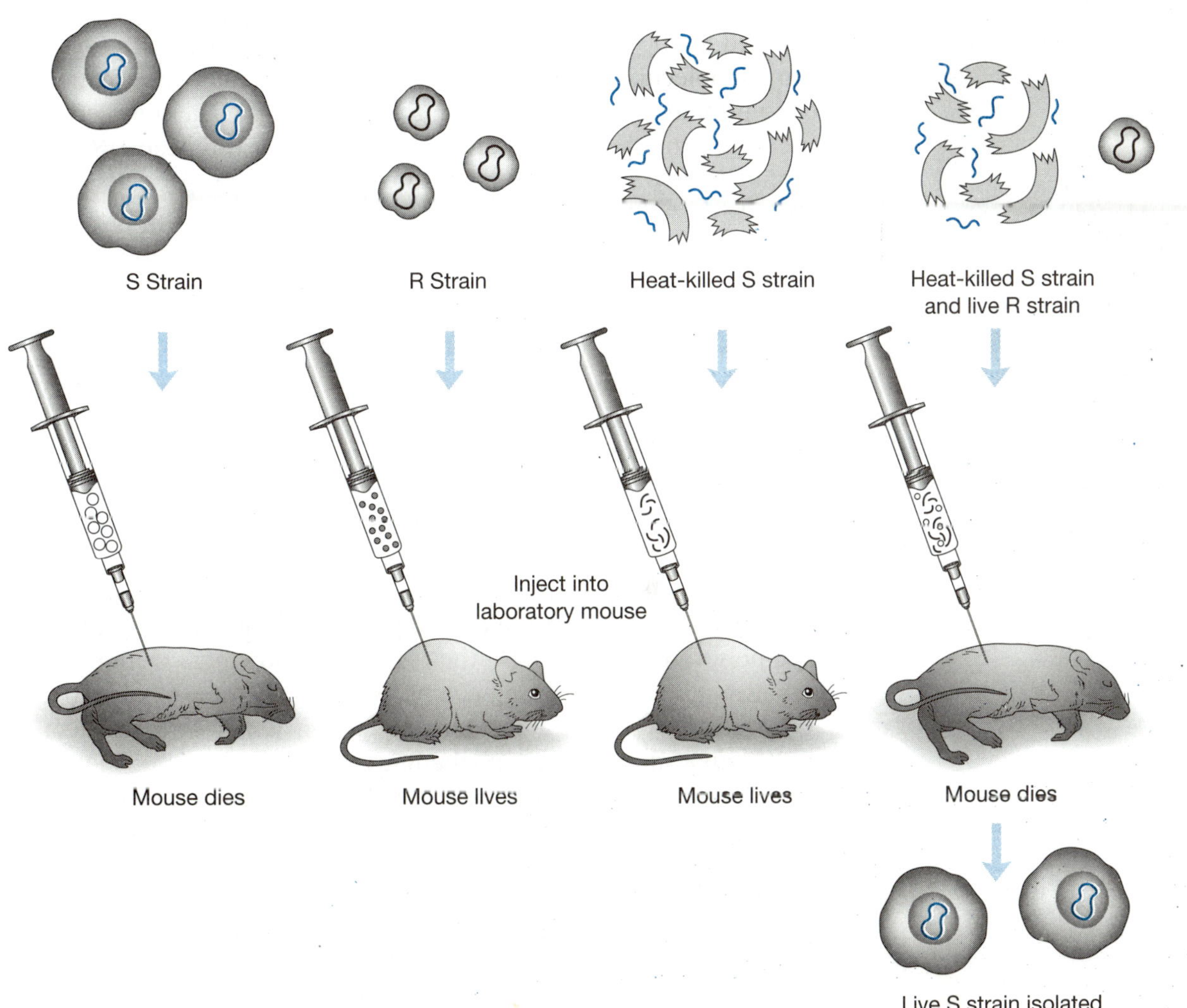

Griffith's Transformation Experiment with Smooth (S) and Rough (R) Strains of Pneumococcus, 1928

Maclyn McCarty, ca. 1936 (Courtesy of The Rockefeller Archive Center.)

Phoebus Levene, ca. 1915 (Courtesy of The Rockefeller Archive Center.)

ry, Colin MacLeod, showed that an extract containing only the polysaccharide coat could not transform R strains, disproving the belief that the coat itself was passed between strains during transformation.

In 1944, Avery, MacLeod, and Maclyn McCarty reported that they had purified the transforming principle. Analysis of its molecular composition and weight indicated that their highly active fraction was primarily DNA. Several enzyme tests were conclusive. Transforming activity was unaffected by treatment with trypsin and chymotrypsin (which digest protein) and ribonuclease (RNase, which digests RNA). However, deoxyribonuclease (DNase) destroyed all transforming activity. Their work was meticulous, and Avery's interpretation was clear: "The inducing substance has been likened to a gene, and the capsular antigen which is produced in response to it has been regarded as a gene product." Therefore, the gene is composed of DNA. (In retrospect, we now know that synthesis of the capsular antigen requires several genes, not a single gene as originally implied by Avery.)

Avery's work should have immediately focused attention on DNA as the molecule of heredity, but it did not. Like many scientific advances, his conclusions did not easily mesh with prevailing dogma—in this case, beliefs about the relative "intelligence" of protein versus DNA molecules. It was agreed that the molecule that functioned as the carrier of heredity must have the capacity to store, presumably in its molecular structure, huge amounts of genetic information. When compared to protein, DNA did not seem to be a very intelligent molecule.

If genetic information was likened to a language, then the DNA language seemed impoverished compared to the protein language. At a basic level, the protein alphabet has 20 letters (the amino acids), whereas DNA has only four letters (the nucleotides). On the face of it, DNA had a more limited combinatorial power—its language had fewer possible words. (Of course, this was before it was known how well a computer can function with a two-letter language.) Furthermore, amino acids were known to articulate into an incredible diversity of large and complex protein molecules that performed numerous important cellular functions, both as enzymes and structural elements. No obvious function for DNA had yet been discovered.

Although DNA was known to be a very large molecule, it was thought to be a repetitious polymer incapable of encoding information. This interpretation was largely the legacy of the "tetranucleotide hypothesis," propounded by Phoebus Levene at The Rockefeller Institute. As a chemist, Levene was influenced by the triumph of polymer chemistry, which had produced wonder materials like polyacetate and polystyrene. Like DNA, these synthetic polymers were composed of small units assembled in very long chains. So it is not hard to see why Levene assumed that DNA was also a regular polymer in which each nucleotide followed another in a monotonous, unchanging pattern.

Common sense and Levene's lingering influence thus prevented many scientists from embracing Avery's conclusion for the better part of a decade. They could not believe that his purified transforming principle had been completely cleansed of protein. DNA might merely be the scaffold to which were attached traces of the true transforming principle—a small amount of protein that had escaped enzyme digestion.

As World War II drew to a close, a remarkable group of scientists coalesced around three men who were primarily responsible for taking the search for the physical basis of heredity to the very fringe of life and nonlife. Max Delbrück, a

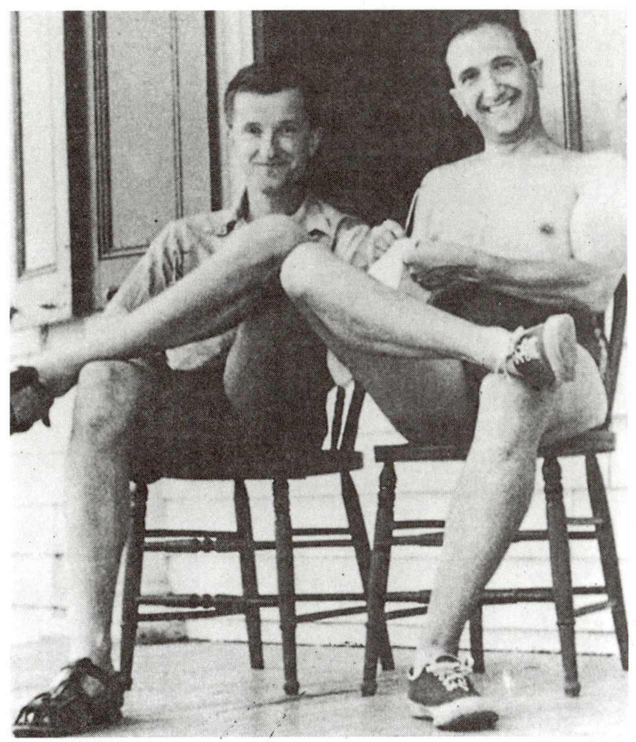

Max Delbrück (Left) and Salvador Luria

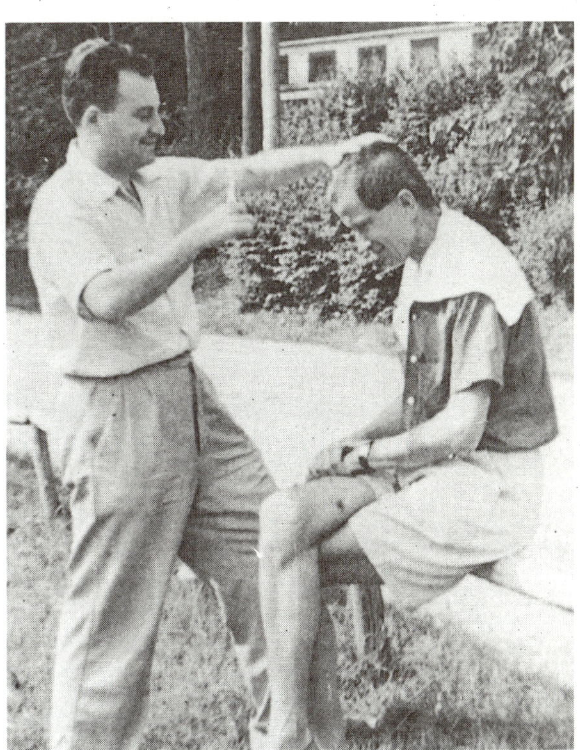

Seymour Benzer Gives Max Delbrück a Haircut

Camaraderie ranked high in 1953 with research among members of the American Phage Group, which convened each summer at Cold Spring Harbor, New York. (Courtesy of Cold Spring Harbor Laboratory Archives.)

German working at Vanderbilt University, and Salvador Luria, an Italian working at Indiana University, had fled to the United States from Nazi Europe; Alfred Hershey was an American working at the Carnegie Institution's Department of Genetics at Cold Spring Harbor, New York.

The object of their research was a group of tiny bacterial viruses, called bacteriophages (or simply, phages). It was known that during infection, phage particles reproduce inside the bacterial cell, which ruptures to release a new generation of viruses. Delbrück, Luria, and Hershey reasoned that the genetic interaction between bacteria and their virus parasites might provide an ideal model system to study the mechanism of heredity.

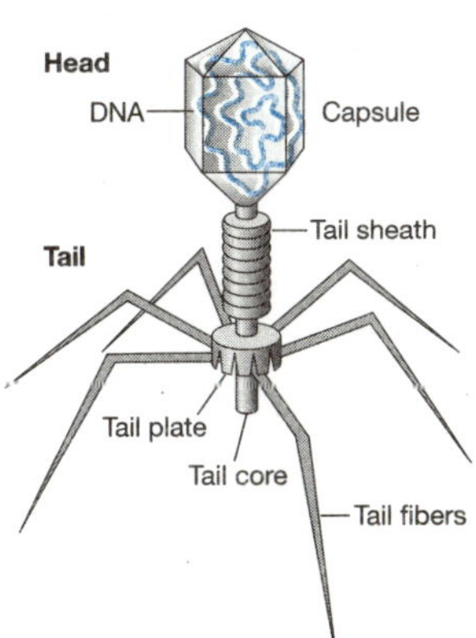

Bacteriophage

The phage particle is essentially a protein capsule surrounding a core of DNA.

Although bacteriophages had been discovered in 1917 by a Canadian, Felix d'Hérelle, Delbrück and Luria were the first to work out quantitative methods for studying the phage life cycle. Key among these was "one-step growth" ensuring that all viruses begin the infection cycle simultaneously. This synchrony was essential to study organisms so minute that they could not be studied individually.

In 1945, Delbrück organized a course at Cold Spring Harbor Laboratory to introduce researchers to the methods that he and Luria had developed. Many historians regard the phage course as the intellectual watershed of molecular biology, for it drew into contact an ever-widening group of scientists, from both the biological and physical sciences, who would hammer out the molecular mechanics of heredity. The phage course, which was taught at Cold Spring Harbor for 26 consecutive years, was a training ground for the first two generations of molecular biologists.

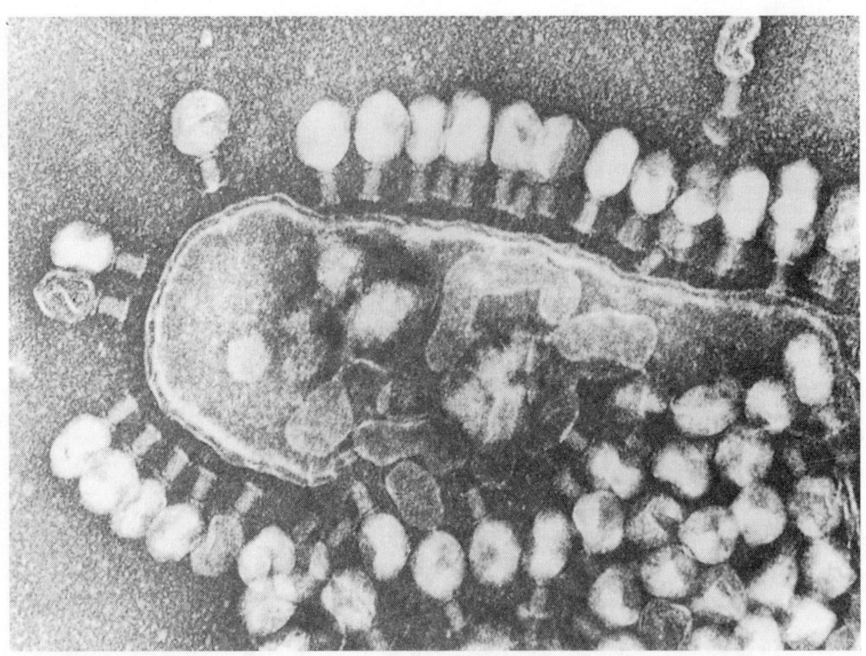

Bacteriophage T4 Infecting a Single *E. coli* Cell
The reproduction of bacterial viruses within *E. coli* provided a simple model to study the molecular basis of heredity. (Courtesy of Lee D. Simon, Photo Researchers, Inc.)

Out of the research on phages came final proof that DNA is the molecule of heredity. The blender experiment, performed by Alfred Hershey and his assistant Martha Chase in 1952, focused attention on DNA in a way that Avery's experiments had not. Their experiment, although regarded as biochemically sloppy, more clearly made the connection between DNA and heredity.

Martha Chase and Alfred Hershey, 1953
(Courtesy of Cold Spring Harbor Laboratory Archives.)

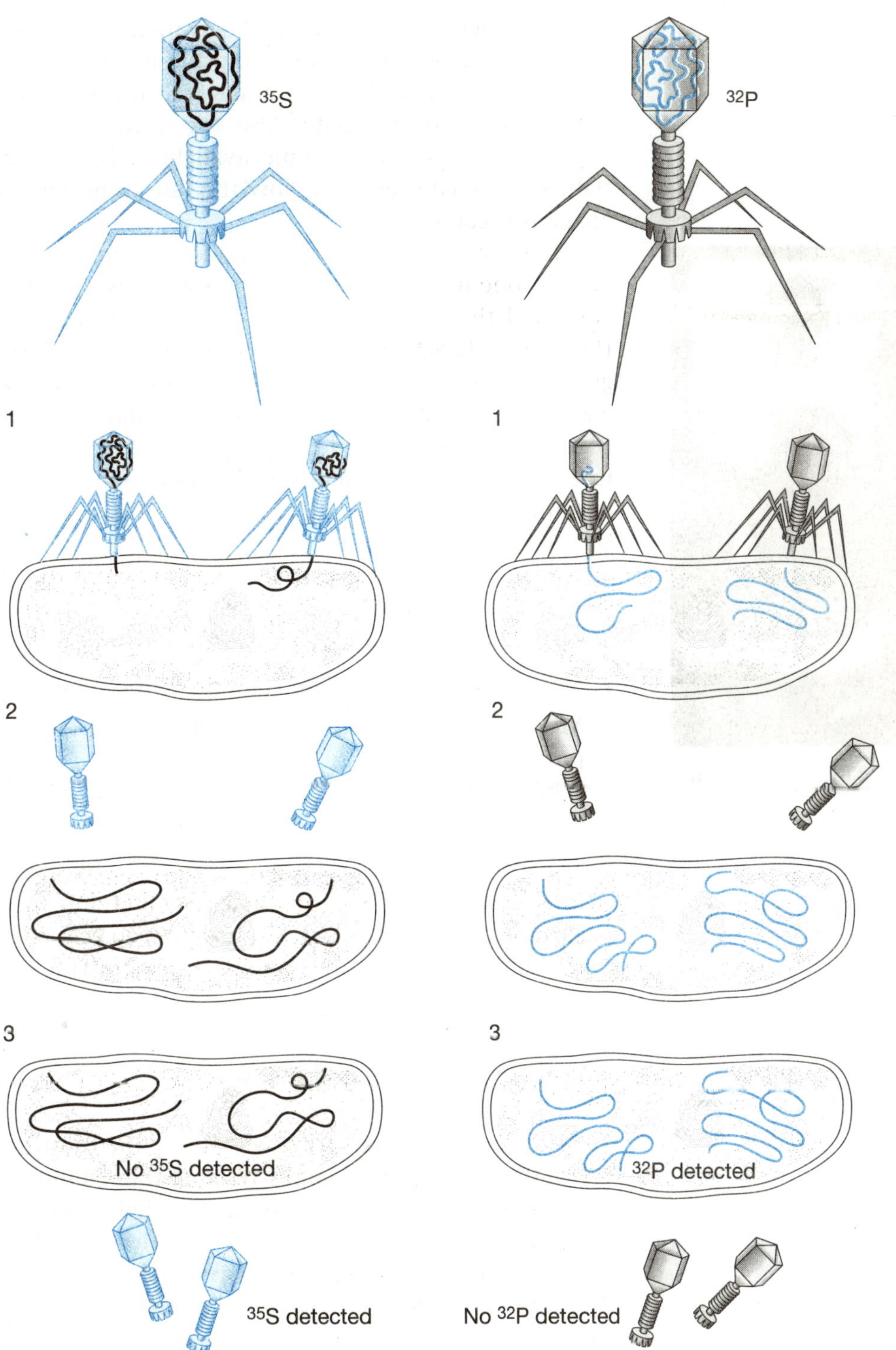

The Hershey-Chase Blender Experiment, 1952

Side-by-side experiments were performed with separate phage cultures in which either the protein capsule was labeled with radioactive sulfur (^{35}S) or the DNA core was labeled with radioactive phosphorus (^{32}P). (*1*) The radioactively labeled phages were allowed to infect bacteria. (*2*) Agitation in a blender dislodged phage particles from bacterial cells. (*3*) Centrifugation pelleted cells, separating them from the phage particles left in the supernatant. Radioactive sulfur was found predominantly in the supernatant. Radioactive phosphorus was found predominantly in the cell fraction, from which arises a new generation of infective phages.

The experimental design took advantage of the unique structure of the bacteriophage, in which an outer capsule of protein surrounds an inner core of DNA. No other organism yet discovered was so perfectly designed to settle the protein versus DNA debate. Also key to the experiment were radioactive isotopes, which had just become available following World War II. Radioactive "labels" allowed them to follow the fate of protein and DNA during bacteriophage replication.

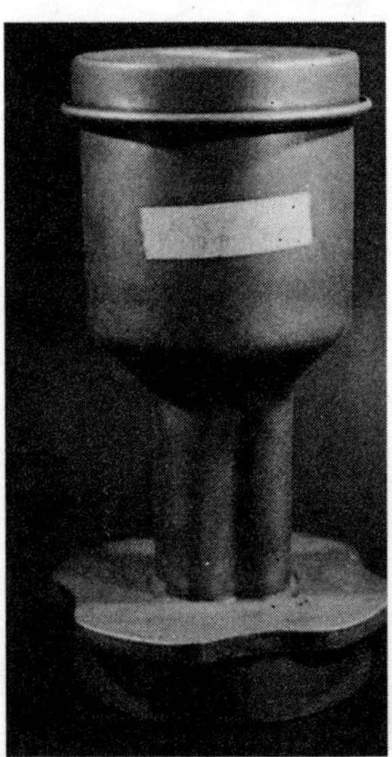
Waring Blender Used in the Hershey-Chase Experiment

Hershey and Chase ran parallel experiments with two populations of phage—one in which the protein capsules were labeled with radioactive sulfur (^{35}S), and the other in which DNA was labeled with radioactive phosphorus (^{32}P). The radioactively labeled phage were allowed to infect cultures of bacteria, giving enough time for the viruses to attach to the bacterial cells. The cultures were then chilled to arrest growth and agitated for several minutes in a Waring Blender, which detached the phage particles from the bacterial cells. The cultures were centrifuged at a speed fast enough to pellet the bacterial cells at the bottom of the tube, leaving the smaller phage particles in the supernatant. The cell pellet and supernatant were then analyzed for the presence of radioactively labeled phage protein or DNA. Hershey gave the following synopsis of the results in a later publication:

- Most of the phage DNA remains with the bacterial cells.
- Most of the phage protein is found in the supernatant fluid.
- Most of the initially infected bacteria (in the cell pellet) remain competent to produce phage.
- If the mechanical stirring is omitted, both protein and DNA sediment with the bacteria.
- The phage protein removed from the cells by stirring consists of more or less intact, empty phage coats, which may therefore be thought of as passive vehicles for the transport of DNA from cell to cell and which, having performed that task, play no further role in phage growth.

WHAT IS THE STRUCTURE OF THE DNA MOLECULE?

Solving the structure of DNA was surely the most important biological discovery of the 20th century. James Watson, trained in the Phage Group as a geneticist, and Francis Crick, a physicist schooled in X-ray crystallography, were the embodiment of the confluence of genetics and physics that led to a detailed understanding of the physical basis of heredity. In their 1953 letter to *Nature*, they assembled pieces of a chemical puzzle that had been accumulating for more than 80 years.

Ironically, DNA was discovered in 1869, only ten years after the publication of Darwin's *Origin of Species* and four years after Mendel's "Experiments in Plant-Hybridization." An enterprising German doctor, Friedrich Miescher, isolated a substance he called "nuclein," from the large nuclei of white blood cells. His source of cells was pus from soiled surgical bandages.

By 1900, the basic chemistry of nuclein had been worked out. It was known to be a long molecule composed of three distinct chemical subunits: a five-car-

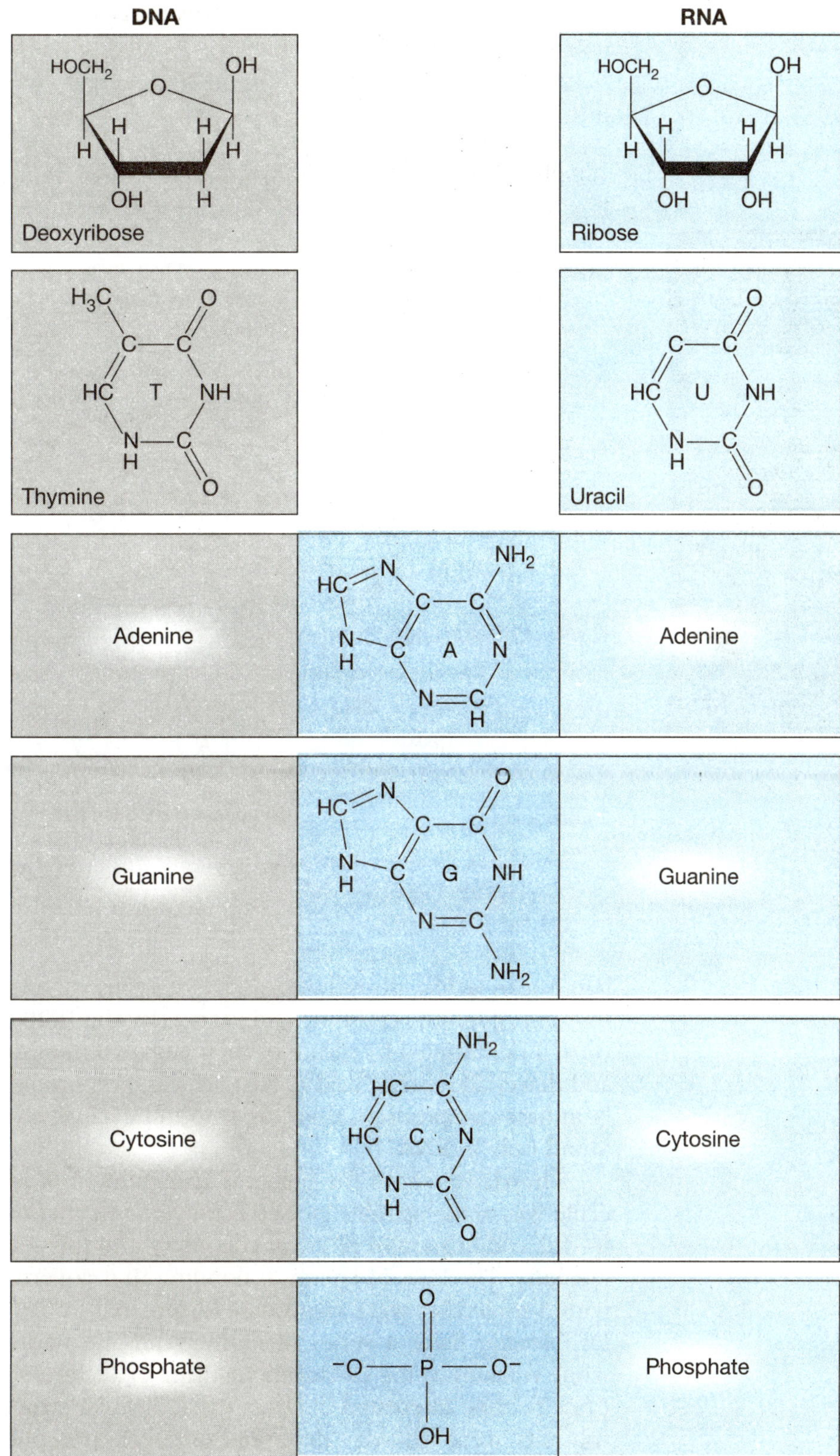

Components of DNA and RNA Molecules
(Art concept developed by Lisa Shoemaker.)

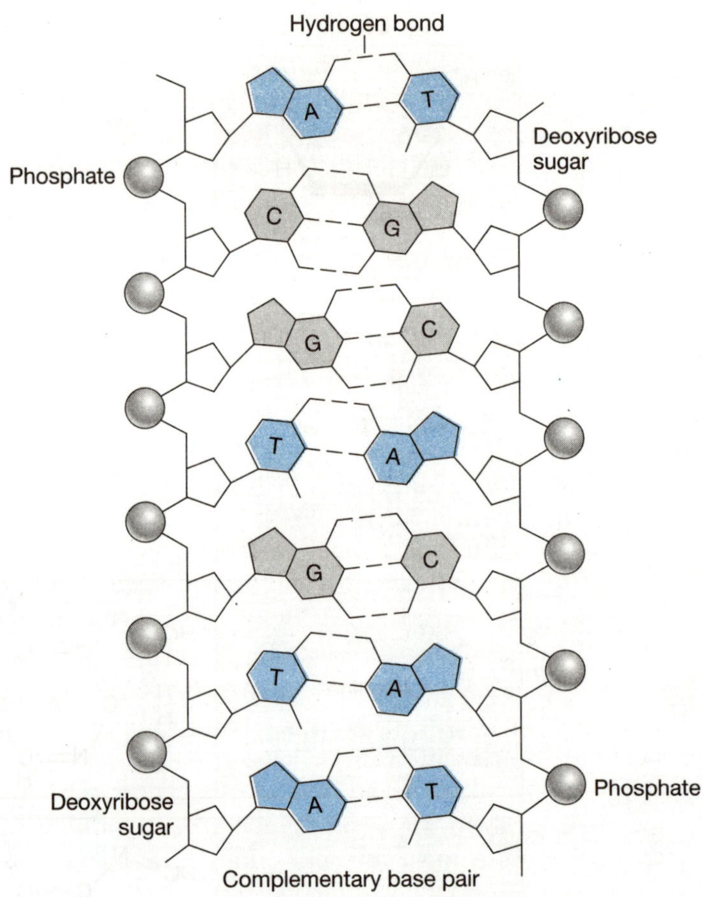

Chemical Structure of DNA

bon sugar, acidic phosphate, and five types of nitrogen-rich bases (adenine, thymine, guanine, cytosine, and uracil). By the 1920s, two forms of nucleic acids were differentiated by virtue of their sugar composition: ribonucleic acid (RNA) and deoxyribonucleic acid (DNA). These forms were also found to differ slightly in base composition; thymine is found exclusively in DNA, whereas uracil is found only in RNA.

Knowledge about the physical arrangement of atoms within the DNA molecule was made possible by two achievements that occurred prior to World War II—the codification of physical chemistry and the development of X-ray crystallography. The laws of chemical bonding that govern the arrangement of atoms within molecules were rigorously formulated by Linus Pauling in *The Nature of the Chemical Bond*, a series of monographs he wrote between 1928 and 1935 while working at the California Institute of Technology. By the time Watson and Crick became interested in DNA, the molecular structures of all of its individual subunits were known—deoxyribose sugar, phosphate, and each of the four nucleotides.

Determining the three-dimensional arrangement of the subunits within the huge DNA macromolecule was beyond the ability of Pauling's laws. This required the use of X-ray crystallography. In 1912, the German physicist Max von Laue discovered that X-rays are diffracted by the regularly arranged atoms of a simple crystal. In the same year, Australian physicist William Henry Bragg and his son, William Lawrence, worked out the mathematical equations needed to interpret diffraction patterns and used them to determine the molecular structure of table salt (sodium chloride). However, it was not until 1934 that John Desmond Bernal, at Cambridge University, obtained the first X-ray photograph of a biologically important molecule—the protein pepsin. This showed that the crystalline structures of giant organic molecules, including DNA, can be studied using X-ray diffraction.

The final three pieces of the DNA puzzle were uncovered only several years prior to the publication of the Watson-Crick *Nature* paper. In 1950, Erwin Chargaff of Columbia University discovered a consistent one-to-one ratio of adenine-to-thymine and guanine-to-cytosine in DNA samples from a variety of organisms. In 1951, Linus Pauling and R.B. Corey obtained precise atomic measurements of a helical polypeptide structure, the α-helix. At about the same time, Maurice Wilkins and Rosalind Franklin obtained sharp X-ray diffraction photographs of DNA. The diffraction patterns strongly suggested a helical molecule with a repeat of 34 angstroms (Å) and a width of 20 Å.

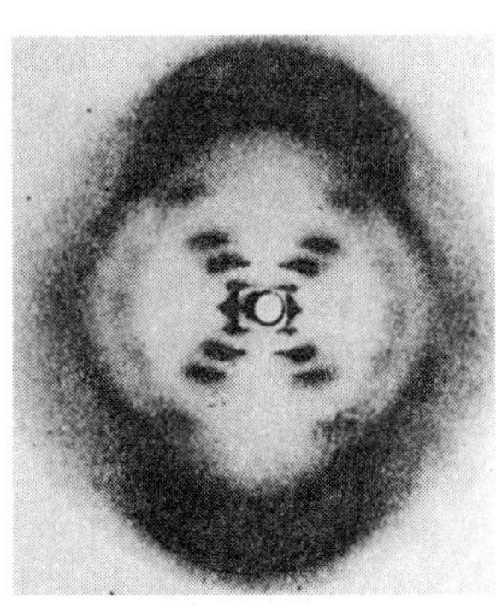

Rosalind Franklin's X-ray Diffraction Photograph of DNA, 1953
(Reprinted, with permission, from R.E. Franklin and R.G. Gosling 1953. Molecular configuration in sodium thymonucleate. *Nature 171:* 740–741; ©Macmillan Magazines, Ltd. Photo courtesy of Cold Spring Harbor Laboratory Archives.)

Watson and Crick were thus faced with the problem of trying to get the DNA subunits to fit in a structure that conformed to known biochemical data and the laws of physical chemistry, yet at the same time functioned as the carrier of heredity. The structure they arrived at, by manipulating paper and then metal models, was elegant in its simplicity. The DNA molecule they proposed is an α helix and resembles a gently twisted ladder. The rails of the ladder, which run in opposite directions (antiparallel), contain alternating units of deoxyribose sugar and phosphate. The planar nucleotides stack tightly on top of one another, forming the rungs of the helical ladder. Each rung is composed of a pair of nucleotides (a base pair) held together by relatively weak hydrogen bonds. Consistent with the 34-Å repeat calculated from the X-ray diffraction data, there are 10 base pairs per turn of the helix, with 3.4 Å between adjacent

Linus Pauling, ca. 1950
(Courtesy of the Archives, California Institute of Technology.)

Rosalind Franklin, 1948
(Courtesy of Anne Sayre.)

Erwin Chargaff, 1947
(Courtesy of Cold Spring Harbor Laboratory Archives.)

Maurice Wilkins, ca. 1955
(Courtesy of Cold Spring Harbor Laboratory Archives.)

base pairs. Of key importance is the complementary relationship between the nucleotides in each pair. In agreement with Chargaff's observation, adenine always pairs with thymine and cytosine always pairs with guanine. Thus, the nucleotide alphabet on one half of the DNA helix determines the alphabet of the other half.

Only weeks before Watson and Crick solved the structure, Linus Pauling published his own structure for DNA—a triple helix with the phosphates positioned on the inside of molecule. In constructing this model, to everyone's astonishment, Pauling had violated basic principles of physical chemistry! Packing the phosphates into the center of the molecule required that he use them in un-ionized form. This is incompatible with DNA functioning as an organic acid, where phosphates liberate hydrogen ions in solution. In their ionized state, the negatively charged phosphates would strongly repel one another, blowing the structure apart.

Why had the greatest physical chemist of his day, and perhaps all time, gotten the structure wrong? Why had Watson and Crick solved the structure ahead of Rosalind Franklin, even though her X-ray diffraction data provided key coordinates? The simplest answer may be that Watson, as the only biologist of the group, had the clearest insight of what a molecule must be to function as a gene. For the others, DNA was primarily an intellectual challenge—a large biomolecule whose structure needed to be solved. For Watson, DNA was the molecule of life.

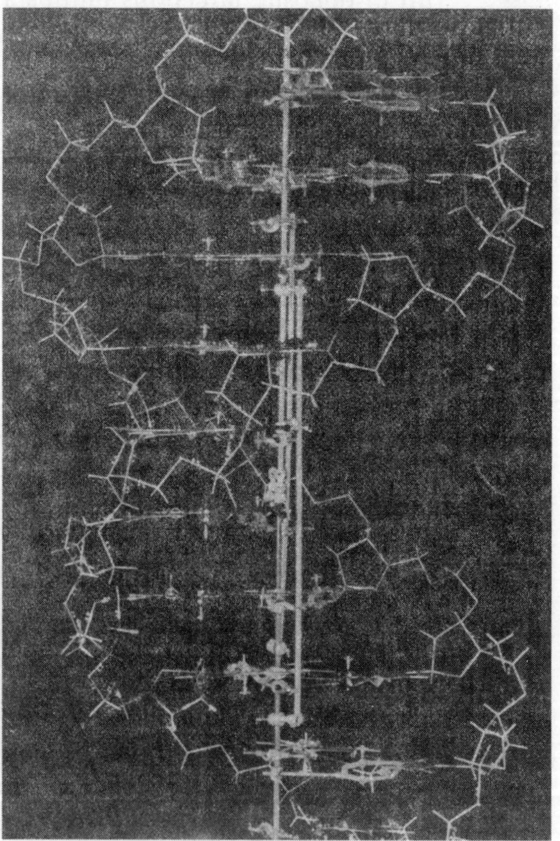

The 6-foot-tall Metal DNA Model Made by Watson and Crick in 1953
(Courtesy of the James D. Watson Special Collection. Cold Spring Harbor Laboratory Archives. From Watson J.D. 1968. *The Double Helix.* Atheneum Press, New York.)

REFERENCES

Garland A. 1978. *Thomas Hunt Morgan: The man and his science.* Princeton University Press, New Jersey.

Cairns J., Stent G.S., and Watson J.D., eds. 1992. *Phage and the origins of molecular biology,* expanded edition. Cold Spring Harbor Laboratory Press, Cold Spring Harbor, New York.

Carlson E.A. 1981. *Genes, radiation, and society: The life and work of H.J. Muller.* Cornell University Press, Ithaca, New York.

Comfort N. 2001. *The tangled field: Barbara McClintock's search for the patterns of genetic control.* Harvard University Press, Cambridge.

Crick F. 1988. *What mad pursuit.* Basic Books, New York.

Fischer E.P. and Lipson C. 1988. *Thinking about science: Max Delbrück and the origins of molecular biology.* W.W. Norton, New York.

Judson H.F. 1979. *The eighth day of creation.* Simon and Schuster, New York.

Korey K., ed. 1984. *The essential Darwin.* Little, Brown and Company, Boston.

McCarty M. 1985. *The transforming principle: Discovering that genes are made of DNA.* W.W. Norton, New York.

Micklos D., ed. 1999. DNA from the beginning (http://www.dnaftb.org). Dolan DNA Learning Center, Cold Spring Harbor, New York.

Perutz M. 1998. *I wish I'd made you angry earlier.* Cold Spring Harbor Laboratory Press, Cold Spring Harbor, New York.

Sayre A. 1975. *Rosalind Franklin and DNA.* W.W. Norton, New York.

Schrödinger E. 1992. *What is Life? The physical aspect of the living cell with mind and matter & autobiographical sketches,* reprint edition. Cambridge University Press, New York.

Sturtevant A.H. 2001. *A history of genetics.* Cold Spring Harbor Laboratory Press, Cold Spring Harbor, New York.

Watson J.D. 1980. *The double helix: A personal account of the discovery of the structure of DNA.* W.W. Norton, New York.

CHAPTER 2

How We Learned the Function of DNA

UNCOVERING THE STRUCTURE OF DNA ushered in an exciting and intense era of biological research. Knowledge of the structure gave obvious clues about how DNA functions as the genetic material. The genetic material must be replicated and passed down to each new cell and new generation, as well as serve as the instructions for protein synthesis. Each complementary strand could be the template for the synthesis of the other strand, which made it clear how each DNA molecule could be copied into two new DNA molecules and so on. But how do these processes work? During the next several years, these and other questions were answered by the ingenious studies of a group of talented scientists. Their discoveries opened up the field of molecular biology and ultimately led to the dawn of the biotechnology era.

HOW DOES DNA STRUCTURE DESCRIBE REPLICATION?

The Watson-Crick structure made it immediately apparent how DNA functions as life's information molecule. This "pretty molecule," as James Watson informally described DNA, embodied the organizing thesis of molecular biology—understanding the structure of a molecule gives clues to its biological function. The Watson Crick structure could neatly explain how the DNA molecule precisely replicates during the cell cycle, so that each daughter cell receives an identical set of hereditary instructions. The structure showed that the DNA molecule is made up of two strands in opposite orientation (5′ to 3′). Each base along the sugar phosphate backbone pairs with a base on the opposite strand, with the pairs held together by hydrogen bonds. Thus, the information in one DNA strand predicts the information in the other strand. To replicate this molecule, the hydrogen bonds holding the complementary nucleotides together could break, allowing the DNA ladder to unzip. Then each complementary half could serve as the template for synthesizing the other half molecule. The end result would be two identical DNA molecules, one to be passed on to each daughter cell.

Experimental support for this hypothesis came in 1958. Matthew Meselson and Frank Stahl, at the California Institute of Technology, devised an innovative experiment proving that DNA replicates by a semiconservative mechanism. One parental DNA strand is conserved in each new generation of DNA molecules, acting as the template for the synthesis of a new complementary strand. In other

Frank Stahl, ca. 1953 Matthew Meselson, 1958

Meselson and Stahl provided experimental evidence for the semiconservative replication of DNA. (Courtesy of Cold Spring Harbor Laboratory Archives; California Institute of Technology Archives.)

words, each newly synthesized DNA molecule is composed of one "old" (conserved) strand and one "new" strand.

Their experiment relied on the ability to differentiate the densities of two isotopes of nitrogen, "heavy" ^{15}N and "light" ^{14}N. *Escherichia coli* bacteria were first grown for several generations in a nutrient medium where ^{15}N was the only source of nitrogen. This heavy nitrogen was incorporated into the nucleotides of newly synthesized bacterial DNA. After 14 generations, the bacteria were abruptly switched into media containing only ^{14}N and allowed to grow for two generations. During this time, newly synthesized DNA would incorporate the light nitrogen.

Samples of cells were taken from the generation before the switch to light nitrogen (1) and from the two generations following the switch (2 and 3). DNA extracted from the cells was added to a solution containing cesium chloride, which forms a density gradient when centrifuged for many hours at high speed. DNA molecules of different densities settle during the centrifugation to form discrete bands. Each DNA band "floats" at a point where its density exactly equals that of the cesium chloride gradient.

The results were picture perfect. The gradient of DNA from generation 1 contained a single high-density band—all the DNA molecules from the parent generation contained two strands made of heavy nitrogen ($^{15}N/^{15}N$). The gradient of DNA from generation 2 contained a single band of intermediate density. This indicated that each daughter DNA molecule was composed of one heavy strand (from the parent DNA) and one complementary strand made of light nitrogen. The gradient of DNA from generation 3 contained two bands, one of intermediate density (in the same position as generation 2) and one of light density. DNA in the intermediate-density band was composed of one heavy strand from generation 2 and a new, light complementary strand. DNA in the light-density band was composed of two light strands, one inherited from generation 2 and one new complementary strand. This study proved that replication occurs by a semiconservative mechanism.

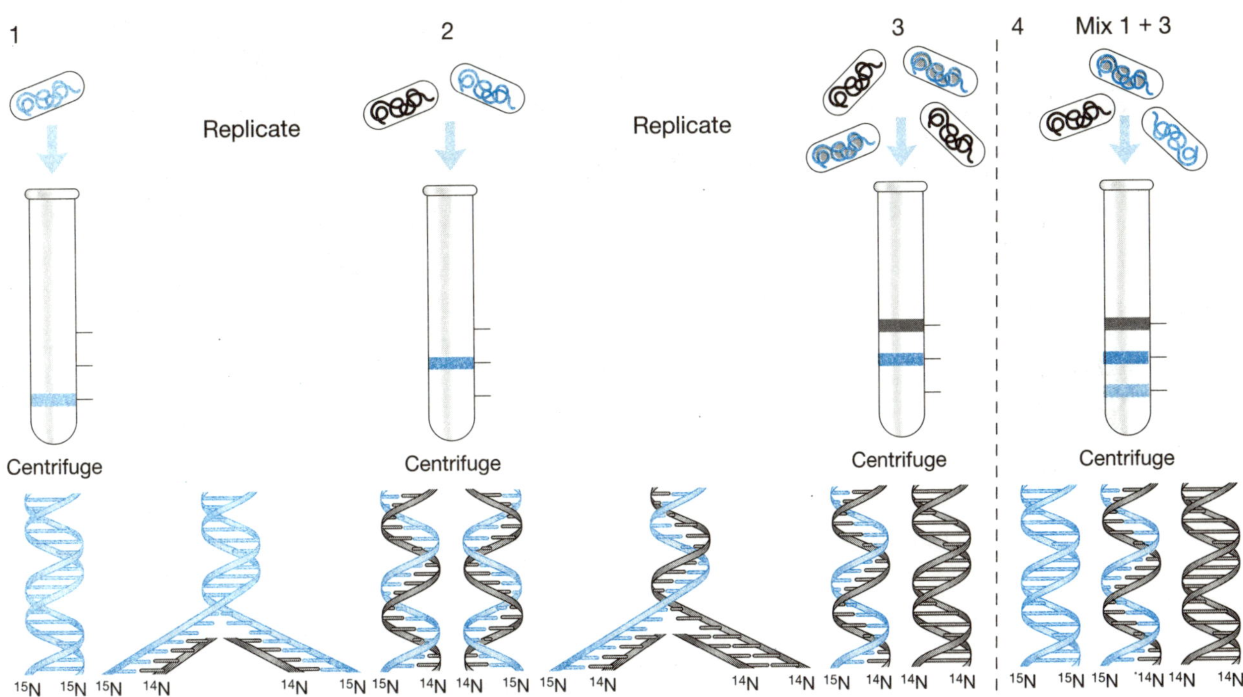

The Meselson-Stahl Experiment Showing Semiconservative Replication of DNA, 1958
(*1*) Following density-gradient centrifugation, DNA isolated from bacteria grown for many generations in a medium containing heavy nitrogen (^{15}N) forms a single band of high density. (*2*) DNA isolated from the generation following a switch to ^{14}N-containing medium shows a single, medium-density band. (*3*) In the subsequent generation, the DNA segregates into a medium-density and a low-density band. (*4*) Mixing DNA from cultures 1 and 3 compares the three density bands.

HOW DNA IS COPIED OVER AND OVER

Arthur Kornberg, ca. 1965 (Courtesy of Stanford University.)

Our understanding of the mechanism of DNA replication began with biochemical studies, using purified or partially purified cellular components in test tube reactions (in vitro or in glass). The work began in 1960 when Arthur Kornberg, from Washington University, published his work on the purification of an enzyme from *E. coli* that was capable of synthesizing DNA. Kornberg had isolated DNA polymerase I (pol I). In this initial study, he demonstrated that in vitro DNA synthesis required only (1) polymerase I, (2) all four deoxynucleotide triphosphates (dNTPs), (3) a double-stranded DNA template, and (4) Mg^{++}.

As it turned out, DNA polymerase I is not the primary DNA polymerase involved in DNA synthesis. John Cairns of Cold Spring Harbor Laboratory found that *E. coli* with a mutation in the gene that codes for polymerase I are sick but still able to replicate their DNA. It was subsequently found that another polymerase, DNA polymerase III (pol III), is the major DNA polymerase in *E. coli* and that DNA polymerase I has a somewhat secondary role.

Why DNA Replication Always Adds Nucleotides onto the 3′OH of the Growing Strand

If we examine the structure of DNA, we find that each deoxyribose sugar is composed of five carbons, labeled 1′ to 5′. The phosphates form covalent bonds

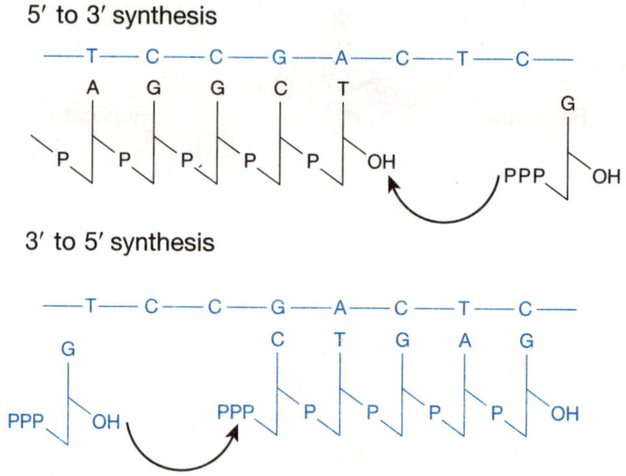

5′ to 3′ synthesis

—T—C—C—G—A—C—T—C—

A G G C T G

P P P P P OH PPP OH

3′ to 5′ synthesis

—T—C—C—G—A—C—T—C—

G C T G A G

PPP OH PPP P P P P OH

DNA Synthesis Is Always 5´ to 3´

(*Top*) DNA synthesis is depicted, in which the incoming nucleotide triphosphate is added to the 3´OH of the existing DNA strand. This is 5´ to 3´ synthesis. (*Bottom*) In 3´ to 5´ synthesis, a nucleotide triphosphate is added to the 5´ triphosphate of the existing DNA strand. In both processes, the cleavage of the triphosphate to a monophosphate produces the energy necessary to form the phosphodiester bond. However, despite the fact that synthesis in either direction is energetically feasible, 3´ to 5´ synthesis is never seen in nature.

between the 3´ and 5´ carbons of adjacent sugars, linking them together. Because two adjacent sugars are connected via oxygens in a single phosphate to form esters, this is termed phosphodiester linkage. In every organism ever studied, the incoming nucleotide is always added onto the 3´ carbon during DNA synthesis. Thus, we refer to DNA synthesis as occurring 5´ to 3´. Although it is theoretically possible to add the nucleotide onto the 5´ carbon and synthesize DNA 3´ to 5´, this does not happen. The reason seems to be that during DNA synthesis, the DNA polymerase makes a mistake about every 10,000 base pairs. The DNA polymerases have "editing" activity, which means that they can retrace to the place the wrong nucleotide was added, remove it, and replace it with the correct one. The energy for putting in nucleotides comes from breaking the phosphodiester bond of the entering nucleotide triphosphate. If synthesis were 3´ to 5´, during DNA editing, the polymerase would have no triphosphate bond to break to create the energy necessary to form a new bond. Evolution has thus solved this potential problem by always adding new nucleotides onto the hydroxy (OH) of the 3´ carbon.

If synthesis is always 5´ to 3´ and DNA is made up of antiparallel strands, running is opposite directions, then how do both strands get replicated? One possibility is to start at one end, copy one strand, and then start at the other end and copy the other strand. But studies have shown that replication of both strands occurs simultaneously and in only one direction. A resolution to the mystery came when Reiji Okazaki at Nagoya University discovered that small, single-stranded DNA fragments form during replication. Okazaki proposed that one strand of the DNA (the "leading" strand) is synthesized continuously in the 5´ to 3´ direction. Synthesis on the second, or "lagging" strand, also occurs in the 5´ to 3´ direction, as required, but in small, discontinuous stretches. These stretches, or Okazaki fragments, are later joined together into a continuous strand.

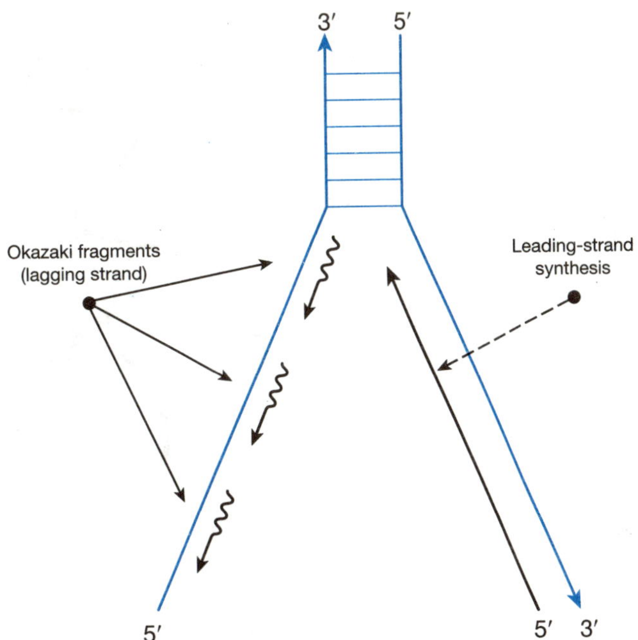

Leading-strand and Lagging-strand Synthesis
For double-stranded DNA to be replicated, the two strands must be locally separated to allow access by DNA polymerase. DNA synthesis of the strand oriented with its 5′ end toward the branch point is synthesized continuously; this is called *leading-strand synthesis*. However, synthesis of the opposite strand occurs as a discontinuous process. A primase adds RNA primers every few hundred nucleotides. DNA polymerase uses these primers to synthesize back to the next primer, generating short DNA segments called Okazaki fragments. RNase H removes the RNA primer, and DNA polymerase fills in the resulting gap with nucleotides. The Okazaki fragments are then joined together. This discontinuous process of DNA synthesis is called *lagging-strand synthesis*.

DNA Polymerases Need Primers and Other Components

It became evident that DNA polymerase can only add nucleotides onto the 3′ end of an existing DNA fragment. This would seem to create a "Catch-22." If one always needs a preexisting piece of DNA to add onto, where does the first piece come from? The answer is that DNA synthesis is initiated when the enzyme primase synthesizes a small piece of RNA, called an RNA primer. Primase is an RNA polymerase, and, like all RNA polymerases, it does not need a primer to initiate synthesis. Another enzyme called RNase H later removes the RNA primer made by primase, which is then replaced with DNA. RNase H specifically degrades RNA that is hybridized to the DNA, and DNA polymerase I fills in the gap with DNA (explaining the role of DNA polymerase I).

Several other components are required for DNA synthesis. For example, for the polymerase to gain access to the DNA, the helix must be opened up and unwound. This is done by an enzyme called DNA helicase, which uses ATP for the energy to carry out this task. But the helicase itself creates a problem. As it unwinds the helix, it creates overwound DNA and eventually knots up the molecule. (To understand this concept, twist two pieces of string into a helix. Hold each end and have someone put a finger between the strands. As the person moves a finger toward one end, the strands become looser behind and tighter

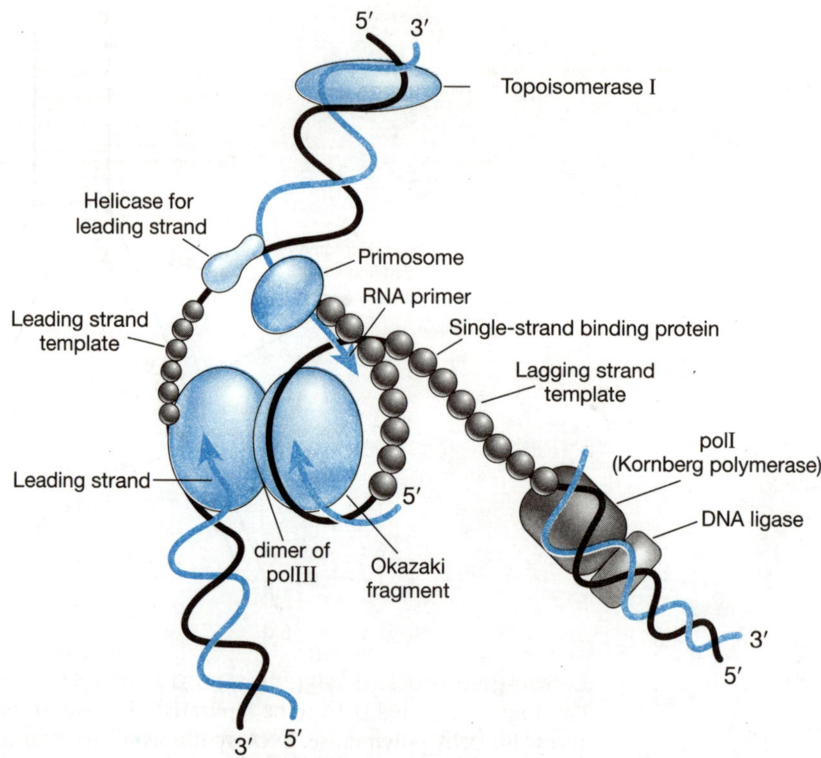

Components of DNA Synthesis

ahead of the finger. This is exactly what happens when DNA is unwound by a DNA helicase.) To deal with this problem, another enzyme, called topoisomerase I, cuts one strand of the overwound DNA, allows it to unwind by one turn, and then reseals the ends. It does this repeatedly to relieve the stress on the DNA created by the helicase.

During replication, the DNA must be unzipped in order to be copied. However, base pairing drives the single strands of DNA to re-form double-stranded DNA. A protein that specifically binds to single-stranded DNA, called single-strand-binding protein (SSB), blocks this process. Finally, in lagging-strand synthesis, the Okazaki fragments must be joined together. This is accomplished by a joining enzyme called ligase.

DNA Synthesis Is Initiated at a Specific Sequence Called an Origin of Replication

Various studies led to the realization that DNA replication originates at a specific sequence called an origin of replication (*ori*). In a simple organism, there may be a single *ori*; most organisms have many origins of replication where DNA synthesis is initiated. Studies in eukaryotes have shown that protein complexes form at the *ori* and that these proteins interact with proteins of the replication machinery. In *E. coli*, the single *ori* is called *oriC*.

HOW ARE THE ENDS OF CHROMOSOMES REPLICATED?

In eukaryotes, replication is even more complicated, but the basic process is the same. The major difference is that the bacterial genome is a circle, whereas the

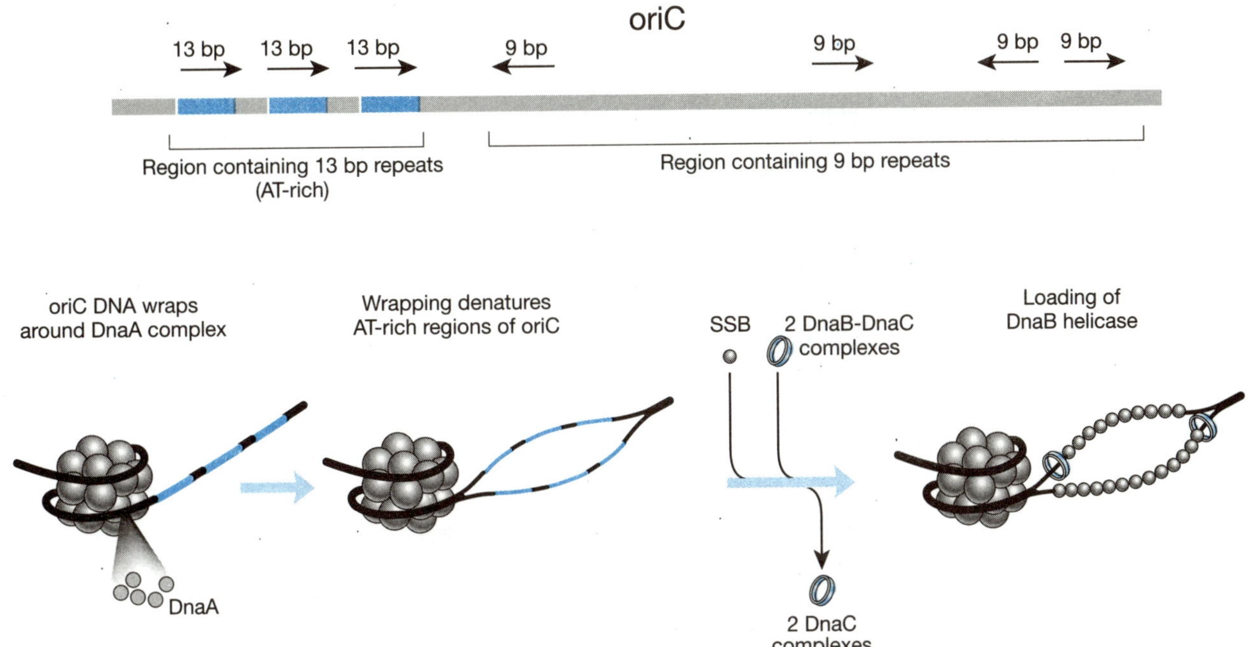

Initiation of DNA Replication at OriC

The *oriC* in *E. coli* is composed of two regions, one containing four repeats of 9 bp and the other containing three repeats of 13 AT-rich base pairs. As the cell grows and nears the time for cell division, the level of DnaA protein accumulates until it is high enough to form a multiprotein complex that binds to *oriC*. This binding leads to unwinding of the adjacent AT-rich repeat region, creating single strands. Single-strand-binding protein (SSB) blocks the strands from reforming double-stranded DNA. Simultaneously, two DnaB-DnaC complexes enter, and DnaC is released from DnaB, activating its helicase activity. Next, two DnaB-DnaC complexes bind to the single-stranded DNA. DnaC, which only helps load DnaB, is then released. Loading of DnaB activates its DNA helicase activity (helicases unwind double-stranded DNA). DnaB tracks ahead of the DNA polymerase, creating single-stranded DNA to be replicated. DNA synthesis now progresses in both directions.

eukaryotic chromosome is linear. The ends of the chromosome are called the telomere and contain many repeat sequences. These ends create a problem during DNA synthesis because there is no way to create a primer to complete synthesis of the lagging strand. A special DNA polymerase called telomerase solves this problem. Telomerase is a type of reverse transcriptase, which means that it makes DNA using RNA as a template. Telomerase carries an RNA molecule that is able to base pair with the unreplicated end of the chromosome, creating a template for the telomerase. Telomerase copies the RNA into DNA and then repeats the process over and over, creating the repeat sequences at the ends of chromosomes. However, telomerase is only found in germ cells, and so the repeat sequences of the telomeres in somatic cells shorten every time the cell replicates its DNA. This shortening is thought to be a kind of biological clock, as shortening telomeres are associated with aging and cancer.

HOW DOES INFORMATION GET FROM DNA TO PROTEINS?

Replication was the simple half of the DNA structure-function question. The more difficult question was how the DNA molecule—a very large molecule but one composed of only four different nucleotides (adenine, cytosine, guanine, and thymine)—could function as the template for the synthesis of so many dif-

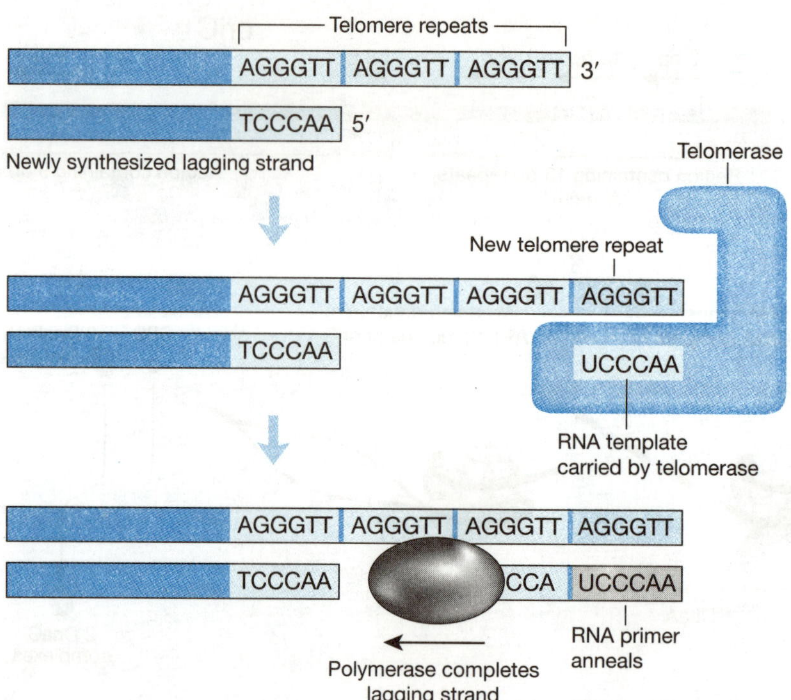

Model of Telomerase Function

In linear DNA, a problem arises in synthesizing the very end of the DNA strand: There is no primer to complete the lagging-strand synthesis. Telomerase solves this problem by adding new telomere repeat sequences to the end of the leading strand, using an RNA it carries as a primer. A different RNA primer then anneals to this added sequence and acts as a primer to complete the lagging-strand synthesis.

ferent proteins? Are the nucleotides a type of code for protein synthesis? How is the code read? It was known that protein synthesis occurs in association with ribosomes, but how does this process work?

In the beginning, much of what was determined about DNA and molecular biology was discovered from work in bacteria and bacteriophages. However, the realization that proteins cannot be made directly from DNA came from the study of eukaryotes. It was recognized that in eukaryotes, DNA resides within the nucleus, whereas protein synthesis occurs in the cytoplasm (bacteria of course do not have a nucleus). It was reasoned that genetic information must be transported from the DNA in the nucleus to the cytoplasm through some intermediate.

Paul Zamecnik
(Courtesy of the National Library of Medicine.)

In the mid 1950s, work primarily in the laboratory of Paul Zamecnik established that protein synthesis in rat liver cells takes place in the cytoplasm, on large molecules composed of RNA and protein (ribonucleoproteins) called microsomes. Today we call them ribosomes. Zamecnik performed most of these studies using cell-free extracts, which are made by breaking open cells and separating the soluble cytoplasmic components from the nuclear and nonsoluble materials, such as the cell membrane and mitochondria. His studies with these soluble extracts established the ribosome as the major cellular component required for protein synthesis. From these and other studies developed the one gene–one ribosome–one protein hypothesis stating that each gene specifies the synthesis of one ribosome, which then produces a particular protein.

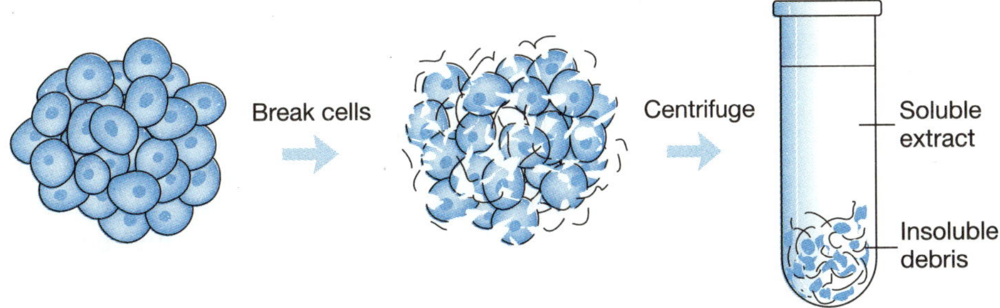

Cell-free Extract
Paul Zamecnik made a cell-free extract proficient for protein synthesis by breaking open cells and removing the insoluble debris by centrifugation. This is much like the process by which cell-free extracts are made today.

By 1960, many scientists began to doubt this model, and attention turned to the existence of some information molecule that would carry the instructions for protein synthesis out of the nucleus to the ribosome. At this time, Benjamin Hall and Sol Spiegelman at the University of Illinois carried out an experiment demonstrating that the DNA sequence is complementary to the RNA sequence. Hall and Spiegelman infected bacteria with bacteriophage T2. They then isolated T2 RNA and DNA from the infected bacteria cells. They denatured the T2 DNA into single strands by heating, mixed it with T2 RNA in 300 mM NaCl, and then slowly cooled the DNA/RNA mixture. The sample was centrifuged on a CsCl density gradient, along with single-stranded RNA and double-stranded DNA. DNA is less dense than RNA, so it settles near the top of the gradient, whereas RNA is denser and settles near the bottom of the gradient. They also found a band of an intermediate density formed by DNA-RNA heteroduplexes. The DNA-RNA hybrid formed because of base pairing between the T2 DNA and its RNA, demonstrating that these molecules are complementary. This study suggested that RNA is an exact copy of the DNA sequence.

At the same time, two studies that would finally dispell the one gene–one ribosome hypothesis were being conducted by Sydney Brenner, François Jacob, and Matthew Meselson and independently by François Gros, James Watson, and

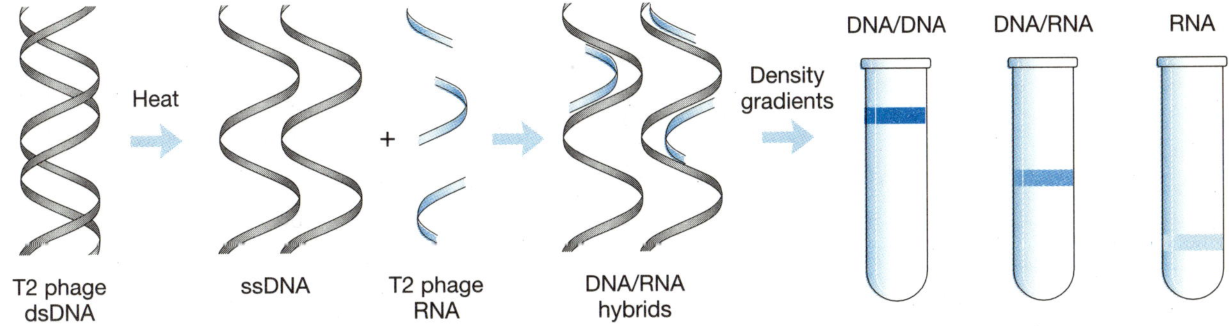

RNA Is a Direct Copy of DNA
In the studies of Benjamin Hall and Sol Spiegelman, T2 phage double-stranded DNA (dsDNA) was isolated and boiled to denature it to single-stranded DNA (ssDNA). The ssDNA was mixed with T2 phage RNA and allowed to hybridize. On CsCl gradients, the DNA-RNA mixture equilibrated to a location that showed its density was intermediate between that of DNA and RNA. Thus, Hall and Spiegelman demonstrated that the DNA and RNA from T2 phage form hybrid molecules and that RNA is copied from DNA.

Walter Gilbert. They demonstrated that each gene does not specify the synthesis of a ribosome, but rather DNA is the template for the synthesis of a messenger RNA (mRNA) molecule that carries the information from the nucleus to the ribosomes. Brenner, Jacob, and Meselson used T4 phage-infected *E. coli* as a model system. It was known that within minutes after T4 infection, *E. coli* synthesize only T4-specific RNA and proteins. These investigators reasoned that any new synthesis of RNA after infection would be specific for T4 phage. They first grew the bacteria for many generations in media containing the "heavy" isotopes ^{15}N and ^{13}C, which labeled all of the cell components. Upon infection with T4 phage, they then switched the cells to "light" media, containing ^{14}N and ^{12}C, so that all ribosomes synthesized after infection were labeled with the light isotopes. They isolated ribosomes from the infected cells and separated them on CsCl gradients. Ribosomes migrate in two peaks on CsCl gradients—a large 70S peak made up of the complete ribosome and a smaller peak made up of the 30S and 50S subunits which together are precursors of the 70S ribosome. When they analyzed the gradients, they found only heavy-isotope-labeled 70S and 30/50S peaks. There was no evidence of newly synthesized, light-isotope-labeled ribosomes.

A second experiment demonstrated that new RNA is synthesized following T4 infection. The researchers added radioactive uracil (^{14}C-labeled uracil) to the light media given just after phage infection to label newly synthesized RNA. (Remember that RNA uses uracil instead of thymine and thus ^{14}C-labeled uracil is specifically incorporated into RNA.) As in the previous experiment, ribosomes were isolated on a CsCl gradient. They found that newly synthesized RNA associated with the 70S heavy ribosome peak of the complete ribosome, where protein synthesis occurs.

A "pulse chase" experiment showed that this newly synthesized RNA is very unstable and rapidly degraded, unlike ribosomes, which are stable for many cell generations. ^{14}C-labeled uracil was added to the light media for a short time (the "pulse") just at the time of phage infection to make highly labeled, or "hot," RNA. Then, a hundredfold excess of unlabeled uracil was added (the "chase") so that RNA synthesized afterward would be unlabeled, or barely labeled ("cold").

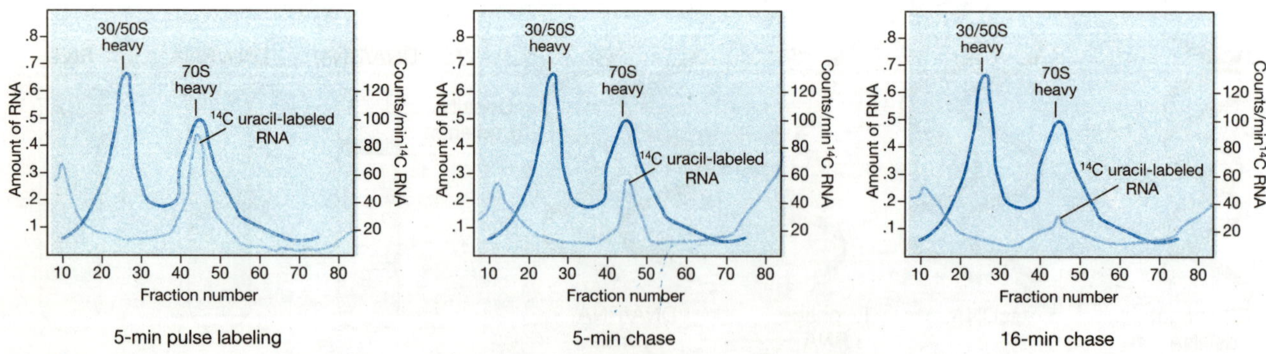

Newly Synthesized mRNA Associates with the 70S Ribosome and Is Short Lived

In this "pulse-chase" experiment, ^{14}C-labeled uracil was briefly added (the "pulse") to a culture of *E. coli* immediately after T4 infection. ^{14}C is a radioactive isotope that can be detected in a scintillation counter. The newly synthesized RNA immediately associates with the 70S ribosome. Upon addition of 100-fold excess of unlabeled uracil to dilute out the labeled uracil (the "chase"), the labeled RNA rapidly disappears from the 70S ribosome. This demonstrated that a short-lived RNA molecule is synthesized from the T4 DNA, sending its information or message to the ribosome for protein synthesis.

After the initial pulse, a very hot peak of ^{14}C-labeled RNA associates with the 70S ribosome. However, RNA samples taken after the addition of cold uracil were found to be progressively less hot, and by 16 minutes after the chase, almost no labeled RNA was detected. Thus, the hot RNA had turned over and was rapidly replaced by unlabeled RNA. This experiment demonstrated that mRNA exists for a very short time within the cell—on the order of a few minutes.

HOW DO CELLS MAKE RNA?

In 1955, Marianne Grunberg-Manago and Severo Ochoa at the University of Illinois described their finding of an enzyme that catalyzes the synthesis of RNA. The enzyme, however, had some properties that seemed to argue against it being RNA polymerase, the enzyme responsible for *transcribing* information from DNA into RNA. This was primarily because the enzyme could link rNTPs into poly rNMPs in the absence of the DNA template. Some time later, it was discovered that Grunberg-Manago and Ochoa had actually isolated polynucleotide phosphorylase, which usually catalyzes the breakdown of RNA, not its synthesis.

In 1960, Jerard Hurwitz and Samual Weiss independently identified the first "true" RNA polymerase. This enzyme requires a DNA template and adds nucleotides to make an RNA. RNA polymerase from *E. coli* has been shown to be composed of five proteins, with a total molecular mass of nearly 500,000 daltons.

In eukaryotes, there are three different RNA polymerases (I, II, and III), each of which synthesizes a different type of RNA molecule. RNA polymerase I makes ribosomal RNA (rRNA), RNA polymerase II synthesizes messenger RNA (mRNA), and RNA polymerase III synthesizes small RNA molecules, including transfer RNA (tRNA) and one rRNA called 5S RNA.

How does RNA polymerase know where to start and stop transcribing DNA into RNA? How does it know where a gene begins and ends? These and other questions about the regulation of transcription are answered in Chapter 3.

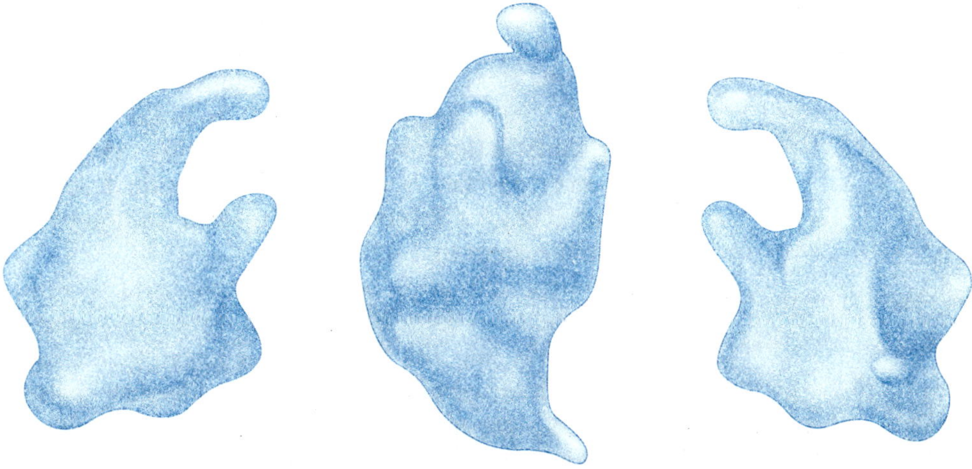

Three Views of *E. coli* RNA Polymerase Holoenzyme
(Redrawn, with permission, from Darst et al. 1989.)

PROTEINS ARE LINEAR POLYMERS MADE UP OF 20 DIFFERENT AMINO ACIDS

Theodor Schwann
(Courtesy of the National Library of Medicine.)

Our understanding of protein structure and function began long before we understood DNA. Interest in proteins probably stemmed from studies of cells and tissues using microscopes. From the beginning of cell biology, scientists were reductionist in their thinking. They were constantly trying to reduce living things into more and more basic forms. This approach to biology continues today. The microscope allowed scientists to deduce that all organisms are made up of basic building blocks called cells. However, cells were not the simplest structures; microscopy revealed that within cells are sub-cellular components (organelles such as mitochondria). Clearly, organelles could be reduced further into the components which make them up, but here microscopists hit a wall. They could not see anything smaller than an organelle. Thus, it was reasoned that cells and tissues needed to be taken apart to uncover the smaller structures that compose them. The histologist Theodor Schwann, who developed the "cell theory" stating that cells are the basic component of all living things, was also the person who identified the first enzyme, pepsin. Schwann isolated pepsin from an animal stomach and showed that it is responsible for the digestion of food. What he did not know at the time is that pepsin specifically digests protein.

The isolation of pepsin spawned a fertile period of biochemical research during which many biological processes mediated by enzymes were studied. Other proteins, such as albumin, which is abundant in blood, were isolated and characterized. In 1839, Gerhardus Johannes Mulder analyzed several purified proteins and determined their chemical components to be primarily carbon, nitrogen, hydrogen, oxygen, and trace amounts of sulfur and phosphate. Other studies on processes such as fermentation in yeast led to the development of various methods of purifying enzymes. With purified enzymes available, more detailed studies of the enzymes were possible, and it was soon recognized that proteins are composed of simpler molecules, amino acids. As the name implies, amino acid structure consists of an amino group (NH_2) and an acidic carboxyl group (COOH). It was also discovered that proteins could be broken down into amino acids by mild heating in dilute acid. In 1820, the French chemist Henri Braconnot used this method to identify the first amino acid glycine in gelatin. During the next 90 years, more amino acids were discovered. It was not until the early 1900s that all 20 different naturally occurring amino acids were known and their structures worked out. All amino acids were found to share a common core structure and a side group (also called an R group) that gives each its unique character. Proteins thus have the combined properties of the amino acids that make them up.

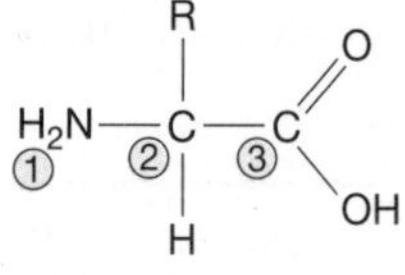

Primary Structure of Amino Acids
(*1*) Amino group; (*2*) α-carbon; (*3*) carboxyl group. R is the variable structure that is unique to each amino acid.

Some amino acids, such as alanine, contain only hydrocarbon (CH) groups within the side chain. Hydrocarbon is not very reactive nor does it like water, so alanine falls into the nonpolar or hydrophobic category of amino acids. The polar or hydrophilic amino acids have side chains that interact with water. Within this group are amino acids with reactive side chains, meaning they can form covalent bonds with other molecules. These are called the charged amino acids. Cysteine has the additional property of forming covalent bonds with other cysteine residues. The side group of this amino acid contains an SH group and the bond formed between two cysteines is a disulfide bond (S–S). Disulfide

$^{-}OOC\text{-}C(H)(NH_3^+)\text{-}CH_2\text{-}SH + HS\text{-}CH_2\text{-}C(H)(NH_3^+)\text{-}COO^-$

$^{-}OOC\text{-}C(H)(NH_3^+)\text{-}CH_2\text{-}S = S\text{-}CH_2\text{-}C(H)(NH_3^+)\text{-}COO^-$

Formation of a Disulfide Bond between Cysteines

Polar R Groups (Hydrophilic)

Glycine (Gly or G)

Serine (Ser or S)

Threonine (Thr or T)

Cysteine (Cys or C)

Tyrosine (Tyr or Y)

Asparagine (Asn or N)

Glutamine (Gln or Q)

Basic

Lysine (Lys or K)

Arginine (Arg or R)

Histidine (His or H)

Acidic

Aspartic acid (Asp or D)

Glutamic acid (Glu or E)

Nonpolar R Groups (Hydrophobic)

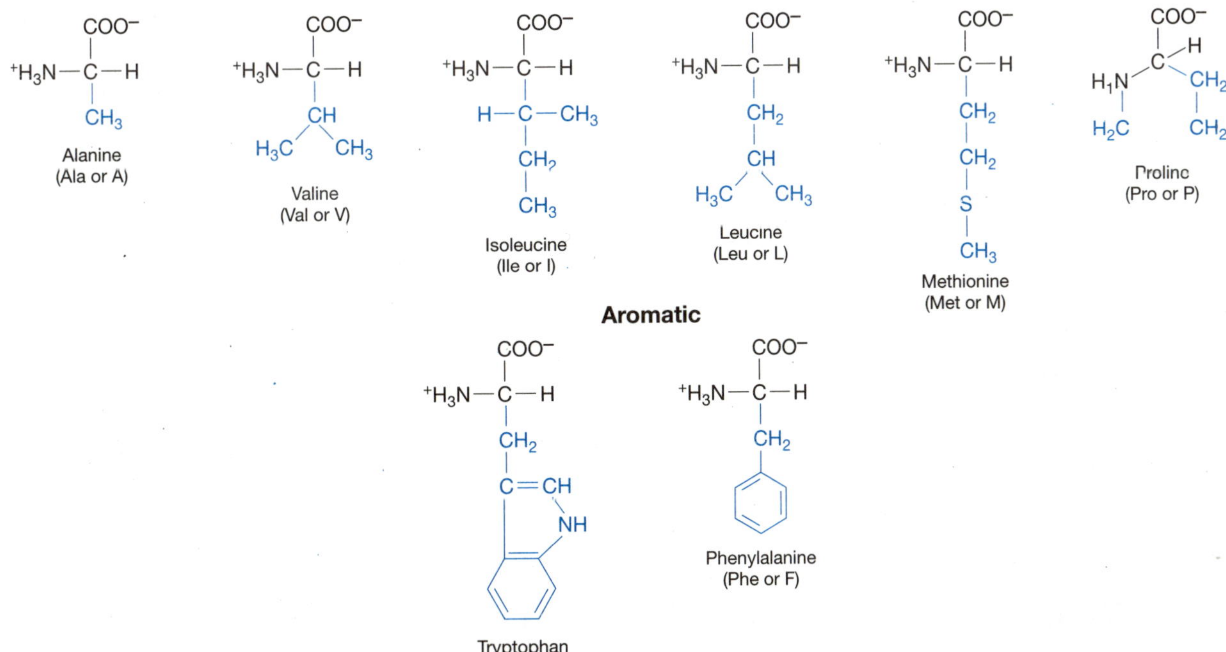

Structures of the 20 Amino Acids from which Proteins Are Synthesized
The side chains (*blue*) determine the characteristic properties of each amino acid.

Peptide Bond

The peptide bond links amino acids into a polypeptide or protein. The carboxyl group of one amino acid joins to the amino group of a new amino acid on the growing peptide chain. One molecule of water is lost, making this a condensation reaction.

bonds are often vital to maintaining a protein's three-dimensional structure or its linkage to other proteins.

In 1901, the biochemist Hermann Emil Fischer discovered how amino acids are linked together in a linear chain to make peptides and proteins. The amino group links to the carboxyl group, forming a peptide bond. Proteins are therefore linear molecules made up of amino acids linked by peptide bonds. Proteins always have an NH_2 group at one end (called the N or amino terminus) and a COOH at the other (called the C or carboxyl terminus). Indeed, it makes sense for proteins to be linear molecules, since the information for their synthesis is included in a linear DNA molecule.

H. Emil Fischer, ca. 1904 (Courtesy of the National Library of Medicine.)

Proteins became a central focus of biochemical study, and even today, we consider proteins the most important molecules of life. Proteins can be structural as in the case of collagen, which gives strength to bone and tendons. Other proteins are enzymes, which catalyze biochemical reactions. Enzymes are generally very specific in that they only work on a particular substrate. For example, the enzyme hexokinase binds to one molecule of the sugar glucose and adds a single molecule of phosphate (PO_4) to it. Hexokinase will not work on any other sugar, even sugars that are nearly identical to glucose. This specificity of an enzyme binding only to its specific substrate is referred to as the "lock and key" model, proposed by Fischer in 1894.

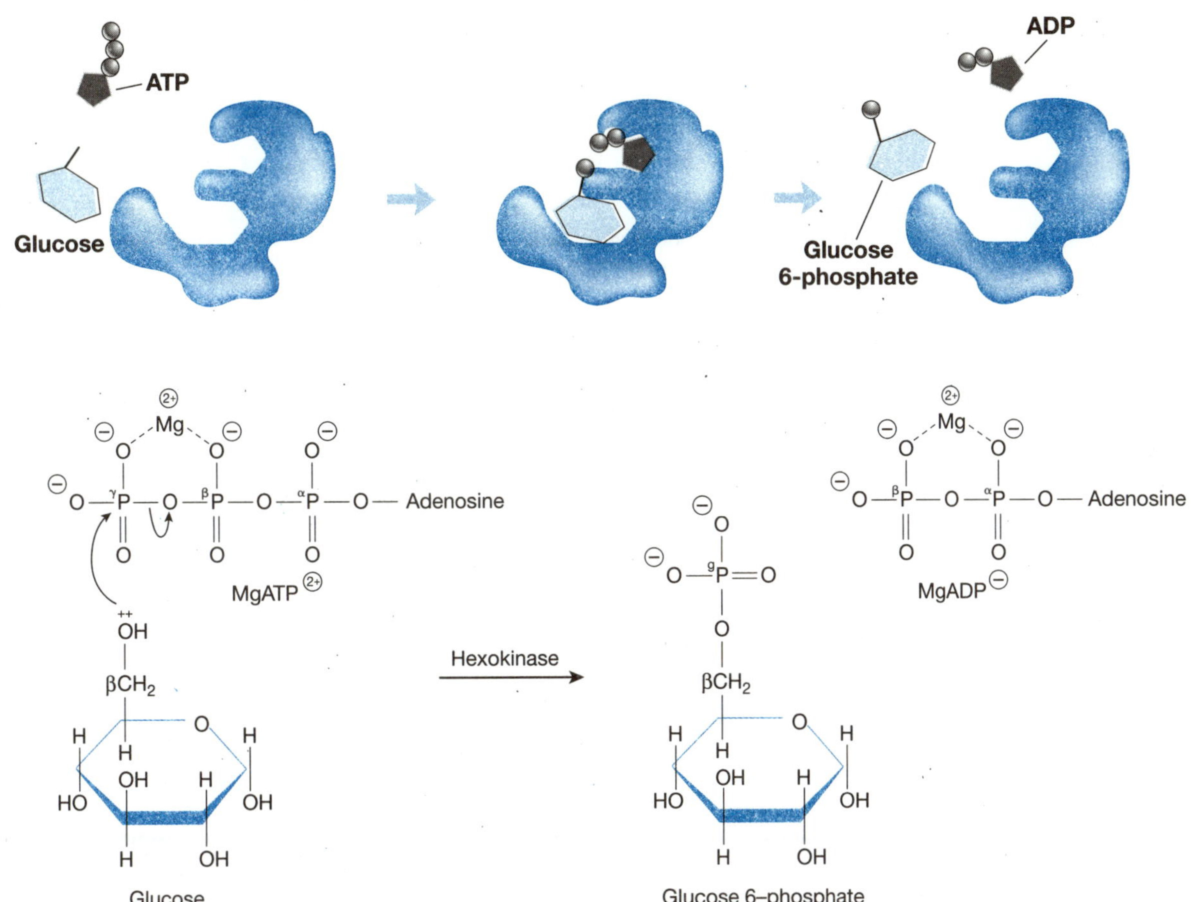

Mechanism of the Enzyme Hexokinase

Hexokinase binds to one molecule of ATP and one molecule of glucose. By putting these two compounds in close proximity, the enzyme achieves a direct nucleophilic attack of the glucose hydroxyl group on the terminal phosphate of the ATP. The divalent cation Mg^{++} also participates in the reaction.

THE PRIMARY STRUCTURE OF THE PROTEIN INSULIN IS SOLVED

By the early 1900s, scientists had established that proteins are composed of amino acids joined together into linear molecules by peptide bonds. The properties of each protein, it was surmised, arose through the combined properties of the amino acids that made them up. In addition, it was clear that not only the composition of amino acids, but also the arrangement (or the order) in which the amino acids are linked gives each protein its unique properties.

After obtaining a Ph.D. in biochemistry from Cambridge University in 1943, Frederick Sanger turned his attention to the problem of protein sequencing. Sanger's first contribution was to develop a method to determine the terminal amino acid of a protein using the dye 1-fluoro-2,4-dinitrobenzene (FDNB). Treatment with FDNB labels the amino-terminal amino group at the end of each protein. Hydrolysis of peptide bonds with acid breaks the protein down into its constituent amino acids. The amino-terminal amino acid is the only one labeled with the FDNB and can be readily identified.

FDNB + Amino-terminal acid of protein → DNP bond to protein + HF

Sanger's FDNB Reaction

FDNB (1-fluoro-2,4-dinitrobenzene) attacks the free amino group at the end of the peptide. This leads to the formation of a dinitrophenyl-amino acid (DNP-aa), which is yellow. The color change can be detected by a spectrophotometer.

When Sanger applied his method to insulin, he found two amino-terminal amino acids: glycine and phenylalanine. This established that insulin is made up of two separate peptide chains. Sanger used mild acid to break the disulfide bonds, holding them together, so that he could study each separately. By this time, he was able to sequence short peptides. To obtain the complete sequence of the A and B peptides, he used proteases (enzymes that cleave proteins) and chemicals to break the peptides into many smaller overlapping fragments. He

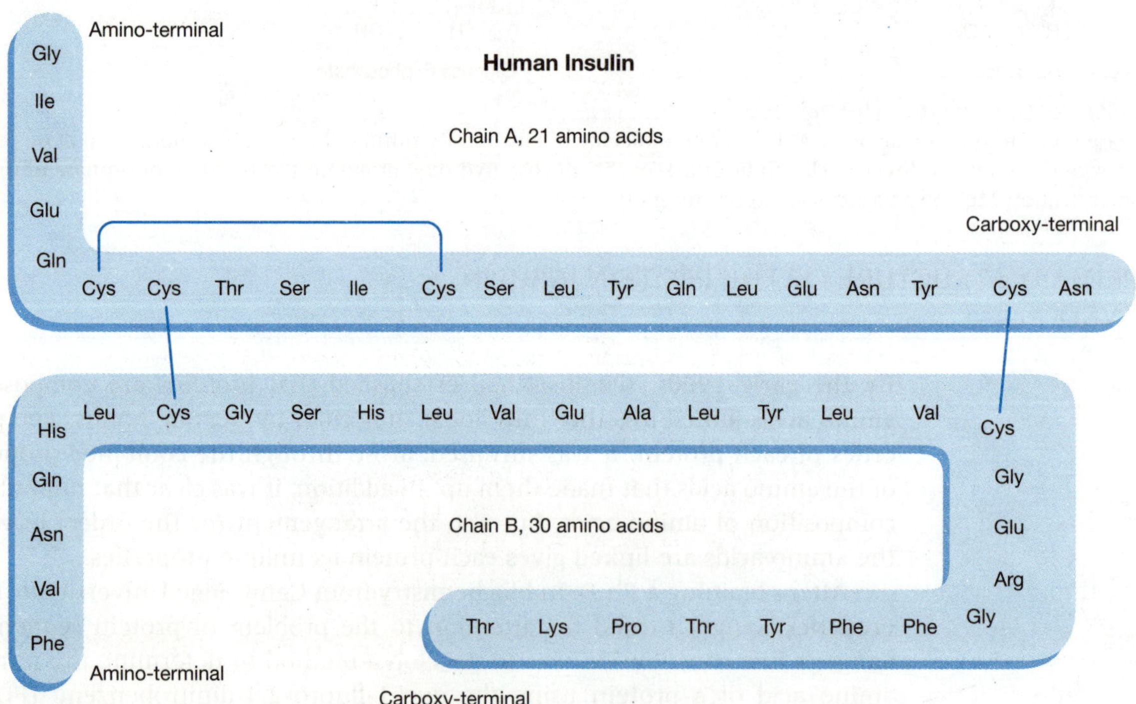

Amino Acid Sequence and Structure of Insulin

Human insulin is made of two peptides (A chain and B chain) joined together by disulfide bonds formed between cysteines. Its sequence is nearly identical to that of bovine insulin sequenced by Frederick Sanger.

then determined the sequence of each fragment. By identifying overlaps, he was able to order the fragments and eventually determine the complete sequence in 1953. Later, Sanger developed methods for sequencing both RNA and DNA. Sanger DNA sequencing is still the primary method in use today.

HOW ARE PROTEINS MADE?

As the studies above demonstrated, the information for making a protein is provided by a gene and that information is transcribed into a messenger RNA, which then associates with the ribosome where protein synthesis occurs. But what happens then? How is the information in the DNA actually stored, and how is it translated into a protein? The answer lies in the fact that DNA is a linear information molecule, which is *transcribed* into a linear mRNA molecule, which then is *translated* into a linear protein sequence. Thus, it makes sense that the linear order of nucleotides in DNA is a code for the genetic information.

A clear explanation of this came to light following years of experimentation. We will first address the issue of the code. There are 20 different amino acids, whereas DNA is made up of only four different nucleotides (abbreviated A, C, G, and T). For the nucleotide sequence to be a code for amino acids, it is necessary to read the nucleotides in groups. The minimum number of nucleotides per group needed to code for 20 different amino acids is three. Francis Crick and Sydney Brenner referred to this nucleotide triplet as a codon.

Robert Holley worked on a class of small RNA molecules (70–80 nucleotides) that today we call transfer RNAs or tRNAs. Intriguingly, tRNAs were found with

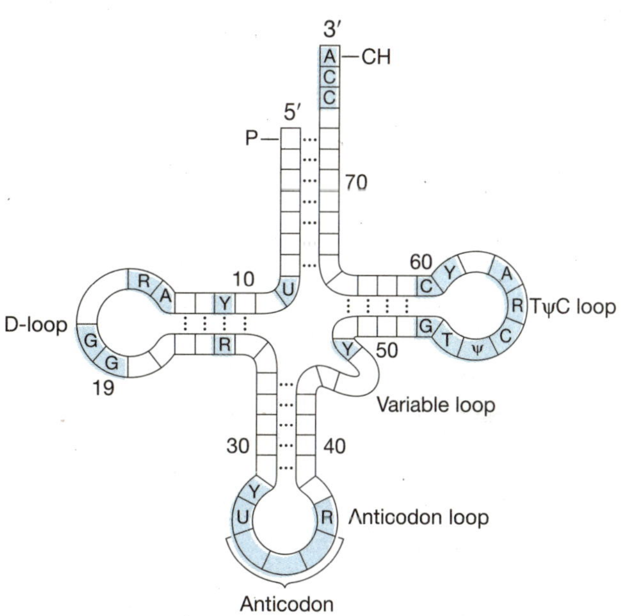

Structure of a Typical tRNA
All tRNAs have the same cloverleaf shape, and the sequence between different tRNAs is fairly highly conserved. The highlighted nucleotides, except for the anticodon, are conserved.

2nd position of codon

1st position of codon (5′ terminus)	U	C	A	G	3rd position of codon (3′ terminus)
U	UUU Phe UUC Phe UUA Leu UUG Leu	UCU Ser UCC Ser UCA Ser UCG Ser	UAU Tyr UAC Tyr UAA Stop (Ochre) UAG Stop (Amber)	UGU Cys UGC Cys UGA Stop UGG Trp	U C A G
C	CUU Leu CUC Leu CUA Leu CUG Leu	CCU Pro CCC Pro CCA Pro CCG Pro	CAU His CAC His CAA Gln CAG Gln	CGU Arg CGC Arg CGA Arg CGG Arg	U C A G
A	AUU Ile AUC Ile AUA Ile AUG Met	ACU Thr ACC Thr ACA Thr ACG Thr	AAU Asn AAC Asn AAA Lys AAG Lys	AGU Ser AGC Ser AGA Arg AGG Arg	U C A G
G	GUU Val GUC Val GUA Val GUG Val	GCU Ala GCC Ala GCA Ala GCG Ala	GAU Asp GAC Asp GAA Glu GAG Glu	GGU Gly GGC Gly GGA Gly GGG Gly	U C A G

The Genetic Code

amino acids covalently attached to their 3′ end. Paul Zamecnik and Mahlon Hoagland discovered a class of enzymes called aminoacyl tRNA synthetases, which were responsible for attaching a specific amino acid to a specific tRNA. Holley and others sequenced the tRNAs, which were all found to share a common three-dimensional structure. Sequence analysis indicated that each tRNA contained a unique three-nucleotide-long sequence in a loop structure. This sequence, today called the anticodon, provided the required specificity for each tRNA (bearing its specific amino acid) to align with the corresponding codon in the mRNA.

By 1966, the laboratories of Marshall Nirenberg and Gobind Khorana broke the genetic code. Nirenberg added to cell-free extracts a synthetic RNA made up of all uracil (making codons UUU-UUU-UUU...), which produced a protein made up solely of the amino acid phenylalanine. This demonstrated that the sequence UUU coded for the amino acid phenylalanine.

Khorana was particularly good at creating synthetic RNA molecules. During the next few years, all possible mRNA combinations were tried, yielding a complete genetic "dictionary" for the translation of mRNA into amino acids. Since nearly all proteins begin with the amino acid methionine (Met), its codon (AUG) represents the "start" signal for protein synthesis. Three codons for which there are no naturally occurring tRNAs—UAA, UAG, and UGA—are "stop" signals that terminate translation.

Interestingly, only two amino acids, methionine and tryptophan, are specified by a single codon; all other amino acids are specified by two or more differ-

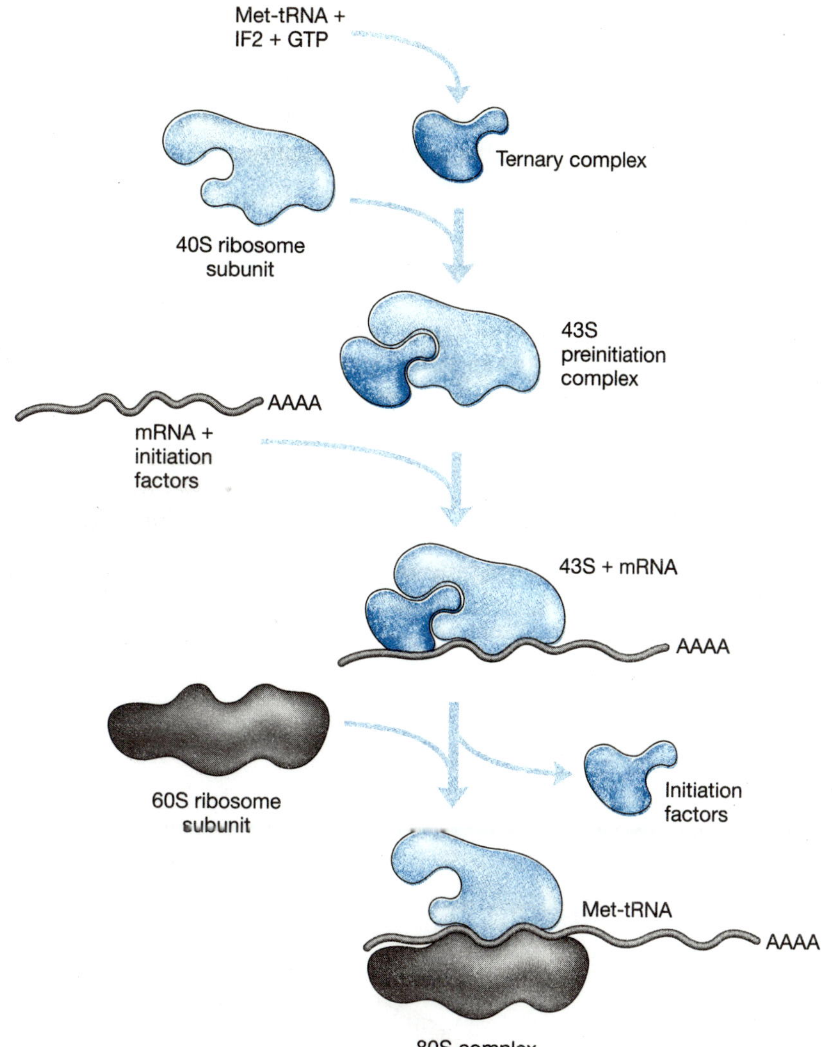

Initiation of Protein Synthesis in Eukaryotes

Initiation factors (IFs), together with the mRNA, the Met-tRNA, and the 40S and 60S ribosome subunits, join to form the 80S initiation complex. This process occurs in multiple steps, beginning with the formation of the ternary complex from the initiating Met-tRNA, IF2, and GTP. The ternary complex binds to the 40S ribosome subunit, along with other IFs and an mRNA, to form the 43S preinitiation complex. Finally, several IFs and GTP are released as the 60S ribosome subunit joins with the 43S mRNA, forming the 80S initiation complex.

ent codons. Because of this redundancy (or more typically referred to as degeneracy or wobble), single-base changes in DNA often do not change the amino acid in the protein. For example, any codon beginning GG, regardless of the nucleotide in the third position (GGU, GGC, GGA, or GGG), specifies the amino acid glycine. So changing the last base of such codons would have no effect on the amino acid sequence of the protein. Changes in the DNA sequence that do lead to a change in the encoded protein can have dramatic effects by causing human disease.

Now we know the answer to how proteins are made. An mRNA is made from the DNA. The mRNA associates with the ribosome where protein synthesis will take place. Next, tRNAs bring in amino acids, aligning with the mRNA by base pairing their anticodons with the codons of the mRNA. The process is very similar in bacteria and eukaryotes.

WHAT IS THE PROCESS OF SYNTHESIZING A PROTEIN?

Translation, the process of building a protein from the RNA message, is usually described in three steps: initiation, elongation, and termination. The initiating event actually occurs away from the ribosomes and mRNA. Essentially all protein synthesis on mRNA starts with the triplet AUG, which codes for methionine. A specific tRNA carries the Met that is needed to initiate translation. This amino acid, called formyl-Met or f-Met, is modified, and the special tRNA for initiation is called f-Met-tRNA. The f-Met-tRNA binds to initiation factor 2 (IF2) and a molecule of GTP (together called the ternary complex). The ternary complex associates with the small 40S subunit of the ribosome, creating the preini-

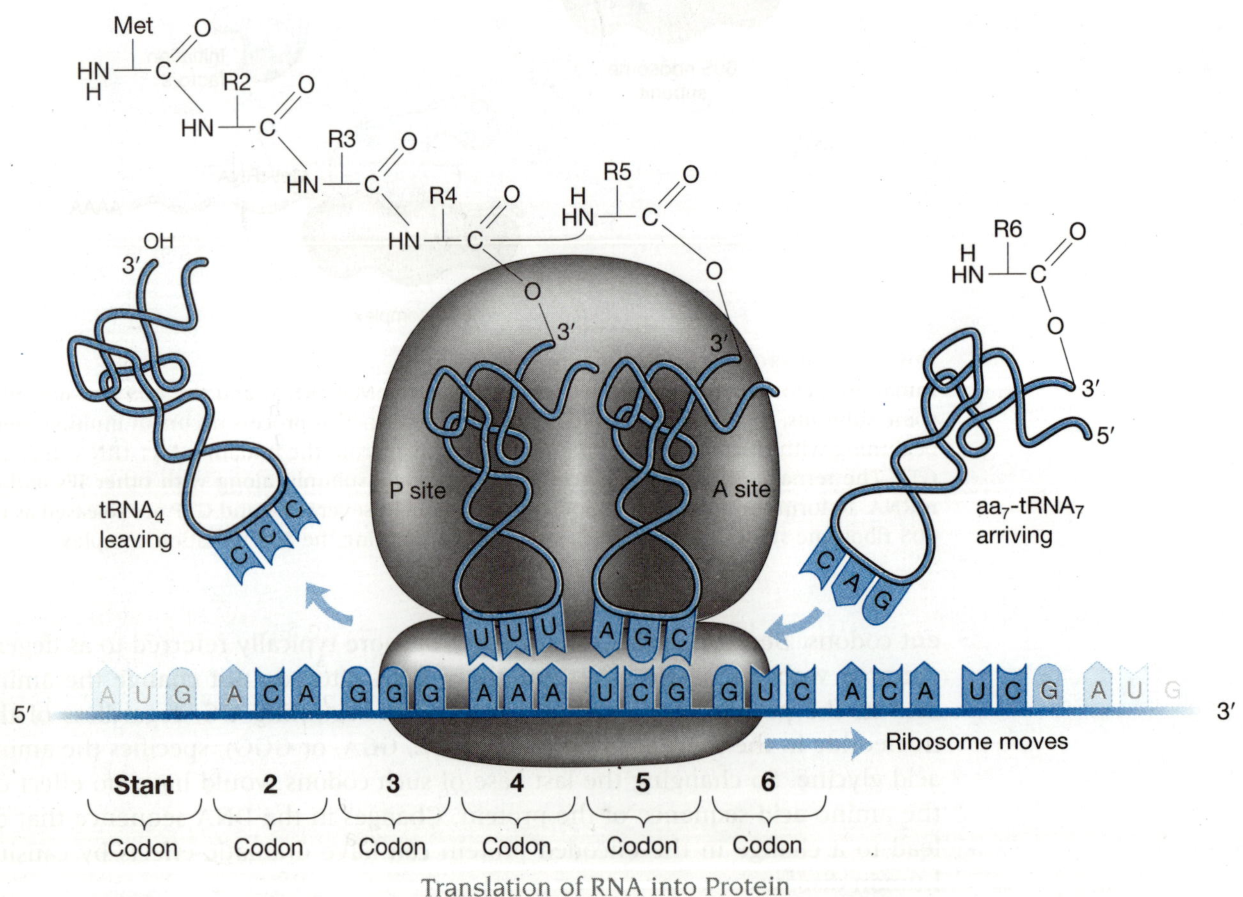

Translation of RNA into Protein

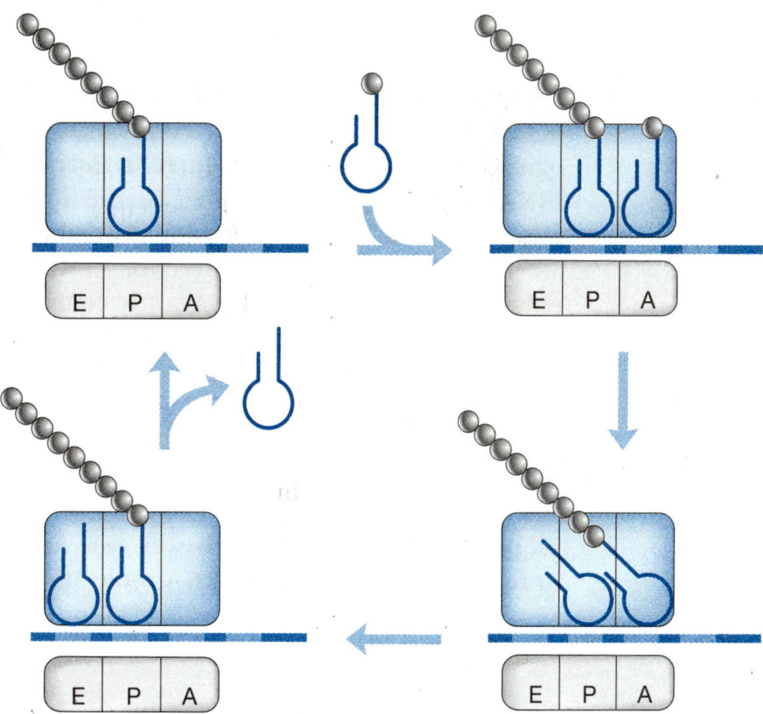

Elongation Phase of Protein Synthesis

During protein synthesis, there are three positions along the mRNA ribosome complex called the peptidyl site (P), aminoacyl site (A), and exit site (E). In each cycle, the growing peptide chain is attached to the tRNA occupying the P site. The incoming charged tRNA enters into the A site. Formation of a peptide bond between the new amino acid and release of the peptide from its tRNA occur simultaneously. The new tRNA, now with the growing peptide chain attached, slides over into the P site, and the old tRNA, which had been in the P site, slides into the E site where it is released. Then the cycle starts over again.

tiation complex. The mRNA meanwhile associates with several initiation factors and then with the preinitiation complex. The preinitiation complex enters the mRNA at the 5′ end and slides down to the initiation AUG where the large ribosomal subunit (60S) associates with the other factors, creating the complete ribosome called the initiation complex. This ends the initiation phase.

In the elongation phase, amino acids are added in succession to create a polypeptide chain. There are three positions on the ribosome. The first is the P (peptidyl) site where the growing protein resides, held in the ribosome by a tRNA. A new tRNA enters at the aminoacyl site accompanied by elongation factor Tu bound to a molecule of GTP (EF-Tu-GTP). (EF-Tu ensures that the proper tRNA enters the A site.) After EF-Tu is released, the new amino acid is added to the growing peptide chain via a peptide bond. This releases the polypeptide from the previous tRNA in the P site. Elongation factor G (EF-G) enters with a GTP molecule. The whole complex slides down to the next triplet codon with hydrolysis of GTP and release of EF-G. The tRNA in the P site now moves to the E (exit) site for release from the ribosome. The tRNA in the A site now moves into the P site, carrying the growing protein with it. The A site is now empty, allowing the next tRNA to enter, and the process continues.

When a stop codon is reached, protein synthesis ends and termination begins. Termination requires several termination, or releasing, factors that recognize a stop codon—any of three triplets for which there are no corresponding tRNAs. GTP is hydrolyzed to release the final amino acid from its tRNA, the protein is released, and the ribosome dissembles.

There is one major difference between translation in bacteria and in eukaryotes. Eukaryotic mRNAs only code for single proteins. Bacterial mRNAs may encode more than one protein, with the information for each arranged in tandem. This is called a polycistronic mRNA. A problem is thus created for bacteria: How is synthesis initiated for the second or third protein of a polycistronic mRNA? The answer is that each mRNA contains a specific ribosome-binding sequence, called a Pribnow box, adjacent to the start site for each protein. This site signals the ribosome to enter and begin translation at the next AUG sequence. In eukaryotes, however, it is sufficient for the preinitiation complex to recognize the end of the RNA molecule to begin protein synthesis.

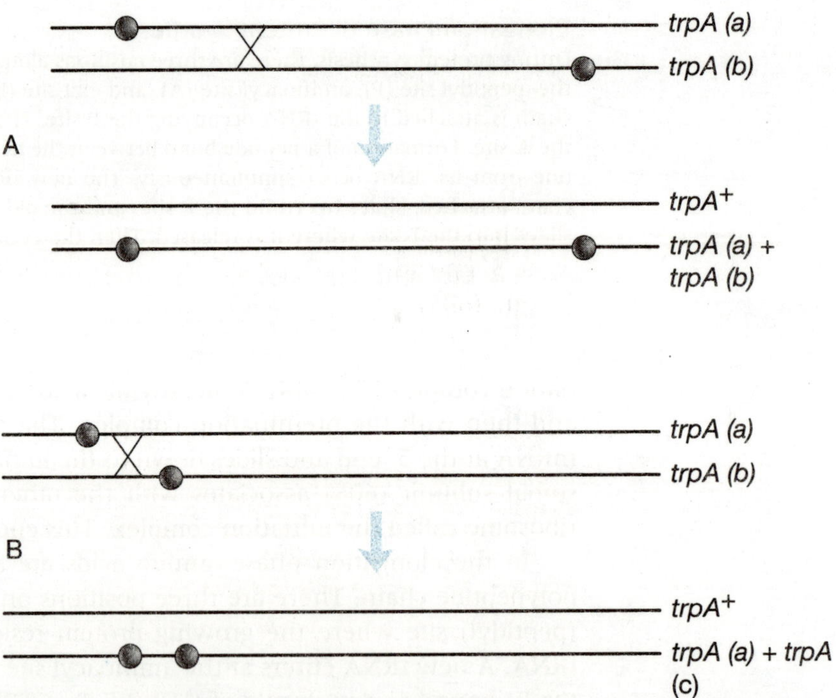

Recombination Complementation
Charles Yanofsky used recombination and complementation to map the positions of the *trpA* mutation. In *A*, the mutations (*dots*) in the two *trpA* genes are far apart, so the probability of a recombination event between them is high. A recombination event will form one wild-type and one double-mutant gene. In *B*, the mutations are close together, and the chances of a recombination event are much lower. By measuring the frequency with which any two mutations recombine, the relative distances between them can be determined.

HOW DO WE KNOW THAT THE INFORMATION IN A GENE IS DIRECTLY TRANSLATED INTO A PROTEIN?

Charles Yanofsky
(Courtesy of the National Library of Medicine.)

Charles Yanofsky designed a set of experiments which proved that the DNA sequence is colinear with the protein sequence. In other words, the first DNA codon produces the first amino acid, the second codon produces the second amino acid, and so on. Yanofsky worked with an *E. coli* gene called *trpA*, which codes for an enzyme involved in synthesizing the amino acid tryptophan. Yanofsky began his studies by making a large collection of *E. coli* mutants that were unable to grow on media lacking tryptophan and, therefore, carried mutations in a gene required for making the amino acid.

Several genes are needed to synthesize tryptophan, so Yanofsky needed to determine which strains had mutations in the *trpA* gene. He took a known *trpA* mutant and crossed it with each of the mutant strains. DNA from two bacteria cells can combine in a single cell, making it diploid. If the unknown mutant being tested had a mutation in the *trpA* gene, then the diploid combination would still lack a functional *trpA*. This new combination would thus be unable to grow on media lacking tryptophan. However, if the mutation was in a different gene involved in tryptophan biosynthesis, then the DNA from each cell could complement the other's mutation, and the cell could now grow on media lacking tryptophan. This is called a complementation assay (discussed further in Chapter 3).

Using this method, Yanofsky identified many strains containing different mutant *trpA* genes. He then crossed each *trpA* mutant with every other *trpA* mutant and measured the frequency of complementation via the ability to grow on media lacking tryptophan. Because both strains contained mutant *trpA* genes, the only way that a cross could produce a functional *trpA* gene was if the two mutant genes recombined to form a functional gene. As we learned in Chapter 1, the frequency of recombination is a measure of distance along the chromosome, and so the further apart the mutations, the higher the chance of recom-

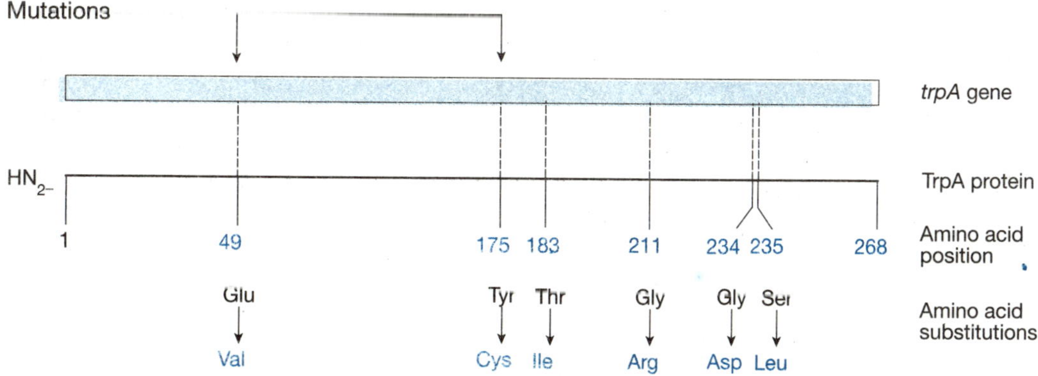

Colinearity between DNA and Protein Sequence
Charles Yanofsky made a precise map of multiple mutations in the *trpA* gene and sequenced the proteins of each mutant. This scheme shows that the positions of the mutated nucleotides in DNA align with the amino acid changes found in each corresponding mutant protein.

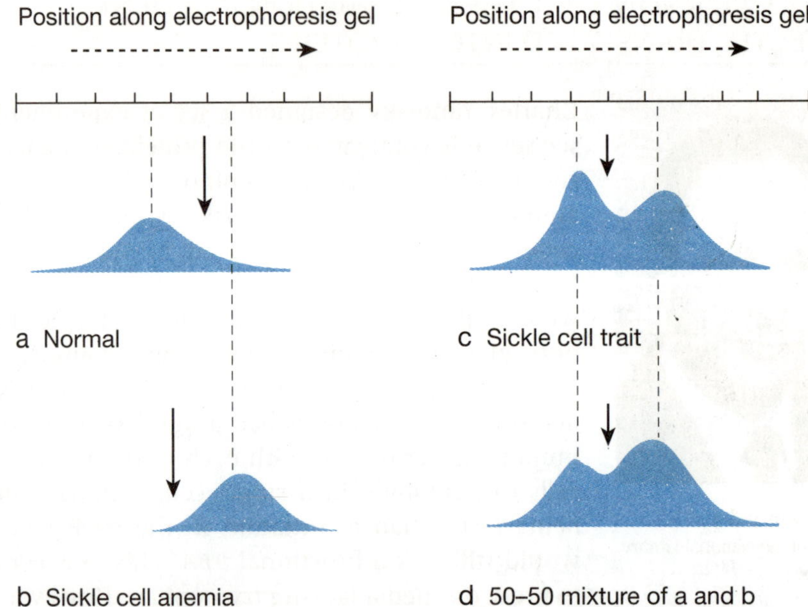

Electrophoretic Difference between Normal and Sickle Cell Hemoglobin
Hemoglobin from normal (*a*), sickle cell anemia (*b*), and sickle cell trait (*c*) individuals was separated by electrophoresis. Note the difference in mobility between the normal and sickle cell hemoglobin and the presence of both forms in the individual with the sickle cell trait. (Redrawn, with permission, from Washington University Biology Department at http://www.nslc.wustl.edu/courses/Bio296A/allen/sicklecell/part2/molecular.html.)

bination between them. From these studies and others, Yanofsky created a genetic map of the positions of each *trpA* mutation. Yanofsky then isolated the mutant protein from every *trpA* mutant strain and determined its amino acid sequence. He found that every mutation in the DNA correlated with an amino acid change at the same relative position in the protein. This study demonstrated the colinearity between genes and proteins.

In 1949, Linus Pauling and Vernon Ingram carried out studies on sickle cell anemia. This genetic disease, which primarily affects individuals of African descent, is caused by a mutation in hemoglobin, the major protein of blood. Genetic studies had shown that this is a recessive disorder, with individuals only being affected if they have two mutated copies of the gene. Thus, heterozygote parents who have one good copy of the gene and one mutated copy of the gene are "carriers" of the sickle cell trait. These individuals are mildly affected. Pauling used a method called electrophoresis, which separates molecules on the basis of their electrical charge, to show that the hemoglobin of normal individuals (Hb) is structurally different from the hemoglobin of sickle cell patients (Hb_s). He further found that the heterozygous carriers have both forms of the protein (Hb and Hb_s)—the same pattern obtained by mixing the hemoglobin of a normal patient and a sickle cell patient. This was the first time that biochemical analysis was used to visualize the difference in properties between a normal and mutant protein.

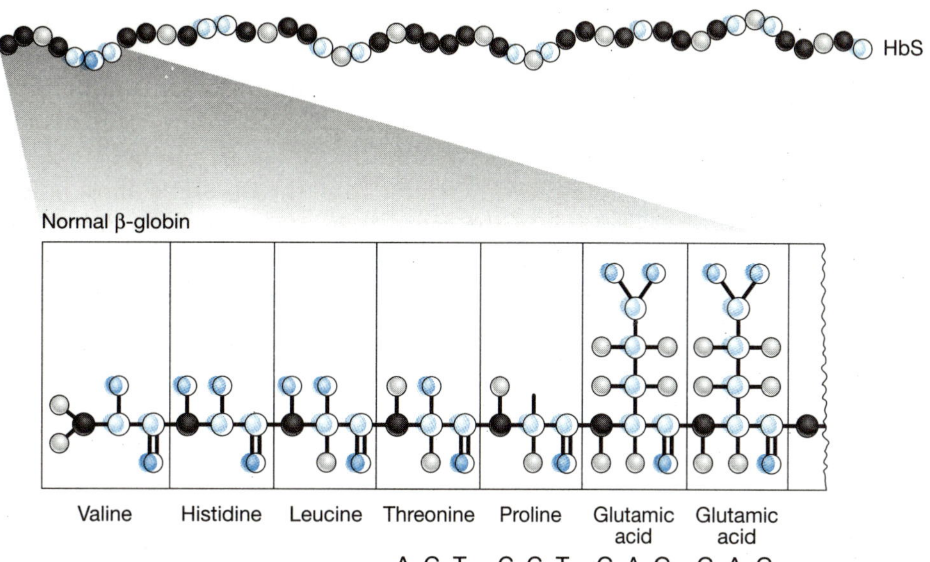

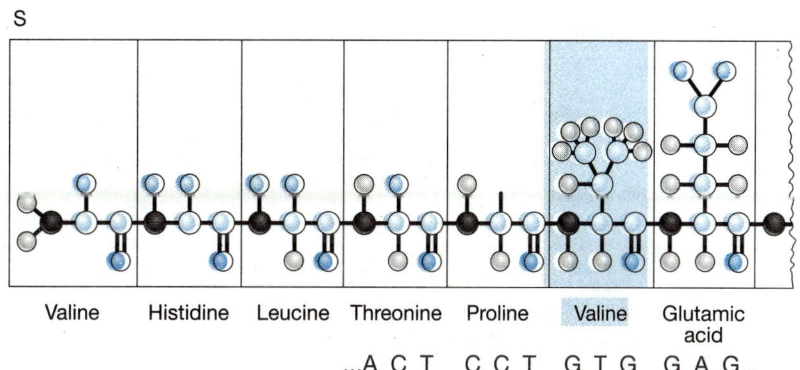

Sickle Cell Anemia Is Caused by a Single-amino-acid Change in β-globin
A valine replaces a glutamic acid at position 6 on the β-globin chain.

Several years later, Pauling and Ingram digested Hb_s and Hb hemoglobin with the enzyme trypsin, which cleaves proteins at specific sites (next to lysines and arginines). They then used paper electrophoresis to separate the fragments generated from the trypsin digest. The pattern of Hb_s hemoglobin was identical to Hb hemoglobin, except for a single spot. This indicated that the change in the hemoglobin protein was limited to a single part or peptide within the whole protein.

In 1957, Ingram separated the α and β chains of hemoglobin and identified the exact mutation in the β-globin chain by protein sequencing. It was a single change of a glutamic acid for a valine at the sixth amino acid in the chain. This change of a charged polar amino acid (glutamic acid) for an uncharged nonpolar amino acid (valine) explains the difference in electrophoretic mobility seen between normal and sickle cell hemoglobin.

PROTEINS ARE ALSO MODIFIED AFTER TRANSLATION

The synthesis of proteins on ribosomes is not necessarily the end of their elaboration into functional molecules. Many proteins are further modified after protein synthesis within the cell or in another tissue.

Following release from the ribosome, most polypeptides in eukaryotes undergo any of the more than 100 known modifications to amino acid side groups. These modifications regulate the protein's biological activity and may be permanent or reversible. For example, permanent addition of acyl groups (acylation) or of a coenzyme, such as biotin or lipoic acid, is necessary for the action of many enzymes. Within the endoplasmic reticulum (ER) and Golgi apparatus, modifica-

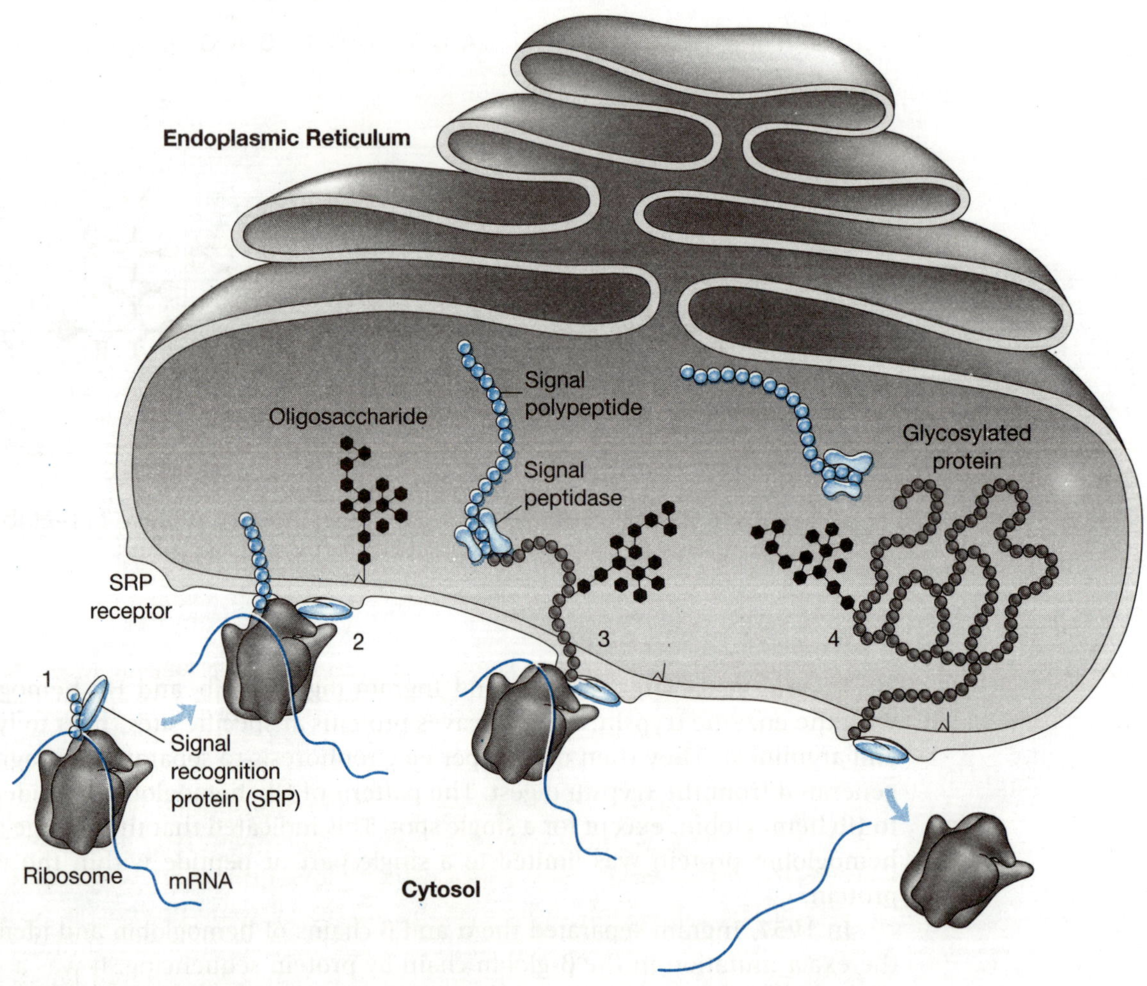

Posttranslational Modification of a Secreted Protein

(*1*) Signal recognition protein (SRP) binds signal polypeptide, which is translated at a free ribosome following initiation of protein synthesis. (*2*) SRP binds its receptor on the ER membrane. (*3*) Protein is translated into the ER lumen, where an oligosaccharide side group is added. (*4*) Signal peptidase cleaves the signal peptide to yield a mature, glycosylated protein.

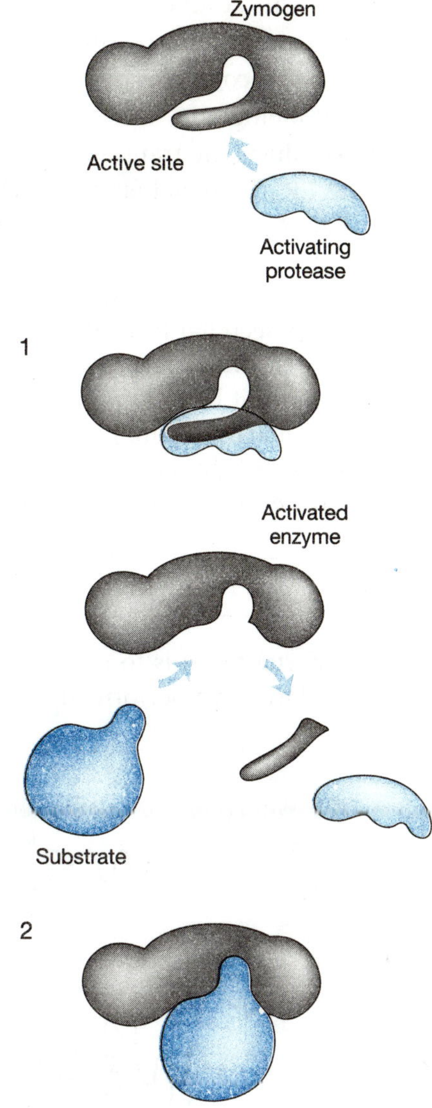

Extracellular Modification of a Zymogen
(*1*) The activating protease cleaves a portion of the zymogen, (*2*) making the active site accessible to bind to its specific substrate in a "lock and key" fashion.

tions are made to proteins that enable them to be transported within the cell or externally secreted. Membrane proteins are modified with fatty acid side chains, whereas secreted proteins have added carbohydrate groups (glycosylation).

Within the cytoplasm, the enzymatic activity of some proteins is regulated by the addition or removal of a phosphate group(s) at a specific serine, threonine, or tyrosine residue(s). Phosphorylation is reversible; protein kinases add phosphate groups, whereas phosphatases remove them. Protein kinases are intensely studied because oncoproteins, the protein products of cancer-causing genes (oncogenes), often exhibit phosphorylating activity. (For more details, see Chapter 7.)

Proteins destined for the ER contain a signal peptide of 16–30 amino acids at their amino terminus. During protein synthesis, the signal peptide is bound by the signal recognition particle (SRP), which, in turn, binds to an SRP receptor in the ER membrane. This anchors the ribosome to the ER, where the signal peptide conducts the translating protein through the membrane. Within the lumen of the ER, the signal peptide is cleaved by a signal peptidase.

Secreted proteins, including many proteolytic enzymes, are often exported from the cell in nonfunctional forms. These pre-enzymes or zymogens are then modified, as needed, to produce the biologically active form. For example, trypsin is secreted from the pancreas as inactive trypsinogen. Proteolytic cleavage of a segment of the trypsinogen polypeptide, making it an active protein, occurs after it has been secreted into the digestive system. Insulin also is secreted in an inactive form, proinsulin, which is activated upon cleavage by a trypsin-like enzyme. This allows insulin to be deployed to the circulatory system, ready to activate in response to a simple signal.

HOW ARE PROTEINS DEGRADED?

As important as it is to make proteins, it is equally important to get rid of them when they have performed their function. This is true of enzymes whose par-

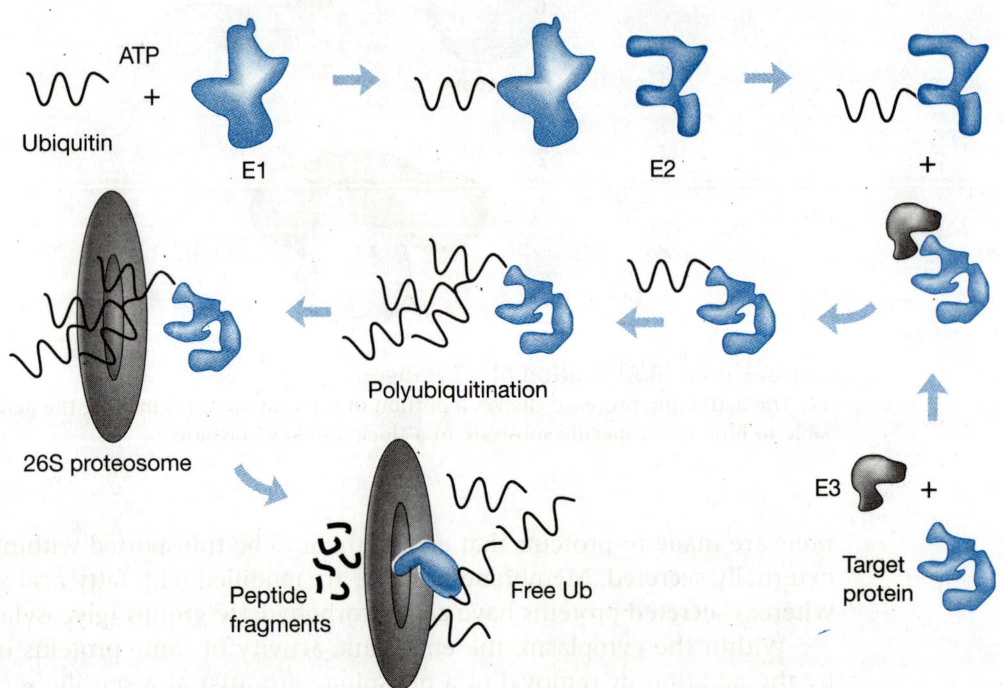

Ubiquitin-Proteosome Proteolytic Pathway

Ubiquitin is activated by the binding of its carboxy-terminal glycine to the ubiquitin-activating enzyme (E1). The activated ubiquitin is then transferred to the ubiquitin-conjugating enzyme (E2). Another enzyme, ubiquitin-protein ligase (E3), binds to the target protein. The E2-Ub interacts with the E3-target protein, and ubiquitin forms an isopeptide bond with the target protein. Multiple ubiquitin molecules are added to the target protein. When at least four ubiquitin molecules have been added, the target protein is recognized by the 26S proteosome. The ubiquitin molecules are released, whereas the target protein is degraded into small peptides.

ticular activity is only wanted at certain times—for example, proteins that regulate cellular processes at certain points of the cell cycle. We already mentioned enzymes that cleave proteins, called proteases, which can randomly degrade proteins. A regulated process of protein degradation occurs via the ubiquitin-proteosome proteolytic pathway.

In 1976, Gideon Goldstein discovered a protein required for the development of white blood cells. This 76-amino-acid protein was named ubiquitin because it is found in all cells of organisms from yeast to humans. Many proteins are targeted for degradation by the ubiquitin-proteosome proteolytic pathway. Several ubiquitin molecules (four seems to be the minimum) bind to a targeted protein, a process mediated by several other proteins. "Ubiquitination" marks the protein so that it is recognized by a ring-shaped protein complex, the 26S proteosome, which then degrades the targeted protein into small peptide fragments. Among the many proteins regulated by the ubiquitin-proteosome proteolytic pathway is p53, the major tumor suppressor protein.

THE CENTRAL DOGMA

This chapter summarized some of the important studies that demonstrated how DNA functions as the genetic material. These studies established the "central dogma," stated first by Francis Crick, which said that genetic information stored in DNA flows through RNA to proteins. The genetic information carried in DNA is a code. This information is first transcribed into the language of RNA, which (in eukaryotes) carries the information to the cytoplasm. In the cytoplasm, the mRNA associates with ribosomes, where the code is translated into proteins. Proteins can then be modified or degraded during posttranslational modification.

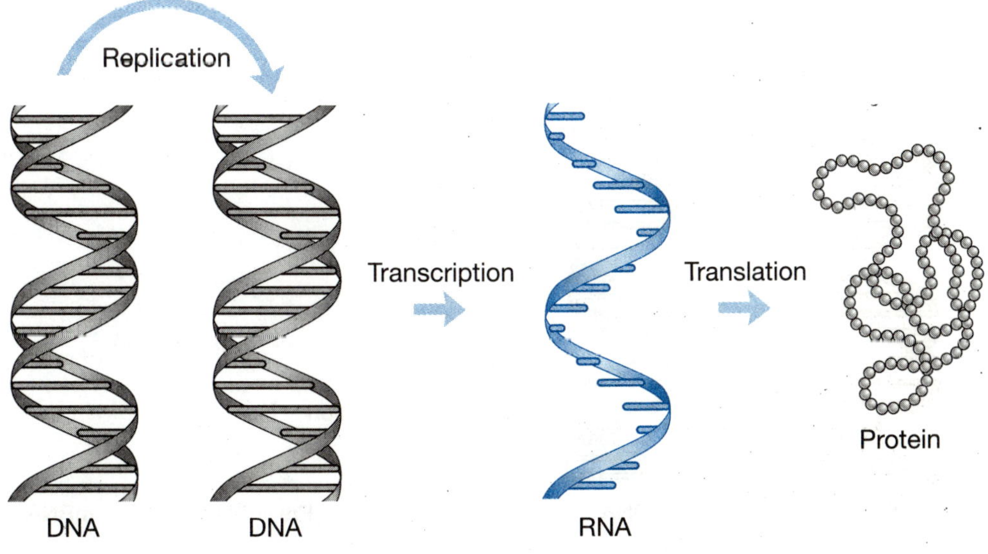

Crick's Central Dogma, 1957

Significant amendments and refinements have been made to Crick's central dogma since the cracking of the genetic code. It is now known that the flow of genetic information is not strictly one-way. Although DNA is, by and large, a very stable molecule, the classic concept of a gene as a unique entity is becoming less and less useful. A single gene may, through rearrangement and biochemical editing, be responsible for the production of many different proteins. Some of the amendments listed below are discussed in subsequent chapters.

- Some organisms do not store their genetic information in DNA but rather use RNA. There are several examples of this in viruses. Prior to replication and protein synthesis, their genetic information is converted to DNA by the enzyme reverse transcriptase. By making DNA from RNA, these organisms represent a reverse flow of information from the central dogma.
- Genes are not immutably fixed on the chromosomes. Transposable genetic elements move about from one chromosomal spot to another and may act as molecular switches to regulate gene expression. Physical rearrangement of genes permits a vast array of proteins to be produced from a relatively limited amount of DNA code.
- DNA sequence and protein sequence are not entirely colinear. The RNA transcript is often extensively "edited" prior to protein synthesis. Interspersed within eukaryotic genes are DNA sequences that do not code for protein. These *introns* are transcribed as part of the primary mRNA (called the pre-mRNA) but are excised by a process called RNA splicing. Furthermore, differential splicing may create different "mature" mRNA molecules that produce different proteins.
- Depending on where translation begins, a single DNA sequence can be read in several overlapping "reading frames" of codons, which are then translated into different proteins. This is usually only seen in viruses where the genomes are very compressed.

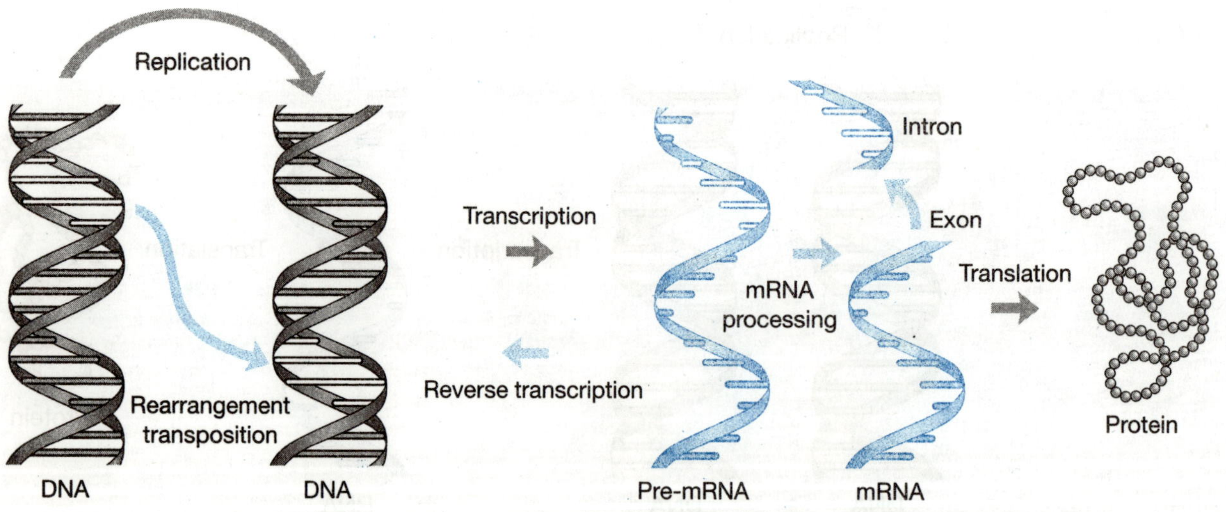

Amendments to the Central Dogma

- Some RNA molecules have been shown to have catalytic activity, acting like primitive enzymes. It has been proposed that RNA molecules may have functioned like proteins before there was protein synthesis in what has been described as the "RNA world."

REFERENCES

Alberts B., Bray D., Johnson A., Lewis J., Raff M., Roberts K., and Walter P. 1998. *Essential cell biology: An introduction to the molecular biology of the cell.* Garland Publishing, New York.

Allen G.E. 1978. *Life science in the twentieth century.* Wiley, New York and Cambridge University Press, United Kingdom.

Cairns J., Stent G., and Watson J., eds. 1992. *Phage and the origins of molecular biology.* Cold Spring Harbor Laboratory Press, Cold Spring Harbor, New York.

Darst S.A., Kubalek E.W., and Kornberg R.D. 1989. Three-dimensional structure of *Escherichia coli* RNA polymerase holoenzyme determined by electron crystallography. *Nature* **340:** 730–732.

Freemont P.S. 2000. Ubiquitination: RING for destruction? *Curr. Biol.* **10:** R84–R87.

Hubscher U., Nasheuer H.P., and Syvaoja J.E. 2000. Eukaryotic DNA polymerases, a growing family. *Trends Biochem. Sci.* **25:** 143–147.

Jacob F. 1995. *The statue within: An autobiography.* Cold Spring Harbor Laboratory Press, Cold Spring Harbor, New York.

Judson H.F. 1979. *The eighth day of creation: The makers of the revolution in biology.* H. Simon Schuster, New York.

Lafontaine D.L.J. and Tollervey D. 2001. The function and synthesis of ribosomes. *Nat. Rev. Mol. Cell Biol.* **2:** 514–520.

Leatherwood J. 1998. Emerging mechanisms of eukaryotic DNA replication initiation. *Curr. Opin. Cell Biol.* **10:** 742–748.

Micklos D., ed. 1999. DNA from the beginning (http://www.dnaftb.org). Dolan DNA Learning Center, Cold Spring Harbor, New York.

CHAPTER 3

How We Learned How Genes Are Regulated

THE EXPERIMENTS DESCRIBED IN CHAPTER 2 LED TO AN UNDERSTANDING of the nature of the gene—revealing the basic mechanisms of DNA replication, transcription, and translation. Researchers learned that a copy of the information in each gene is made in the form of a messenger RNA and that mRNA carries the coded message to the ribosomes, where it is decoded to synthesize a protein.

Several major questions remained unanswered in these experiments: How is the information in each gene accessed and how is this process regulated? We already know that the first step involves making an mRNA copy of the gene. But why, and how, is a specific gene transcribed into an mRNA at a specific moment or in a particular cell? In addition, some proteins are needed in greater quantity than others, and thus the amount of protein synthesis also must be regulated. These important and complex questions have been central to molecular biology for the past 35 years, and we still have much to learn.

The *lac* Operon

The study of gene regulation was initiated by the landmark work in the late 1950s and early 1960s of François Jacob and Jacques Monod at the Pasteur Institute in Paris. These researchers set out to determine how lactose metabolism genes are regulated. Lactose is a disaccharide made up of two simple sugars (glucose and galactose). Providing *E. coli* lactose as a carbon source, in the absence of glucose, rapidly induces the production of three proteins not normally present in the bacterial cell: (1) β-galactosidase splits lactose into glucose and galactose; (2) permease acts at the cell membrane to allow lactose into the cells; and (3) transacetylase, whose function is not directly related to the usage of lactose. By breaking lactose down into glucose and galactose, the cell obtains a supply of glucose for energy.

As it turns out, the genes for β-galactosidase, permease, and transacetylase, *lacZ, lacY,* and *lacA*, respectively, lie in tandem on the *E. coli* chromosome and are transcribed into a single mRNA. Although this situation is not found in eukaryotes, it is common in bacteria, and such gene clusters are called *operons*. An mRNA coding for multiple proteins is known as a *polycistronic* mRNA. The set of genes encoding proteins that metabolize lactose is called the lactose operon, or simply the *lac* operon. Jacob and Monod performed a series of genetic experiments to determine how these genes are switched on in the presence of lactose.

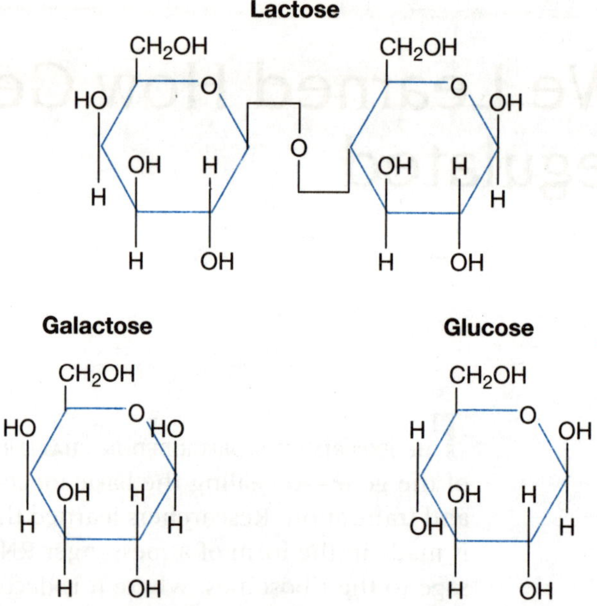

Conversion of Lactose into Glucose and Galactose

Lactose is broken down into its composite sugars, galactose and glucose. Glucose is the universal food source of cells. Glucose is ultimately broken down into carbon dioxide and water during several metabolic processes releasing its stored energy, forming ATP molecules. A total of 38 ATP molecules can be produced from one molecule of glucose. Galactose is converted into glucose by other enzymes and also used in glycolysis.

One experiment involved making partial diploids of the *lac* genes. *E. coli* contains a single circular chromosome and is thus a haploid organism. However, *E. coli* can contain other smaller chromosomes or plasmids. The F plasmid (fertility) is a sort of sex chromosome, containing genes encoding proteins needed to assemble an appendage called a pilus, which forms a bridge between mating *E. coli* cells. The F plasmid integrates into the donor chromosome and carries it across the pilus to the recipient cell. It can carry across all or part of the donor genome, making the recipient cell diploid for all or part of the genome. This is essentially sexual reproduction for bacteria. The F plasmid can be engineered so that it carries particular genes from the *E. coli* genome (this engineered plasmid is referred to as F′). The F′ plasmid makes the recipient cell diploid for only those genes carried on the F′ plasmid, and the cell becomes a partial diploid.

Jacob and Monod constructed a series of partial diploid strains where the F′ plasmid and genomic DNA contained wild-type (lac^+) or mutant (lac^-) versions of the genes in the *lac operon*. These strains were then studied for their ability to be regulated by lactose and metabolize it. From these studies, Jacob and Monod grouped mutations into different classes. The simplest class of lac^- genomic mutants were those strains able to regain the ability to metabolize lactose if a wild-type version of the gene were present on the F′ plasmid. This simple complementation by a gene on the F′ plasmid demonstrated that these mutations

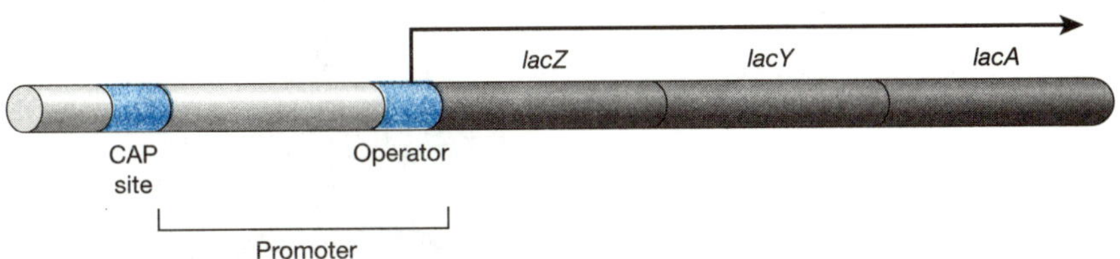

The *lac* Operon

The *lacZ* gene is transcribed in a single mRNA along with two other genes, *lacY* and *lacA*. *lacZ* encodes for β-galactosidase, *lacY* encodes the permease that brings lactose into the cell, and *lacA* encodes an acetylase that is believed to detoxify thiogalactosides, which, along with lactose, are transported into the cell by *lacY*. The promoter spans about 60 bp, and the CAP (catabolite activator protein) site and the operator (the Lac repressor-binding site) are about 20 bp each. The operator lies within the promoter, and the CAP site lies just upstream of the promoter. The drawing is not to scale: The *lacZ* gene, for example, is about 3500 bp long. The entire element shown is called an "operon." (Reprinted, with permission, from Ptashne and Gann 2002.)

François Jacob
(Courtesy of Cold Spring Harbor Laboratory Library Archives.)

Jacques Monod
(Courtesy of Cold Spring Harbor Laboratory Library Archives.)

were recessive, following predictable Mendelian rules of inheritance. Such a *lac*⁻ mutation was in either the *lacZ* or *lacY* gene.

Jacob and Monod recognized other classes of mutations that affected the regulation of the *lacZ* and *lacY* genes. In one class, Jacob and Monod identified mutants that expressed *lacZ* or *lacY* independently of the presence of lactose, called constitutive expression. These mutations fell into two types. One acted in a recessive fashion, meaning that the presence of a wild-type gene on the F′ plasmid restored lactose-dependent expression. Jacob and Monod referred to these as *trans* mutations because they could assert their influence across a distance (from the F′ plasmid to the genomic DNA). It turns out that this class of mutations were within the *lac repressor* (or *lacI*) gene which codes for a protein that binds to the DNA, blocking expression of *lacZ* and *lacY*. Thus, the Lac repressor protein is an inhibitor of lac^{+} expression. The second type of constitutive mutations could not be complemented by the presence of a normal gene on the F′ plasmid. These "dominant-like" mutations were said to act in *cis* because they only worked on the genes of the genome that they were physically associated with. These mutations were found to be in the *lacO* region (or *operator*) which is the region on the DNA near the start of the *lacZ* gene where the Lac repressor protein binds. Mutations in this sequence presumably prevented Lac repressor binding, leading to constitutive expression of *lacZ* and *lacY*.

The final class of mutations also acted in *cis*, but these mutations were unable to express genes of the *lac* operon. These mutations were outside of the *lacO* region. They were found to be mutations in the *lacP* region which we now know as the promoter. This is the sequence recognized by RNA polymerase as the place to associate with the DNA and start synthesizing RNA. Thus, the arrangement of the *lac* operon is as follows. The first gene in the sequence is the *lac repressor* which constitutively expresses the Lac repressor protein. Lac repressor binds to *lacO* which lies adjacent to *lacP*. Binding of the Lac repressor to *lacO* blocks RNA polymerase from moving down the DNA from its docking position at *lacP*. When lactose is present, it binds to the Lac repressor protein. This leads to a change in

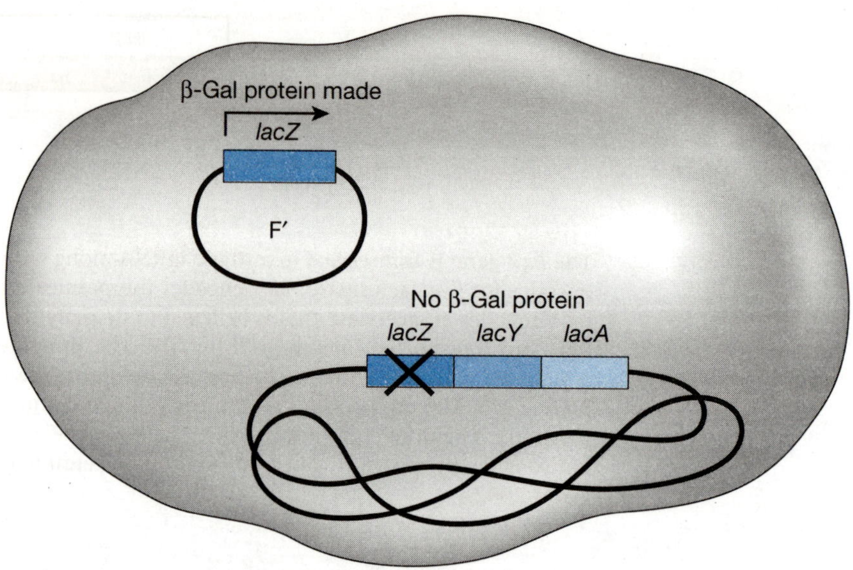

Complementation

The endogenous copy of *lacZ* found in *E. coli* is defective, and it cannot make the β-galactosidase protein needed to process lactose. However, the F′ plasmid contains a wild-type copy of the *lacZ* gene. Using this gene, the cell can make a functional β-galactosidase protein and grow on lactose. The wild-type gene on the F′ plasmid is said to "complement" the mutation in the *E. coli*.

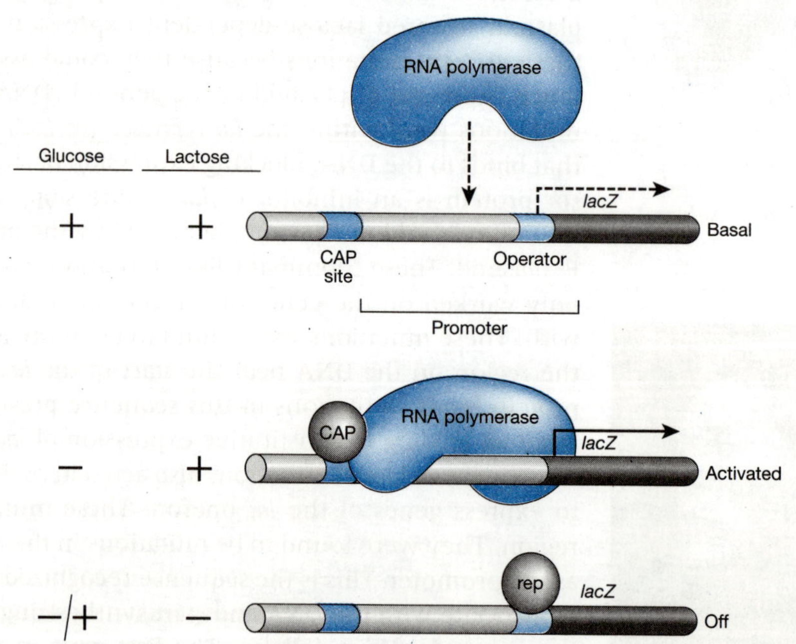

Signals from the Environment Control the Binding of Factors at the *lac* Gene

When glucose is absent, the CAP protein can bind, activating transcription. When lactose is present, it prevents the binding of the Lac repressor protein, allowing transcription to occur. (Reprinted, with permission, from Ptashne and Gann 2002.)

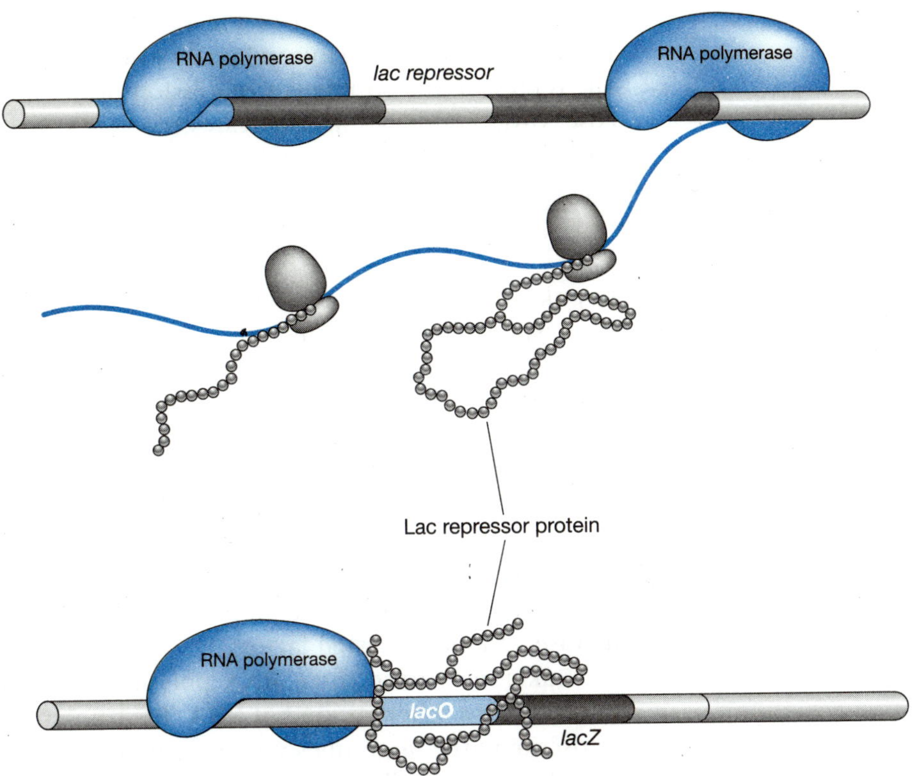

cis and *trans* Elements
The *lac repressor* (*lacI*) gene can be anywhere in the genome to act on its target (*lacO*) because it produces a protein which diffuses to the *lacO* site where it regulates the expression of the *lacZ* gene. Therefore, the *lac repressor* gene acts in *trans*. *lacO* is a sequence to which the Lac repressor binds, blocking the RNA polymerase from moving down from the promoter. It does not produce a protein and mutations in it only affect the downstream *lacZ* gene. Its function is completely dependent on its position and so it is referred to as a *cis* element.

the structure of the Lac repressor, leading to its release from the *lacO* DNA sequence. Once the Lac repressor is released, the RNA polymerase is free to transcribe the genes of the *lac* operon. Although this description of negatively regulated genes was a landmark in our understanding of gene regulation and introduced the concept of a promoter which is ubiquitous to all genes, positive regulation at the promoter is the more common form of gene regulation, particularly in eukaryotes.

In addition to being controlled by Lac repressor, the *lac* genes are controlled by another regulatory protein, called CAP. This additional regulator stimulates (activates) expression of the genes. With one surface, CAP binds to a site upstream of the promoter and, with another surface, CAP interacts with polymerase. Instead of blocking the polymerase from binding to the promoter, it assists binding and so increases transcription. This exemplifies positive regulation of gene expression (activation).

The *lac* genes are only expressed efficiently when lactose is present, as mentioned above, and glucose (a more efficient energy source) is absent. Lactose controls the DNA binding of the Lac repressor, whereas glucose controls the DNA binding of CAP. Thus, when lactose is present, it blocks the binding of the Lac repressor and so relieves repression of the *lac* genes. In the absence of glucose, CAP binds DNA and activates the genes.

GENE REGULATION IN EUKARYOTES

In eukaryotes, gene regulation is somewhat more complicated than in bacteria. For instance, in eukaryotes, RNA polymerase cannot bind to the promoter by itself; a group of other proteins must bind with it. The DNA sequences involved in regulating expression are far more extensive, with regulatory proteins binding not only near the promoter, but also to sites further away (called enhancers). Furthermore, in eukaryotes, DNA is complexed with several types of proteins, in a structure called *chromatin* (see Chapter 6). The presence of this chromatin structure can prevent regulatory proteins and the transcription machinery from gaining access to the DNA. Chromatin can be chemically modified by enzymes found in the cell. Many of these modifications help the binding of RNA polymerase and its associated factors.

Eukaryotes have three RNA polymerases. Each polymerase has a role in transcribing a different kind of RNA. RNA polymerase I (pol I) is responsible for transcribing the large ribosomal precursor RNA (45S rRNA, which is processed into the three smaller 18S, 28S, and 5.5S rRNAs). RNA polymerase II (pol II) is responsible for transcribing primarily mRNAs and a few specialized RNA molecules. RNA polymerase III (pol III) transcribes tRNAs, 5S RNAs, and other small RNAs. We will focus on RNA polymerase II, a molecule made up several protein subunits.

Initiation of Transcription

The promoter region of a eukaryotic gene can be divided into distinct parts. The first part is the TATA box, which is located approximately 25–30 bp upstream of the start point of transcription. The TATA box is so named because of its consensus sequence, which is TATAAAA, although small variations in this sequence are found. A number of factors that associate with pol II bind to, or around, the TATA box. For example, TFIID is a complex of many subunits, including the protein that binds specifically to the TATA box, called TATA-binding protein (TBP). The TBP is essential for the initiation of transcription and forms a central part of the *preinitiation complex*. TBP binds in the narrow (minor) groove of DNA at the TATA box and bends the DNA. This forces the helix to open slightly, probably allowing better access to RNA polymerase. The rest of TFIID is made up of *T*BP-*a*ssociated *f*actors (TAFs); TAFs contain some nine proteins, the composition of which may vary, with different promoters.

Transcriptional Regulation

Some regulatory proteins (called transcription factors) bind close to the promoter, whereas others bind to more distant enhancer elements. Transcription

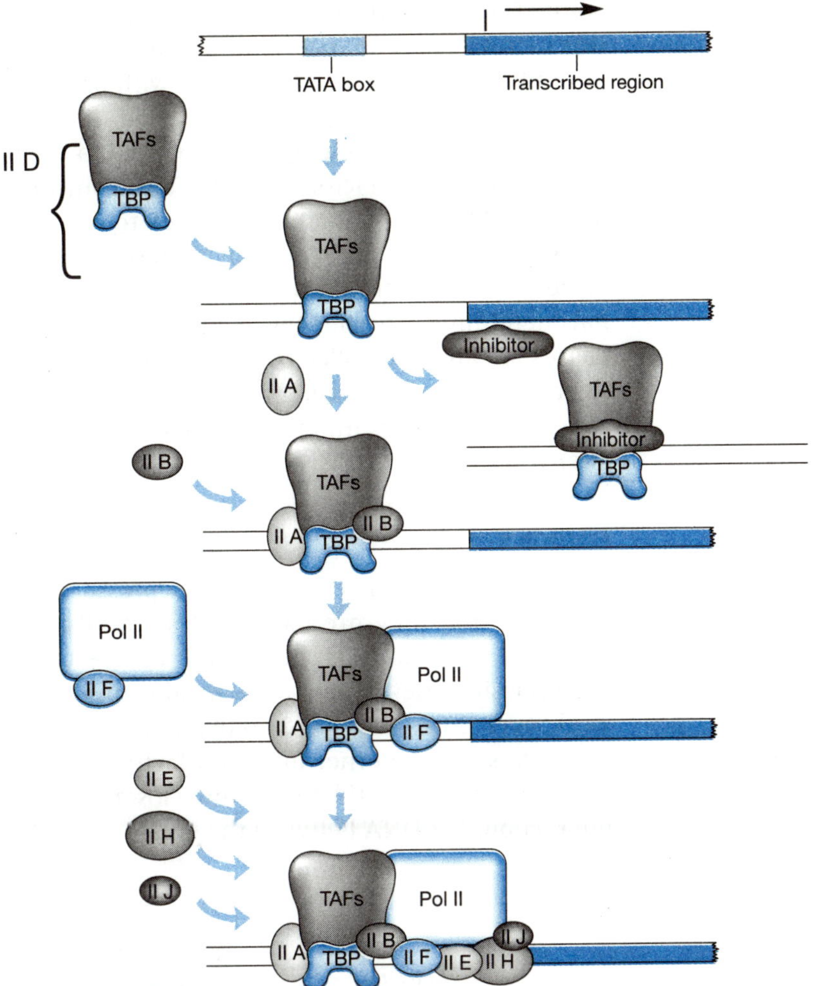

Binding at the TATA Box

A number of transcription factors (TFs) participate to form a preinitiation complex for essentially all genes transcribed by RNA polymerase II. The process begins with the binding of the basal transcription factor TFIID, composed of several transcription-associated factors (TAFs) and the TATA-binding protein. TFIID binds the inhibitor, which is then displaced by the arrival of additional transcription factors and RNA polymerase II.

factors can act positively to promote transcription (*activators*) or negatively to reduce transcription (*repressors*), just as was seen in the bacterial *lac* operon. The regulation of a specific gene requires the binding of multiple transcription factors. For example, the GC box is a common element in eukaryotic promoters; its consensus sequence is GGGCGG (as is true with the TATA box, variations of this sequence also exist). The GC box may be present in one or more copies, which can be located between 40 and 100 bp upstream of the start point of transcription. This element is bound by the transcriptional activator Sp1. The CCAAT box (consensus sequence GGCCAATCT, with some variations seen) is also often found between 40 and 100 bp upstream of the transcription start point. The CCAAT box element is bound by the transcriptional activator C/EBP (for CCAAT-box/enhancer-binding protein).

Eukaryotic activators bind DNA with one surface and interact with the transcription factors associated with pol II at the TATA box with another surface. In this way, activators help polymerase bind to the promoter and initiate transcription. Transcription factors probably also control the rate at which pol II releases from the TATA box initiation complex to carry out transcription.

We saw at the *lac* operon in bacteria that the regulators (e.g., CAP and Lac repressor) are controlled by signals (glucose and lactose, respectively). These signals determine whether the regulators bind DNA. In eukaryotes, transcription factors are also controlled by signals, such as growth factors, but in this case, the signals control their target proteins in a variety of ways, including entry of the transcription factors into the nucleus, and the activity of the transcription factor after it is bound to DNA.

Enhancers

Enhancers also bind activators and repressors, but differ from the promoter elements in that they are located at some distance from the gene. In addition, enhancer sequences can be moved relative to the promoter, or even flipped in the opposite orientation, without affecting their function. How do enhancers work over great distances? Activators bound to enhancers interact with RNA polymerase or the other initiation factors at the promoter, and so activate in a manner similar to activators bound closer to the promoter. To accommodate this interaction, the DNA between the enhancer and the promoter must form a loop.

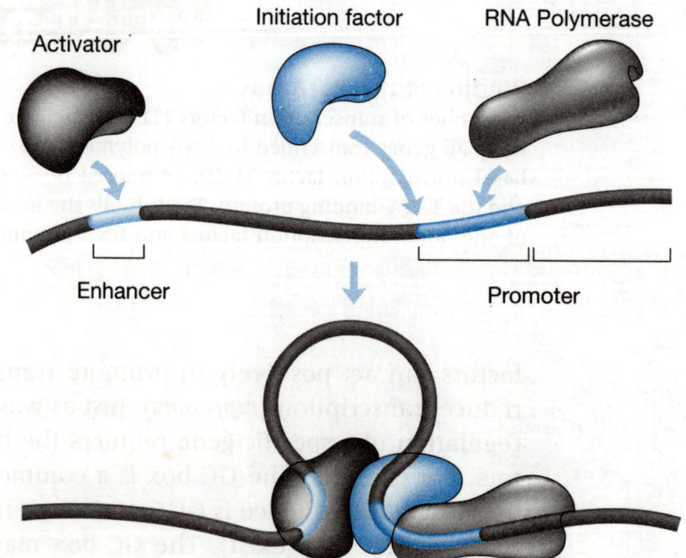

How Enhancers Act at a Distance
Studies have shown that activator proteins bound to the enhancer physically interact with proteins bound to the promoter region through DNA looping. This physical interaction allows the enhancer-binding proteins to exert their effect on the promoter, although it can be a great distance away on the DNA.

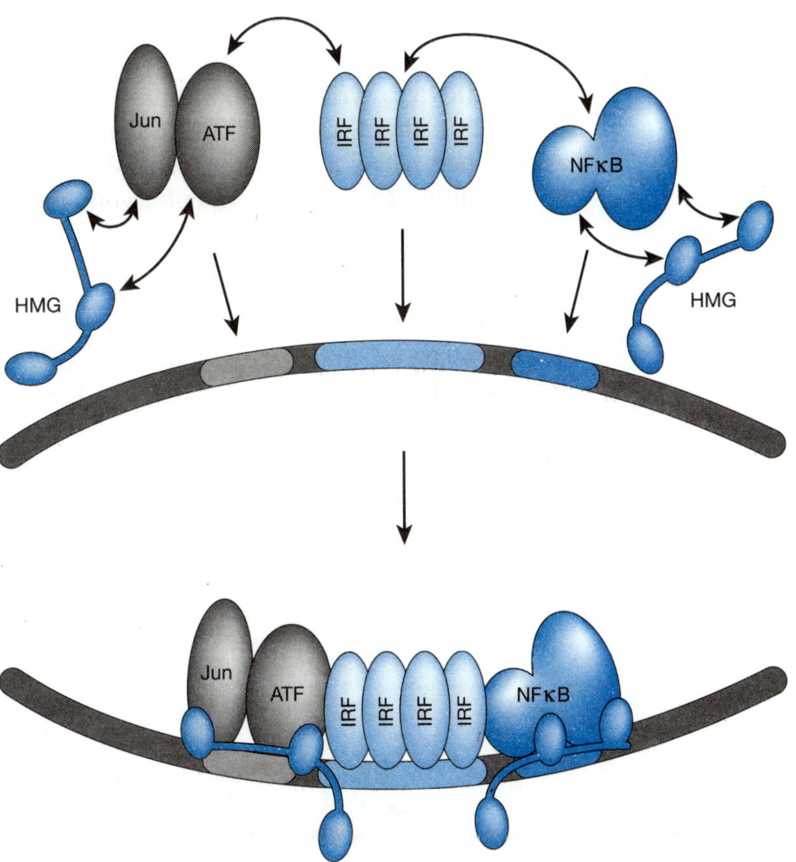

The Human Interferon-β Enhanceosome

Three factors, Jun/ATF, IRF, and NF-κB interact with each other and bind cooperatively to the interferon-β enhancer. HMG-I also interacts with these proteins and bends the DNA of the enhancer in a manner that helps the factors bind. (Reprinted, with permission, from Ptashne and Gann 2002.)

Signal Integration at Enhancers

Activators bound at an enhancer can work together to switch on a gene, which is an example of "signal integration." With the *lac* genes in *E. coli*, we saw how two regulators work together—a repressor and an activator—to ensure that the genes are only expressed efficiently when two signals are received. Many eukaryotic enhancers coordinate the input from multiple regulators as well. One example is the enhancer that controls the human *interferon*-β gene; the product of this gene helps cells fight viral infection. When a virus infects a cell, it triggers three different activators: NF-κB, IRF, and JUN/AP1. These activators bind to sites adjacent to one another upstream of the *interferon*-β promoter. Each activator helps the other activators bind, and they only bind if all three are present; this is called cooperative binding, and in this case, it works in two ways. First, the activators interact with each other, enhancing their ability to bind DNA. Second, an additional protein, HMG-I, binds within the enhancer, bending the DNA in a way that helps the activators bind and interact with each other.

RNA PROCESSING

As an mRNA is synthesized, it is also being further processed. A "cap" structure is added to the 5′ end. Following completion of transcription, a long string of adenosines, called a poly(A) tail, is added at the 3′ end. This processing partially determines the life span of eukaryotic mRNA.

Creating a 5′ Cap Structure

The mRNA cap is composed of a guanosine nucleotide covalently attached to the 5′ phosphate of the first nucleotide of the mRNA. The guanine base is then methylated at position N7. Further methylation of the first encoded nucleotide of the mRNA can also occur. This process involves three separate enzymatic reactions, which may be catalyzed by the same polypeptides or by different polypeptides, depending on the organism:

- When mRNA is synthesized, it initially always contains a triphosphate on its 5′ end. Two of the three phosphates are removed by RNA 5′-triphosphatase, creating a monophosphate.

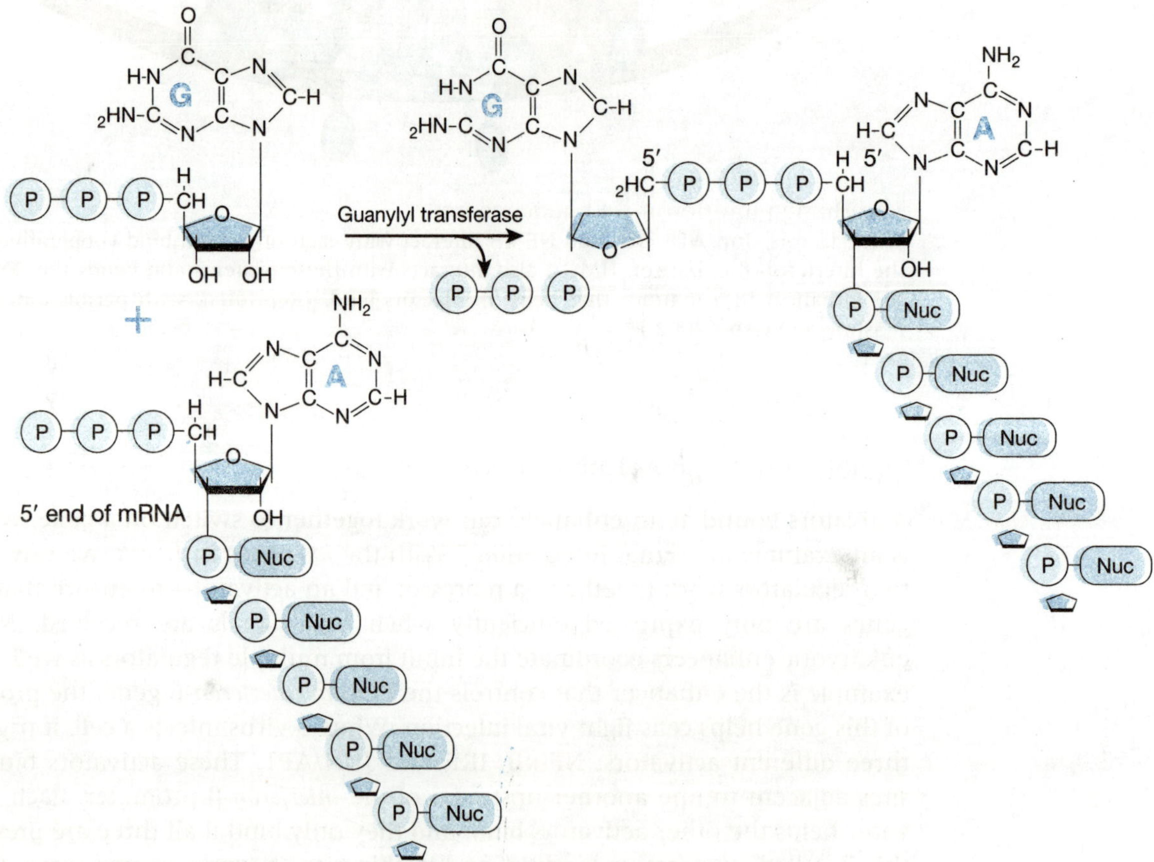

Addition of the 5′ Cap Structure to mRNA
The 5′ cap on mRNAs is formed when a guanylyl transferase joins a molecule of GTP to the 5′ end of a newly synthesized mRNA which still contains a triphosphate. Two phosphates are removed from the mRNA and one phosphate is removed from the GTP. Once joined, the guanine is methylated. The adenosine can also be methylated.

- Next, a guanosine diphosphate (GDP) is brought in and ligated to the remaining 5′ monophosphate of the mRNA by mRNA guanylyl transferase. This is also called capping enzyme.
- Finally, N7 of the guanine base is methylated by RNA guanine-7-methyltransferase.

The cap structure not only stabilizes the mRNA, but also is involved in mRNA splicing, export of mRNA from the nucleus, and recognition of mRNA by the translation machinery in the initiation of protein synthesis.

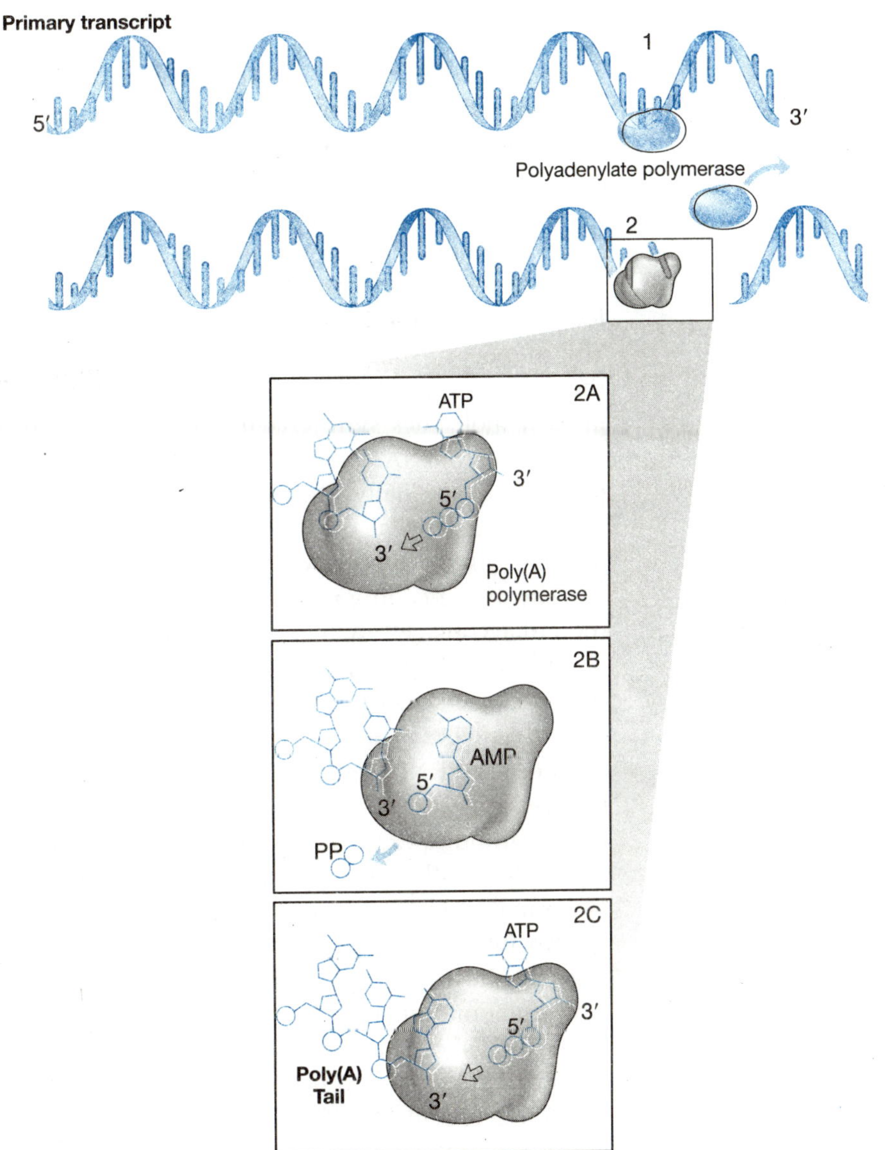

Addition of the 3′ Poly(A) Tail to Pre-mRNA

(*1*) Polyadenylate polymerase cleaves adjacent to the sequence AAUAAA near the 3′ end of the pre-mRNA. (*2A–C*) Polyadenylate polymerase catalyzes the addition of some 200 adenosine residues to the 3′ end.

Creating a Poly(A) Tail

At the 3′ end of essentially all mRNA molecules exists a run of adenosines called a poly(A) tail. A specific mRNA sequence, AAUAAA (again, variations of this sequence exist), is recognized by the enzyme polyadenylate polymerase, an endonuclease that cleaves the primary transcript approximately 11–30 bases 3′ of this sequence element. A stretch of 200–250 adenosines (in mammalian cells) is then added to the 3′ end of the mRNA by the polyadenylate polymerase, creating the poly(A) tail. The poly(A) tail stabilizes the mRNA and is an extremely useful feature for scientists in isolating and working with mRNA.

mRNA Stability and Turnover

In Chapter 2, mRNA was described as a short-lived messenger that is rapidly turned over. This makes sense, because as long as an mRNA is around, it can be used to make protein. However, the cell may only need limited production of that particular protein. Thus, the number of protein molecules produced by a single mRNA is dependent on, among other factors, the length of time an mRNA exists in the cell before being degraded or "turned over." The stability of a specific mRNA is determined by measuring the time it takes for half the population of transcripts to be degraded—termed the mRNA half-life. For example, the mRNAs that code for α- and β-globin have very long half-lives, on the order of hours or even days. The mRNAs of some regulatory proteins, on the other hand, have half-lives on the order of seconds. Other factors being equal, the longer the half-life, the more protein molecules synthesized from a given mRNA transcript. One factor that determines mRNA half-life is the poly(A) tail. It has been observed that the poly(A) tail gradually shortens (degrades) as the mRNA "ages" in the cytoplasm. This appears to be a contributing factor to mRNA half-life, since mRNA molecules degrade when the poly(A) tail shortens to a particular length.

The 3′-noncoding region of mRNA can also be involved in mRNA stability in the eukaryotic cell. A sequence rich in adenine and uracil (AU) residues is frequently found in the 3′-noncoding region of mRNAs from genes that are transiently expressed. This sequence appears to be a signal for the selective degradation of mRNAs. The AU-rich sequence is recognized by a 3′ nuclease, which degrades the RNA molecule from its 3′ end. This effect has been demonstrated by inserting an AU-rich sequence in the 3′-noncoding region of the β-*globin* gene. Although the β-*globin* transcript is normally very stable, engineered β-*globin* mRNAs with the 3′ AU sequence degrade rapidly and accumulate at only 3% of normal levels.

An intriguing experiment with the *fos* oncogene, a gene mutated in some cancers, shows that altering regulation of mRNA half-life can have devastating developmental effects on a cell. The mRNA of the normal cellular gene (c-*fos*) has a very short half-life. The same gene carried in a retrovirus (v-*fos*) causes cancer in infected organisms. v-*fos* mRNA is considerably more stable and has a much longer half-life. There is convincing evidence that the prolonged presence of the Fos protein contributes to malignant transformation of a normal cell into a cancerous one. The rapid degradation of the normal c-*fos* mRNA is attributed to the presence of a 67-bp AU-rich sequence in the 3′-noncoding region. The mRNA of v-*fos* differs from that of c-*fos* in several regions, including the absence

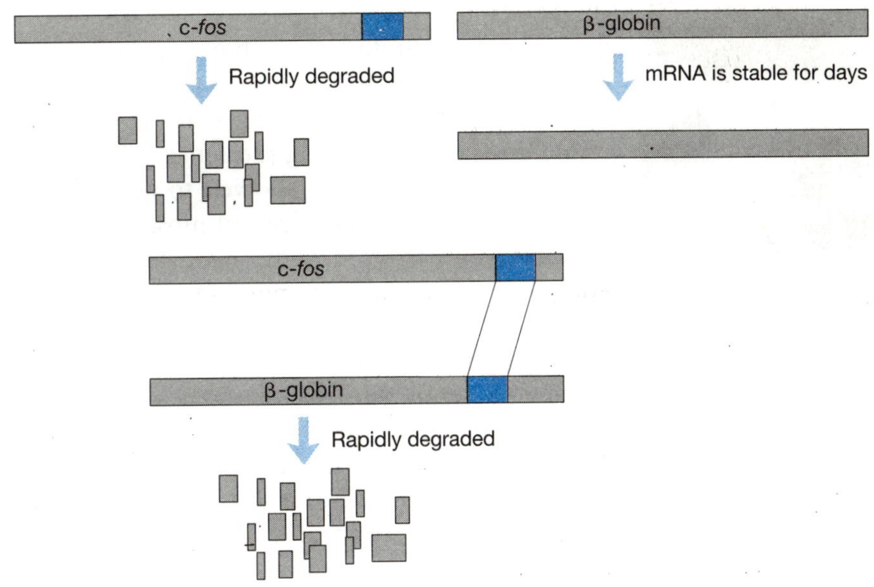

mRNA Turnover

Although c-*fos* mRNA turns over very rapidly, on the order of seconds or minutes, the β-*globin* mRNA has been shown to be stable for days. One explanation for this difference is the presence of an AU-rich sequence in the 3´-untranslated region of the c-*fos* gene which is a target for degradation. This is demonstrated by the finding that placing this AU-rich sequence into the 3´-untranslated region of the normally stable globin genes greatly decreases its stability.

of the AU-rich sequence. It was found that the normal c-*fos* gene could be made oncogenic by increasing its transcription and deleting the 3´ AU-rich sequence, thus making it resemble the 3´ region of v-*fos*. In addition, the c-*fos* gene contains a sequence within its coding region called mCRD, which also has a major role in degrading the c-*fos* mRNA. mCRD-dependent degradation occurs through interaction with the poly(A) tail of the mRNA.

RNA Splicing

In 1976, Jerry Lingrel, Jeffrey Ross, and Charles Weissman, whose laboratories were studying β-*globin* mRNA, reported that they had discovered a β-*globin* precursor RNA in the nucleus that was larger than β-*globin* mRNA. They found that this precursor was then processed into the mature β-*globin* mRNA found in the cytoplasm. This seemed to explain the earlier finding of James Darnell and others that the nucleus is filled with large RNA molecules, called heterogeneous nuclear RNA or hnRNA.

At the same time, studies in adenovirus, a human DNA virus, also indicated the existence of precursor RNAs, or pre-mRNA. A group headed by Richard Roberts at Cold Spring Harbor Laboratory and, independently, a second group headed by Philip Sharp at the Massachusetts Institute of Technology made an amazing finding. These groups created heteroduplex molecules by hybridizing specific adenovirus mRNAs to single-stranded adenovirus DNA. They examined these molecules by electron microscopy and found that the mRNA hybridized to several discontinuous regions of the DNA, that is, the DNA looped out in long

Richard Roberts
(Courtesy of Cold Spring Harbor Laboratory Archives.)

Philip Sharp
(Courtesy of Cold Spring Harbor Laboratory Archives.)

single strands between the regions where RNA hybridized to DNA. Their explanation of these findings was that genes (and pre-mRNAs) contain one or many blocks of sequence, called *introns*, that are not represented in the mature mRNA. A process called RNA splicing edits out these introns and joins together the remaining pieces of coding sequence, called *exons*, to form the messenger RNA. What Roberts and Sharp showed to be true in the adenovirus genome turned out to be true in the genomes of most eukaryotes.

Other investigators went on to show that introns are excised in a two-step process of sequential *trans*-esterification reactions. In the first reaction, the cleavage of the 5′ intron/exon junction occurs simultaneously with the formation of a new phosphodiester bond between the phosphate at the 5′ end of the intron and an adenine residue just within the 3′ end of the intron. This unusual phosphodiester bond is formed between the 5′ phosphate and the 2′ hydroxyl group of the sugar. The resulting structure has the appearance of a branched lariat. In the second step, another *trans*-esterification reaction occurs, in which the phosphodiester bond at the 3′ intron/exon junction is cleaved and the two cleaved exons are joined. The net result of these two reactions is that two exons are joined together and the intervening intron is released as a branched lariat structure.

What are the components of the machinery that are responsible for RNA splicing? Five small U-rich RNAs, designated U1, U2, U4, U5, and U6, are abundant in the nuclei of mammalian cells and, along with more than 60 proteins, make up of the cellular splicing machinery. These small nuclear RNAs (snRNAs) range in size from 107 to 187 nucleotides. Several observations led to the suggestion that snRNAs assist in the splicing reaction. First, a short consensus sequence at the 5′ end of introns (CAG|GUAAGU) was found to be complementary to a sequence near the 5′ end of the U1 snRNA. Second, snRNAs were found to be associated with hnRNAs in nuclear extracts. Each snRNA associates in the nucleus with at least 10 proteins and, in some cases, more than 20 proteins to form small nuclear ribonucleoprotein (snRNP) particles.

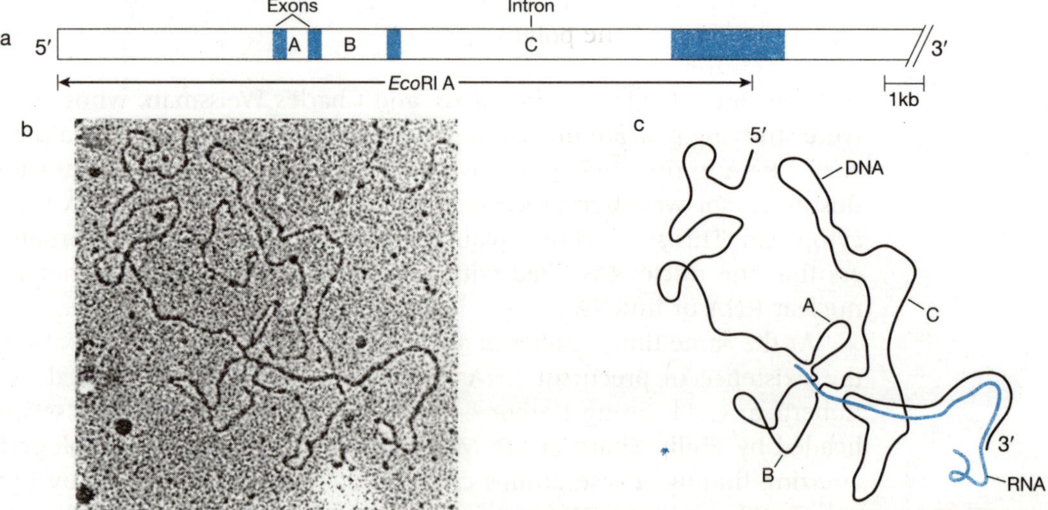

Genes Contain Introns That Are Spliced Out of mRNAs

RNA splicing was first identified in adenovirus. Heteroduplex molecules are formed between an adenovirus mRNA and its DNA. The molecules were examined by electron microscopy as seen in *b*. The drawing in *c* makes clear that there are some regions of DNA (introns) that are not included in the mRNA. (*Bottom:* Reprinted, with permission, from Berget S.M., Moore C., and Sharp P.A. 1977. Spliced Segments at the 5′ terminus of adenovirus 2 late mRNA. *Proc. Natl. Acad. Sci. 74:* 3171–3175.)

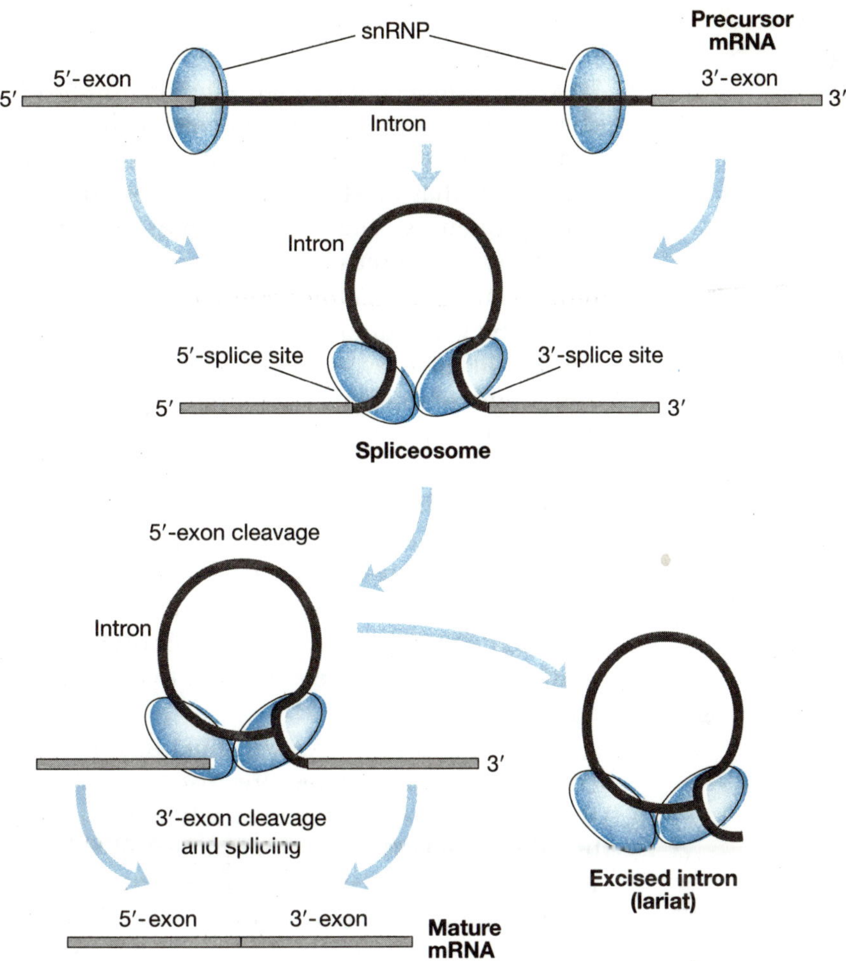

RNA Splicing

The 5′ intron/exon junction is cleaved simultaneously with the formation of an unusual 5′-2′ phosphodiester bond between the phosphate at the 5′ end of the intron and an adenine residue just within the 3′ end of the intron. The resulting structure has the appearance of a branched lariat. The 3′ intron/exon junction is then cleaved and the two cleaved exons are joined. The net result of these two reactions is that two exons are joined together and the intervening intron is released as a lariat structure.

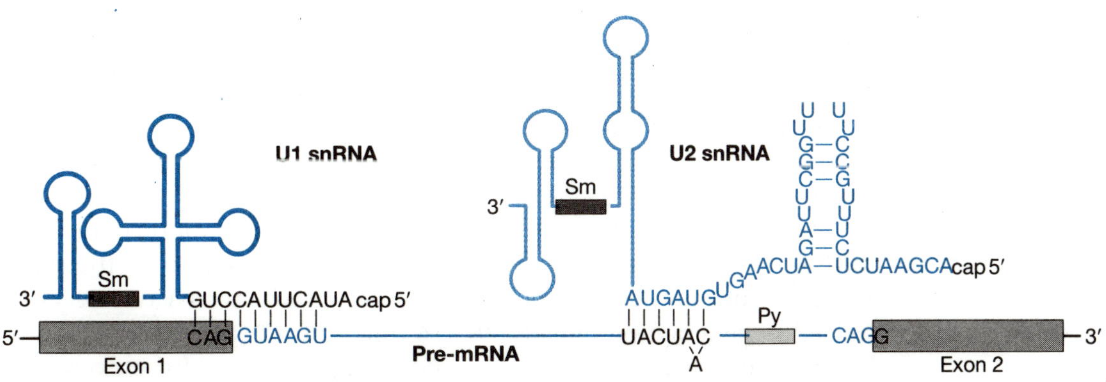

How snRNAs Work

Two snRNAs are shown forming partial hybrids with a pre-mRNA. U1 is forming a hybrid at the junction of the 5′ exon and U2 is forming a partial hybrid with a sequence near the 3′ exon.

ALTERNATIVE SPLICING

This new understanding of the organization of eukaryotic genes radically changed our view of the genome and led scientists to realize that it is much more complex than originally thought. As more and more genes were discovered, it was found that most eukaryotic genes consist of more intron sequences than exon sequences. The β-*globin* gene, discussed above, has a relatively simple structure, with three exons separated by two introns. Thus, the splicing of the β-*globin* gene is relatively straightforward compared with other genes, such as *FBN1*, the gene mutated in Marfan syndrome. Marfan syndrome was first recognized in 1896 by the French physician Antonin Bernard Jean Marfan, in a young girl who he was treating. The disease results in tall individuals, with abnormally long limbs, elastic joints, spinal column deformities, an arched palate, and crowned teeth. In addition, patients suffer from myopia and cardiovascular deformities, including problems with the heart valves and with the aorta. Individuals with Marfan syndrome often die prematurely as a result of these cardiovascular irregularities. (Interestingly, Abraham Lincoln is suspected to have had Marfan syndrome, which was not recognized as a disease during his lifetime.) The *FBN1* gene codes for fibrillin-1, a large protein (nearly 3500 amino acids long) that provides structure to connective tissue. The gene is 110 kb long and contains 65 exons.

One of the largest known human genes is *Titin*, which has more than 175 exons. With such a large precursor mRNA, it seems that it would be easy to splice together the wrong exons. In mammalian cells, intron removal appears to be largely dependent on specific features of individual introns, including how well the splice signal sequences match consensus splice site sequences, intron length, and other regulatory sequences located within the pre-mRNA. Splicing occurs cotranscriptionally, that is, while transcription is still in progress. One model suggests that some genes maintain the correct order of splicing by joining exons almost as soon as they are synthesized.

Recent studies on the human genome estimated that the number of genes in the genome is between 30,000 and 40,000. This number was surprising because the estimated number of different human proteins is on the order of 100,000 to 150,000. It is now thought that these gene estimates are low and that the actual number is more than 50,000. However, we already know that the number of genes alone does not tell the whole story about the complexity of genetic information. This is because of a phenomenon known as alternative RNA splicing.

With some proteins, multiple variations can exist at different times or in different tissues. In the example of the α and β globins discussed later in this chapter, this is accomplished by having different genes encoding each variation. However, there are many examples of multiple forms of a protein arising from a single gene. In fact, early estimates based on analysis of human chromosome 22 mRNAs indicate that about 60% of genes make more than one form of mRNA. A striking example of alternative splicing is seen in the gene for the Down's syndrome cell-adhesion molecule (*Dscam*) which can be alternatively spliced to generate potentially 38,016 different mRNAs. These variations occur because, during splicing, there is differential selection of the exons in the precursor mRNA for forming a mature mRNA transcript.

ACAUUUGCUUCUGACACAACUGUGUUCACUAGCAACCUCAAACAGACACCAUGGUGCACCUGACUCCUGAGGAGAAC
(met) val his leu thr pro glu glu lys
1 5

UCUGCCGUUACUGCCCUGUGGGGCAAGGUGAACGUGGAUGAAGUUGGUGGUGAGGCCCUGGGCAGGUUGGUAUCAAG
ser ala val thr ala leu trp gly lys val asn val asp glu val gly gly glu ala leu gly arg
10 15 20 25 30

GUUACAAGACAGGUUUAAGGAGACCAAUAGAAACUGGGCAUGUGGAGACAGAGAAGACUCUUGGGUUUCUGAUAGGC

ACUGACUCUCUCUGCCUAUUGGUCUAUUUUCCCACCCUU**AGGCUGCUGGUGGUCUACCCUUGGACCCAGAGGUUCUUU**
leu leu val val tyr pro trp thr gln arg phe phe
31 35 40

GAGUCCUUUGGGGAUCUGUCCACUCCUGAUGCUGUUAUGGGCAACCCUAAGGUGAAGGCUCAUGGCAAGAAAGUG
glu ser phe gly asp leu ser thr pro asp ala val met gly asn pro lys val lys ala his gly lys lys val
45 50 55 60 65

CUCGGUGCCUUUAGUGAUGGCCUGGCUCACCUGGACAACCUCAAGGGCACCUUUGCCACACUGAGUGAGCUGCAC
leu gly ala phe ser asp gly leu ala his leu asp asn leu lys gly thr phe ala thr leu ser glu leu his
70 75 80 85 90

UGUGACAAGCUGCACGUGGAUCCUGAGAACUUCAGGGUGAGUCUAUGGGACCCUUGAUGUUUUCUUUCCCCUUCUUU
cys asp lys leu his val asp pro glu asn phe arg
95 100 104

UCUAUGGUUAAGUUCAUGUCAUAGGAAGGGGAGAAGUAACAGGGUACAGUUUAGAAUGGGAAACAGACGAAUGAUUG

CAUCAGUGUGGAAGUCUCAGGAUCGUUUUAGUUUCUUUUAUUUGCUGUUCAUAACAAUUGUGUAUAACAAAAGGAAAU

AUCUCUGAGAUACAUUAAGUAACUAAAAAAAAACUUUACACAGUCUGCCUAGUACAUUACUAUUUGGAAUAUAUGUG

UGCUUAUUUGCAUAUUCAUAAUCUCCCUACUUUAUUUUCUUUUAUUUUAAUUGAUACAUAAUCAUUAUACAUAUUUAUG

GGUUAAAGUGUAAUGUUUUAAUAUGUGUACACAUAUUGACCAAAUCAGGGUAAUUUUGCAUUUGUAAUUUUAAAAAAU

GCUUUCUUCUUUUAAUAUACUUUUUUGUUAUCUUAUUUCUAAUACUUUCCCUAAUCUCUUUCUUUCAGGGCAAUAAUGA

UACAAUGUAUCAUGCCUCUUUGCACCAUUCUAAAGAAUAACAGUGAUAAUUUCUGGGUUAAGGCAAUAGCAAUAUUU

CUGCAUAUAAAUAUUUCUGCAUAUAAAUUGUAACUGAUGUAAGAGGUUUCAUAUUGCUAAUAGCAGCUACAAUCCAG

CUACCAUUCUGCUUUUAUUUUAUGGUUGGGAUAAGGCUGGAUUAUUCUGAGUCCAAGCUAGGCCCUUUUGCUAAUCAU

GUUCAUACCUCUUAUCUUCCUCCCAC**AGCUCCUGGGCAACGUGCUGGUCUGUGUGCUGGCCCAUCACUUUGGCAAA**
leu leu gly asn val leu val cys val leu ala his his phe gly lys
105 110 115 120

GAAUUCACCCCACCAGUGCAGGCUGCCUAUCAGAAAGUGGUGGCUGGUGUGGCUAAUGCCCUGGCCCACAAGUAU
glu phe thr pro pro val gln ala ala tyr gln lys val val ala gly val ala asn ala leu ala his lys tyr
125 130 135 140 145

CACUAAGCUCGCUUUCUUGCUGUCCAAUUUCUAUUAAAGGUUCCUUUGUUCCCUAAGUCCAACUACUAAACUGGGGG
his **stop**

AUAUUAUGAAGGGCCUUGAGCAUCUGGAUUCUGCCUAAUAAAAAACAUUUAUUUUCAUUGC

β-globin Gene Structure

The *β-globin* gene sequence is shown. Intron sequences are shown in black and exon sequences are in blue.

Another example of a gene that produces multiple mRNAs codes for the actin-binding protein tropomyosin, which is expressed in all cells. There are four *tropomyosin* (*TM*) genes: *TM-α*, *TM-β*, *TM-4*, *TM-5*. However, more than 25 different isoforms of tropomyosin are known to exist. At least nine isoforms of TM-α have been identified, and recent data suggest that several more isoforms may exist. Studies have shown that the expression of specific isoforms is specific both for the tissue that expresses the gene and for the stage of development of the organism.

The regulation of alternative splicing is not completely understood, but information on a number of splicing factors that control alternative RNA splicing is beginning to accumulate. Polypyrimidine-tract-binding protein (PTB) binds to the polypyrimidine tract of precursor mRNAs, a 10–40-nucleotide tract of pyrimidines (Us and Cs) present in most mammalian introns. PTB exists as three isoforms, which themselves arise from alternative RNA splicing: PTB1, PTB2, and PTB4. These proteins have been shown to have a direct role in alternative splicing of *TM*-α. Exon 3 of *TM*-α is not present in smooth muscle. Binding of PTB4 enhances the exclusion of exon 3 in *TM*-α, whereas binding of PTB1 seems to promote the inclusion of exon 3. Thus, PTB4 is expressed in smooth muscle and PTB1 is not. There are many known splicing factors that have a role in the processing of mRNAs from a wide variety of genes.

Possibly the most intriguing aspect of alternative RNA splicing has yet to be understood. Alternative splicing occurs at extremely high levels in nervous tissue and has been shown to have a major role in neurotransmission, learning, and memory. Many proteins that regulate alternative splicing have been shown to be active in nerve cells. In addition, extracellular stimuli such as changes in Ca^{++} are believed to alter splicing patterns of RNAs in nerve cells. One possible explanation is that the extracellular signal activates a series of events that lead to the modification of splicing regulatory proteins, which changes the pattern of splicing of targeted pre-mRNAs. The role of alternative RNA splicing in the nervous system is an active area of study and is likely to be central to our understanding of memory, learning, and even our emotions.

The association of unusual alternative splicing patterns with some neurological and psychiatric diseases highlights the importance of alternative splicing in

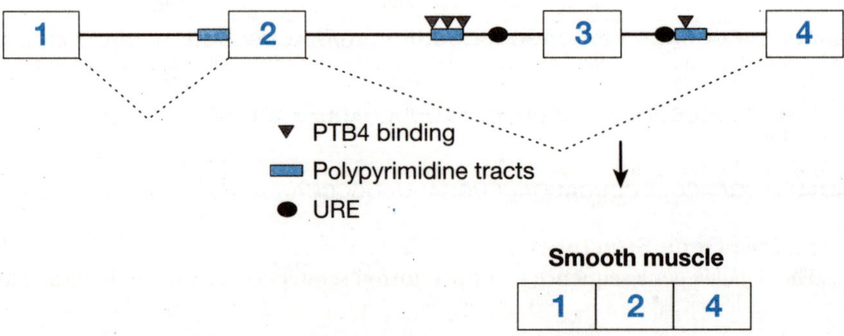

Alternative RNA Splicing of α-tropomyosin

In smooth muscle, exon 3 of α-tropomyosin pre-mRNA is removed during splicing. This is controlled by binding of the polypyrimidine-tract-binding protein (PTB) which binds to polypyrimidine tracts (shown as small boxes). PTB4 binding enhances the exclusion of exon 3. The upstream regulatory element (URE) is a CUG repeat sequence that also controls splicing. (Adapted, with permission, from Wollerton M.C., Gooding C., Robinson F., Brown E.C., Jackson R.J., and Smith C.W. 2001. Differential alternative splicing of isoforms of polypyridime tract binding protein (PTB). *RNA 7:* 819–832.)

Splicing Factors

Protein	Organism	Known target genes	Inhibition (–) or activation (+) of splicing
Sxl	*Drosophila*	*sxl, tra, msl-2*	–
PSI	*Drosophila*	gene for P-element transposase	–
hrp48	*Drosophila*	gene for P-element transposase	–
hnRNPA1	mammalian	HIV *tat*, FGF receptor hnRNPA1 (autoregulation)	–
hnRNP-F	mammalian	*src*	+
hnRNP-H	mammalian	*src*	+
hnRNP-H	mammalian	TM-β	–
PTB (hnRNP-I)	mammalian	*src* and genes for TM-α and TM-β, α-actinin, $GABA_A$ γ2 receptor	–
PTB (hnRNP-I)	mammalian	gene for calcitonin/CGRP	+
CUG-BP (hNAB50)	mammalian	genes for cardiac troponin-T, CLCB(?), NMDA(?)	+
KSRP	mammalian	*src*	+
Tra	*Drosophila*	*dsx, fru*	+
Tra-2	*Drosophila*	*dsx, fru*	+
Tra-2	*Drosophila*	*tra-2*	–
RBP1 (d9G8?)	*Drosophila*	*dsx*	+
dSRp30 (dsF2?)	*Drosophila*	*dsx*	+
RSF1	*Drosophila*	not known	–
SWAP	*Drosophila*	gene for SWAP	–
SRp20	mammalian	gene for SRp20 gene for calcitonin	+
SRp30a/SF2/ASF	mammalian	genes for bovine growth hormone, fibronectin gene for TM-β gene for CD45	+
SRp30/SF2/ASF	mammalian	gene for SRp20 gene for RSV gene for adenovirus L1 transcription unit	–
SC35	mammalian	gene for TM-β	–
9G8	mammalian	gene for fibronectin	+
SRp40	mammalian	gene for cTnT gene for fibronectin	+
SRp55	mammalian	gene for TnT gene for CD45	+

(Adapted, with permission, from Smith C.W.J. and Valcárcel J. 2000. Alternative pre-mRNA splicing: The logic of combinatorial control. *Trends Biochem. Sci. 25:* 381–388.)

nervous tissue. For example, in amyotrophic lateral sclerosis (ALS, also known as Lou Gehrig's disease), neurons die as a result of accumulating toxic levels of the neurotransmitter glutamate. This accumulation is due to reduced levels of excitory amino acid transporter 2 (EAAT2), which rids the cell of excess glutamate. Studies of ALS patients have revealed that there is no defect in the *EAAT2* gene or in its transcription, and researchers were left to find another explanation for this disease. Recent studies revealed that at autopsy, many ALS patients showed the presence of abnormally spliced *EAAT2* mRNAs, which could account for reduced *EAAT2* activity and explain the cause of the disease. Altered

RNA splicing patterns have also been reported in schizophrenia. The γ aminobutyric acid type A ($GABA_A$) receptor has two splice variants, γ2L and γ2S, with γ2S being the dominant form in normal brain. In the postmortem brains of schizophrenia patients, the ratio of γ2L to γ2S is greater, indicating a shift in splicing patterns. Abnormal patterns of two other alternatively spliced mRNAs, those of neural cell adhesion protein and NMDAR1 receptor, have also been seen in postmortem brains of schizophrenia patients.

Gene mutations that disrupt splicing are now believed to have a role in many human diseases. Investigating these diseases has become an important area of splicing research. As splicing mechanisms are beginning to be understood, approaches to correct defects therapeutically are beginning to look promising.

TRANSLATIONAL CONTROL

Protein levels can also be regulated at the translation step. There is no better example of this than the regulation of proteins that utilize iron (Fe^{++}/Fe^{+++}). Iron is tightly regulated due to the difficulty in obtaining it. In addition, it has toxic side effects, so cells limit the amount of iron which they possess. The toxicity of iron is largely due to its ability to create reactive oxygen molecules, which can destroy tissues and cells by oxidation. It was recognized that the synthesis of iron-using proteins (such as ferritin) is diminished when iron levels are low. At the same time, there is increased synthesis of proteins involved in increasing iron levels in cells (such as the transferrin receptor which transports iron into cells). The opposite situation arises when iron levels are high.

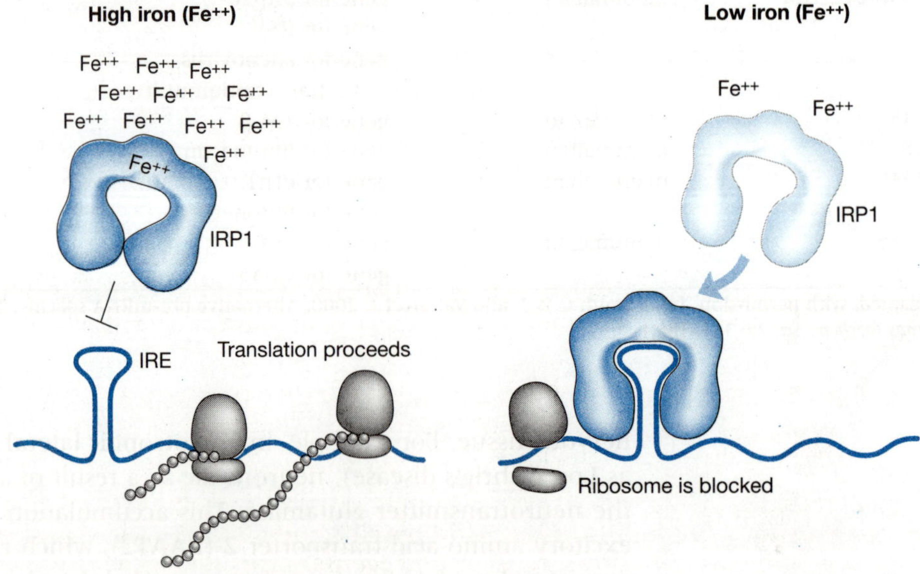

Regulation of Translation of mRNA Containing an IRE by IRP Binding

The iron responsive protein (IRP1) binds iron (Fe^{++}) when iron concentrations are high. When iron is bound to the IRP1, it assumes a closed configuration. In this state, IRP1 cannot bind to the iron response element (IRE) of the mRNA, thus allowing for protein synthesis (shown at left). However, when iron is low, it is not bound to IRP1, leading to its taking on an open conformation. This open form of IRP1 binds to the IRE of the mRNA blocking the ribosome from translating the mRNA into protein.

Studies showed that the mRNAs for these proteins contain *stem loop* structures that bind to a specific protein. The stem loop structure was named the iron response element (IRE), and the protein that binds to IREs has been named the IRE-binding protein (IRP1). Proteins whose synthesis is reduced in low iron and elevated in high iron, like ferritin, have IREs at the 5′ end of their mRNAs. Studies showed that binding of IRP1 to these 5′ IREs prevents the binding of the mRNA to the ribosome, blocking protein synthesis. On the other hand, proteins that increase cellular iron levels have IREs at their 3′ end. Binding of the IRP1 to 3′ IREs increases the stability of the mRNA so that it exists for a longer time before being degraded, which results in being able to make more protein. At high iron levels, the IRP1 stops binding to the IRE, which leads to increased ferritin protein synthesis and decreased transferrin receptor levels.

How does the IRP1 control its binding to IREs according to iron level? It turns out that IRP1 is itself an iron-binding protein and so functions as a sensor of cellular iron concentrations. Binding of iron to IRP1, as with any chemical reaction, is in an equilibrium. When iron is plentiful in the cell, IRP1 binds iron, but when iron levels are low, IRP1 is more likely to exist without bound iron. Iron-bound IRP1 exists in a closed conformation and cannot bind to an IRE; when IRP1 loses its iron, it is converted into an open configuration and is able to bind to IREs. Once iron levels increase again, IRP1 again binds iron and no longer binds to IREs.

TRANSPOSABLE ELEMENTS: MAIZE

From the founding of genetics in the first years of the 20th century, many biologists were intrigued by variegated, or mosaic, pigmentation. In the late 1940s and early 1950s, Barbara McClintock, of the Carnegie Department of Genetics at Cold Spring Harbor Laboratory, developed a daring hypothesis about the genetic basis of the striking color variations in the leaves and kernels of Indian corn (maize). She proposed that the movement of genetic "controlling elements" from one chromosomal location to another results in unstable mutations in color-forming genes.

One such transposable, or movable, genetic element is Dissociator (*Ds*), which causes mutations when it inserts into an active gene locus. For example, assume that we begin with an embryo (developing seed) in which the *Ds* element has inserted into a color gene, thereby disrupting pigment production. Division of these progenitor cells results in a group of unpigmented cells, each of which inherits the mutated gene. Subsequently, the *Ds* element moves out of the locus, which restores gene function, giving rise to groups of normally pigmented cells. In this case, the size of pigmented sectors within a nonpigmented background indicates the timing of *Ds* transpositions that restored gene function. Transposition events early in development would give rise to broad stripes or large patches of pigmented tissue, whereas events occurring late in development would give rise to narrow stripes or small speckles. To complicate analysis further, the transposition of *Ds* is controlled by a second element, Activator (*Ac*).

Barbara McClintock, 1951
(Courtesy of Cold Spring Harbor Laboratory Archives.)

McClintock's theory of transposition contradicted long-held dogma that genes were immutably fixed along the length of the chromosomes, and her the-

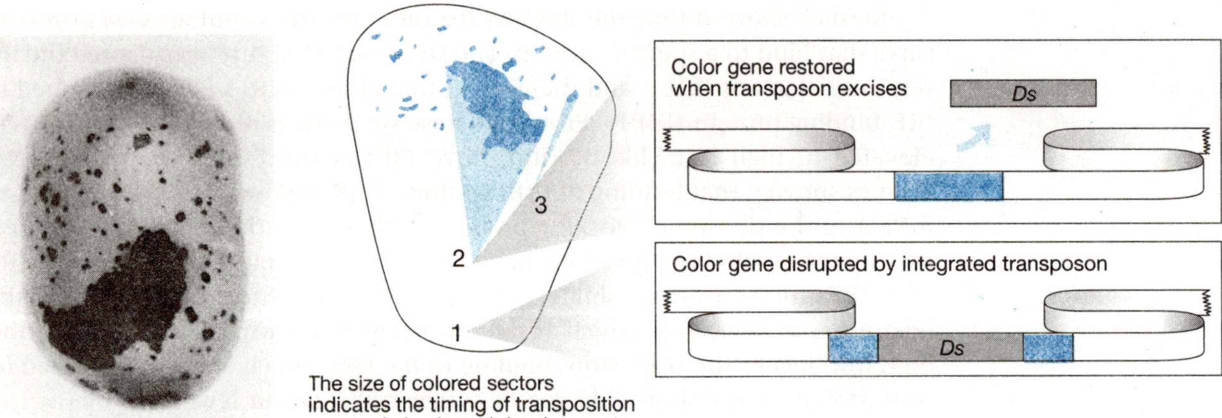

The Phenotypic Effect of *Ds* Transposition Events in a Single Corn Kernel
(*1*) At the beginning of kernel development, the *Ds* transposon is inserted into a gene encoding kernel color, giving rise to a region of colorless tissue. (*2*) *Ds* transposition in a cell early in development leads to a large region of colored tissue. (*3*) *Ds* transposition later in development leads to a smaller region of colored tissue. (Courtesy of B. McClintock.)

ory was not widely accepted for more than 20 years. With the advent of recombinant DNA techniques, her ideas were confirmed. Furthermore, it has been shown that transposition is not merely a peculiarity of maize, but it is a widespread genetic phenomenon common in both prokaryotic and eukaryotic cells. Transposition is now accepted as a major means to activate and organize gene expression during development. However, McClintock also suggested that induction of gene instability by transposable elements may provide a mechanism to reorganize the genome rapidly in response to stress and thus may have an important role in generating diversity.

GENE REARRANGEMENTS

Simple Switches in *Salmonella*, Trypanosomes, and Yeast

An extremely simple recombination event in the *Salmonella* bacterium (which causes food poisoning and other diseases) works exactly like a toggle switch. The inversion of a central 1000-bp sequence activates either of two flanking genes that code for surface antigens. This genetic "flip-flop" brings one of the flanking regions under the control of a promoter positioned at one end of the reversible region; each gene is transcribed in the opposite direction. Antigenic switching allows *Salmonella* to evade the immune response of its hosts. When host antibodies are produced against one surface antigen, an inversion switch allows a clone to arise that produces the alternate surface antigen. Similar simple switches are found in several bacteriophages.

A more involved sort of antigenic switching is practiced by trypanosomes, the unicellular flagellates that cause African sleeping sickness. Trypanosomes possess a repertoire of up to several thousand different genes encoding variable surface glycoproteins (VSGs). Antigenic switching is accomplished when a duplicate copy of one of the many VSG genes is made and shuttled from its storage area into a single active site, where it displaces a previously active VSG gene. The duplicated VSG gene is then expressed under the control of several promoters resident in the active site. In this way, only one surface antigen is produced at a time.

A switching mechanism that represents a hybrid of the two systems described so far exerts even more fundamental control of development in the brewer's yeast, *Saccharomyces cerevisiae.* Here, transposition is the master switch that sets into motion complex developmental programs that regulate the two most critical aspects of the yeast life cycle—expression of mating type and alteration of haploid and diploid generations. As in *Salmonella,* the yeast mechanism activates one of two alternate genes. As in the trypanosomes, yeast switching involves expression of a duplicate gene at an active locus.

Haploid yeast cells exist as one of two mating types (**a** or α). During the haploid phase of their life cycle, either mating type can divide mitotically to give rise to a *mixed population* composed of *both* mating types. This means that a change in the expression of a single (haploid) genome gives rise to a "mating-type switch," which can occur at nearly every mitotic division. Haploid **a** and α cells can fuse to form diploid cells. Under conditions of nutrient depletion, a diploid cell avoids starvation by undergoing meiosis to form four spores.

In the mid 1970s, Jeffrey Strathern, James Hicks, and Ira Herskowitz at the University of Oregon discovered that mating-type switching is controlled by three gene loci, or "cassettes," on yeast chromosome 3. A central mating-type locus (*MAT*) is flanked by unexpressed ("silent") copies of α and **a** cassettes. In the **a** mating-type, a duplicate **a** cassette is inserted, and expressed, at the *MAT* locus. During a mating-type switch, the α cassette is duplicated at its silent locus and then transposes into the *MAT* locus to displace the **a** gene.

Proteins expressed at the *MAT* locus, in turn, control expression of a number of cell-specific genes whose proteins regulate differential metabolic pathways in **a**, α, and diploid (**a**/α) cells. For example, **a** and α cells prepare for mating by each secreting a pheromone that binds to a specific receptor on the surface of the opposite mating type. Entirely different sets of genes are induced in diploid cells that initiate meiosis and sporulation.

Antibody Diversity

A far more complex series of gene rearrangements enables the human immune system to produce antibodies that recognize and destroy foreign particles and disease-causing pathogens. These and other substances that elicit an immune response are called antigens. A multitude of antigenic molecules are present on the surfaces of viruses, bacteria, dust, pollen, and other foreign particles that may enter the body. Living pathogens also secrete numerous antigens as toxins or by-products of their metabolism. Although antigens are most often proteins, polysaccharides and nucleic acids can also elicit an immune response. Exposure to an antigen stimulates B lymphocytes found throughout our bodies in extracellular spaces, in spleen, and in lymph nodes. B lymphocytes produce immunoglobulin proteins (antibodies) that bind to the antigen, facilitating its destruction. Moreover, several different immunoglobulins may be produced that recognize different molecular features (epitopes) of a single antigen. For proteins, an epitope consists of only several amino acids. Over time, only those B lymphocytes that produce antibodies having the strongest interactions with the antigen receive necessary survival and proliferation factors.

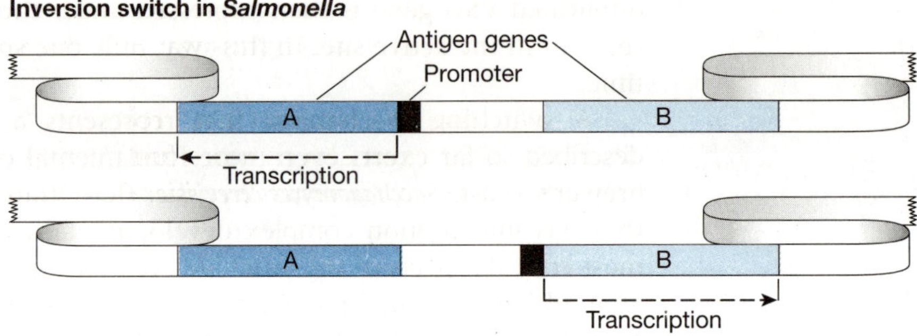

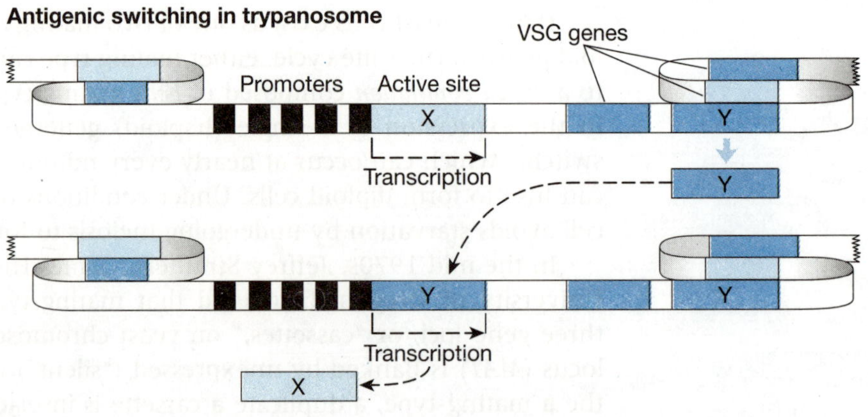

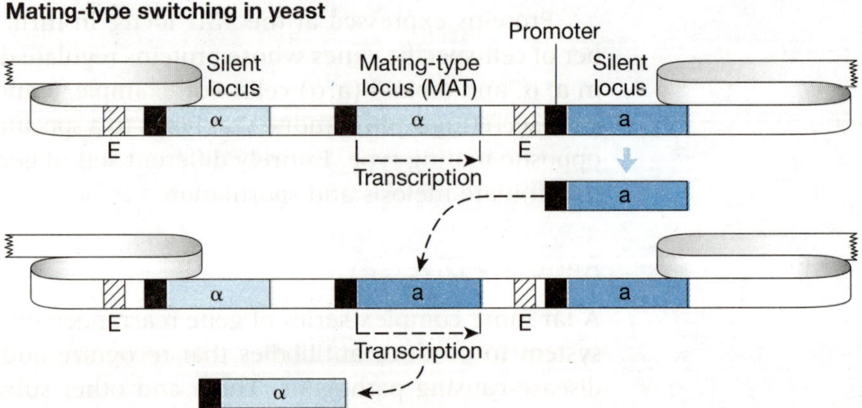

Simple Genetic Switches

In the *Salmonella* bacterium, a promoter lies between the genes encoding two different surface antigens. Inversion of a 1000-bp sequence "flip-flops" the promoter to alternately transcribe either gene. The flagellate *Trypanosome* possesses several hundred genes encoding variable surface glycoproteins (VSGs). A transposed VSG gene is expressed at the active site under the control of five resident promoters. The yeast *Saccharomyces* stores unexpressed copies of **a**- and α-mating-type genes at silent loci, where they are under negative control of the regulatory element E. Transposition of either the **a** or α gene releases it from negative control by the E element, allowing it to be expressed at the active mating-type locus (*MAT*).

Susumu Tonegawa
(Photo by Betsy Cullen Photography ©2001.)

Obviously, the set of potential antigens, which is essentially infinite, is greater than the total number of genes in the entire genome (~50,000). Because it seemed impossible that the genome could contain preexisting genes for every potential antibody, controversy long raged about the mechanism through which B lymphocytes can synthesize an almost infinite diversity of antibody molecules. In the late 1970s and early 1980s, Susumu Tonegawa of the Massachusetts Institute of Technology was key among researchers who cloned immunoglobulin genes and probed their organization. The controversy was settled by showing that recombining a relatively small number of chromosomal DNA segments, which make up the immunoglobulin locus, creates the diversity of functional antibody genes.

There are five classes of antibodies, each varying in structure and function, characterized by the part of the protein known as the heavy-chain constant region. They are IgG, IgE, IgA, IgD, and IgM. IgG is the most common and abundant antibody. IgE is responsible for allergic reactions and IgA is found mostly in mucosal secretions. IgD is rarely found freely circulating in the blood, but it is found on the surface of B lymphocytes and helps them recognize antigens. IgM is the antibody that typically appears first in an immunological response to antigen exposure, but this is ultimately replaced by IgG. As elucidated in 1969 by Gerald Edelman at The Rockefeller University, an IgG antibody molecule consists of four polypeptide chains: two identical heavy (H) chains of about 440 amino acids each and two identical light (L) chains of about 220 amino acids. The amino acid sequences of the carboxyl ends of both heavy and light chains show little variation from antibody to antibody; they are termed constant (C) regions: C_H and C_L. In contrast, the amino ends of both types of chains show great variation in amino acid sequence and are termed variable (V) regions: V_H and V_L. The four chains form a Y-shaped molecule with all variable regions oriented at the tips of the arms, where antigen binding occurs. Moreover, the variable region of each chain contains hypervariable regions that give antibodies more variability and specificity for binding to its antigen. Like IgG, IgE is composed of two heavy

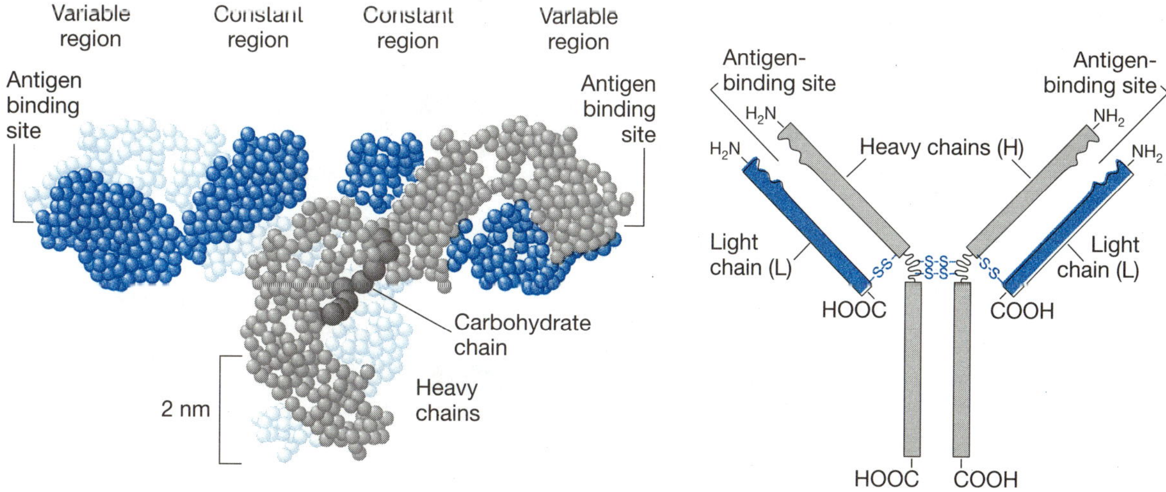

Models of an Antibody Molecule

Each sphere (at left) represents a single amino acid. (Redrawn, with permission, after Selverton E.W. et al. 1977. *Proc. Natl. Acad. Sci. 74:* 5140. From Alberts B. et al. 1989. *Molecular biology of the cell*, pp. 1013, 1022. Garland Publishing, New York.)

chains and two light chains. IgA is composed of either four heavy and light chains or six of each. IgM is the largest antibody, composed of ten heavy and light chains. Although IgA and IgM are larger, they maintain the same basic structure of the other antibodies; they simply appear as multiple Y structures linked side by side.

In humans, a single locus of genes for immunoglobulin heavy chains is located on chromosome 14; separate light-chain loci λ and κ are found on chromosomes 2 and 22. The approximate number of immunoglobulin genes in these loci, as well as the basic scheme of their rearrangement, was discovered by hybridizing cDNA probes of antibody mRNA to genomic DNA. Constant and variable regions are coded by different sets of genes, widely separated from each other on the chromosome. The human κ variable gene region is made up of several hundred variable (*V*) gene segments, some 30 diversity (*D*) gene segments, and 5 joining (*J*) gene segments. *V* gene segments make up the bulk of the variable domain, with *J* gene segments coding for 4–6 amino acids and *D* gene segments coding for 2–13 amino acids. These gene segments are joined together in a diverse variety of combinations. Each gene segment is flanked by a recombination signal

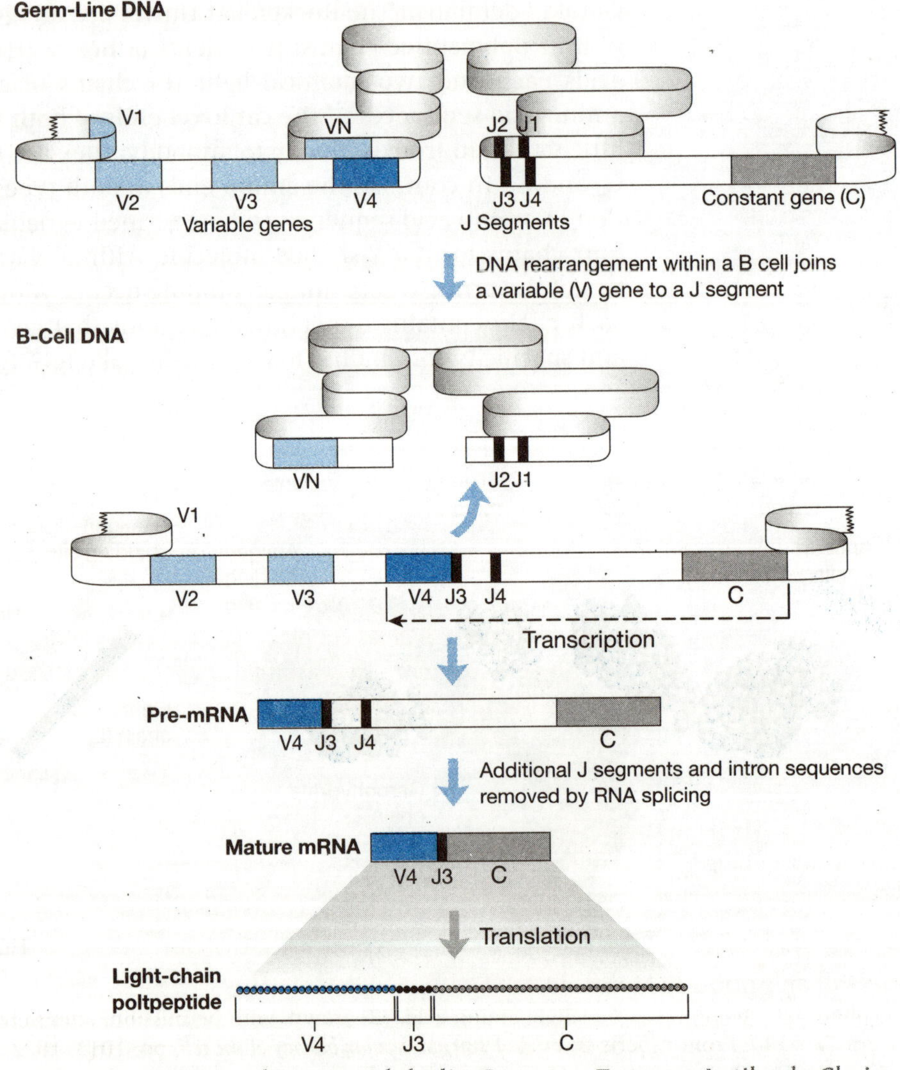

Rearrangement of Immunoglobulin Genes to Form an Antibody Chain

sequence (RSS), which is cleaved by a specific nuclease encoded by the *RAG1* and *RAG2* genes. Taken together, the following molecular genetic events are easily capable of generating a nearly infinite number of different antibodies:

1. The creation of the heavy-chain variable gene occurs by DNA rearrangements that randomly join a *V, D,* and *J* gene (*V*(*D*)*J* recombination). The first step in the process is joining one *J* gene to one *D* gene, creating a *DJ* fusion gene. Next, one of the *V* genes fuses with the *DJ* gene, creating the V_H gene. The V_H sequence lies upstream of the C_H gene. The creation of the V_H gene stimulates its transcription, creating a pre-mRNA that includes the downstream C_H gene. RNA splicing subsequently joins the V_H and C_H domains. Light-chain genes are linked through a single recombination event termed *V-J* joining, during which one of hundreds of *V* genes join with any of four *J* segments to create a V_L gene. V_L and a downstream C_L gene are transcribed together and fused during RNA splicing. *V*(*D*)*J* and *V-J* joining are the major source of antibody diversity.
2. *V-J* and *V*(*D*)*J* joining is intentionally imprecise, leading to further diversity. The junctions that form between *V, D,* and *J* genes during recombination do not occur at precise sites, so that the final V_H or V_L gene can gain or lose information for one or a few amino acids. Specific enzymes are used to randomly add or delete nucleotides in these junctions, and as a result, only one in three rearrangements is functional (two of the three result in frameshifts). Only lymphocytes can afford to be so wasteful because of their tremendous proliferative capacity.
3. Point mutations in the fully assembled V_H genes, termed somatic hypermutation, are another source of variation. Recent studies have shown that a cytosine deaminase (which changes cytosine into uracil that can be mistaken for thymine by the DNA polymerase creating a mutation) is involved in creating these mutations in the variable region. Through the process of somatic hypermutation, any nucleotide can be substituted for another. These changes have the potential to alter the genetic code of the gene.
4. Each V_H gene can be linked to a different C_H gene, which further affects how the antibody functions, e.g., whether it be membrane-bound or freely circulating. C_H genes also undergo gene rearrangement through a process known as class switch recombination (CSR).

T cells, another major immunological cell type, also make antigen-binding proteins called T-cell receptors (TCRs), which do not circulate but stay attached to the cell surface. The variable domains of TCRs form much like those of antibodies. Unlike immunoglobulin genes, TCR genes do not undergo somatic hypermutation.

TISSUE-SPECIFIC GENE REGULATION

Obviously, what makes a particular cell type or tissue type unique is the particular proteins that it synthesizes. In a multicellular organism, nearly every somatic cell possesses the exact same DNA content, so the diversity of cell types is largely a function of which genes are transcribed and translated in that particular cell. Some genes are expressed in all cell types such as actin, which is a major part of the cytoskeleton. Such genes have been referred to as "housekeeping genes." The

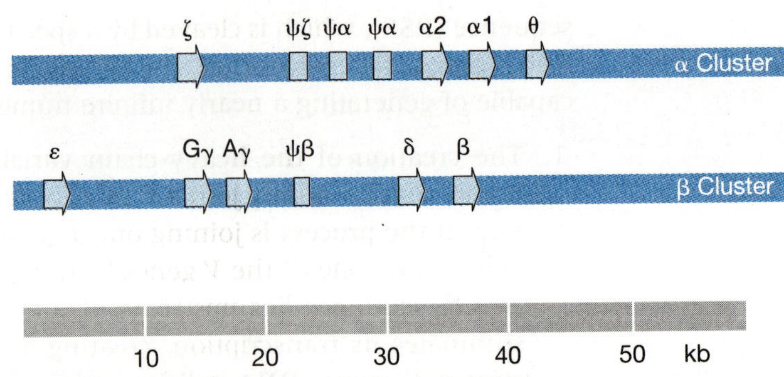

Stage of Development at which Each Hb Gene Is Expressed

Embryonic (<8 weeks)	$\zeta_2\,\varepsilon_2\,\zeta_2\,\gamma_2\,\alpha_2\,\varepsilon_2$
Fetal (3–9 months)	$\alpha_2\,\gamma_2$
Adult (from birth)	$\alpha_2\,\delta_2\,\alpha_2\,\beta_2$

Organization of the Globin Genes
Shown here are the relative positions of the α-like and β-like globin genes. The order of expression during development is the same as the order in which they are arranged in the genome. Mixed in with the functional genes are several pseudogenes (ψ). These genes all contain fatal mutations and are not expressed.

particular set of genes expressed in a given cell depends on the signals received by cell and the activators and repressors that function within it.

Hemoglobin is one example of a cell-specific protein that is also differentially expressed during development. Hemoglobin plays a role in the transport of O_2 and CO_2 to and from cells, and it is synthesized exclusively in erythrocytes (red blood cells). Hemoglobin is made up of four peptides and a central heme group, which contains iron and is the functional part of this protein. Different forms of hemoglobin are expressed at different times of development, making the regulation of the genes encoding the hemoglobin proteins a complex process.

In human adults, the four hemoglobin proteins are composed of two α-globin and two β-globin molecules. There is also a minor form of hemoglobin (constituting 2–3% of total hemoglobin) that contains two α globins and two δ-globin chains. In early embryos, there is a different hemoglobin protein. This hemoglobin, called embryonic hemoglobin, is composed of two ζ chains and two ε chains. At 8 weeks following fertilization, the genes producing embryonic hemoglobin are turned off and two new genes are expressed, making a new hemoglobin composed of two α chains and two γ chains, called fetal hemoglobin. (There are actually two forms of γ-globin, γ^A and γ^G.) Finally, the adult form of two α chains and two β chains switches on at birth. All of the globin genes are found in two gene clusters, the α-like globin cluster on chromosome 11 and the β-like globin cluster on chromosome 16. Together, they represent a gene family. The arrangement of these genes parallels the order of their expression during development.

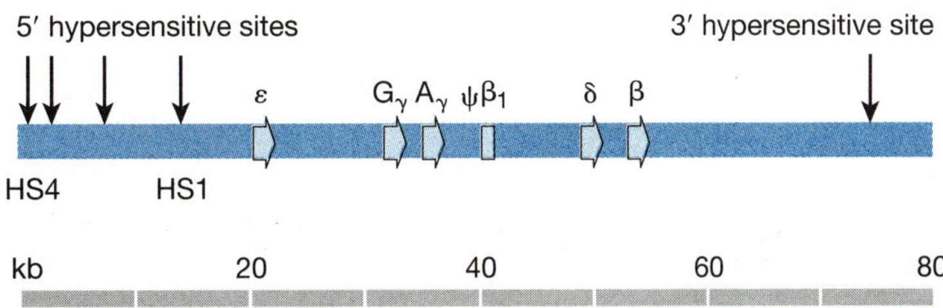

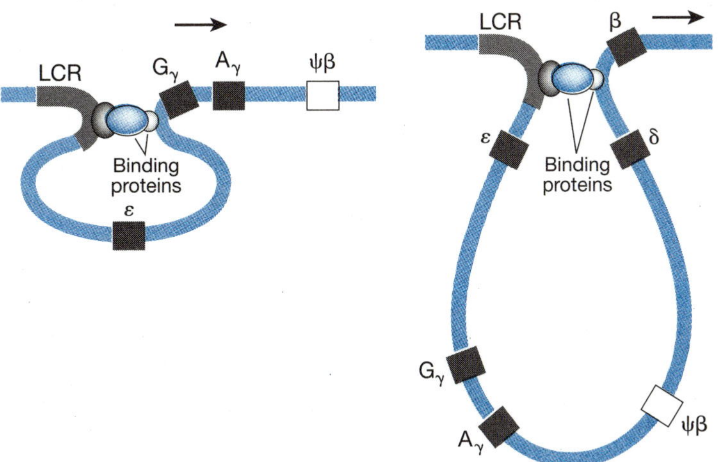

Hypersensitive Sites in the β-*globin* Gene

As shown in the top diagram, experiments identified positions upstream and downstream from the globin gene cluster (marked by arrows) that are hypersensitive to DNase treatment. These sites turned out to be regulatory regions of the globin gene cluster referred to as the locus control region (LCR). In the bottom diagram, a looping model is shown to demonstrate how the LCR might regulate expression of specific globin genes. (*Top:* Adapted, with permission, from Townes T.M. and Behringer R.R. 1990. Human globin locus activation region (LAR): Role in temporal control. *Trends Genet. 6:* 219–223; *bottom:* redrawn, with permission, from http:// bssv01.lancs.ac.uk/ADS/BIOS336/336L9.html.)

The mechanism for regulating expression of these genes has been the subject of intensive investigation during the past 20 years. Our understanding has come from both laboratory investigations and studies of mutated forms of these genes, which cause a variety of diseases involving hemoglobin known as thalassemias. There are many different forms of thalassemia; some arise by mutations within the part(s) of the globin genes that code for proteins. Other thalassemias arise by mutations lying outside the coding regions of the globin genes, presumably within regulatory regions. Laboratory studies identified regions in the chromosome, particularly 5′ to the β-*globin* gene cluster, that were potential sites of gene regulation. A region of chromatin (DNA bound to specific types of proteins, as it exists in cells) that is 5′ to the β-*globin* gene cluster was shown to have four sites hypersensitive to degradation by treatment with DNase. This sensitivity is presumably due to binding of transcription factors to these regions of the chromatin. These hypersensitive sites are thousands of base pairs away from the coding sequences of the β-*globin* gene cluster.

Studies of the genes of patients suffering from a particular form of thalassemia called γδβ-thalassemia were particularly informative in understanding the regulation of the β-*globin* gene cluster. Patients with this disorder are unable to make γ-, δ-, or β-globin. Analyzing the DNA of these patients revealed that many had a normal complement of the genes in the β-*globin* cluster, but lacked the sequences corresponding to the hypersensitive sites mentioned above. These DNase hypersensitive sites are collectively called the locus controlling region (LCR). The LCR is equivalent to an enhancer, binding to regulatory proteins. These regulatory proteins, in turn, interact with specific transcription factors that bind to the promoter regions which lie before each individual globin gene.

COORDINATE EXPRESSION OF GENES

As in the example of the *lac* operon, there are many groups of genes that must be expressed in a coordinate fashion. Eukaryotes do not make polycistronic

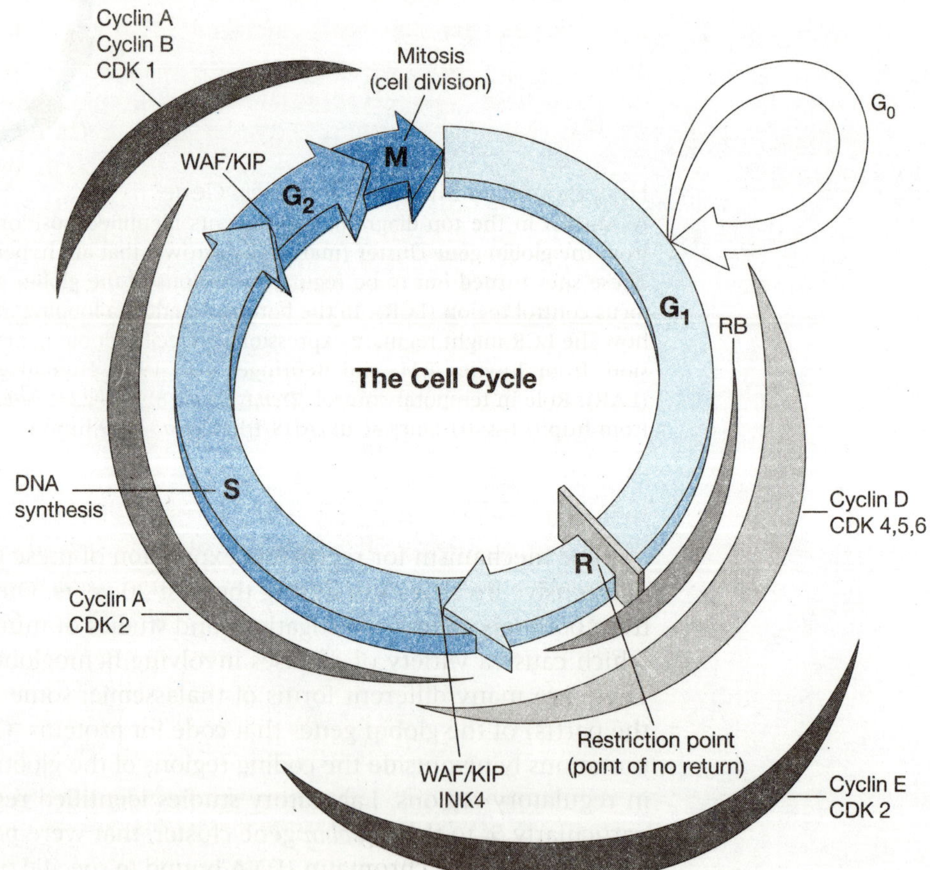

The Cell Cycle

The cell cycle is the eukaryotic cell's regular rhythm of rest and division. G_0 is a resting phase, when the cell is quiescent. During G_1, the cell grows. In the S phase, the cell replicates its DNA (synthesis). In G_2, the cell prepares for division, which occurs during the M phase. Molecules that regulate each phase and transition of the cell cycle are also shown.

mRNAs, so to express multiple genes in a coordinated fashion, the gene expression must be highly organized and share common regulatory elements.

The Cell Cycle

A network of genes are involved in controlling the eukaryotic cell's regular rhythm of rest and division, which is called the cell cycle. All cells cycle through four different stages:

- G_1 (gap 1): The cell grows and replenishes its resources.
- S (synthesis): The cell must synthesize DNA in preparation for cell division.
- G_2 (gap 2): The cell makes cytoplasmic components necessary for cell division.
- M (mitosis): The cell divides, and the whole cycle repeats.

Mature, fully differentiated cells exit the cell cycle into a quiescent (resting) phase called G_0. The cell cycle is largely governed by cyclin proteins—labeled A, B, C, D, and E—whose concentrations peak at different times. Each cyclin, in turn, reaches its threshold level and signals the cell to move into the next phase of the cell cycle. For example, D cyclins accumulate and peak during G_1 and drive the cell past the Restriction Point. This Restriction Point is essentially the "point of no return" at which the cell is committed to complete the cell cycle. Synthesis of D cyclins is stimulated by a number of growth factors.

The cyclins exert control over the cell cycle by activating specific sets of cyclin-dependent protein kinases (CDKs), which interact with numerous molecules to drive cells through the cell cycle. Although CDKs are up-regulated by cyclins, they are down-regulated by CDK inhibitors (CKIs). For example, members of the INK4 family bind to G_1 CDKs, preventing them from interacting with D-type cyclins; members of the WAF/KIP family bind CDKs needed for the G_1-S and G_2-M transition.

The cell cycle must be rigorously controlled to ensure that each daughter cell receives a single set of high-quality chromosomes. Thus, movement from each of the gap phases is regulated by checkpoint control genes, whose proteins "sense" key parameters. For example, it is critical that the genome be duplicated exactly once during S phase. If mitosis commences before synthesis is complete, the daughter cells receive incomplete sets of chromosomes. Checkpoint genes also have a major role in controlling progression through the cell cycle when cells encounter DNA damage. When DNA is damaged in cells following exposure to a reagent that either modifies the bases of the DNA (like UV radiation) or causes a break in the DNA backbone (as is done by X-rays), DNA repair proteins are recruited to the site of DNA damage to repair it. If this damage occurs in G_1 and its repair is not completed, checkpoint proteins prevent cells from entering S phase. This is because DNA synthesis will arrest at the damaged base. If this situation is not remedied, cells will ultimately die when they try to segregate their chromosomes during mitosis. DNA damage occurring in cells after commitment to S phase or when cells have already entered S phase activates a different set of checkpoint genes whose products prevent exit from S phase prior to completion of DNA synthesis. Similarly, checkpoint genes exist that block cells from entering mitosis if DNA damage occurs in G_2, again until DNA repair is completed.

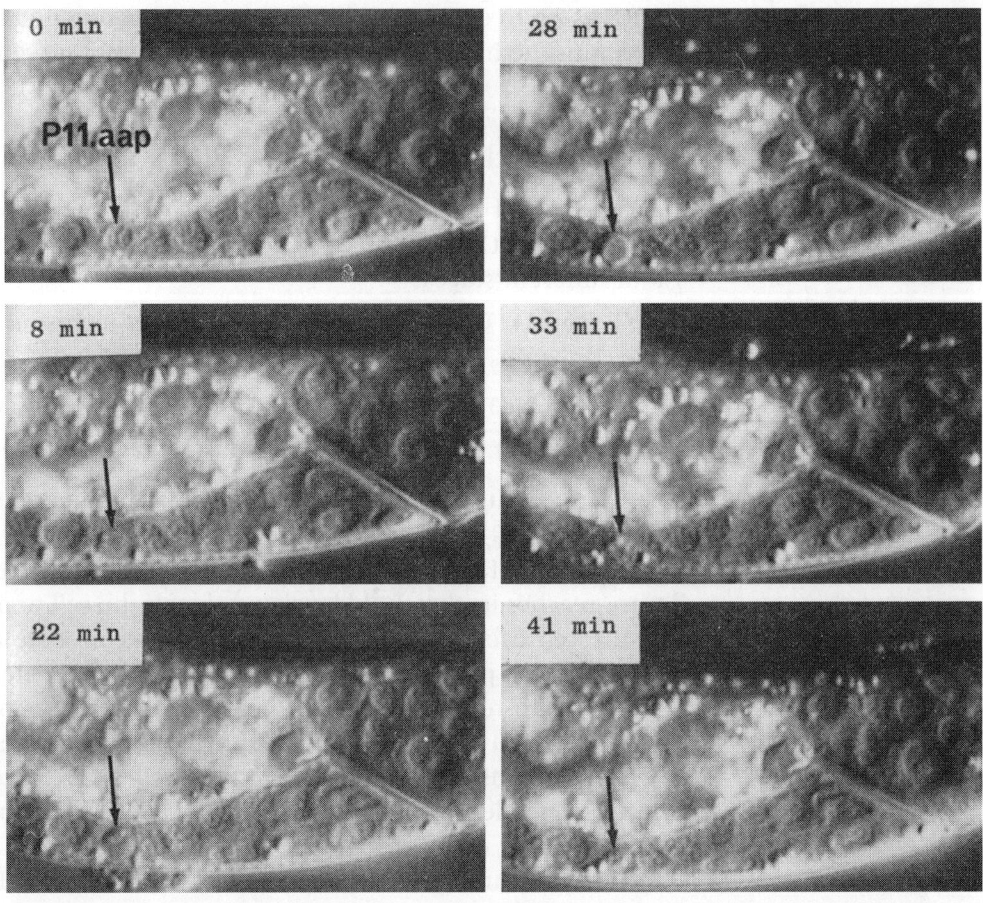

Cell Death

Sequential photographs of a developing *C. elegans*. The cell labeled P11.aap undergoes apoptosis. (Reprinted, with permission, from Sulston J.E. and Horvitz H.R. 1977. Post-embryonic cell lineages of the nematode *Caenorhabditis elegans*. *Dev. Biol. 78:* 110–156.)

Leland Hartwell

The first checkpoint control gene, *rad9*, was identified in the 1970s by Leland Hartwell in his studies of cell division cycle (*cdc*) mutants in yeast at the University of Wisconsin. He found that several mutants were unable to complete the cell cycle due to defects in enzymes involved in the DNA replication machinery. One of these, *cdc9*, has a defective DNA ligase, needed to knit together the single-stranded DNA fragments (Okazaki fragments) created during "lagging-strand" synthesis. This results in incomplete chromosomes and numerous DNA fragments. Another mutant, *rad9*, carries on with cell division despite exposure to ionizing radiation. Radiation, like ligase deficiency, produces double-stranded breaks in the DNA molecule. Hartwell found that double mutants (*rad9* and *cdc9*) die quickly after replicating a fragmented genome. This showed that the role of *rad9* is to sense DNA damage or incomplete synthesis and to halt cells from completing the cell cycle until the chromosomes are intact. It is believed that *BRCA1*, a gene mutated in familial breast cancer, may be the human equivalent of *rad9*.

Apoptosis Sculpts the Developing Limb in Chickens
The limb bud grows out into a paddle-shaped structure. The shaded areas undergo apoptosis, leading to the separation of individual digits.

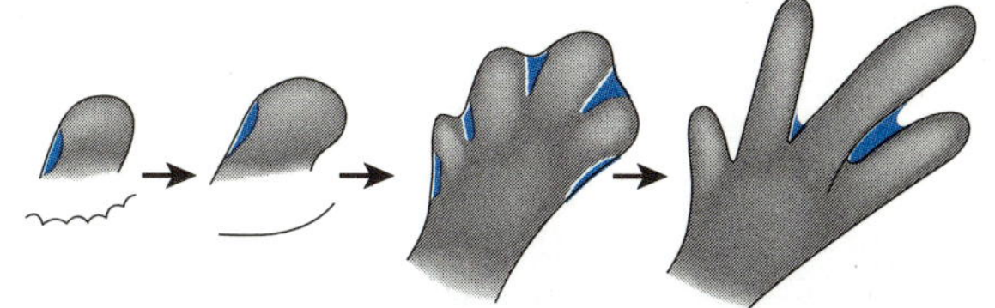

Apoptosis

H. Robert Horvitz
(Photo by Donna Coveney.)

Apoptosis, or programmed cell death, balances mitosis to ensure that higher organisms maintain the proper number of healthy cells. In the nematode *C. elegans,* apoptosis trims the total of 1090 cells generated during development down to 959 cells in the adult worm. Because the lineage and location of every *C. elegans* cell are known, the elimination of any cell by apoptosis can be easily monitored by microscopic observation. By the same token, a failure of apoptosis leads to the observable "resurrection" of additional cells. Like the cell cycle, apoptosis is regulated by a complex network of genes.

Using *C. elegans,* Robert Horvitz and Michael Hengartner at the Massachusetts Institute of Technology identified key genes involved in cell death (*ced*). One gene, *ced-3,* encodes a protease that helps accomplish apoptosis by degrading the cellular proteins. In *ced-3* mutants, cells that normally die, do not; they survive and often assume the function of their sister cells. Another gene, *ced-9,* encodes a protein that interacts with *ced-3,* blocking its protease activity. Sequence comparison showed that the human homolog of *ced-3* is *caspase-9,* whereas *bcl-2* is the human counterpart of *ced-9*. These genes are so closely related that a copy of human *bcl-2* can protect against cell death in *C. elegans* whose own *ced-9* is deleted.

Michael Hengartner
(Courtesy of Cold Spring Harbor Laboratory Archives.)

Apoptosis is not unique to *C. elegans*. Cell death is used to clear away transitory structures during development, such as the early tissues that contribute to the growing kidney in higher vertebrates. Many tissues are sculpted by apoptosis. In vertebrates, the limb buds start as a paddle-shaped structure. Cells in between digits undergo apoptosis and are cleared away, leaving the fingers and toes. Apoptosis also contributes to the correct formation of heart loops and to the organization of cells in the nervous system. As discussed in Chapter 7, alterations in apoptosis can contribute to cancer.

Body Axis Formation

The development of an organism—from a fertilized egg through embryonic and juvenile stages to adulthood—requires the coordinated expression of sets of genes at the proper times and in the proper places. Studies of several bizarre mutations in the fruit fly, *Drosophila,* provided keys to understanding the molecular basis of large-scale developmental plans. During larval development, embryonic genes express proteins that set up the orientation and define the body segments of the fly embryo. Then, during metamorphosis, homeotic genes act in coordination to produce unique appendages in the thorax (legs and wings) and head (antenna).

Christiane Nüsslein-Volhard
(Courtesy of Cold Spring Harbor Laboratory Archives.)

Eric Wieschaus
(Reprinted, with permission, The Nobel Foundation.)

Edward B. Lewis
(Reprinted, with permission, The Nobel Foundation.)

In the late 1970s, Christiane Nüsslein-Volhard and Eric Wieschaus, at the European Molecular Biology Laboratory, identified key genes that guide the spatial development of segments in the *Drosophila* embryo. Soon after fertilization, several sets of genes are differentially expressed to establish anterior-posterior (i.e., head-to-tail) and dorsal-ventral (i.e., back to abdomen) orientation, which guides the development of bilaterally symmetrical parts. For example, the concentration of the bicoid protein decreases from anterior to posterior, whereas the concentration of the Dorsal protein grades from ventral to dorsal. The Torso protein is concentrated at the anterior and posterior ends, whereas oskar and nanos proteins are concentrated exclusively at the posterior end. The relative mix and concentration of these proteins activate three types of genes (Gap, pair rule, and segment polarity genes), which work together to determine the identity of body segments.

Gap genes guide the differentiation of segments along the head-to-tail axis, leading to head, thoracic, and abdominal regions. For example, *Krüppel* is a Gap gene expressed mainly in the thoracic segments, and thus *Krüppel* mutants are missing those segments. Gap proteins, in turn, control expression of pair rule genes that define each segment. For example, *fushi tarazu* (*ftz*) is expressed in the boundaries between segments, so *ftz* mutants are missing every other segment. *engrailed* is an example of a segment polarity gene, which regulate the anterior/posterior orientation within each segment. In *engrailed* mutants, the posterior end of each segment is a mirror image the anterior end.

With work that began in the 1950s at the California Institute of Technology, Edward Lewis had a key role in understanding how the *Drosophila* body plan is determined. He studied how homeotic genes control the specialization of body parts within the segments. Homeosis describes the process in which one body part becomes like another by assuming its identity. For example, in the homeotic mutant *Ultrabithorax* (*Ubx*), the halteres (a type of small modified wing that helps the fly balance) of the third thoracic segment have been converted into a second pair of wings, mimicking the normal situation in the second thoracic segment. In the *Antennapedia* (*Antp*) mutant, antennae in the head segment are converted into an extra pair of legs, normally found in the second thoracic segment.

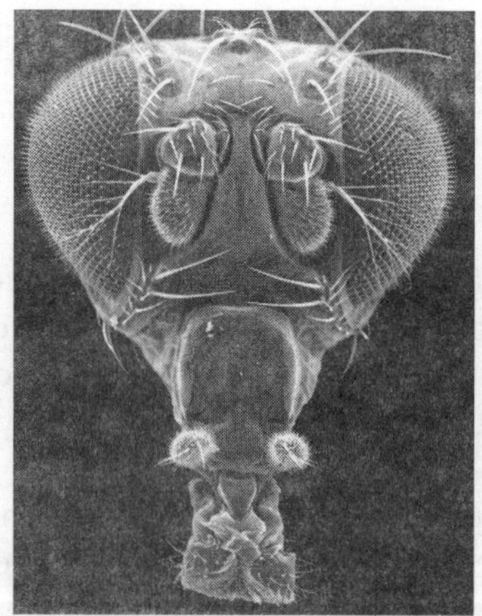

Antennapedia Mutation
(*Left*) Wild-type *Drosophila* head. (*Right*) *Antennapedia* mutant fly. Note the presence of legs where antennae normally form. (Reprinted, with permission, courtesy of F.R. Turner, Indiana University, http://flybase.bio.indiana.edu.)

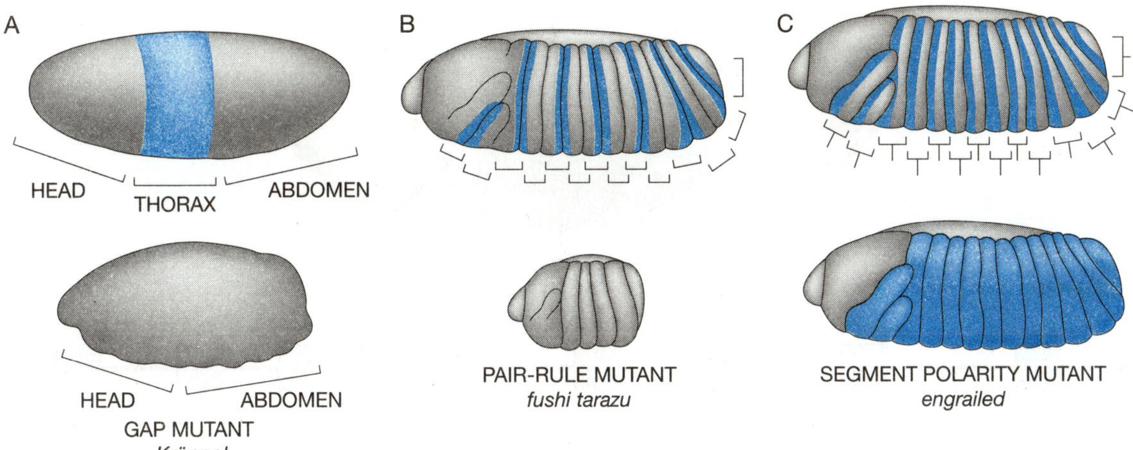

Three Types of Body Patterning Mutations in *Drosophila*

(*A*) Gap genes control the differentiation along the head to tail axis. In this case, a *Krüppel* mutation leads to the loss of the thoracic segments. (*B*) Pair-rule genes define each segment. A mutation in *fushi tarazu* leads to the loss of every other segment. (*C*) Segment polarity genes regulate the anterior/posterior axis within each segment. Each segment of *Drosophila* normally has an anterior and a posterior half. Mutations in *engrailed* lead to flies where every segment is entirely anterior.

Homeotic genes are found in two clusters in *Drosophila*. The bithorax cluster includes *Ultrabithorax* and two other genes, whereas the *Antennapedia* cluster includes five genes. These genes are involved in controlling the specialization of segments and are laid out on the chromosome in the same spatial order they are expressed during development. Genes at the 3′ end of a complex are expressed earlier, and toward the head of the animal, whereas genes at the 5′ end of the complex act later and more posteriorly. This arrangement, in which the physical order of the genes reflects the physical order of their activity, is called "colinearity." Homeotic control is not just limited to fruit flies. Vertebrates have

Extensive Homology in Amino Acid Sequences of Five Homeodomains

	1																			20
Mouse *MO* 10	Ser	Lys	Arg	Gly	Arg	Thr	Ala	Tyr	Thr	Arg	Pro	Gln	Leu	Val	Glu	Leu	Glu	Lys	Glu	Phe
Frog *MM3*	Arg	Lys	Arg	Gly	Arg	Gln	Thr	Tyr	Thr	Arg	Tyr	Gln	Thr	Leu	Glu	Leu	Glu	Lys	Glu	Phe
Antennapedia	Arg	Lys	Arg	Gly	Arg	Gln	Thr	Tyr	Thr	Arg	Tyr	Gln	Thr	Leu	Glu	Leu	Glu	Lys	Glu	Phe
Fushi tarazu	Ser	Lys	Arg	Thr	Arg	Gln	Thr	Tyr	Thr	Arg	Tyr	Gln	Thr	Leu	Glu	Leu	Glu	Lys	Glu	Phe
Ultrabithorax	Arg	Arg	Arg	Gly	Arg	Gln	Thr	Tyr	Thr	Arg	Tyr	Gln	Thr	Leu	Glu	Leu	Glu	Lys	Glu	Phe

	21																			40
Mouse *MO*-10	His	Phe	Asn	Arg	Tyr	Leu	Met	Arg	Pro	Arg	Arg	Val	Glu	Met	Ala	Asn	Leu	Leu	Asn	Leu
Frog *MM3*	His	Phe	Asn	Arg	Tyr	Leu	Thr	Arg	Arg	Arg	Arg	Ile	Glu	Ile	Ala	His	Val	Leu	Cys	Leu
Antennapedia	His	Phe	Asn	Arg	Tyr	Leu	Thr	Arg	Arg	Arg	Arg	Ile	Glu	Ile	Ala	His	Ala	Leu	Cys	Leu
Fushi tarazu	His	Phe	Asn	Arg	Tyr	Ile	Thr	Arg	Arg	Arg	Arg	Ile	Asp	Ile	Ala	Asn	Ala	Leu	Ser	Leu
Ultrabithorax	His	Thr	Asn	His	Tyr	Leu	Thr	Arg	Arg	Arg	Arg	Ile	Glu	Met	Ala	Tyr	Ala	Leu	Cys	Leu

	41																			60
Mouse *MO*-10	Thr	Glu	Arg	Gln	Ile	Lys	Ile	Trp	Phe	Gln	Asn	Arg	Arg	Met	Lys	Tyr	Lys	Lys	Asp	Gln
Frog *MM3*	Thr	Glu	Arg	Gln	Ile	Lys	Ile	Trp	Phe	Gln	Asn	Arg	Arg	Met	Lys	Trp	Lys	Lys	Glu	Asn
Antennapedia	Thr	Glu	Arg	Gln	Ile	Lys	Ile	Trp	Phe	Gln	Asn	Arg	Arg	Met	Lys	Trp	Lys	Lys	Glu	Asn
Fushi tarazu	Ser	Glu	Arg	Gln	Ile	Lys	Ile	Trp	Phe	Gln	Asn	Arg	Arg	Met	Lys	Ser	Lys	Lys	Asp	Arg
Ultrabithorax	Thr	Glu	Arg	Gln	Ile	Lys	Ile	Trp	Phe	Gln	Asn	Arg	Arg	Met	Lys	Leu	Lys	Lys	Glu	Ile

From W.J. Gehring. 1985. *Sci. Am. 253/4:* 159.

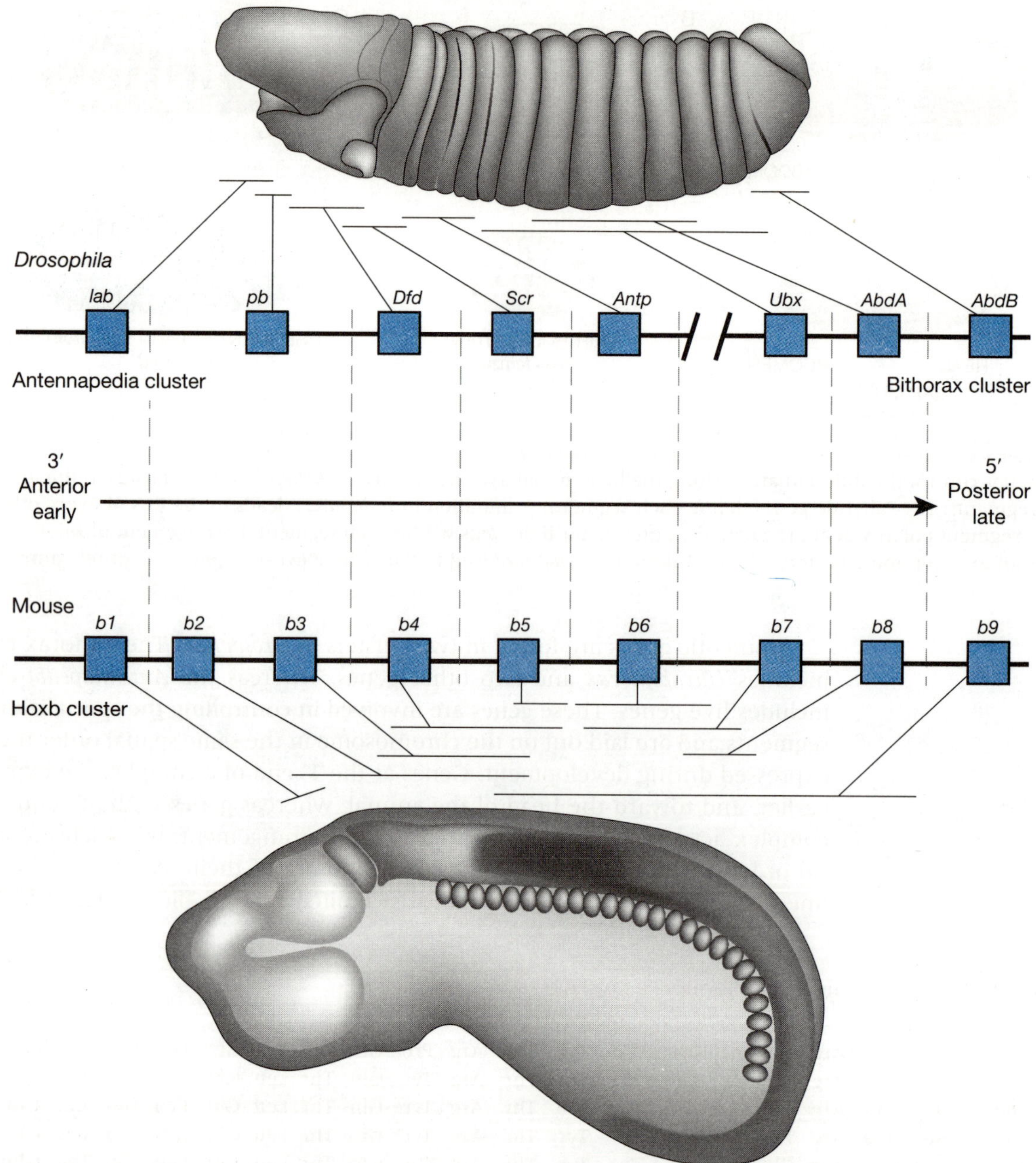

Comparison of Homeotic Gene Clusters in *Drosophila* and Mouse

The *Drosophila Antennapedia* and *Bithorax* clusters are shown aligned with the mouse *Hoxb* cluster (mice have four such clusters, *Hoxa, Hoxb, Hoxc,* and *Hoxd*). The relative position of expression for each gene along the body axis is indicated.

homeotic genes, called *Hox* genes, which are arranged in clusters and expressed in the same colinear manner as in *Drosophila*.

Sequence analysis showed that homeotic genes from *Drosophila* and vertebrate animals share a 180-nucleotide region, called the homeobox. The homeobox encodes a DNA-binding region that is structurally related to other transcription factors. Homeotic proteins have been shown to have a role in transcriptional regulation, activating or repressing other genes, which is likely to explain how they function to regulate development. Homeobox-containing genes have been found in every eukaryotic species tested, from hydra to humans.

There is still much to be understood as to how these families of genes work together to provide the information that determines the position and identity of regions within a developing embryo.

REFERENCES

Bier E. 2000. *The coiled spring: How life begins.* Cold Spring Harbor Laboratory Press, Cold Spring Harbor, New York.

Carey M.F. and Smale S.T. 2000. *Transcriptional regulation in eukaryotes: Concepts, strategies, and techniques.* Cold Spring Harbor Laboratory Press, Cold Spring Harbor, New York.

Comfort N. 2001. *The tangled field: Barbara McClintock's search for the patterns of genetic control.* Harvard University Press, Cambridge.

Fox-Keller E. 1983. *A feeling for the organism: The life and work of Barbara McClintock.* W.H. Freeman, New York.

Holland S., LeBacqz K., and Zoloth L., eds. 2001. *The Human embryonic stem cell debate.* MIT Press, Cambridge.

Jacob F. 1995. *The statue within: An autobiography.* Cold Spring Harbor Laboratory Press, Cold Spring Harbor, New York.

Krainer A.R., ed. 1997. *Eukaryotic mRNA processing.* IRL Press, New York.

Lewin B. 1999. *Genes VII.* Oxford University Press, United Kingdom.

Lewis J.D. and Tollervey D. 2000. Like attracts like: Getting RNA processing together in the nucleus. *Science* **288:** 1385–1389.

McClintock B. 1987. *The discovery and characterization of transposable elements: The collected papers of Barbara McClintock.* Garland Publishing, New York.

Micklos D., ed. 1999. DNA from the beginning (http://www.dnaftb.org). Dolan DNA Learning Center, Cold Spring Harbor, New York.

Müller-Hill B. 1996. *The* lac *Operon: A short history of a genetic paradigm.* Walter de Gruyter, New York.

Niehrs C. and Pollet N. 1999. Synexpression groups in eukaryotes. *Nature* **402:** 483–487.

Nissim-Rafinia M. and Kerem B. 2002. Splicing regulation as a potential genetic modifier. *Trends Genet.* **18:** 123–127.

Ptashne M. and Gann A. 2001. *Genes & signals.* Cold Spring Harbor Laboratory Press, Cold Spring Harbor, New York.

Reilly P.R. 2000. *Abraham Lincoln's DNA and other adventures in genetics.* Cold Spring Harbor Laboratory Press, Cold Spring Harbor, New York.

Sonenberg N., Hershey J.W.B., and Mathews M.B., eds. 2000. *Translational control of gene expression.* Cold Spring Harbor Laboratory Press, Cold Spring Harbor, New York.

Tonegawa S. 1985. The molecules of the immune system. *Sci. Am.* **253:** 122–131.

CHAPTER 4

Basic Tools and Techniques of DNA Science

Mr. DNA from *Jurassic Park* (Courtesy, with permission, ©Universal City Studios, Inc. and Amblin Entertainment, Inc.)

IN *JURASSIC PARK*, FICTIONAL SCIENTISTS ARE ABLE to fill an island near Costa Rica with newly created dinosaurs. In the movie, a video is shown featuring an animated "Mr. DNA" who explains that the dinosaurs were made through the miracle of genetic cloning. Mr. DNA tells his audience that "Dino DNA" was extracted from mosquitoes that had dined on dinosaur blood just before being encased in amber, millions of years ago. The dinosaur DNA was inserted into unfertilized crocodile ova and the ova were placed into eggs, from which cloned dinosaurs hatched.

Although the concept of bringing extinct species back to life is intriguing, *Jurassic Park* is a work of fiction. The techniques described in the movie are based on real technologies in use today, but their application to generate dinosaurs is improbable, and certainly impossible with our current abilities. A great deal of effort has gone into isolating ancient DNA from insects trapped in amber with very little success.

The scientific community and the world, however, were taken by surprise in 1997 when Dr. Ian Wilmut and his team from the Roslin Institute in Scotland announced that they had cloned a sheep named Dolly. The Scottish scientists

Dolly, the First Cloned Sheep, Gave Birth to Her First Lamb in April of 1998
(Courtesy of Roddy Field, Photographer, The Roslin Institute.)

removed the nucleus from an unfertilized egg and fused this egg with the nucleus of an adult cell taken from a sheep's udder. The fused egg was grown in culture and then transplanted into a surrogate mother. This was the first report of animal cloning using DNA from an adult animal. There are still questions about the Dolly experiment. Whether Dolly is truly as healthy as a normal sheep is unknown and whether she is a true clone is unclear. More recently, cloning by this nuclear transplantation method has been achieved in mice, cattle, and cats.

Taken in its pure sense, the term cloning refers to the reproduction of daughter cells by fission or mitotic division. During these processes, DNA from a parent cell is replicated, and identical sets of genetic information are passed on to daughter cells. Successive generations of cells in turn divide, giving rise to a population of genetically identical *clones*, all derived from a single ancestral cell.

Gene cloning takes advantage of the natural ability of cells to duplicate DNA by replication. Briefly, the gene of interest is inserted into a carrier DNA molecule, termed a vector. The vector, with its gene insert, is introduced (transformed) into the appropriate host cell, where it is duplicated by the host cell's replication machinery. Subsequent mitosis of the host creates a population of "clones," each containing the gene of interest. From this population, the gene insert can be harvested in quantities needed for research.

RESTRICTION ENDONUCLEASES

Restriction endonucleases are possibly the most powerful tools in biotechnology. They allow scientists to precisely cut DNA in a predictable and reproducible manner. In the early 1950s, Salvador Luria and Giuseppe Bertani of the University of Illinois and Jean Weigle of the California Institute of Technology found evidence of a primitive immune system in bacteria. They observed that certain strains of the bacterium *Escherichia coli* are resistant to infection by various bacteriophages. The phenomenon seemed to be a property of the bacterial cell, which is able to *restrict* the growth and replication of certain phages. In 1962, Werner Arber, at the University of Geneva, provided the first evidence that the resistant bacterium possesses an enzyme system that selectively recognizes and destroys foreign DNA while modifying its own chromosomal DNA to prevent self-destruction.

Several years later, Arber and his associate Stuart Linn, as well as Matthew Meselson and Robert Yuan at Harvard University, isolated *E. coli* extracts that efficiently cleaved phage DNA. These extracts contained the first known *restriction endonucleases*, enzymes that recognize and cut a specific sequence of DNA. Restriction endonucleases are members of a larger group of enzymes called nucleases, which generally break the phosphodiester bonds that link adjacent nucleotides in DNA. *Endo*nucleases cleave DNA at internal positions, whereas *exo*nucleases progressively digest from the ends of DNA molecules.

In addition to an endonuclease cutting activity, some of these enzymes also possess a modification activity that is protective to the host. It was later realized that the enzyme protects bacterial host DNA from digestion by adding methyl groups to a nucleotide within the sequence recognized by the restriction enzyme. This modification of the DNA blocks the restriction enzyme from recognizing its sequence-specific binding sites in the host. Although typically both strands of host DNA are methylated, bacterial DNA is protected from digestion even when only one strand is methylated (hemimethylated). This is important

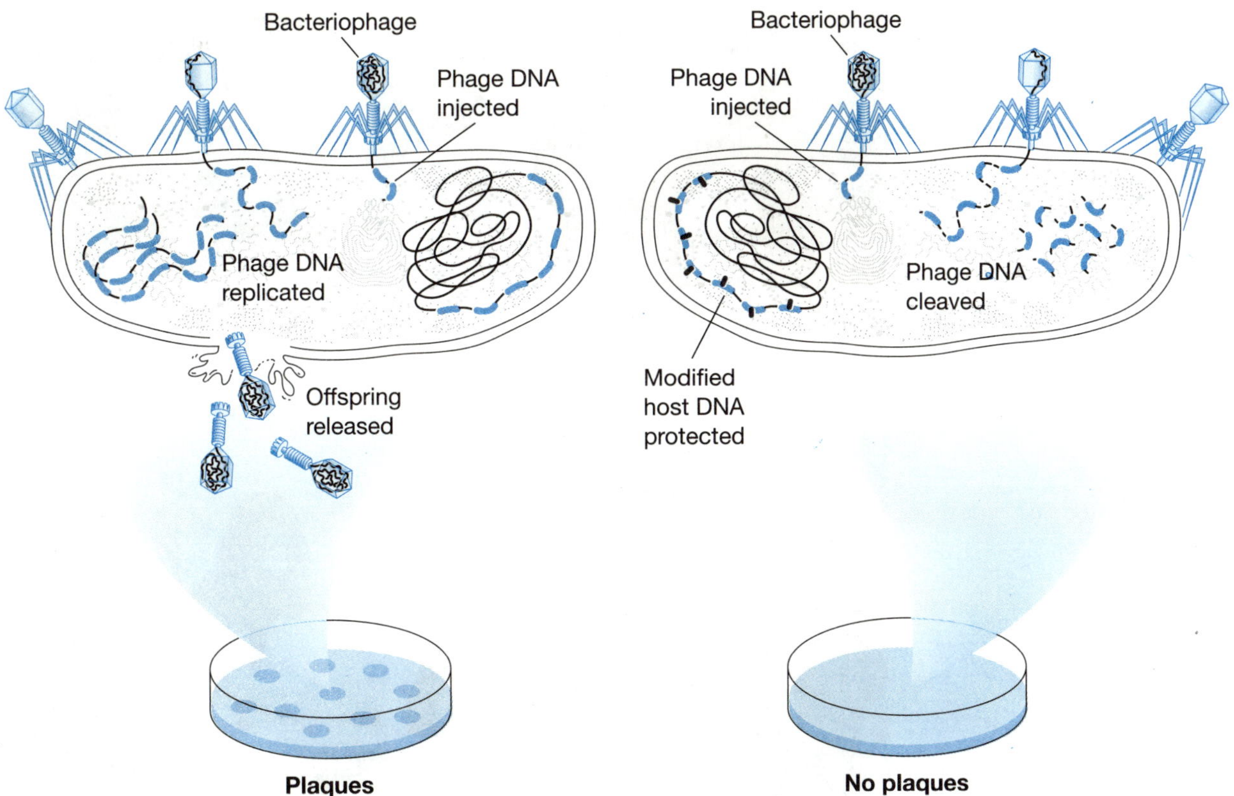

Host-controlled Restriction
Bacteriophage infects strain at left, producing clear plaques in the bacterial lawn. Resistant strain at right possesses a restriction enzyme that cleaves incoming phage DNA, as well as a modifying enzyme that methylates its own DNA to protect it from cleavage. (Art concept developed by Lisa Shoemaker.)

because following DNA replication, the newly synthesized DNA strand is not immediately methylated. Because hemimethylated DNA is protected, the methylated parental strand protects the host chromosome while the new daughter strand becomes methylated.

The particular restriction enzymes isolated by Arber, Linn, Meselson, and Yuan were not of practical value as tools for manipulating DNA. Although these enzymes recognize specific nucleotide sequences, they cut the DNA molecule at positions that may be thousands of nucleotides distant from the recognition site. The phenomenon of restriction modification was therefore of little use as a tool until 1970, when Hamilton Smith and his student Kent Wilcox, at Johns Hopkins University, isolated a new restriction endonuclease from *Haemophilus influenzae*. The restriction activity of this enzyme, named *Hin*dII, differed from previously discovered endonucleases in two important ways. First, the enzyme showed restriction activity but no modification activity. Second, *Hin*dII cleaved DNA predictably by cutting at a precise position *within* its recognition sequence, not at a distance.

There are three major classes of restriction endonucleases: Types I, II, and III. Type I and III enzymes have both restriction and modification activity, and they cut DNA at sites outside of their recognition sequences. Using ATP for energy, the enzymes move along the DNA molecule from recognition site to cleavage

Hamilton Smith (left) and Daniel Nathans, 1978

Paul Berg, 1980

Herbert Boyer, 1977

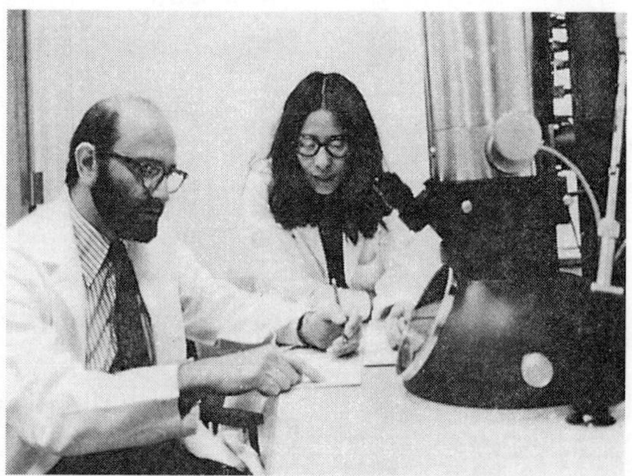

Stanley Cohen and Annie Chang, 1973

Key Figures in the Construction of the First Recombinant DNA Molecules
(Courtesy of [top] Smith and Nathans, The Baltimore Sun; Berg, News and Publication Service, Stanford University. [Bottom] Boyer, Paul Conklin Academy Forum; Cohen and Chang, News and Publication Service, Stanford University.)

site. The lack of predictability in cutting and the requirement of ATP make type I and III restriction enzymes of little practical use.

Type II restriction endonucleases are ideal as tools for manipulating DNA molecules for several reasons: (1) Each has restriction activity but no modification activity; (2) each cuts in a predictable and consistent manner, at a site within or adjacent to the recognition sequence; and (3) type II enzymes require only the magnesium ion (Mg^{++}) as a cofactor; ATP is not needed. Type II restriction enzymes have been isolated from a huge number of organisms. More than 200 restriction enzymes are commercially available, enabling researchers to cut DNA at more than 100 different recognition sequences. The names for restriction endonucleases reflect their origin:

*Eco*RI E = genus *Escherichia*
co = species *coli*
R = strain RY13
I = first endonuclease identified

Type II restriction endonucleases recognize sequences that are generally four to eight nucleotides in length. These sequences are usually palindromic—that is, they read the same on opposite strands. For example, the recognition sequence for *Bam*HI is

5´GGATCC3´
3´CCTAGG5 ´

The two DNA strands read the same 5´-3´ (read the top strand left to right and the bottom strand right to left). When *Bam*HI cuts the DNA, the result is two fragments with the following ends:

5´G GATCC3´
3´CCTAG G5´

The cut site within the recognition sequence, like the sequence itself, is always the same for type II restriction enzymes. Type II restriction enzymes cleave their restriction sites in one of three possible ways, depending on the enzyme: (1) They can cleave in the middle of the site directly opposite each other on the two strands, creating a "blunt ended" molecule; (2) they can cleave 3´ of center leaving either 2 or 4 single-stranded bases that extend from the 5´ end; and (3) they can cleave 5´ of center, leaving either 2 or 4 bases that extend from the 3´ end. The single-strand-overhanging ends from the latter two scenarios are complementary to any end cut by the same enzyme. This complementarity of the single-strand ends gives restriction enzyme cut sites "sticky ends" that are important for gene cloning. (A subcategory, type IIS cleaves up to 20 nucleotides away from their recognition sequence, making them of little use to molecular biologists.)

Structural studies of various restriction enzymes bound to their DNA recognition sequence have been carried out, giving us a fairly clear understanding of how type II restriction endonucleases function. *Eco*RI acts as a homodimer (two *Eco*RI molecules acting together). The two *Eco*RI molecules align at the recognition site in opposite orientation on the DNA and work by accessing the bases on the inside of the DNA molecule. Specific amino acids within *Eco*RI form hydrogen bonds with the DNA recognition sequence, GAATTC. Once bound, other residues within the restriction enzyme catalyze a hydrolysis reaction that uses water to break the phosphodiester linkage on each strand of the DNA helix within the recognition sequence. The DNA molecule is cut in two, with a phosphate group at the 5´ end and a hydroxyl group at the 3´ end. By working as a dimer, the enzyme is able to cut the DNA simultaneously on both strands.

Specificities of Some Typical Endonucleases

Source	Enzyme	Recognition sequence	Number of cleavage sites		
			λ	Adenovirus-2	SV40
Bacillus amyloliquefaciens H	*Bam*HI	G¬GATCC	5	3	1
Bacillus globigii	*Bgl*II	A¬GATCT	6	12	0
Escherichia coli RY13	*Eco*RI	G¬A$\overset{*}{A}$TTC	5	5	1
Escherichia coli R245	*Eco*RII	¬C$\overset{*}{C}$TGG	>35	>35	16
Haemophilus aegyptius	*Hae*III	GG¬CC	>50	>50	19
Haemophilus influenzae R_d	*Hind*II	GTPy¬Pu$\overset{*}{A}$C	34	>20	7
Haemophilus influenzae R_d	*Hind*III	$\overset{*}{A}$¬AGCTT	6	11	6
Haemophilus parainfluenzae	*Hpa*II	C¬CGG	>50	>50	1
Nocardia otitidis-caviarum	*Not*I	GC¬GGCCGC	0	7	0
Providencia stuartii 164	*Pst*I	CTGCA¬G	18	25	2
Serratia marcescens S_b	*Sma*I	CCC¬GGG	3	12	0

Recognition sequences are written 5´ to 3´. Only one strand is represented. The arrows indicate cleavage sites. Pu (purine) denotes that either A or G will be recognized. Py (pyrimidine) denotes that either C or T will be recognized. Asterisks represent positions where bases can be methylated.

How frequent a particular restriction cut site occurs is based on the length of the enzyme recognition sequence. Assuming that the four nucleotides (A, C, T, and G) are distributed randomly and with equal distribution within a DNA molecule, then any 4-nucleotide combination will occur, on average, every 256 nucleotides (4 x 4 x 4 x 4), and any 6-nucleotide sequence will occur every 4096 nucleotides (4 x 4 x 4 x 4 x 4 x 4). Restriction endonucleases recognize 8 nucleotides, with a calculated frequency of occurrence of (4 x 4 x 4 x 4 x 4 x 4 x 4 x 4) or once every 65,536 base pairs. The actual frequency of cutting by a restriction enzyme is typically different from this calculated number, because DNA sequence is not randomly arranged. For example, the restriction enzyme *Not*I recognizes the sequence GCGGCCGC, an 8-base recognition sequence that would be predicted to occur every 65,536 base pairs. In mammalian genomes, *Not*I cutting frequency is actually approximately once per 1,000,000 base pairs because sequences rich in GC content occur infrequently (as discussed in Chapter 6).

Daniel Nathans, a colleague of Hamilton Smith at Johns Hopkins University, was the first to show the broad applicability of restriction endonucleases coupled

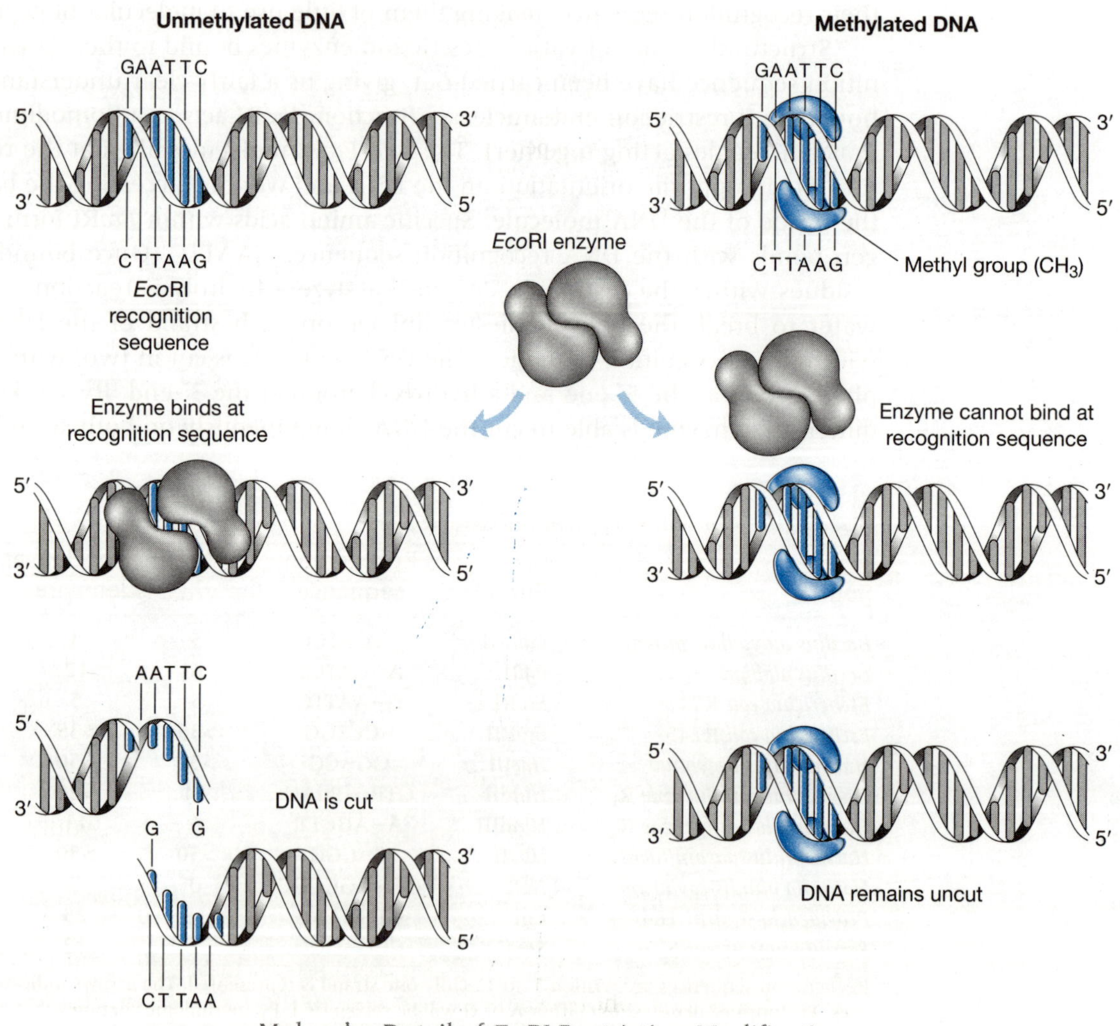

Molecular Detail of *Eco*RI Restriction-Modification
(Art concept developed by Lisa Shoemaker.)

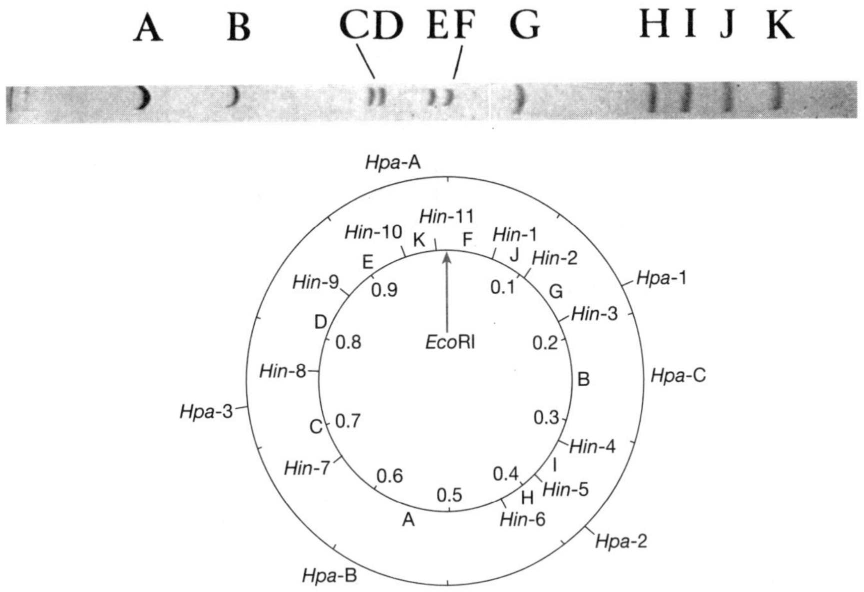

Cleavage Map of the SV40 Genome

(*Top*) DNA fragments generated from SV40 DNA digested with *Hin*dII restriction endonuclease separated by electrophoresis. (Reprinted, with permission, from Danna K. and Nathans D. 1971. Specific cleavage of simian virus 40 DNA by restriction endonuclease of *Hemophilus influenzae*. *Proc. Natl. Acad. Sci. 68:* 2913–2917.) (*Bottom*) A cleavage map of the SV40 genome. Fragments created by *Hin*dII digestion are labeled A to K, corresponding to fragments seen on the gel above. (Reprinted, with permission, from Danna K., Sack G.H., and Nathans D. 1973. Studies of simian virus 40 DNA. VII. A cleavage map of the SV40 genome. *J. Mol. Biol. 78:* 363–376.)

with gel electrophoresis. Nathans used Smith's *Hin*dII enzyme to cut the purified DNA of a small virus that infects monkeys (simian virus 40 or SV40). Reactions where restriction enzymes are used to cut DNA samples are called *digests*. Nathans then used gel electrophoresis to separate the *Hin*dII fragments of SV40 by size. He deduced the order of the fragments (and corresponding restriction sites) in the 5000-base-pair circular chromosome, creating a *restriction map* that was then related to the existing genetic map of SV40. Similar use of restriction enzyme digests to create restriction maps played an integral part in mapping the human genome.

GEL ELECTROPHORESIS

DNA fragments of different sizes, such as those resulting from digestion with restriction endonucleases, were originally separated by a laborious method called velocity-sedimentation ultracentrifugation. DNA samples are spun in a high-speed centrifuge through a gradient of salt or sugar that would separate DNA fragments by size, largest to smallest, top to bottom.

In 1970, Daniel Nathans used polyacrylamide gel electrophoresis as a simple and rapid means to separate DNA fragments. Whereas centrifugation uses gravitational force to separate molecules, electrophoresis means literally *to carry with electricity*. Gel electrophoresis takes advantage of the fact that, as an organic acid, DNA is negatively charged. DNA owes its acidity to phosphate groups that alternate with deoxyribose to form the rails of the double helix ladder. In solution, at neutral pH, negatively charged oxygens radiate from phosphates on the outside of the DNA molecule. When placed in an electric field, DNA molecules are attracted toward the positive pole (anode) and repelled from the negative pole (cathode).

During electrophoresis, DNA fragments sort by size in the polyacrylamide gel. The porous gel matrix acts as a molecular sieve through which smaller molecules can move more easily than larger ones; thus, the distance moved by a DNA fragment is inversely proportional to its molecular weight. In a given period of time, smaller restriction fragments migrate relatively far from the origin compared to larger fragments. Because of its small pore size, polyacrylamide efficiently separates small DNA fragments of up to 1000 nucleotides. However, this level of resolution was inappropriate for isolating gene-sized fragments of several thousand nucleotides.

Although less time consuming than centrifugation, early polyacrylamide gel electrophoresis was still labor intensive and required the use of radioactively labeled DNA fragments. Following electrophoresis, the polyacrylamide gel was cut into many bands, and the amount of radioactivity in each slice was determined in a scintillation counter. The pattern of radioactivity was used to reconstruct the pattern of DNA bands in the gel.

A research team at Cold Spring Harbor Laboratory, led by Joseph Sambrook, introduced two important refinements to DNA electrophoresis that made possible rapid analysis of DNA restriction patterns. First, they replaced polyacrylamide with agarose, a highly purified form of agar. An agarose matrix can efficiently separate larger DNA fragments ranging in size from 100 nucleotides to more than 50,000 nucleotides. DNA fragments in different size ranges can be separated by adjusting the agarose concentration. A low concentration (down to 0.3%) produces a loose gel that separates larger fragments, whereas a high concentration (up to 2%) produces a stiff gel that resolves small fragments.

Second, they used a fluorescent dye, ethidium bromide, to stain DNA bands in agarose gels. Following a brief staining step, the fragment pattern is viewed directly under ultraviolet (UV) light. This technique is extremely sensitive; as little as 5 ng (0.000000005 g) of DNA can be detected. Thus, it is not difficult to understand why ethidium bromide staining quickly replaced radioactive labeling for the routine analysis of DNA restriction patterns.

Currently used methods are identical to those published by the Cold Spring Harbor team in 1973. Molten agarose is poured into a casting tray in which a plastic or Plexiglas comb is suspended. As it cools, the agarose hardens to form

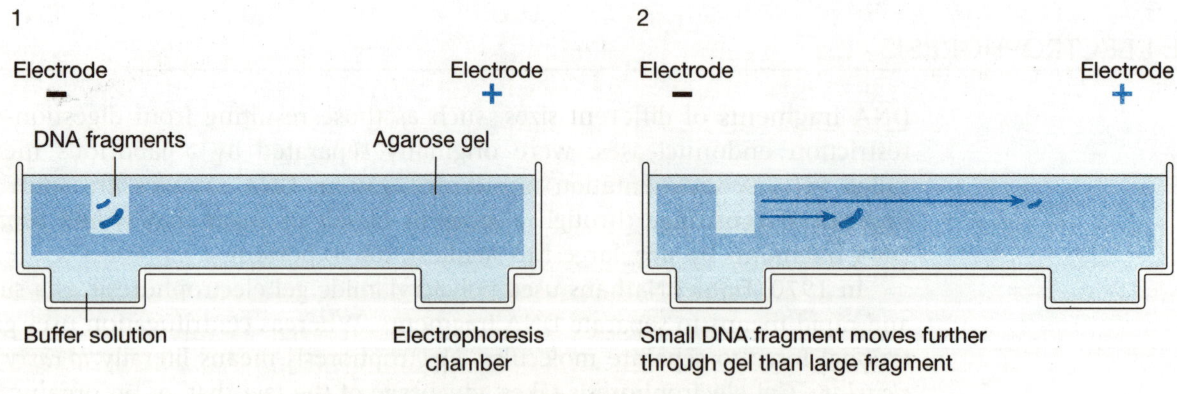

Agarose Gel Electrophoresis of DNA Fragments
(Art concept developed by Lisa Shoemaker.)

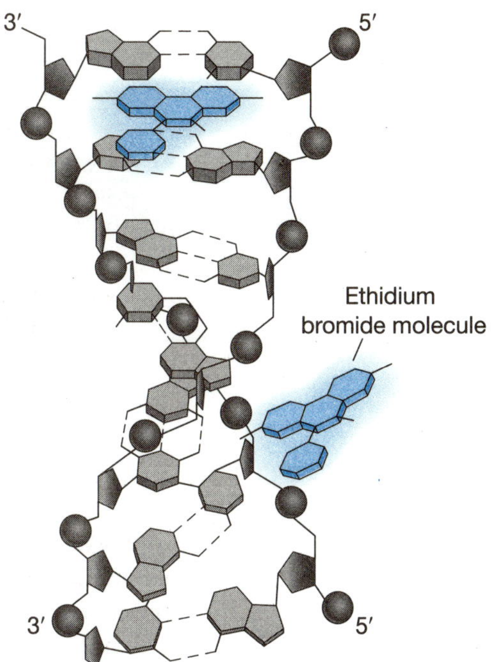

Intercalation of Ethidium Bromide into DNA Helix

a Jell-O-like substance consisting of a dense network of cross-linked molecules. The solidified gel slab is immersed in a chamber filled with buffer solution, which contains ions needed to conduct electricity. When the gel solidifies, the comb is removed, leaving a number of wells into which DNA samples are loaded. Just prior to loading, the digested DNA is mixed with a loading solution that consists of sucrose and one or more visible dyes. The dense sucrose solution weights the DNA sample, helping it to sink when loaded into a well.

Current supplied through electrodes at either end of the chamber creates an electric field across the gel. The negatively charged DNA fragments move from the wells into the gel, migrating through the pores in the matrix toward the positive pole. The negatively charged dye molecules do not interact with the DNA, but migrate independently toward the positive pole. For example, the commonly used marker bromophenol blue migrates at a rate equivalent to a DNA fragment of approximately 300 nucleotides (in a 1% gel). The visible movement of the dye allows one to monitor the relative migration of the unseen DNA bands.

Following electrophoresis, the gel is soaked in a dilute solution of ethidium bromide. The stain diffuses throughout the gel, becoming highly concentrated in regions where it binds to DNA fragments. (Alternatively, ethidium bromide is incorporated into the gel and buffer prior to beginning electrophoresis.) A planar molecule, the ethidium bromide intercalates between the stacked nucleotides of the DNA helix, staining DNA bands in the gel.

The stained gel is then exposed to medium-wavelength UV light. The DNA/ethidium bromide complex strongly absorbs UV light at 300 nm, retains some of the energy, and reemits visible light in the orange range at 590 nm. Under UV illumination, the stained restriction fragments appear as fluorescent orange bands in the gel. It is important to understand that a band of DNA seen in a gel is not a single DNA molecule. Rather, the band is a collection of millions of identical DNA molecules, all of the same nucleotide length.

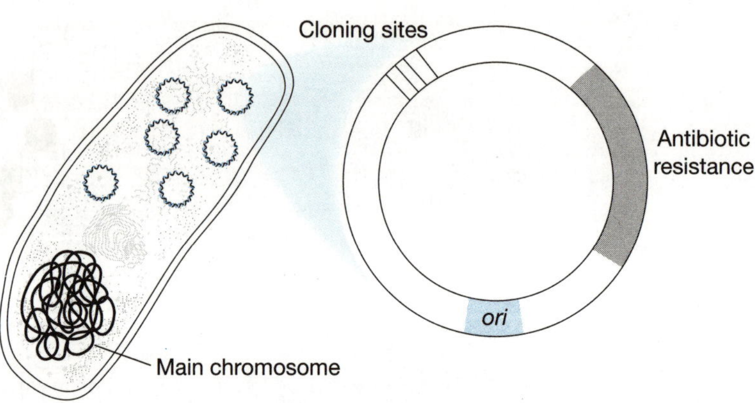

Escherichia coli with Plasmids

THE PLASMID VECTOR

In medical terminology, a vector is an organism that carries a pathogen from one host organism to another. In molecular biology, a vector is a DNA molecule that is used as a vehicle to carry foreign DNA sequences into a host cell. Plasmids are the simplest bacterial vectors. Ranging in length from 1,000 to 200,000 base pairs, they are circular DNA molecules that exist separately from the main bacterial chromosome. As explained below, molecular biologists can alter plasmids to carry specific gene sequences. These plasmids can be used as tools to propagate large quantities of a given DNA sequence or as a means of getting a foreign gene expressed in a host cell.

Propagation of Plasmids

The use of plasmids to propagate large quantities of specific gene sequences takes advantage of the rapid growth rate of bacterial cells. If a plasmid is placed in a host cell, and it is duplicated with each cell division (and each daughter cell receives a copy), many copies of the plasmid will be replicated in culture after several generations.

To be propagated through successive bacterial generations, the plasmid vector must contain a specific DNA sequence, called the origin of replication (*ori*),

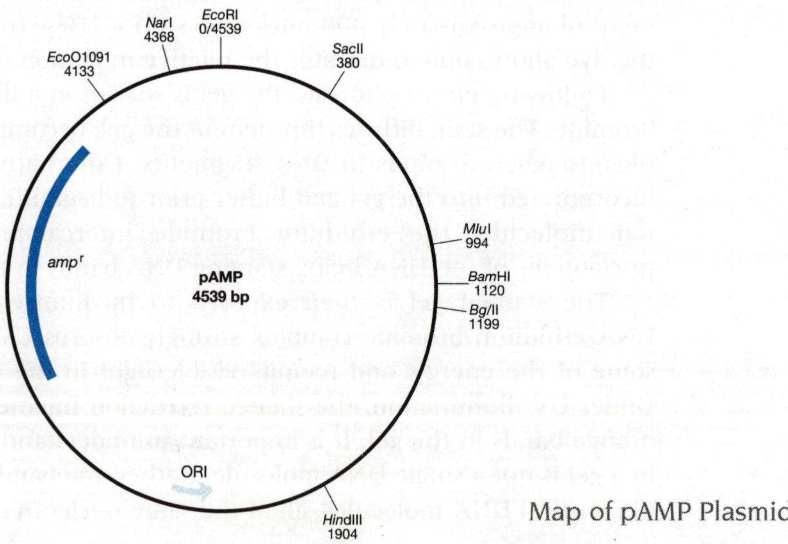

Map of pAMP Plasmid

which allows it to be replicated within the host cell. *E. coli* DNA polymerase and other proteins required to initiate DNA synthesis bind to the *ori*.

Plasmids can be divided into two broad groups, according to how tightly their replication is regulated. Plasmids under "stringent control" replicate once per cell division, along with the main bacterial chromosome. "Relaxed" plasmids replicate their DNA autonomously throughout the cell cycle of the bacteria and accumulate in hundreds of copies per cell. Relaxed plasmids are useful for amplifying large amounts of cloned DNA.

Selectable Markers

When using plasmids in experiments, it is very important to be able to distinguish between cells that actually contain the plasmid and cells that do not. Because of this, most plasmids contain genes that encode selectable marker proteins. The most common type of selectable marker is antibiotic resistance. When

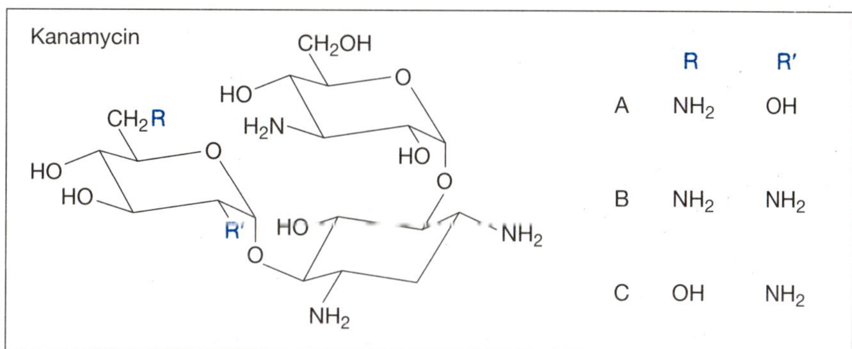

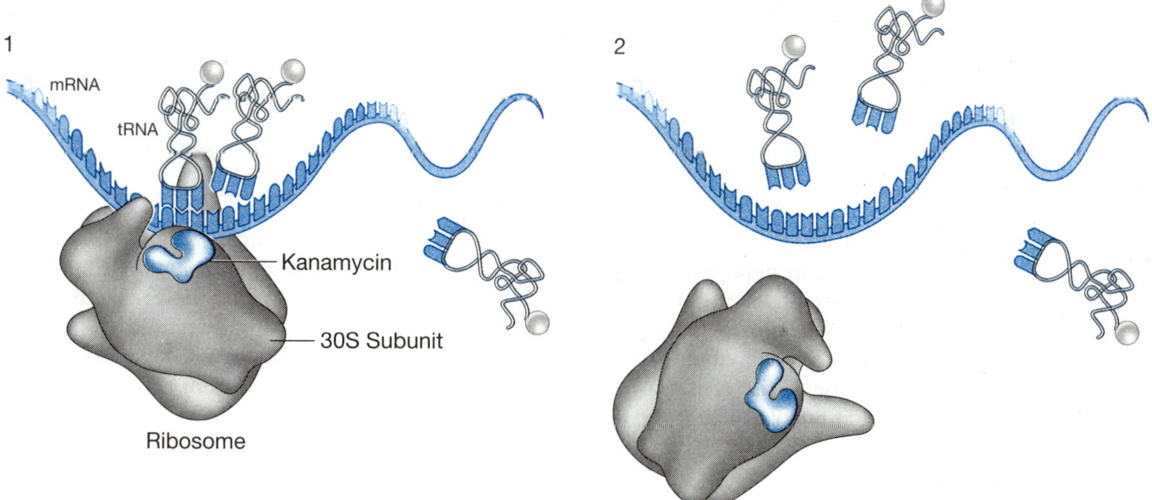

Kanamycin Action

Kanamycin poisons protein synthesis by irreversibly binding the 30S subunit of the ribosome. (*1*) The kanamycin/ribosome complex initiates protein synthesis by binding mRNA and the first tRNA. (*2*) However, the second tRNA is not bound, and the ribosome-mRNA complex dissociates. The pKAN resistance protein, from the aminoglycoside family, is a phosphotransferase that adds phosphate groups to the kanamycin molecule, thus blocking its ability to bind the ribosome. Other resistance proteins from this family can be acetyltransferases, which add acetyl groups to the antibiotic. Three forms of the phosphotransferase (A, B, or C) vary by the group present at R and R´, as indicated in the figure.

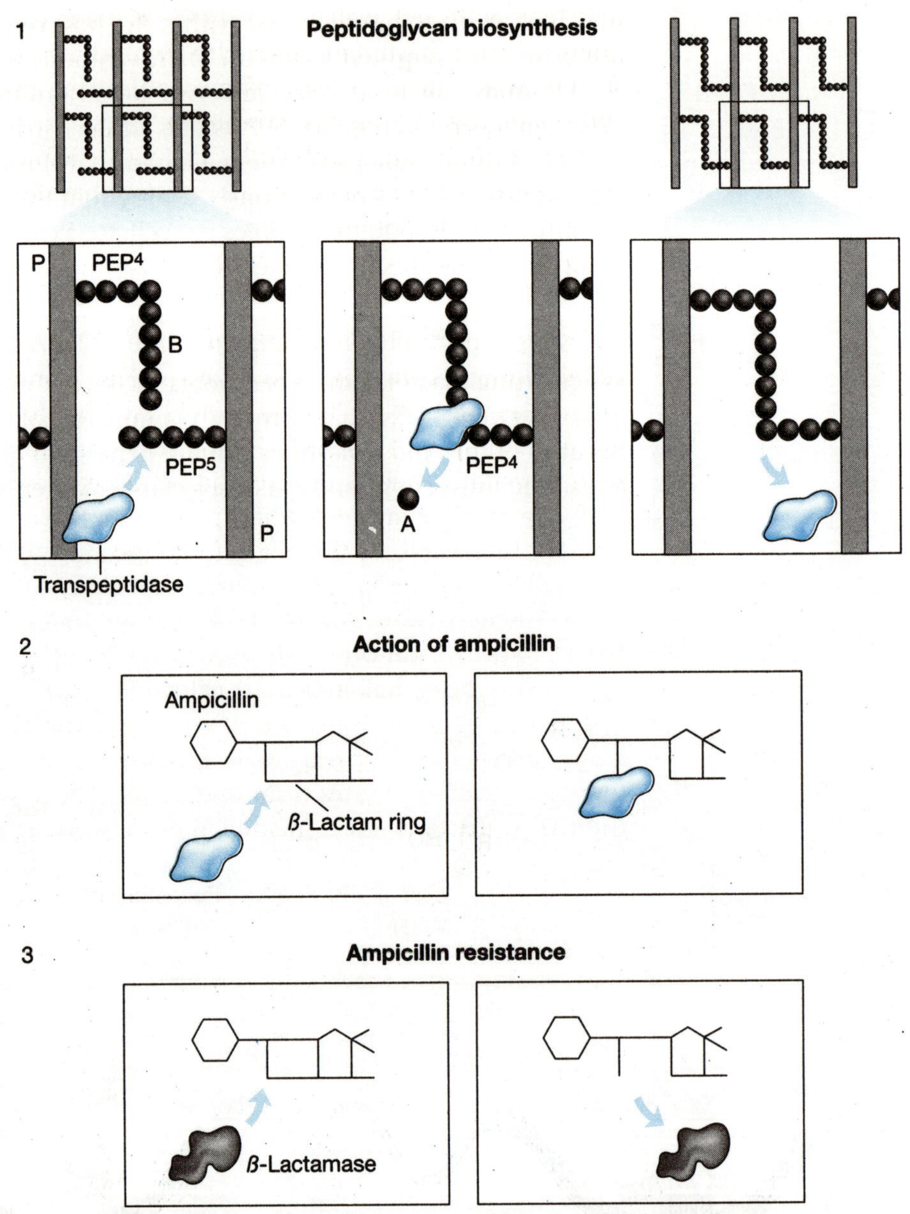

Peptidoglycan Biosynthesis, Ampicillin Action, and Ampicillin Resistance
(*1*) A transpeptidase removes an alanine residue (A) from a pentapeptide (PEP^5). The resulting tetrapeptide (PEP^4) is joined to the peptide bridge (B) to cross-link two adjacent polysaccharide chains (P). (*2*) The β-lactam ring of ampicillin structurally mimics the peptide bridge and irreversibly binds the transpeptidase, making it unavailable for peptidoglycan synthesis. (*3*) The pAMP resistance protein, β-lactamase, cleaves the β-lactam ring of ampicillin, making it unable to bind the transpeptidase.

the host cells are grown in media that contains an antibiotic, only cells that contain a plasmid with resistance to that antibiotic can survive. By killing off the nonplasmid-containing cells, the investigator is left with a pure population of cells that all carry the plasmid. Antibiotics function by several different mechanisms. For example, members of the penicillin family (including ampicillin) interfere with cell wall biosynthesis. Kanamycin, tetracycline, and chloramphenicol arrest bacterial cell growth by blocking various steps in protein synthesis. Likewise, there are various mechanisms of antibiotic resistance. The ampicillin, kanamycin, and chloramphenicol resistance genes produce proteins that inactivate their target antibiotics through chemical modification. The tetracycline resistance gene specifies an enzyme that prevents transport of the antibiotic through the cell membrane.

Inserting New Genes into Plasmids

The simple principle of cutting and pasting DNA fragments is the backbone of gene cloning technology. A key step of gene cloning is to put a gene of interest into a plasmid vector. Plasmid vectors are therefore designed to contain one or more "cloning sites." These cloning sites are a series of restriction recognition sequences for a variety of restriction endonucleases called a polylinker. Cutting with a restriction enzyme opens up the circular plasmid DNA and allows an insert DNA to be spliced in. Combining two pieces of DNA creates a new piece containing both, called a *recombinant DNA molecule.*

As discussed above, many restriction enzymes create a single-stranded overhang, a sticky end, when they cut DNA. This sticky end can form hydrogen bonds with the complementary nucleotides in the overhang of another fragment generated by the same restriction enzyme. Hydrogen bonding of the sticky ends is not sufficient to form a stable molecule, and associations between complementary ends constantly form and break. This transient interaction does hold the two restriction fragments together long enough for an enzyme called DNA

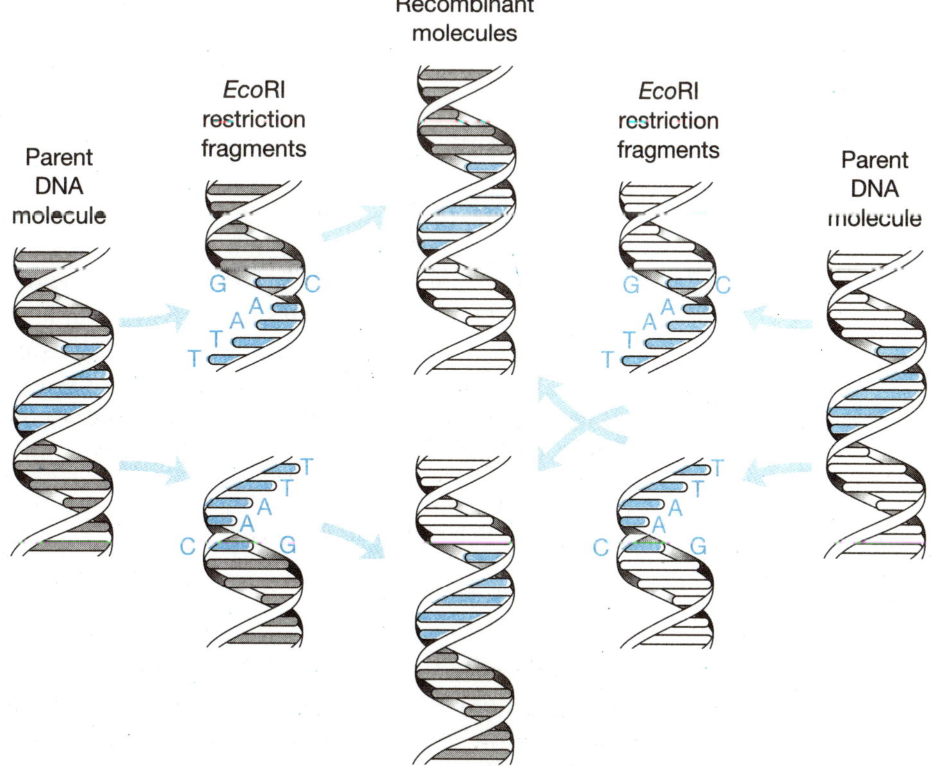

Joining Complementary "Sticky Ends" of *Eco*RI Restriction Fragments

ligase to re-form phosphodiester bonds between adjacent nucleotides. Two pieces of DNA that have been cut with the same restriction enzyme can be pasted together, or *ligated,* this way. During the ligation reaction, an ester linkage is formed between the terminal phosphate of the 5′ overhang of one fragment and the adjacent deoxyribose ring at the 3′ end of the second fragment. This is accompanied by the loss of one molecule of water, making ligation an example of a condensation reaction. Ligation of two restriction fragments covalently links the deoxyribose-phosphate rails of the two fragments into a stable double helix.

DNA pieces can be ligated into vectors using different approaches. DNA containing a gene of interest can be digested with a restriction enzyme producing a small number restriction fragments. This mixture of fragments can be ligated in one reaction into a plasmid vector, and the resulting clones analyzed to find a plasmid containing the specific desired DNA insert. Ideally, a gene is isolated on a restriction fragment created by two different endonucleases that cleave on either side of the target region and that generate distinctive sticky ends. Because the overhangs are different, they are noncomplementary and cannot hydrogen bond to each other. This prevents any fragment from rejoining its own ends and encourages recombination between the insert fragment and a plasmid vector that has been opened up with the same two enzymes. Such "directional cloning," using two different enzymes, produces clones reliably and fixes the orientation of the inserted fragment in the plasmid.

THE HOST CELL: THE BACTERIUM *ESCHERICHIA COLI*

The manipulation of DNA and creation of recombinant DNA molecules described so far takes place in the test tube. Ultimately, the propagation of a DNA sequence must take place inside a living cell. Thus, transformation—the cellular uptake and expression of DNA in a bacterium—is crucial to the research process. The following elements are required to make the transformation process efficient and controllable enough to be of general use for introducing foreign genes into living cells: (1) a suitable host organism in which to insert the gene, (2) a self-replicating vector to carry the gene into the host organism, and (3) a means of selecting for host cells that have taken up the gene.

The bacterium *E. coli* has become the most widely used organism in molecular biology because it provides a relatively simple and well-understood genetic environment in which to isolate foreign DNA. Its primary genetic complement is contained on a single chromosome of approximately 4.6 million base pairs, making it only 1/600th the size of a haploid set of human chromosomes (the human genome). The *E. coli* genome has been completely sequenced and more than 4000 genes identified.

Because the genetic code is nearly universal, *E. coli* can accept foreign DNA derived from any organism. The DNA of a bacterium, a human, a corn plant, or a fruit fly is constructed of the same four nucleotides (adenosine, cytosine, guanosine, and thymidine), is assembled in the same structure, and is replicated by the same basic mechanism. Each organism transcribes DNA into messenger RNA, which is in turn translated into proteins according to the genetic code. A foreign gene inside *E. coli* is replicated, and in some cases translated, in exactly the same manner as the native bacterial DNA. *E. coli* "sees" foreign DNA as its own.

Under the best of circumstances, the uptake of a specific foreign gene is a relatively rare occurrence and is thus most easily accomplished in a large population of organisms that are reproducing rapidly. *E. coli* is an ideal genetic organism in this regard.

A recombinant plasmid is biologically amplified when a transformed bacterium replicates by binary fission to create a clone of identical daughter cells. Under favorable conditions, *E. coli* replicates once every 22 minutes, giving rise to 30 generations and more than 1 billion cells in 11 hours. This number of cells can be contained in a single milliliter of culture solution. Moreover, since each bacterium can carry up to several hundred copies of a cloned gene, the foreign DNA sequence is potentially amplified by a factor of several hundred billion.

E. coli is a constituent of the normal bacterial fauna that inhabits the human colon, where it absorbs digested food materials. Thus, it grows best with incubation at 37°C in a culture medium that approximates the nutrients available in the human digestive tract. An example of such a medium is Luria-Bertani (LB) broth, which contains carbohydrates, amino acids, nucleotide phosphates, salts, and vitamins derived from yeast extract and milk protein.

Bacterial growth falls into several distinct phases. During *lag phase*, cells adjust to the nutrient environment and gear up for rapid proliferation; little or no cellular replication takes place. During *logarithmic* (*log*) *phase*, the culture grows exponentially and the cell number doubles every 22 minutes. During *stationary phase*, the cell number remains constant as new cells are produced at the same rate as old cells die. After an extended period, the culture enters *death phase*; the number of viable cells decreases as nutrients deplete and wastes accumulate.

Masses of bacterial cells are grown in a suspension culture; shaking provides aeration and keeps cells suspended in the medium. To isolate individual colonies, cells are spread on the surface of LB agar plates. Although the individual cells are invisible to the naked eye, after plating onto solid medium, each cell divides to form a visible colony of identical daughter cells in 12–24 hours.

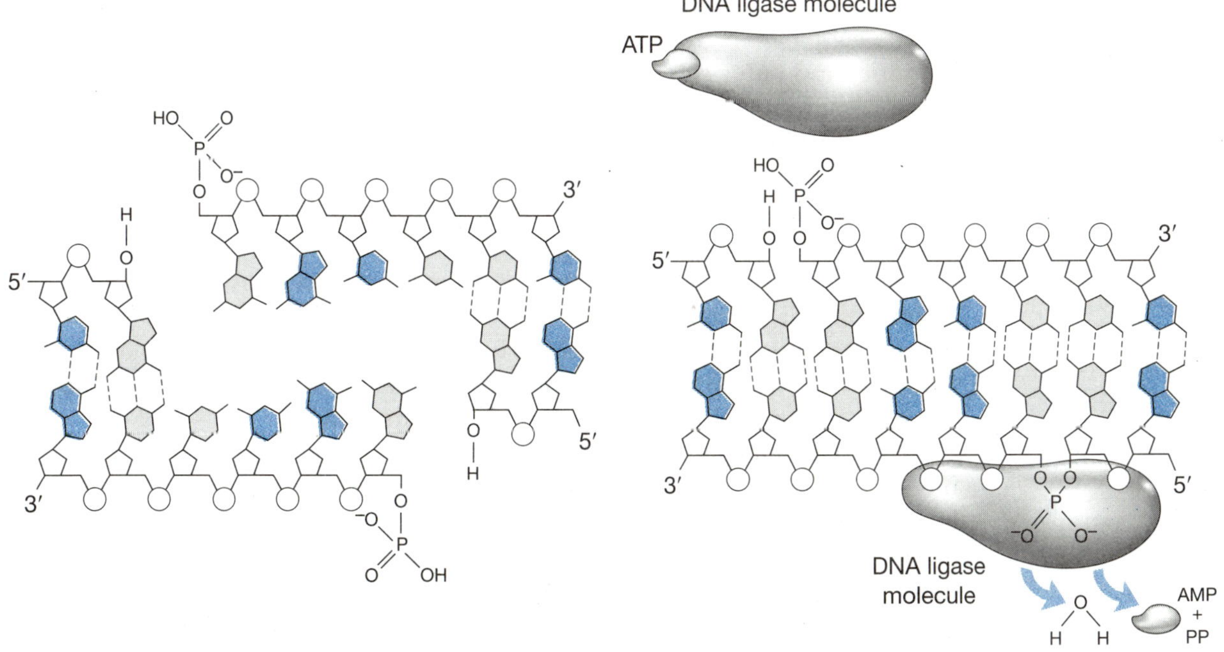

Molecular Detail of DNA Ligation by T4 Ligase

Hydrogen bonding between complementary nucleotides aligns *Bam*HI fragments while ligase reforms phosphodiester bonds on each side of the DNA molecule.

TRANSFORMATION

The phenomenon of transformation, which provided a key clue to understanding the molecular basis of the gene, also provided a tool for manipulating the genetic makeup of living things. The natural transformation described by Griffith and Avery is an exceedingly rare event. However, in 1970, Morton Mandel and Akiko Higa found that *E. coli* becomes markedly *competent* for transformation by foreign DNA when cells are suspended in cold calcium chloride solution and subjected to a brief heat shock at 42°C. They also found that cells arrested in early- to mid-log growth can be rendered more competent than can cells in other stages of growth.

Their calcium chloride procedure, which is still in wide use, yields transformation efficiencies of 10^5 to 10^7 transformants per microgram of plasmid DNA. (Transformation efficiency is generally expressed as the number of transformed cells that would be obtained from a microgram of intact plasmid DNA.) However, there is nothing particularly magical about calcium (Ca^{++}) ions.

Subsequent research showed that treatment with other divalent cations, such as magnesium (Mg^{++}), manganese (Mn^{++}), and barium (Ba^{++}), produces comparable or even greater transformation efficiencies. A protocol using a reducing agent and a complex transformation buffer composed of a mixture of positive ions (Ca^{++}, Mn^{++}, K^{+}, and Co^{+++}) achieves efficiencies of up to 10^9 transformants per microgram of plasmid DNA.

Under conditions that yield high-efficiency transformation (10^9 transformants per microgram of plasmid), approximately 1/10th of all viable cells are rendered competent. At an efficiency of 10^5 transformants per microgram of plasmid, only 1/100,000th of cells become competent. Both size and conformation of the DNA molecule affect transformation efficiency. Small plasmids are more readily taken up than larger ones, although no preferred size cutoff is evident. Linear DNA fragments transform at a negligible rate, in part because they are susceptible to degradation by exonucleases present in the periplasm between the inner and outer cell membrane of *E. coli*.

Because transformation is limited to a subset of cells that are competent, increasing the number of available plasmids does not increase the probability that a cell will be transformed. A suspension of competent cells becomes saturated at a very low concentration of DNA, roughly 0.1–0.2 μg per milliliter of mid-log cells (10^8 cells). Increasing plasmid concentration beyond this point decreases transformation efficiency, because the excess DNA does not transform additional cells. Competent cells will, however, readily take up more than one plasmid under saturating conditions. Experiments have shown that when equal amounts of two different plasmids are added under saturating conditions, 70–90% of all transformed cells contain both plasmids.

The exact mechanism of plasmid DNA uptake by competent *E. coli* cells is unknown. Unlike salts and small organic molecules such as glucose, DNA molecules are too large to diffuse or be readily transported through the cell membrane. Some bacteria possess membrane proteins that recognize DNA and facilitate the absorption of short DNA sequences derived from related species. However, *E. coli* appears not to have evolved such an uptake mechanism.

One hypothesis is that DNA molecules pass through any of several hundred channels formed at *adhesion zones*, where the outer and inner cell membranes

1 Source DNA containing gene of interest (GOI) and vector digested with *Eco*RI and *Bam*HI

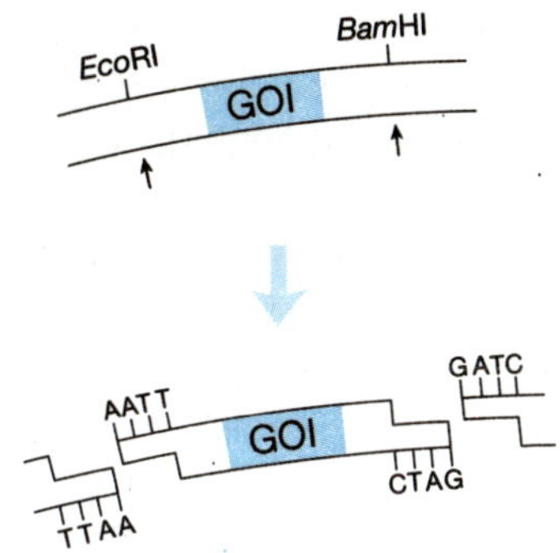

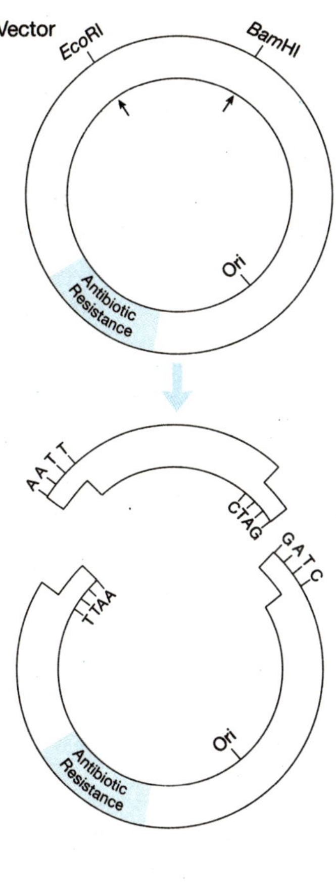

2 Separate fragments on agarose gel

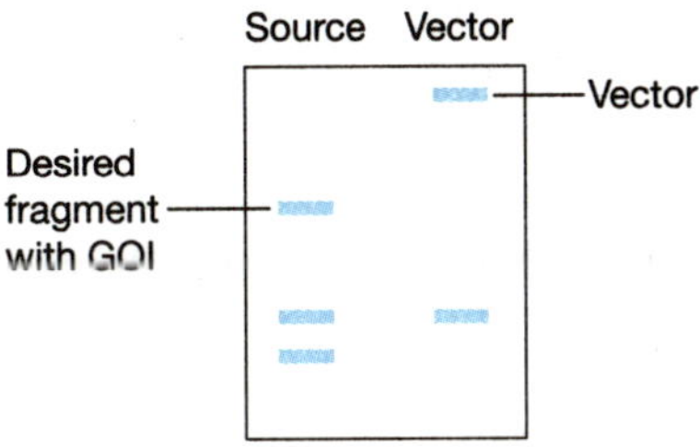

3 Elute fragments from gel and mix together with DNA ligase

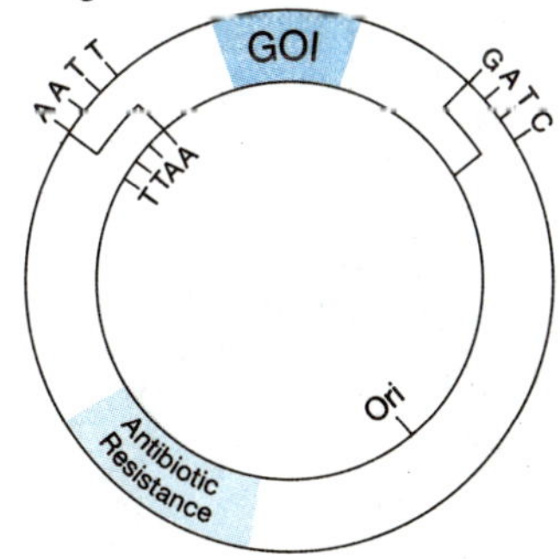

Directionally cloned recombinant plasmid

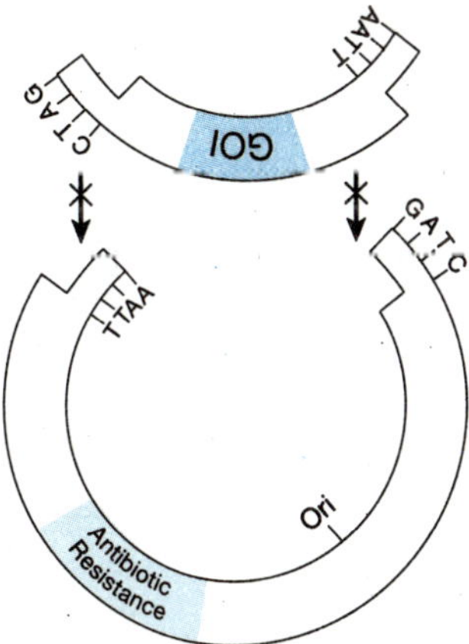

Insert fragment has two different sticky ends, so it can join the vector in only one orientation

Directional Cloning

are fused to pores in the bacterial cell wall. The fact that adhesion zones are only present in growing cells is consistent with the observation that cells in logarithmic growth can be rendered most competent for plasmid uptake. However, acidic phosphates of the DNA helix are negatively charged, as are a proportion of the phospholipids composing the cell membranes and lining the membrane pore. Thus, electrostatic repulsion between anions may effectively block the movement of DNA through the adhesion zones.

Analysis of this ionic interaction produces a plausible hypothesis for DNA uptake in bacteria. Treatment of the cells at 0□C crystallizes the fluid cell membrane, stabilizing the distribution of charged phosphates. The cations in a transformation solution (Ca^{++}, Mn^{++}, K^{+}, Co^{+++}) form complexes with exposed phosphate groups, shielding the negative charges. With this ionic shield in place, a plasmid molecule can then move through the adhesion zone. Heat shock complements this chemical process, perhaps by creating a thermal imbalance on either side of the *E. coli* membrane that physically helps to pump DNA through the adhesion zone.

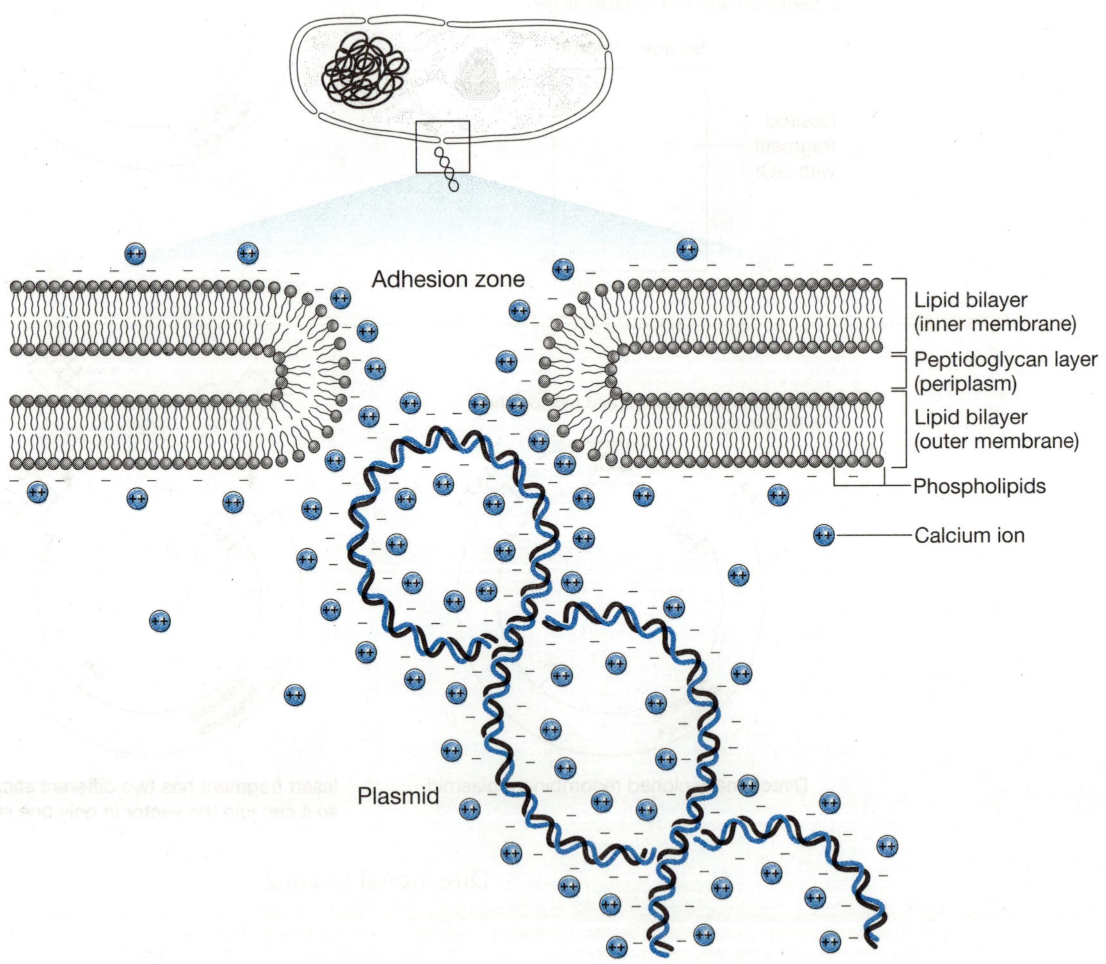

Proposed Molecular Mechanism of DNA Transformation of *E. coli*
Calcium ions (++) complex with negatively charged oxygens (–) to shield DNA phosphates from phospholipids at the adhesion zone.

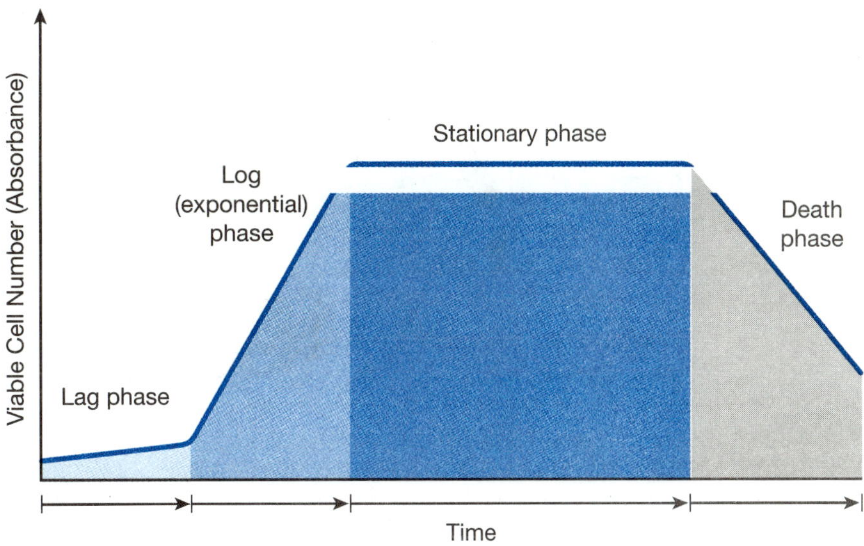

Escherichia coli Growth Curve

It is difficult to successfully transform large DNA plasmids. Although there is no exact size cutoff, plasmids of greater than 15–20 kb are not efficiently transformed into competent *E. coli.*

ELECTROPORATION

To insert large circular DNA molecules into host cells, a method called electroporation is used. With this method, cells are grown to late log phase and are prepared simply by washing them thoroughly in ion-free water. The cells and the vector are placed into a small chamber whose two sides are made of metal that serve as electrodes. A pulse of electricity is then applied across the chamber. The current causes a change in the electrical properties of the cells and the DNA enters. There is no theoretical limit to the size of DNA that is taken up in this method, and pieces up to 1 million base pairs have been successfully electroporated into cells. Electroporation has thus effectively eliminated the previous barrier of the size of DNA that can be cloned. Furthermore, electroporation can be performed with a wide variety of host cells. Electroporation is even used to insert DNA into cells of living animals, such as developing fish, frog, and chicken eggs. In these cases, a solution containing the plasmid is injected into an area of the animal next to the target cells. Electrodes are then placed into the tissue surrounding the solution and the target cells and a current is generated. The DNA enters into the target cells, creating a small population within the living organism that can be assayed for the effects of the foreign DNA.

ISOLATION AND ANALYSIS OF RECOMBINANT PLASMIDS

The growth of colonies on antibiotic medium provides *phenotypic* evidence that cells have been transformed. To confirm this at the *genotypic* level, plasmid DNA is isolated from transformants. Restriction analysis of the purified plasmid DNA,

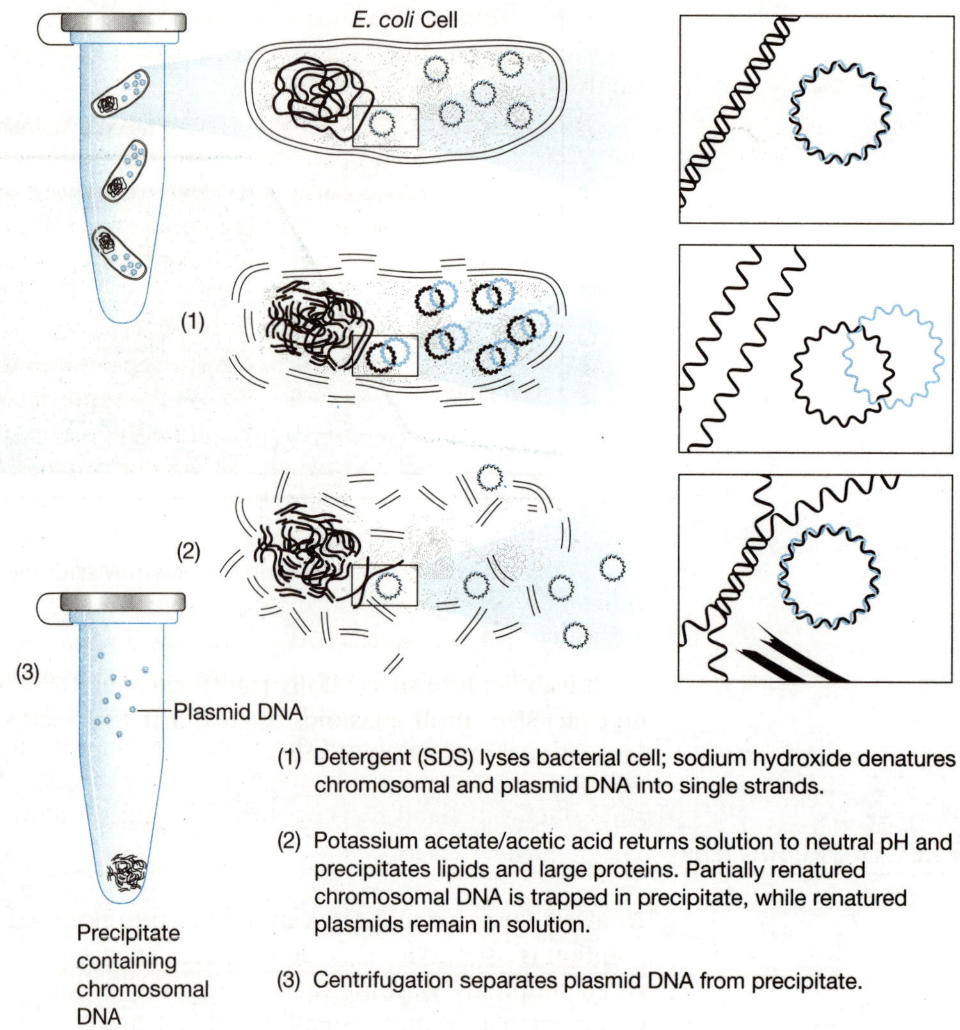

Plasmid Isolation

along with the original DNAs used to make the recombinant, presents proof of its genetic identity. The small size and circular structure of plasmids make it relatively easy to separate them from the host cell's chromosomal DNA.

DNA Miniprep Procedure

A rapid method for making a small preparation of purified plasmid DNA from culture volumes as low as 1 ml is called a *miniprep*. Transformed cells from an antibiotic-resistant colony are grown to stationary phase in an overnight suspension culture. The cells are collected by centrifugation and resuspended in a buffered solution of glucose and ethylenediaminetetraacetic acid (EDTA), which binds divalent cations (such as Mg^{++} and Ca^{++}) necessary for cell membrane stability.

The resuspended cells are then treated with a mixture of SDS and sodium hydroxide. SDS, an ionic detergent, dissolves the phospholipid and protein components of the cellular membrane. This lyses the membrane, releasing the cell contents. Sodium hydroxide denatures both plasmid and chromosomal DNAs into single strands. The chromosomal DNA separates completely into individual strands; however, the single-stranded plasmid loops remain linked together like interlocked rings.

Subsequent treatment with potassium acetate and acetic acid forms an insoluble precipitate of SDS/lipid/protein and neutralizes the sodium hydroxide from the previous step. At neutral pH, DNA renatures. In the miniprep, the long strands of chromosomal DNA only partially renature and become trapped in the SDS/lipid/protein precipitate. The linked, single-stranded plasmid DNA completely renatures into double-stranded molecules that remain in solution and largely escape entrapment in the precipitate.

The precipitate is pelleted by centrifugation and discarded, leaving the plasmid DNA (as well as RNA molecules) in the supernatant. Ethanol or isopropanol is added to the supernatant to precipitate the plasmid DNA out of solution. The plasmid DNA is pelleted by centrifugation, washed with ethanol, dried, and resuspended in a small volume of buffer. Subsequent treatment with RNase destroys RNA, leaving relatively clean plasmid DNA.

Samples of the isolated miniprep plasmid and the original plasmids are typically cut with the same restriction enzymes used to make the recombinant constructions. The digested DNAs are run on an agarose gel and the fragment patterns are compared. The bands containing the cloned gene and the vector backbone should line up perfectly with the corresponding fragments from the original DNA. Each fragment of the same size will have migrated the same distance. Larger-scale preparations, called *maxipreps*, can also be made with larger volumes of cultures and reagents to yield a greater amount of the plasmid DNA.

THE FIRST RECOMBINANT DNA MOLECULES

In the early 1970s, the techniques described above came together to yield the first recombinant DNA molecules. These pioneering experiments set the stage for a revolution in biological research.

In 1972, Paul Berg's group, at Stanford University, worked out the "tailing" method of joining DNA molecules, which was modeled after the sticky ends found at the chromosome ends of the bacteriophage λ. Using the restriction

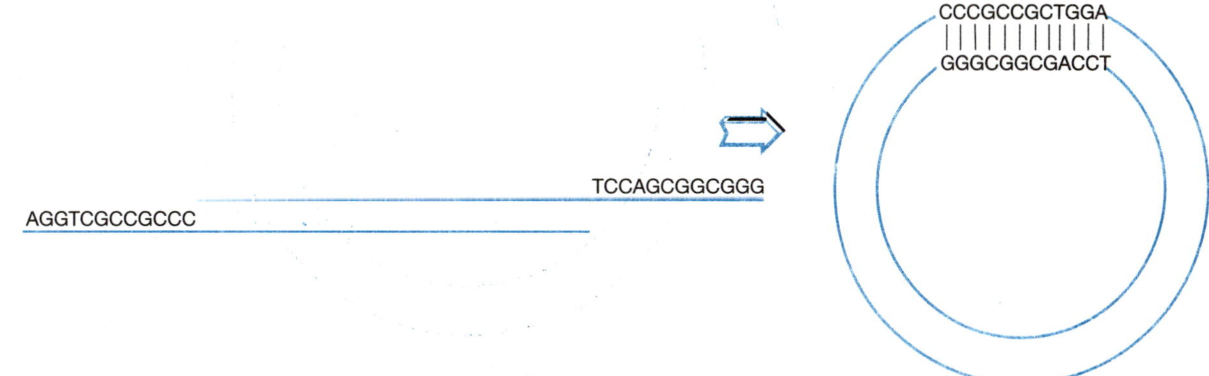

Lambda *cos* Sites

The complementary base pairs of the single-stranded ends of bacteriophage λ (*cos* sites) can form a circle.

enzyme *Eco*RI, Berg's lab cut the circular SV40 chromosome as well as a plasmid derived from *E. coli*. The restriction enzyme cut each molecule in one place, opening the DNA loops to produce linear strands. A single-stranded "tail" of 50–100 adenine residues was added to the ends of the SV40 DNA using the enzyme terminal transferase. A tail of thymine residues was added to the *E. coli* plasmid DNA by the same method. When the two DNAs were mixed together, the complementary adenine and thymine tailed molecules base-paired to form an uninterrupted circular DNA molecule. Two enzymes completed the job: DNA polymerase filled in single-stranded gaps, and DNA ligase sealed the junction points between the SV40 and plasmid fragments. Thus, the first known recombinant DNA molecule was created.

Berg's experiment provided two key pieces of the recombinant DNA puzzle. He showed that a restriction enzyme can be used to cut DNA in a predictable manner and that DNA fragments from different organisms can be joined together. In 1973, Stanley Cohen and Annie Chang, from Stanford University, carried the Berg experiment a step further. They showed that a recombinant DNA molecule could be maintained and replicated within *E. coli*.

Cohen and Chang constructed a plasmid called pSC101 that contained several important features: (1) a unique restriction site, allowing for the plasmid to be cut at a single location by the restriction enzyme *Eco*RI; (2) an origin of replication; and (3) a gene coding for resistance to the antibiotic tetracycline, to allow for selection of bacteria incorporating the plasmid. Cohen and Chang transformed pSC101 into *E. coli* cells. The transformed cells were then spread on nutrient agar plates containing tetracycline; the appearance of colonies of bacteria with resistance to the antibiotic showed that the plasmid had been taken up and expressed.

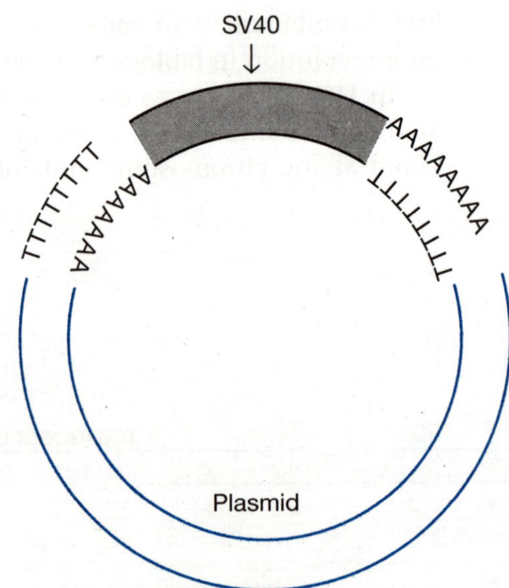

A/T Tailing
Berg added a poly(A) tail to the SV40 and a poly(T) tail to the plasmid molecule to create sticky ends. It would later be shown that many restriction enzymes cut, leaving single-stranded sticky ends.

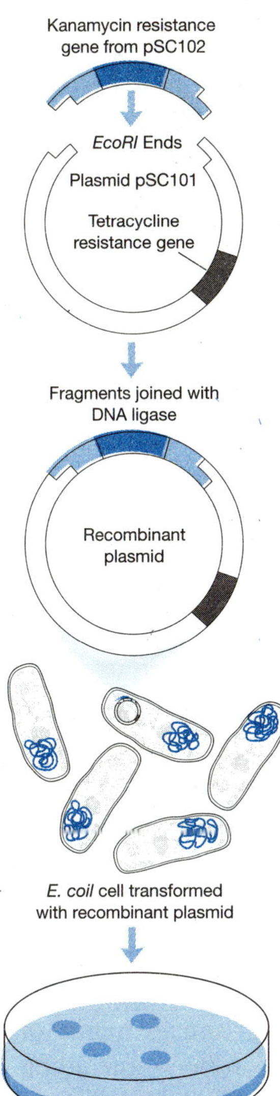

The Boyer-Cohen-Chang Experiment, 1973

The next step in the recombinant DNA experiment was to construct a recombinant plasmid and insert it into *E. coli*. A second plasmid, pSC102, was constructed, which contained an antibiotic resistance gene for kanamycin flanked by *Eco*RI sites. Herb Boyer's laboratory, at the University of California, San Francisco, had demonstrated that *Eco*RI creates its own complementary sticky ends when it cuts, making Berg's tailing method unnecessary for recombinant plasmid construction. So pSC101 and pSC102 were simply cut with *Eco*RI, mixed together, and rejoined with DNA ligase. The resulting recombinant plasmid was transformed into *E. coli* cells, which were plated on media containing both tetracycline and kanamycin. The appearance of colonies with resistance to both tetracycline and kanamycin could be due to two different possibilities: (1) Resistant *E. coli* cells carried a copy of both pSC101 and pSC102 or (2) resistant *E. coli* cells carried one new recombinant plasmid that contained genes for resistance to both antibiotics. Plasmid DNAs isolated from several resistant colonies were tested by restriction enzyme digestion and gel electrophoresis. Plasmid DNA from these colonies was digested with *Eco*RI. Where two plasmids were present, the gel showed three bands, corresponding to pSC101, and two pieces of pSC102 (the kanamycin insert and the rest of the plasmid). But, some colonies that contained only one plasmid were also found, as the gel showed only two bands, corresponding to pSC101 and the kanamycin insert from pSC102. This new plasmid, called pSC105, confirmed that a recombinant DNA molecule had been successfully created and introduced into living bacterial cells.

Shortly thereafter, Cohen and Boyer teamed up to produce a recombinant molecule containing DNA from two different species. To accomplish this, they spliced a gene encoding a ribosomal RNA from the frog *Xenopus laevis* into the plasmid pSC101 and transformed the recombined DNA into *E. coli*. Some of the colonies resistant to tetracycline were found to contain plasmid with the frog gene inserted. The recombinant DNA techniques developed by Berg, Cohen, Chang, and Boyer revolutionized the frontier of genetic research. These first recombinant DNA experiments established techniques that are still critical to molecular biologists today.

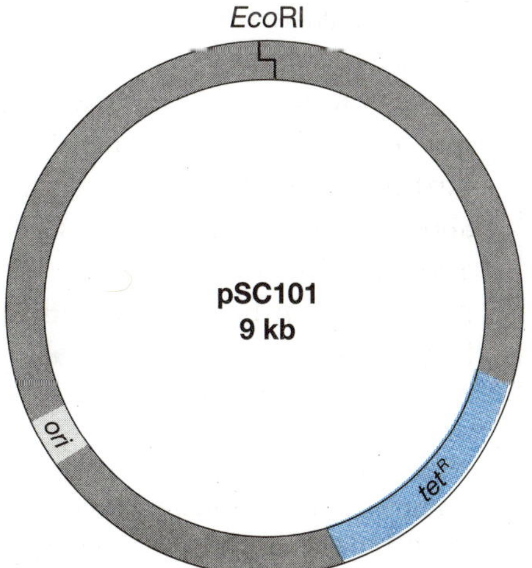

pSC101

This plasmid contains the three essential components for a vector. (*1*) An origin of replication (*ori*); (2) a selectable marker (tetracycline resistance tet^R); and (3) a novel restriction site or sites (*Eco*RI).

It Is Important to Pick the Right Model System

MODEL SYSTEMS

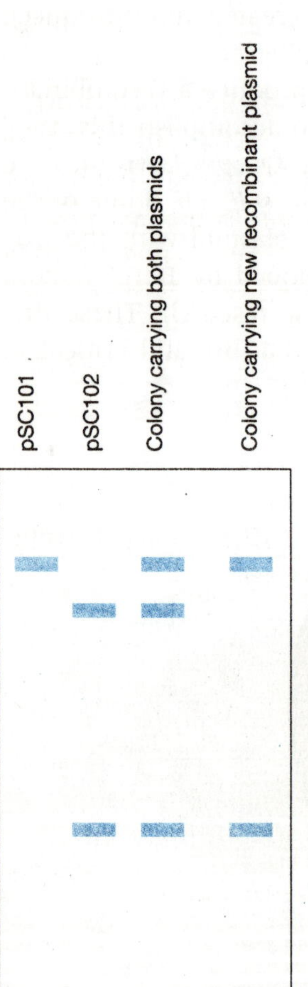

Biomedical research is propelled by funding from pharmaceutical companies and government agencies, such as the U.S. National Institutes of Health—with the clear objective of improving human health. Although this demands knowledge of how molecular processes work in human beings, experimenting on people presents many practical and ethical difficulties. Thus, biologists often employ model systems that mimic elements of human physiology and biochemistry.

The theory behind the use of model systems dates to 1926, when the Dutch microbiologist Albert Jan Kluyver proposed the "unity of biochemistry," emphasizing the similarity of enzymatic reactions found in a variety of organisms. Basic life processes are, indeed, conserved between species, and elements of what is learned from one organism may be extrapolated to others—including humans. Single-celled organisms, such as bacteria and yeast, allow biologists to examine basic cell biochemistry, without the complexity posed by multicellular tissues and organs.

By the same token, however, the metabolism of a single-celled organism cannot explain every process needed to manage a multicellular existence. Furthermore, every organism has its own idiosyncrasies. So one must be cautious about over-extending the results from experiments conducted in a model system, and choosing an appropriate model depends on the question being asked. Each model system has its own advantages and disadvantages, so many laboratories find it useful to perform parallel experiments in two or more different systems.

Proof That Boyer-Cohen-Chang Experiment Created a Recombinant DNA Molecule
Samples of the original plasmids (pSC101 and pSC102) and plasmids isolated from tetracycline/kanamycin-resistant colonies were digested with *Eco*RI, and the resulting DNA fragments were separated on an agarose gel. pSC101 was cut once and generated one fragment. pSC102 was cut twice and generated two fragments. Colonies containing both plasmids generated three fragments. Colonies containing the new recombinant plasmid pSC105 generated two fragments, one corresponding to the pSC101 plasmid and one corresponding to the insert fragment from pSC102.

E. coli

In the 1920s and 1930s, studies demonstrating the uniformity of biosynthetic pathways led many biochemists to study pathways in microorganisms. Microorganisms are easy and inexpensive to maintain, and large quantities of cultured cells can be easily obtained. The microorganism of choice became *E. coli*, first discovered in 1855. It is a normal inhabitant of the human gut and is the dominant species of bacteria present in feces. *E. coli* functions in the gut to prevent the growth of other harmful types of bacteria, as well as to provide certain vitamins. Pathogenic strains of *E. coli* often are associated with sickness or even death. We have all almost certainly heard of *E. coli* outbreaks from uncooked food or unsanitary conditions where individuals become severely ill or even die. It is important to understand that these harmful *E. coli* strains are different from the nonpathogenic strain that we work with in the laboratory or that are typically found in our gut.

Today, *E. coli* and other bacteria are key models for studying and understanding metabolism—biochemical reactions that produce energy (catabolism) and the synthesis of cellular components from simpler molecules that require energy (anabolism). Anabolism is a multistep process in which multiple enzymes work in a chain-wise fashion to produce a final product. The biosynthesis of any molecule is blocked if any enzyme in the "pathway" is inhibited or lost. Scientists use strains of *E. coli* that carry mutations in different genes in these pathways, allowing each specific step to be studied.

E. coli continues to be a workhorse in molecular biology and biochemistry. Of course, not everything that occurs in this bacterium is directly translated into higher organisms. Likewise, there are many processes in eukaryotes that do not even occur in bacteria or are much more complicated than in prokaryotes. To study events unique to eukaryotes, researchers utilize a series of model organisms of increasing complexity.

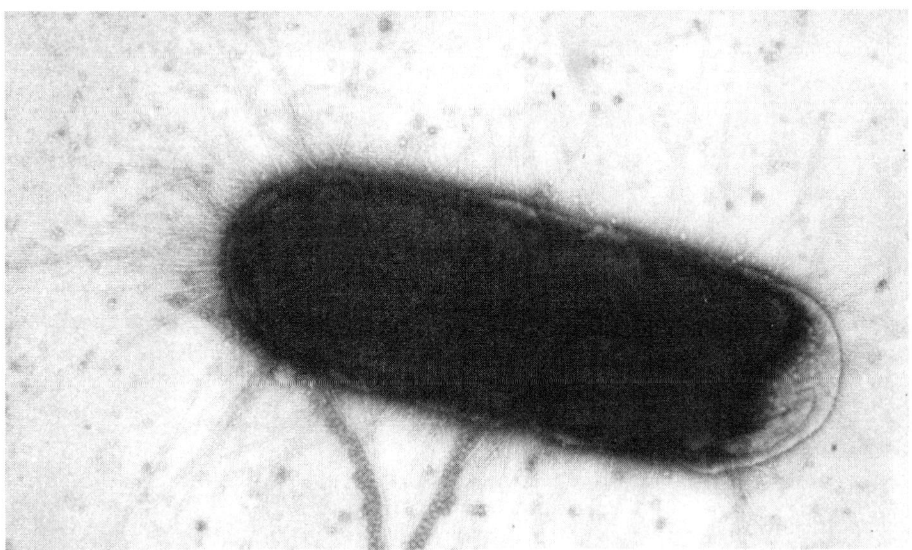

The Bacterium *Escherichia coli*

(Photo courtesy of Lucien Caro.) (Reprinted, with permission, from Miller J.H. 1972. *Experiments in molecular genetics*. Cold Spring Harbor Laboratory, Cold Spring Harbor, New York.)

Yeast

Yeast are a great organism in which to study many molecular processes, particularly those that differ between eukaryotes and prokaryotes. Yeast have the advantage over other eukaryotes of being single-celled organisms, making genetic studies exceedingly easy to carry out. Yeast share many properties with *E. coli.* They are haploid (meaning they carry only one copy of each chromosome), making genetic studies straightforward, and they are single-celled. Yeast cells grow very rapidly (2-hour doubling time) using inexpensive nutrients and equipment, and large quantities of material can be obtained from a small culture. At the same time, like in all eukaryotes, yeast DNA is located in the nucleus; they have mitochondria; and they have a distinct cell cycle. Scientists study two major species of yeast: *Saccharomyces cerevisiae* and *Schizosaccharomyces pombe. S. cerevisiae* reproduces by budding, whereas *S. pombe* divides by fission.

Many important biological discoveries have been made in yeast. An example of one such breakthrough aptly illustrates the power of yeast as a model organism. *S. cerevisiae* has multiple chromosomes and multiple origins of replica-

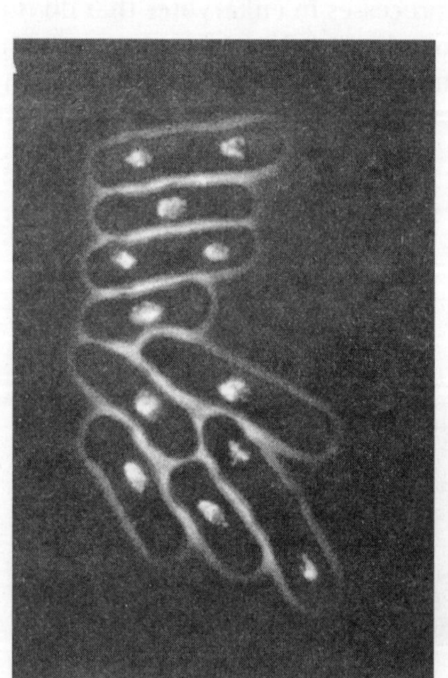

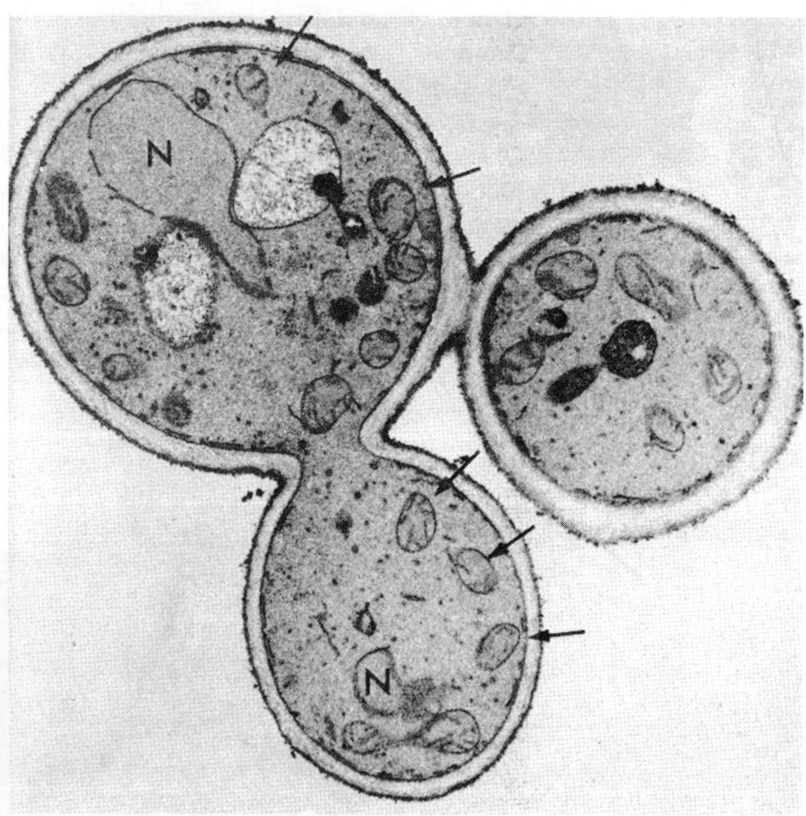

The Yeast *Schizosaccharomyces pombe* The Yeast *Saccharomyces cerevisiae*

(*Left:* Reprinted, with permission, from Alfa C., Fantes P., Hyams J., McLeod M., and Warbrick E. 1993. *Experiments with fission yeast: A laboratory course manual.* Cold Spring Harbor Laboratory Press, Cold Spring Harbor, New York.) (*Right:* Reprinted, with permission, from Strathern J.N., Jones E.W., and Broach J.R. 1981. *The molecular biology of the yeast* Saccharomyces: *Life cycle and inheritance.* Cold Spring Harbor Laboratory, Cold Spring Harbor, New York.)

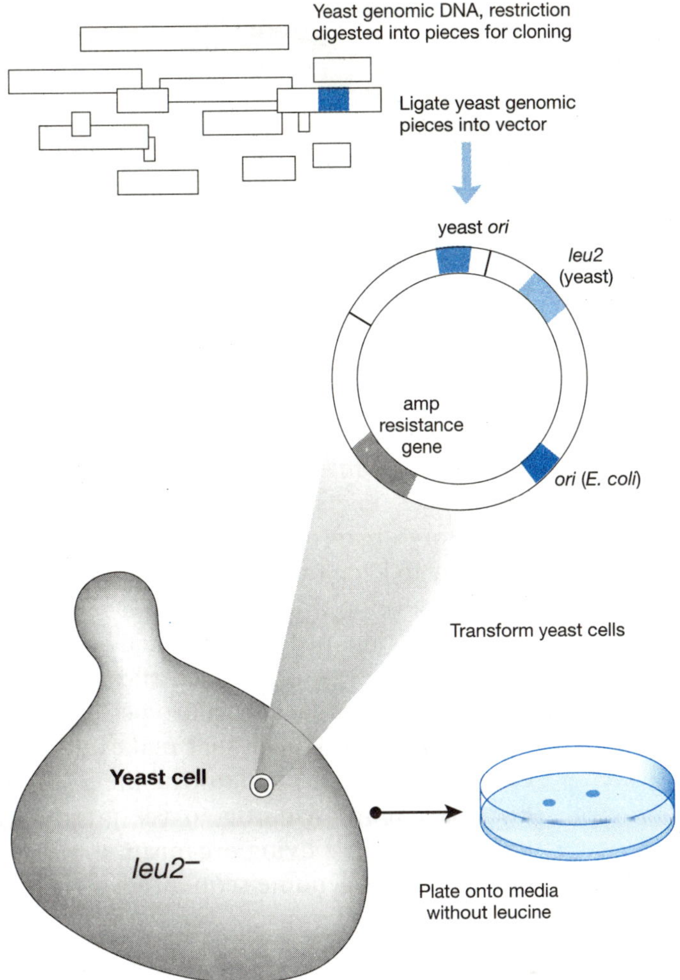

Cloning the Yeast Origin of Replication

The entire genome of yeast was cut into small restriction fragments, which were individually cloned at random into plasmid vectors. The vector contained an antibiotic resistance marker and an *E. coli* origin of replication (*ori*), and thus recombinant plasmids could be grown and selected for in *E. coli* cells. The plasmid also contained the yeast *leu2* gene, which allows yeast to survive in media lacking leucine. The recombinant plasmids were transformed into yeast cells that had a mutated version of the *leu2* gene. Growth on leucine-deficient media selected for yeast cells that carried the *leu*$^+$. However, the *E. coli ori* does not function in yeast. Any yeast colonies that grew on leucine-deficient media had to contain a yeast *ori* fragment in their recombinant plasmid.

Bruce Stillman

tion within each chromosome, unlike bacteria which have one chromosome and one origin of replication. Thus, yeast is a considerably better model system for studying DNA replication in higher eukaryotes. Landmark work in yeast genetics, in the laboratory of Bruce Stillman at Cold Spring Harbor Laboratory, led to the isolation of the first origin of replication in a eukaryotic cell. The discovery of replication origins in yeast was critical to the understanding of DNA replication initiation in all eukaryotes. DNA replication remains an active area of research and many scientists continue to choose yeast as the model system for studying this basic cellular process.

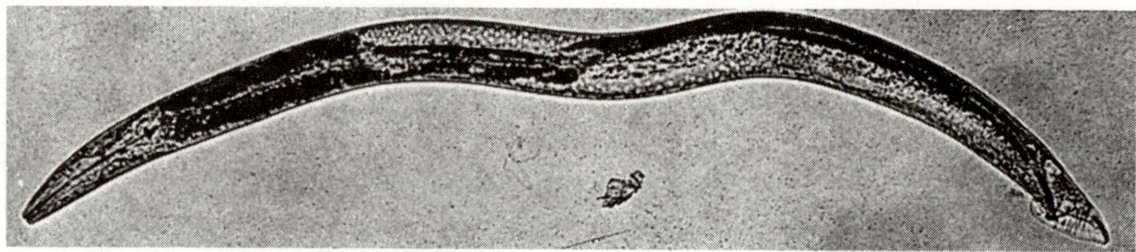

The Nematode *Caenorhabditis elegans*
(Reprinted, with permission, from Sulston J.E. and Horvitz H.R. 1977. Postembryonic cell lineages of the nematode, *Caenorhabditis elegans*. *Dev. Biol. 56:* 110–156.)

Nematode Worm

The nematode *Caenorhabditis elegans* was introduced as a model system in the 1960s by Sydney Brenner at the Medical Research Council Laboratory in Cambridge, England. *C. elegans* has many characteristics that make it an ideal laboratory organism. It is small (10,000 worms can be raised on one Petri dish) and easy to keep (it feeds on bacteria, such as *E. coli*). *C. elegans* takes about three days to develop from an egg to a reproducing adult, which is about the same amount of time needed for genetic crosses in yeast. Unlike yeast, *C. elegans* is a multicellular organism, which allows scientists to view the complex interactions between cells. In fact, a wild-type *C. elegans* is made up of 959 cells. Because it is transparent, each cell that makes up a *C. elegans* has been identified and can easily be traced throughout the life of the worm. Understanding the lineage of every cell allows scientists to ask questions about how genes influence individual cells within a living organism. *C. elegans* was the first multicellular eukaryote to have its genome sequenced.

Fruit Fly

Many important genetic discoveries have been achieved using the fruit fly, *Drosophila melanogaster*. But why the fruit fly? Why not higher mammals? Genetic studies on fruit flies grew out of the identification of mutants. These mutations were in traits that were easily visualized in the fly, such as eye color or wing shape. Such mutations are very rare, occurring on the order of once in a million flies or so. But two flies can produce hundreds of offspring in a few days, so mutations can be found in due time with a little patience. Mammals, at most, produce a few young at a time, breeding only every few weeks to months or years. Thus, we are not likely to find useful mutants very readily. The rapid growth of *Drosophila* also makes it a good model organism to study. The fruit fly develops from egg to adult in about 12 days. Mammalian development is measured in months or years. Finally, a million fruit flies can fit inside a large box and can be fed for a few dollars a day. This is certainly not the case with mammals.

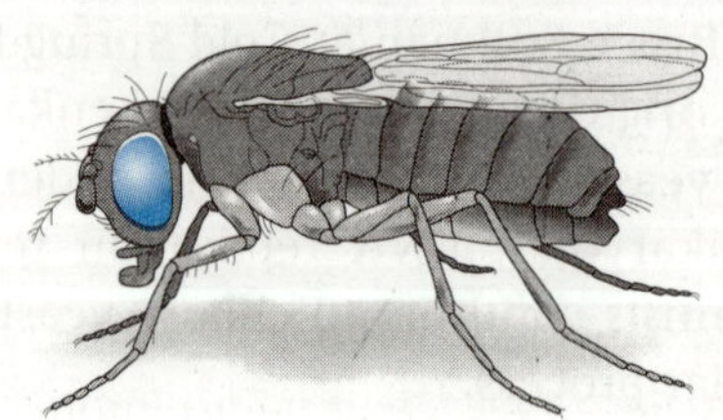

Drosophila melanogaster
(Reprinted, with permission, from Bate M. and Martinez-Arias A. 1993. *The development of* Drosophila melanogaster. Cold Spring Harbor Laboratory Press, Cold Spring Harbor, New York.)

The groundbreaking work of Christiane Nüsslein-Volhard and Eric Wieschaus, at the European Molecular Biology Laboratory (discussed in Chapter 3), revolutionized the field of developmental genetics. They found key genes involved in the embryonic development of *Drosophila* using a large-scale *genetic screen*. Mutations in *Drosophila* genes occur spontaneously, but only at a very low rate. Nüsslein-Volhard and Wieschaus took advantage of chemical mutagenesis to increase the frequency of mutations. After mutagenizing a large number of flies with a chemical that would damage approximately half the *Drosophila* genes at random, they then had the arduous task of examining *Drosophila* embryos resulting from genetic crosses of the mutagenized strains. After a year of work, they identified more than 100 genes that affect the normal formation of the *Drosophila* body, leading to our current understanding of how organisms specify a body axis.

The innovative screening method devised by Nüsslein-Volhard and Wieschaus has been used to identify genes that govern a variety of processes in *Drosophila*. Like many of the genetic techniques first developed in *Drosophila*, their screening method has been adapted for use in many other model organisms.

Zebrafish

In the late 1980s, Nüsslein-Volhard turned her attention toward the zebrafish (*Danio rerio*), a model system pioneered by Charles Kimmel at the University of Oregon. Although *Drosophila* had proven to be a tremendous model organism, scientists were looking for a system that was more closely related to higher organisms, but still had many of the positive features of the fruit fly. Zebrafish, like *Drosophila*, reproduce fairly rapidly and in high numbers. It is also possible to keep large quantities of zebrafish in a relatively small laboratory space. In contrast to *Drosophila*, zebrafish are vertebrates, and their developmental processes are much more similar to those of humans. Zebrafish therefore provide a useful bridge between insects such as *Drosophila* and more complex mammalian organisms. Another benefit of zebrafish is that their eggs are clear. Scientists can place a developing zebrafish egg under a microscope and watch as changes occur. In 1996, Nüsslein-Volhard and collaborators published the results of the first large-scale genetic screen of zebrafish, identifying 1200 mutants. These mutant strains are providing valuable clues to the genetic machinery of higher vertebrates.

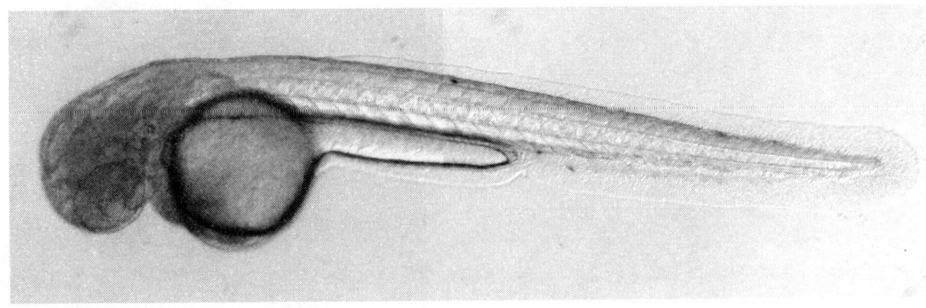

The Zebrafish *Danio rerio* Embryo
(Courtesy of Reinhard Köster, California Institute of Technology.)

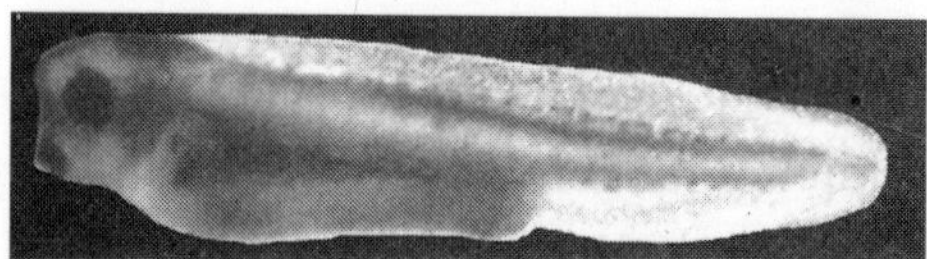

Xenopus laevis Embryo
(Reprinted, with permission, from Larabell C.A., Torres M., Rowning B.A., Yost C., Miller J.R., Wu M., Kimelman D., and Moon R.T. 1997. Establishment of the dorsoventral axis in *Xenopus* embryos is presaged by early asymmetries in beta-catenin that are modulated by the Wnt signaling pathway. *J. Cell Biol. 136:* 1123–1136.)

Hilde Mangold
(Reprinted, with permission, from Hamburger V. 1988. *The heritage of experimental embryology: Hans Spemann and the organizer.* Oxford University Press, New York.)

Amphibians

Amphibian embryos, especially the African clawed frog (*Xenopus laevis*), have a long history as a model system for studies in the field of developmental biology. *Xenopus* eggs are very large when compared with eggs from species like zebrafish or mouse, which makes them ideal for experimental manipulation.

In the 1920s, Hilde Mangold in the laboratory of Hans Spemann at the Kaiser Wilhelm Institute in Dahlem, Germany, performed a very important experiment. Using an amphibian, the common striped newt (*Triton taeniatus*), Mangold transplanted small regions from one embryo (the donor) to a second embryo (the host). Transplanting a specific region from the dorsal side of the donor to the ventral side of the host yielded a surprising result: a two-headed embryo with a complete second axis opposite the normal one. This small region became known as the "Spemann Organizer," since the cells it contains instruct the surrounding cells to organize the embryo's axis. Similar organizing centers have since been found in many different vertebrate species, and the molecular signals that they use to instruct developing cells are the focus of intense research.

Chicken

The egg of the chicken (*Gallus gallus*) is also a laboratory staple in the study of embryonic development. Like the frog embryo, the relatively large size and accessibility of the chicken embryo make it an ideal system for observing growth and differentiation. They are evolutionarily closer to human embryos than the

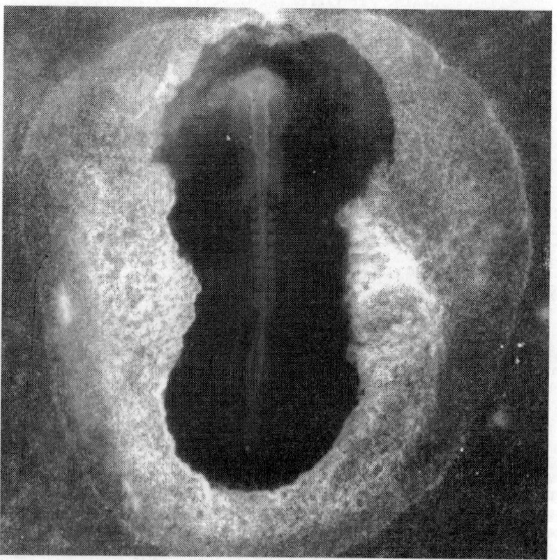

Chicken Embryo
(Courtesy of Paul Kulesa, California Institute of Technology.)

systems discussed above, and the external development of chicken embryos makes them much easier to access than internally developing mammalian embryos. The researcher needs only to cut a small hole in the chicken egg to visualize and manipulate the developing embryo. After experiments, tape can be placed over the hole, and the egg will continue to develop. Contrast this with the surgery needed to access a mouse embryo, and it is easy to understand why many scientists prefer to work with chickens when they can. One drawback to the use of the chicken as a laboratory animal is that many of the techniques for genetic manipulations discovered in other species have not adapted well to the chicken. In addition, chickens require a considerable amount of space to be maintained, so raising multiple generations in a laboratory is not feasible.

Mouse

Studies in mice, like studies in any model organism, have both advantages and disadvantages. Of course, mice are vertebrates, and, because they are mammals, they are more like humans than any of the models discussed thus far. However, mice are more expensive to maintain, and they reproduce more slowly than flies or yeast. Because mice have a physiology similar to that of humans, and go through similar stages of development, they are becoming increasingly valuable as models for many human diseases.

Many laboratory strains of mice have been inbred to have a homogeneous genome. By starting with genetically identical mice, researchers are better able to conclude that an observed response is due to the experimental treatment, rather than individual differences in the mice. One of the most useful techniques available for use with laboratory mice is the ability to directly manipulate their genes. Transgenic mouse techniques (discussed in detail in the Chapter 5) allow scientists to introduce genes randomly into a genome as well as "knock out" the function of specific genes within the genome. These alterations are passed on to the experimental mouse's offspring, providing a stock of animals to study.

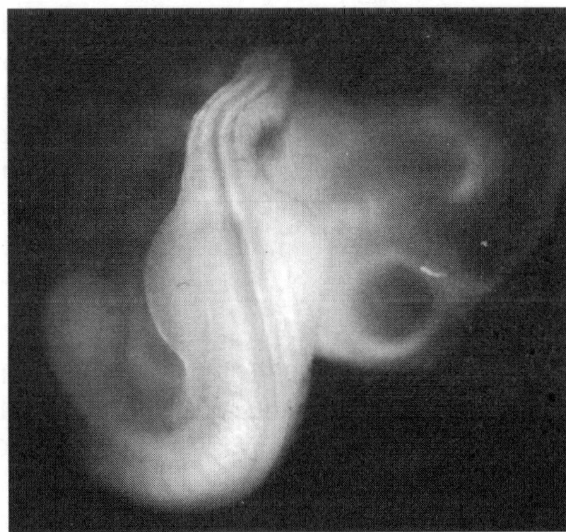

Mouse Embryo
(Courtesy of David Kremers, California Institute of Technology.)

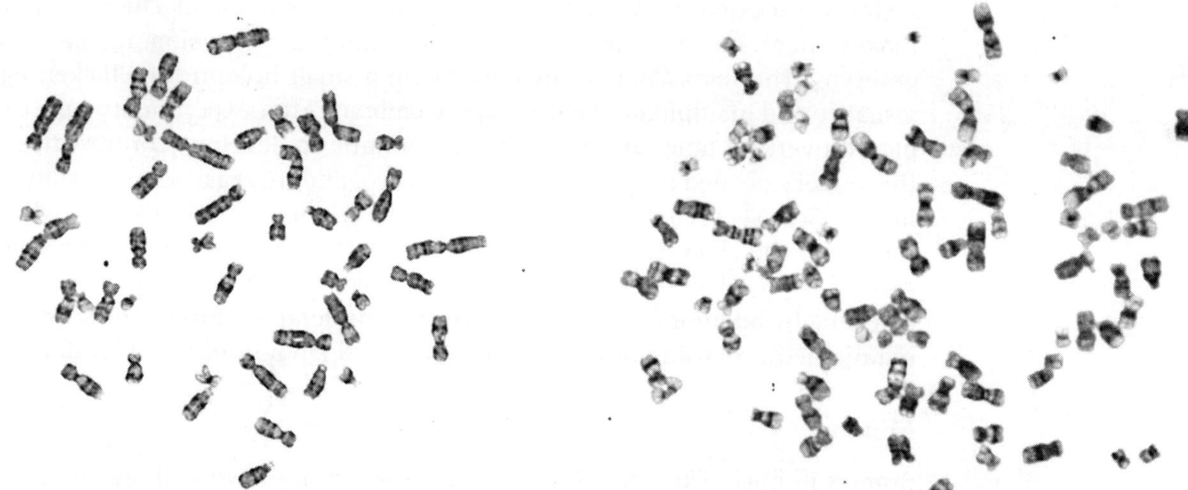

Giemsa-stained chromosomes from a normal cell

Giemsa-stained chromosomes from a cell that is polyploid

Giemsa-stained Chromosomes from Normal and Polyploid Cells
(Courtesy of Dorothy Warburton, Columbia University.)

Human Cell Culture

Mammalian cell culture provides an ethical and noncontroversial means to study the biochemistry of human cells. In human cell culture experiments, referred to as in vitro experiments, cells derived from human blood or tissue are grown in media either on plates or in suspension, much like bacteria or yeast cells. "Primary cultures" are cells derived directly from a living tissue. The advantage of primary cultures is that the cells are believed to behave like normal human cells do in vivo, or in the organism. The disadvantage of primary cultures is that they only grow for a short period of time (several generations) and then stop. This tendency to have a limited number of cell divisions has led most researchers to use "immortalized cultures," which grow indefinitely.

The process of making cells immortal, however, causes them to behave partially like a cancer cell. For example, immortalized cells often have abnormal genomes with extra (a state called polyploidy) or missing (a state called aneuploidy) copies of chromosomes. But despite the shortcomings, much has been learned about cell function and disease using immortalized cells.

Plants

The Department of Agriculture's 2.3-billion-dollar research budget request for fiscal year 2003 is less than one-tenth that of the National Institutes of Health's (27.2 billion dollars). Plant research lags far behind mammalian research, in part due to bias toward human medicine and in part due to inherent difficulties of working with plants.

Plants grow slowly and have relatively long generation times. Plant cells are surrounded by a "wooden box," the cellulose outer wall, which makes gene transfer more difficult than in animal cells. Indeed, reliable transformation sys-

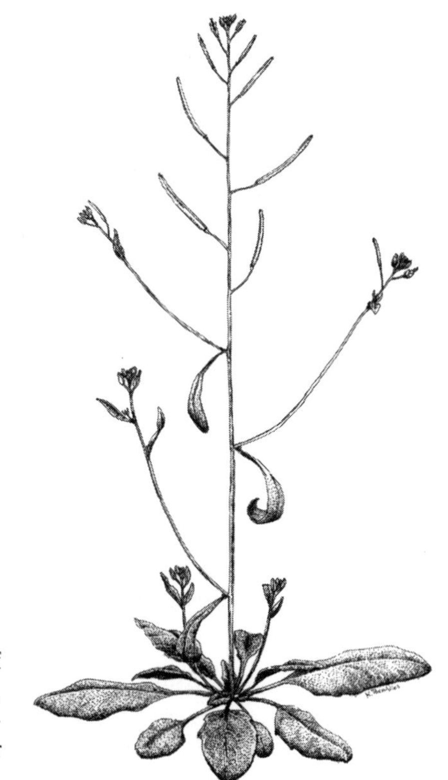

Arabidopsis thaliana
(Courtesy of Kirsten Bomblies, University of Wisconsin, Madison. Reprinted, with permission, from Weigel D. and Glazebrook J. 2002. Arabidopsis: *A laboratory manual.* Cold Spring Harbor Laboratory Press, Cold Spring Harbor, New York.)

tems for monocot plants were only developed in the 1990s. Plants have surprisingly large genomes; at 15 billion base pairs, the haploid corn genome is about five times larger than the human genome. Plant genomes are also complex: Polyploid chromosomes confuse genetic analysis, and high proportions of repetitive DNA complicate sequencing.

Prior to the molecular era, model plants were of agricultural or horticultural importance, including corn, tomato, pea, tobacco, petunia, and snapdragon. Corn (*Zea mays*) has been a model plant since the early days of genetics. Molecular analysis was accelerated with the adaptation, in the 1980s, of Barbara McClintock's *Ac/Ds* transposon system as a research tool to produce new mutations in corn. In this system, the *Ds* transposon randomly inserts into a new site in the corn genome, "tagging" a resident gene for retrieval. In the mid 1990s, Cold Spring Harbor Laboratory researcher Robert Martienssen successfully adapted the *Ac/Ds* system for tagging genes in *Arabidopsis thaliana*, the model plant in genomic research.

Although *Arabidopsis* was first used experimentally in 1907, only during the past 20 years has it gained widespread acceptance as *the model* for flowering plants. Unlike corn and other domestic plants that have been intensely manipulated by humans for generations, *Arabidopsis* is essentially a wild-type plant with thousands of close relatives among the flowering plants. It is small, easy to grow, and goes through its life cycle—from seed to seed—in about eight weeks. Unlike many plants, it is easy to transform (by simply spraying flowers with a solution of tumor bacterium containing the transgene). *Arabidopsis* is self-fertile and a single individual produces thousands of seeds, making it easier to screen for rare mutants or transgenes.

At about 117 million base pairs, *Arabidopsis* has one of the smallest genomes of any flowering plant, with relatively little repetitive DNA. Its entire genome was sequenced in 2000, and work is ongoing to annotate the functions of its 25,500 genes. With more than 50,000 mapped polymorphisms, a large collection of characterized mutants, whole genome microarrays, and numerous Internet data resources, the *Arabidopsis* research community has virtually everything needed to launch a "postgenome" exploration of the molecular details of plant development and physiology.

REFERENCES

Bier E. 2000. *The coiled spring: How life begins.* Cold Spring Harbor Laboratory Press, Cold Spring Harbor, New York.

Cherfas J. 1982. *Man-made life.* Pantheon Books, New York.

Kornberg A. 1989. *For the love of enzymes.* Harvard University Press, Cambridge, Massachusetts.

Lewin B. 1999. *Genes VII.* Oxford University Press, New York.

Neidhardt F.C., ed. 1987. Escherichia coli *and* Salmonella typhimurium: *Cellular and molecular biology,* volumes I and II. American Society for Microbiology, Washington, D.C.

Micklos D., ed. 1999. DNA from the beginning (http://www.dnaftb.org). Dolan DNA Learning Center, Cold Spring Harbor, New York.

Novick R.P. 1980. Plasmids. *Sci. Am.* **243:** 103–127.

Sambrook J. and Russell D. 2001. *Molecular cloning: A laboratory manual,* 3rd edition. Cold Spring Harbor Laboratory Press, Cold Spring Harbor, New York.

Shapiro J.A. 1988. Bacteria as multicellular organisms. *Sci. Am.* **258:** 82–89.

Sylvester E.J. and Klotz L.C. 1987. *The gene age.* Charles Scribner's Sons, New York.

Watson J.D., Tooze J., and Kurtz D.T. 1983. *Recombinant DNA,* 2nd edition. Scientific American Books, New York.

Wilmut I. 1998. Cloning for medicine. *Sci. Am.* **279:** 58–63.

CHAPTER 5

Methods for Finding and Expressing Important Genes

PRIOR TO THE ADVENT OF BIOTECHNOLOGY, molecular biology could be separated into two fields: biochemistry and molecular genetics. Biochemistry is the study of the chemistry of living systems. Molecular genetics is a field that uses simple organisms such as bacteria and bacteriophage to understand genetics at its most basic level, such as seen in the work of Alfred Hershey, Salvador Luria, and Charles Yanofsky. Although these two fields of study are still very much alive today, molecular biology borrows heavily from each and, combined with biotechnology, creates a new field.

The growing questions facing biochemists and molecular geneticists in the late 1960s and early 1970s focused on the regulation of genes. From the work of François Jacob and Jacques Monod, scientists understood that there are regions of genes called promoters that regulate gene expression. Molecular geneticists knew that genes are located on chromosomes and inferred that a different gene encodes each protein. But how does the whole system work? What does a promoter look like? What is the organization of genes? Are there any surprises in the structure of genes? The only way to answer these questions was to isolate individual genes.

In the early 1970s, biochemists were using new techniques to purify and study an increasing number of proteins. Biochemical pathways involved in every conceivable cellular process were being worked out, including the synthesis of amino acids, nucleotides, and fatty acids. Pathways for the production of energy molecules, such as glycolysis and the Krebs cycle were being determined. At the same time, molecular genetics was making progress in understanding the relationship between DNA and proteins. At the heart of all of these studies was the gene.

The attention of many biochemists and molecular geneticists was captured by the idea of purifying genes for further study. The initial approach was to isolate abundant genes like those coding for ribosomal RNA (rDNA), which were known to exist in hundreds or thousands of copies per cell. But rDNA was likely to be different than the genes that coded for proteins. So what gene should be studied first?

The average atomic mass of one base pair is 635 daltons (a dalton is 1/12 the mass of a carbon atom)

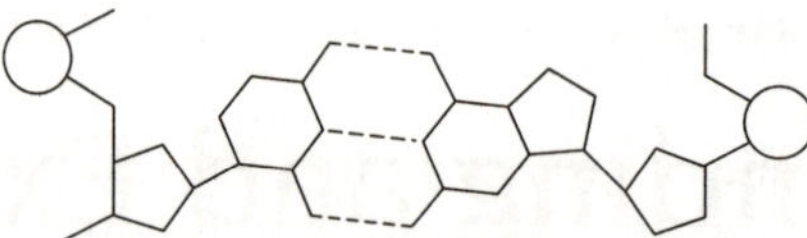

The *β-globin* gene is 2000 bp in length

So, the atomic mass of the *β-globin* gene is:

2000 bp
×
635 daltons/bp

= 1.27×10^6 daltons

The average atomic mass of one base pair is 635 daltons (a dalton is 1/12 the mass of a carbon atom)

The *β-globin* gene is 2000 bp in length

So, the atomic mass of the *β-globin* gene is:

2000 bp
×
635 daltons/bp

= 1.27×10^6 daltons

Mass of *β-globin* gene in an adult human

There are two copies of the *β-globin* gene per cell

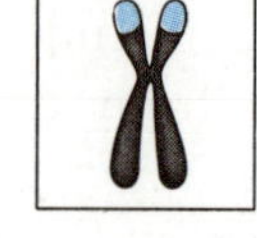

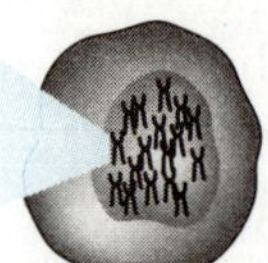

There are 10^{14} cells per individual

So the total atomic mass of *β-globin* DNA per individual is:

1.27×10^6 daltons/gene
×
2 genes/cell
×
10^{14} cells/individual

= 2.54×10^{20} daltons

If there are 6.02×10^{23} daltons per gram, then:

$$\frac{2.54 \times 10^{20} \text{ daltons}}{6.02 \times 10^{23} \text{ daltons/gram}}$$

= 0.00042 grams

= 0.42 mg

= 420 μg *β-globin* DNA

Mass of *β-globin* gene in a liter of *E. coli*

There are 500 copies of the *β-globin* gene (on plasmids) per cell

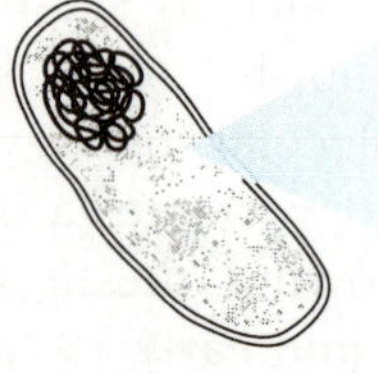

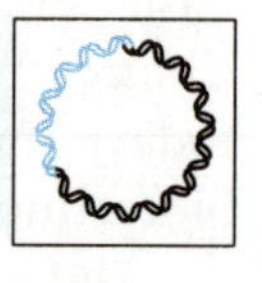

There are 5×10^{11} cells per liter

So the total atomic mass of *β-globin* DNA per liter is:

1.27×10^6 daltons/gene
×
500 genes/cell
×
5×10^{11} cells/liter

= 3.175×10^{20} daltons

If there are 6.02×10^{23} daltons per gram, then:

$$\frac{3.175 \times 10^{20} \text{ daltons}}{6.02 \times 10^{23} \text{ daltons/gram}}$$

= 0.000527 grams

= 0.527 mg

= 527 μg *β-globin* DNA

Mass of β-*globin* DNA in Adult Human vs. 1-liter Culture of *E. coli* Carrying the β-*globin* Gene on Plasmid

THE BIOCHEMIST'S AND MOLECULAR GENETICIST'S PROBLEM

The magnitude of growth that has taken place in molecular biology and genetics since those first cloning experiments of Paul Berg, Stanley Cohen, Herb Boyer, and Annie Chang is hard to grasp. To appreciate these experiments and to better understand the basis of today's technology, consider how the first specific genes were cloned and isolated. Obviously, the initial cloning of genes was a sort of training ground where methods and techniques were tested. Not surprisingly, the first genes isolated were genes whose products are abundant, well-characterized, and potentially easy to isolate. The genes coding for hemoglobin were among the first mammalian genes to be isolated using recombinant DNA technology.

Philip Leder

The research design used by Philip Leder and his co-workers at Harvard University to isolate the mouse β-*globin* gene from genomic DNA illustrates how molecular cloning elegantly solves two critical problems of biochemical analysis: (1) separating the gene of interest from the mass of genomic DNA and (2) purifying enough DNA for analysis. Shortly after Leder's work, Tom Maniatis, at Harvard and the California Institute of Technology, isolated human globin genes using a slightly different protocol.

Hemoglobin is the oxygen-carrying molecule of red blood cells. It is made up of four proteins, two α-globin and two β-globin polypeptides, surrounding a central heme group. Although the two globin polypeptides are similar in structure, each is encoded by a different gene. The human β-globin protein is 147 amino acids long. Predicting the size of the gene from the protein is difficult. In addition to 441 base pairs (bp) encoding amino acids (remember that there are three nucleotides per codon), the β-*globin* genes include a 5′-untranslated region, stretching from the promoter to the start codon (51 bp), a 3′-untranslated region, stretching from the stop codon to the poly(A) signal (134 bp), plus two introns (1374 bp). Thus, the β-*globin* gene is actually 2000 bp in length. Since a base pair has an average molecular mass of 635 daltons, the β-*globin* gene has a total mass of about 1.27×10^6 daltons. Multiplying this figure by the number of copies of the gene per cell (2) and the number of cells in the adult human body (100 trillion, or 1×10^{14}), we calculate a total mass of 0.00042 g (0.42 mg or 420 μg) of β-*globin* DNA per individual. Clearly, attempting to pull human genes straight out of cells is not the best way to go. Biochemical analysis might require on the order of 1 mg of DNA, which seems small considering that one teaspoon of sugar weighs approximately 4500 mg. However, it would take the equivalent cell mass of 2.5 humans to extract 1 mg of β-*globin* DNA.

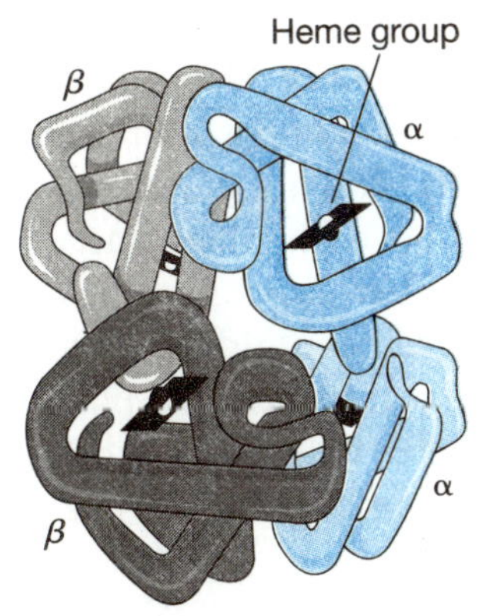

Hemoglobin
(Redrawn, with permission, from Rawn J.D. 1989. *Biochemistry*, p. 125. Neil Patterson Publishers, Burlington, North Carolina.)

Even if this difficult and onerous task was successfully completed, the problem of separating the β-*globin* gene from the rest of the DNA would be nearly impossible, because the β-*globin* gene represents 1/1,500,000 of the 3 billion (3×10^9) base pairs of DNA that comprise a haploid set of human chromosomes (the haploid human genome). Thus, the isolation procedure from the equivalent of 2.5 humans would yield a total mass of 1.5 kg of genomic DNA, of which only 1 mg would be β-*globin* DNA. Because DNA is a polymer of only four repeating subunits, for all intents and purposes, any 2000-bp fragment appears biochemically identical to any other 2000-bp piece. The problem is even worse than searching for the proverbial needle in a haystack, where at least the needle looks different from the surrounding hay.

THE MOLECULAR BIOLOGIST'S SOLUTION: CREATING AND SCREENING A LIBRARY

Instead of trying to fish one gene out of the mass of genomic DNA, molecular biologists came up with a novel strategy. The idea is to physically separate each gene in the genome and then rapidly screen these individual pieces to pick out the desired one.

The DNA Library

We start with the creation of a genomic library, given this name because, like a library, it contains a large quantity of items that are easily searchable. To create this library, we begin by isolating genomic DNA (since Leder was interested in isolating mouse globin genes, he used mouse cells and made a mouse genomic library). Cells are isolated and lysed in an aqueous solution. Proteins, lipids, and other molecules are eliminated by organic extraction in which these molecules are solubilized but in which DNA is insoluble. Trace organic solvents are removed from the aqueous phase by precipitating the DNA using ethanol or isopropanol. The precipitated DNA is solubilized in an aqueous buffer, such as TE (10 mM Tris-HCl, 1 mM EDTA).

A sample of the human genomic DNA is cleaved with a restriction endonuclease, which cuts on average once every 4000 nucleotides and generates approximately 750,000 DNA fragments. These genomic restriction fragments are mixed with plasmid molecules that have been digested with the same restriction enzyme. The plasmid vector and the genomic DNA fragments are covalently joined with DNA ligase (as described in Chapter 4). The collection of 750,000 recombinant plasmids creates a genomic library. The recombinant plasmids are then transformed into competent *Escherichia coli* cells. In theory, each cell takes up a different plasmid carrying a different fragment of human DNA. The transformed bacteria are diluted and spread onto plates as single cells. Each cell reproduces itself until a visible colony forms. Each colony contains millions of identical cells, each with many copies of a unique recombinant plasmid.

The problem of separating the gene of interest from all of the other genomic DNA has now been solved. On average, 1 of every 750,000 bacterial colonies should contain an intact β-*globin* gene. Methods for screening the colonies to identify those containing the β-*globin* gene are discussed later in this chapter.

The problem of purifying enough DNA is now also easily solved. Only about 50 μg of genomic DNA is needed to construct the genomic library, and this amount can be isolated from less than 1 g of tissue or cells growing in culture. Once the bacterial colony containing the β-*globin* gene has been identified, large amounts of plasmid can be prepared from liquid cultures; 1 liter of *E. coli* culture at stationary phase contains approximately 5×10^{11} cells, each containing up to 500 copies of the recombinant plasmid. Multiplying the total number of plasmids per liter times the mass of the β-*globin* gene (1.27×10^{6} daltons), we calculate that up to 527 μg of β-*globin* DNA can be extracted from each liter of bacterial culture.

Better Vectors Make Better Libraries

In the "plasmid library" described above, we estimated that the entire human genome could be represented by approximately 750,000 clones. In reality, a library of 750,000 clones would represent only about 90% of the genome, due

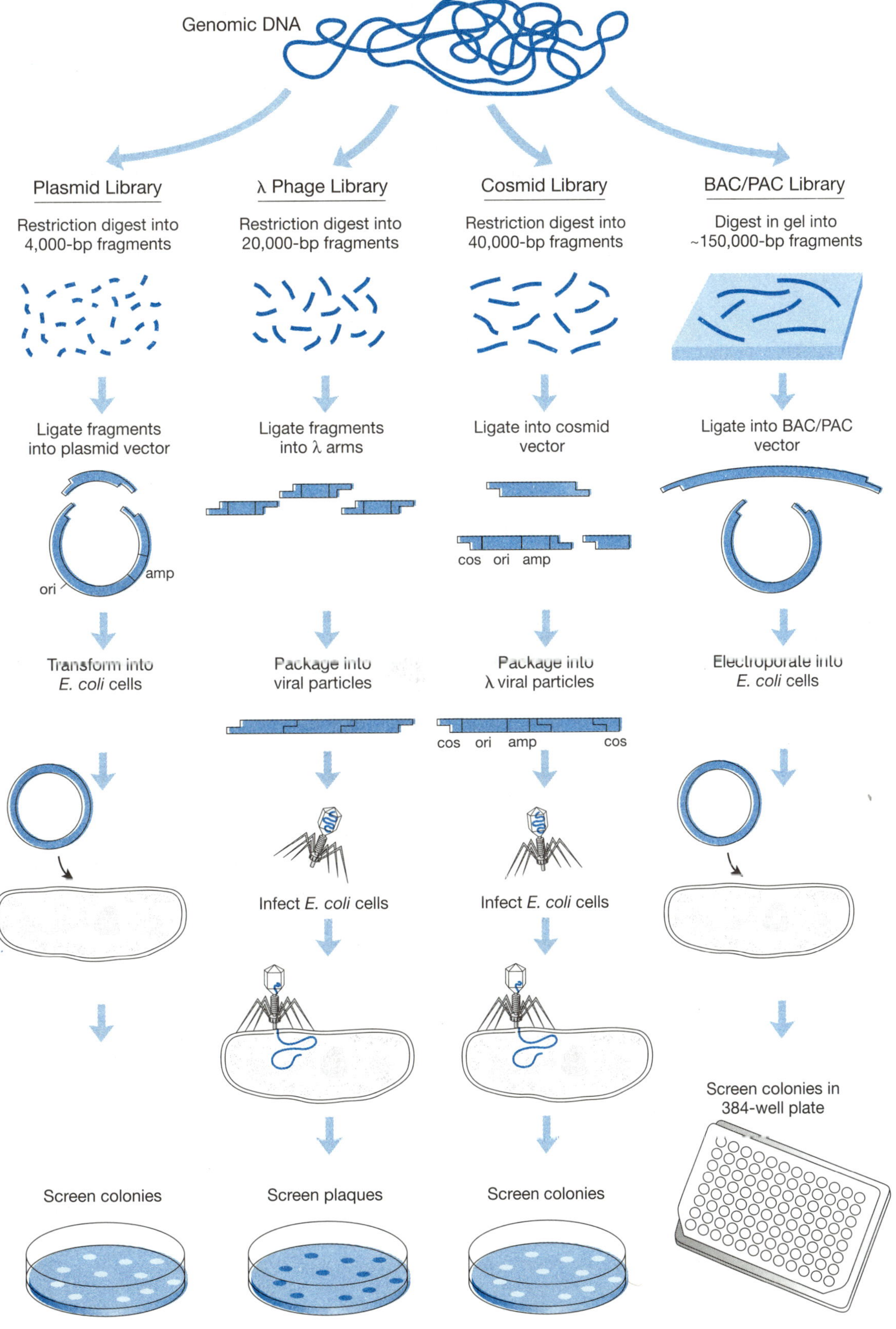

Making a Genomic Library

to the probability associated with any particular piece of DNA becoming ligated into a plasmid. Consequently, to increase the chance that every gene is represented, a library must contain molecules equivalent to several times the mass of the genome. This means that a plasmid library that truly represents the entire genome must be on the order of several million colonies. A library of this size creates problems. One problem is that the process of screening it to find a specific gene means screening millions of colonies. Another is that many genes are much larger than the 4000-bp average size of DNA fragments in this library. For example, the *FBN1* gene discussed in Chapter 3 is 110,000 bp (or 110 kilobases or kb) long. If the average piece of DNA in the library is 4000 bp, then the *FBN1* gene would be split up into at least 28 different fragments and their corresponding colonies. Since the point is to isolate an entire gene, various vector systems were developed that allowed for cloning larger DNA fragments.

Bacteriophage λ. The first vectors developed to clone larger DNA fragments for libraries were not plasmids; instead, they were modified forms of bacteriophage λ (see Chapter 1). The natural life cycle of phages makes them ideal vectors for molecular biologists. A phage infects cells by transferring its genetic material into the cell, where it replicates, producing hundreds of new copies of the viral genome. Several modified versions of bacteriophage λ were developed as cloning vectors in which viral genes could be replaced with an insert of foreign DNA.

The genome of bacteriophage λ is approximately 50 kb long, and it contains more than 40 genes. The protein capsule of the bacteriophage λ has a tight constraint on the amount of DNA that will fit inside of it (~55 kb). There is no room to add extra material. Fortunately, a number of nonessential genes lie in the middle of the λ genome, whereas the essential genes lie at the ends of the genome. The nonessential genes can be removed via restriction digests and fragments of genomic or other DNA can be ligated in. The DNA insert of choice is ligated in between the end pieces of the λ genome, called the λ "arms." The cohesive ("sticky") ends of bacteriophage λ, called *cos* sites, hybridize to one another, creating a long concatemer made up of multiple recombinant λ DNA molecules. When these concatemers are mixed in a test tube with protein extracts from bacteriophage λ, the concatemers are cleaved into individual λ molecules and efficiently packaged into new phage viruses.

Total genomic DNA is carefully isolated to prevent mechanical shearing into small pieces. The genomic DNA is digested by a restriction enzyme that is also present in the phage vector as a cloning site. Unlike the restriction digests described for the plasmid library, here the DNA is briefly incubated with a limited amount of enzyme. This "partial digest" produces a range of fragment sizes, considerably larger than the 4000-bp average-sized pieces that would result

Cloning Vectors and Minimum Number of Clones Needed to Represent the Haploid Human Genome

Vector	Average insert (bp)	Clones to represent genome
Plasmid	4,000	750,000
Lambda	20,000	150,000
Cosmid	40,000	75,000
Bacterial artificial chromosome (BAC)	150,000	20,000

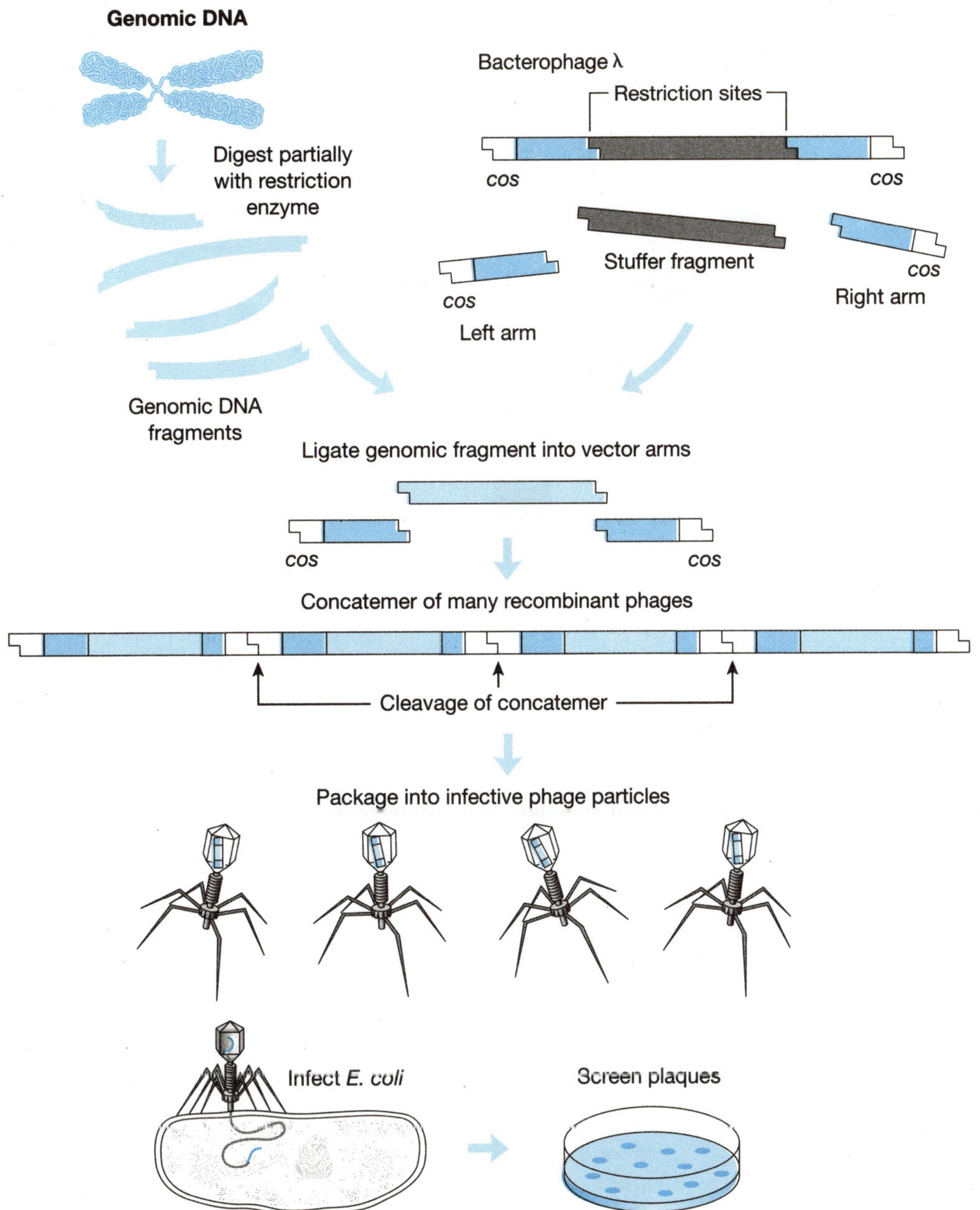

Generating a Genomic Library with Bacteriophage λ Vector

from a complete digest by a typical restriction enzyme that recognizes six nucleotides. (Remember from Chapter 4 that the likelihood of such a site occurring can be calculated $4 \times 4 \times 4 \times 4 \times 4 \times 4 = 4056$.) Ultimately, about 20-kb-sized fragments are selected for cloning. Originally, these 20-kb fragments were isolated by centrifuging the DNA in a sucrose gradient where different-sized fragments would settle at different levels. More recently, clonable fragments are isolated by separating the restriction fragments in an agarose gel and then cutting out a gel slice that contains DNA of the appropriate size, as determined by comparison to a DNA size marker. The DNA is extracted from the agarose slice and

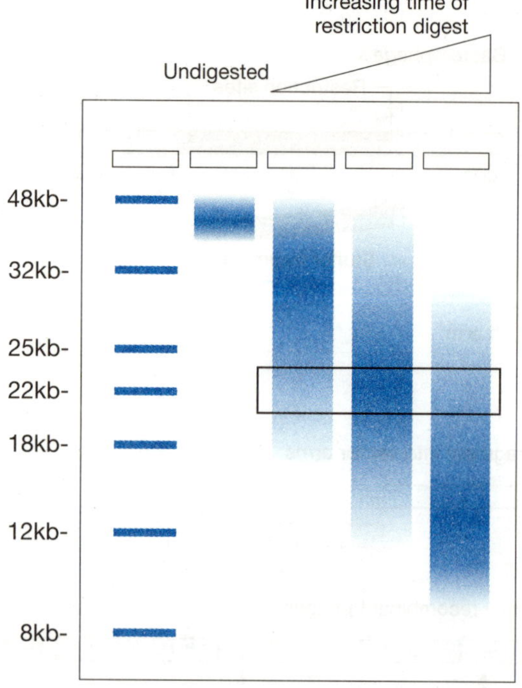

Size-selecting Genomic DNA for a Bacteriophage λ Library
Total genomic DNA is partially digested with an appropriate restriction enzyme and separated on an agarose gel. The first lane is a DNA marker showing DNA fragments of known sizes. The second lane is undigested DNA. The last three lanes are partially digested DNA incubated for increasing amounts of time. The box represents the region of the gel containing appropriately sized DNA fragments, which are used to make the λ library.

ligated with the phage DNA. Following ligation of the λ vector with genomic DNA, the ligation mixture is added to extracts containing the proteins needed to form a complete phage particle. Once "packaged," the library is used to infect *E. coli*.

Typically, *E. coli* and the bacteriophage λ library are briefly incubated together to allow the phage to infect the bacteria. There are many times more bacteria than phage. The mixture is spread on top of agar plates. The phage injects its genetic material into the *E. coli* cell, which is then replicated many times. The phage also encodes the capsule proteins needed to package new viral particles, which are then assembled within the cell. The newly replicated viral DNA is packaged into these new viral particles. The cell bursts (or lyses), recombinant viruses are released to infect neighboring uninfected cells, and the process is repeated. As the bacterial cells continue to lyse, a clear spot is created on the surface of the plate, called a plaque. Each plaque originates from a single infected

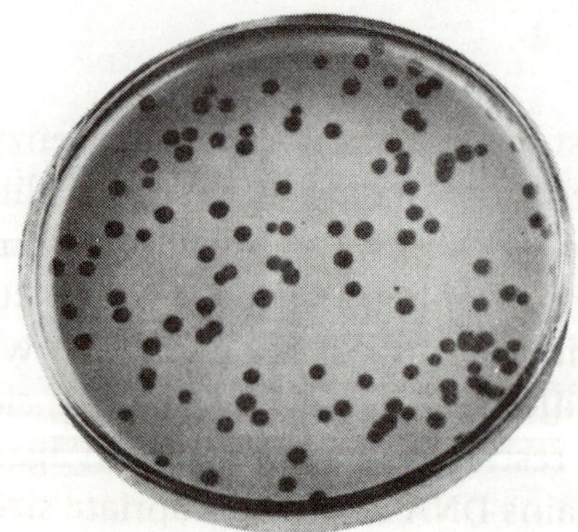

Phage Plaques on a Lawn of *E. coli* Bacteria
(Reprinted, with permission, from Stent G.S. 1963. *Molecular biology of bacterial viruses*, p. 41. W.H. Freeman, San Francisco.)

cell. The ratio of phage to bacterial cells in the original culture is adjusted so that several thousand plaques form on the plate. Each plaque can be isolated, and the viral DNA separated from the cellular debris.

Two features of λ vectors advanced the construction of genomic DNA libraries. First, λ vectors typically can accommodate inserts averaging 20,000 bp (20 kb); therefore, one-fifth as many recombinant molecules are required to represent the entire human genome. This makes the job of screening the library much more manageable. Second, introducing recombinant DNA into *E. coli* by phage infection is many times more efficient than transforming *E. coli* with plasmids.

Cosmids. Cosmids, which can be used to clone DNA fragments averaging 40,000 bp, were the next advance in vector development. These hybrid vectors are essentially plas**mids** that contain the bacteriophage λ **cos** sites. To create a cosmid, a λ DNA fragment containing the *cos* site is ligated into a plasmid containing an origin of replication and a selectable marker. Genomic DNA is digested into 40-kb pieces and ligated into the cosmid vectors. The vectors are then

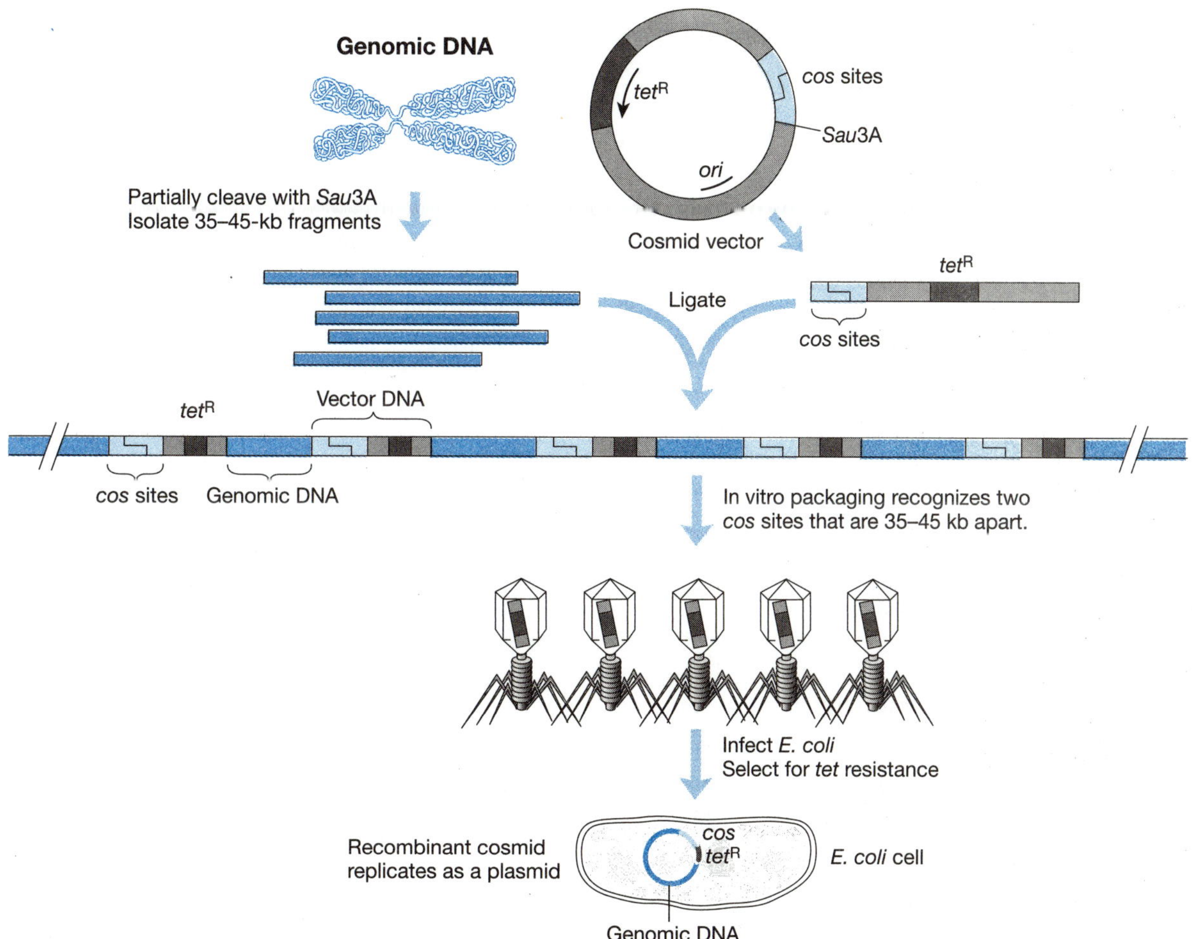

Cosmid Cloning

In this strategy, tandemly ligated cosmids with inserts are formed that will be reduced to single cosmids during packaging into phage particles.

mixed with bacteriophage λ protein extracts, which recognize the *cos* site and package the cosmid vectors into phage particles, creating infectious viruses. These recombinant viruses are used to infect *E. coli* cells. Since all of the viral genes have been removed from the cosmid, new λ viruses are not replicated—the infected cells do not lyse and form plaques. Instead, the cosmid replicates like a plasmid within the bacterial cell, and infected cells grow into normal bacterial colonies on selective media. The cosmids can easily be isolated by the standard techniques for isolating plasmids (discussed in Chapter 4).

BACs and PACs. Cosmids are limited as vectors by the amount of DNA that can physically fit inside the phage capsule. By the same token, the chemical transformation method of Mandel and Higa are limited to relatively small plasmids. Electroporation, popularized in the 1990s, provided a method to directly introduce really large (100,000–300,000 bp) DNA molecules directly into *E. coli*. DNA molecules of up to 1 million bp have been electroporated, and there is no theoretical limit to the size of the molecule taken up by this method. Electroporation also eliminated the preparation of elaborate phage packaging extracts or highly competent *E. coli* cells. Electroporation changed our strategy in vector design. Now vectors used to clone large DNA fragments need only contain a bacterial origin of replication and a selectable marker.

Two types of cloning vectors emerged that would allow for cloning of 100–300-kb DNA fragments: BACs (bacterial artificial chromosomes) and PACs (P1-derived artificial chromosomes). As the name implies, BACs contain an *E. coli* origin of replication from the F episome (described in Chapter 3). PACs contain the bacteriophage P1 origin of replication.

pBACe3.6 is a typical BAC vector. The selectable marker that allows identification of cells carrying the BAC is the *SacBII* gene, which allows for selective growth on sucrose. BAC libraries containing the entire human genome are now

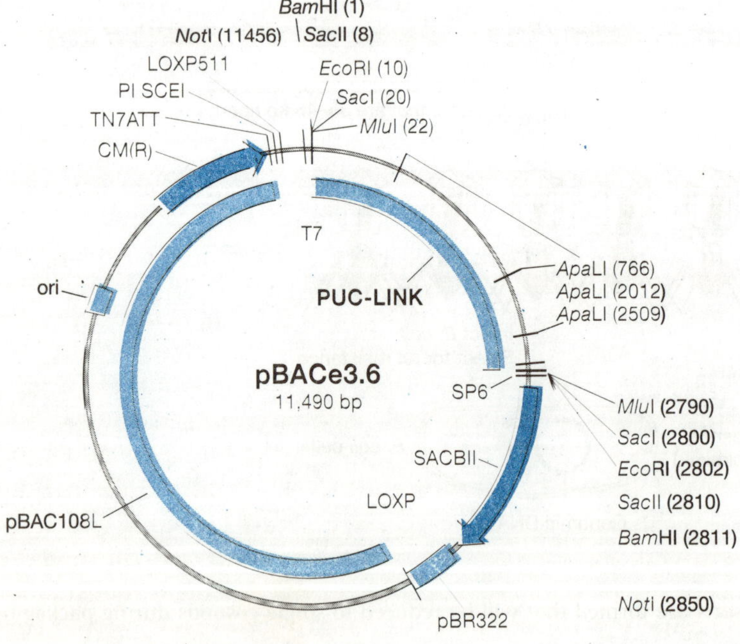

Map of BAC Plasmid pBACe3.6
The *SacBII* gene codes for levansucrose, which converts sucrose to levan that is toxic to host cells. The genomic insert is placed into the middle of *SacBII*, disrupting it, and preventing production of the full levansucrose protein. Plasmids lacking inserts will produce the levansucrose protein, failing to grow on media containing sucrose.

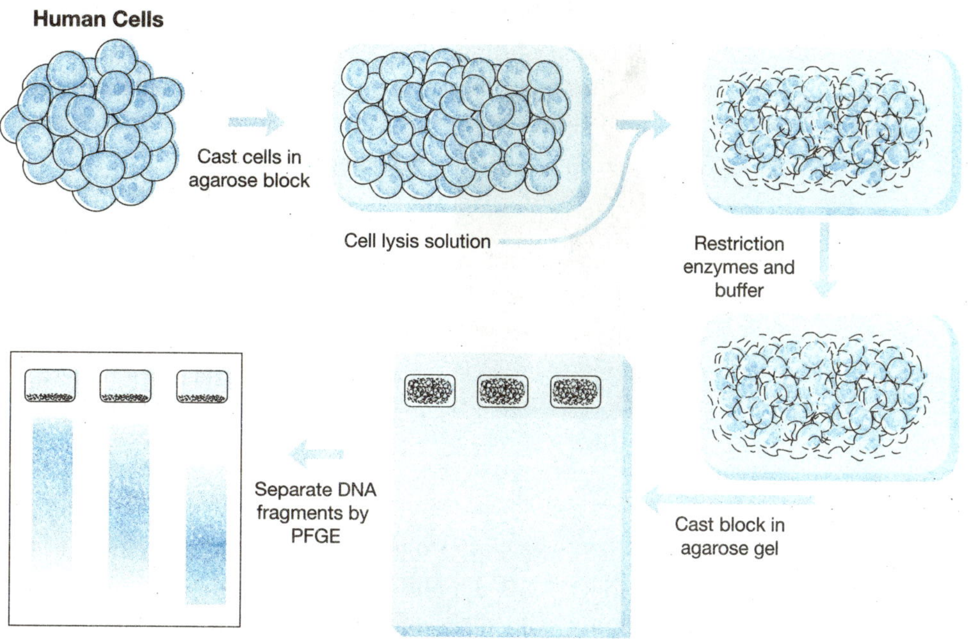

Isolating Large-sized DNA for BAC Cloning

commercially available. They typically contain enough clones for ten times coverage of the human genome, which statistically represents 99% of the genome.

Standard methods of DNA isolation shear DNA into fragments smaller than 100 kb. Thus, construction of BAC and PAC "megabase" libraries, requiring DNA fragments of 100–300 kb, demands careful manipulation. To minimize DNA breakage, cell samples are embedded in small agarose blocks, each having a volume of approximately 100 μl. Reactions to release and digest the genomic DNA are then performed "in situ" by diffusing reagents into the agarose block. First, cells are lysed with detergent and proteases. After a wash, restriction enzyme and buffer are diffused into the block. Following incubation, EDTA is diffused in to halt the restriction reaction to produce a partial digest. A larger agarose gel is cast around the reaction block and restriction fragments are separated by electrophoresis.

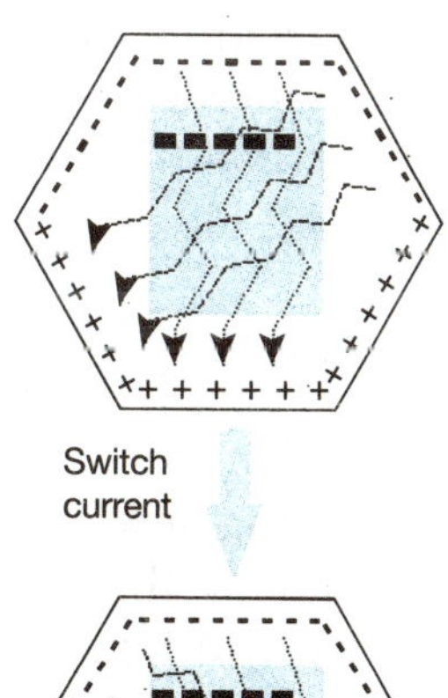

PFGE
Switching the current between the side electrodes allows for greater separation of large DNA fragments.

Conventional (fixed-field) electrophoresis cannot separate DNA fragments of greater than 50,000 bp. However, several newer techniques can separate fragments up to 10 million base pairs in length. In pulse-field gel electrophoresis (PFGE), pioneered by David Schwartz and Charles Cantor at Columbia University, current is alternated between several pairs of opposing electrodes. The electrophoresis box is set up in a hexagon, with electrodes at the top and bottom of the gel and two sets of electrodes on either side of the gel at a 120□ angle. Current is applied constantly between the top electrode and the bottom electrode. Current applied from the side electrodes is switched back and forth between the two sets. This alters the path of the DNA as it migrates through the gel, allowing for better separation of large fragments. In field-inversion or orthogonal-field-alternation gel electrophoresis (OFAGE), a longer forward cur-

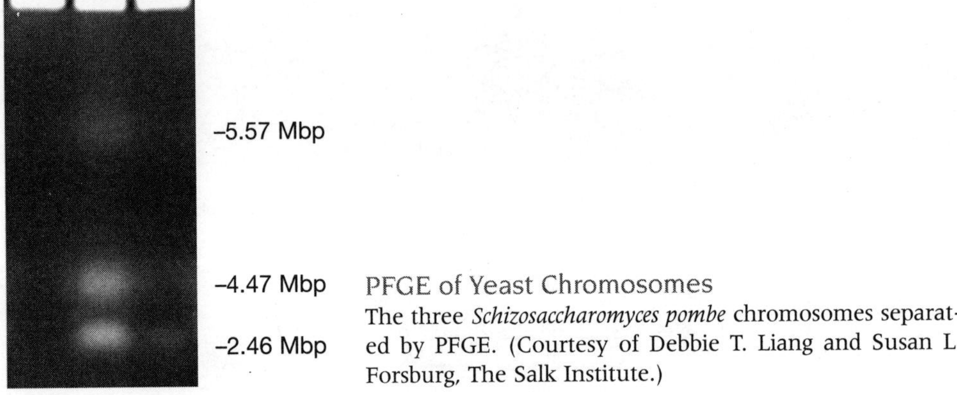

PFGE of Yeast Chromosomes
The three *Schizosaccharomyces pombe* chromosomes separated by PFGE. (Courtesy of Debbie T. Liang and Susan L. Forsburg, The Salk Institute.)

rent is alternated with a shorter reverse current. The resolving ability of PFGE and OFAGE depends on the length of time the two alternating currents are applied. Gel slices containing DNA fragments of the appropriate size are isolated gently and ligated into BAC or PAC vectors. PFGE has even been applied to separate whole yeast chromosomes, which can be millions of bp in length.

The enzymes used for megabase cloning cut 15–75 times less frequently than would be predicted by the random occurrence probabilities of their recognition sequences. For example, *Not*I recognizes an 8-bp sequence and would be expected to produce restriction fragments averaging 65,536 bp (4^8). In fact, it produces fragments averaging 1 million base pairs. *Mlu*I, *Nru*I, and *Pvu*I each recognize six-nucleotide sequences with random occurrence probabilities of 4096 nucleotides (4^6), yet each produces restriction fragments averaging 300,000 nucleotides. This is because the distribution of nucleotide sequences in the human genome is not random. The recognition sequences of all megabase-cutting enzymes contain one or more CG dinucleotides, a sequence that turns out to be very rare in mammalian genomes.

Screening DNA Libraries

We have discussed how difficult it is to distinguish one DNA fragment from another biochemically. Because the relative abundance of adenine, thymine, guanine, and cytosine is constant within the DNA polymer, every DNA fragment has more or less the same chemical composition. This was one of the facts that encouraged scientists during the 1940s and early 1950s to believe that DNA was incapable of being the genetic material. However, to work creatively with DNA, one must think in terms of the exact sequence of base pairs along the length of a particular DNA fragment. A genomic library is screened on the basis of two attributes of the DNA molecule: (1) the uniqueness of even relatively short nucleotide sequences and (2) the complementary binding of nucleotide pairs within the sequence.

Consider the length of a chain of nucleotides needed in order to be a unique sequence of human DNA—a sequence that statistically would be found only once in the entire human genome. To be unique, the sequence must have a probability of occurring less than once in 3 billion nucleotides.

Occurrence Probabilities of DNA Sequences of Increasing Numbers of Nucleotides

Nucleotides	Occurrence Probability (1/)
1	4
2	16
3	64
4	256
5	1,024
6	4,096
7	16,384
8	65,536
9	262,144
10	1,048,576
11	4,194,304
12	16,777,216
13	67,108,864
14	268,435,456
15	1,073,741,824
16	4,294,967,296

For the purpose of this discussion, we will adopt the view that, on the whole, the four bases are randomly distributed in a DNA molecule. (For the sake of simplicity, a DNA sequence usually takes into account only the nucleotide arrangement on one strand, since the opposite complementary strand can then be predicted.)

The probability of any DNA sequence occurring can be calculated by taking $1/4^n$, where 4 equals the number of different nucleotides and n equals the total number of nucleotides in that sequence. For example, starting at any nucleotide position along a DNA strand, there is a one in four (1/4) chance that it will be adenine. There is also a one in four (1/4) chance that the adjacent nucleotide will be cytosine. Therefore, the combined chance of encountering the dinucleotide sequence adenine-cytosine (AC) is $1/4 \times 1/4 = 1/4^2 = 1/16$. Every other dinucleotide sequence would also have an occurrence probability of 1 in 16, AA, TT, GG, CC, AT, AG, TA, TG, TC, GA, GT, GC, CA, CT, CG.

A simple exercise with a calculator shows that the chance of encountering a specific 10-nucleotide sequence is $1/4^{10}$, or 1 in 1,048,576. Statistically, only 3000 copies of a specific 10-nucleotide sequence should be found in the human genome. A 15-nucleotide sequence has an occurrence probability of $1/4^{15}$, or 1 in 1,073,741,824, whereas a 16-nucleotide sequence has an occurrence probability of $1/4^{16}$, or 1 in 4,294,967,296. A 15-nucleotide sequence should occur three times in the human genome, whereas a 16-nucleotide sequence should occur only once.

This calculation suggests a general rule that a sequence of 16 nucleotides or longer is likely to be unique in the human genome. However, we must bear in mind that repetitive DNA sequences comprise a significant percentage of human chromosomes. Repetitive sequences, found throughout the genome, are important in human genetic studies described in Chapter 8. Obviously, a 16-nucleotide sequence from *within* a repetitive element would be common to each of the repeated units. Some longer repetitive sequences include the hundreds of copies of the ribosomal RNA genes, but, in general, repetitive sequences are short—ranging from two to a few hundred nucleotides in length.

Hybridization
The radioactive probe (black) finds its complementary sequence in the genomic DNA (blue) and the bases hydrogen bond, forming a duplex molecule.

Identifying Specific DNA Sequences

In 1961, Sol Spiegelman and Benjamin Hall discovered that single-stranded DNA will hydrogen bond to its complementary RNA sequence to form a stable, double-stranded (duplex) molecule. The same bonding can be accomplished between single-stranded DNA molecules as well. This complementary base pairing provides a powerful tool to probe for unique DNA sequences in a genomic library.

Incubation of double-stranded DNA at a temperature above 90°C, at a pH of greater than 10.5, or with various organic compounds (such as urea or formamide) disrupts the hydrogen bonds between base pairs, causing the complementary strands to dissociate. This process is called denaturation. Under proper conditions of salt, temperature, and pH, the two single-stranded molecules can be renatured to re-form the original duplex DNA molecule. This process of complementary single-stranded molecules aligning and forming double-stranded molecules is known as *hybridization*. (Annealing is the term applied to the hybridization of short DNA sequences.)

Under reaction conditions of "high stringency," stable DNA duplexes form only when complementary base pairing is essentially perfect along the entire length of the DNA strands. Under conditions of "low stringency," partial hybridization occurs between strands that have lesser degrees of complementarity. Both conditions are useful in library screening.

Hybridization is the method used to identify and isolate a colony or plaque containing a specific gene from a library. The idea is to create a "probe," usually a radioactively labeled single-stranded DNA fragment, that contains a sequence complementary to the gene of interest and then to incubate it with the denatured genomic DNA from the plasmid or phage library. The probe is added in great excess, so it is more likely that the gene will hybridize to the probe, rather than its original complementary strand.

Making a Probe for β-*globin*

To make a probe for a gene of interest requires a purified DNA sequence complementary to that gene. This creates a "Catch 22," because in order to identify your target gene, you need to already have a piece of that gene. So how does one purify a complementary sequence from which to make the probe? Returning to our original example, we can see how Philip Leder and his co-workers overcame this apparently unsolvable problem in cloning the mouse β-*globin* gene.

Although the chromosome region containing a gene of interest looks more or less identical to every other region, mRNAs within a cell are somewhat unique. Each mRNA is a separate molecule, and the size of the mRNA molecule reflects the size of the gene. Thus, different mRNAs can be separated by size. mRNA is complementary to the DNA sequence of the gene from which it is transcribed, and it will readily hybridize to this DNA sequence. Leder reasoned that he might take advantage of hybridization, in combination with the unique properties of red blood cells and eukaryotic mRNA, to make a probe for the β-*globin* gene.

Red blood cells can be thought of as bags of hemoglobin because they are filled with this protein and little else. The nucleated precursors of red blood cells, reticulocytes, thus essentially function as hemoglobin factories. Whereas most mammalian cells contain relatively small amounts of thousands of different kinds of mRNA molecules, α- and β-*globin* mRNAs account for the major portion of all mRNA molecules in reticulocytes. Reticulocytes are not normally found in the bloodstream, but can comprise more than 50% of circulating blood cells in mice with induced anemia. Thus, reticulocytes were the raw materials from which Leder and co-workers created a probe for the β-*globin* gene.

Isolating Globin mRNA. As described in Chapter 3, all but a few eukaryotic mRNAs contain a long stretch of 100–200 adenine residues at their 3´ ends. This poly(A) tail makes it possible to isolate mRNA in a single step using oligo(dT)-cellulose affinity chromatography. In this protocol, a column is filled with cellulose particles, to which are attached short synthetic sequences (10–20 nucleotides) of deoxythymine (dT). The stretch of T residues will hybridize to the stretch of As in the poly(A) tail of mRNA under high-salt conditions. This allows for the capture

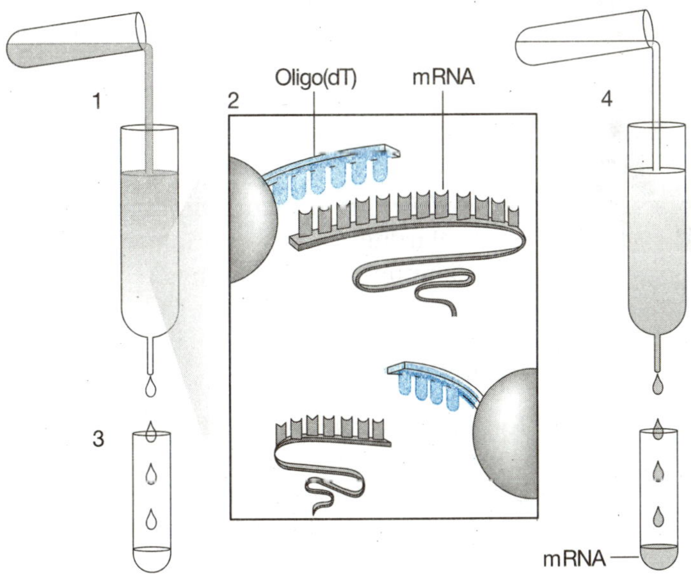

Purifying Globin mRNA

(*1*) Lysed reticulocytes in high-salt buffer are poured through a column of oligo(dT)-cellulose. (*2*) Poly(A) tails of mRNA bind to oligo(dT) molecules on the cellulose beads. (*3*) Non-mRNA molecules pass through the column. (*4*) Wash with low-salt buffer releases mRNA from the column.

of poly(A) mRNA as it binds to the column. When a reticulocyte lysate is passed through the oligo(dT)-cellulose column in a high-salt buffer containing sodium chloride, the poly(A) tails of the mRNA bind to the complementary T residues of the column. mRNA molecules are thus retained on the column while all other molecules pass through. The salt is then rinsed from the column, which releases the bound mRNA. Since α- and β-*globin* represent 98% of the mRNAs contained in reticulocytes, no further purification is necessary.

Tom Maniatis
(Courtesy of Kris Snibbe, Harvard News Office.)

Making a cDNA Probe from mRNA. Chapter 2 discussed the central dogma of molecular biology—DNA is transcribed to make RNA that is translated to make protein. However, recall that there were some exceptions to this rule. To make a probe for library screening, we can take advantage of one of these exceptions to make DNA from RNA. For this, we need a special DNA polymerase called RNA-dependent DNA polymerase, or reverse transcriptase (RT).

RT is found in *retroviruses*, which store their genetic material as RNA. When a retrovirus infects a cell, it must convert its RNA genome into DNA to begin its life cycle. The retrovirus RT uses RNA as a template to build a new complementary strand of DNA (this new strand is called "cDNA"), which is converted into a double-stranded DNA molecule.

In the mid 1970s, several groups discovered methods for using RT in the test tube to make cDNA from mRNA. The methods in use today derive from work done at Harvard by Tom Maniatis, Argiris Efstratiadis, Fotis Kafatos, and Allan Maxam. Using RT in the test tube, a radioactively labeled cDNA probe can be created from the purified globin mRNA. First, a short "primer" composed of oligo(dT) is hybridized to the 3′ poly(A) tail of the purified mRNA. Like all DNA polymerases, RT initiates DNA synthesis at the primer; it then builds a DNA chain that incorporates a complementary nucleotide for each position on the mRNA molecule. As is true for all polymerases (both DNA and RNA polymerases), RT synthesizes 5′ to 3′—always adding a new nucleotide to the 3′ OH of the previously added nucleotide. The synthesis of the cDNA probe takes place in a reaction mixture containing RT and the four deoxynucleotide triphosphates, one or more of which contain ^{32}P (a radioactive isotope of phosphorus).

Argiris Efstratiadis

Leder used this reaction to produce ^{32}P-labeled cDNA probes from the α- and β-*globin* mRNA that he isolated from reticulocytes. Because mRNAs for both α- and β-*globin* are present, the library screen identified colonies containing both genes. Further characterization of the isolated clones allowed him to distinguish between the two globin genes.

Screening

To isolate the mouse β-*globin* gene, Leder and his colleagues screened a bacteriophage λ library. At the time, these were the libraries containing the largest DNA inserts, but today, investigators most likely start with a BAC or PAC library. The screening of either bacteriophage λ or plasmid-like libraries is essentially the same so we should consider the cosmid library as an example. When screening a library, individual cells are not screened for the gene, but rather, we screen colonies (the clump of cells growing on a solid surface that arise from a single cell as it grows).

The first step is to grow up colonies. We begin by placing a circular nylon or nitrocellulose filter onto an agar Petri dish containing the appropriate nutrients

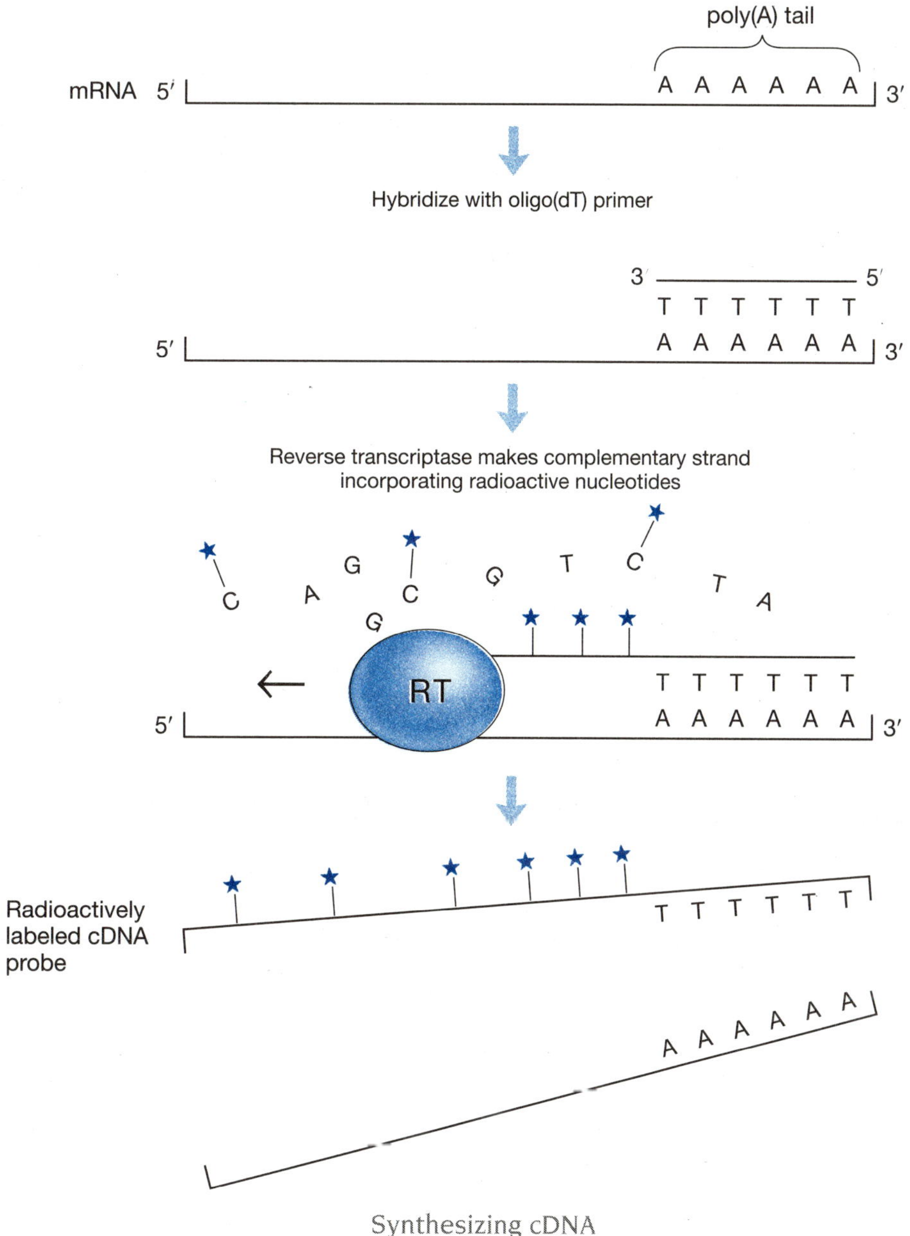

Synthesizing cDNA

and the selection marker, for example, ampicillin. A suspension of cells from the library is plated onto the filter and spread out evenly. The cells containing the library must be diluted and spread out enough on the filter surface so that there is a minimum of colony overlap (remember that we will later want to isolate one particular colony). Typically, several thousand cells are plated and multiple filters are prepared in order to screen enough inserts to cover the genome several times over. After overnight incubation, colonies appear on the filters.

The process of library screening requires destroying the cells on the filter. To obtain a viable colony after screening, a duplicate set of filters must be made by replica plating each filter. This is accomplished by placing a fresh filter over each colony-containing filter. Registration marks are made by poking small holes through

Plate cells onto filter overlayed on a agar plate

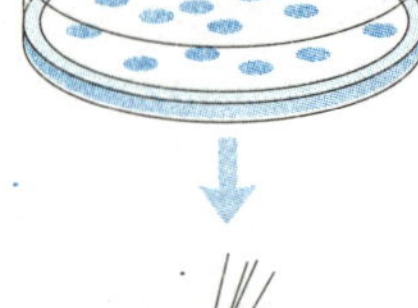

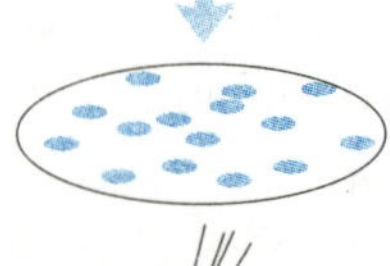

Make replica filter

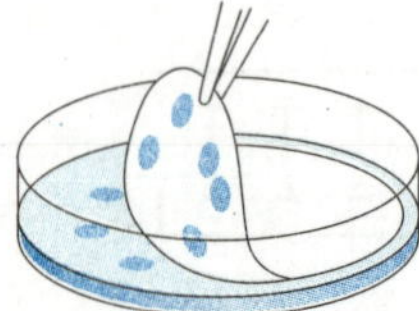

Remove replica filter

Lyse cells; denature DNA and hybridize to radiolabeled probe in sealed plastic bag

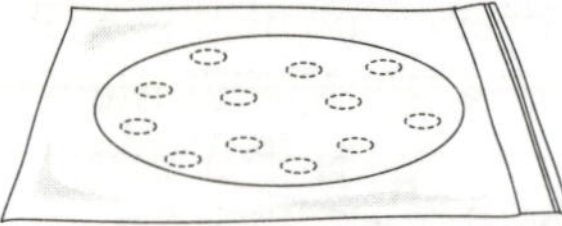

Overlay filter with X-ray film

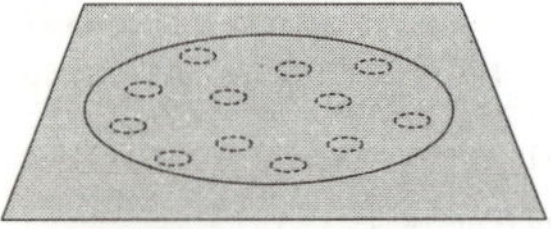

Developed autoradiogram reveals location of colony to which probe hybridized

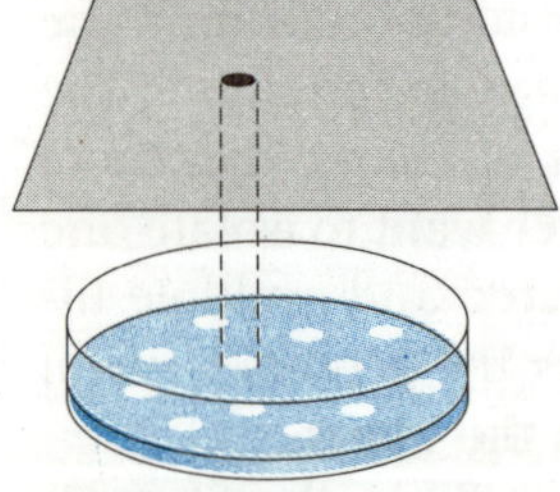

Screening a Cosmid Library
An autoradiogram indicates locations of colonies on master plate to which a radioactive probe hybridized.

the edges of the filters, so that the filters can be realigned later, and then they are separated. Some cells from each colony are transferred to the new filter. The new filter is placed on a fresh plate and both plates are incubated until full-size colonies appear. (This is an adaptation of a technique for making replica bacterial colonies published in 1952 by Joshua and Esther Lederberg, at the University of Wisconsin.)

One replica set of filters is left on the Petri dish and refrigerated. The other replica filter is screened. The filters are soaked in an alkaline solution of sodium hydroxide, which simultaneously lyses the cells and denatures the DNA. The filters are then soaked in buffer at neutral pH. The DNA remains denatured, and the single-stranded DNA is fixed in place on the nylon/nitrocellulose by baking the filters under vacuum at 80°C.

The dried filters are then screened with the probe (^{32}P-labeled cDNA) to detect the presence of the gene of interest. To accomplish this, the filters are placed in a plastic sealable bag with the probe, containing hybridization buffer (buffer, salt, and usually proteins such as BSA and some polymers such as ficol and polyvinyl pyrimidol that coat the filter, blocking the probe from nonspecific binding), under conditions that encourage hybridization of complementary sequences: neutral pH, presence of sodium ions (which neutralize the negative charges along the DNA backbone), and elevated temperature (65°C, which untangles the DNA strands). During an incubation of 12–24 hours, the probe should hybridize to the DNA from only those colonies containing all or part of the gene of interest. Remember that *E. coli* cells are used, so the presence of the bacterial DNA should not be a hindrance for screening—the probe for a unique genomic sequence should not hybridize to the bacterial DNA.

Following hybridization, the filters are washed to remove excess probe, blotted dry, and covered in plastic wrap. The covered filters are tightly sandwiched against a sheet of X-ray film, which is exposed for several hours. The β particles released during the decay of the radioactive ^{32}P in the probe expose the film. Following development, the exposed area appears as a dark spot on the X-ray film. Realignment of reference marks on the film, filter, and culture plate matches an exposed spot to its corresponding colony on one of the duplicate filters that has been stored in the refrigerator. This colony thus contains DNA sequences complementary to the probe and therefore contains the gene of interest. The colony is picked from the plate and can be grown in quantity for isolation of the cosmid DNA and further analysis.

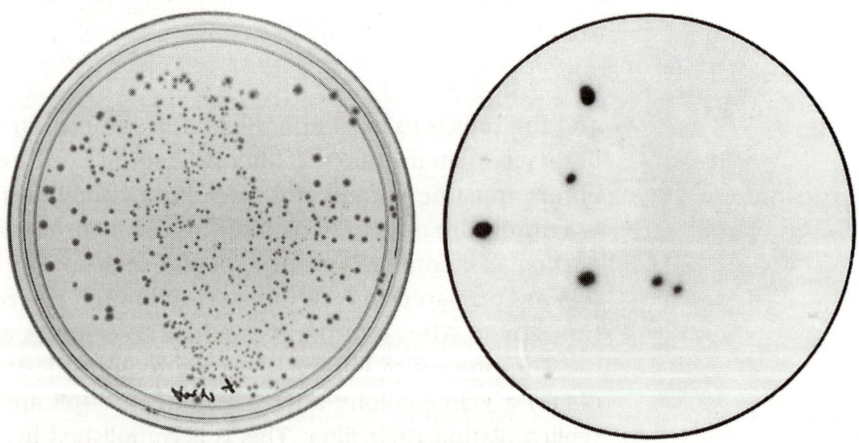

Colony Hybridization
LB plate with transformed colonies (*left*). Filter with positive clones (*right*).

Other Strategies for Screening

Once a gene is isolated from one organism, it can often serve as a probe to isolate the equivalent genes from different organisms (orthologs). For example, the mouse and rabbit globin genes were used as probes to isolate the human globin genes. Many genes also exist in "families" of similar genes. For example, in Chapter 3, we discussed the *Hox* genes, each of which contains a highly conserved sequence called the "homeobox." Conserved sequences from within the homeobox were used as probes to find other *Hox* family members. When a non-identical sequence is used as a probe to detect another gene, the process is called "cross-hybridization." Less stringent conditions (lower temperature and higher salt concentration) are used in cross-hybridization screening, allowing hydrogen bonding to occur despite a few mismatches between probe and target sequence.

Synthetic Oligonucleotides. Progress in protein biochemistry meant that, by the start of the gene cloning era, many proteins had been fully or partially sequenced. The advent of automated DNA synthesizers, in the early 1980s, made it feasible to use virtually any partial amino acid sequence to produce short oligonucleotide probes (oligo = a few) for the corresponding gene.

Since each amino acid is encoded by a triplet codon, a six-amino-acid sequence can be used to derive an 18-nucleotide DNA probe, which should hybridize to a unique sequence in the human genome. (Remember the calculation for the length of a unique sequence?) However, most amino acids are encoded by more than one triplet. Because of this degeneracy, or "wobble," of the genetic code, the exact DNA sequence cannot be derived precisely from an amino acid sequence.

This problem can be circumvented by selecting a string of amino acids with the least degeneracy—ideally, those encoded by only one or two different codons. Then, oligonucleotides are synthesized that represent every possible codon combination. This probe mixture is radioactively labeled using ^{32}P ATP and polynucleotide kinase, which transfers the γ phosphate of the ATP to the 5′ end of the oligonucleotide. The probe is used to screen a genomic or cDNA library, and the complete amino acid sequence of the protein is predicted from the DNA sequence of the cloned gene.

BAC and PAC Library Screening. Because BAC and PAC libraries contain such large pieces of DNA, many fewer clones need to be screened to find a particular gene. These libraries are not typically screened by spreading cells onto a culture plate. Instead, individual BAC or PAC colonies are first propagated in 384-well microtiter dishes. A mechanical robot arm spots the colonies onto a filter at a high density. The filter is then screened using standard hybridization methods. The positions of positive colonies on the filter are correlated with specific wells in the microtiter dish.

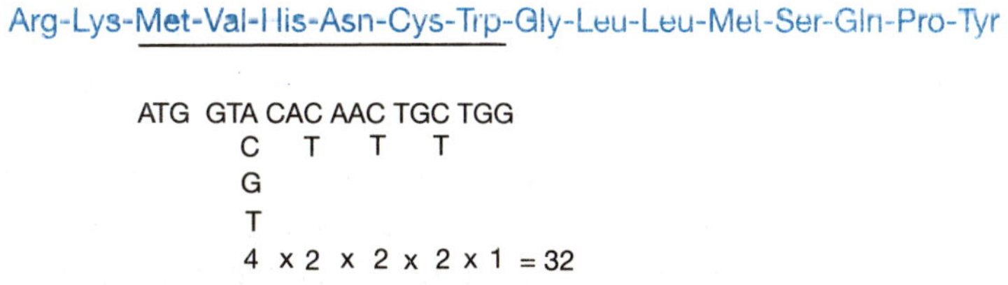

Arg-Lys-Met-Val-His-Asn-Cys-Trp-Gly-Leu-Leu-Met-Ser-Gln-Pro-Tyr

```
ATG GTA CAC AAC TGC TGG
      C   T   T   T
      G
      T
      4 x 2 x 2 x 2 x 1 = 32
```

Choosing the Best Sequence for a Probe

In the hypothetical protein sequence above, we scan for a region where the amino acids have the least degeneracy, or wobble. The underlined sequence has few degenerate amino acids, and a mixture of 32 probes will account for every possible coding sequence.

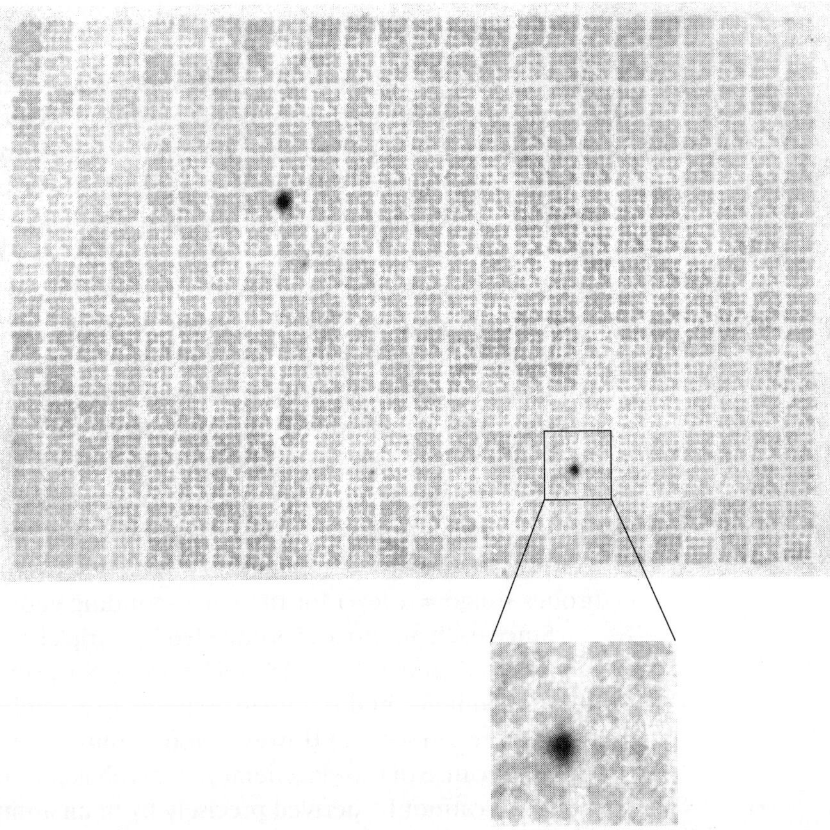

PAC Library Screened for the *CRAF* Gene
Colonies were spotted onto the filter using a robot, allowing for a high-density filter. To visualize all of the colonies, the weak isotope ^{35}S was used to make labeled *E. coli* DNA, which when hybridized to the filters, creates the gray background spots. Each square contains 16 colonies and there are 16 x 24 squares or a total of 6144 colonies. If each PAC contains inserts of 150,000 to 300,000 bp, then this single filter contains between 1 and 2 billion base pairs of DNA or 33–66% of the human genome. The probe was labeled with ^{32}P to screen for the *CRAF* gene. Two positive colonies are seen as dark spots. (Courtesy of Sergey Kalachikov and T. Conrad Gilliam, Columbia University.)

Once a genomic clone is identified as containing a target gene, it must be located within the very large BAC or PAC. The easiest approach is to digest the BAC with a restriction endonuclease and to "subclone" these fragments into another vector, such as a plasmid or bacteriophage λ. This mini-library is rescreened with the same probe used to identify the BAC. Today, because DNA sequencing is so rapid and inexpensive, subclones are usually directly sequenced to identify the target gene.

cDNA Libraries. Up until now, we have described cloning genomic DNA. Remember that a genomic DNA sequence typically contains many times as much intron sequence as exon sequence. This means that a single gene can be spread out over a large distance, making it harder to isolate and analyze. Another problem with screening a genomic library is that the library contains all of the genes present in the organism's genome. Often, a scientist wishes to look for a specific gene that is active in a particular tissue or cell type.

cDNA libraries provide solutions to both of these problems. cDNAs are made from mRNAs, which, when processed, do not contain any introns. This means that any cDNA isolated will only contain coding sequences for that gene, which reduces the amount of DNA one has to isolate and analyze. In addition, a cDNA

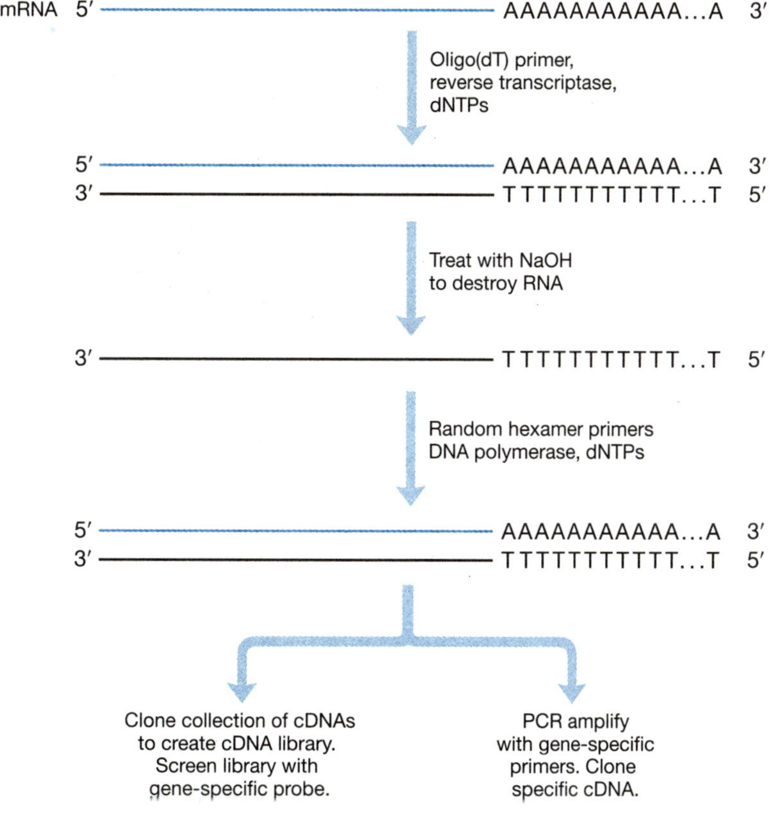

Making a Double-stranded cDNA for Cloning

library represents only those genes that are expressed in a particular cell type, reducing the number of genes to be screened.

To produce a cDNA library, total mRNA is isolated from the tissue or cells of interest using oligo(dT) affinity chromatography, and cDNA copies are synthesized using oligo(dT) to prime the RT, as described earlier. This synthesis produces a DNA-RNA hybrid molecule. The next step is to synthesize the second DNA strand to make double-stranded cDNA (ds cDNA). DNA synthesis requires a primer. Ideally, we would like to use a primer that begins at the most 5′ end of the mRNA in order to synthesize the whole molecule, as was accomplished at the 3′ end using oligo(dT). However, it is very difficult to begin priming at the very 5′ end of the mRNA, since each mRNA has its own unique sequence in this region. Instead, we must rely on a method called random priming. After completely removing the mRNA, the remaining cDNA strand is incubated with random hexamers—oligonucleotides representing every possible six-nucleotide sequence. The primers bind and prime from sequences within the cDNA, typically resulting in double-stranded cDNAs that lack the 5′ end.

The resulting double-stranded cDNA is prepared for ligation into a plasmid or a phage vector by adding linkers to the ends using T4 DNA ligase. Linkers are short double-stranded oligonucleotides containing a restriction recognition sequence. For example, a linker with the sequence GAATTC contains the recognition site for *Eco*RI. If desired, prior to the addition of the linkers, the double-stranded cDNA can be incubated with *Eco*RI methylase, blocking any *Eco*RI sites within the cDNA itself from cleavage during the next step. (In directional cloning

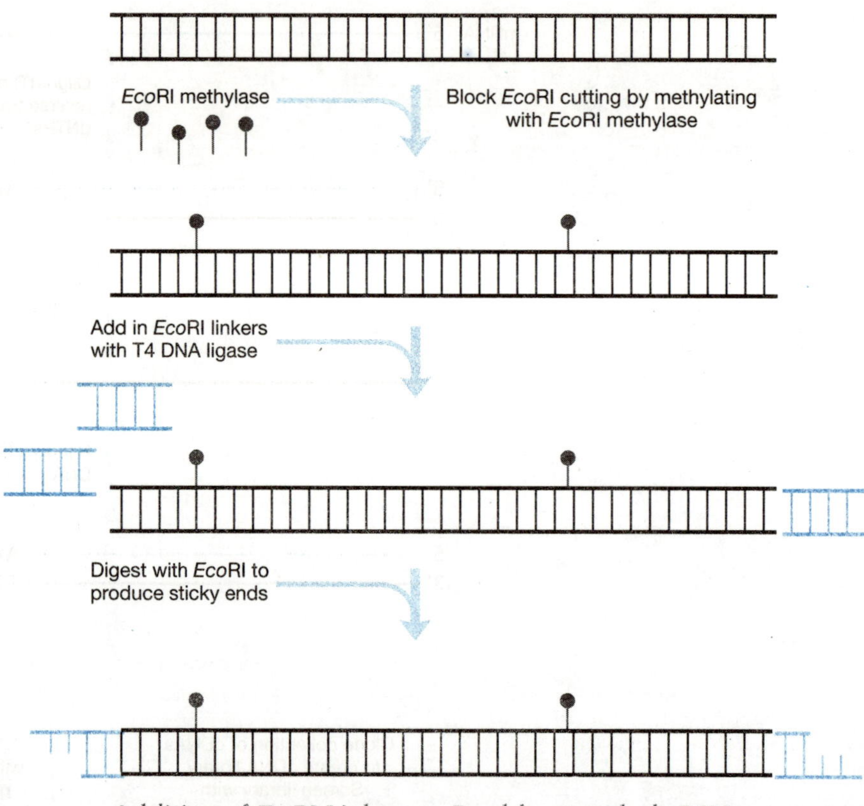

Addition of *Eco*RI Linkers to Double-stranded cDNA

strategies, a different linker is added to each end of the double-stranded cDNA, creating a different restriction endonuclease site at either end.)

Digestion with the restriction enzyme *Eco*RI cuts the linkers and creates sticky ends for ligating cDNAs into *Eco*RI-digested vectors. The resulting recombinant plasmids or phage are transformed into or used to infect *E. coli*, which are spread onto agar plates. Antibiotic in the medium selects for colonies that express a resistance gene from the recombinant vector. The presence of phage containing cells is seen through the formation of plaques. Either way, the plates now contain a cDNA library that can be screened in the same way as a genomic DNA library.

Computer-based Screening. With the genome sequences of many organisms complete or nearly complete, it is not generally necessary to isolate a cDNA from a library. A gene of interest can be identified by screening a genome database. Using this information, primers specific to the 5′ and 3′ end of the gene can be made. The 3′ primer is annealed to mRNA isolated from cells that express the gene of interest, and a cDNA is made using RT. Next, the 5′ primer is added and second-strand synthesis is carried out using DNA polymerase from *E. coli*, producing a specific double-stranded cDNA. This approach has one weakness. If the mRNA is present in low abundance, it may be difficult to get enough double-stranded cDNA with which to work. The alternative is to reverse-transcribe mRNA to make the cDNA and then add the 5′ primer and use the polymerase chain reaction (PCR) to amplify the cDNA into as many copies as are necessary (see Chapter 6). This method is called reverse transcriptase PCR (RT-PCR).

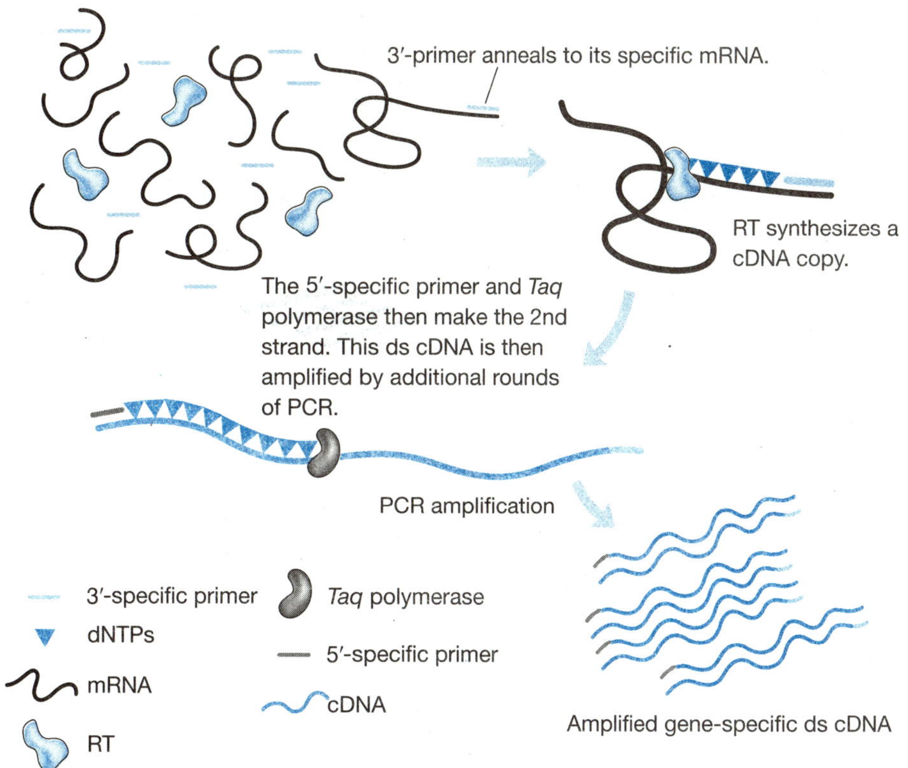

Reverse Transcriptase–Polymerase Chain Reaction

Working with Cloned Genes

Once a gene has been isolated from a library, a world of possibilities opens up for analysis.

Southern Hybridization. Southern hybridization (or blotting), named after its discoverer Ed Southern, can be used to study the organization of genomic DNA. As with library screening, Southern blotting is based on the ability of a DNA strand to seek out and bind to its complementary sequence.

Total DNA is isolated and digested in separate reactions with one or several restriction enzymes. The digested DNA is then separated by electrophoresis on an agarose gel. Following electrophoresis, the DNA in the gel is denatured to create single strands and then transferred to a nylon or nitrocellulose filter by capillary action. A single-stranded radioactive probe is incubated with the filter under conditions that promote hybridization. The filter is washed and exposed to X-ray film. As in plaque or colony screening, the radioactive probe hybridizes to its complementary DNA sequence on the filter, and a band on the film mirrors the position of this sequence on the agarose gel.

PCR has largely supplanted Southern hybridization for the analysis of relatively short DNA fragments, including forensic and identity testing. However, blotting is still used to study large-scale DNA arrangements and to genotype plants and animals that have undergone germ-line manipulation to introduce a new gene (transgene). In the latter case, genomic DNA is isolated and hybridized with a probe to show the presence or absence of the transgene.

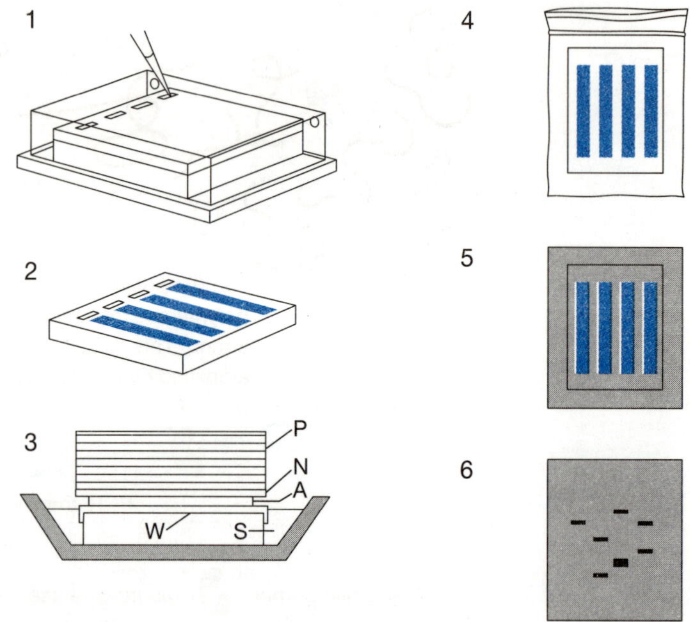

Southern Blotting

(*1*) Load digested DNA into agarose gel and separate by electrophoresis; (*2*) visualize DNA, and denature in gel; (*3*) transfer DNA to nitrocellulose filter by capillary action (P = paper towels, N = nitrocellulose, A = agarose gel, W = wick, S = high-salt solution); (*4*) hybridize radioactive probe to filter, which, if stained, would be a mirror replica of the gel; (*5*) wash filter, and overlay with X-ray film; (*6*) developed autoradiogram reveals location of bands to which probe binds. (Adapted from art concept by Lisa Shoemaker.)

mRNA Analysis

All cells in the body (except the germ cells) carry the same set of genes, but only a subset of those genes is active in a particular cell or tissue. For example, nerve cells need different structural proteins and enzymes than kidney cells. To understand which genes have roles in which cells, we need to see which genes are being transcribed into mRNA ("expressed"). Once a gene is cloned out of a library, it can be used as a probe in the analysis of mRNA.

Northern Hybridization. Northern blotting (given this name as a word-play on Southern blotting) is a method to detect mRNA and to determine whether a gene of interest is expressed in a specific tissue or cell type. Northern blotting basically follows the same protocol as Southern blotting. mRNA is isolated from a particular tissue, separated by electrophoresis on a denaturing gel, and blotted onto nitrocellulose or nylon filters. Radioactive probes for the mRNA are made by transcribing the gene of interest in the *opposite* direction. (Remember that the probe must be complementary to hybridize to the mRNA, so the strand that normally does not make the mRNA is used to make the probe.)

Northern blots can show which genes are active or inactive in particular cell types. They can also show changes in the levels of gene activity following a particular treatment or perturbation. However, northern blots are not particularly sensitive and may not detect a gene transcribed at a low level. Northern blots also require a fairly large quantity of mRNA, which may be difficult to obtain from a small tissue, such as the embryonic kidney.

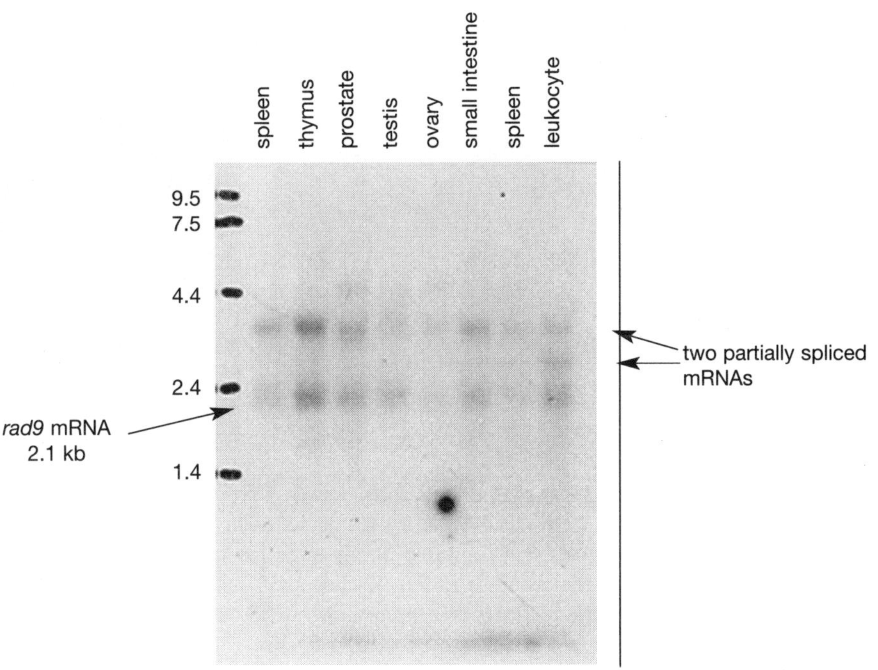

Northern Blot
This northern blot shows the relative amounts of the human *rad9* mRNA in various tissues (listed above each lane). A size marker was also run, and the positions of these mRNAs are shown on the left in kilobases. Fully spliced *rad9* mRNA of 2.1 kb is seen at various intensities in each lane. A partially spliced mRNA is also seen in each lane. The last lane shows another partially spliced mRNA of intermediate size. (Courtesy of Kevin Hopkins and Howard Lieberman, Columbia University).

In Situ Hybridization. In situ ("in place") hybridization is a more sensitive method for visualizing gene activity directly in fixed cells or tissues. It was developed independently by two groups in 1969: Joseph Gall and Mary Lou Pardue at Yale, and Max Birnstiel and Ken Jones at the Institute of Molecular Biology in Zurich and the MRC in Edinburgh. Rather than extracting mRNA from cells, the mRNA is left in place and the entire cell, tissue, or embryo is fixed using paraformaldehyde. The fixative prevents the breakdown of molecules and holds them in place for analysis.

In early uses of this technique, the tissues were cut into very thin sections and adhered onto microscope slides. Radioactive probes were used to look for gene activity directly on the tissue sections. Probe-containing solution was added to each slide, and after an incubation, the slides were then washed to remove any unbound probe. Instead of exposing the slides to photographic film, the slides themselves were dipped in a photographic emulsion. The slides were given some time to expose the emulsion, and then were developed using the same chemicals used to develop photographic film.

Some of the drawbacks of this technique are the safety issues involved in using radioactive materials, and the need for sectioned tissues. Sectioning is time-consuming, and it is difficult to obtain an overall picture of gene expression throughout an organism from small sections. These problems are nicely addressed by in situ hybridization of whole-mount specimens with nonradioactive probes.

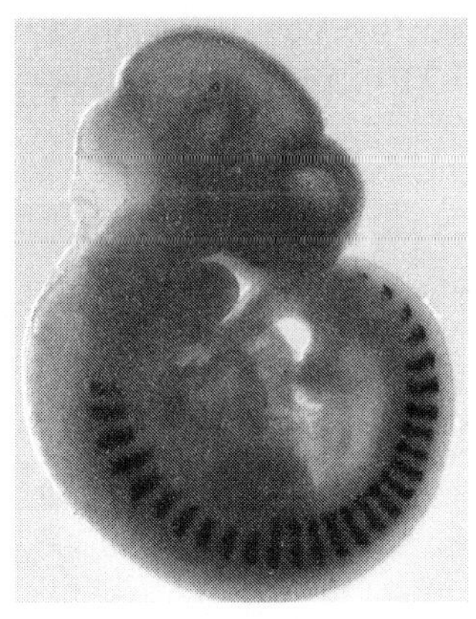

In Situ Hybridization of a Mouse Embryo with a *Myogenin* Probe
The probe has been visualized with the BCIP/NPT indicator. Expression of the gene is shown as the dark patches in the myotome of the somites, which will develop into muscle. (Courtesy of Deborah Chapman, University of Pittsburgh.)

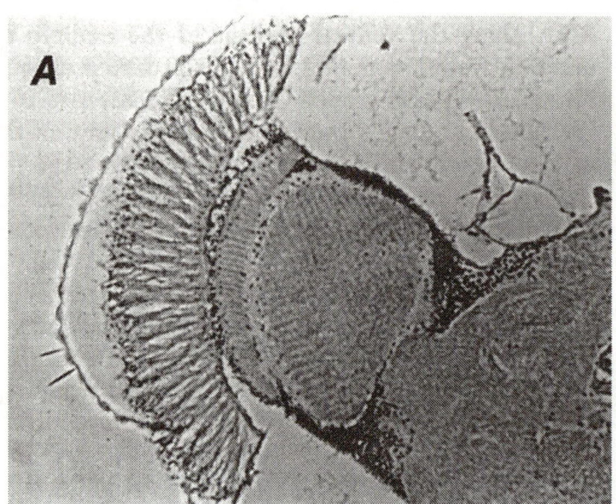

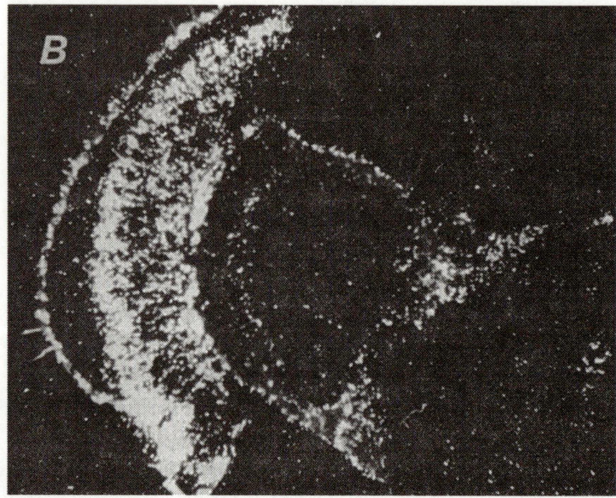

In Situ Hybridization of a Tissue Section of an Adult *Drosophila* Eye

A radioactive probe was used to detect mRNA for the *period* gene. (*A*) Eye tissues are shown with bright-field microscopy; (*B*) dark-field microscopy reveals the presence of silver grains from the photographic emulsion in cells where *period* is expressed. (Reprinted, with permission, from Liu X., Lorenz L., Yu Q., Hall J.C., and Rosbash M. 1988. Spatial and temporal expression of the *period* gene in *Drosophila melanogaster. Genes Dev. 2:* 228–238.)

Instead of labeling probes with radioactive nucleotides, probes are linked to substrates that fluoresce or produce a colored precipitate (such as biotin, digoxygenin, and alkaline phosphatase). When a fluorescent substrate is used for a label, the probe bound to mRNA in the tissue can be visualized *directly*, through the use of a fluorescence microscope. Probes labeled with alkaline phosphatase, biotin, or digoxygenin must be visualized *indirectly*.

Alkaline phosphatase reacts with BCIP/NPT (5-bromo-4-chloro-3-indolyl phosphate/nitroblue tetrazolium chloride) to form an insoluble purple precipitate, so this reporter chemical is added directly to the tissue for visualization of bound mRNA. Digoxygenin-labeled probes are visualized through the use of antidigoxygenin antibodies with an attached reporter enzyme such as alkaline

Oxidation of BCIP and Reduction of NBT in the BCIP/NPT Indicator Reaction

(Reprinted, with permission, from Kessler C. 1991. The digoxigenin:anti-digoxigenin (DIG) technology—A survey on the concept and realization of a novel bioanalytical indicator system. *Mol. Cell. Probes 5:* 161–205.)

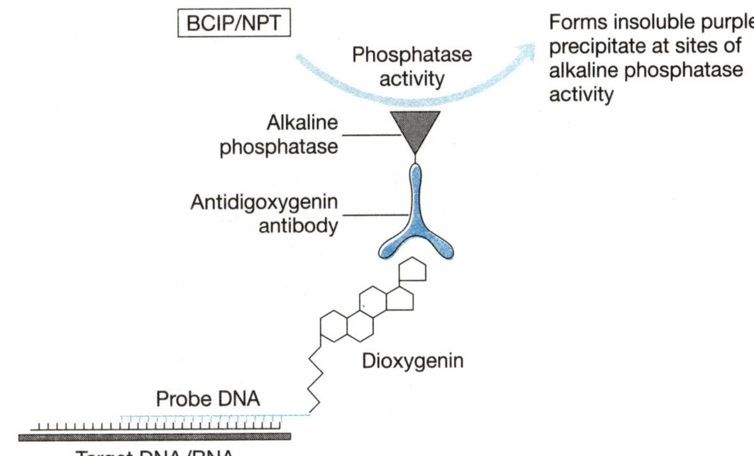

Detection of Digoxygenin-labeled Nucleic Acid Probes with BCIP/NPT (Reprinted, with permission, from Sambrook J. and Russell D. 2001. *Molecular cloning: A laboratory manual,* 3rd edition. Cold Spring Harbor Laboratory Press, Cold Spring Harbor, New York.)

phosphatase. Again, BCIP/NTP is added, which forms a colored precipitate. Biotin is a water-soluble vitamin that binds with high affinity to streptavidin, a basic glycoprotein found in egg white. Streptavidin also is attached to a reporter enzyme for visualization.

Nonradioactive probes have many advantages. Although radioactive probes decay over a short time, nonradioactive probes can be stored frozen over long periods with no loss in activity. Nonradioactive probes have supplanted radioactive ones for labeling blots and in situ hybridization of tissue sections on slides. In addition, they can be used for in situ hybridization of whole tissues or small organisms, eliminating the need for sectioning and providing a clearer overall three-dimensional picture of gene expression.

With careful preparation, one probe can penetrate intact embryos and small organisms, such as *C. elegans*. Dissected tissues are fixed in paraformaldehyde and treated with a proteinase and a detergent to help break down cellular structures and membranes, allowing better penetration of the probe. After incubation with the probe, the tissues are washed and then soaked in a solution containing the substrate needed for visualization. After observing expression in the whole tissue, sectioning can then be performed to look at expression on a cellular level.

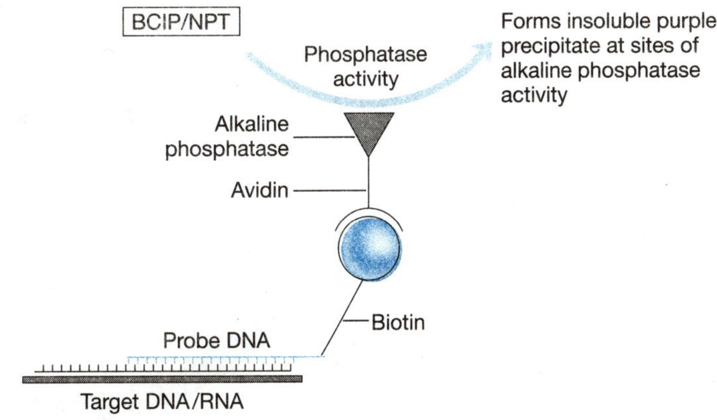

Detection of Biotinylated Nucleic Acid Probes with BCIP/NPT

Protein Analysis

While DNA provides the instructions, and mRNA is the messenger molecule, proteins are the effector molecules of the cell. To understand the structural and enzymatic processes of the cell, we must understand the protein molecules that carry out these functions. DNA cloning technology has given us a means for investigating proteins.

Making Proteins by Expressing Foreign Genes in E. coli *and Other Systems.* One of the great achievements of DNA cloning is the ability to produce large amounts of a specific protein by expressing specific genes at high levels in *E. coli* or other cells. To express a human protein in *E. coli,* we start with the isolation of a full-length double-stranded cDNA. The next step is to insert the double-stranded cDNA into an expression vector, which is tailored to function in *E. coli*—regardless of the original source of the gene. To express the protein in *E. coli,* the cDNA must be transcribed from an *E. coli* promoter. In addition, all bacterial genes contain a sequence recognized by the ribosome for translation known as the Shine-Delgarno sequence (AGGAGG). To obtain the highest level of protein production, we select a "strong" promoter, such as the promoter from bacteriophage T7. The T7 promoter requires the presence of functional RNA polymerase from T7 phage, and thus the host cell used for protein expression must contain the T7 RNA polymerase gene. In a standard expression vector, the Shine-Delgarno sequence is placed downstream (on the 3′ side) from the T7 promoter. Next to the Shine-Delgarno sequence are an ATG translation initiation sequence and, finally, a polylinker (a series of side-by-side restriction recognition sequences for cloning in our cDNA).

In addition to these elements for protein expression, the vector may also contain tags, which are sequences that facilitate the isolation of the expressed protein. Tags are usually short DNA sequences that encode a specific peptide

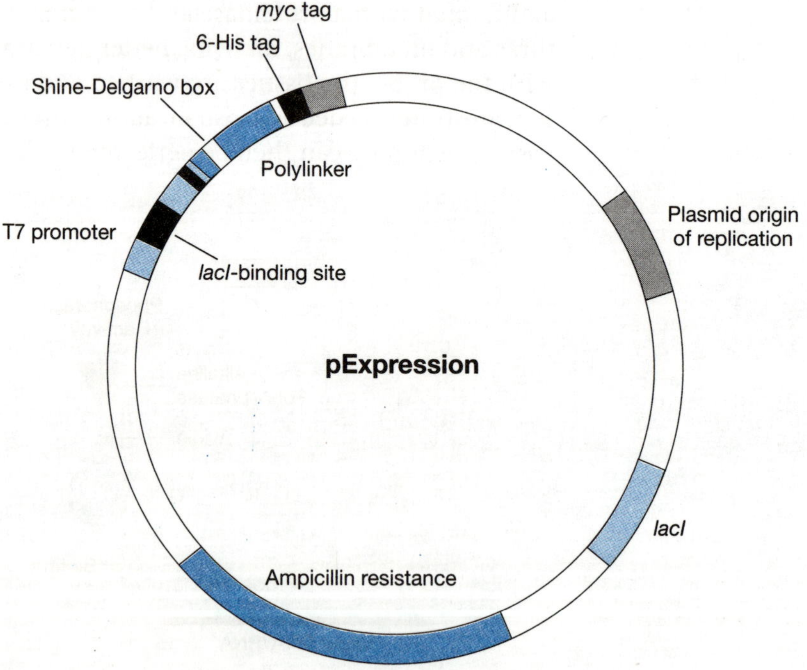

A Generic Expression Plasmid and Its Key Elements

that is translated as part of the expressed protein. One type of tag binds to a specific compound that can be immobilized on a bead and used to affinity-purify the expressed protein. For example, a 6-His tag is a sequence of six consecutive histidines that creates a site that binds nickel or other metals. Extract from cells expressing a 6-His-tagged protein is passed over a column containing nickel. The 6-His protein binds to the column and the other proteins pass through. The 6-His-tagged protein is eluted from the column with imidizol which mimics histidine and competes for nickel binding, thus releasing the protein.

Another type of tag is an epitope—a peptide sequence that is recognized and bound by a specific antibody. One epitope tag is the *myc* tag, a 10-amino-acid sequence (EQKLISEEDL) from the *myc* oncogene. This tag is readily bound by anti-*myc* antibodies.

Many proteins are toxic to the host cell, especially when expressed at high levels. One way to safeguard against this problem is to create an inducible promoter so that the protein is not expressed until the cells are grown to an appro-

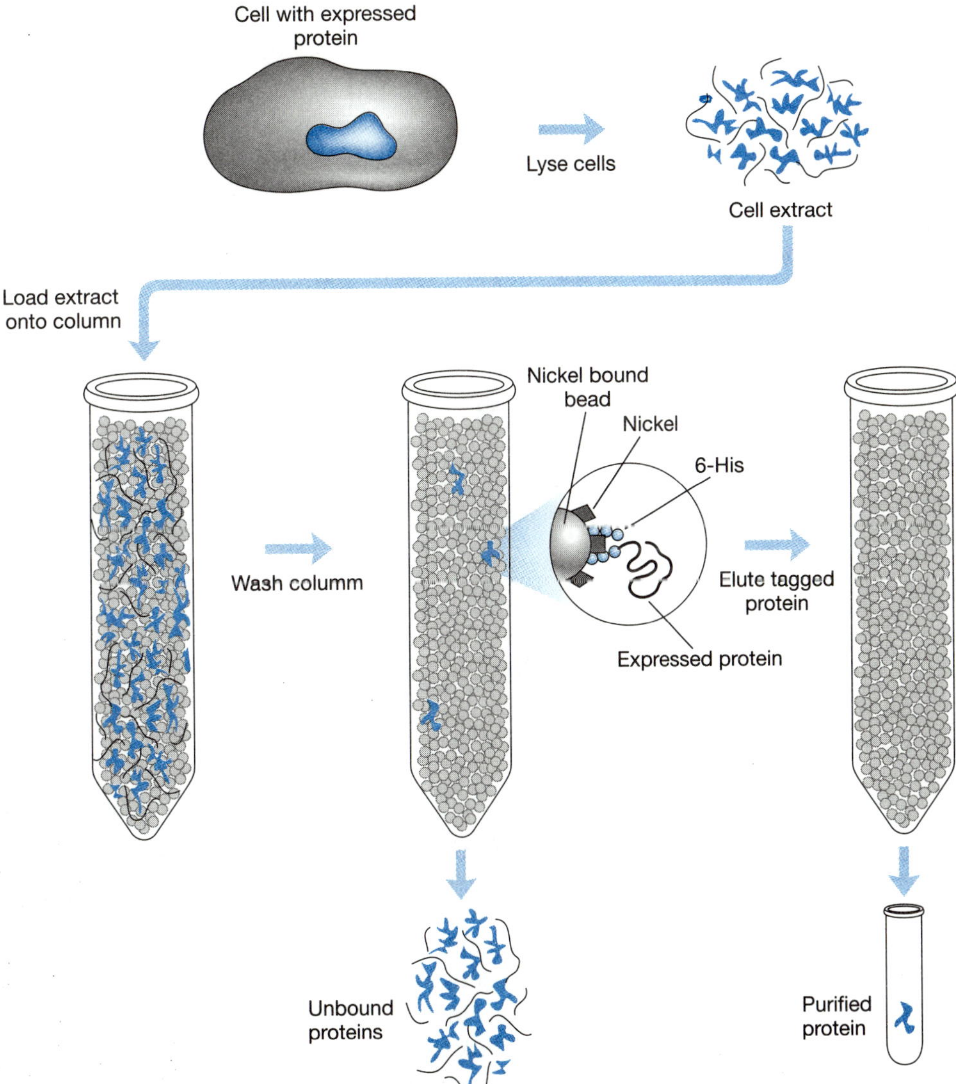

Purifying 6-His-tagged Protein on a Nickel Column

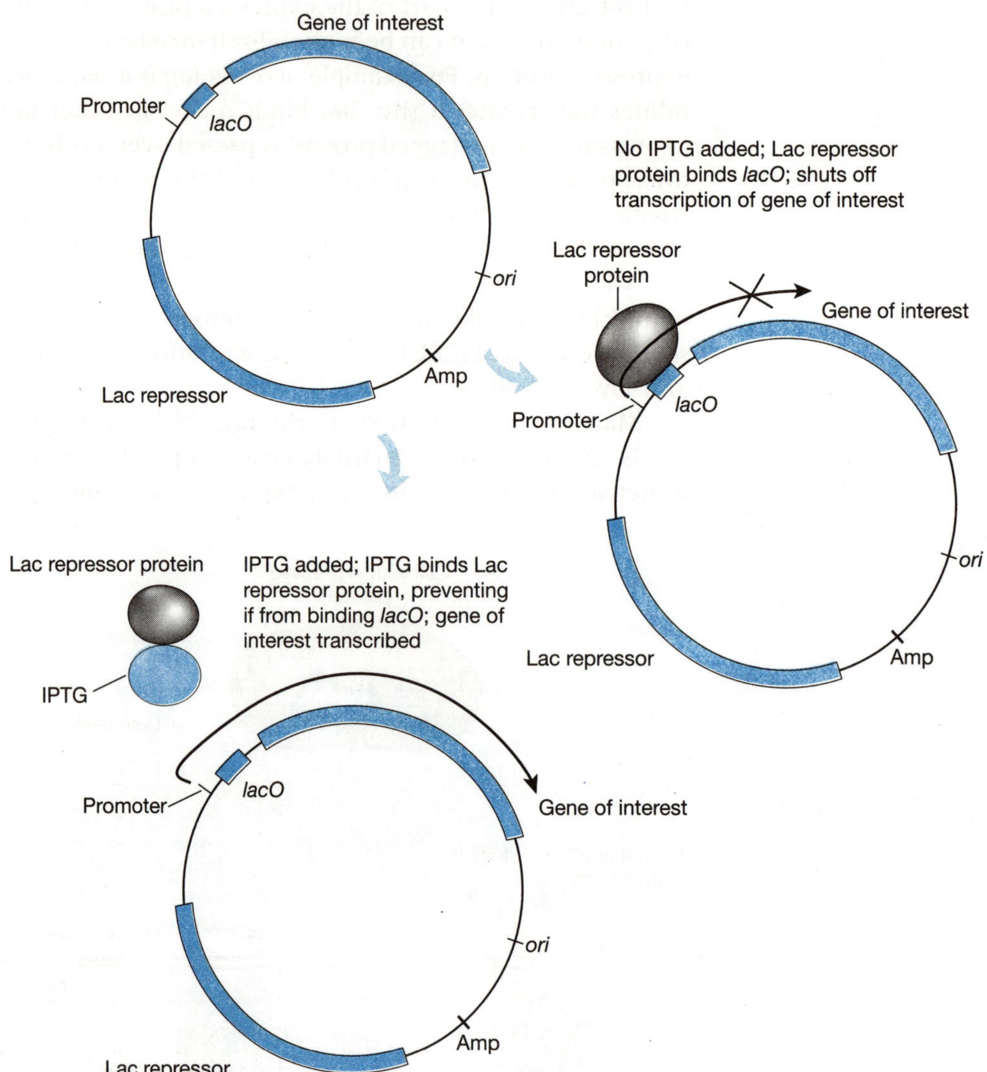

Expressing a Protein Using an Inducible Promoter

priate density. Typically, an inducible promoter is created by inserting a *lacO* sequence downstream from the promoter. A *lac repressor* gene is present either in the plasmid or in the genome of the host cell and expresses Lac repressor protein, which binds the *lacO* sequence and prevents transcription. The addition of the lactose analog IPTG (isopropyl-β-D-thiogalactopyranoside), inhibits Lac repressor protein and initiates transcription of the cloned gene. After induction, cells are collected and lysed, and the protein is purified chromatographically, using a tagged sequence.

Proteins can also be expressed in yeast or mammalian cells, provided the vector contains an appropriate species-specific promoter sequence. In addition, there is a consensus sequence at the translation start site known as the Kozak sequence that includes the ATG initiation sequence ATG (GCCACCATGG). Expressing a eukaryotic gene in a eukaryotic cell system ensures that the proper posttranslational modifications will be made to ensure a biologically active polypeptide.

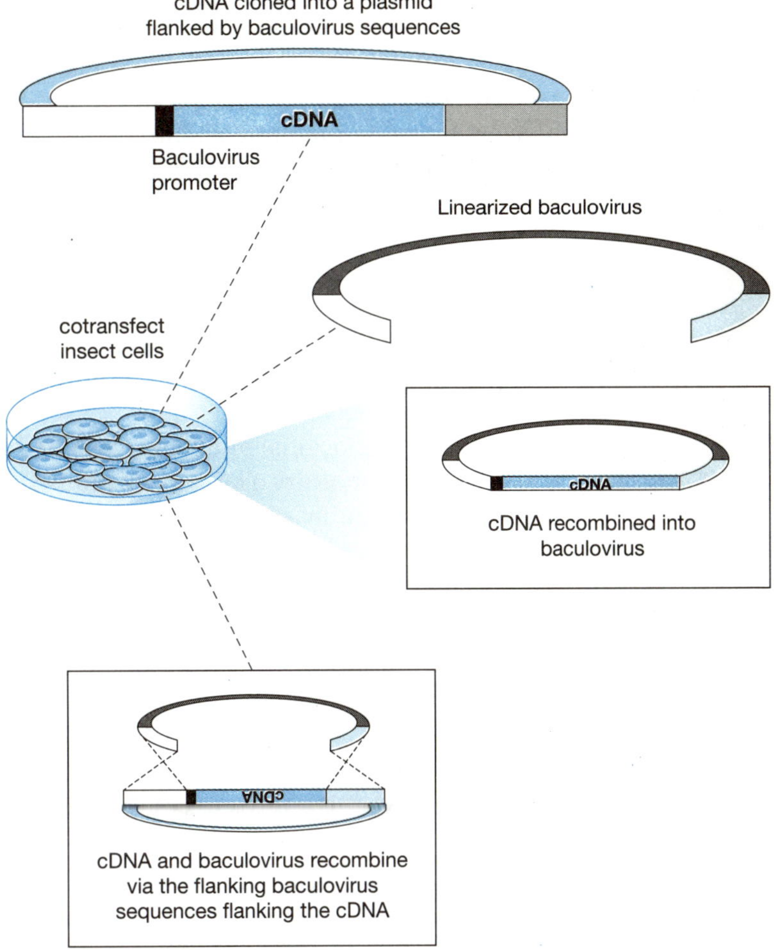

The cDNA Gene to be Expressed Is Introduced into the Baculovirus by Recombination

Another favored expression system uses insect cells, which expresses many proteins at high levels. In addition, mammalian proteins expressed in insect cells are usually correctly folded and modified. In this system, an insect virus, commonly called baculovirus, from *Autographa californica* nucleopolyhedrovirus (AcNPV) is used as the host. Recombinant baculoviruses are usually obtained from a natural host, the larva of *Trichoplusia ni*, a moth common throughout North America. The DNA sequence of the protein to be expressed is cloned into a plasmid, with flanking baculovirus sequences. Linearized baculovirus DNA and the plasmid are cotransfected into cells where the inserted sequence recombines with the baculovirus, creating a recombinant virus. This virus is then propagated and used to infect cells to produce large amounts of protein.

Using Antibodies and Marker Proteins

Antibodies are formed as part of the immune response to the presence of foreign substances (antigens). Exposure to an antigen stimulates B lymphocytes to produce antibodies (immunoglobulins) that bind to the antigen and facilitate its destruction. Each antibody recognizes a different surface feature (an epitope) of the antigen molecule consisting of as few as several amino acids. A clone of lymphocyte cells secretes one antibody specific for one epitope.

César Milstein, ca. 1976 (Courtesy of Cold Spring Harbor Laboratory Archives.)

For years, scientists have produced antibodies directed against specific proteins by injecting the proteins into experimental animals such as mice and rabbits. The antibody-rich serum collected from an immunized animal contains a mixture of antibodies that are directed against numerous epitopes of the injected protein, as well as against other antigens to which the animal has been exposed. Thus, serum used from an immunized animal is referred to as *polyclonal* antibody. Working at the British Medical Research Council (MRC) Laboratory of Molecular Biology, in 1976, César Milstein developed hybridoma technology that provided a means to make pure preparations of *monoclonal* antibodies with single-epitope specificity.

To make monoclonal antibodies against human actin, for example, the protein is used to immunize a laboratory mouse. After a few weeks, when the mouse has mounted a sufficient immune response, it is sacrificed, and B-lymphocyte cells from its spleen are mixed with cultured mouse myeloma cells. The myeloma cells used are immortalized cells, derived from a mouse bone cancer, that live essentially forever in culture. The addition of polyethylene glycol (PEG) encourages fusion of the two cell types to form *hybridomas*. The myeloma confers on the

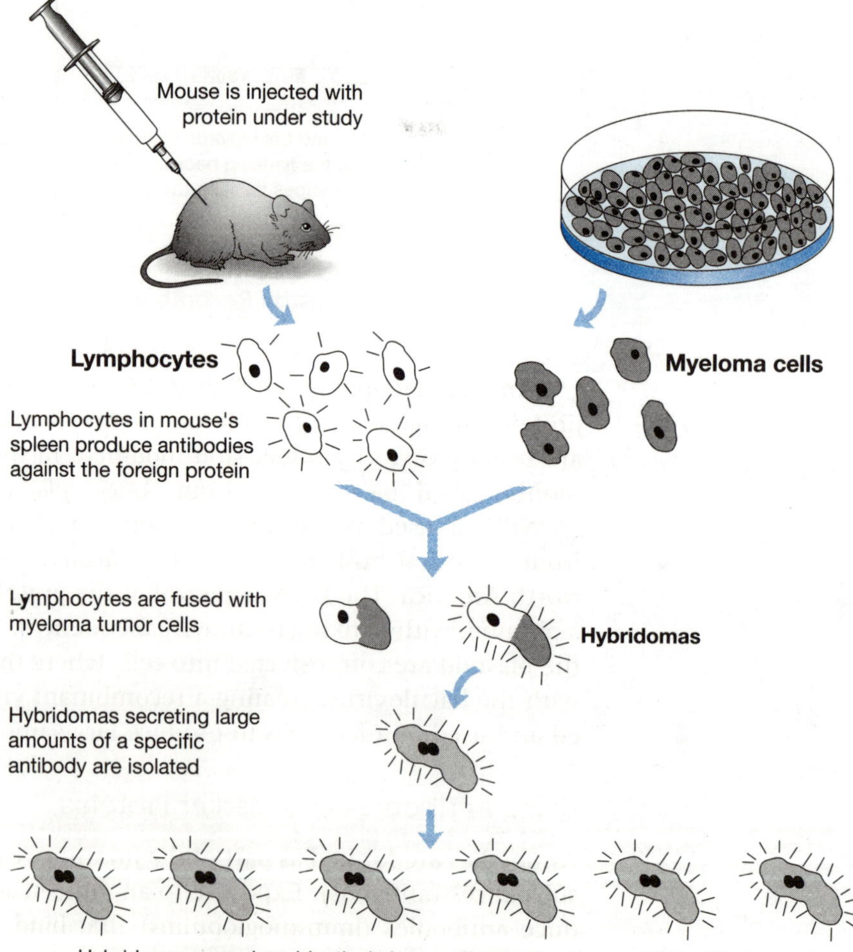

Making Monoclonal Antibodies

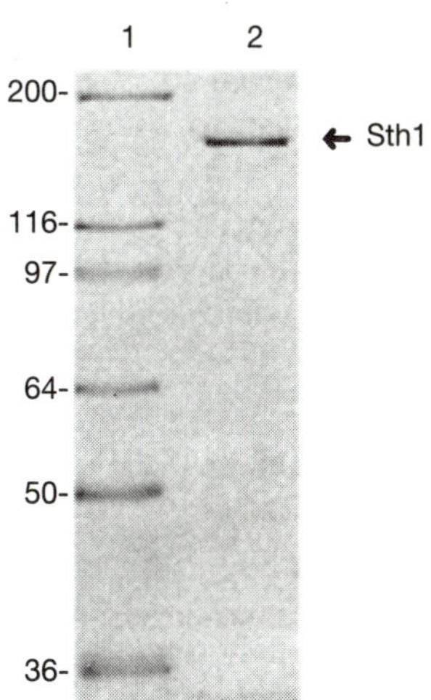

Western Blot
The band in lane 2 shows the presence of the Sth1 protein in yeast cells. In this case, a His-tagged Sth1 protein was produced at high levels using an expression vector. Lane 1 shows molecular weight markers. (Reprinted, with permission, from Saha A., Wittmeyer J., and Cairns B.R. 2002. Chromatin remodeling by RSC involves ATP-dependent DNA translocation. *Genes Dev. 16:* 2120–2134.)

hybridoma the capacity for unlimited reproduction in culture, whereas the B lymphocyte confers the ability to synthesize a specific anti-actin antibody.

The cell mixture is propagated in a selective medium in which only fused cells can survive. The fused cells are then diluted to isolate individual hybridomas, which multiply to produce a pure colony of clones. This collection of hybridomas is screened to identify clones that produce the antibody of interest—by their reaction with the antigen. Positive clones are further screened to identify those producing the high-affinity antibodies which bind strongly to the antigen. The progeny clones of the single ancestral hybridoma all produce identical antibodies of the same specificity, hence the term monoclonal antibody. In this way, antibodies against specific epitopes of an antigen can be purified.

Uses for Antibodies. Antibodies have many uses in scientific research, as well as for medical treatments. One example is the production of an antibody that recognizes cell-surface proteins present in tumor cells, such as the monoclonal antibody known as herceptin. This antibody specifically recognizes HER2 (human epidermal growth factor receptor 2) found on the surface of many breast cancer cells. The antibody binds to the cell, blocking the signal for cell growth and providing a new tool to treat breast cancer.

Methods for Studying Proteins: Western Blots and Immunocytochemistry. Western blots (named to continue the series after Southern and northern blots were developed) utilize the specific binding of antibodies to proteins, much like Southern and northern blots utilize the specific hybridization of probes made from complementary nucleotide sequences. First, proteins are separated by gel electrophoresis and transferred to a nylon or nitrocellulose filter. The filter is then incubated with an IgG antibody that specifically binds to a protein of interest. This "primary" antibody may be a monoclonal or a polyclonal antibody, typically made in a mouse or rabbit. Excess primary antibody is washed away, and a "secondary" antibody is added that binds to a constant region of all antibodies produced by a species. (For example, rabbit IgG can be injected into goats and used to generate goat antibodies that will recognize all rabbit IgG molecules.)

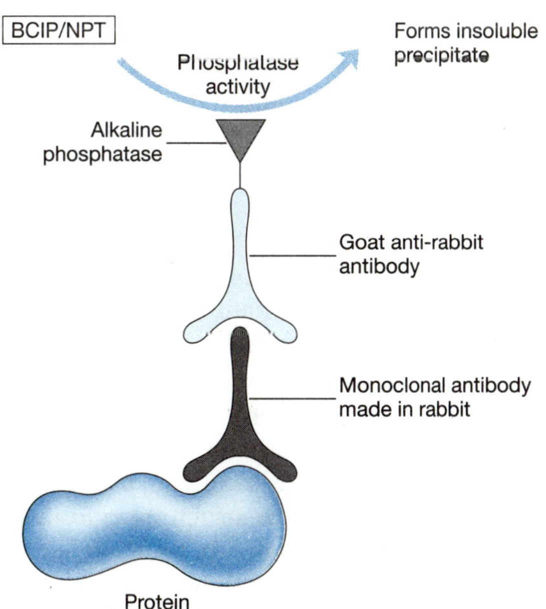

Detection of Antibody Probes with BCIP/NPT

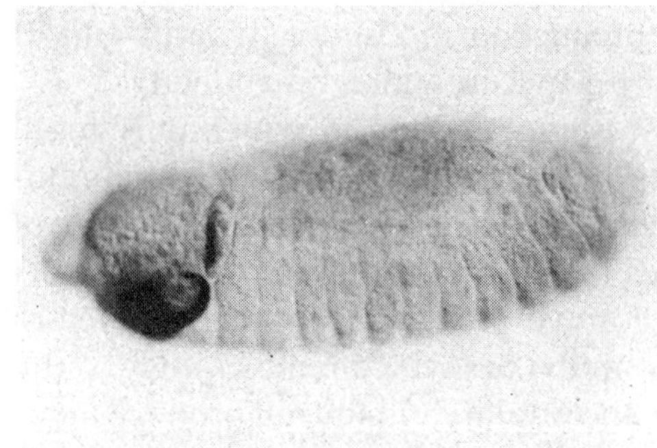

Immunocytochemistry Used to Locate the Deformed Protein in a *Drosophila* Embryo

Staining reveals the presence of antibodies against Deformed binding to the protein in cells in the maxillary segment of the developing head. (Reprinted, with permission, from Jack T., Regulski M., and McGinnis W. 1988. Pair-rule segmentation genes regulate the expression of the homeotic selector gene, *Deformed. Genes Dev. 2:* 635–651.)

This secondary antibody is covalently fused to a reporter enzyme, usually alkaline phosphatase (AP) or horseradish peroxidase (HRP). Excess secondary antibodies are washed from the filter, and an appropriate substrate is added that will react with the reporter enzyme to produce a detectable color or fluorescent compound.

Immunocytochemistry is the protein equivalent of in situ hybridization. In this procedure, antibodies are introduced into fixed tissues and cells. These primary antibodies are then reacted with secondary antibodies containing a reporter enzyme, which identifies the location of the protein of interest within a tissue or a cell.

Green Fluorescent Protein (GFP). The methods described thus far provide static assays for gene and protein activity. Most of the compounds used for visualization, whether radioactive or nonradioactive, are toxic to living cells. The methods for getting the probes into cells are often destructive as well. Because of this, much of our knowledge about where proteins are translated and where they localize in the cell comes from studies of fixed, dead tissues. To study changes in expression and activity over time, a nondestructive, nontoxic method for visualizing gene activity is needed. Ideally, this method would consist of a naturally occurring protein that can be expressed in many cell types and provide a simple method for indicating its presence.

The jellyfish *Aequorea victoria* provided a solution to this dilemma. In the early 1960s, Osamu Shimomura and Frank Johnson at Princeton University collected specimens of jellyfish in studies to understand their "bioluminescence," One of the compounds they discovered was named green fluorescent protein (GFP) because it glowed bright green under UV light. Many years later, in 1992, Douglas Prasher at the Woods Hole Oceanographic Institute cloned a cDNA for GFP. In 1994, Prasher and Martin Chalfie at Columbia University were the first to realize the potential for use of GFP as a reporter molecule.

GFP is a naturally occurring protein of 238 amino acids. The coding sequence for GFP can easily be cloned into a variety of vectors, and GFP maintains its fluorescence when expressed in cells from different species. The expression of GFP does not appear to have toxic effects, so it can be followed in living cells over time. The protein can even be directed to specific subcellular regions, allowing observation of individual parts of the cell, such as mitochondria, cell membranes, nuclei, or the structural cytoskeleton. Fluorescent molecules from other species have subsequently been cloned, including YFP (yellow fluorescent protein), CFP (cyan, or blue fluorescent protein), and RFP (red fluorescent protein).

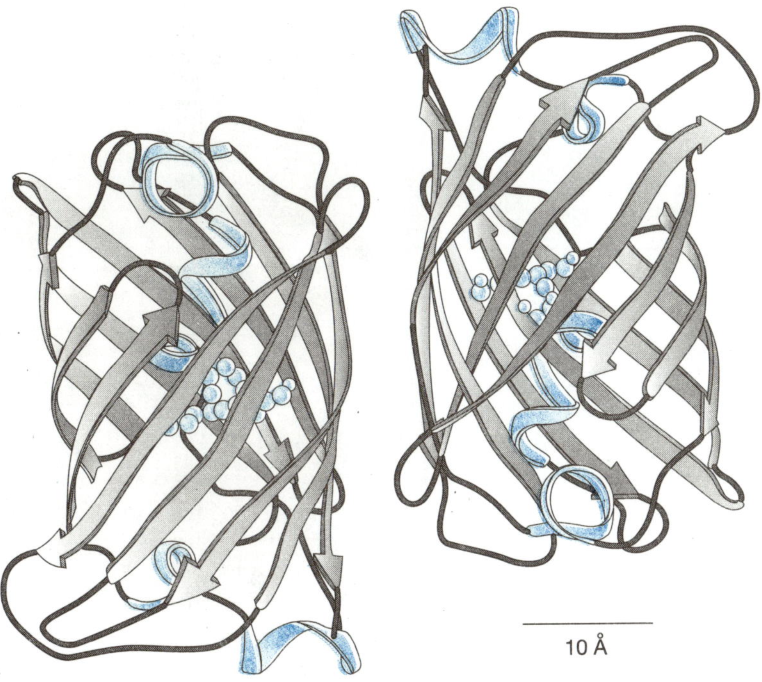

Structure of Green Fluorescent Protein
GFP forms a cylindrical structure called a "β-can." The fluorophore is in the center of the can. (Adapted, with permission, from Yang F., Moss L.G., and Phillips G.N. Jr. 1996. The molecular structure of green fluorescent protein. *Nat. Biotechnol. 14:* 1246–1251.)

GFP has many uses in biological studies. Inserting a simple GFP expression vector allows one to mark a particular cell with a visual signal. For example, a GFP-containing expression vector can be inserted into nerve cells in the developing zebrafish via electroporation. Under a fluorescence microscope, these cells can now be observed as they grow and migrate throughout the body. A recom-

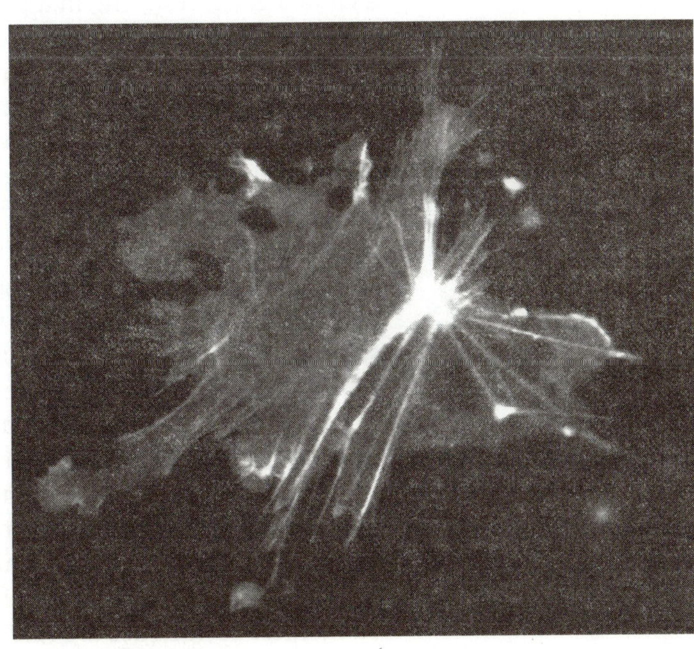

GFP in a Living Neural Crest Cell in Culture
In this case, GFP has been localized to actin, which is a key structural component of the cell. (Courtesy of Andrew Ewald, California Institute of Technology.)

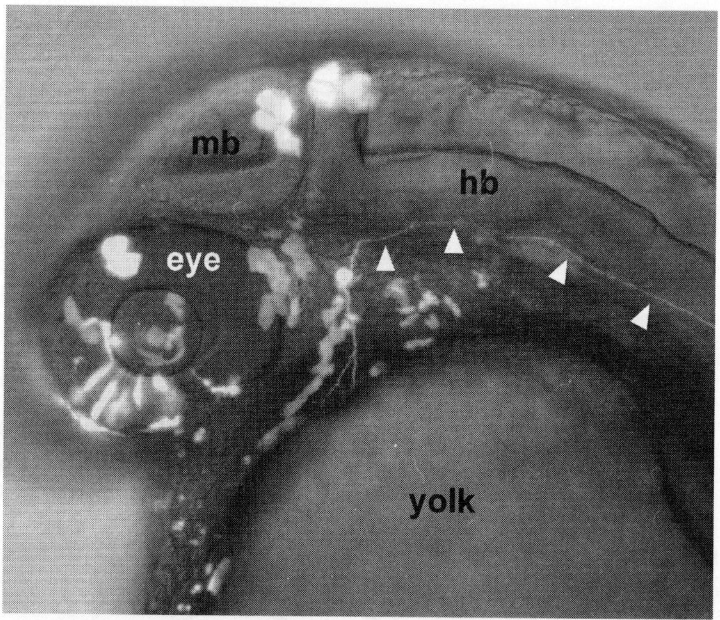

GFP in a Living Zebrafish Embryo
GFP was introduced into individual cells to visualize cell morphology and cell behavior during development. This enables axons to be followed as they emerge from developing neurons and project into other body regions for communication purposes (arrowheads). (mb) Midbrain; (hb) hindbrain. (Courtesy of Reinhard Köster, California Institute of Technology.)

binant DNA construct can be made that combines the regulatory regions for a gene of interest with the coding region of GFP. Insertion of this construct into a living cell provides a tool for watching changes in gene expression over time. Since GFP is under the control of the gene of interest's regulatory sequences, GFP will only be made when and where that gene's product is normally made. In addition, the GFP-coding sequence can be recombined with the sequence of a cloned gene, resulting in a fusion protein. The gene for this fusion protein can now be introduced into living cells. The localization, turnover, and intracellular associations of this GFP fusion protein can provide valuable clues about the function of the native protein.

Putting It All Together: Transgenic Animals

Molecular genetics is a reductionist science. By breaking the larger questions down into simpler, more straightforward pieces, molecular biologists have been able to learn a great deal about the machinery of life. However, a broader understanding cannot be gleaned until these smaller pieces are reassembled into a whole. When an isolated molecule is observed in a test tube, one cannot fully understand how it behaves as part of a large, interacting system. Because of this, it is necessary to take the lessons learned in vitro and see how they hold up in vivo—in a living organism. Chapter 4 discussed methods of creating recombinant DNA molecules and expressing them in single-celled organisms, such as bacteria and yeast. These techniques have been adapted for use in higher, more complex organisms. Animals and plants that carry new genetic material inserted by these techniques are referred to as "transgenic." Most model systems have

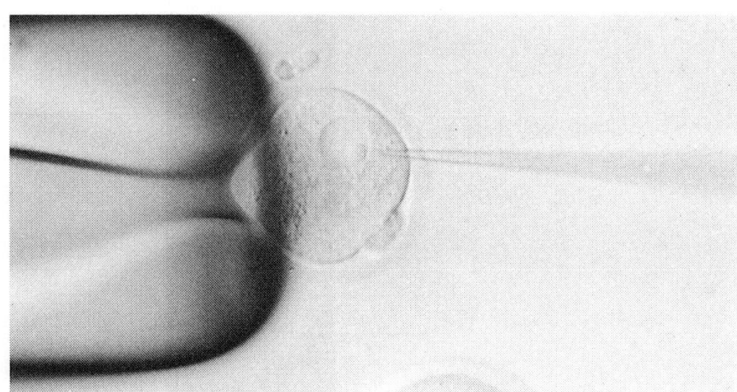

Microinjection of a One-celled Mouse Embryo
(Courtesy of Mark Steinhelper.)

proven amenable to carrying engineered transgenes. Mice have proven to be the transgenic animal of choice, since they are relatively easy to rear and provide a valuable model for studying human disease.

Transgenic Mice. The groundwork for making transgenic mice was laid long before the invention of recombinant DNA techniques. Methods for isolating early mouse embryos called "blastocysts" were developed in the 1950s. Anne McLaren, working in Edinburgh, optimized methods for transplanting these isolated embryos back into foster mothers, where they complete development. In 1966, Teh Ping Lin at the University of California, San Francisco, showed that mouse eggs can survive having small quantities of liquid injected into them with a fine glass needle. Rudolph Jaenisch at the Salk Institute and Beatrice Mintz at the Fox Chase Cancer Center used these methods to make the first "transgenic" mouse in 1974. They injected purified DNA from a simian virus (SV40) into mouse blastocysts. When the resulting mice were raised to adulthood, SV40 DNA could be detected in their tissues, suggesting that the injected DNA had integrated into their genome.

Jon Gordon and Frank Ruddle at Yale University introduced a cloned gene into fertilized mouse eggs via microinjection in 1980 and showed stable integration of the injected gene into somatic cells. Within a year, several other groups used this technique to prove that transgenes not only integrate into the host's genome, but also are expressed and passed on to offspring through germ cells. The integration into germ cells is particularly important, because it allows one to mate mice to produce large numbers of offspring carrying a particular transgene.

Transgenic mice have become a valuable tool for studying the activity of normal and mutated genes in a complex living organism. Scientists could microinject foreign genes and see their effects on the resulting mice. Mutated versions of normal mouse genes could also be tested. The use of DNA microinjection to make transgenic animals has since been successfully applied to many other species—including fish, birds, pigs, sheep, rats, and rabbits.

One drawback to this method is the random integration of the transgenic DNA in the host genome. It is not possible to predict where or how many copies of the transgene will integrate. Sometimes, a transgene may integrate in the middle of an endogenous gene, disrupting it; other times, the foreign DNA may integrate next to a powerful regulatory element, coming under its control. These issues make interpretation of transgenic mouse experiments more difficult. Is the

Chimeric Mouse
The mouse on the left is a chimera. This can be seen by the presence of dark fur derived from the injected ES cells. (Courtesy of Andras Nagy, Mount Sinai Hospital, Toronto.)

Martin Evans

Matthew Kaufman

Gail Martin

effect seen due to the protein coded by the transgene or is it due to the alteration of another gene when the transgene integrated? Although microinjection is a useful method for introducing genes into an embryo (gain-of-function experiments), it does not allow for removing, or turning off, resident genes (loss-of-function experiments). This requires methods for gene targeting.

Chimeras, ES Cells, and Homologous Recombination. The Chimera is a mythical animal, made up of parts of a lion, a goat, and a snake. The word chimera is now used to refer to an animal made from two or more distinct embryonic sources. Cells from different embryos can be combined, resulting in one mixed embryo. If a chimera is made by combining the embryos from an albino mouse and one with darkly colored hair, the resulting pup will be a mix of the two. Areas of the pup derived from the albino embryo cells will be white, whereas areas from the dark embryo cells will be dark. In 1968, Richard Gardner showed that chimeras can be made by injecting isolated cells into host blastocysts.

In 1981, Martin Evans and Matthew Kaufman at Cambridge University and Gail Martin at the University of California, San Francisco, independently produced lines of mouse embryonic stem (ES) cells directly from cultured blastocysts. These undifferentiated cells, which are derived from the inner cell mass of a developing blastocyst, are *pluripotent*—that is, they have the ability to develop and differentiate into a variety of tissue and cell types. If injected into mouse blastocysts, ES cells are readily taken up and contribute to the development of many tissues, especially germ cells. A key benefit of ES cells is that they can be grown in cell culture like bacteria or yeast, allowing for genomic manipulation and use of selectable markers. Working with Evans and Kaufman, Elizabeth Robertson and Allan Bradley showed that ES cells can be genetically manipulated using retroviruses and that these transformed cells could be introduced into mice and passed on through generations.

In 1987, Mario Capecchi, at the University of Utah, and Oliver Smithies, at the University of North Carolina, independently developed "gene targeting" as a means to precisely introduce novel genetic changes into mouse ES cells. They took advantage of homologous recombination, the mechanism through which chromatids "cross over" to exchange like (homologous) regions during meiosis. Although homologous recombination is common in germ cells, it occurs at a very low frequency in somatic (body) cells. Thus, a foreign DNA sequence usually integrates randomly into a host genome, but it occasionally integrates at a homologous sequence. Smithies and Capecchi pioneered means to select and enrich for cells in which homologous recombination deletes a specific gene (termed a "knock-out") or adds an engineered copy of a gene ("a knock-in").

Making A Knock-out Mouse. To create a mouse knock-out, a targeting vector is constructed by inserting a portion of the target gene into plasmid backbone, along with a positive selection marker (*neomycin-resistance* gene, *neo*) and a nega-tive selection marker (the *thymidine kinase* gene, *tk*, from the herpes simplex virus). The flanking portions of the transgene maintain strict homology with the genomic target sequence; however, deletions in the central portion of the target sequence are designed to abolish expression of a functional protein. The *neo* gene is inserted *within the target sequence* (this alone is often enough to disable the gene), and the *tk* gene is added to a nonhomologous region *outside the target sequence*.

The targeting vector is transformed into *E. coli* cells and amplified in culture. The purified vector is then cut once with a restriction enzyme, and the linearized plasmid is transformed into cultured ES cells by electroporation. A two-stage selection then identifies ES cells that have correctly integrated the transgene. Only cells that have taken up the targeting construct containing the *neo* resistance gene survive an initial positive selection with the neomycin analog G418. Then, negative selection with the antiviral drug gancyclovir kills cells that have randomly integrated the *tk* gene along with the transgene. During targeted insertion, the homologous flanking regions of vector and the host chromosome align and exchange, excluding the nonhomologous *tk* gene. Thus, only cells with a homologously inserted transgene survive gancyclovir, and continued selection enriches for these ES cells.

In the next step, blastocysts are harvested from a pregnant mouse. A number of ES cells containing the targeted transgene are injected into the blastocyst, where they develop along with the host cells. After a brief period in culture, several chimeric blastocysts are implanted into a surrogate mother. About half of the surrogate mothers will give birth to chimeric offspring, which are a patchwork of cells derived from the host blastocyst and ES cells.

A chimeric pup is, in fact, derived from four parents, whose coat colors are carefully selected to visibly show their contributions. ES cells are usually derived from mating black parents, and blastocysts are the product of two albino parents. The transgenic chimeras have black and white patches of hair, and they can be quickly assessed by the amount of black hair their coats contain. The more black, the larger the contribution of the ES cells.

Producing a chimeric pup is only the first step of creating a gene knock-out. Some of the chimeric pups have sex cells derived from the ES cells. These are from black mice and so will have a gene encoding black coat color. Breeding the chimeras to albino mice produces some black mice with the ES genome in their germ plasm. Because the chimeric mouse produces two types of sex cells with the black coat gene—one with the transgene and one without—black mice are then screened to show which carry the *neo* gene insert. Although Southern blots were used for genotyping in original experiments, PCR is more widely used today.

Each of the positive mice is heterozygous for the gene knock-out, because the targeting construct was integrated into only one of the paired chromosomes during the initial manipulation of the ES cells. However, if two heterozygous mice are mated, 25% of their offspring will be homozygous for the knock-out. This is termed a null mutation. These homozygous mice are then screened for bio-

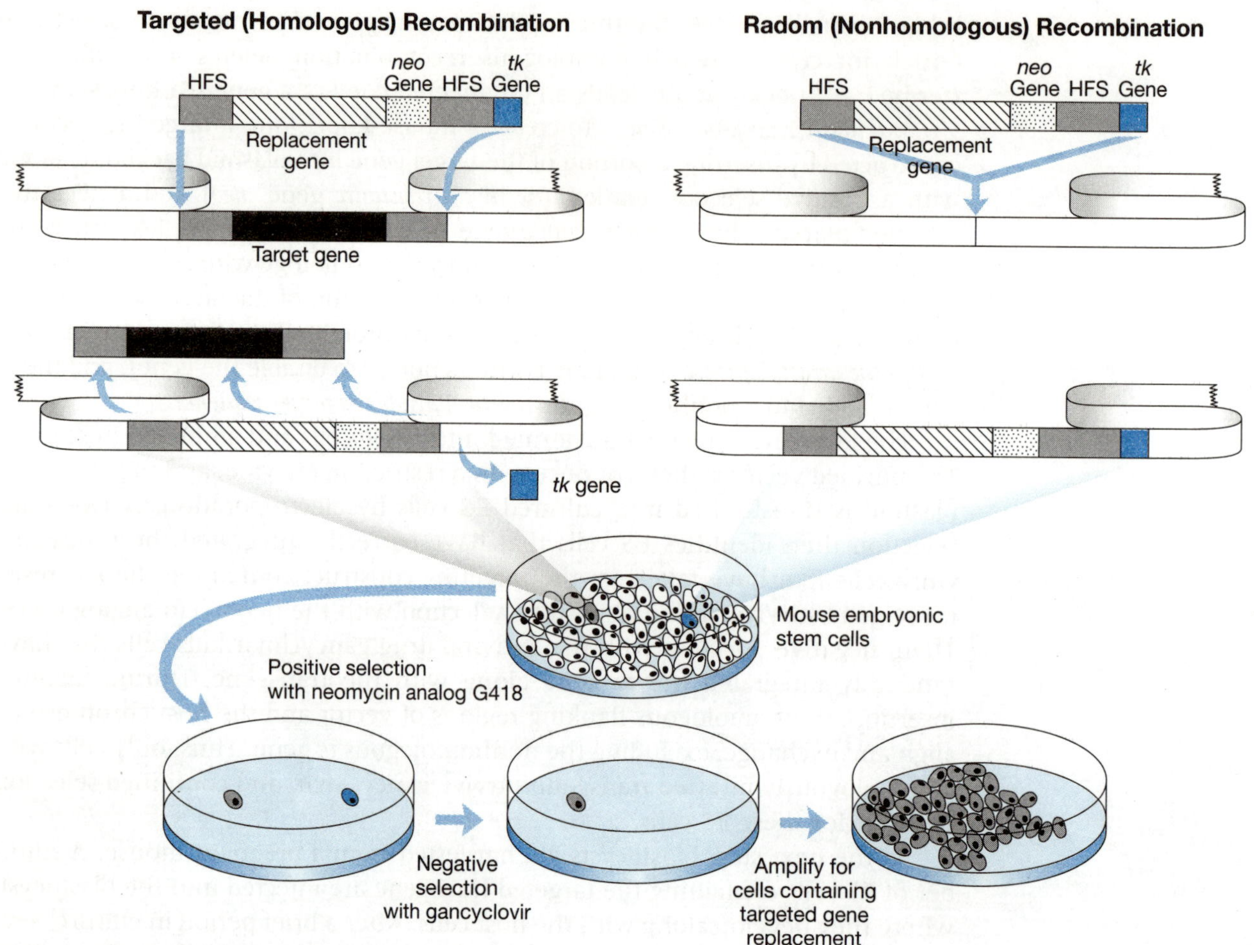

Selection of Correctly Targeted ES Cells

The replacement gene construct contains two marker genes, *neomycin resistance* (*neo*) and *thymidine kinase* (*tk*) and is bordered by the homologous flanking sequence (HFS). Following electroporation, the replacement gene almost always integrates randomly into the genome (at *right*). These cells integrate the *tk* gene and are sensitive to gancyclovir. Rarely, the replacement gene integrates specifically at the site of its cellular homolog, replacing the endogenous gene (at *left*). These cells do not integrate the *tk* gene. Although every electroporated cell is neomycin-resistant, only those where the gene recombines homologously are gancyclovir-resistant.

chemical, developmental, anatomical, or behavioral differences (heterozygotes may have intermediate phenotypes).

In addition to the *neo* gene, scientists often find it useful to knock-in a marker protein, like GFP. This gives a second tool for analysis from one experiment. Sequences coding for GFP can be inserted so they are regulated by elements that normally control expression of the endogenous gene. Mice that are heterozygous for the transgene will still produce the normal protein from the one copy that is left and produce GFP from the transgene. Cells where the gene is active can now be observed through fluorescence microscopy.

Example of a Knock-out Mouse Study. In the early 1990s, Argiris Efstratiadis and Elizabeth Robertson at Columbia University used mouse knock-outs to study the effects of insulin-like growth factors (IGFs). Mice homozygous for the *IGF2* knock-out transgene showed a startling phenotype. Their growth was severely retarded, and they reached only 60% of the size of their normal littermates.

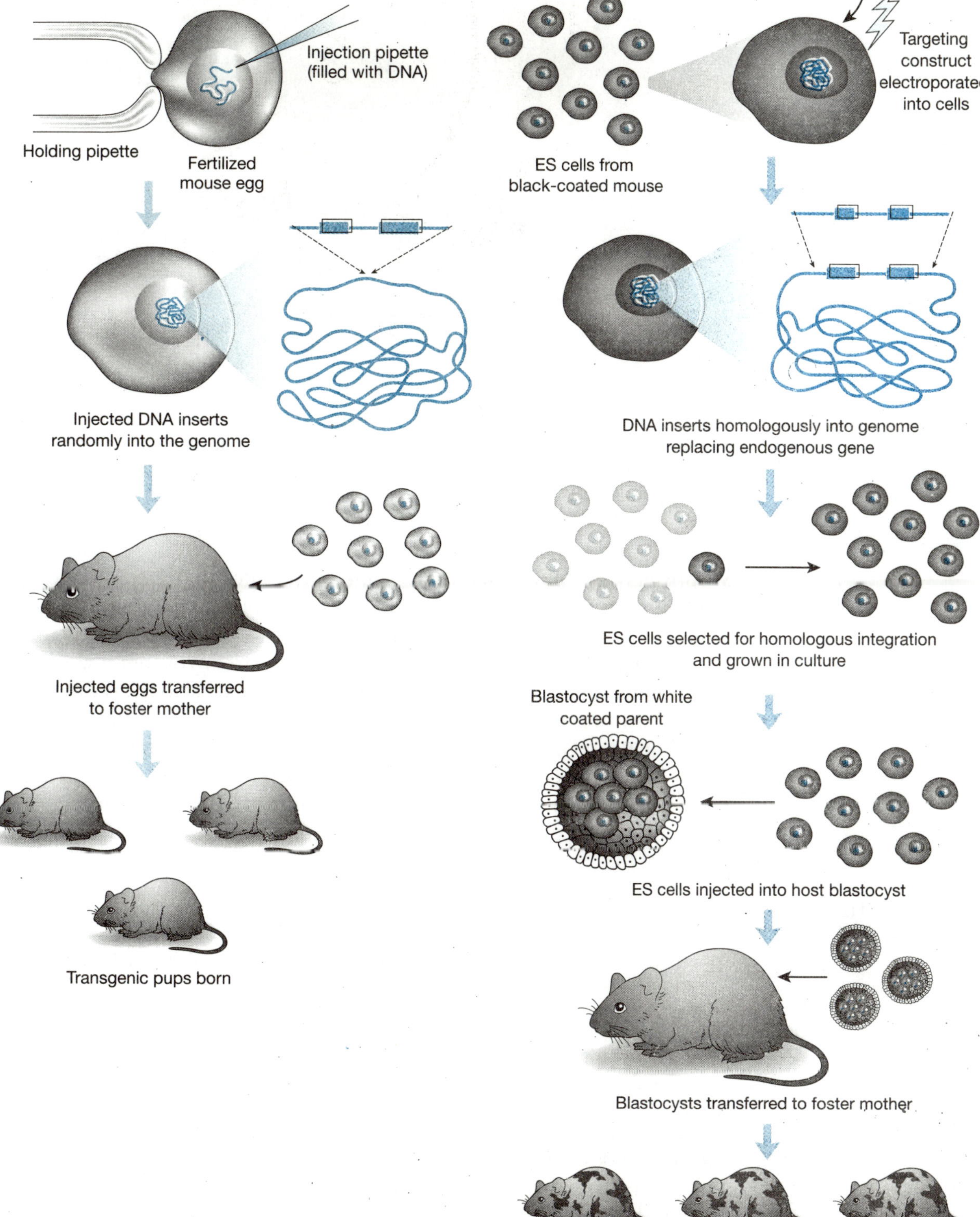

Making a Transgenic Mouse

Transgenic Dwarf Mouse Lacking the *IGF2* Gene
The smaller mouse on the bottom is a heterozygote, inheriting paternally a null *Igf2* allele, and the normal-sized mouse on the top is a wild-type littermate. (Courtesy of Argiris Efstratiadis, Columbia University.)

Surprisingly, in later generations, some of the mice that were heterozygous for the transgene also showed the growth-retarded phenotype. If the transgene was inherited from the male parent, the mouse was small, but if it was inherited from the female parent, the mouse was normal-sized. This phenomenon, where the parental source of a gene makes a difference in its activity, is called "imprinting." Thus, one experiment informed us not only about what the *IGF2* gene does, but also something about how it is regulated, and that there are differences in the activity of genes inherited from each parent. A series of later experiments knocking out related genes have shown that the *IGF* gene family is a major regulator of growth during mammalian development.

FOR FURTHER READING

Alberts B., Johnson A., Lewis J., Raff M., Roberts K., and Walter P. 2002. *Molecular biology of the cell,* 4th Edition. Garland Publishing, New York.

Lodish H., Berk A., Zipursky S.L., Matsudaira P., Baltimore D., and Darnell J. 2000. *Molecular cell biology,* 4th Edition. W.H. Freeman, New York.

Micklos D., ed. 1999. DNA from the beginning (http://www.dnaftb.org). Dolan DNA Learning Center.

Nagy A., Gertenstein M., Vintersten K., and Behringer R.R. 2003. *Manipulating the mouse embryo,* 3rd Edition. Cold Spring Harbor Laboratory, Cold Spring Harbor, New York.

Sambrook J. and Russell D. 2001. *Molecular cloning: A laboratory manual,* 3rd. Edition. Cold Spring Harbor Laboratory Press, Cold Spring Harbor, New York.

Spector D.L., Goldman R., and Leinwand L., eds. 1998. *Cells: A laboratory manual,* Volumes I–III. Cold Spring Harbor Laboratory Press, Cold Spring Harbor, New York.

Watson J.D., Hopkins N.H., Roberts J.W., Steitz J.A., and Weiner A.M. 1987. *Molecular biology of the gene,* 4th Edition. Benjamin/Cummings, Menlo Park, California.

CHAPTER 6

Modern Methods for Analyzing Whole Genomes

THE PUBLICATION OF THE FIRST DRAFT OF THE DNA SEQUENCE of the human genome in February 2001 was the culmination of the biological revolution that began with Watson and Crick's discovery of the DNA structure half a century earlier. The Human Genome Project is the most ambitious experiment in the history of biological science. The enormity of the task comes into perspective when one considers the information needed to encode and perpetuate the human species.

The genome is an organism's entire genetic endowment—the complete nucleotide sequence of DNA comprising all of its chromosomes. The human genome is thus contained in the 23 pairs of large chromosomes in the nucleus of a human cell—22 pairs of autosomes and 1 pair of sex chromosomes (X and Y)—*plus* one tiny chromosome contained in the mitochondrion. The unique set of chromosomes that sets the parameters of each human life comes together at the moment of conception. One member of each pair of nuclear chromosomes (plus the mitochondrial chromosome) is provided by the mother's egg; the other member of each pair is provided by the father's sperm.

J. Craig Venter and Francis Collins
Leaders of the private and public sequencing efforts at a press conference announcing the completion of the draft sequence of the human genome. (Photo courtesy of A/P WIDE WORLD PHOTOS; Rick Bowmer.)

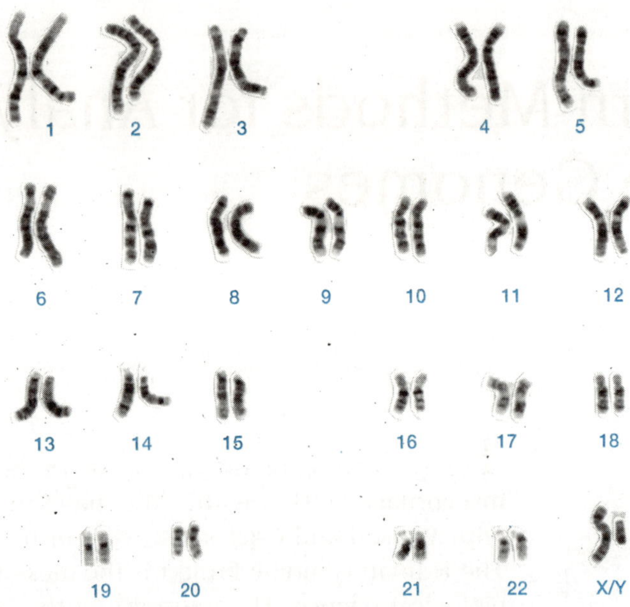

Human Karyotype

The human genome sequence consists of the 22 autosomal chromosomes, the X and Y chromosomes, and the mitochondrial DNA. The mitochondrial DNA is very small in comparison to the chromosomes and is not shown in this picture. (Reprinted, with permission, from Witherly J.L., Perry G.P., and Leja D.L. 2001. *An A to Z of DNA science*. Cold Spring Harbor Laboratory Press, Cold Spring Harbor, New York.)

With the exception of the X/Y pair in males, each chromosome pair has the same linear arrangement of genes. Thus, for the sake of simplicity, biologists are concerned only with the *haploid* genome composed of a single set of 25 chromosomes: the 22 autosomes, the X and Y chromosomes, and the mitochondrial chromosome. Each chromosome is a complex package containing a single, unbroken strand of DNA. The bulk of the chromosome structure is contributed by protein molecules (histones) that form a scaffold around which the DNA strand is wound, like coils of rope around a capstan. Thus, it is useful to think of the human genome as a set of very long DNA molecules, one corresponding to each chromosome. The 25 DNA molecules of the haploid human genome contain approximately 3.1 billion (3.1×10^9) nucleotides.

The magnitude of the Human Genome Project becomes apparent when we imagine a "genomic book of life," in which each nucleotide is represented by a single letter (A, T, C, or G). Using the typography of this text as an example—without spaces, punctuation, or illustrations—each page would contain 3675 nucleotides. At this rate, the entire sequence of the human genome would fill 843,500 pages. If contained in volumes about the length of this text (approximately 600 pages), the genomic book of life would comprise a stack of 1406 volumes. At 3.5 cm per volume, the stack would measure 49.21 meters—the height of a 15-story building!

Although the Human Genome Project began as a purely academic effort, it soon drew the attention of the business sector. The project pitted a publicly funded consortium led by five major academic research centers in the United States and England against an upstart company, Celera, the spin-off of the major producer of automated sequencing machines. The leaders of the public and pri-

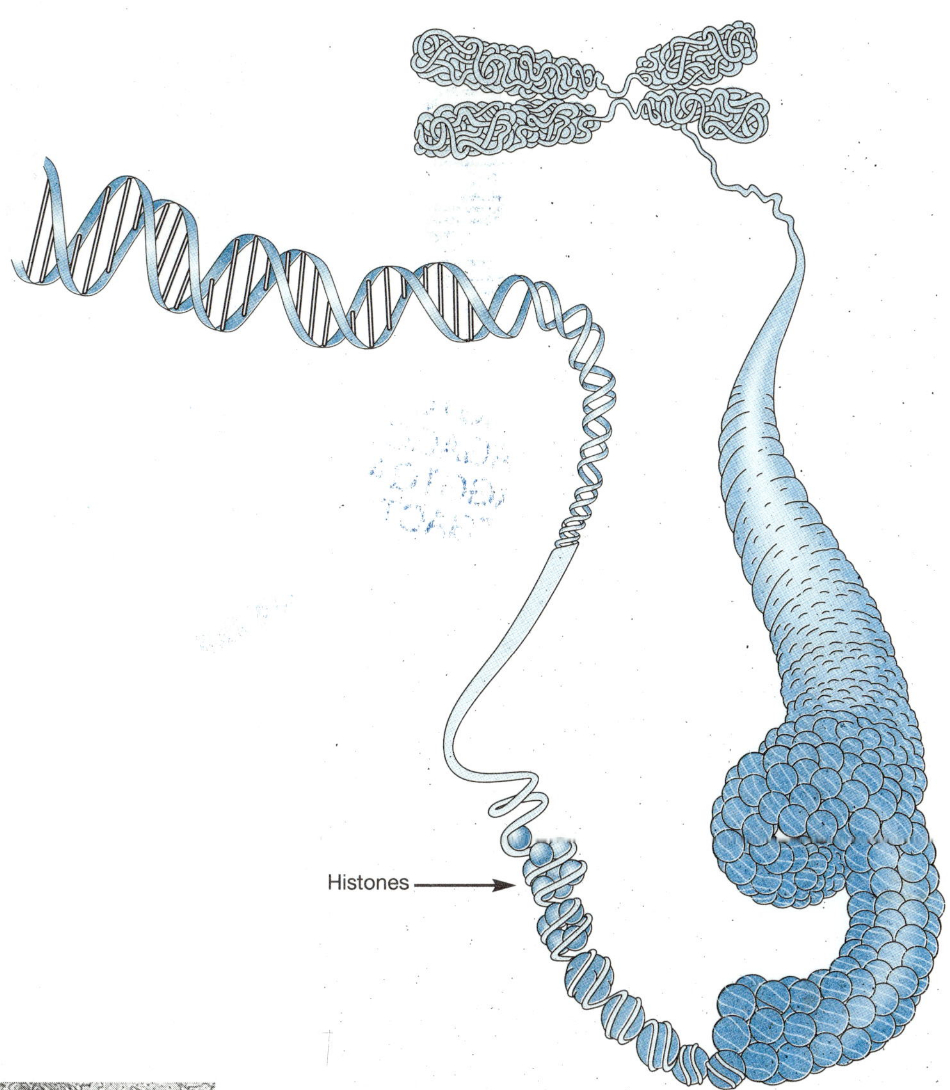

Chromatin
Complex folding around histone proteins allows the approximately 6 feet of DNA in each human cell to be compacted into the 46 nuclear chromosomes.

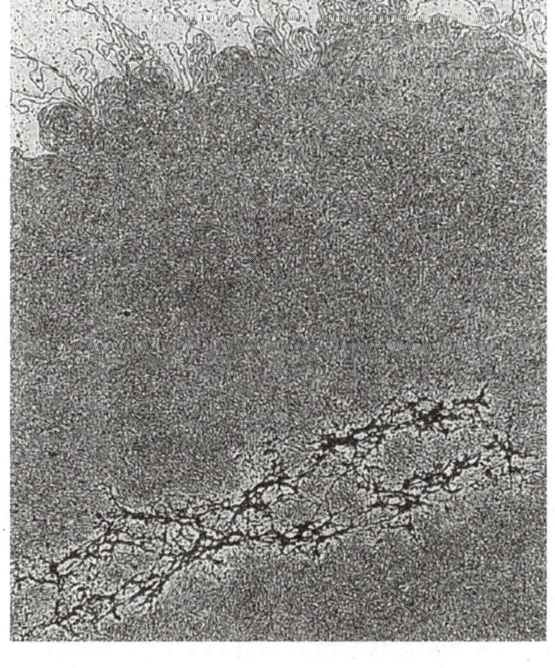

Electron Micrograph of a Histone-depleted Chromosome
The entire protein scaffold takes the shape of a chromosome with DNA spilling all around. (Reprinted, with permission, from Moran L.A., Scrimgeour K.G., Horton H.R., Ochs R.S., and Rawn J.D. 1994. *Biochemistry*. Neil Patterson Publishers/Prentice Hall, Englewood Cliffs, New Jersey. Courtesy of U.K. Laemmli.)

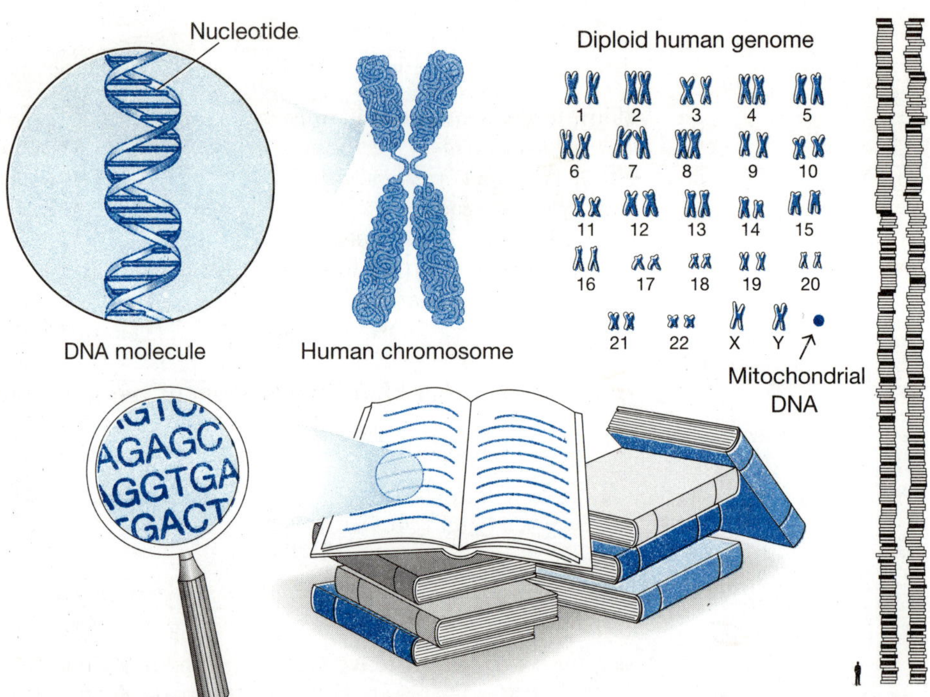

The Genomic Book of Life

If each nucleotide of the human genome is equated to a single letter in this textbook and the nucleotide sequence is written without paces, punctuation, illustrations, or tables, then 3675 nucleotides would fill one page of type. At this rate, the DNA sequence of the haploid human genome (the 22 autosomes, the X and Y chromosomes, and the mitochondrial chromosome) would fill 843,500 pages. If contained in volumes the size of this textbook, the genomic book of life would comprise a stack 49.21 meters tall. Twice this amount would be needed to represent the genetic information of the full diploid genome.

vate ventures sniped at each other—with some even casting various players as forces of "good" and "evil." The competition to sequence the human genome thus took on the aspect of an America's Cup sailing race—teams with different influential backers using different technologies, each betting the other that it would be the first to wrest victory in the face of nature.

Before the draft sequences were made available, most scientists estimated that the human genome contained 100,000 genes, with some estimates as high as 150,000 genes. So it was a surprise to almost everyone when the announcement came that two independent sequencing efforts had arrived at a number hovering around 35,000 genes. This raised an obvious question: How can humans get along with less than twice as many genes as a microscopic nematode (with 19,099 genes). Since this initial announcement, some studies have increased the number of estimated genes in the human genome to about 50,000. Estimates of the total mass of protein-coding sequence have been similarly reduced from 5% of the genome to less than 1.5%. Although knowledge of the extent of the noncoding portion of the human genome has increased during the past 30 years, the fact that nearly 99% of the human genome does not code for genes is unsettling to many.

The general public and most scientists are focusing on the genes revealed by the Human Genome Project. Naming and knowing all of our genes will be a great step forward in understanding how our cells work, and exactly what goes

Sizes of Sequenced Genomes

Species	Base pairs	Number of genes
Completely sequenced genomes		
Mycoplasma genitalium (bacterium)	578,000	400
Haemophilus influenzae (bacterium)	1,830,138	1,738
Escherichia coli (bacterium)	4,639,221	4,377
Saccharomyces cerevisiae (budding yeast)	12×10^6	5,885
Caenorhabditis elegans (nematode)	95.5×10^6	19,820
Drosophila melanogaster (fruit fly)	1.8×10^8	13,601
Arabidopsis thaliana (mustard plant)	1.17×10^8	25,498
Genomes sequenced but total number of genes still unknown[a]		
Mus musculus (mouse)	3×10^9	30,000
Homo sapiens (human)	3.3×10^9	30,000–50,000
Oryza sativa (rice)	4.3×10^8	30,000–63,000

[a]The number of genes is estimated.

wrong in various diseases. The causative gene has been identified for only a fraction of the several thousand known genetic diseases of humans. Mapping and sequencing these disease genes will facilitate diagnosis, and detailed biochemical study of their protein products should lead to improved therapies. Making human life healthier and, presumably, happier are the practical outcomes of the Human Genome Project. However, the extent of noncoding DNA suggests that, just as meaning sometimes lies between the words of a poem, some of the meaning of our genome lies in the "junk" between our genes.

In this chapter, we trace the development of the Human Genome Project, from a controversial idea to a worldwide collaboration involving thousands of scientists.We then review the laboratory methods for determining and assembling DNA sequences, computer methods for finding and comparing genes, and new microarray technologies for analyzing thousands of genes at once. We conclude with a discussion of the basic structure of the human genome, how we make do with so relatively few genes, and why so little of our DNA encodes protein.

GENESIS OF THE PROJECT

The possibility of a coordinated effort to sequence the human genome was first raised at a meeting held at the University of California at Santa Cruz in 1985. Although the idea intrigued many, the proposition had few strong supporters initially. Detractors continued to raise the same set of objections well after the formal initiation of the project. Most arguments against a large-scale sequencing project centered on several key questions:

- ***What will it do to biology?*** Throughout most of its history, biology had been centered on individual investigators and their small group of students, junior research fellows, and lab technicians. A typical molecular biology group in the 1980s averaged a dozen or so individuals; only the most established scientists had larger groups. A decentralized genome project would shift effort away from the numerous individual investigators toward large consortia. At the time, when the lines between academic and commercial biological research were still clearly drawn, a massive effort smacked of industrialized "big science." In short, many thought the genome project would change the "soul" of biological research.

- *How will the project be funded?* Reaching for numbers to express the cost and duration of a human genome project, the common estimate came to be 3 billion dollars (1 dollar per nucleotide) over a 15-year period. Competition for research grants was keen enough in the mid 1980s, and thus many biologists were concerned that a massive genome project would divert funding away from established programs and away from single investigators.
- *Is it real science?* Although the information from a genome project would certainly be interesting, the actual sequencing was seen as monotonous technical work. For some, this seemed below the aspirations of rigorous, hypothesis-driven science. Sydney Brenner famously quipped that large-scale DNA sequencing was a job best suited for convicts. In fact, the Sanger Centre, which would become the largest academic sequencing facility, initially resorted to recruiting lab help in local grocery stores around Cambridge, England.
- *Why sequence it all?* By the 1980s, all of the major types of noncoding DNA had been identified, and it was well known that they comprised the majority of the human genome. Many scientists could see no reason to sequence the whole genome. Why not merely sequence out from known genes to link them to others located within the same region of a chromosome? The debate between "head-to-tail" (telomere-to-telomere) sequencing and "cream skimming" (focusing on genes) was never fully resolved. It resurfaced at two key points, divided the research community, and ultimately fueled the "race" to sequence the human genome.

As the biological community debated the pros and cons of a genome project funded by the National Institutes of Health (NIH), another federal agency had completed its own review and decided to move ahead. In 1986, when the Department of Energy (DOE) announced its intention to start its own Human Genome Initiative, with funding of 5.3 million dollars in fiscal year 1987, the move puzzled many. Why would the DOE be interested in sequencing the human genome? Ostensibly, the DOE is primarily concerned with the civil and military uses of atomic energy. It was founded to manage the system of national laboratories used to manufacture nuclear weapons. From its inception, it had engaged in biological research, focusing, by necessity, on the effects of and responses to ionizing radiation. Studying radiation-induced mutations and their repair led to a focus on the molecular genetics of humans. Thus, there was a biological rationale for the DOE's interest in sequencing the human genome.

The DOE had little concern about the issues debated by biologists. It was experienced in administering huge collaborative projects with big budgets and long time horizons. Billions of dollars and decades of time could be devoted to planning and constructing a single particle accelerator to be shared among thousands of researchers. The DOE's huge budget could easily accommodate a move into genomics, and the work could be housed at several of the national laboratories with strong biological programs. Because many of its projects relied on the collection of large-scale data sets, head-to-toe sequencing did not seem particularly daunting or tedious. Moreover, the DOE had an existing computer infrastructure for storing and sharing the copious amounts of sequence data that would be generated.

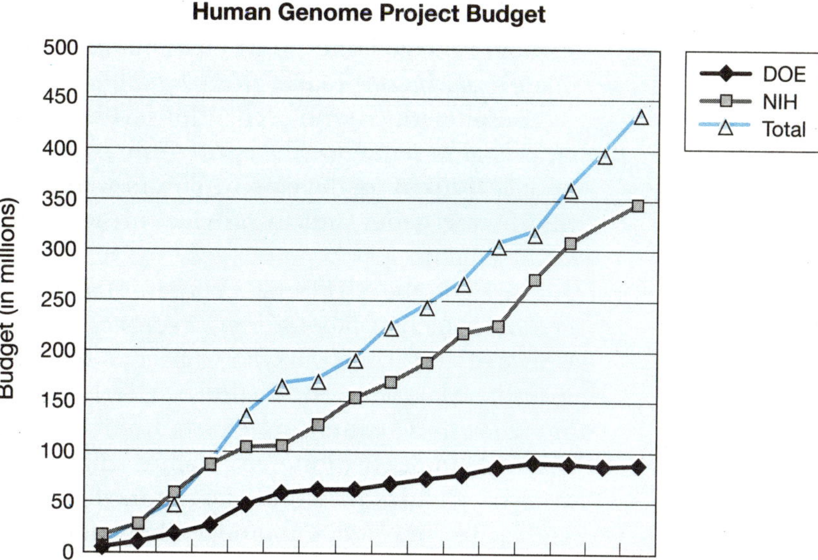

U.S. Government Funding for the Human Genome Project

The DOE's stunning move left biologists to ponder the ramifications of allowing an agency controlled by physicists to lead what promised to be the biological project of the century. After the 1988 publication of a National Research Council report endorsing genome sequencing, biologists set aside their differences, and the NIH announced its own genome project in March. That same year, the NIH and the DOE signed a memorandum to coordinate their sequencing efforts.

In 1988, James Watson was selected to lead the NIH sequencing effort. He spent the next several years commuting between Washington, D.C. and Long Island, New York, where he continued as the director of Cold Spring Harbor Laboratory. As one of the cofounders of the DNA structure that had launched the DNA revolution, Watson brought scientific stature, intuition, and visibility to the project. Foreseeing the many implications of obtaining the script of human life, Watson decreed that at least 3% of the research budget be devoted to examining the *e*thical, *l*egal, and *s*ocial *i*mplications (ELSI) of genome research. The ELSI program was an unprecedented experiment in exploring the interaction of science and society.

James D. Watson
(Courtesy of the James D. Watson Collection, Cold Spring Harbor Laboratory Archives.)

Early on, it was determined that sequencing the human genome should be paralleled by sequencing the genomes of several research organisms, including the bacterium *Escherichia coli*, the yeast *Saccharomyces cerevisiae*, the nematode worm *Caenorhabditis elegans*, the fruit fly *Drosophila melanogaster*, and the mouse *Mus musculus*. Moving back and forth between sequences from humans and those of research organisms would provide insights at several levels. On one level, genes previously identified in the research organisms can be used to identify corresponding human genes. On another level, the function of human genes and their protein products can be readily tested in these experimental systems. Support for sequencing the genomes of *S. cerevisiae* and *C. elegans*, as well as targeted regions of the mammalian genomes, provided the first tests of large-scale sequencing. Separate projects were also initiated (although not NIH-funded) to sequence the genomes of several important plant species, including the mustard plant *Arabidopsis thaliana*, rice (*Oryza sativa*), and corn (*Zea mays*).

Despite critics who argued for focusing on expressed genes, Watson set the clear goal to sequence the entire human genome from head to tail. Initially, the project focused on developing chromosome maps and markers needed to identify disease genes and to provide anchor points for specifying the location of DNA sequences.

In 1995, the NIH funded pilot human genome sequencing centers at a number of academic laboratories, eventually focusing on centers at Washington University (St. Louis), Baylor College of Medicine (Houston), and the Whitehead Institute/Massachusetts Institute of Technology (Cambridge). The DOE eventually coalesced sequencing efforts from several national laboratories at its new Joint Genome Institute in Walnut Hills, California. In the mid 1990s, the Wellcome Trust, the world's largest medical charity, established its own sequencing facility outside Cambridge, England. Named after Nobel Laureate Fred Sanger, it became the largest academic sequencing center. These five sequencing centers of the public genome project became known as the "G5" and ultimately contributed 80% of the data used to assemble the draft sequence. Fifteen other major sequencing centers in the United States, Japan, France, Germany, and China cooperated in sequencing specific human chromosomes.

Work on the model organisms and at the major centers developed key technologies that greatly accelerated the speed with which DNA could be sequenced and assembled, ultimately reducing the cost of sequencing from about 1.00 dollar per finished base pair down to 10 cents. These technologies included (1) automated methods for processing cloned DNA, (2) biochemical and analytical systems for rapid DNA sequencing, (3) computer systems for organizing and distributing DNA sequence data, and (4) algorithms for assessing sequence fidelity, assembling DNA sequences, and identifying genes.

CHROMOSOME MAPS AND MARKERS

Constructing high-resolution maps of each chromosome was an important objective at the outset of the genome project. Like road maps, human chromosome maps can be constructed with varying degrees of resolution. *Genetic maps* describe the locations of genes and their patterns of inheritance. *Genetic linkage maps*, like those first constructed for *Drosophila* by Thomas Hunt Morgan and his colleagues in the early 1900s (see Chapter 1), have been useful in making low-resolution schemes of whole chromosomes. These gave rise to the classical measure of genetic distance, the centiMorgan (cM).

The cM is a measure of the frequency of chromosome recombination during gamete formation. Recall that during meiosis, the paired chromosomes align and "cross over" to exchange DNA segments. One cM is defined as the distance between two genes or markers such that they separate during meiosis 1% of the time. At the start of the genome project, a cM in genetic distance was estimated to equal about 1 million nucleotides in physical distance. The draft sequence of the human genome produced by Celera showed an average of 1.22 million nucleotides (megabases, Mb) per cM.

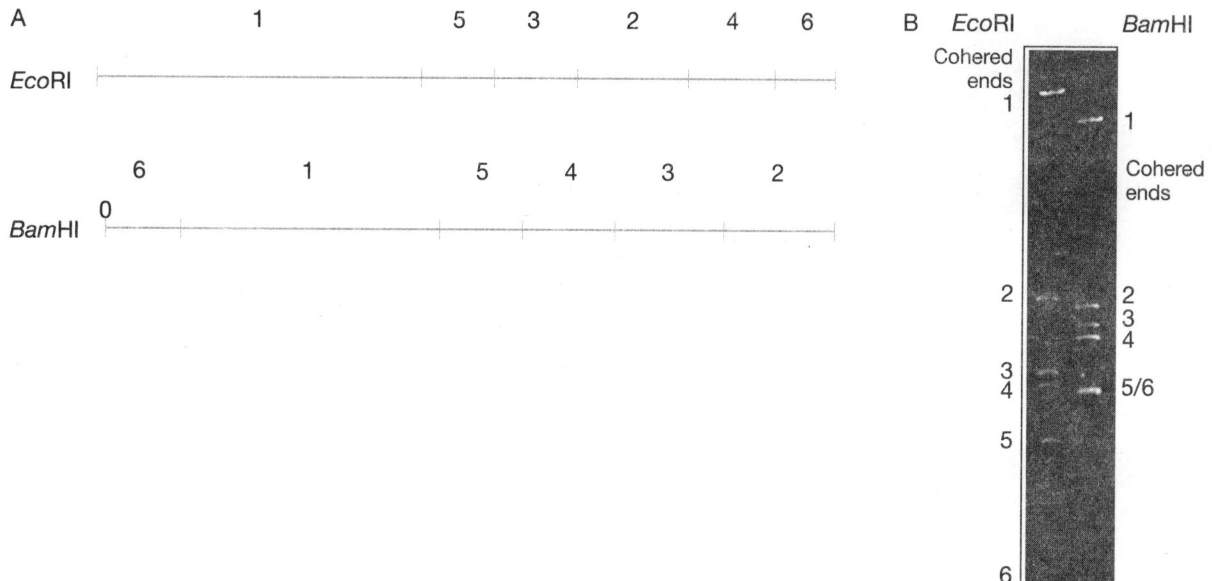

Restriction Map of Bacteriophage λ DNA

(*A*) Maps showing the restriction sites for the enzymes *Eco*RI and *Bam*HI in λ DNA. Vertical bars represent sites of enzyme cutting. (*B*) An electrophoretic gel separation of *Eco*RI and *Bam*HI enzyme digests of λ DNA. Numbers indicate the corresponding fragment shown on the map. The DNA has not been separated by electrophoresis sufficiently long enough to resolve bands 5 and 6 in the *Bam*HI digest. The bands labeled "Cohered ends" represent the two terminal fragments, joined by the cohesive ends (*cos* sites) of λ DNA. (Reprinted, with permission, from Snyder L.A., Freifelder D., and Hartl D.L. 1985. *General Genetics*. Jones and Bartlett, Boston.)

Early experiments in the 1970s created maps of the cutting sites of restriction enzymes along the length of viral chromosomes. These were the first *physical maps*, with each site marking the occurrence of a structural characteristic of the DNA molecule itself. Although restriction enzymes had proved to be useful for mapping small chromosomes, they cut too frequently to be of use in constructing large-scale maps of the human chromosomes. Eventually, this problem was overcome by using restriction enzymes that cut infrequently. Pulse-field gel electrophoresis, a method of separating very large pieces of DNA, was used to analyze these large fragments.

Identifying population variations in the cutting patterns of restriction enzymes provided a means to make the first physical maps of entire human chromosomes. By the mid 1980s, it was realized that point mutations occasionally create or delete restriction enzyme recognition sites along the length of a chromosome. Also, the addition of short tandem repeats can increase the distance between enzyme cut sites. Each mutation or tandem repeat variation is a different allele of the DNA molecule. Different alleles produce different-sized restriction fragments, or restriction-fragment-length polymorphisms (RFLPs), whose lengths can be measured in nucleotides.

Classical genetics had used genes as markers to define discrete locations (loci) on chromosomes. An allele was defined as an alternative form of a gene, with respect to a specific detectable phenotype. Restriction enzyme cutting sites provided a new definition of loci as a specific DNA sequence. RFLPs extended

Maynard Olson
(Courtesy of CSHL Archives.)

Leroy Hood

Charles Cantor

David Botstein
(Courtesy of CSHL Archives.)

the definition of an allele to mean an alternative form of DNA—a different length DNA molecule. Although RFLPs were first used in identity testing and DNA diagnosis (see Chapter 8), they became extremely useful to chromosome mappers, even in the absence of any phenotypic relationship. Measuring the crossover frequency between RFLP markers provided the first means to relate genetic distance (measured in cMs) to physical distance (measured in nucleotides). The first linkage map of the human genome, incorporating 403 RFLP markers, was published in 1987 by Helen Donis-Keller and colleagues at Collaborative Research, Inc. in Boston, Massachusetts.

In 1989, Maynard Olson (Washington University), Leroy Hood (California Institute of Technology), Charles Cantor (Lawrence Berkeley Laboratory), and David Botstein (Genentech, Inc.) provided a common language of DNA markers with which scientists could easily exchange sequence information and obtain probes for known regions of the genome. In retrospect, what they proposed seemed obvious—that a DNA sequence *itself* can act as a chromosome marker! Any single-copy sequence of 200–500 nucleotides—a sequence-tagged site (STS)—can be used as physical landmark on a chromosome.

Each RFLP or STS marker must be localized to a specific chromosome position by using a cloned fragment of DNA. *Contiguous* (*contig*) maps align the ends of adjacent DNA fragments, thus stringing together continuous sequences. Contig maps orient physical markers on adjacent fragments, which provide points of reference spaced at regular intervals throughout the genome. By the late 1980s, Ray White and his associates at the University of Utah had established a set of RFLP markers spanning the chromosomes at approximately 10-cM intervals. An early goal of the genome project was to put more markers on the chromosome maps and decrease the intervals between them. By the publication of the rough draft sequence, the resolution of the physical map had increased 100-fold—with 30,000 distinct markers at intervals of approximately 0.1 cM (100,000 bp).

As the quality of these genome maps increased, they became valuable tools for identifying genes involved in human disease. Previously, finding a disease gene was an arduous task involving years of intense work. For example, it took Mark Skolnick at the University of Utah 4 years (1990–1994) to identify and locate the breast-cancer-related gene, *BRCA1*. Improved maps helped Richard Wooster at the Haddow Laboratories Institute of Cancer Research to clone *BRCA2* in a matter of months.

STSs represented the final step in the evolution of DNA markers. Genes, restriction recognition sites, and RFLPs all represent functional properties associated with DNA. STSs are "pure markers" in that they can be mapped without knowledge of genes or restriction enzymes. But the practical and widespread use of STS markers did require two technologies that emerged in the 1980s: polymerase chain reaction and automated DNA sequencing.

Polymerase Chain Reaction

Polymerase chain reaction (PCR), discovered in 1985 by Kary Mullis at Cetus Corporation, proved key to the genome project in several ways. First, it provided a simple means to obtain DNA markers and to genotype chromosome fragments, without the need of exchanging clones. Second, it provided a new chemistry that allowed DNA sequencing to be automated.

Kary Mullis
(Courtesy of CSHL Archives.)

Until the advent of PCR, the only way to purify a DNA sequence was through biological amplification in cultured cells. To obtain enough pure sample of a given DNA sequence, it had to be cloned into a vector and grown in a bacterial or yeast culture. Thus, at the beginning of the genome project, there was no alternative but to maintain reference cultures of bacteria or yeast, each transformed with a fragment of a human chromosome identified by a specific STS or RFLP marker. The prospect of maintaining a mammoth central bank of clones and dispersing these among research groups worldwide was daunting. PCR provided a simple means of isolating pure quantities of specific DNA sequences. PCR uses enzymatic amplification to increase the copy number of any DNA fragment of up to 6000 bp in length. Using data available from an STS database, a researcher anywhere in the world can use PCR to assay a DNA fragment for a particular STS in several hours.

PCR is based on the phenomenon of primer extension by DNA polymerase, discovered in the 1960s (see Chapter 2). Synthesis of each DNA strand is

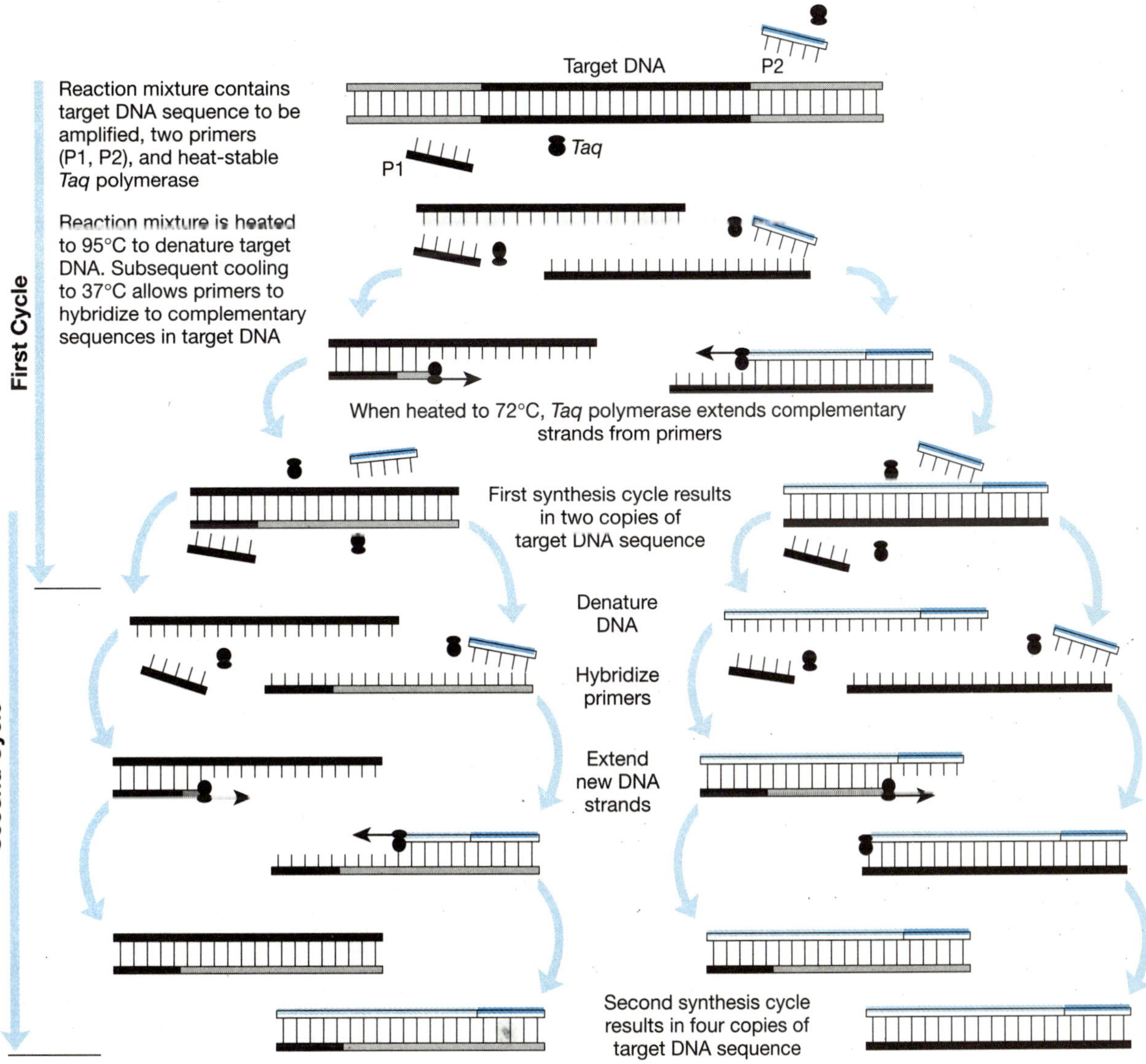

First Cycle

Target DNA

P2

Reaction mixture contains target DNA sequence to be amplified, two primers (P1, P2), and heat-stable *Taq* polymerase

P1

Taq

Reaction mixture is heated to 95°C to denature target DNA. Subsequent cooling to 37°C allows primers to hybridize to complementary sequences in target DNA

When heated to 72°C, *Taq* polymerase extends complementary strands from primers

First synthesis cycle results in two copies of target DNA sequence

Second Cycle

Denature DNA

Hybridize primers

Extend new DNA strands

Second synthesis cycle results in four copies of target DNA sequence

DNA Amplification Using Polymerase Chain Reaction

"primed" by a short double-stranded stretch. Working from this primer, the polymerase adds nucleotides complementary to the template DNA, extending the double-stranded region. First, a pair of DNA oligonucleotide (oligo meaning a few) primers approximately 20 nucleotides in length are synthesized that bracket the "target" region to be amplified. The primers are designed to anneal to complementary DNA sequences at the 5′ end of each strand of the DNA molecule. The two primers are mixed in excess with a DNA sample containing the target sequence, along with a DNA polymerase. The cofactor magnesium (Mg^{++}) and the four deoxyribonucleoside triphosphates (dNTPs) are also provided. The reaction mixture is then taken through multiple synthesis cycles consisting of the following:

1. ***Denaturing:*** Heating to near boiling (94°C) denatures the target DNA and creates a set of single-stranded templates. Heating increases the kinetic energy of the DNA molecule to a point that it is greater than the energy needed to maintain hydrogen bonds between base pairs, and the double-stranded DNA separates into single strands.
2. ***Annealing:*** Cooling to approximately 65°C encourages oligonucleotide primers to anneal to their complementary sequences on the single-stranded templates. The optimum annealing temperature varies according to the proportion of AT to GC base pairs in the primer sequence. Because the primers are added in excess and are short, they will anneal to their long target sequences before the two original strands can come back together.
3. ***Extension:*** Heating to 72°C provides the optimum temperature for the DNA polymerase to extend from the oligonucleotide primer. The polymerase synthesizes a second strand complementary to the original template.

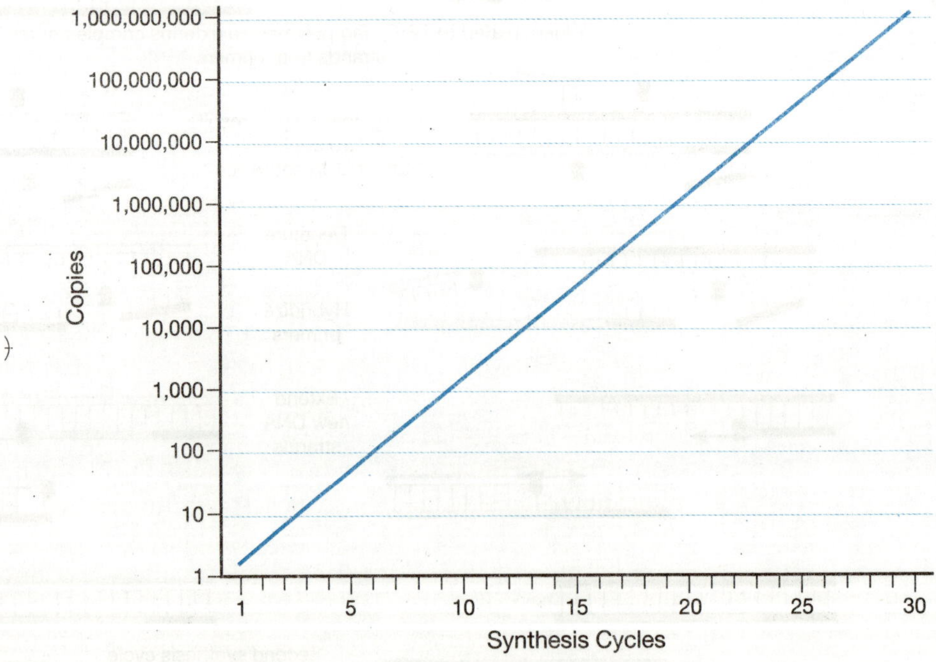

Polymerase Chain Reaction Theoretical Amplification

During each synthesis cycle of approximately 2 minutes, the number of copies of the target DNA molecule is doubled. *Twenty-five* rounds of synthesis theoretically produce 1,000,000-fold amplification of the target sequence in as little as 1 hour.

The high temperatures used in PCR required using a heat-stable DNA polymerase. In the late 1980s, the first heat-stable DNA polymerase had been isolated from a cultured sample of the bacterium *Thermus aquaticus*, originally obtained in the 1960s from the Mushroom Pool in Yellowstone National Park. This bacterium is adapted for life in the 80°C waters of the thermal pool, and its DNA polymerase (called *Taq* polymerase) maintains activity throughout repeated exposure to near-boiling temperatures, making it ideal for PCR. Volcanic ocean vents became another prominent source for organisms adapted to extreme conditions (extremophiles) with heat-stable polymerases.

Temperature cycling was initially achieved by moving reaction tubes back and forth between several water baths with fixed temperatures. This tedium was relieved by the advent of automated DNA thermal cyclers in which cycling temperatures and times are entered into a microcomputer controller that raises or lowers the temperature in a heat block. This automated the process, as the tubes containing the reaction stay in one place while they are repeatedly cycled through the PCR steps. The combination of heat-stable polymerases and automated cycling brought rapid DNA analysis within reach of virtually any scientist or student.

Polyacrylamide Gel Electrophoresis

Sequencing reactions loaded into denaturing gel

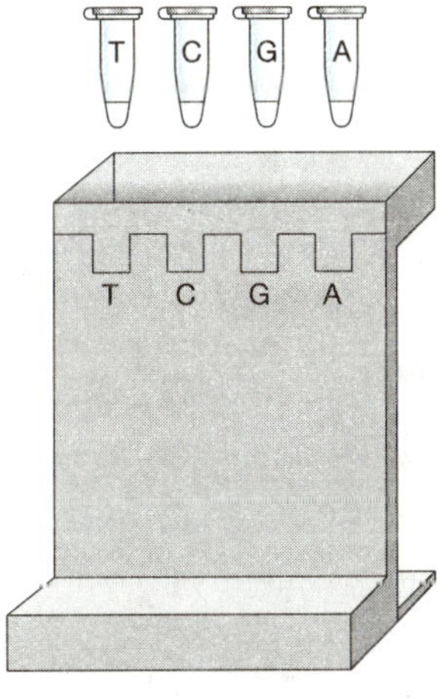

Autoradiogram of Gel

Lanes read in sequence from bottom to top of gel

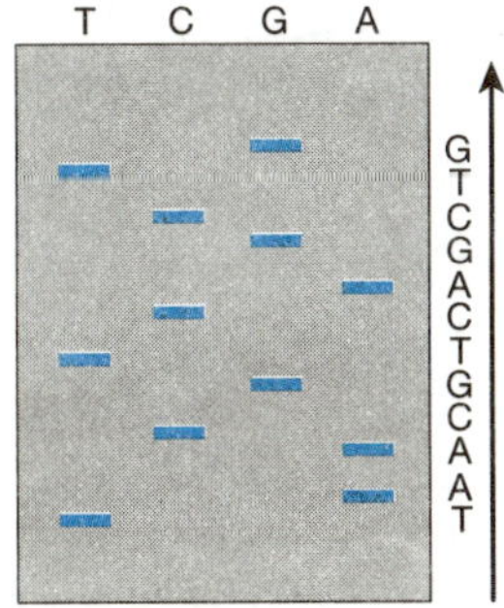

Dideoxy Method for Sequencing DNA

DNA Sequencing

At the start of the genome project, manual sequencing of DNA was time- and labor-intensive. A major thrust of the genome project was to develop automated sequencing technologies that ultimately would increase sequencing speed to 500,000 or more nucleotides per day at a cost of pennies per nucleotide.

Two methods for sequencing DNA were developed independently in 1977. The chemical cleavage method of Allan Maxam and Walter Gilbert, at Harvard University, involves the specific chemical modification of each different nucleotide. This alteration allows subsequent cleavage at a specific site and identifies the nucleotide present at that position. However, modern DNA sequencing is based entirely on the chain termination method, developed by Fred Sanger at the Medical Research Council's Laboratory of Molecular Biology in Cambridge, England.

The Sanger method is based on two facts of DNA synthesis. First, in the presence of the four dNTPs, DNA polymerase will initiate synthesis of a new strand of DNA when a short DNA primer is hybridized to a single-stranded DNA template. Second, if dideoxynucleotide triphosphates (didNTPs) are included in the reaction, DNA elongation will stop when a didNTP is incorporated. This occurs because didNTPs lack a 3′ hydroxyl group (–OH), which is necessary to form the phosphodiester linkage that joins adjacent nucleotides.

In the dideoxy sequencing protocol, four reaction tubes (A,T,C,G) are set up. Each of the reactions contains a DNA template, a primer sequence, DNA polymerase, and the four deoxynucleotide triphosphates (dATP, dTTP, dCTP, and dGTP, one of which is radioactively labeled). A single type of didNTP is added to each of the four reactions—did*A*TP (to tube A), did*T*TP (tube T), did*C*TP (tube C), or did*G*TP (tube G).

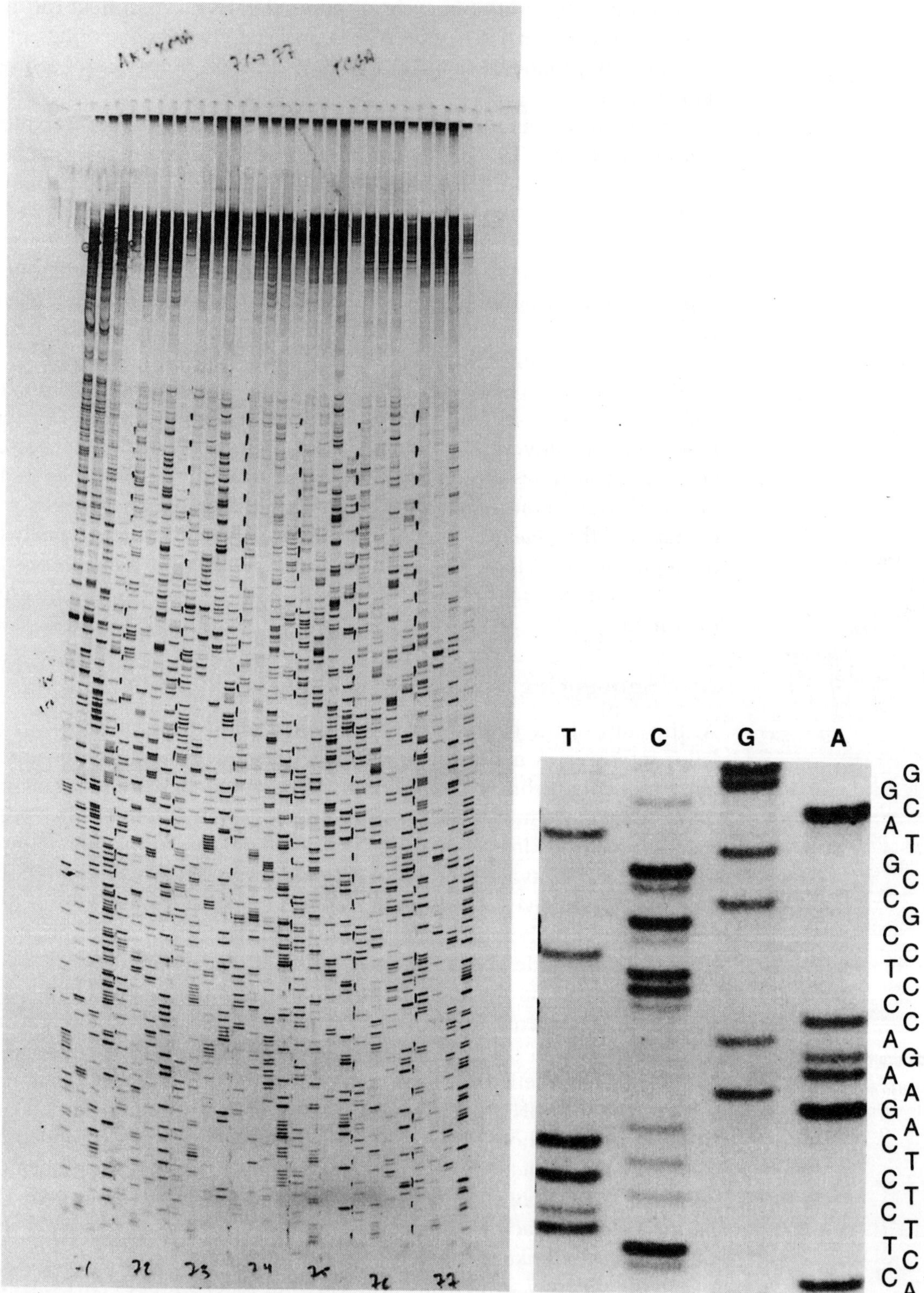

Autoradiogram of Dideoxy Sequencing Gel

Seven separate clones containing restriction fragments were sequenced in side-by-side reactions and separated by electrophoresis on a sequencing gel. Each set of four columns (T, C, G, A reactions) on this autoradiogram are the sequence data from a different clone. A portion of the sequence data for clone 5 is shown in detail. This experiment was performed in the laboratory of Fred Sanger, who developed the dideoxy sequencing method. (Courtesy of W. Herr, Cold Spring Harbor Laboratory.)

Working from the primer, the polymerase randomly adds dNTPs or didNTPs that are complementary to the DNA template. The ratio of dNTPs to didNTPs in the reaction is adjusted so that a didNTP is incorporated into the elongating DNA chain approximately once every 100 nucleotides. Each time didNTP is incorporated, synthesis stops, and a DNA strand of a discrete size is generated. After replication, there are millions of copies of the template DNA sequence terminated at each nucleotide position.

In Reaction A, for example, for every 100 thymine residues encountered on the DNA template, the polymerase will incorporate 99 units of the complementary nucleotide deoxyadenosine (remember that in DNA synthesis, A complements T in the opposite strand). In 1 instance out of 100, the polymerase will incorporate dideoxyadenosine, halting synthesis of the strand at that point. In other words, the DNA chain will be terminated with the thymine position in 1% of all template molecules. Because millions of DNA molecules are present, Reaction A results in a collection of DNA fragments in which synthesis has been arrested to mirror the position of each thymine residue in the template DNA.

When the reactions are complete, formamide is added to denature the newly synthesized strands from the template DNA. Each reaction is loaded into a different lane of a polyacrylamide gel containing urea, which prevents the DNA strands from forming secondary structures during electrophoresis. The fragments migrate through the gel according to size, eventually resolving to form a "ladder" of bands, each composed of DNA molecules differing in length by a single nucleotide.

Following electrophoresis, the gel is placed in contact with X-ray film. The radioactively labeled nucleotides that were incorporated expose the X-ray film, producing a series of bands, indicating the length of fragments generated in the A, T, C, and G reactions. The gel is then "read" from bottom to top, beginning with the smallest DNA fragment, then scanning across the lanes to identify each successively larger fragment. Optical scanners were eventually introduced to read gels, because manual sequencing was far too slow to accomplish the goal of sequencing the human genome rapidly and accurately.

Automated sequencing was made possible by dye chemistry developed by Lee Hood and Lloyd Smith at the California Institute of Technology. In 1986, they paired a different fluorescent dye with each of the four didNTP reactions. The four sequencing reactions were added to a single lane of a sequencing gel, and the fluorescent labels were detected as the terminated fragments passed an Argon laser aimed at the bottom of gel. When struck by the laser light, each fluorescent terminator emits a colored light of a characteristic wavelength, which is then interpreted by the computer software as an A (green), T (red), C (blue), or G (yellow) at that position.

Hood then collaborated with Mike Hunkapiller, at Applied Biosystems, Inc. (ABI), to produce the first commercial instrument to read sequences from dye-labeled fragments. The ABI Model 370, first marketed in 1987, employed a polyacrylamide slab gel to resolve a ladder of DNA fragments. However, rather than hand or optical gel reading, it employed a laser to detect the fluorescently labeled nucleotides. The sequencer incorporated a computer program that builds a simulated gel image of colored DNA bands as they pass a scanning laser during electrophoresis. The final output took the form of an electropherogram, showing colored peaks corresponding to each nucleotide position. The ABI 370 DNA Sequencer, equipped with a 16-lane polyacrylamide gel, had the capacity to sequence as many as 20,000 nucleotides per day. Increasing the number of lanes to 32, 48, or ultimately 96 brought daily output to 80,000 nucleotides or more.

By allowing all four nucleotides to be analyzed in a single lane, Hood's fluorescent chemistry quadrupled the output of sequencing gels. Parallel improvements in DNA preparation further increased output. By the late 1980s, thermal stable polymerases (such as *Taq)* and DNA thermal cyclers were pressed into service to automate dye labeling, a hybrid method that became known as cycle sequencing. In 1987, Dupont introduced "dye terminators," which attach a different fluorescent dye directly to each of the four terminator nucleotides (didATP, didTTP, didCTP, or didGTP). This allowed all four nucleotides to be labeled simultaneously in a single reaction, streamlined sample preparation, and effectively quadrupled a technician's daily sequencing output.

The final phase in the development of high-throughput sequencing came by replacing the polyacrylamide slab gel with multiple capillary gels, whose feasibility was first demonstrated in 1990. New automated sequencing machines linked a 96-capillary array with a robot mechanism capable of automatically reloading samples up to 12 times per day. The capillary machines eliminated the time-consuming elements of pouring and loading sequencing gels, reducing human intervention to maintaining reagent levels and loading microtiter plates into the autoloader. As the race to complete the first draft of the human genome heated up, each of the major sequencing centers had become automated factories equipped with 30 or more of these sequencers, each churning out up to 400,000 nucleotides of sequence per day. The speed of sequencing had increased 400-fold since the start of the project.

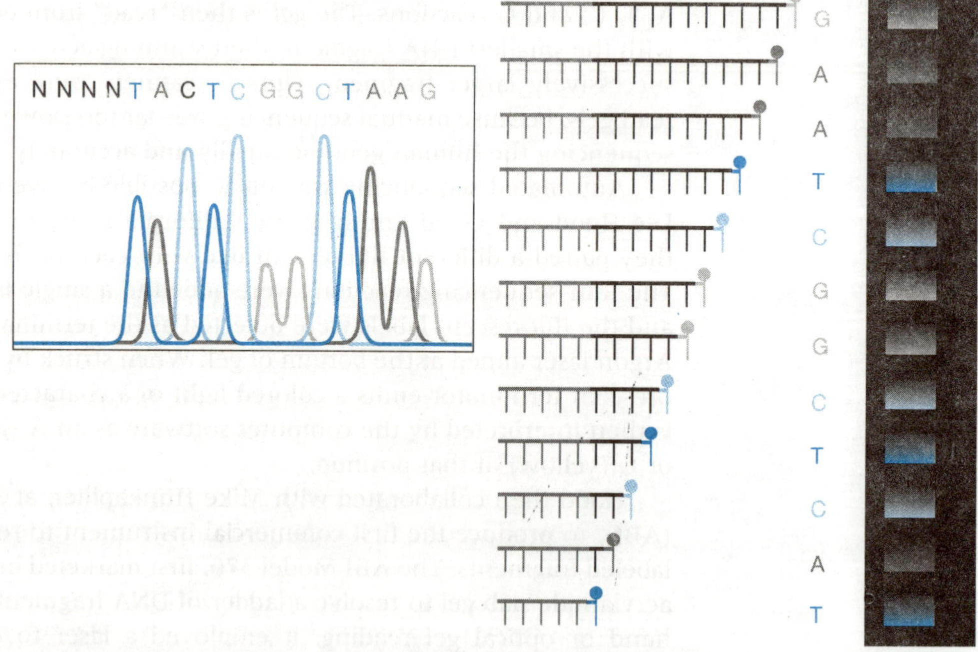

Sample Output from an Automated DNA Sequencer
(Left) Electropherogram showing peaks corresponding to each nucleotide position. *(Middle)* Representation of the DNA fragments created in the sequencing reaction that correspond to each peak. *(Right)* Representation of the actual sequencing gel for this fragment.

HIERARCHICAL SHOTGUN CLONE LIBRARIES

The NIH-DOE projects embraced a top-down strategy of creating an ordered library of cloned DNA molecules, starting with large-scale maps and working down to individual sequences. This involved cloning large pieces of human chromosomal DNA into vector molecules and then using STSs, RFLPs, and restriction sites to orient the large fragments into contiguous chromosome regions, or *contigs*. The larger inserts were then subcloned into sequencing vectors, and the sequenced bits reassembled. By linking together adjacent sequences and maintaining reference markers between each level, investigators can move up and down this hierarchy of ordered clones. In this way, ever-larger contiguous stretches of DNA sequence can be united using the map coordinates.

Two major cloning vectors were available at the start of the genome project: bacteriophage λ, which accept inserts averaging 20,000 nucleotides, and cosmids, with inserts averaging 40,000 nucleotides. To reduce the number of fragments needed to form a contig map, it seemed imperative to create new vectors capable of holding much larger inserts. Moving away from these systems required the development of artificial chromosomes that could be propagated inside living cells. Yeast artificial chromosomes (YACs), developed in the late 1980s by David Burke and Maynard Olson, at the Washington University in St. Louis, could accept inserts of up to 500,000 bp (500 kb). Although they were widely used in mapping the *C. elegans* and yeast genomes, YACs proved inherently unstable. Rearrangements and deletions make it difficult to ensure that the insert being sequenced is complete and in the same order as on the chromosome.

Bacterial artificial chromosomes (BACs), developed by Mel Simon at the California Institute of Technology, ultimately proved to be the workhorses of the genome project. Although they held smaller inserts, ranging from 50 to 250 kb, they proved to be extremely stable and amenable to large-scale automation. Simon and Pieter de Jong, at Roswell Park Cancer Institute in Buffalo, New York, constructed extensive BAC libraries upon which much of the public sequencing was based.

To construct BAC libraries, human genomic DNA is randomly cleaved with restriction enzymes to produce fragments averaging 150,000 nucleotides. This can be achieved with rare-cutting enzymes, such as *Mlu*I, *Nru*I, and *Pvu*I, whose recognition sequences are rare in mammalian genomes. Alternatively, a partial digest of genomic DNA can be achieved by exposing it to a mixture of *Eco*RI and *Eco*RI methylase. At the proper ratio, the methylase protects enough restriction sites so that the endonuclease cuts infrequently, producing very large restriction fragments (see Chapter 4). Either method has the effect of shredding the genome into bits, much like a shotgun shreds material at close range. Thus, assembling libraries for sequencing by this method became known as shotgun sequencing.

Individual BACs were identified by their unique *Hin*dIII restriction patterns, termed fingerprint clones. A computer algorithm analyzed shared bands in fingerprint clones, converting overlaps into a contig map. Some BACs were anchored to existing chromosome maps with STSs and RFLPs, or by sequencing several hundred nucleotides at each end ("BAC ends"). Using this information, a set of BAC clones could be identified whose sequence and marker patterns

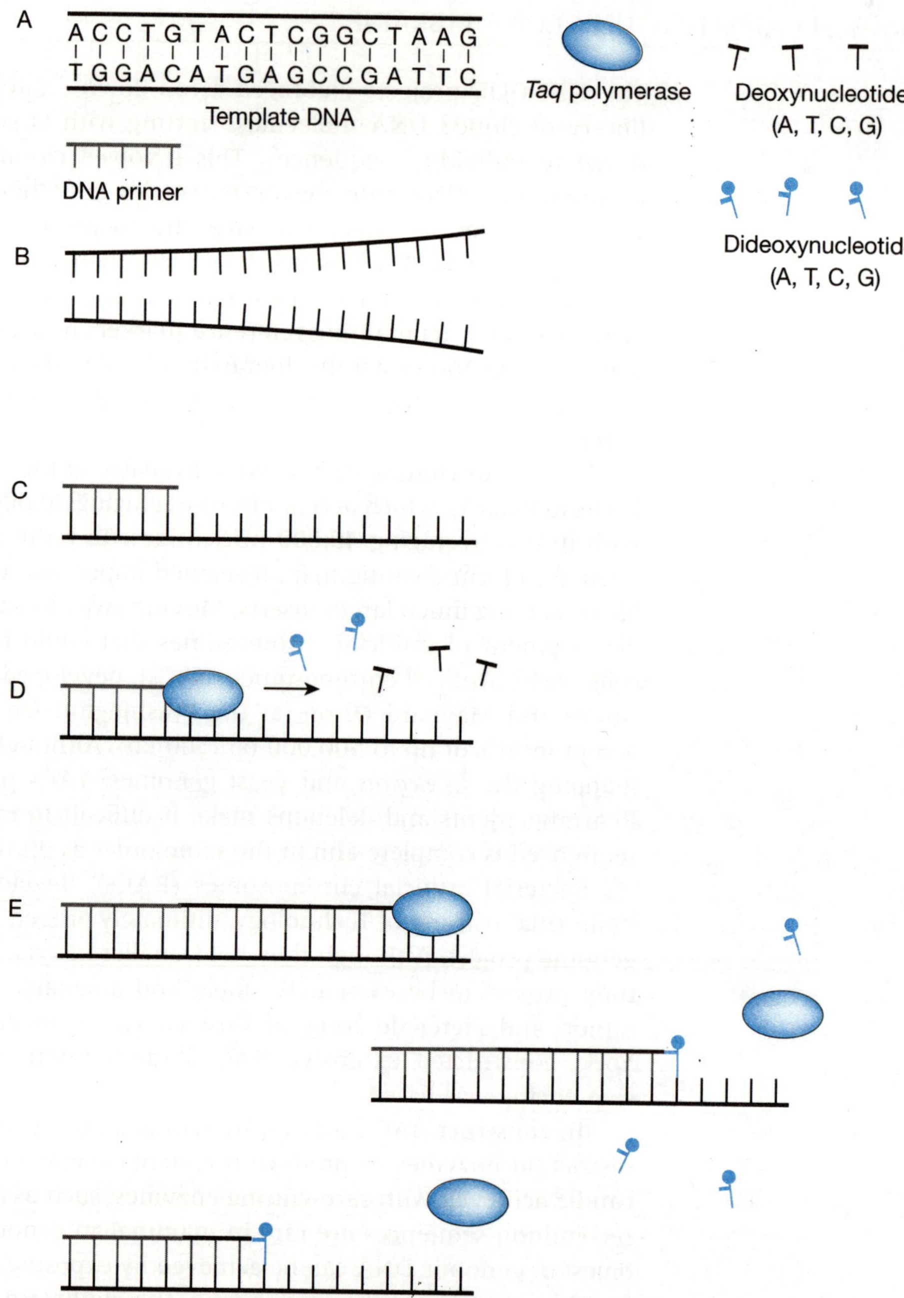

Cycle Sequencing
(*A*) Double-stranded template DNA is added to a test tube with a single-stranded DNA primer, *Taq* polymerase, deoxynucleotides (A, T, C, and G), and fluorescently labeled dideoxynucleotides (A, T, C, and G, each with a different colored label). The primer is designed to anneal to one end of the region to be sequenced. (*B*) The mixture is heated, and the double-stranded template DNA separates. (*C*) The mixture is cooled, and the primer anneals to the now single-stranded DNA template. (*D*) *Taq* polymerase binds and creates a new second strand, incorporating deoxynucleotides from the solution. (*E*) When a dideoxynucleotide is incorporated, extension of the new strand ceases. This cycle is then repeated many times.

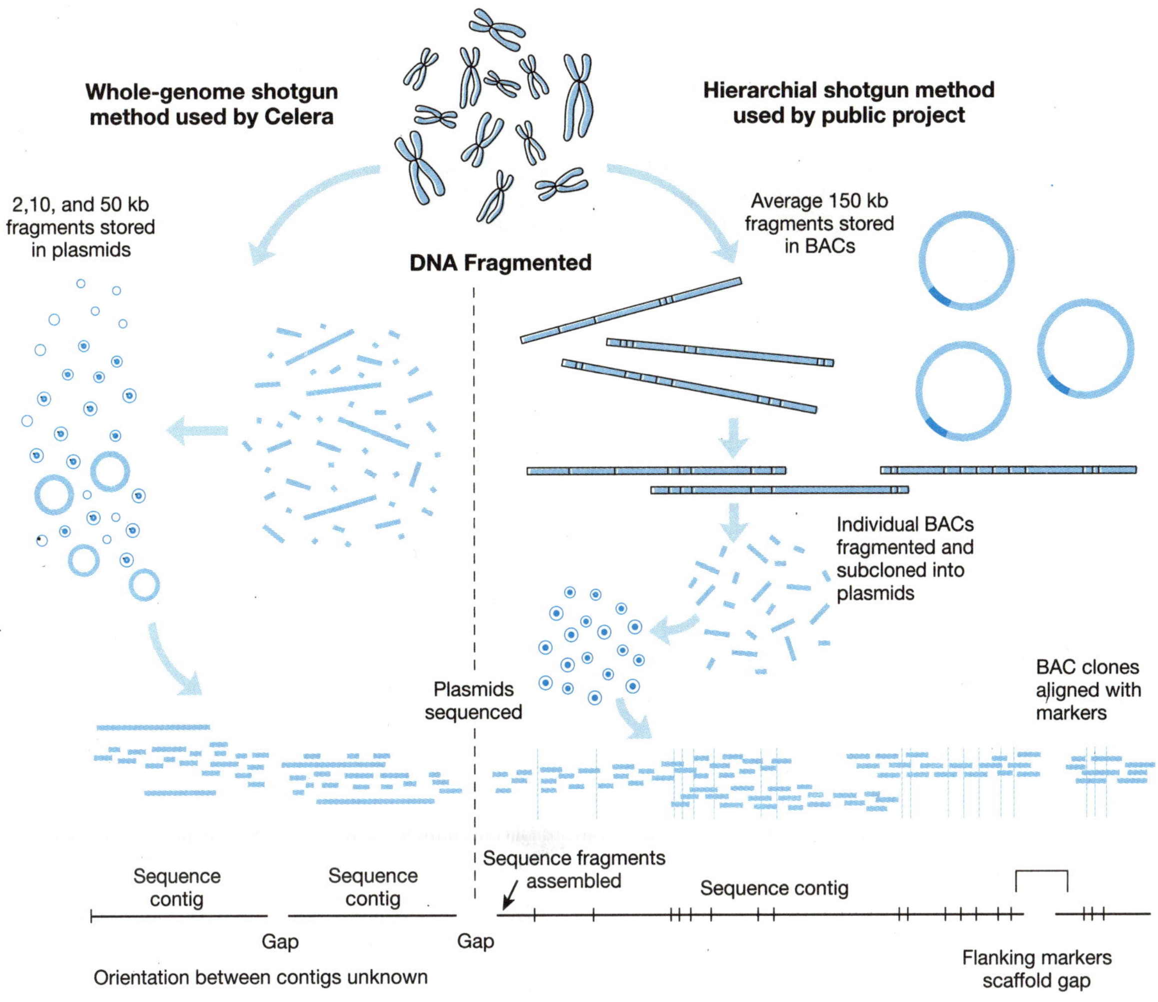

Sequencing the Human Genome

overlapped to cover large contiguous chromosome regions. Ideally, a single "tiling path" would span an entire chromosome. However, the first draft of the human genome had numerous gaps or breaks. In many cases, adjacent contigs could be oriented with respect to each other using BAC ends, plasmid ends, known mRNAs, and map information to achieve the highest order of large-scale integration called "scaffolds." As the name implies, a scaffold provides a framework to link adjacent contigs through unsequenced chromosome regions.

To produce the final sequence, BACs were selected from within the tiling path. Each BAC was restriction digested to produce fragments averaging 1500 nucleotides, and these pieces were cloned into sequencing vectors. Taking into account the length of the source BAC clone, enough sequencing clones were selected that each nucleotide in the BAC clone would be represented four to eight times. Because of the significant overlap, sequence was generated from one or both ends of a clone, but the central portion usually was not sequenced entirely.

The largest sequencing centers functioned like factories, with robots automatically picking clones, purifying DNA samples, setting up sequencing reac-

Total Human Sequence Deposited in GenBank

Sequencing center	Finished human sequence (kb)
The Sanger Centre	284,353
Washington University Genome Sequencing Center	175,279
US DOE Joint Genome Institute	78,486
Baylor College of Medicine Human Genome Sequencing Center	53,418
Genoscope	48,808
Whitehead Institute, Center for Genome Research	46,560
Department of Genome Analysis, Institute of Molecular Biotechnology	17,788
RIKEN Genomic Sciences Center	16,971
University of Washington Genome Center	14,692
Keio University	13,058
Multimegabase Sequencing Center; Institute for Systems Biology	9,676
University of Oklahoma Advanced Center for Genome Technology	9,155
The Stanford Human Genome Center and Department of Genetics	9,121
University of Texas Southwestern Medical Center at Dallas	7,028
GTC Sequencing Center	7,014
Beijing Genomics Institute/Human Genome Center	6,297
Stanford Genome Technology Center	3,530
Max Planck Institute for Molecular Genetics	2,940
GBF–German Research Centre for Biotechnology	2,338
Cold Spring Harbor Laboratory Lita Annenberg Hazen Genome Center	2,104
Other	35,911
Total	842,027

Adapted, with permission, from the International Human Genome Sequencing Consortium (2001. Initial sequencing and analysis of the human genome. *Nature 409:* 860–921). HTGS, high-throughput genome sequence.

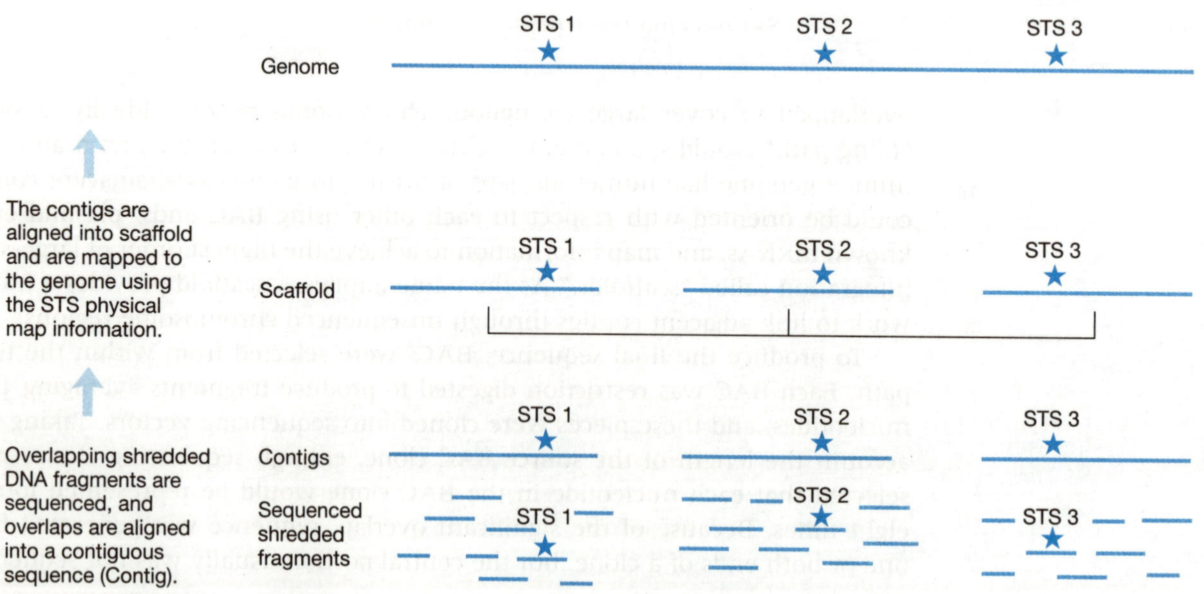

Anatomy of a Whole-genome Assembly

(Adapted, with permission, from Venter J.C., Adams M.D., Myers E.W., et al. 2001. The sequence of the human genome. *Science 291:* 1304–1351.)

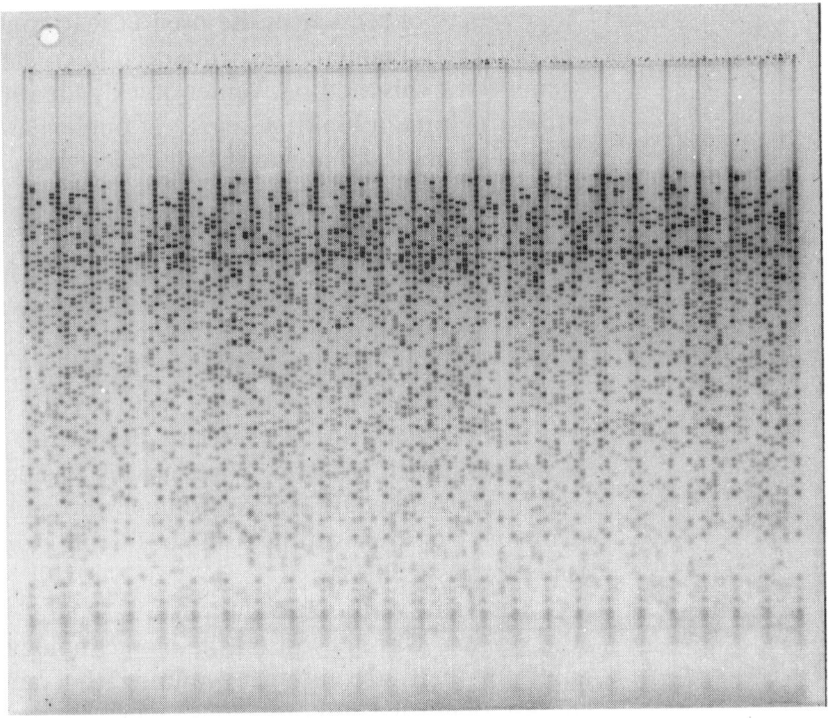

Gel Photo of BAC Fingerprint Clone Panel
(Courtesy of James McPherson, School of Medicine, Washington University, St. Louis.)

tions, and passing the samples on to automated capillary sequencers. Working around the clock, these centers could process more than 200,000 sequencing reactions per day. The raw sequences were processed and assembled using PHRED and PHRAP software written in 1995 by Phil Green and Brent Ewing of the University of Washington. According to rules established at a meeting of public genome sequencers in Bermuda in 1996, all assembled sequences of greater than 1000 bp (1 kb) were deposited in public databases within 24 hours. Using map information provided by BAC fingerprints, a program called GigAssembler strung together sequences from local assemblies of nearly 5000 contigs to produce a nearly continuous genome sequence, dubbed "the golden path." The code for GigAssembler was written in 4 weeks by Jim Kent, a graduate student at the University of California, Santa Cruz.

EXPRESSED SEQUENCE TAGS

Early DNA sequencing efforts quickly showed that cDNAs, corresponding to transcribed mRNAs (see Chapter 5), would be extremely important to verify genes predicted in genomic sequence. The expressed sequence tag (EST) method, developed in 1991 by J. Craig Venter, at the National Institute of Mental Health, provided a high-throughput method to identify the DNA-coding regions of thousands of transcribed genes. He started with a human cDNA library, created by reverse transcribing mRNA from brain tissue, and cloned the resulting cDNAs into phage particles. Using a nonspecific primer, which bound to a variety of sequences, he used PCR to randomly amplify 150–400-bp fragments. Each fragment was sequenced to create a unique "tag" for an expressed gene. Using this method, Venter quickly identified more than 3000 unique ESTs from the human brain. A search of public databases showed that only 17% of brain ESTs matched previously identified genes; 83% of the ESTs potentially represented unknown genes.

The announcement that the NIH was seeking a patent on Venter's sequenced-tagged genes raised a furor among academic researchers. Many were shocked at the implications of patenting a fragment of a gene whose identity and biological function were unknown. This move seemed to be more appropriate to a drug company than to the federal government. NIH genome office chief James Watson went so far as to say the venture was "sheer lunacy." The rift this caused with NIH director Bernardine Healy, in addition to other disagreements, led Watson to resign his position in the genome project. This vacancy was filled by Francis Collins, who had established his reputation at the University of Michigan by using positional cloning to identify the gene responsible for cystic fibrosis.

The value of the EST approach in separating the "wheat from the chaff" of the genome was not lost on the biotechnology business sector. Several companies got their start in "gene mining" by rapidly creating EST libraries from dozens of normal and cancerous tissues. By the late 1990s, these companies had collections of 50,000–100,000 unique ESTs, which theoretically represented the majority of genes in the human genome. They, in turn, applied for patents on hundreds of ESTs of potential use for diagnostics and pharmaceuticals.

Although the NIH eventually withdrew its patent application for Venter's ESTs, it raised issues that remain at the heart of gene patenting. To receive a patent, an invention must be novel and useful. An EST may pass the patent requirement for novelty. However, without complete sequence information, it is virtually impossible to demonstrate usefulness. To date, the US Patent Office has not allowed patents on parts of genes of unknown biological function. Under some circumstances, it has allowed patents on ESTs for human genes whose function can be inferred by homology with a previously described gene. So far, no EST patents have been tested in court, so it remains unclear whether they will hold up.

WHOLE-GENOME SHOTGUN SEQUENCING

Venter's EST method was important on a philosophical level. By demonstrating a simple, scalable method to identify expressed genes—to the exclusion of noncoding DNA—it cast doubt on the genome project's "head-to-toe" method. At a new institution, The Institute for Genomic Research (TIGR), Venter set out to show that the algorithm used to analyze ESTs could be used to assemble an entire genome from shotgun clones, without the need for first constructing chromosome maps of the sort that had consumed the first phase of the public genome project.

Turned down for support by the NIH genome project in 1995, Venter's team nevertheless used whole-genome shotgun sequencing to determine the entire 1.8 million base pair (mega base pair or Mbp) genome of the bacterium

Haemophilus influenzae. The TIGR group went on to use the whole-genome shotgun method to sequence several additional microbial genomes.

By 1997, only about 5% of the human genome had been fully sequenced. James Weber, of Marshfield Medical Research Foundation, and Edward Myers, of the University of Arizona, proposed that the whole-genome shotgun method be used as a means to speed human sequencing. Their plan was based on sequencing both ends of short (0.4–1.2 kb) and long (5–20 kb) human inserts cloned in *E. coli*. "Paired-end reads" of short inserts overlap to provide continuous sequence information, whereas end reads of long inserts provide information on spacing and orientation that improve sequence assembly, especially in sequences containing interspersed repetitive elements. However, most scientists doubted that the method could be scaled up to the needed 150-fold size over microbial genome sequencing. In addition, microbial genomes have virtually no repetitive DNA, and many were skeptical that assembly algorithms could deal with the large amount of repetitive DNA in a eukaryotic genome.

Genome researchers were thus shocked when, in May 1998, Venter announced that he was leaving TIGR to head a new company called Celera. A subsidiary of Applied Biosystems, Celera would use the newest capillary sequencers to sequence the human genome in 3 years at a cost of 300 million dollars. The public project responded with an accelerated timetable to produce a working draft of the human genome in 2001. The Wellcome Trust doubled support of the public project to 330 million dollars, assuming responsibility to sequence a third of the genome at its Sanger Centre. By the end of 1998, John Sulston, of the Sanger Centre, and Robert Waterston, of Washington University, reported the first genomic sequence of a multicellular eukaryotic organism, the nematode *C. elegans*.

To test the feasibility of the whole-genome assembly on a eukaryotic organism, Celera teamed with Gerald Rubin and the Berkeley *Drosophila* Project, sequencing the 120-Mbp euchromatin portion of the fruit fly genome in one year. Although the sequence did not cover the entire genome, it was considered to be *finished*—the main regions where genes exist had been sequenced. Repeat regions were not sequenced. Several large gaps remained and several anomalous arrangement issues were not resolved. The *Drosophila* sequence is planned to be completed, with all gaps filled in and all sequences confirmed to a very low error rate, by the end of 2002.

The issue of what is considered a *finished* sequence is very important and depends on what a scientist wants to do with the sequence data. For example, to identify a disease gene, a high-quality map with gene location and order is sufficient. In contrast, studies on molecular evolution (how DNA sequences change between species over time) require much more accurate sequences, and sequences from noncoding regions are also important.

From the beginning of the genome project, scientists realized the importance of obtaining accurate DNA sequences. Even a single-nucleotide mistake can have a major impact on the interpretation of a DNA sequence. For example, an insertion or deletion of a nucleotide within a coding sequence (exon) shifts the reading frame and predicts an erroneous amino acid sequence from that point on through the end of that coding region. Remember also that the genome sequence for each individual (aside from identical twins) is unique. High-fidelity sequence is needed to ascertain the bona fide sequence differences between individuals.

At the outset of the genome project, sequencing accuracy averaged about 99%, or approximately one error per 100 nucleotides sequenced. Most authorities thought it would be necessary to achieve a consistent sequencing accuracy of 99.9%, which translates into one error per 1000 nucleotides. Assuming that the average exon is 200 nucleotides in length, even at this rate, a sequencing error is likely to occur in one exon in five. As it turned out, the exponential increase in sequencing speed and decrease in sequencing cost were accompanied by an increase in sequencing accuracy. The lesson was the same as we have seen in mass production of consumer goods—the greater the mechanization, the less the human error.

FINISHING THE HUMAN GENOME SEQUENCE

Armed with experience from *Drosophila* sequencing and assembly, Celera began work on the human genome. The bulk of whole-genome shotgun sequencing was completed in a 10-month period, from September 1999 to June 2000. As with the public project, the Celera project began with DNA obtained with informed consent from multiple anonymous donors of varied sex and ethnic backgrounds. Also like the public program, Celera reported sequencing a mix of DNAs from an anonymous pool. However, Craig Venter subsequently revealed that the majority of DNA sequenced by Celera had, in fact, been his own.

Plasmid libraries were created by forcing a solution of DNA through a syringe and cloning the physically sheared fragments in three size classes: 2 kbp, 10 kbp, and 50 kbp. Sequencing the ends of each plasmid insert produced 27.27 million sequence reads averaging 543 nucleotides, a total of 14.8 billion base pairs. A total of 4.4 billion base pairs of BAC sequence data from the public genome project was "shredded" by computer to produce "faux" reads of 550 bp each. The combined data set was assembled by two methods: (1) a whole-genome shotgun assembly to generate contigs and scaffolds de novo by matching overlaps between individual 550-bp reads and (2) a "compartmentalized" assembly using the public genome map to anchor local assemblies of the combined data. Thus, having access to the sequences and chromosome maps generated by the public project greatly facilitated Celera's efforts.

During the final stretch of sequencing in 2000 that led to the publication of the human genome, Celera with a staff of 65 persons achieved a sequencing throughput of 175,000 samples and 90 million base pairs per day. Working virtually nonstop, each automated sequencing machine required only approximately 15 minutes of human intervention per day. The publicly funded centers together produced a similar output. Together, the two competing genome projects produced more than 2000 nucleotides of raw sequence per second, completing the equivalent of an entire human genome in less than 3 weeks. In the end, better than 95% of the sequences were produced at an accuracy of at least 99.9%.

BIOINFORMATICS: FINDING THE INFORMATION IN OUR GENES

After the raw sequence data were assembled into contiguous regions came the task of analyzing information encoded by the sequence. Genes are identified in raw data by comparing them to known genes or by searching for telltale gene features. In comparative methods, an algorithm such as BLAST (Basic Local Alignment Search Tool) attempts to align a query sequence with all sequences

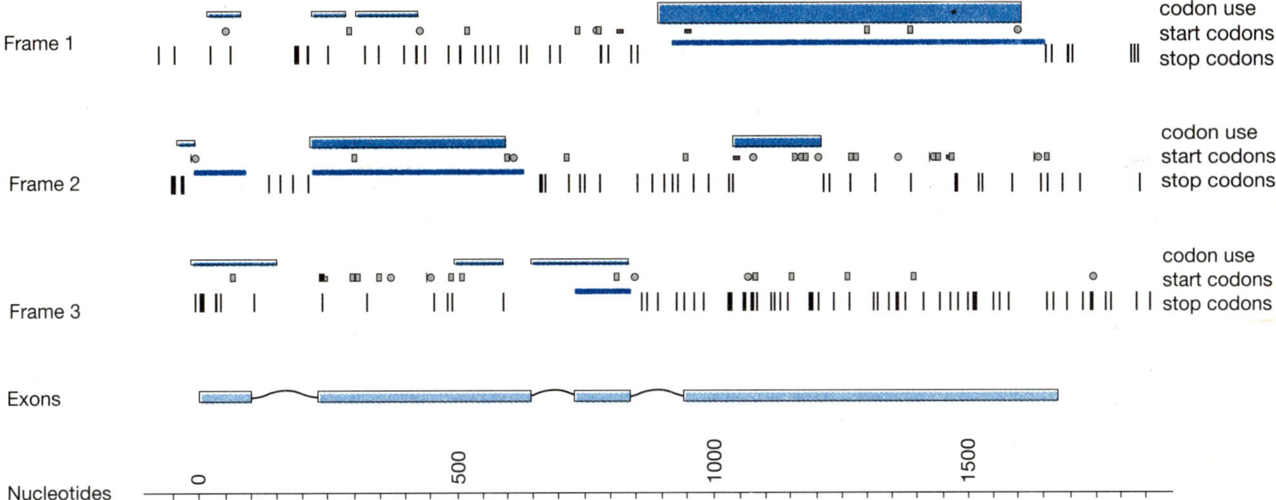

Ab Initio Gene Prediction in Three Reading Frames

Gene-finding algorithms search for signs of coding exons in the three reading frames available on each strand of the DNA molecule. Codon use identifies sequence patterns that are consistent with known biases in the use of specific codons in different organisms. The absence of stop codons defines an open reading frame—and a potential exon. A start codon followed by an open reading frame defines the first exon. The gene map (at bottom) shows that the first two exons are in Frame 2, followed by one exon each in Frame 3 and Frame 1.

in the database by introducing the fewest mismatches and gaps. A "hit" of statistical significance strongly suggests shared structural or biochemical properties. Direct evidence that a DNA sequence is transcribed (and therefore a coding sequence) is provided by a match to a known EST or mRNA sequence. Indirect evidence for a sequence being a gene is based on homology with other known genes or proteins, in humans and in other species. Here, the growing data from parallel sequencing of model species—notably *E. coli, S. cerevisiae, C. elegans, S. pombe, Drosophila,* mouse, and *Arabidopsis*—provide strong databases of previously identified genes.

Comparative methods like those described above identified about 60% of the genes found so far in the first draft of the human genome. The remaining 40% were identified *ab initio*—from the beginning—by searching for structure in the genome sequence. In some cases, computer algorithms searched relatively large "windows" of sequence for broad patterns associated with regions containing genes, such as "CpG islands," which contain a relatively high proportion of CG dinucleotides. These have been associated with transcriptional regulatory sequences. In other cases, the search focused on specific sequences that define various functional elements of genes.

The classic method of identifying a putative gene is looking for the triplet codons that specify a string of amino acids. An open reading frame is a stretch of hundreds to thousands of nucleotides that is not interrupted by a stop codon (TAA, TAG, TGA). If DNA sequences are read in triplets (remember three nucleotides per codon), a stop codon should arise once every 21 triplets on average. Remember there are 64 possible triplets 4 x 4 x 4 = 64. Three of these are stop codons, which is an average of 1 every 21 codons (or 63 nucleotides). This method works rather well with bacterial genes, which almost always begin with

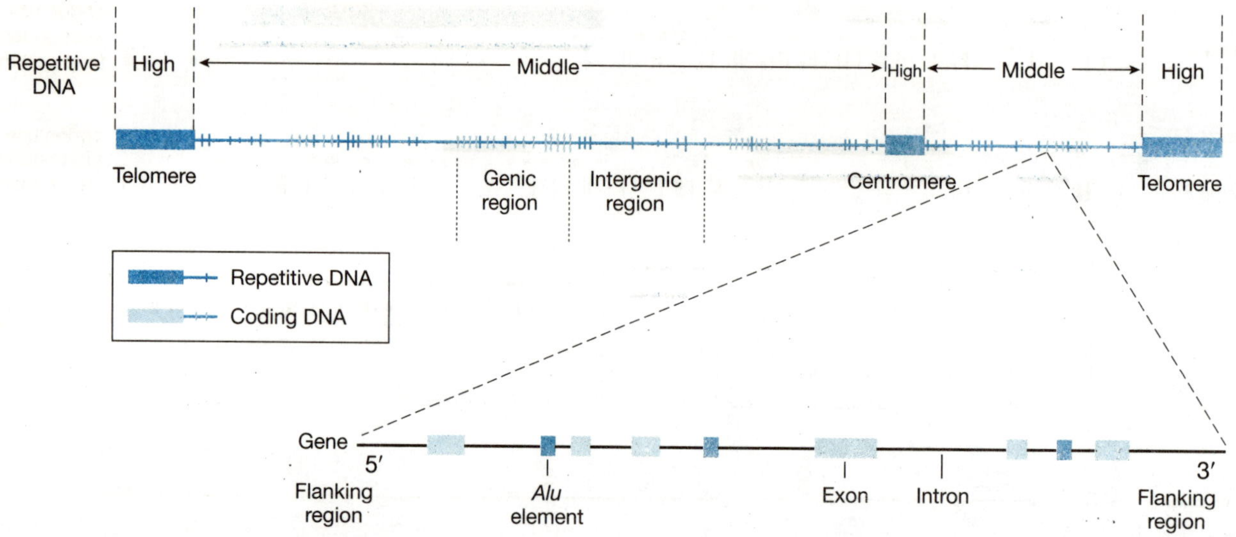

Structure of a Human Chromosome

Highly repetitive DNA includes microsatellites (2–5 bp) and minisatellites (6–50 bp). The telomere repeat is an example of a minisatellite. Middle repetitive DNA includes SINEs (75–500 bp), LINEs (up to 6.5 kb), and ribosomal RNA genes. *Alu* is an example of a SINE.

a start codon (ATG) and run continuously in-frame to their ends. It can be complicated by the fact that prokaryotes and viruses at times have overlapping genes using two or more of the six reading frames in any given region of double-stranded DNA, as viruses, in particular, have compact genomes.

Eukaryotic genes infrequently overlap and rarely occupy positions on opposite strands of the DNA molecule. However, the presence of introns makes open reading frames a far less reliable predictor of genes. Consider the "average" human gene identified by the genome project, with eight exons of about 135 bp in length, and seven introns of about 2200 bp. In introns, stop codons are of no consequence and are encountered at about the rate predicted by chance alone. The open reading frame of an exon can extend into an intron, making it virtually impossible to determine where an exon actually ends or begins. Thus, a eukaryotic gene typically is composed of a number of relatively short coding sequences that are separated by numerous introns.

```
Exon                         Intron                                      Exon

~~cgatgccaccgaggcatcag | GTttgtagacgcggcctaac~~~~~ctctttccattgtgtttcAG | tgcacccactctggacggtc~~
~~atttggaagcagtatgttg  | GTaagctgcatattgaaat~~~~~tgtgcttctcggtctctcAG  | actgaatttggaagcagta~~
~~gccacgcgtaccccgcagag | GTgagtgccagctggggctc~~~~~ctcactgcctttatttccAG | gaattagtggtgggcccttg~~
~~tgccggaaataaaacataag | GTaatcacttgctcgggtgt~~~~~ctcctttgtctgccatgcAG | ctcggggcccatcatcagac~~
~~tgataggcgtgtttttaaag | GTaagtagagcctcaaattt~~~~~ctccgtctctcaccctgcAG | ctgcatttagttcactggaa~~
~~gctcaccaccccgccgctga | GTgagccgtgcccgtaactg~~~~~ctggctgccctctattctAG | gacgcgcgggcggccatgcg~~
~~aggaaagaactgtgaattag | GTaagaggttcagagtggct~~~~~taacttcttcatatctccAG | atggagatcagtgtgagtcc~~
~~tgaagaagaaccatgaggag | GTgagtgccagctggggctc~~~~~ctcactgcctttatttccAG | gtatgagacggaggtgtccc~~
                        ^                                              ^
                   Splice site                                    Splice site
```

Intron/Exon Splice Junction Consensus Sequences

Intron-exon boundaries are defined by consensus sequences, with four invariant nucleotides (GT and AG) defining either side of the intron. A consensus sequence within the intron is also used in prediction. Even the best algorithms make many false-positive predictions of splice junctions.

A number of gene elements have characteristic "consensus sequences." Although they are not identical, consensus sequences share several invariable nucleotides in combination with a number of variable nucleotides. For example, GT and AG nucleotides flank each end of an intron, but the intron/exon boundary is further specified by a larger consensus sequence at the splice site and a more diffuse signal within the intron. The "TATA box," with the consensus sequence of TATAAAA, marks the binding site for RNA polymerase in a majority of eukaryotic promoters. The ends of genes are marked by a polyadenylation signal, AATAAA, which orchestrates the addition of the poly(A) tail found on most eukaryotic mRNAs. Used in combination with open reading frames, these consensus signals can help delineate many genes.

Taken together, the various functional elements of a gene decrease the randomness of DNA sequence and create a bias toward the use of certain nucleotide combinations. Although these overall sequence characteristics cannot be readily discerned by simple rules, hidden Markov models (HMMs) incorporate statistical information about splice sites, coding bias, and exon and intron length from known genes to identify new genes.

In the worm and fly, HMMs can correctly identify about 90% of individual exons and can correctly identify every exon in about 40% of genes. However, these figures drop to 70% and 20% in human DNA. Gene prediction programs are rather good at identifying internal exons, which have two splice junctions (left and right) adjacent to two introns. The first and last exons are frequently missed because they have only half of the sequence information used in prediction. Furthermore, exon prediction generates a large number of false positives. Consensus splice-site sequences are very common in introns, making "pseudo-exons" much more common than bona fide ones.

BEYOND THE HUMAN GENOME

DNA Microarrays

Now that scientists have turned over a catalog of tens of thousands of human genes, the challenge remains to understand how these genes work together. Although every nucleated cell (except the germ cells) possesses the same set of genes, different cells express different sets of genes at different levels. Understanding differences in gene expression is vital to understanding the development and the function of different cell types.

Pat Brown

The emergence of DNA arrays, in the mid 1990s, provided the means to analyze the expression patterns of thousands of genes at a time. Pat Brown, at Stanford University, developed an array based on cDNA libraries in common use by molecular biologists. First, a glass slide is coated with polylysine, and then different cDNAs are spotted at discrete positions onto the slide. The spotting can be done with a set of needle-like pins or even with an inkjet printer! The negatively charged DNA molecules form ionic bonds to the positively charged polylysine substrate, holding them firmly in position in the microarray. The finished microarray contains immobilized probes representing thousands of genes from a single cell type or from a single species.

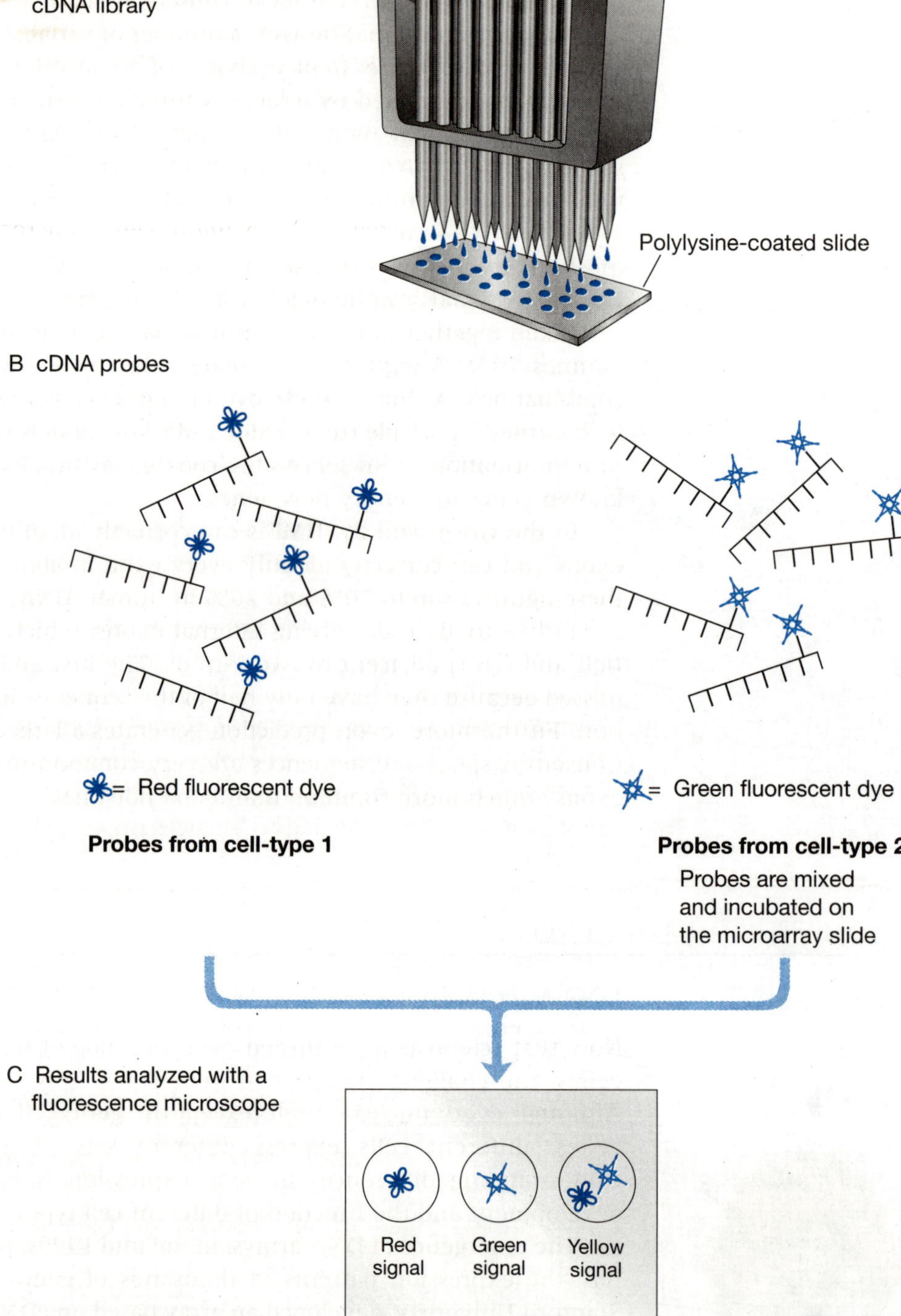

Microarray Experiment

(*A*) A cDNA library, representing the entire genome, is prepared and spotted onto a polylysine-coated slide. (*B*) cDNA probes are constructed from two cell types. Probes from cell-type 1 are labeled with a red fluorescent dye, and probes from cell-type 2 are labeled with a green fluorescent dye. The probes are mixed together and incubated on the DNA microarray slide. (*C*) The microarray is washed, removing any probe that has not specifically hybridized. A red fluorescent spot shows a gene on the array that is expressed only in cell-type 1. A green fluorescent spot shows a gene on the array that is expressed only in cell-type 2. A yellow spot shows a gene where both probes have hybridized, and it is expressed in both cell types.

In a typical experiment using microarrays, mRNA is isolated from the experimental and control cells to be compared—for example, tumor versus normal cells, mitotic versus quiescent cells, or cells from different tissues. cDNAs are made from these mRNAs (for details of this process, see Chapter 5). The cDNAs from each cell type are then labeled with either a green or red fluorescent dye, of the same types used for automated DNA sequencing. The labeled DNAs are incubated with the microarray, where they hybridize to positions containing their complementary sequences. Unbound DNAs are washed away, and the microarray is imaged under a fluorescence microscope. Red or green signals indicate genes that are differentially expressed in the two populations of cells, and the intensity of the signal indicates the level of expression.

In one early experiment, a "chip" containing 17,856 cDNAs expressed by the lymph nodes was used to compare diffuse large B-cell lymphomas from different cancer patients. Two different expression patterns—which correlated with different clinical outcomes—were found in tumors that could not be distinguished by microscopic examination (for further details on this experiment, see Chapter 7).

The company Affymetrix took a different approach to the construction of DNA microarray construction. Combining micro-photolithography borrowed from computer chip manufacture and combinatorial chemistry from the pharmaceutical industry, Affymetrix patented an industrial method to produce high-quality DNA microarrays. Rather than attaching cDNA probes to the array, oligonucleotides are built anew at individual positions on a quartz wafer using light-directed chemical synthesis. Each wafer may yield 50–400 GeneChips, depending on the number of probes in the microarray.

To make a GeneChip, the wafer is first coated with a linker molecule. Each linker molecule is attached to a single nucleotide with a protecting group that blocks polymerization, in the same manner that a dideoxynucleotide terminates a growing nucleotide chain. The protector group is sensitive to light (photolabile) and is released on exposure to ultraviolet (UV) light. A filter mask is placed between the wafer and the UV light source so that only specific positions are exposed to the light and become deprotected. A new nucleotide is then added to the chain at these deprotected positions. A new protecting group is added to these positions at the end of each synthesis step and the process starts over. A computer program controls the process, and a wafer containing a wide variety of oligonucleotides can be built up in 50–100 synthesis steps. The oligonucleotide arrays are considerably more accurate and versatile than the printed cDNA arrays. Because they are synthesized, the probes can be based on sequence information from Genbank and other public databases.

Functional Analysis

Although microarrays will help us understand large-scale coordinated gene expression, ultimately, we need to understand the biochemical roles of the proteins encoded by the approximately 30,000–50,000 genes that comprise the human genome. About 60% of known human genes have been named or assigned to classes, by virtue of a sequence similarity to genes previously identified in other organisms. Sequence comparison is useful in determining evolutionary relationships and classifying a gene by type, but it says little about the exact function of the expressed protein or the pathways in which it participates.

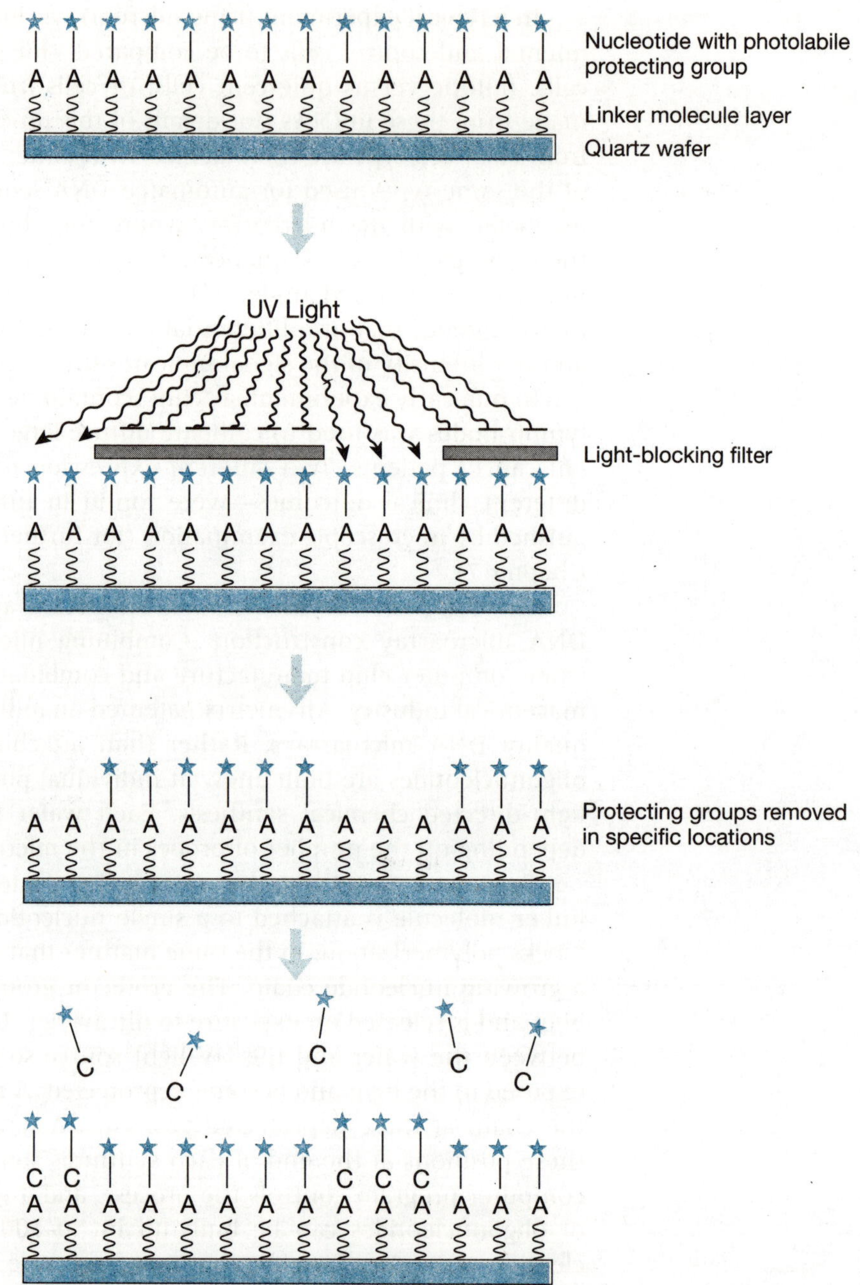

Making a Gene Chip

A quartz wafer is coated with a linker molecule and a solution of a nucleotide is added. Each linker molecule binds to one nucleotide, adhering it to the wafer. Each nucleotide carries a photolabile protecting group. With this in place, no other nucleotides can be added to each chain. A light-blocking filter is placed above the wafer, with specific areas that allow the passage of UV light. The UV light removes the protecting group from any nucleotide it reaches. A second nucleotide, with its own protecting group, is added to the wafer. The second nucleotide can only bind to nucleotides where the protecting group has been removed. The process is repeated, as more specific nucleotides are added to specific chains.

For example, identifying an unknown gene as a protein kinase suggests that the gene is generally involved in signal transduction, but it does not tell us in which signaling pathway it functions or which proteins it actually phosphorylates.

The huge amounts of data generated by high-throughput methods to sequence genomes and to analyze gene expression may lead us to believe that the work of biology is nearly done. Nothing could be further from the truth. By all accounts, the functional analysis of the genes discovered by the human and other genome projects will be much more painstaking than their sequencing. Traditionally, gene function has been deduced by comparing the phenotypes of mutant and wild-type organisms. Then, in the manner of Beadle and Tatum's *Neurospora* experiment (see Chapter 1), multiple mutants can be used to discern the roles of specific proteins in a biochemical or signaling pathway. Modern methods for isolating genes and transforming them back into cells is a continuation of this classic approach.

Research on the nematode *C. elegans* aptly illustrates where we stand in converting sequence data into a functional understanding of genes. The *C. elegans* genome was sequenced in 1998, and the organism has been intensively studied for 30 years. During this period, numerous mutant strains have been identified, and the developmental lineage of each cell has been determined. Yet, even in this superb experimental organism, fewer than 10% of its 19,000 genes have been functionally analyzed. The functional analysis of human genes becomes even more difficult with the obvious restriction that humans cannot be directly manipulated in experiments.

To date, human disease genes constitute the largest identified and functionally analyzed groups of genes. In these cases, the disease gene provides the equivalent of a "mutant" phenotype that can be compared with normal function. The Online Mendelian Inheritance in Man (OMIM) database lists 13,700 genes involved in human disease, but only 1,100 of these have been described phenotypically, and only several hundred have been cloned for detailed functional studies.

Many biologists have used the mouse—with its ease of care, rapid reproduction, and close DNA similarity—as a model in which to explore the function of key human genes and biochemical pathways. During the past 15 years, powerful methods have been developed that allow researchers to introduce new genes (transgenes) into the mouse germ line, giving rise to stable strains of mice that can be used in ongoing functional studies (discussed in detail in Chapter 5). Some scientists believe that creating legions of transgenic mice will be the best way to functionally analyze human genes. This amounts to carrying Beadle and Tatum's "one gene–one protein" experiments to a massive scale.

THE STRUCTURE AND MEANING OF THE HUMAN GENOME

The sequences published by the public and private genome projects provide an overview of the structure of the human genome. Genes span about 25% of the human chromosomes, with coding sequence accounting for a little over 1% and *intragenic* noncoding sequence (introns) accounting for about 24%. *Intergenic* noncoding regions, located between genes, comprise about 75% of the genome.

About 20–30% of genes are clustered in "CpG islands" that comprise less than 10% of the genome. Gene "deserts," tracts of 500,000 or more nucleototides without genes, comprise about 20% of the human genome. The published sequences exclude the human heterochromatin regions at the centromere and telomeres, which are believed to have few functioning genes.

Various repeated DNA sequences—ranging from dinucleotides to functional and nonfunctional genes to chromosome regions—comprise at least two thirds of the human genome. The extent of repetitive DNA in eukaryotic genomes first became apparent in hybridization studies conducted by Roy Britten and David Kohne in the 1950s and 1960s at the Carnegie Institution of Washington. They sheared DNA, denatured it at high temperature, and then measured the rate at which the single-stranded fragments reassociate (renature) with complementary sequences to form duplex DNA. Reassociation was measured by passing the DNA through a hydroxyapatite column and comparing the amounts of double-stranded DNA bound by the column versus the flowthrough of single-stranded DNA. The fraction of reassociated DNA was then plotted versus the log of the product of DNA concentration and time (C_0t).

Reassociation is a function of DNA complexity. The more complex the DNA—the greater the number of unique sequences—the longer it takes to "find" complementary matches. In contrast to bacterial DNA, which shows a single reassociation curve, eukaryotic DNA shows three distinct components. Two components of eukaryotic DNA reassociate more quickly than bacterial DNA, and the fastest component renatured nearly as quickly as poly(A) with poly(U). The fastest renaturing component is composed of highly repetitive DNA, which, like poly(A), finds complementary matches very quickly. The second component is composed of "middle" repetitive sequences. Radioactive mRNA hybridizes strongly with the slow DNA fraction of unique sequences, indicating that only a minority of eukaryotic DNA encodes protein.

Highly repetitive DNA is composed primarily of very short tandem repeats, several to thousands of repeated units lined up head to tail, like the cars of a train. Tandem repeats are classified by size of the repeated unit: microsatellites (2–5), minisatellites (6–50), and satellites (up to several hundred). Micro- and minisatellites are concentrated in and around heterochromatin regions, includ-

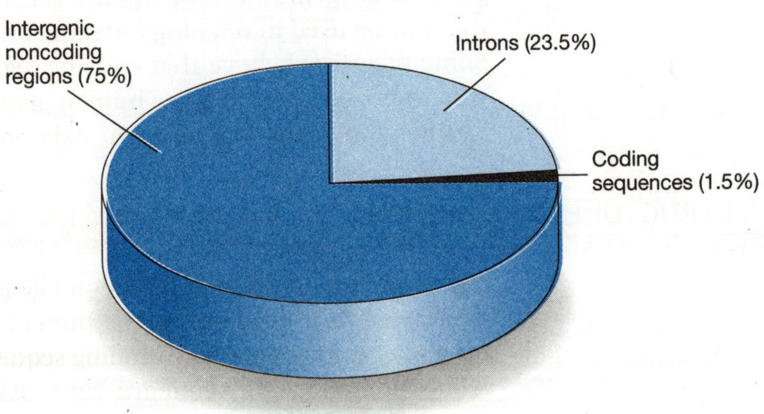

Composition of the Genome

ing the centromere and telomere. Repeat DNA located at the centromere may aid in chromatid pairing during cell division, whereas the telomere repeat TTAGGG protects the free chromosome ends from degradation.

It is thought that short DNA repeats arise when the DNA polymerase "stutters" during chromosome replication, losing its place among a string of repeats and adding an extra unit from time to time. The repeated regions thus increase in size over the long course of evolution. Inherently unstable, triplet microsatellites are associated with a class of neuromuscular and behavioral disorders, including Huntington's disease, Fragile X mental retardation, and several ataxias. The number of micro- and minisatellite repeats can vary at a particular chromosome locus,

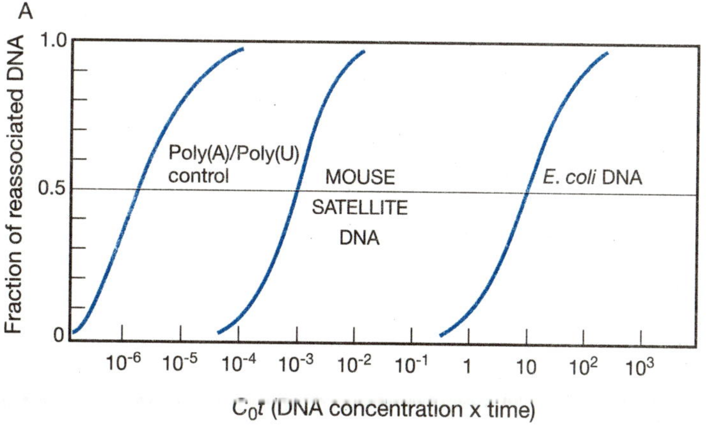

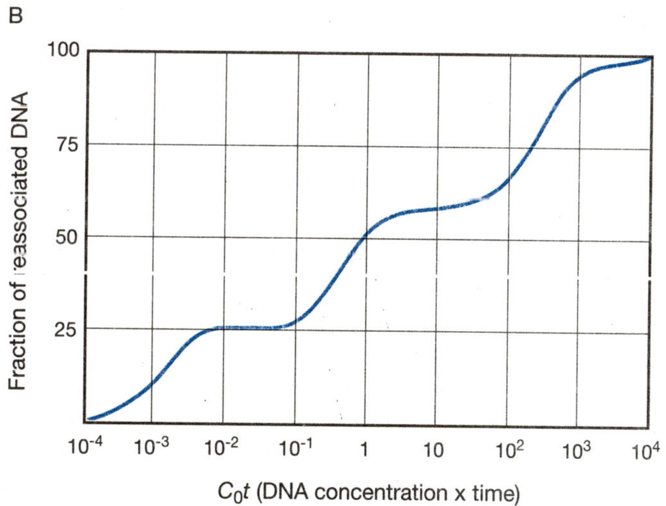

C_0t Graphs Showing Reassociation of DNA

(*A*) Strands of poly(A) and poly(U) were synthesized and used as a control. These strands reassociate very quickly. DNA from *E. coli* takes much longer to reassociate because repetitive sequences are not present. Mouse satellite DNA contains many repeated sequences, and thus, it reassociates faster than *E. coli*. (*B*) The average eukaryotic genome has a reassociation curve that looks like this. The first part of the curve is the fast component and represents highly repetitive DNA. The second part of the curve is the intermediate component where moderately repetitive DNA reassociates. The third part of the curve is the slow component, containing mainly single genes.

creating a length polymorphism that can be detected by gel electrophoresis. Minisatellite polymorphisms are generally termed VNTRs (variable number of tandem repeats), whereas microsatellite polymorphisms are termed STRs (short tandem repeats). STR polymorphisms are readily "scored" on a DNA sequencing gel and have supplanted VNTRs in human identification and disease linkage studies (see Chapter 8).

Middle repetitive DNA repeats are found outside the centromeres. Short interspersed elements (SINEs) are 75–500 bp in length, whereas long interspersed elements (LINEs) are up to 6.5 kb long. These repeats are not due to duplication errors. Included in this group are genes that encode ribosomal RNA (rDNA), which exist in several hundred copies in the human genome. However, the vast majority of middle repetitive DNA is derived from transposable elements (so-called jumping genes) that have copied themselves and moved from chromosome to chromosome. Transposable elements comprise more than half of the human genome. An additional 10% of the human genome is involved in large-scale duplications of greater than 5 kb. Many of these duplications are *pseudogenes,* nonfunctional copies of bona fide genes that may be located on entirely different chromosomes. Pseudogenes, along with other types of repetitive DNA, greatly confound sequence assembly.

WHY SO LITTLE GENE INFORMATION AND WHY SO MUCH "JUNK?"

Some fraction of noncoding DNA is far from junk. Noncoding regions between genes contain promoter and enhancer sequences that regulate when, where, and to what level mRNA is synthesized, which in turn determines how much of a specific protein is produced (see Chapter 3). This coordinated activity of genes is especially important during embryonic growth and development. Introns contain information that directs RNA splicing, including alternative splicing through which a diversity of proteins is produced by a single gene. There is increasing evidence that the majority of human genes generate two or more transcripts by alternative splicing, effectively doubling the information in the genome. Thus, gene number does not necessarily reflect the true extent or complexity of an organism's protein complement.

The noncoding DNA that surrounds promoters, enhancers, and splice signals can be explained as the necessary evolutionary "grist" from which these gene regulators emerge. Short tandem repeats might be dismissed as mere mistakes in the complex machinery needed for replicating DNA. The transposable elements are more difficult to explain.

Two transposable elements—a LINE called L1 and a SINE called *Alu*—are the most frequent gene-size sequences in the human genome. About 1 million copies of *Alu* and 500,000 copies of L1 comprise about 27% of human DNA by weight. Each of these hundreds of thousands of copies arose from an individual "jump" at some point in human evolution. L1 transposons came into the genome about 150 million years ago, so we share these sequences with many vertebrates, including mice. *Alu* is only about 65 million years old, young enough that its jumping is confined to primates, the "monkey" branch of the evolutionary tree. Several thousand *Alu*s are found only in humans, so they have made the jump in the past 6–7 million years, after humans diverged from a common ancestor with chimps. *Alu*s are dispersed throughout the genome.

L1 carries a gene for the enzyme reverse transcriptase (RT), which converts L1 RNA into a mobile DNA copy. This same enzyme enables retroviruses, such as human immunodeficiency virus (HIV), to insert into positions on the human chromosomes from which they cause infection. Current thinking holds that the retroviruses borrowed the *RT* gene from transposons in the host genome. As the largest endogenous source of RT, L1 is largely responsible for a creation of processed pseudogenes. These pseudogenes arise when an mRNA is reversed-transcribed into a cDNA, integrated into the chromosome, and subsequently mutated to become nonfunctional.

The L1 RT enzyme has an additional function that has not been found in retroviral RT enzymes. L1 makes staggered nicks on each side of the DNA molecule, thus providing a site into which a reverse-transcribed DNA copy can integrate at a new chromosome position. The initial nick is made at the consensus sequence AATTTT, but the second nick is unpredictable. Thus, the jumping about that is mediated by L1 RT lies somewhere between predictability and randomness.

With no functional genes at all, *Alu* is a defective transposon. It must be mobilized for transposition using RT from another source. *Alu* is thought to rely on the RT enzyme produced by L1, making it a parasite in a molecular symbiosis that puts a point on the Jonathan Swift poem:

So, naturalists observe, a flea
Hath smaller fleas that on him prey;
And these have smaller still to bite 'em:
And so proceed ad infinitum.

Biologists struggle to find the meaning of transposons that take so much space in the human genome. Some believe transposons have been successful because they have enabled the genomes they inhabit to compete more successfully on the battleground of evolution. *Alu* reached a jumping peak about 40 million years ago, with a rate of perhaps one new jump in each newborn primate. This period roughly coincided with the evolutionary "radiation" that led to the development of forerunners of modern branches of the primate family. *Alu* accumulates in gene-rich regions, and transcription is activated under stressful conditions. These facts have led some to speculate that *Alu* transposition has had a positive role in primate evolution, creating gene variations that provided a selective advantage for evolving primates as they adapted to changes in the earth's environment.

Other biologists believe that the success of *Alu* and other transposons rests entirely in their own reproductive ability. *Alu*, which makes no protein, may exist solely for its own replication. This fits Richard Dawkins' moniker of "selfish DNA." Continuing on this line of thought, one might craft a definition of life from the viewpoint of DNA: *the perpetuation and amplification of a DNA sequence through time*. By this view, *Alu* is a supremely successful life form, with a million copies of itself perpetuated in each of the billions of humans and primates alive today.

Of course, most people would not like to entertain the possibility that our genomes are merely vessels for the reproduction of some selfish and fidgety DNA. Those so inclined might find some cheer knowing that some transposons have become as extinct as the dinosaurs. The *Alu*/L1 symbiosis replaced an earlier and virtually identical symbiosis between sequences called Mer and L2.

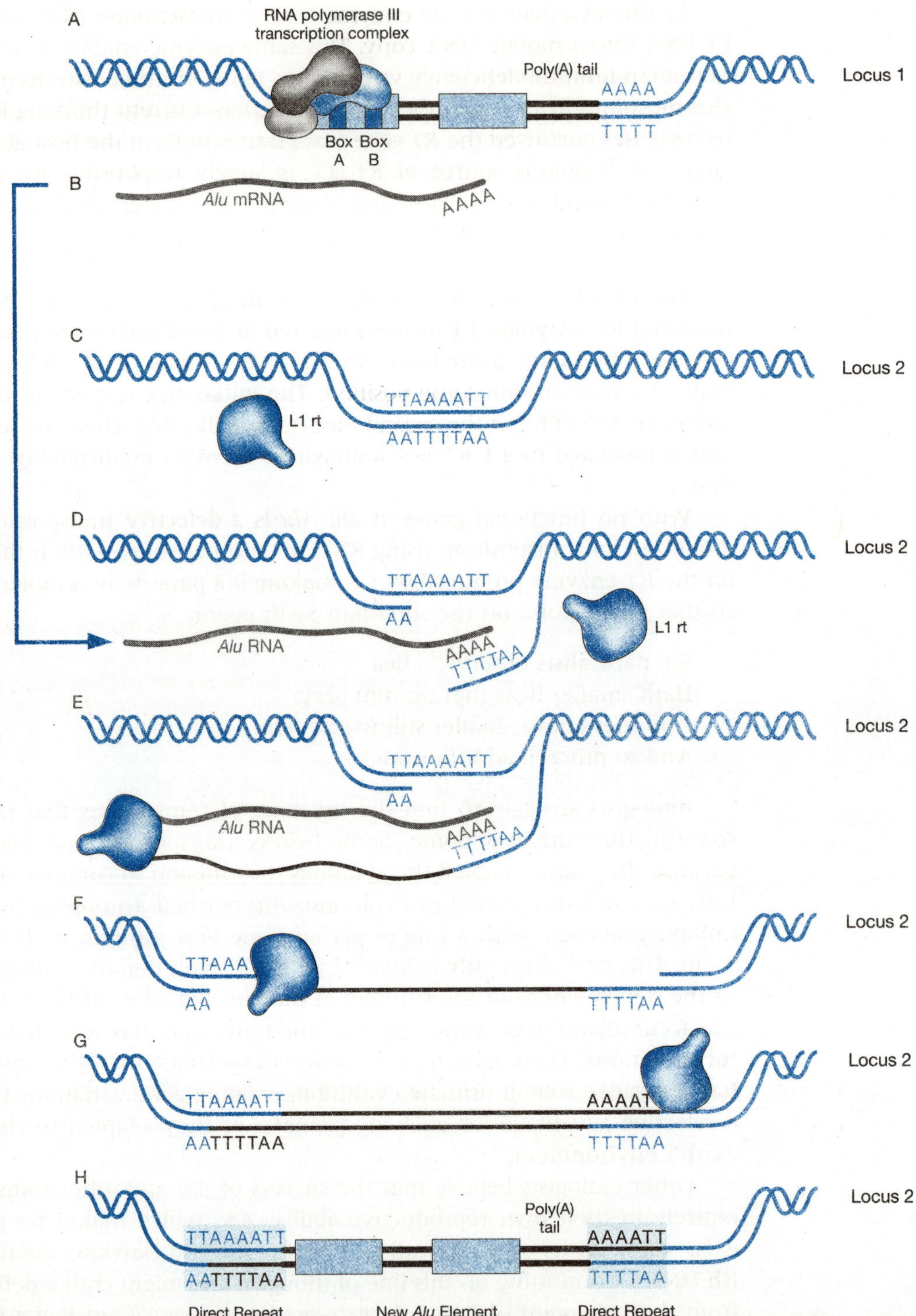

Model for *Alu* Transposition

(*A*) The *Alu* element at Locus 1 contains two G-C rich regions (*blue*). Box A and Box B contain sequences recognized by transcription factors and RNA polymerase. (*B*) *Alu* element is transcribed into mRNA. (*C*) L1 reverse transcriptase (RT) nicks one strand of DNA at Locus 2 at the consensus sequence AATTTT. (*D*) *Alu* mRNA transposes to Locus 2, where its poly(A) tail hydrogen bonds to Ts at the nick site. (*E*) L1 RT binds to this primed region and reverse transcribes the *Alu* mRNA into DNA. (*F*) L1 RT nicks the opposite DNA strand a short distance from the initial site. (*G*) L1 RT uses flanking regions as a primer to synthesize a second complementary strand, and any remaining nicks are sealed by DNA ligase. (*H*) This results in a new *Alu* element integrated at Locus 2. The transposition creates direct repeats on either side of the element.

Although they have not jumped for at least 100 million years, Mer and L2 "fossils" still litter about 5% of the human genome.

All told, various types of transposons make up more than half of the human genetic endowment. If possession is nine-tenths of the law, one might ask "Who is in charge of this genome, anyway?" The truth is, we will probably never know for certain whether transposons are part of the plan or just little fleas playing dice with our DNA.

FOR FURTHER READING

Celera Web Site: http://www.celera.com/

Dieffenbach C.W. and Dveksler G.S. 1995. *PCR primer: A laboratory manual.* Cold Spring Harbor Laboratory Press, Cold Spring Harbor, New York.

DOE Web Site: http://www.ornl.gov/hgmis/

Human Genome Project Information: http://www.ornl.gov/hgmis/

Kevles D.J. and Hood L. 1993. *Code of codes: Scientific and social issues in the human genome project.* Harvard University Press, Cambridge, Massachusetts.

Micklos D., ed. 1999. DNA from the beginning (http://www.dnaftb.org). Dolan DNA Learning Center, Cold Spring Harbor, New York.

NCBI Web Site: http://www.ncbi.nlm.nih.gov/

National Center for Biotechnology Information: http://www.ncbi.nlm.nih.gov/

Ridley M. 2000. *Genome.* Harper Collins, New York.

Sambrook J. and Russell D. 2001. *Molecular cloning: A laboratory manual,* 3rd edition. Cold Spring Harbor Laboratory Press, Cold Spring Harbor, New York.

Sanger Institute, The Wellcome Trust: http://www.sanger.ac.uk

The Human Genome. 2001. *Nature* **409:** 745–964.

The Human Genome. 2001. *Science* **291:** No. 5507, 1145–1434.

CHAPTER 7

The DNA Science of Cancer

On December 23, 1971, United States President Richard M. Nixon signed the National Cancer Act, initiating the so-called "war on cancer." It was the culmination of the growth of federal sponsorship of medical research that had been accelerating in the United States since World War II. During the ensuing 30 years, 46 billion dollars in research support funneled through the National Cancer Institute stimulated an unprecedented expansion of basic biological research. The increased funding fell on a biological research community enormously excited by its new understanding of the role of DNA in directing the machinery of life. With a basic molecular description of DNA replication and protein synthesis in hand, scientists were now anxious to apply what they had learned to study cancer.

By the late 1960s, sophisticated methods had been developed to grow mammalian and human cell lines in culture dishes in the manner of bacteria. It was also found that certain animal viruses, known as tumor viruses, can "transform" cultured cells into cancer cells. It thus became possible to mimic cancer in a Petri dish and to perform controlled experiments on the origin and progression of malignancy. Bacteriophages had provided a means to probe the molecular genetics of bacterial cells; now the tumor virus made possible a similar approach to mammalian cells. The potential power of the tumor virus system prompted a migration of former phage biologists into the field of basic cancer research.

Research on the molecular genetic basis of cancer came of age precisely a decade after passage of the National Cancer Act. In the fall of 1981, research teams at Cold Spring Harbor Laboratory, the Massachusetts Institute of Technology, and the Sidney Farber Cancer Center of Harvard University announced that they had tracked the origin of human cancer directly to the genetic material. Working independently, each group employed the tools of recombinant DNA to isolate oncogenes—specific genes that induce cultured cells to become malignant. Understanding the cellular function of proteins encoded by oncogenes led to an enormous growth in our understanding of how metabolism is perturbed in cancer cells. Techniques developed to dissect the molecular genetics of cancer cells were translated into quick progress in the fight against human immunodeficiency virus (HIV) and provided a foundation for the Human Genome Project.

In the years following the National Cancer Act, improved diagnosis and treatment have increased the 5-year survival rate from 50% to 60%. Death rates for childhood cancers and cancers of the stomach, uterus, and colon have dropped dramatically. However, death rates from lymphoma, lung, prostate, liver, and brain cancers have risen markedly. Although cancer is still not conquered, many believe that even difficult cancers will come under control with a new generation of diagnostics and treatments based on a detailed understanding of the inner workings of tumor cells.

WHAT IS CANCER?

Cancer is perhaps the most perplexing and ubiquitous group of diseases. Forms of cancer strike every part of the body. "Liquid tumors," such as leukemias and lymphomas, affect the white cells of the blood. "Solid tumors," such as sarcomas and carcinomas, are found in various organs and structural tissues. Although seemingly different diseases, each manifestation of cancer is characterized by the uncontrolled growth and spread of abnormal cells. Cancer cells proliferate without regard for the regulatory mechanisms that restrain cell reproduction, crowding out their normal counterparts.

Scientists have always looked to the cell's replicative machinery for a hint about the common basis of all cancers. This line of reasoning led to DNA, for errors in cell replication are ultimately errors in DNA replication. To understand the origin of malignancy, one must focus on the somatic (body) cells that make up tissues and organs. Most cancers are thought to arise when DNA changes within a *single* somatic cell lead to loss of growth control and rapid proliferation. Thus, tumor cells are progeny, or clones, derived from a single aberrant ancestor. This is the "somaclonal" theory of cancer formation.

All cancers can be thought of as genetic diseases, in the sense that they originate with changes to the genetic material, DNA. There is increasing evidence that a number of cancers are also genetic diseases in the sense that a predisposition to the condition is inherited. There are familial, or inherited, forms of several cancers, including neuroblastoma, colon carcinoma, and breast cancer. Still, inherited cancers probably only account for 5–10% of all cases.

Although cancer affects people of all nationalities, cancer "hot spots"—regions of the world with high cancer rates—give clues about cancer causes, as well as prevention. Epidemiologists attempt to find environmental, dietary, cultural, or lifestyle factors that are common to the regions where cancers are most frequent.

- ***Lung cancer*** is directly related to smoking. The highest rates are found in developed countries, where people can afford the luxury of a cigarette habit. Lung cancer cases rose in men following World War II, but have fallen in recent years. Unfortunately, smoking—and lung cancer—are on the rise in women. Lung cancer is the leading cause of cancer deaths in the United States, but deaths would drop 85% if no one smoked.
- ***Liver cancer*** is related to hepatitis B virus and also to aflatoxin—a powerful mutagen released by mold on peanuts and other foodstuffs. High rates of liver cancer roughly coincide with areas of hepatitis infection and where foodstuffs are improperly stored under damp conditions.
- ***Stomach cancer*** appears to be strongly influenced by diet. The high rates found in Asia are puzzling, but are thought to be related to pickled and smoked foods prevalent in the diet.

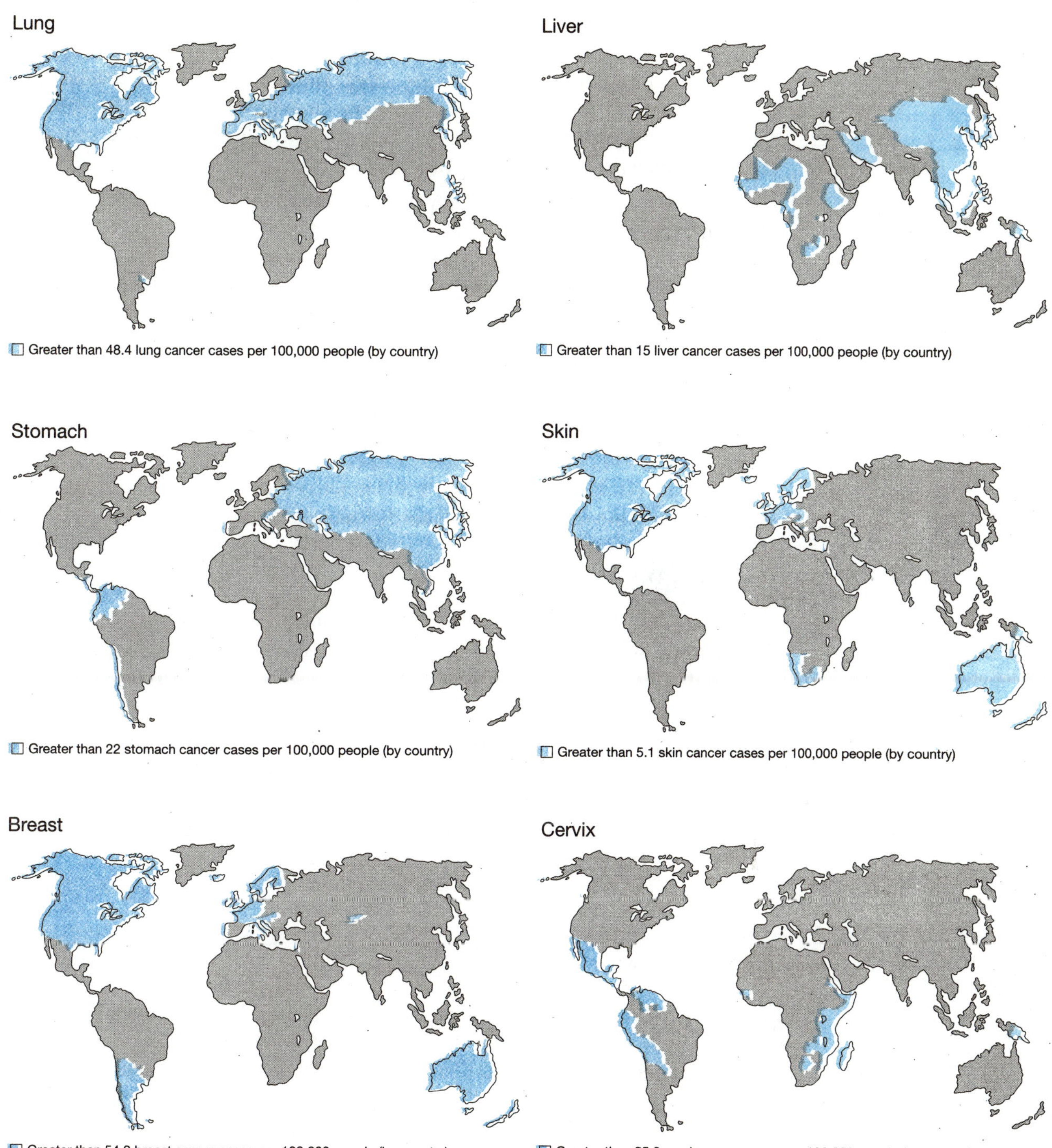

Maps of Cancer Incidence

(Adapted, with permission, from Ferlay J., Bray F., Pisani P., and Parkin D.M. 2001. Globocan 2000: Cancer incidence, mortality, and prevalence worldwide. International Agency for Research on Cancer [IARC], France.)

- *Skin cancers* are mostly caused by sunlight exposure. Pigments in dark or tanned skin absorb some of the ultraviolet light that is responsible for the DNA mutations behind skin cancers. High rates of deadly melanomas are found mainly in the middle latitudes, where large populations of fair-skinned Caucasian people live. The highest rates are found in dry, sunny climates, where people can spend a lot of time outdoors.
- *Breast cancer* "hot spots" are suspected by many people to be caused by the environment. However, a woman's lifetime exposure to estrogen (either produced by the body or taken as a drug) is probably the best predictor of cancer risk. This is roughly measured as the length of time between the onset of menstruation and first childbirth. Breast cancer rates are highest in affluent areas, where good nutrition decreases the average age at which girls reach puberty and where working women are most likely to postpone childbirth. A high-fat diet may also contribute to breast cancer.
- *Cervical cancer* is related to genital warts caused by papillomavirus. It is one of the most preventable cancers, since precancerous cells can be detected by a Pap smear test. Thus, areas of high incidence correlate with underdeveloped regions with poor routine health care for women.

CANCER IN THE UNITED STATES

- About 1.25 million new cases of cancer are diagnosed each year.
- Approximately 550,000 die of cancer each year—about one person every minute.
- About 25% of all deaths are due to cancer.
- Treatment of cancer equals about 5% of all health care costs.
- Half of all men and one third of all women in the United States will develop some form of cancer during their lifetimes.
- More than 9 million people alive today have a history of cancer, including those currently under treatment and those considered "cured."

TUMOR VIRUSES AS CANCER MODELS

Close study of bacteriophage replication within *Escherichia coli* during the 1950s and 1960s revealed many basic mechanisms of gene regulation. Thus, when molecular geneticists moved into basic cancer research, it was logical that they should turn again to simple viral systems for clues to the basic mechanisms of oncogenesis. The viruses that came into use belong to a group known as tumor viruses, which cause malignancies in various vertebrate animals. Tumor viruses share striking similarities with bacteriophages:

- Both types of viruses have small genomes.
- Both are essentially protein capsules surrounding a core of nucleic acid.
- Both can be propagated in cultured cells.
- Both display a lysogenic phase, during which the viral genome integrates into the host-cell chromosome, followed by an infective (lytic) phase, during which virus replication occurs.

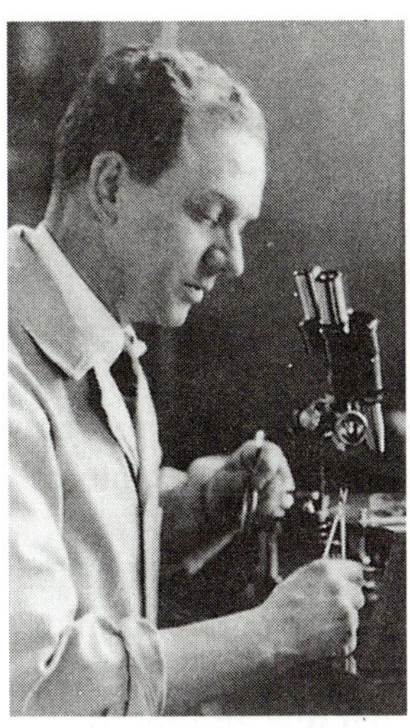

Peyton Rous, ca. 1911
(Courtesy of The Rockefeller University Archive Center.)

- Both may transduce host-cell DNA sequences. During lysogeny, the integrated virus may pick up host-cell DNA sequences and subsequently transmit them to other cells during infection.

Several of the most intensively studied tumor viruses are Rous sarcoma virus (RSV), adenovirus, polyomavirus, and simian virus 40 (SV40). The closely related polyomavirus and SV40 have ring-shaped genomes the size of bacterial plasmids (5292 and 5243 bp, respectively). At 36,000 bp, adenovirus is smaller than the bacteriophage λ (48,000 bp) and RSV (60,000 bp).

The relationship between viruses and cancer was first established by Peyton Rous in 1911. He found that in chickens, a filterable extract from sarcomas (connective tissue tumors) can be injected to induce tumors in healthy chickens. The infective agent was later shown to be a virus whose genome is composed of RNA. RSV belongs to the class of RNA viruses called retroviruses, which integrate into host genomes by using a reverse transcriptase to convert their genetic material from RNA to DNA.

The first DNA tumor viruses—rabbit fibromavirus and papillomavirus—were isolated by Richard E. Shope during the 1930s. He also demonstrated a marked

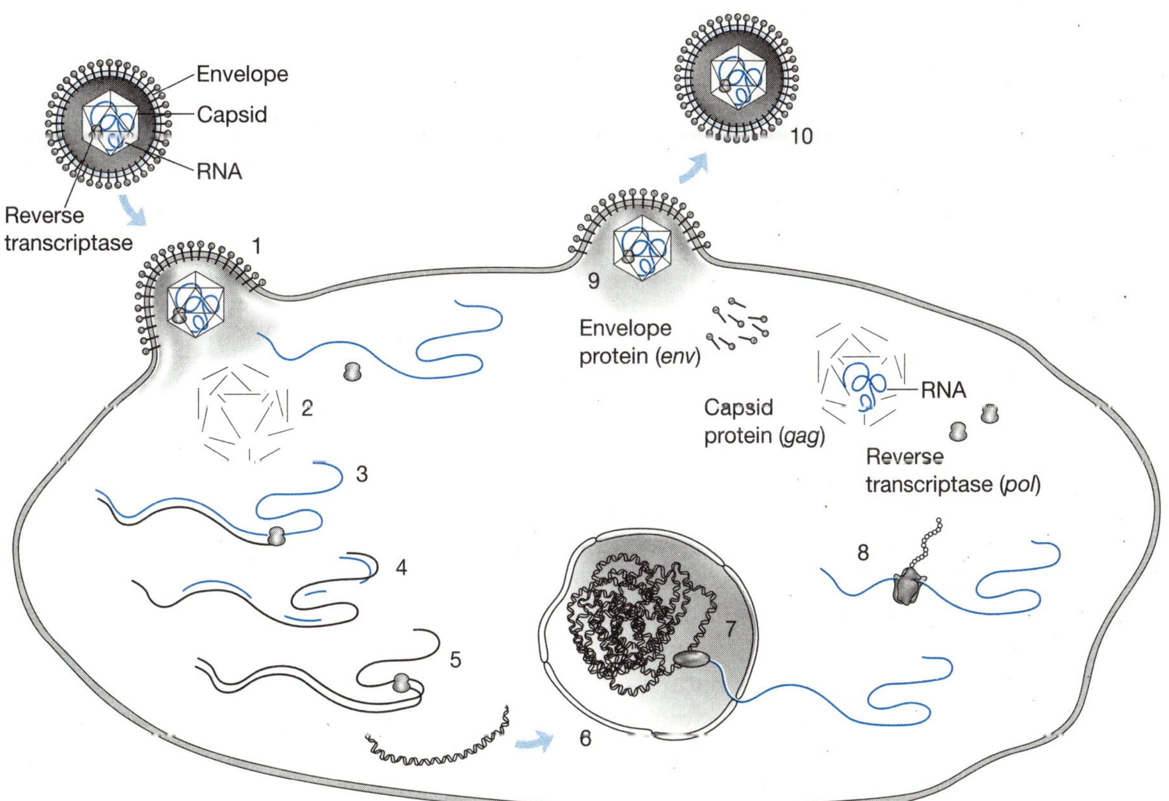

Life Cycle of Rous Sarcoma Virus
(*1*) Viral envelope fuses with host-cell membrane. (*2*) Capsid uncoats. (*3*) Reverse transcriptase synthesizes complementary DNA strand from RNA template. (*4*) RNase H degrades RNA strand. (*5*) Reverse transcriptase synthesizes second DNA strand. (*6*) DNA double helix integrates into host-cell chromosome. (*7*) Host-cell RNA polymerase transcribes viral genes into RNA. (*8*) mRNA translated into viral proteins. (*9*) RNA genome and reverse transcriptase packaged into capsid. (*10*) Viral particle buds from host-cell membrane.

species-specific effect: papillomavirus, which causes only benign warts in wild rabbits, induces malignant tumors in domestic rabbits. Thus, tumor viruses often have normal, or permissive, hosts in which they cause only mild disease, and nonpermissive hosts in which they cause tumors.

The genetic mechanism of tumor induction by DNA and RNA viruses was not amenable to serious research until the development of in vitro systems that allow mammalian cells to be studied in isolation. Increasingly sophisticated techniques were developed in the 1950s and 1960s for growing pure cultures of mammalian cells in defined media containing amino acids, vitamins, salts, glucose, and animal serum. Particularly important was the isolation of stable cell lines that can be propagated indefinitely in culture.

Most primary cells—those taken directly from a living organism—are difficult to maintain in culture. They have a limited life span; for example, primary cells from humans typically undergo about 50 divisions before dying. However, a number of "immortal" cell types have been discovered that live and reproduce essentially forever in culture. Usually, cell lines are derived from tumor cells, whereas others have arisen from mutations that occur in previously established cultured cells. Cell lines that have come into common use include the human cervix HeLa cell, rat embryo fibroblast (REF), Chinese hamster ovary (CHO), and mouse NIH-3T3 fibroblast. When grown in culture, noncancerous cells display four important characteristics:

- *Flattened morphology.* Cultured cells, such as fibroblasts, usually have a flattened shape.
- *Anchorage-dependent growth.* The cells can only grow when adhered to a solid support; the bottom of the culture plate provides the needed substrate for attachment. Noncancerous cells fail to grow when suspended in soft agar.
- *Growth in confluent monolayer.* New cells are only reproduced to the point of spreading evenly across the culture dish to form an uninterrupted layer that is only one to several cells thick.
- *Contact inhibition.* Replication and growth ceases when cells come in physical contact with their neighbors.

The characteristics of points three and four above are analogous to organized tissues, where cells grow only enough to form and maintain specific structures. For example, when a tissue is injured, only enough new cells are reproduced to replace those that were killed in the injured area, and then growth ceases.

By the mid 1960s, it was established that tumor virus infection of cultured cells can take either of two markedly different courses, which correspond to those previously observed in bacteriophage infection of *E. coli*. During lytic infection, new viral particles are replicated, and host cells are lysed. In contrast, during lysogenic-like infection, the virus integrates into the host's DNA and virtually "disappears"—there is no evidence of virus replication or cell lysis. However, lysogenic-like infection of cultured cells by tumor viruses *does* induce dramatic changes in cell morphology and growth. This parallels the transformation process in pneumococcus described by Fred Griffith and Oswald Avery, so the term transformation was also applied to cultured cells that have incorporated tumor virus DNA. Transformation thus took on the expanded meaning of in vitro changes in cultured cells that are the "test tube" equivalent to cancer in a living animal.

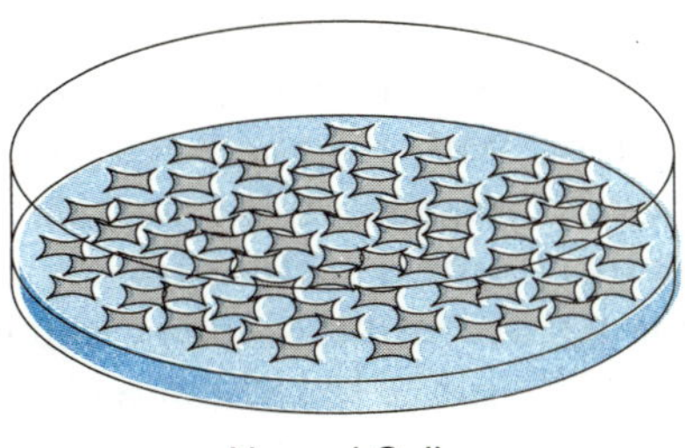

Normal Cells

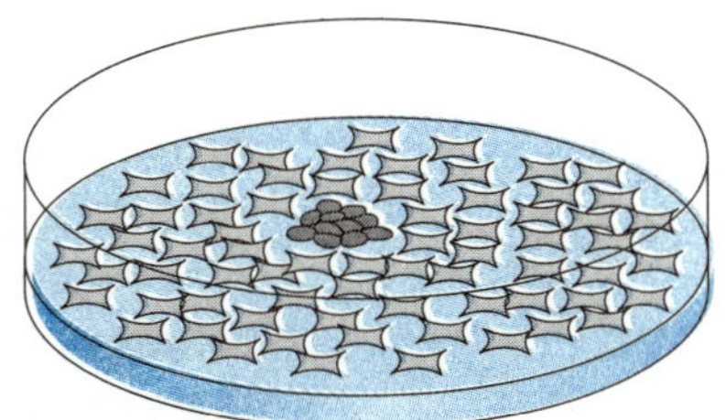

Transformed Cells

Contact-inhibited cells grow in confluent monolayer

Dense foci of transformed cells are not contact inhibited

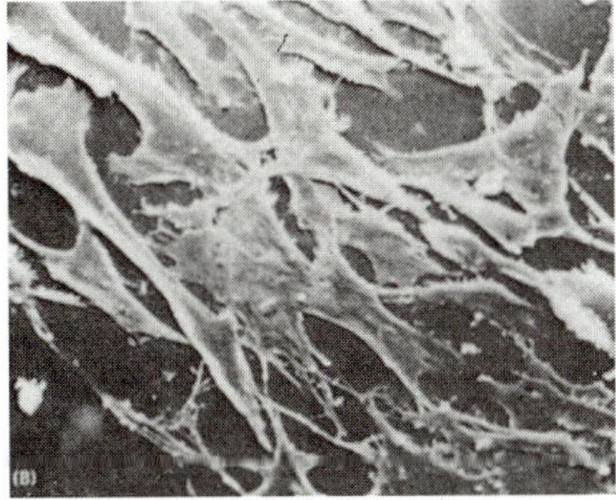

Flattened morphology

Rounded morphology

Characteristics of Normal and Transformed Cells
(Courtesy of L.B. Chen, Dana Farber Cancer Institute.)

Cultured cells transformed by tumor viruses look and act strikingly different from their noncancerous counterparts. They display a rounded, globular morphology. They are not contact-inhibited—cells derived from an original transformant pile up on one another to form a clump called a focus (plural, *foci*). The dense focus of transformed cells is easily spotted against the confluent monolayer of noncancerous cells. Transformed cells are anchorage-independent and will continue growing when transferred to soft agar. This focus assay allowed the transforming ability of various viruses to be tested in a variety of different cell lines. As was the case for the Shope papillomavirus, viruses that cause only benign disease in their permissive hosts may cause malignant transformation of the cultured cells of nonpermissive species. For example, adenovirus causes flu-like infections in humans; it is not associated with malignancy of any type. However, adenovirus transforms many rodent cell lines, including 3T3 mouse fibroblasts. Although this is a contrived situation, transformation of cultured cells has proven to be an extremely convenient, and accurate, model of many steps in oncogenesis.

The advent of restriction enzymes and recombinant DNA made possible the detailed analysis of viral transformation. Restriction mapping of genomic DNA isolated from transformed cells showed that tumor virus DNA integrates randomly into the host chromosome as a "provirus." The DNA genomes of adenovirus, polyomavirus, and SV40 DNA are directly integrated, whereas the RNA genome of a retrovirus is first reverse-transcribed into a DNA copy. In either case, the provirus DNA is transcribed along with the host-cell DNA to make messenger RNA (mRNA), which is then translated into polypeptides at the host-cell ribosomes.

Deletion analysis was coupled with restriction mapping to show the organization of tumor virus DNAs. In this strategy, a portion of the viral genome is deleted, and the mutated virus is tested for replication, capsule formation, and transforming ability. A change in activity can then be correlated to a specific DNA fragment, which is then mapped with respect to other functional sequences.

The tumor virus genome is typically composed of only a handful of genes that fall into two categories. One set of genes codes for structural proteins that form the viral particle within the host cell. A second set of proteins regulates the expression of both the viral genes and those of the host cell. Some of the genes from this latter class are oncogenes that code for proteins which bring about malignant transformation.

Adenovirus contains two oncogenes, *E1A* and *E1B* (E is for early), which are located in a region of the genome that is transcribed in the *early* phase of infection. SV40 and polyomavirus each contains two analogous oncogenes called tumor (T) antigens, confusingly designated *large T antigen* and *small t antigen* in SV40, and *middle T antigen* and *large T antigen* in polyomavirus. RSV contains a single oncogene, viral *src* (v-*src*).

THE ORIGIN OF VIRAL ONCOGENES: TRANSDUCTION

J. Michael Bishop (*Left*) and Harold Varmus, 1989 (Courtesy of the University of California, San Francisco.)

A detailed understanding of the evolutionary origin of viral oncogenes came in the late 1970s from studies of the v-*src* gene. J. Michael Bishop and Harold Varmus, of the University of California at San Francisco, used a DNA copy of v-*src* to probe the cellular DNA of various animals. A closely related homolog, cellular *src* (c-*src*), was found in the DNA from normal chickens, fish, mammals, humans, and even *Drosophila*. Subsequently, other viral oncogenes were shown to have related cellular forms that have been conserved from species to species during the course of evolution.

In the early 1980s, Hidesaburo Hanafusa, at The Rockefeller University, found that retroviruses with an incomplete *src* oncogene can still produce tumors in animals. Viruses subsequently isolated from the malignant tumors had reassembled a complete *src* gene, indicating that they had become oncogenic by "capturing" the missing portion of the *src* gene from the host-cell DNA.

The work of Bishop, Varmus, and Hanafusa showed that retroviral oncogenes are modified cellular genes captured from the genomes of their vertebrate hosts. During provirus integration, all or part of the coding region of a cellular gene may be integrated within the sequences of the viral genome. Thus, the virus acts like a transducing phage, removing a cellular gene as it excises from the host DNA. The cellular gene is then packaged, along with viral sequences, into an infectious virus particle.

Oncogenes, like c-*src*, encode regulatory proteins that are essential for normal cell proliferation. However, overexpression of the altered v-*src* gene apparently disrupts normal growth control. Consistent with this hypothesis, Bishop and Varmus found that the *src* gene is expressed at low levels in normal cells and at inappropriately high levels in virally transformed cells. These normal cellular

Hidesaburo Hanafusa
(Courtesy of Cold Spring Harbor Laboratory Archives.)

genes have the potential to become oncogenes through mutation or increased expression.

Overexpression of a viral or cellular oncogene occurs by several mechanisms in cancers caused by retroviruses. Some oncogenic viruses, like RSV, carry a captured cellular oncogene that is under the direct control of a strong viral promoter. The v-*src* gene is under the control of a promoter located at the end of the viral genome, called the LTR (long terminal repeat). Other oncogenic retroviruses carry no oncogene; however, a cellular oncogene adjacent to the site of provirus integration may be overexpressed when it comes under the control of the strong viral promoter. For example, the proviral DNA of avian leukosis virus (ALV, which causes leukemia in chickens) nearly always inserts upstream of c-*myc*, a known cellular oncogene.

Experiments in which the c-*src* gene is overexpressed by placing it in front of the viral LTR show that increased levels of *src* gene product, alone, are not sufficient to cause transformation. Therefore, transformation also appears to require a mutation in the c-*src* gene that arises during the viral infection.

There is evidence that transforming viruses, like RSV, integrate preferentially at "fragile sites" on the host chromosome that are prone to breakage. These regions often correspond to "hot spots of recombination," where exchange of DNA sequences occurs at an unusually high frequency. RSV often integrates into a recombinational hot spot located at the 3′ end of the c-*src* gene. When RSV excises from the host genome and transduces the cellular gene, the 3′ coding sequences are left behind.

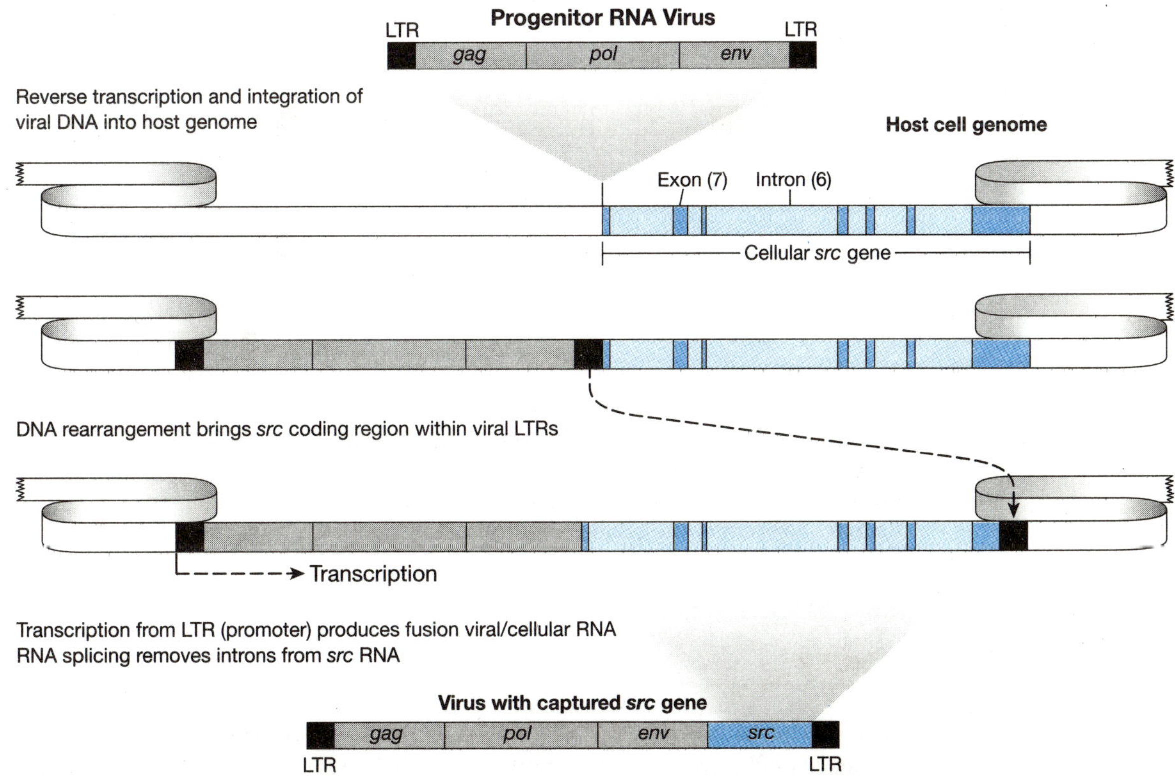

Capture of the Cellular *src* Oncogene by a Progenitor Rous Sarcoma Virus

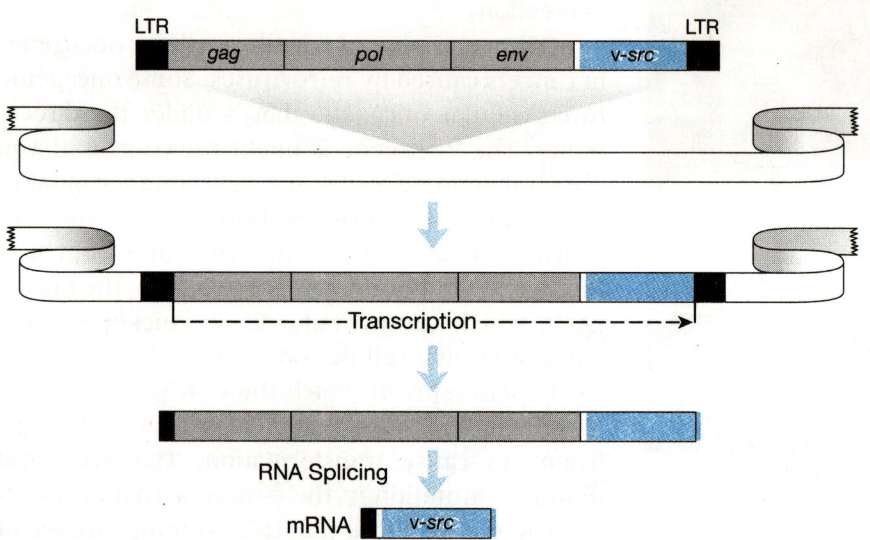

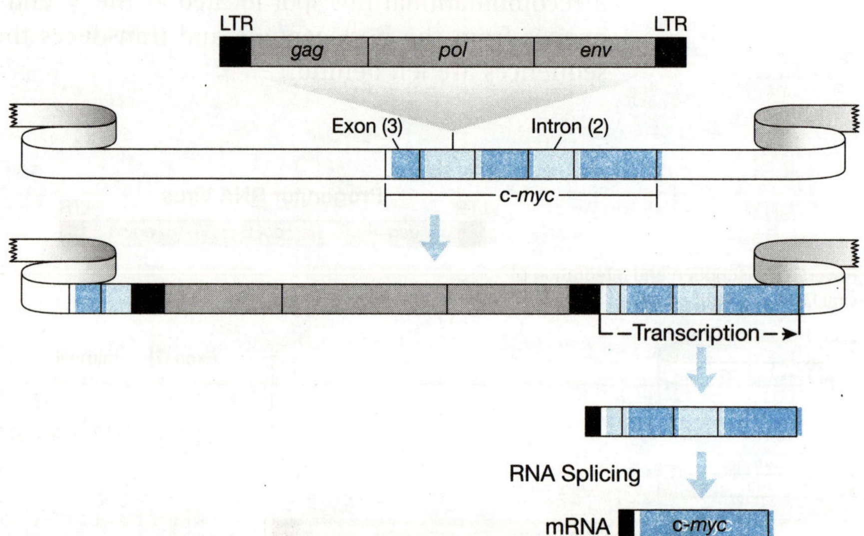

Acute versus Chronic Transformation by Retroviruses

In acute transformation, a tumor virus carrying a resident oncogene integrates into the host chromosome, where the viral oncogene is transcribed from the viral promoter or long terminal repeat (LTR). In chronic transformation, a tumor virus lacking a resident oncogene integrates adjacent to the cellular proto-oncogene, which is transcribed from the viral LTR.

VIRUSES AND HUMAN CANCER

Study of the tumor viruses of various vertebrate species revealed much about the basic mechanisms of oncogenesis, both in vivo and in vitro. However, the relationship between viruses and human cancer remained hypothetical until 1980, when Robert Gallo, at the National Institutes of Health, found that a retro-

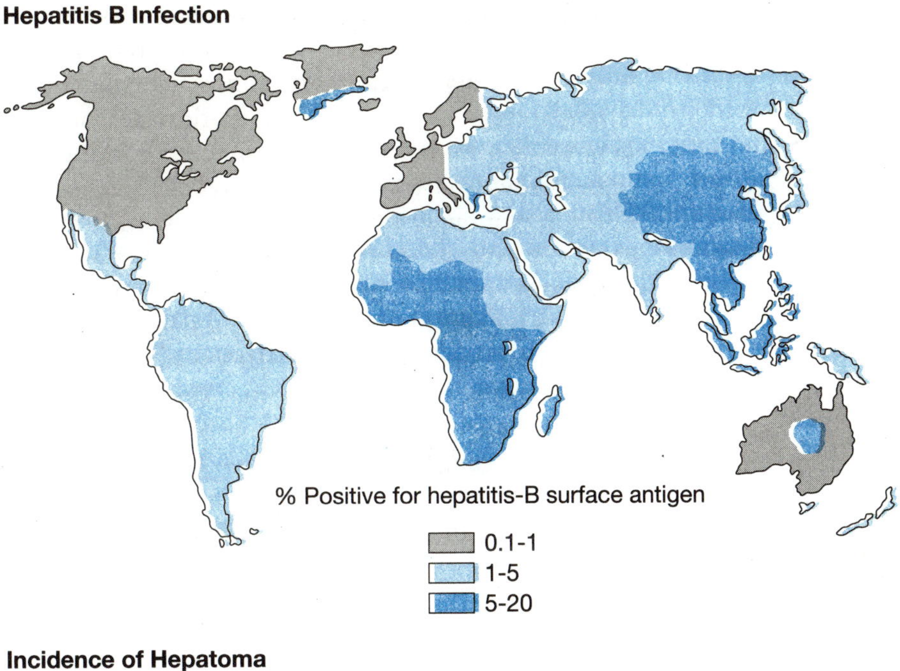

Coincidence of Hepatitis B and Hepatoma (Liver Cancer)
(After P. Maupos and J.L. Melnick. 1981. *Prog. Med. Virol. 27:* 1.)

virus is responsible for a rare form of leukemia. Human T-lymphotrophic virus type I (HTLV-I) acts in a manner analogous to that of ALV by integrating upstream of a cellular regulatory gene. The provirus promoter causes overexpression of the cellular gene coding for T-cell growth factor, which stimulates proliferation of T lymphocytes. In 1983–1984, a close relative of HTLV-I was isolated from patients suffering from acquired immune deficiency syndrome (AIDS). This virus, independently isolated and named lymphadenopathy-associated virus (LAV) by Luc Montagnier and HTLV-III by Gallo, was ultimately named human immunodeficiency virus (HIV).

DNA viruses have now been linked to several human cancers. Exposure to hepatitis-B virus increases several hundredfold the risk of hepatoma, a type of liver cancer that is the most common malignancy worldwide. In addition to a strong correlation between the worldwide coincidence of hepatitis-B virus and hepatoma, viral DNA is found integrated in the genomic DNA of hepatic tumor cells. Epstein-Barr virus (EBV), which causes infectious mononucleosis, produces lymphomas in certain primates and immortalizes human lymphocytes grown in culture. Antigens for EBV and copies of viral DNA are present in the malignant cells of two cancers whose worldwide incidence is restricted—Burkitt's lymphoma (in Africa) and nasopharyngeal cancer (in China).

Human papillomaviruses (HPVs) have been similarly linked to genital cancers, including cervical carcinoma. The Pap smear test, which detects precancerous cell changes of the cervix, has been primarily responsible for a 70% decline in uterine cancer deaths in the United States over the last four decades. However, cervical cancer remains a major cause of female death in underdeveloped nations and among members of lower socioeconomic groups in the developed nations.

MAMMALIAN CELL TRANSFECTION

Richard Axel
(Photo by Don Hamerman.)

Michael Wigler

A new approach to understanding the action of oncogenes was opened up by fine tuning techniques for transferring naked DNA into mammalian cells. Prior to the development of these methods, genetically engineered DNA or RNA viruses were the only means to ferry new genes into cultured mammalian cells. In 1973, Alex van der Eb and his associates at the University of Leiden, The Netherlands, showed that cultured mammalian cells can be transformed with precipitates of calcium phosphate and purified SV40 or adenovirus DNA. Although the mechanism of uptake, or transfection, is still unknown, it is believed that the cells engulf (phagocytize) the calcium/DNA granules.

Several years later, Richard Axel and Michael Wigler, at Columbia University, found that cotransfecting a selectable marker *along with* genomic DNA, cDNA, or a cloned gene makes it possible to readily identify cultured cells that have taken up foreign DNA. The first mammalian selectable marker was the thymidine kinase (*tk*) gene, which allows transformants to use an alternative metabolic pathway to incorporate thymidine into DNA. Because of limitations of the *tk* selection system, other markers have been incorporated into cloning vectors.

In one plasmid vector, SV*neo,* a bacterial gene for neomycin resistance is fused to the SV40 promoter, allowing it to be expressed in mammalian cells. The foreign DNA need not be ligated directly to the SV*neo* plasmid; the two DNAs simply can be mixed together and coprecipitated with calcium phosphate. Cells that phagocytize calcium phosphate granules thus receive a mixture of DNAs and are selected by growth in medium containing the neomycin analog G418, which kills nonresistant eukaryotic cells.

The transfected DNA forms long "concatemers" of several hundred thousand base pairs in which multiple copies of the SV*neo* plasmid are linked together with segments of genomic DNA. The concatemer integrates randomly as a unit into the host-cell chromosome. Since the genomic DNA and the SV*neo* plasmid are often closely linked within the concatemer, plasmid sequences can be used to probe for a gene of interest.

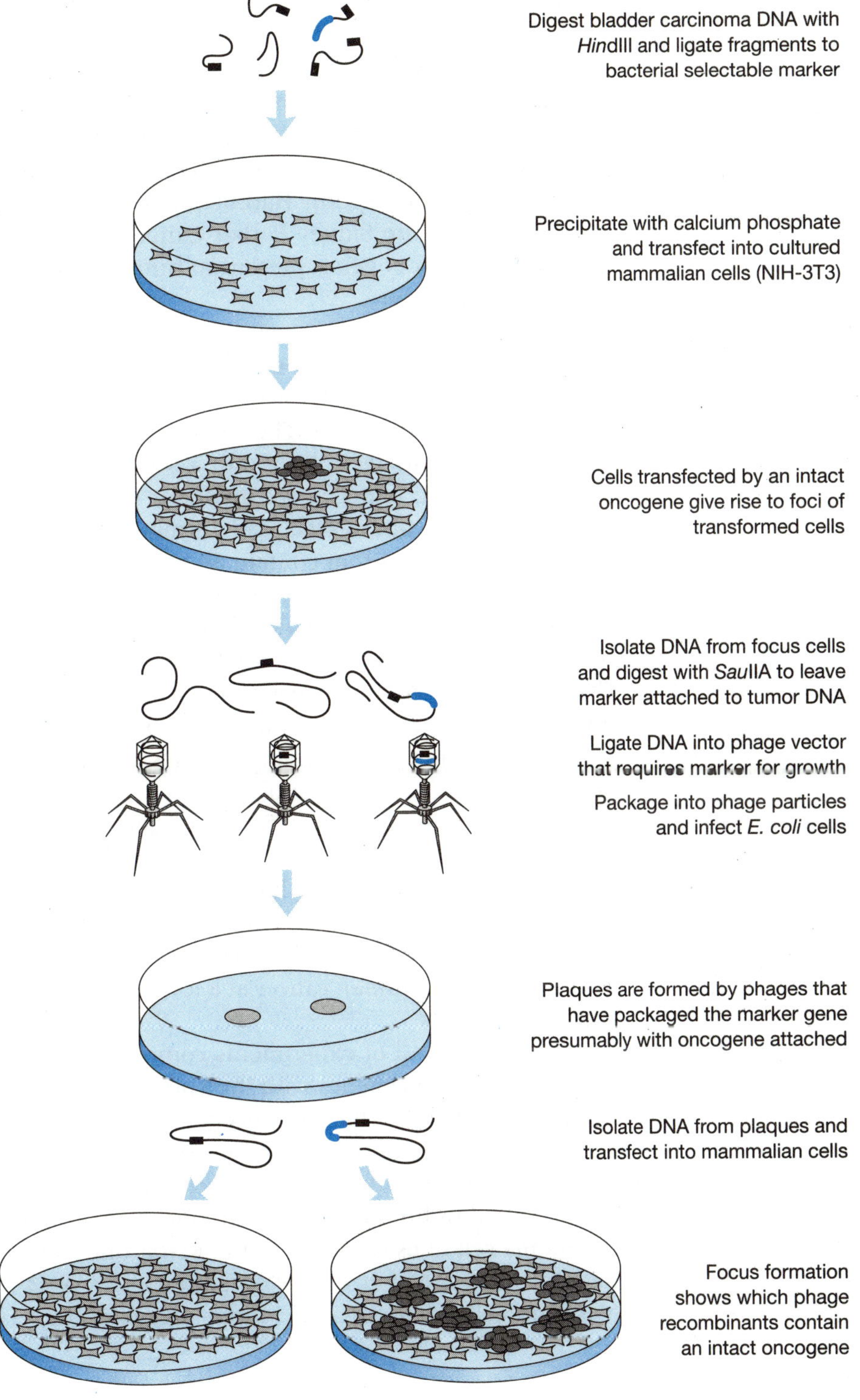

Wigler's Strategy for Isolating the Ha-*ras* Oncogene, 1981
(After R. Weinberg. 1983. *Sci. Am. 249/5:* 138.)

HUMAN ONCOGENES: GOOD GENES GONE BAD

Considering the variety of oncogenes that had been isolated from virally transformed cells and animal tumors, it seemed plausible that cellular oncogenes could be isolated from human cancer cells. Coupling mammalian cell transfection with the focus assay provided the technology for research teams to test this hypothesis at Cold Spring Harbor Laboratory, the Massachusetts Institute of Technology, and the Sidney Farber Cancer Center. Their task was to screen the tens of thousands of genes present in human genomic DNA in the hope of finding a single oncogene.

Although different strategies were employed by each group, the scheme used by Michael Wigler's group at Cold Spring Harbor Laboratory is representative. Genomic DNA was extracted from a cultured cell line (T24) derived from a human bladder carcinoma. The tumor DNA was digested with *Hin*dIII, and the resulting fragments were ligated to a selectable bacterial marker—a suppressor tRNA gene for *E. coli*. The recombinant molecules were then transfected into NIH-3T3 mouse fibroblasts. The formation of foci indicated cells harboring restriction fragments that contain an intact oncogene.

DNA was isolated from cells from several foci and digested with *Sau*IIA, which left the selectable marker linked to the oncogene. This DNA was religated to a mutant λ vector that *requires* the suppressor tRNA marker for growth. The new recombinant phages that formed plaques under selection were reasoned to contain restriction fragments possessing the marker *along with the oncogene*. DNA isolated from these λ plaques was transfected into 3T3 cells; formation of foci confirmed which of the λ constructs contained the oncogene.

Prior to the isolation of human oncogenes, it had been feared that the genetic mechanisms of human cancer might be incomprehensible. Now, human cancer could be thought of as having discrete causes. Oncogenes might provide insights to cut through the confusion of cancer and provide common pathways accessible to experimentation. These revolutionary experiments provided a means to analyze human cancer at a level of precision approaching that achieved with viral models.

The next round of experiments compared the DNA sequences of the newly isolated human oncogenes with the known sequences of cellular and viral genes. In the case of the T24 bladder oncogene, sequence comparison revealed that it is a cellular homolog of the Ha-*ras* oncogene of Harvey sarcoma virus, which causes connective-tissue tumors in rats. Ha-*ras* is related to oncogenes isolated from several different human tumors. An oncogene isolated from both colon and lung carcinoma cell lines is related to the Ki-*ras* oncogene of Kirsten sarcoma virus of mice. A third member of the *ras* gene family, N-*ras*, was isolated from neuroblastoma cells. Although the three *ras* genes have different arrangements of

... Glu Tyr Lys **Ile** Val Val Val Gly Gly Gly Gly Val Gly Lys Ser Ala Leu Thr Ile Gln **Phe** Ile Gln **Ser** **Tyr** Phe ... YEAST
... Glu Tyr Lys **Ile** Val Val Val Gly Gly Gly Gly Val Gly Lys Ser Ala Leu Thr Ile Gln Leu Ile Gln Asn His Phe ... SLIME MOLD
... Glu Tyr Lys Leu Val Val Val Gly **Pro** Gly Gly Val Gly Lys Ser Ala Leu Thr Ile Gln Leu Ile Gln Asn His Phe ... DROSOPHILA
... Glu Tyr Lys Leu Val Val Val Gly Ala Gly Gly Val Gly Lys Ser Ala Leu Thr Ile Gln Leu Ile Gln Asn His Phe ... CHICKEN
... Glu Tyr Lys Leu Val Val Val Gly Ala Gly Gly Val Gly Lys Ser Ala Leu Thr Ile Gln Leu Ile Gln Asn His Phe ... HUMAN

Ras Protein Sequence Comparison between Species
The amino acid sequences are extremely similar, with the few differences shown in bold type.

introns and exons, later sequence analysis showed that each codes for a closely related protein of 189 amino acids.

Homologs of the *ras* genes were identified in a variety of normal human tissues and were found in organisms throughout the evolutionary tree, from yeast to slime molds to *Drosophila* to birds. Experiments later showed that human and yeast *RAS* genes are functionally equivalent. Yeast cells with deleted *RAS* genes fail to germinate; however, insertion of a human *ras* gene complements the missing *RAS* function.

These studies showed that human oncogenes, sometimes referred to as proto-oncogenes, are derived from normal genes with important cellular functions. Their discovery also raised the unsettling proposition that a cancerous potential exists in the proto-oncogenes in each individual's cells. Oncogenes are not unique to cancerous tissue—they are not simply "bad" genes confined to tumors. Instead, they are essential genes whose normal function has in some way been altered. What mechanisms, then, activate the cancerous potential of the human *ras* and other oncogenes?

Close examination of the human Ha-*ras* oncogene provided incredibly precise insight into one mechanism of oncogene activation. The nucleotide sequences of Ha-*ras* genes isolated from carcinoma cells and from normal placental cells were virtually identical. Only a single nucleotide difference was found in the protein coding region of the two genes. The 12th codon of the normal cellular gene reads *GGC* (which codes for the amino acid glycine), whereas the 12th codon of the T24 carcinoma gene reads *GTC* (which codes for valine). Thus, this point mutation in the DNA code results in a glycine-to-valine substitution in the Ras protein. Subsequently, other sites of mutational activation of the *ras* gene family were found—notably at codons 59 and 61—which result in single-amino acid changes. Underscoring the importance of this mutation, it was later shown that a yeast *RAS* gene with an analogous "activating" mutation transforms NIH-3T3 cells in culture.

Some Mutations in *ras* Genes from Viruses and Human Tumor Cell Lines and Resulting Amino Acid Changes

	Codon (Amino Acid)					
	12		59		61	
Ha-*ras*						
normal	GGC	(Gly)	GCC	(Ala)	CAG	(Gln)
Harvey sarcoma virus	**AGA**	**(Arg)**	**ACA**	**(Thr)**	CAA	(Gln)
bladder tumor (T24)	G**T**C	**(Val)**	GCC	(Ala)	CAG	(Gln)
lung tumor (HS242)	GGC	(Gly)	GCC	(Ala)	C**T**G	**(Leu)**
Ki-*ras*						
normal	GGT	(Gly)	GCA	(Ala)	CAA	(Gln)
Kirsten sarcoma virus	**A**GT	**(Ser)**	**A**CA	**(Thr)**	CAA	(Gln)
colon tumor (SW480)	G**T**T	**(Val)**	GCA	(Ala)	CAA	(Gln)
lung tumor (Calu)	**T**GT	**(Cys)**	GCA	(Ala)	CAA	(Gln)
N-*ras*						
normal	GGT	(Gly)	GCT	(Ala)	CAA	(Gln)
neuroblastoma (SK-N-SH)	GGT	(Gly)	GCT	(Ala)	**A**AA	**(Lys)**
malignant melanoma (Mel)	GGT	(Gly)	GCT	(Ala)	**A**AA	**(Lys)**
lung carcinoma (SW1271)	GGT	(Gly)	GCT	(Ala)	C**G**A	**(Arg)**

Because these experiments were carried out with cultured cells, it was not certain how well they related to cancer formation in a living human being. Mariano Barbacid's research team at the National Cancer Institute was first to show a close relationship between human cancer and mutational activation of the *ras* oncogene. Sequence comparison of the Ki-*ras* gene isolated from malignant and normal tissues taken from a patient with lung carcinoma showed that the 12th codon mutation was present only in tumor cells.

TUMOR SUPPRESSOR GENES

The viral and cellular oncogenes discussed so far can all be described as "dominant-acting." The *presence* of a single mutated copy of an oncogene or of increased numbers of normal copies results in a gain of tumor function. In 1986, a fundamentally different type of oncogene was found to be involved with the rare childhood tumor, retinoblastoma. In this cancer, malignancy results from the *absence* of a functional copy of the retinoblastoma (*Rb*) gene, which is there-

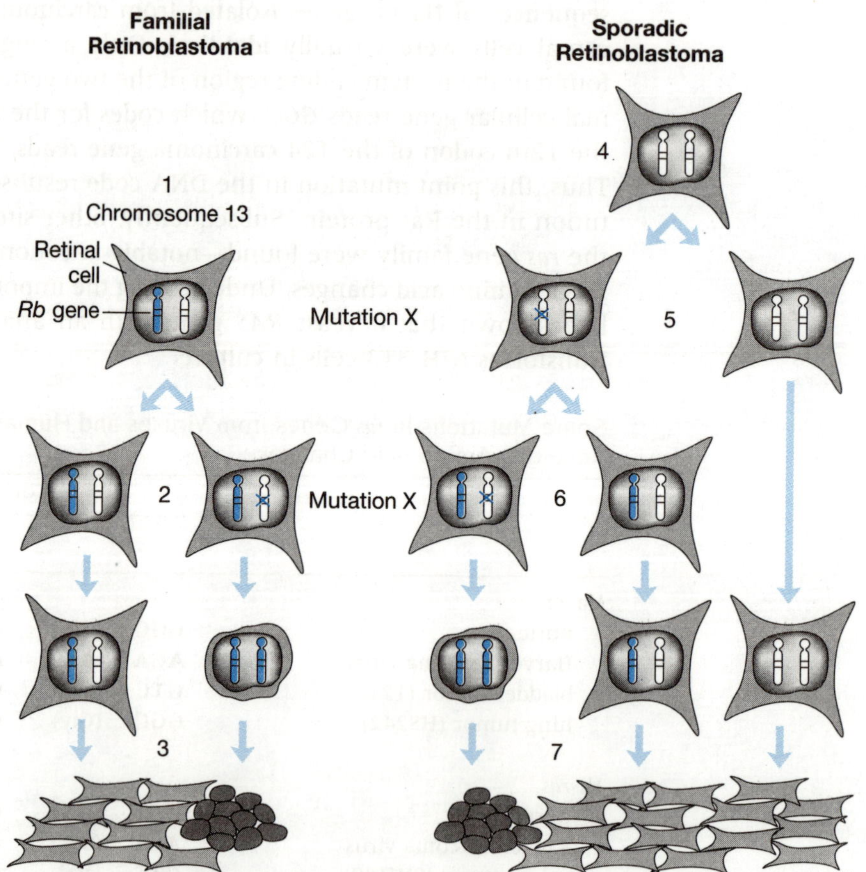

Rb Gene Mutations in Familial and Sporadic Retinoblastoma
In familial retinoblastoma, one normal (*white*) and one mutated *Rb* gene (*blue*) are inherited (*1*). (*2*) Subsequent mutation in *any* retinal cell inactivates remaining *Rb* gene (*blue*), (*3*) leading to loss of growth control in a clone of tumor cells. In sporadic retinoblastoma, two normal *Rb* genes are inherited (*4*). (*5*) First mutation inactivates one copy of *Rb* gene (*blue*); (*6*) subsequent mutation within *same* retinal cell inactivates remaining copy of *Rb* gene (*blue*), (*7*) leading to loss of growth control in a clone of tumor cells.

Alfred Knudson

fore said to be "recessive-acting." *Rb* is a tumor suppressor, because its presence (even in a single copy) inhibits formation of this particular cancer. Whereas the oncogenes described so far encourage cell proliferation, *Rb* suppresses the mitogenic response.

Statistical studies of children with retinoblastoma conducted in the 1970s by Alfred Knudson at the Fox Chase Cancer Center in Philadelphia anticipated the existence of the *Rb* gene and its role as a tumor suppressor. Children who develop the "familial" form of the disease inherit a normal copy of the *Rb* gene from one parent and a mutated copy from the other parent. Subsequently, a spontaneous mutation in a blast cell of the developing retina causes loss of the remaining normal copy and results in tumor cells lacking any functional *Rb* gene. These patients develop multiple tumors in both eyes (bilateral disease). Children with the sporadic form of the disease inherit normal *Rb* genes from each parent and thus are born with two functional *Rb* genes per cell. Cancer arises when two successive somatic mutations *within a single cell* of the developing retina inactivate both copies of the *Rb* gene. Such children present with a single tumor (unilateral disease).

Retinoblastoma patients typically show chromosome deletions and rearrangements of the *Rb* gene on chromosome 13. Heterozygous cells, with one functioning copy and one mutated copy of *Rb*, behave normally. However, deletion of the second copy of the *Rb* gene—referred to as a loss of heterozygosity (LOH)—leads to loss of Rb's tumor protective function. An experiment illustrates this concept. Wilms' tumor of the kidney results from deletions on chromosome 11, where a different tumor suppressor gene, *WT1*, resides. If a normal copy of chromosome 11 is introduced into Wilms' tumor cells, they revert to normal and no longer form tumors.

The *Rb* gene has been implicated in a number of human cancers, indicating that it has a generalized role in negatively regulating cell growth in a variety of tissues. Patients who are cured of familial retinoblastoma carry a germ-line *Rb* mutation in every cell of their body and are at increased risk for developing other tumors by LOH. *Rb* gene deletions or chromosome 13 abnormalities have been found in each of the tumors that commonly affect *Rb* patients later in life, including osteosarcoma (bone cancer), small cell lung cancer, and breast cancer.

The *p53* gene, so named because it encodes a protein with a mass of 53 kD, was first isolated in 1979 from cells transformed by the DNA tumor virus SV40. *p53* was identified bound to large T antigen, the cancer-causing protein of SV40. Originally thought to be a dominant-acting oncogene, the true role of *p53* as a tumor suppressor was uncovered in 1989. Since that time, an "avalanche" of research has found that *p53* malfunction is a common denominator in a majority of human tumors, with hundreds of tumor-associated mutations identified.

UNITING ONCOGENES AND TUMOR SUPPRESSORS

The existence of oncogenes *and* tumor suppressor genes suggests an elaborate system of positive and negative controls that maintains cell growth within normal limits. Taken separately, the oncogenes and tumor suppressors provide alternate pathways of oncogenesis. Mutation of oncogenes stimulates mitogenic pathways, resulting in cell proliferation. Mutation of tumor suppressors results in loss of control of cell proliferation.

Edward Harlow

A discovery in 1988 by Ed Harlow at Cold Spring Harbor Laboratory provided a link between oncogenesis caused by DNA tumor viruses and suppressor oncogenes. What Dr. Harlow's group discovered was that the adenovirus tumor-causing protein E1A actually binds to Rb, blocking its function. This also explained the association between SV40 large T antigen and p53, where again binding to p53 inactivates it, leading to tumorigenesis.

Cervical cancer further illustrates this link between viral oncogenes and cellular tumor suppressors. Virtually all cervical cancer is caused as a consequence of infection by human papillomavirus (HPV), presenting the clearest epidemiological association of any cancer to a causative agent. The Pap smear test in fact detects the abnormal phenotype of HPV-infected cells. Two of the oncoproteins encoded by HPV bind to tumor repressor proteins, leading to their inactivation—E6 binds to p53, and E7 binds to Rb.

COOPERATIVE TRANSFORMATION AND "MULTI-HIT" ONCOGENESIS

Epidemiological studies convincingly show that the incidences of most cancers increase dramatically with age. Although cancers strike people of all ages, 80% of cancers are diagnosed at age 55 or over. For example, the median age of onset for chronic myeloid leukemia (CML) is about 65 years. These data strongly support the hypothesis that malignancy results from an accumulation, over time, of sev-

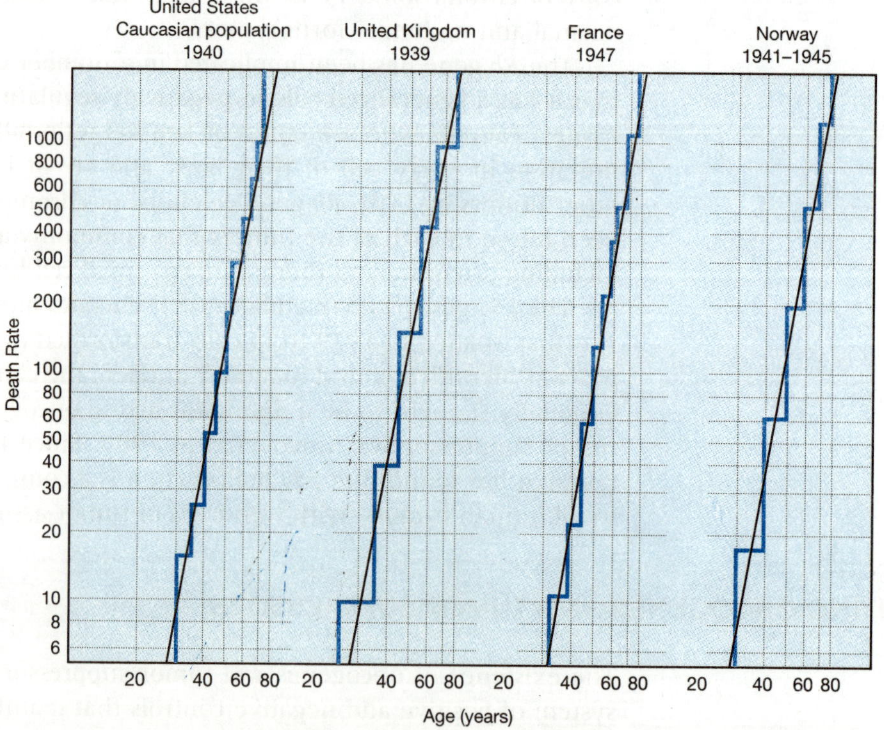

Log:Log Plots of Cancer Deaths in Males (per 100,000) versus Age, Showing a Linear Relationship

(Redrawn, with permission, from Nordling C.E. 1953. A new theory on the cancer-inducing mechanism. *Br. J. Cancer 6:* 68–72.)

eral genetic changes. Knudson's analysis of familial and sporadic retinoblastoma suggested that both forms of the disease entailed two mutation events. In the inherited form, one mutation is inherited in the germ line and the other occurs in a somatic (body) cell of the developing retina. In the sporadic form, both mutations are somatic. This became known as the "two-hit" hypothesis.

The transfection experiments that identified the first human oncogenes suggested, however, that changes in a single dominant-acting oncogene are responsible for cancer formation. Many scientists were uncomfortable with this interpretation, in part because they doubted that the 3T3 cells used in the focus assay from these studies represent normal cells. Unlike primary cells taken directly from a living tissue, 3T3 cells are immortal in culture and occasionally even undergo spontaneous transformation. There was another major inconsistency: Although the *ras* oncogene can efficiently transform 3T3 cells, it fails to transform primary cells. On the other hand, the *myc* oncogene cannot transform 3T3 cells at all.

Again, a lesson was learned from previous work with viral oncogenes. Transfection experiments had shown that the oncogenes of DNA tumor viruses have similar and complementary functions. Adenovirus *E1A* and *E1B* genes and polyomavirus *large T* and *middle T* genes act in a cooperative fashion to transform primary cells. *E1A* and *large T* genes immortalize primary cells, establishing them so that they can grow in culture. *E1B* and *middle T* genes convert immortalized cells to the malignant phenotype. However, acting alone, none of the four oncogenes can transform primary cells.

Extensions of these experiments worked exactly as predicted. When cotransfected with *E1A* or *large T*, *ras* transforms embryonic rat cells to the malignant phenotype. Primary cells transfected with *myc*, in combination with *E1B* or *middle T*, are also transformed. Finally, *myc* and *ras* can cooperate to transform primary cells. The implication was clear. Cooperation between at least two onco-

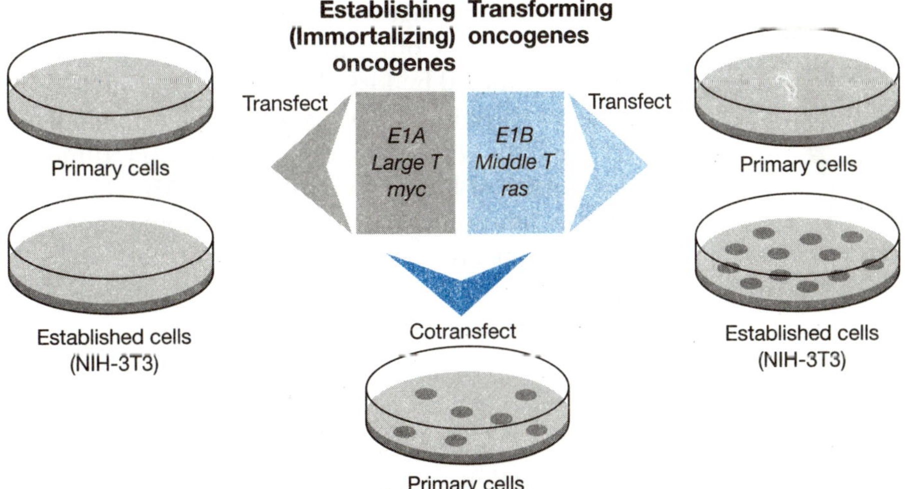

Cooperative Transformation of Primary Cells by Two Oncogenes

Transfected alone, neither immortalizing nor transforming oncogenes transform primary cells. However, cotransfection of both types of oncogenes causes foci of transformed cells. Established cells, such as NIH-3T3, require only a transforming oncogene to form foci.

genes is normally required to transform primary cells in culture. Whereas *ras* is a transforming gene, *myc* is an immortalizing gene. The *ras* oncogene scores positively on the 3T3 focus assay because 3T3 fibroblasts have already been immortalized; they are halfway to the transformed phenotype.

The multi-step model of carcinogenesis is supported by other lines of evidence. Many types of tumors appear to contain two or more independently activated oncogenes. For example, activated forms of both *myc* and *Blym* are found in chicken lymphomas. Chemical carcinogens, such as 3-methylcholanthrene and benzopyrene, immortalize cells and can act cooperatively with transforming oncogenes to transform primary cells. 3-methylcholanthrene also activates the *ras* oncogene in cultured cells and laboratory animals and thus potentially can accomplish two steps of oncogenesis.

Eric Fearon

Experiments with transgenic mice indicate that additional steps, beyond immortalization and transformation, are needed to create tumors. For example, mice carrying a fusion gene of the *insulin* promoter/*large T* antigen show correct tissue-specific expression of large T antigen protein in the β cells of the islets of Langerhans. Although the immortalizing and transforming functions of T antigen induce abnormal proliferation of essentially all β cells, tumors develop in only a small percentage of islets. If expression of T antigen is sufficient for oncogenesis, one would expect to observe progression to tumor in every islet. Similarly, when the c-*myc* oncogene is fused to the promoter/enhancer region of the mouse mammary tumor virus, some transgenic females develop mammary tumors at the time of pregnancy. Hormonal changes associated with lactation are thought to induce sufficient expression of the Myc protein to induce immortalization of all breast cells. However, localized tumors develop in only a small percentage of mammary cells.

Bert Vogelstein

Work in the 1990s by Eric Fearon and Bert Vogelstein at Johns Hopkins Medical School convincingly showed a multi-step model for colorectal cancer involving activation of oncogenes and deletion of tumor suppressors. In this disease, benign polyps (adenomas) increase in size and ultimately progress to fully malignant carcinomas. An analysis of DNA from adenomas and carcinomas at various clinical stages showed that mutations in specific genes accumulate in a relatively specific order during the progression of malignancy. Oncogenesis typically is initiated by loss of both copies of the *APC* (adenomatous polyposis coli) gene on chromosome 5. As in retinoblastoma, patients with familial colon cancer are born with a single copy of the *APC* tumor suppressor, greatly increasing the chance of loss of heterozygosity. Mutational activation of the Ki-*ras* oncogene is followed by deletions in chromosomes 17 and 18 that inactivate the tumor suppressors *p53* and*SMAD2/SMAD4*.

CARCINOGENS, POINT MUTATIONS, AND INDIVIDUALS AT INCREASED RISK

The discussion so far begs the question: What causes the mutations in multistep carcinogenesis? As one would expect, carcinogens can induce key mutations in oncogenes and tumor suppressors. Activating mutations in the 12th codon of *ras* genes are found in a number of rodent tumors (including thyroid,

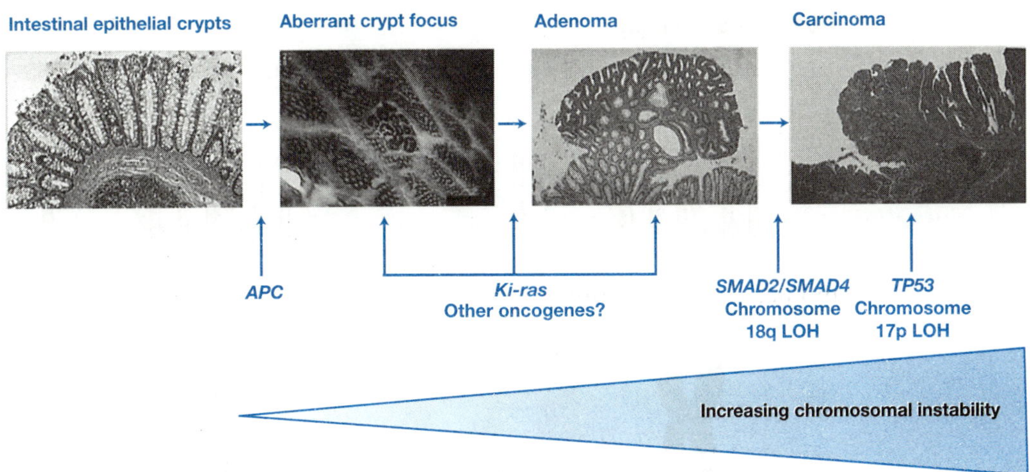

Diagram of Tumor Steps in the Human Intestine and Specific Gene Mutations
(Adapted, with permission of Edward C. Klatt, M.D., Florida State University College of Medicine, and Takayama T. et al. 1998. Aberrant crypt foci of the colon as precursors of adenoma and cancer. *N. Engl. J. Med. 339:* 1277–1284.)

kidney, and lung) induced by benzo[a]pyrene, *N*-bis (2-hydroxypropyl)-nitrosamine (DHPN), and *N*-nitroso-*N*-methylurea (NMU). Coal tar, which contains high concentrations of polycyclic aromatic hydocarbons such as benzo[a]pyrene, primarily induces G to T transversion, which is the activating mutation in the Ki-*ras* gene and a common mutation in the p53 tumor suppressor. G to T transversions are the most common mutations in the *p53* gene of lung cancers, but this type of mutation is infrequent in lung tumors of non-smokers and in other tumors.

Since lung cancer is one of the most preventable cancers, another risk of cigarette smoking bears mentioning. *Cytochrome P450* genes encode metabolic enzymes that can protect against lung cancer by disabling carcinogens in smoke. Polymorphisms that alter the effectiveness of *cytochrome P450*s put some smokers at significantly higher cancer risk than others. For young people with these P450 mutations, cancer can occur at lower levels of exposure to carcinogens. This is in addition to the already higher risk that young people face, because rapid proliferation of lung epithelium increases the chance of mutations during DNA replication.

CHROMOSOME ABNORMALITIES AND GENE AMPLIFICATION

At some point in carcinogenesis, the cancer cell genome often becomes extremely unstable, leading to wholesale chromosome duplication, deletion, and translocation. In 1890, the German scientist David von Hansemann reported that tumor cells have chromosome abnormalities and mitotic defects. In the early 1900s, Theodor Boveri proposed that the addition of specific chromosomes was responsible for cancer. The use of Giemsa and Wright stains, beginning in the 1950s, allowed more accurate counting of chromosomes and showed that cultured tumor cells frequently have many extra chromosomes.

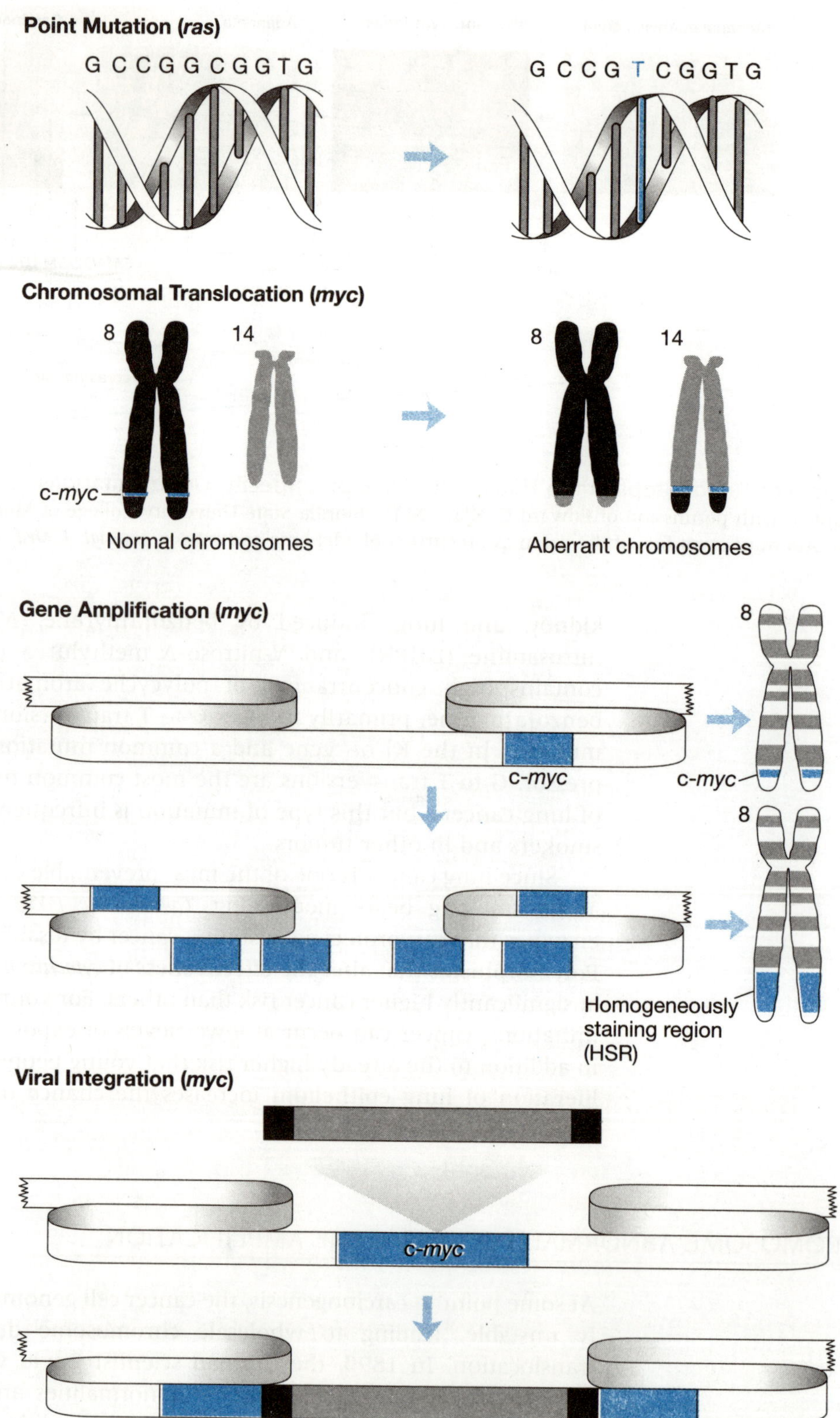

Examples of Molecular Mechanisms of Oncogenesis

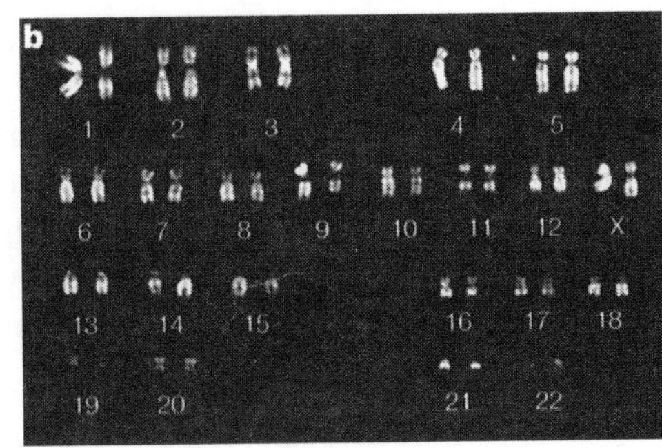

Two Methods of Visualizing Chromosomes
(*a*) Giemsa staining; (*b*) Q-banding. (Adapted, with permission from Rowley J.D. 2001. Chromosome translocations: Dangerous liaisons revisited. *Nat. Rev. Cancer 1:* 245–250.)

In the 1970s, new staining techniques were developed that allowed chromosomes to be identified by distinctive patterns of light and dark bands: Giemsa (G) staining and quinacrine (Q) banding. Studies showed that translocations—exchanges of chromosome pieces between nonhomologous chromosomes—are characteristic of several blood cancers. In 1972, Janet Rowley identified the diagnostic translocation in 95% of CML patients, an exchange between the long arms of chromosome 9 and 22. This recombination results in a much shortened chromosome 22—the so-called Philadelphia (Ph) chromosome—named after the city in which it was first identified in CML patients. In 1976, Lore Zech identified a translocation of the tips of chromosomes 8 and 14 in patients with Burkitt's lymphoma.

During the last 20 years, molecular analysis has pinpointed the genes disrupted by these translocations, making leukemias and lymphomas some of the best characterized malignancies. The Ph translocation occurs in a 5.8-kb region of chromosome 22 known as the breakpoint cluster region (BCR). Translocation fuses the 5′ portion of the resident *BCR* gene with the 3′ portion of the *abl* proto-oncogene on chromosome 9. The truncated *abl* gene present in the *BCR/abl* fusion is analogous to the v-*abl* oncogene of the Abelson mouse leukemia virus, and mice transfected with the human *BCR/abl* fusion develop CML.

Like many oncoproteins, v-Abl is a tyrosine kinase. Although the normal c-Abl protein has little or no tyrosine kinase activity, the BCR/Abl fusion protein is endowed with a tyrosine kinase activity equivalent to v-Abl. The Ph chromosome is also found in some cases of acute lymphoblastic leukemia (ALL), although the translocation produces a different BCR/Abl fusion protein. Consistent with the more aggressive nature of this tumor, ALL cells express an Abl tyrosine kinase with activity greater than that of CML cells.

Activating mutations that change one or more amino acids in an oncoprotein (such as Ras or Fos) or fusion products (such as BCR/Abl) result in *altered* proteins. These are termed *qualitative* changes in protein expression. Mutations that increase or decrease the *amount* of oncoprotein produced are termed *quantitative* changes in protein expression.

Translocations in Burkitt's lymphoma provide examples of mutations that produce quantitative changes in protein expression. In this case, translocation alters transcription of the *myc* oncogene on chromosome 8. Although the translocation from chromosome 8 to 14 is most common, translocations of chromosome 8 to chromosomes 2 and 22 also occur. The role of these translocations in a cancer of the immune system makes good intuitive sense. The breakpoints where *myc* fuses on chromosomes 14, 22, and 2 coincide with three major immunoglobulin loci, where several hundred gene segments actively rearrange to generate the huge variety of antibody genes. The immunoglobulin loci are also extremely active sites of gene transcription. There is evidence for two different mechanisms through which translocation into an immunoglobulin locus may activate transcription of the *myc* oncogene, as indicated by increased abundance of mRNA. In some cases, the translocation appears to *de*regulate the c-*myc* gene by destroying a repressor region that normally keeps the gene "silent" in its position on chromosome 8. In other cases, c-*myc* appears to come under the influence of the immunoglobulin enhancer, which stimulates transcription.

For many years, chromosome abnormalities were thought to be a by-product of oncogenesis. However, the detailed molecular analysis of CML and Burkitt's lymphoma showed that translocation directly contributes to oncogenesis by interfering with key growth regulators. Although translocations occur most frequently in blood tumors, many are associated with solid tumors. About 50 common translocations have been identified, with most associated with disruptions in genes involved in the control of cell division.

Gene amplification, where multiple copies of an oncogene arise, is another means of increasing production of an oncoprotein. Two forms of large-scale DNA duplications are common in tumor cells: numerous duplicate chromosome fragments called double minutes (DMs); or duplicated regions inserted into a chromosome locus, which create a widened band called a homogeneously staining region (HSR). A resident oncogene would, as a matter of course, become amplified during these large-scale duplications. In situ hybridization, where a radioactively labeled *myc* probe is hybridized to intact chromosomes, has confirmed that numerous copies of the N-*myc* oncogene are present in HSRs and DMs from neuroendocrine tumor cells. Since gene copy number tends to correlate with cancer progression, gene amplification may be a secondary result of chromosome "crisis" during tumorigenesis.

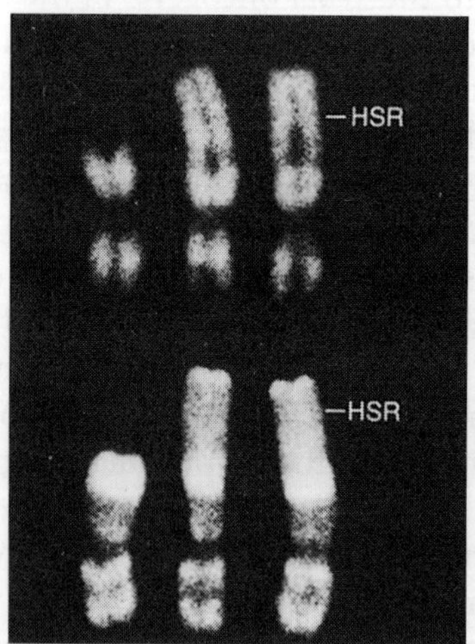

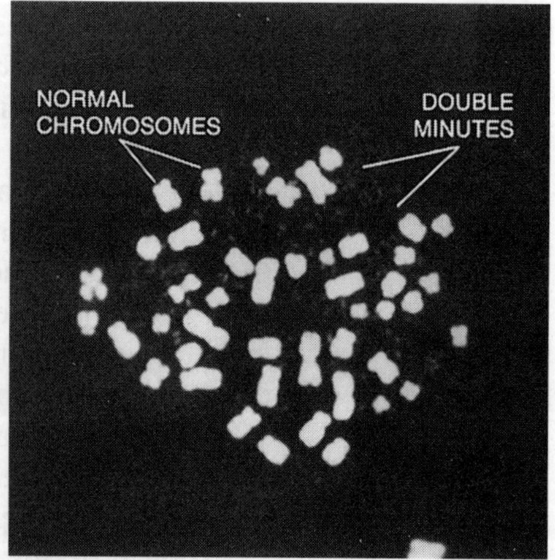

Cytological Evidence of Oncogene Amplification
Amplification of the N-*myc* oncogene results in both homogeneously staining regions (HSRs) and double minutes (DMs) in neuroblastoma cells. (Courtesy of R. Alt, Columbia University.)

SIGNAL TRANSDUCTION: CANCER AS A DISEASE OF COMMUNICATION

Throughout this chapter, cancer has been framed in terms of cellular *proliferation*. To begin to understand the oncoprotein functions in cell biochemistry, it would serve us well to reconsider cancer as a disease of cellular *communication*. The molecular pathway through which growth signals are relayed from outside the cell to the cytoplasm and nucleus is called *signal transduction*. At various points in the transduction pathway, the signal may branch to follow one of several alternative routes. Further along, each of these routes may branch again and again, reaching multiple targets. Thus, by analogy to a series of small waterfalls interacting and spilling into successive pools, the reactions of a transduction pathway are often referred to as a "cascade."

A signal transduction pathway begins with the arrival of a "first messenger" at the cell surface. First messengers (including hormones, growth factors, and insulin-like molecules) communicate between cells and may be produced in a distant part of the body. The first messenger binds to a receptor molecule anchored in the cell membrane. The receptor transduces the chemical signal through the cell membrane to the internal membrane surface, by way of physical changes in its structure that open access to one or more reactive sites on the intracellular side. Receptor dimerization is the most common such physical change. The modified receptor then interacts with "second messengers" that carry the signal into the cytoplasm. Many signals ultimately are carried into the nucleus. Here, signal transducers interact with DNA-binding proteins that attach to regulatory sequences to initiate transcription of specific genes—whose proteins, in turn, regulate cell division and growth.

Mutations in any of the proteins in a transduction pathway can disrupt signals regulating cell proliferation. Thus, oncoproteins have been identified in major steps of many signal transduction pathways. The highly branching nature of transduction pathways provides redundancy and makes it unlikely that a mutation in any single molecule would completely disrupt signaling. Thus, cancer cells typically show mutations in several proteins involved in signal transduction. This is consistent with the fact that the development of most cancers requires multiple mutational "hits."

The first direct link between an oncoprotein and biochemical signal transduction came, in 1983, from studies of the amino acid structure of platelet-derived growth factor (PDGF). This protein is released by platelets to stimulate tissue repair at wound sites. Like other growth factors, it is a powerful mitogen that induces cell growth. A computerized comparison of the amino acid sequence of PDGF with those of other proteins in a data bank revealed that a 104-amino-acid stretch of PDGF is virtually identical to the protein structure of the v-*sis* oncogene of simian sarcoma virus. Further experiments showed that the Sis oncoprotein, produced in virus-transformed cells has properties identical to those of PDGF. This result suggested that the simian retrovirus captured cellular sequences that encode a portion of PDGF or a closely related protein. This potentially explained how v-*sis* acted as an oncogene: Virus-transformed cells carrying v-*sis* secrete a PDGF-like protein that, in turn, stimulates a mitogenic response when it binds to PDGF receptors on the cell surface. Alternatively, internal expression of the Sis oncoprotein might stimulate replication in cell types lacking PDGF receptors.

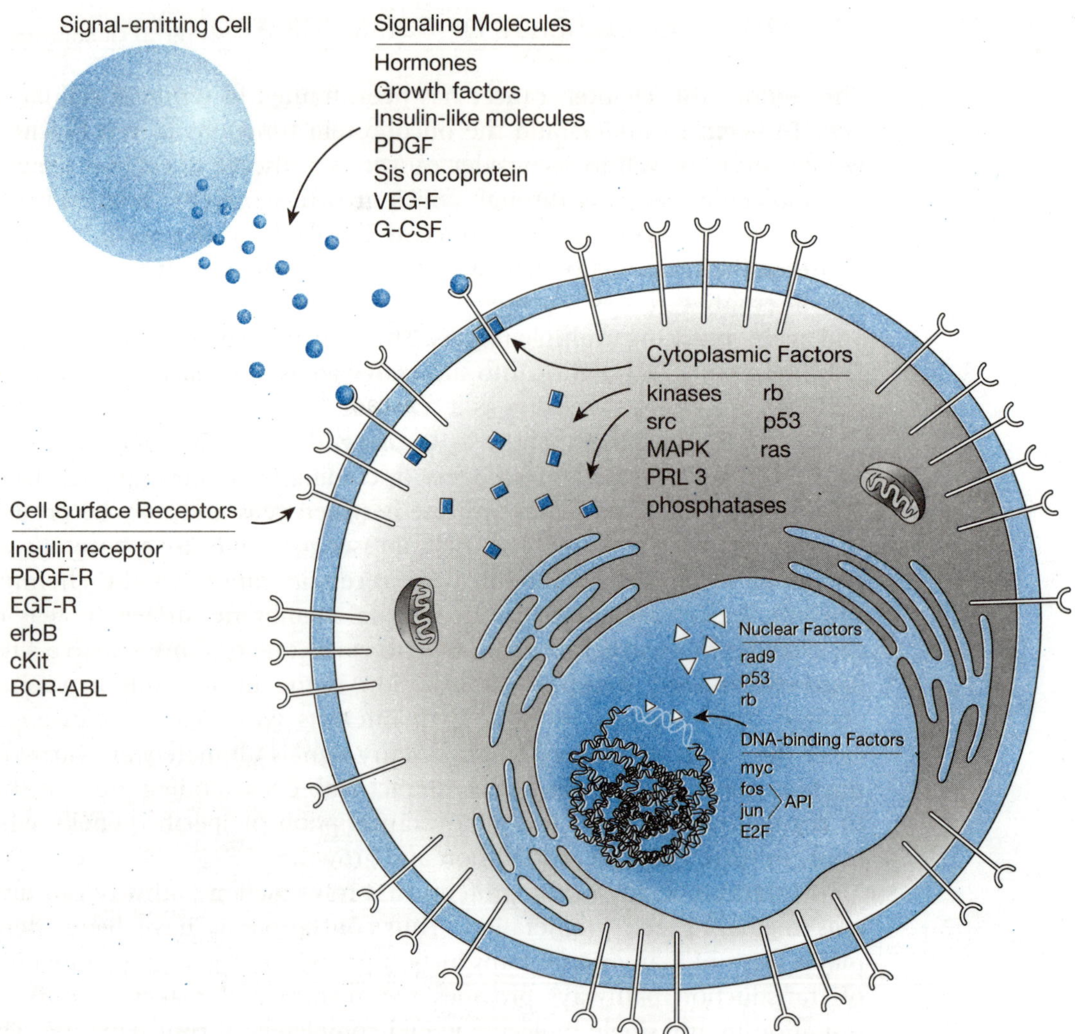

Signal Transduction Proteins That Have Roles in Cancer

Another connection between oncoproteins and growth factors quickly emerged. The amino acid sequence of the cell surface receptor for epidermal growth factor (EGF) shares remarkable homology with the sequence of the *erbB* oncogene of avian erythroblastosis virus. It appears that *erbB* is an example of viral capture of an incomplete cellular gene. The EGF receptor has three domains: (1) An extracellular region projects outside the cell and is the site of EGF binding; (2) a membrane-spanning region anchors the protein in the cell membrane; and (3) an intracellular domain projects into the cytoplasm and interacts with cytoplasmic signaling factors. The ErbB protein shares with the EGF receptor common intracellular and membrane-spanning regions, but it is missing most of the extracellular binding site. It is hypothesized that the loss of binding function leads to deregulation of the receptor and that ErbB is a growth factor receptor stuck in the "ON" position. The human homolog of *erbB* was later identified as *Her2/Neu*.

The relay of a signal is accomplished by a close molecular interaction between two or more molecules, whose reactive surfaces have a unique "lock

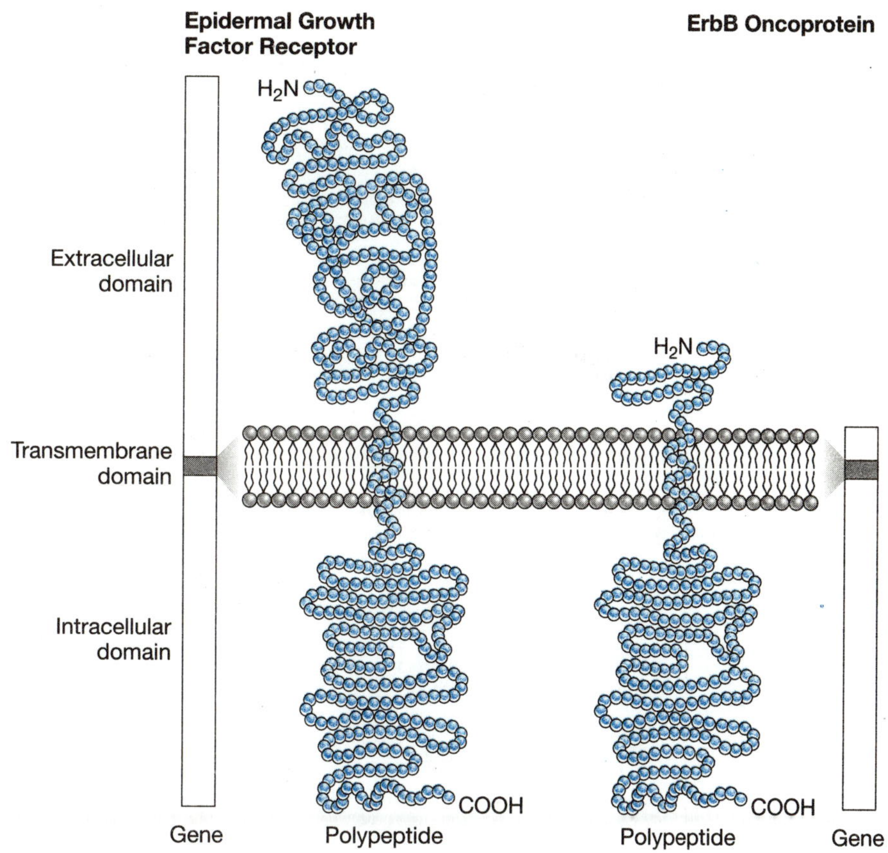

Comparison of Epidermal Growth Factor Receptor and ErbB Oncoprotein

and key" fit. In many cases, interacting molecules are "activated" or "deactivated" to interact by the addition or subtraction of a phosphate group. Thus, protein kinases—which add phosphates to protein molecules—and protein phosphatases—which remove them—are important regulators of signal transduction. For example, a number of cell surface receptors, including those for EGF, PDGF, and insulin, are protein kinases whose activity is regulated by phosphorylation by other protein kinases, and by themselves (autophosphorylation). Although the vast majority of kinases add phosphate groups to serine or threonine residues, many oncogenic kinases exclusively catalyze the addition of phosphate groups to the amino acid tyrosine. This makes sense because protein kinases in the signaling pathway for cell growth are largely tyrosine kinases.

The Src and Ras oncoproteins, which are integrated into the inner cell membrane, represent the second major step in signal transduction. Each accepts a signal from a membrane receptor and then relays it to other targets in the cytoplasm. The *src* gene illustrates the complexity and importance of tyrosine phosphorylation in signal transduction. A tyrosine kinase with many cellular targets, Src itself is activated by tyrosine phophorylation. As mentioned previously, the v-*src* gene is missing 3′ c-*src* sequences, which include a codon for tyrosine. The phosphorylation of this position in the c-Src protein is essential for negative control of its kinase activity. Thus, the absence of this phosphorylated residue in v-Src deregulates its tyrosine phosphorylation activity, enhancing its activity in tumor cells.

```
V-erbB          ------------------------------------------------------------
erbB1   (EGFR)  MRPSGTAGAALLALLAALCPASRALEEKKVCQGTSNKLTQLGTFEDHFLSLQRMFNNCEVVLGNLEITYV
erbB2   (HER2)  ---MELAALCRWGLLLALLPPG--AASTQVCTGTDMKLRLPASPETHLDMLRHLYQGCQVVQGNLELTYL

V-erbB          ------------------------------------------------------------
erbB1   (EGFR)  QRNYDLSFLKTIQEVAGYVLIALNTVERIPLENLQIIRGNMYYENSYALAVLSNYD----------ANKT
erbB2   (HER2)  PTNASLSFLQDIQEVQGYVLIAHNQVRQVPLQRLRIVRGTQLFEDNYALAVLDNGDPLNNTTPVTGASPG

V-erbB          ------------------------------------------------------------
erbB1   (EGFR)  GLKELPMRNLQEILHGAVRFSNNPALCNVESIQWRDIVSSDFLSNMSMDFQNHLGSCQKCDPTCPNGSCW
erbB2   (HER2)  GLRELQLRSLTEILKGGVLIQRNPQLCYQDTILWKDIFHKNNQLALTLIDTNRSRACHPCSPHCKGSRCW

V-erbB          ------------------------------------------------------------
erbB1   (EGFR)  GAGEENCQKLTKIICAQQCSGRCRGKSPSDCCHNQCAAGCTGPRESDCLVCRKFRDEATCKDTCPPLMLY
erbB2   (HER2)  GESSEDCQSLTRTVCAGGCA-RCKGPLPTDCCHEQCAAGCTGPKHSDCLACLHFNHSGICELHCPALVTY

V-erbB          ------------------------------------------------------------
erbB1   (EGFR)  NPTTYQMDVNPEGKYSFGATCVKKCPRNYVVTDHGSCVRACGADSYEM-EEDGVRKCKKCEGPCRKVCNG
erbB2   (HER2)  NTDTFESMPNPEGRYTFGASCVTACPYNYLSTDVGSCTLVCPLHNQEVTAEDGTQRCEKCSKPCARVCYG

V-erbB          ------------------------------------------------------------
erbB1   (EGFR)  IGIGEFKDSLSINATNIKHFKNCTSISGDLHILPVAFRGDSFTHTPPLDPQELDILKTVKEITGFLLIQA
erbB2   (HER2)  LGMEHLREVRAVTSANIQEFAGCKKIFGSLAFLPESFDGDPASNTAPLQPEQLQVFETLEEITGYLYISA

V-erbB          ------------------------------------------------------------
erbB1   (EGFR)  WPENRTDLHAFENLEIIRGRTKQHGQFSLAVVSLNITSLGLRSLKEISDGDVIISGNKNLCYANTINWKK
erbB2   (HER2)  WPDSLPDLSVFQNLQVIRGRILHNGAYSLTLQGLGISWLGLRSLRELGSGLALIHHNTHLCFVHTVPWDQ

V-erbB          ------------------------------------------------------------
erbB1   (EGFR)  LFGTSGQKTKIISNRGENSCKATGQVCHALCSPEGCWGPEPRDCVSCRNVSRGRECVDKCNLLEGEPREF
erbB2   (HER2)  LFRNPHQALLHTANRPEDECVGEGLACHQLCARGHCWGPGPTQCVNCSQFLRGQECVEECRVLQGLPREY

V-erbB          ------------------------------MKCAHFIDGPHCVKACPAGVLGENDTL-VRKYADANAVCQ
erbB1   (EGFR)  VENSECIQCHPECLPQAMNITCTGRGPDNCIQCAHYIDGPHCVKTCPAGVMGENNTL-VWKYADAGHVCH
erbB2   (HER2)  VNARHCLPCHPECQPQNGSVTCFGPEADQCVACAHYKDPPFCVARCPSGVKPDLSYMPIWKFPDEEGACQ

V-erbB          LCHPNCTRGCKGPGLEGCP-NGSKTP--SIAAGVVGGLLCLVVVGLGIGLYLRRR-HIVRKRTLRRLLQE
erbB1   (EGFR)  LCHPNCTYGCTGPGLEGCPTNGPKIP--SIATGMVGALLLLLVVALGIGLFMRRR-HIVRKRTLRRLLQE
erbB2   (HER2)  PCPINCTHSCVDLDDKGCPAEQRASPLTSIISAVVG-ILLVVVLGVVFGILIKRRQQKIRKYTMRRLLQE

V-erbB          RELVEPLTPSGEAPNQAHLRILKETEFKKVKVLGSGAFGTIYKGLWIPEGEKVKIPVAIKELREATSPKA
erbB1   (EGFR)  RELVEPLTPSGEAPNQALLRILKETEFKKIKVLGSGAFGTVYKGLWIPEGEKVKIPVAIKELREATSPKA
erbB2   (HER2)  TELVEPLTPSGAMPNQAQMRILKETELRKVKVLGSGAFGTVYKGIWIPDGENVKIPVAIKVLRENTSPKA

V-erbB          NKEILDEAYVMASVDNPHVCRLLGICLTSTVQLITQLMPYGCLLDYIREHKDNIGSQYLLNWCVQIAKGM
erbB1   (EGFR)  NKEILDEAYVMASVDNPHVCRLLGICLTSTVQLITQLMPFGCLLDYVREHKDNIGSQYLLNWCVQIAKGM
erbB2   (HER2)  NKEILDEAYVMAGVGSPYVSRLLGICLTSTVQLVTQLMPYGCLLDHVRENRGRLGSQDLLNWCMQIAKGM

V-erbB          NYLEERRLVHRDLAARNVLVKTPQHVKITDFGLAKLLGADEKEYHAEGGKVPIKWMALESILHRIYTHQS
erbB1   (EGFR)  NYLEDRRLVHRDLAARNVLVKTPQHVKITDFGLAKLLGAEEKEYHAEGGKVPIKWMALESILHRIYTHQS
erbB2   (HER2)  SYLEDVRLVHRDLAARNVLVKSPNHVKITDFGLARLLDIDETEYHADGGKVPIKWMALESILRRRFTHQS

V-erbB          DVWSYGVTVWELMTFGSKPYDGIPASEISSVLEKGERLPQPPICTIDVYMIMVKCWMIDADSRPKFRELI
erbB1   (EGFR)  DVWSYGVTVWELMTFGSKPYDGIPASEISSILEKGERLPQPPICTIDVYMIMVKCWMIDADSRPKFRELI
erbB2   (HER2)  DVWSYGVTVWELMTFGAKPYDGIPAREIPDLLEKGERLPQPPICTIDVYMIMVKCWMIDSECRPRFRELV

V-erbB          AEFSKMARDPPRYLVIQGDERMHLPSPTDSKFYRTLMEEEDMEDIVDADEYLVPHQGFFN----------
erbB1   (EGFR)  IEFSKMARDPQRYLVIQGDERMHLPSPTDSNFYRALMDEEDMDDVVDADEYLIPQQGFFS----------
erbB2   (HER2)  SEFSRMARDPQRFVVIQ-NEDLGPASPLDSTFYRSLLEDDDMGDLVDAEEYLVPQQGFFCPDPAPGAGGM

V-erbB          ------SPSTS-------------------RTPLLSSLSATSN--NSATNCIDRNG-QGHPVREDSFVQRY
erbB1   (EGFR)  ------SPSTS-------------------RTPLLSSLSATSN--NSTVACIDRNGLQSCPIKEDSFLQRY
erbB2   (HER2)  VHHRHRSSSTRSGGGDLTLGLEPSEEEAPRSPLAPSEGAGSDVFDGDLGMGAAKGLQSLPTHDPSPLQRY

V-erbB          SSDPTGNFLEES----IDDGFLPAPEYVNQLMPKKPSTAMVQNQIYNFISLTAISKLPMDSRYQNSHSTA
erbB1   (EGFR)  SSDPTGALTEDS----IDDTFLPVPEYINQSVPKRP-AGSVQNPVYHNQPLNPAPSR--DPHYQDPHSTA
erbB2   (HER2)  SEDPTVPLPSETDGYVAPLTCSPQPEYVNQPDVRPQPPSPREGPLPAARPAGATLER---PKTLSPGKNG

V-erbB          VDNPEYLN---TNQSPLAKTVFESSPYWIQSGNHQINLDNPDYQQDFLPTSCS-----------------
erbB1   (EGFR)  VGNPEYLN---TVQPTCVNSTFDSPAHWAQKGSHQISLDNPDYQQDFFPKEAKPNGIFKGS-TAENAEYL
erbB2   (HER2)  VVKDVFAFGGAVENPEYLTPQGGAAPQHPPPAFSPAFDNLYYWDQDPPERGAPPSTFKGTPTAENPEYL

V-erbB          ------------
erbB1   (EGFR)  RVAPQSSEFIGA
erbB2   (HER2)  GLDVPV------
```

Amino Acid Alignment of v-ErbB, Human ErbB1 (EGFR), and ErbB2 (HER2)

The extracellular domain is indicated by the single vertical line at the left. The transmembrane domain is boxed. The tyrosine kinase domain is indicated by the double vertical line. v-*erbB* lacks most of the extracellular domain and has several intracellular mutations that result in dimerization and phosphorylation in the absence of growth factor. Herceptin, used in treatment of metastatic breast cancer, is a monoclonal antibody directed against the extracellular domain of *erbB2*. The blue shading shows identical residues, and the gray shading shows conservative changes in amino acids.

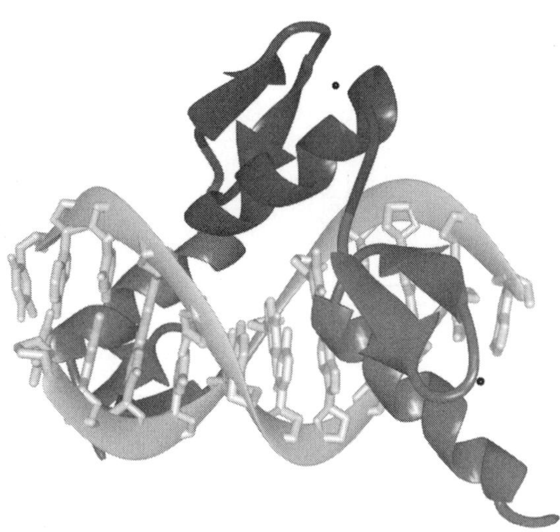

Crystal Structure of Zinc Fingers Bound to DNA
The zinc finger motif is found in many DNA-binding proteins. Each of the three zinc fingers (*black*) from the mouse Zif268 protein contains a zinc ion (small black sphere). The zinc fingers bind to the major groove of the DNA (*light gray*). (Figure created by Dr. Song Tan, Pennsylvania State University, using Midas-Plus software.)

Ras belongs to the group of guanine-nucleotide-binding (G) proteins, which are inactive when bound to GDP and active when bound to GTP. The subsequent hydrolysis of GTP to GDP shuts off the G protein following signal transmission. Mutated Ras oncoproteins lack this hydrolytic activity, essentially leaving them stuck in the ON position. The Ras signal is passed through the cytoplasm, via phosphorylation of several substrates, to a mitogen-activated protein kinase (MAPK), which then activates substrates that transpose to the nucleus.

Like many other pathways, *ras* signal transduction culminates in the activation of transcription factors that interact with gene promoter sequences. Transcription factors have common protein structures through which they interact with DNA, as well as other proteins in the transcription complex. For example, the oncoproteins Myc, Fos, and Jun share the DNA-binding motif known as a "leucine zipper." Repeated leucine residues form an α-helix region from which side chains project, in line, from one side of the protein molecule. The leucine side chains of two such proteins can interdigitate, like teeth of a zipper, to form a stable dimer. The zipped region maintains alignment while free ends of the molecules contact the major groove of the DNA molecule in a "scissors grip." The *ras* signaling pathway ends with the dimerization of Fos and Jun to form the transcription factor known for years as AP1.

THE CELL CYCLE AND APOPTOSIS

Many signal transduction pathways converge on the cell cycle, the eukaryotic cell's regular rhythm of rest and division (see Chapter 3). The two key steps of cell division—DNA synthesis (S phase) and mitosis (M phase)—are separated by two "gaps" (G_1 and G_2), during which the cell prepares for the next phase. Mature, fully differentiated, cells exit the cell cycle into a quiescent (resting) phase called G_0.

Binding of DNA by the Basic Leucine Zipper (bZip) Motif
The yeast Gcn4 transcription factor (*black*) forms long helices that dimerize through the leucine zipper region and interact with DNA (*light gray*) via the basic region of the motif. (Figure created by Dr. Song Tan, Pennsylvania State University, using Midas-Plus software.)

During G_1, the cell receives input from mitogens—growth factors, cytokines, and insulin-like molecules—that prepare it for cell division. About midway through G_1 phase, a restriction (R) point is reached at which stimulation by growth factors is no longer necessary. This is essentially the "point of no return" at which the cell is committed to complete the cell cycle. It has been hypothesized that loss of control of R is a key determinant of cancer. Cells lacking G_1 control may bypass R and enter the cell cycle in the absence of proper mitogen signaling. Cell cycle regulators are frequently mutated in human tumors. Mutations leading to loss of *Rb* and *cyclin-dependent kinase inhibitor* (*CKI*) function or to overexpression of cyclins all have the effect of inappropriately driving cells through the cell cycle.

The tumor suppressors *Rb* and *p53* have emerged as key checkpoint control genes. Their role as master controllers is emphasized by the fact that binding assays show each interacts with more than 50 different cellular proteins, most of which must be coordinately regulated during different phases of the cell cycle. This is consistent with mounting evidence that large protein complexes modulate numerous inputs and outputs to achieve a high degree of coordination in eukaryotic cells. This is the case with the transcriptional machinery, which transcribes DNA; the spliceosome, which splices pre-mRNA; and the Hebbesome, which detects patterns of synaptic activity.

Rb regulates gene expression in G_1 phase. Ablating *Rb* immortalizes cultured cells by deregulating G_1 and driving them through the R point. Rb exerts its reg-

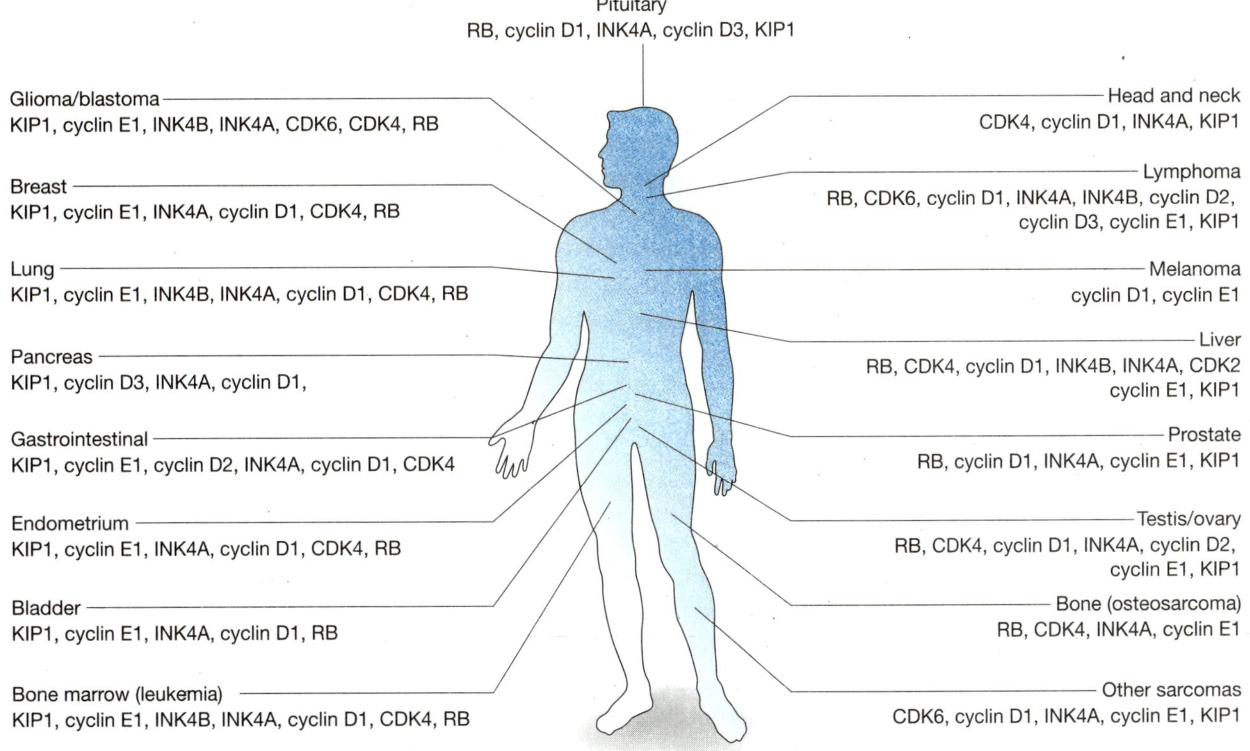

Cell Cycle Regulators and the Types of Cancer in Which They Have Been Found to Act
(Adapted, with permission, from Malumbres M. and Barbacid M. 2001. To cycle or not to cycle: A critical decision in cancer. *Nat. Rev. Cancer 1:* 222–231.)

ulatory function by binding the E2F family of transcription factors, inactivating them during M and G_0 phases. Interestingly, the viral oncoproteins E1A and large T antigen bind Rb at the same site as E2F, releasing E2F to stimulate transcription.

The p53 oncoprotein exerts checkpoint control in both G_1 and G_2 by suppressing replication of cells with DNA damage. p53 receives feedback from DNA repair systems, halting the cell cycle until the damage has been corrected. This surveillance system is extremely sensitive; p53 expression may be triggered by even a single break in a double-stranded DNA molecule. If DNA damage is too extensive, p53 induces apoptosis or "programmed cell death," thus preventing the establishment of a tumor-prone clone of cells with large numbers of mutations. For example, p53 apoptosis is induced in skin cells, where absorption of ultraviolet radiation produces a characteristic covalent linking of adjacent pyrimidines (Cs + Ts) in DNA (cyclobutane pyrimidine dimers [CPDs]). Apoptotic cells of the epidermis slough off, producing the peeling skin that follows a sunburn. Because of these crucial roles in DNA quality control, p53 has been called "the guardian of the genome."

This dual control is exerted in the nucleus, where p53 binds to promoter sequences to activate expression of key proteins involved in the cell cycle, including WAF1 and protein 14-3-3 σ. Entry into S and M phase is blocked by stimulating expression of WAF1, whereas M phase is blocked by stimulating expression of protein 14-3-3 σ. p53 activates transcription of several apoptosis-controlling proteins, including Bax, a member of the *bcl-2* gene family.

Apoptosis can also have a role in cancer. Translocations are seen in human B-cell lymphomas, which juxtapose the *bcl-2* oncogene on chromosome 18 with the immunoglobulin locus on chromosome 14. This situation brings *bcl-2* under the control of strong immunoglobulin enhancers. As immune defenders, B cells typically have short life spans and are eliminated by apoptosis several days after activation. Overexpression of *bcl-2* in cells with a chromosome 18-14 transposition protects B cells from apoptosis, presumably allowing them to linger long enough to acquire additional mutations needed to become fully malignant.

IMPROVING CANCER DIAGNOSIS AND TREATMENT

Radiation and chemotherapy are widely used to treat a variety of cancers. However, these treatments kill healthy cells as well as malignant cells, resulting in unpleasant or dangerous side effects. Aggressive therapy is clearly the only option for patients with highly malignant tumors—those most likely to metastasize or recur. However, in many cases, it is difficult to determine whether a particular tumor is highly malignant. In these situations, standards of medical care often require that patients undergo aggressive therapy. Although this is the prudent thing to do, it means that some patients with low-grade tumors undergo unneeded radiation or chemotherapy.

One early benefit from molecular genetic studies will thus be to correlate specific genetic changes in oncogenes and suppressors with malignancy and prognosis (outcome) and to use this information for better treatment. Colon cancer illustrates the precision with which we are coming to understand the sequence of genetic events that, by degree, convert a normal cell into a malignant one. Similar natural histories of accumulating mutations are being worked out for other important malignancies, including cancers of the stomach, breast, and prostate. It will be especially important to discover the mutational events that correlate with metastasis of tumors.

Gene amplification is one marker of tumor progression, and it can be readily detected by measuring mRNA hybridization to an oncogene probe. This method has identified amplified *myc* oncogenes in numerous human tumors and tumor cell lines, including carcinomas of the colon, lung, and breast; leukemia; retinoblastoma; and neuroblastoma. Studies in the 1980s of untreated neuroblastomas showed that amplification of the N-*myc* oncogene correlates with advancing stages of the disease; 3–300 copies of the *myc* gene were present in tumors from half of patients in advanced stages, whereas no amplification was detected in primary tumors taken from patients in early stages of the disease.

A similar analysis recently found that the tyrosine phosphatase gene *PRL3* is overexpressed in colorectal tumor cells that metastasize to the liver. *PRL3* is located in a region of chromosome 8q that is amplified in colorectal tumors. Like other phosphatases, PRL3 acts in opposition to kinases, removing a phosphate group to "deactivate" one or more proteins in a signal transduction pathway.

At present, many tumors are classified by microscopic examination. Although histological examination can separate tumors into different grades of

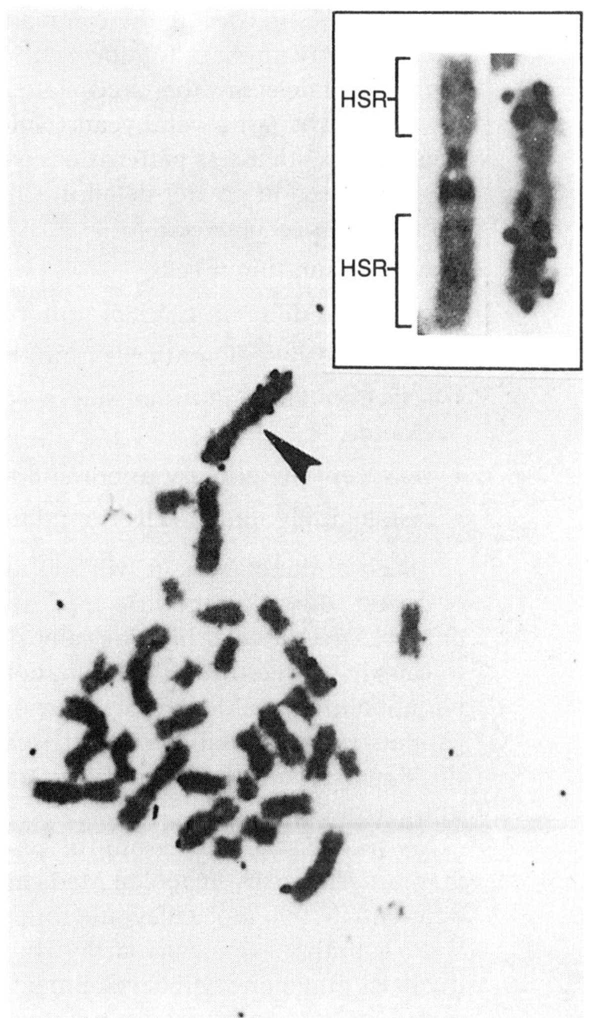

In Situ Hybridization Reveals Oncogene Amplification

Dark grains indicate regions where a radiolabeled probe for the *myc* oncogene hybridizes to homogeneously staining regions (HSRs) of chromosomes from neuroendocrine tumor cells. (*Inset*) Comparison of labeled and unlabeled chromosomes. (Courtesy of J. Bishop and H. Varmus, University of California, San Francisco. From K. Alitalo et al. 1983. *Proc. Natl. Acad. Sci. 80:* 1707.)

increasing severity, it ultimately fails to separate patients according to prognosis or response to therapy. Two carcinomas that appear to be identical under the microscope may, in fact, contain different mutations. One, for example, may have acquired the mutations needed to metastasize, whereas the other has not. Thus, two patients with seemingly identical carcinomas may have different clinical outcomes.

We have discussed some of the changes in signal transduction in cancer cells and how these pathways intersect with those involved in cell replication, DNA repair, and apoptosis. Although we only touched on some of the genes and proteins involved in these interlinked processes, the extreme complexity of this system of checks and balances can be sensed. Now imagine the ways in which each of these pathways, in turn, interacts with webs of kinases, phosphatases, enzymes, and substrates to execute the commands to divide, repair, or die. Add to that the fact that many tumor cells undergo a "crisis" of chromosome translocations, duplications, and deletions that precipitate large-scale disruptions in gene expression.

Within the last several years, biologists have turned to DNA microarrays or "gene chips," to assess patterns of gene expression in tumor cells. This technology (discussed in greater detail in Chapter 6) allows a comparison of levels of gene expression between

- normal and tumor cells,
- cells from different types of tumors,
- histologically different cells from the same tumor,
- cells from different tumor stages—precancerous dysplasia, adenoma, and carcinoma,
- cells from the primary tumor and metastatic cells, and
- histologically similar cells from different patients.

Each of these analyses will add to our understanding of the natural history of cancer. However, the first application of microarray technology has been to differentiate between histologically similar tumors—those that remain difficult to classify by microscopic examination. This is the case with diffuse large B-cell lymphoma (DLBCL), one of the most common blood tumors. About 40% of patients respond well to current treatments and survive long term; however, there is no reliable way to differentiate this group from patients whose cancer is untreatable.

A multidisciplinary group of researchers headed by Ash Alizadeh at the Stanford University School of Medicine used a "lymphochip" DNA microarray to differentiate two molecular subgroups of DLBCL. RNAs from 40 patient tumors were hybridized to a microarray containing 18,000 cDNAs representing genes involved in lymphocyte development. Two contrasting patterns of gene expression correlated with two distinct stages in B-cell differentiation. One resembled the expression profile of B cells from germinal centers in the lymph nodes; the other resembled activated B cells in the bloodstream. Three quarters of the patients with germinal center B-cell-like expression survived, compared to only 32% with activated B-cell-like expression.

Because of its key position at the interface between cell division and cell death, determining *p53* status has become critical in cancer prognosis. Most p53 mutations in human tumors abolish the protein's ability to bind DNA and, so, activate transcription of various controllers of the cell cycle and apoptosis. Mutations found in a number of solid tumors—breast, colon, lung, esophagus, and head and neck—correlate with shorter patient survival. There is mounting evidence that *p53* status predicts response to cancer treatments that rely on tumor killing via apoptosis. Chemotherapeutic agents, including cisplatin, doxorubicin, etoposide, and mitomycin C, induce apoptosis by damaging DNA in dividing cells. As described earlier, upon "sensing" irreparable DNA damage, p53 stalls the cell cycle and induces apoptosis. Thus, tumors with inactivating mutations in p53 tend to respond poorly to chemotherapy and to relapse quickly.

Because p53 inactivation is the most common genetic aberration shared across tumor types, strategies to rescue p53 function hold out great promise. Restoring p53 function could render many types of tumors more susceptible to traditional treatments. A number of gene therapy protocols are attempting to replace mutant *p53* genes with functioning genes. However, pharmaceutical companies see greater promise in screening chemical libraries in hopes of finding small molecules that stabilize the p53 protein by directly binding to a mutated site to restore a reactive conformation. "Rational drug design" goes a step beyond high-throughput, but random, screening methods. The rational approach uses X-ray structural data on the mutated p53 protein to *design* small molecules that bind to the mutated p53 protein, effectively "resurfacing" it to restore function to key reactive sites.

Understanding the action of traditional treatments may make them more effective. For example, 5-fluorouracil (5-FU), a chemotherapeutic treatment effective against colon cancer, illustrates that some agents may utilize p53 to both induce and sense DNA damage. *Ferredoxin reductase* (*FR*), one of the relatively few genes induced by 5-FU, codes for an electron-transferring molecule of the mitochondrion. The *FR* promoter contains a binding site for p53, and so may be directly activated by p53. When present in excess, FR releases electrons, which may then generate reactive oxygen species that, in turn, have also been implicated in apoptosis.

Chemotherapy targets rapidly dividing cells, both cancerous and healthy. For example, the killing of normal cells of the blood and digestive tract results in the common side effects of fatigue and nausea. Cytokines, natural stimulators of cell division, are used increasingly to stimulate the replenishment of blood cell types depleted by cancer treatment. Granulocyte colony stimulating factor (G-CSF), among the first therapeutics made from a cloned human gene, stimulates the proliferation and differentiation of granulocyte stem cells and improves function of mature neutrophils. Similarly, erythropoietin helps cancer patients overcome debilitating anemia by stimulating redevelopment of red blood cells.

Our growing knowledge of signal transduction pathways—and points at which they are perturbed in cancer cells—is beginning to revolutionize cancer treatment. The first use of these data has been the development of antibodies that bind to a mutated growth factor or cell surface receptor, blocking its action and "turning off" the signal cascade. About a dozen therapeutic antibodies have been approved or are undergoing testing—targeting epidermal growth factor receptor (ERB2), vascular endothelial growth factor (VEGF), cell adhesion molecules (Cams), and cell surface antigens found on specific types of tumor cells.

The results of the first generation of antibody therapeutics have been less than spectacular. Relatively low response rates have kept them as mainly drugs of last resort for terminally ill patients. For example, the prototype for this group, Herceptin, targets HER2/NEU (ERB2) and was approved for use in the United States in 1998. However, only 15% of patients responded to Herceptin treatment for metastatic breast cancer in Phase III clinical trials. Response rates improve dramatically when Herceptin is given in combination with chemotherapy, but at the cost of increased toxicity. Antibodies armed with toxins or radionuclides have been approved for use in leukemia and lymphoma, and they promise improved response for other tumors.

The use of antibodies as therapeutics is limited by two facts: First, being proteins, they must be given by infusion. Second, being relatively large molecules, they do not readily enter cells and so must target cell surface features. Thus, pharmaceutical companies are actively searching for small-molecule drugs that

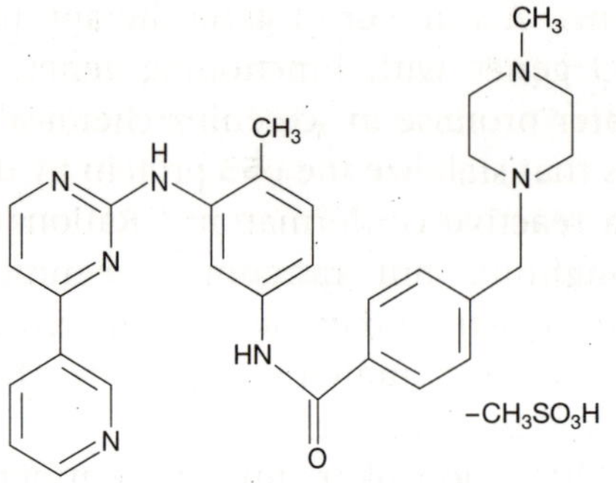

Chemical Structure of Gleevec™ (Novartis Pharmaceuticals)

remain active when taken orally and that diffuse easily into tumor cells. In the past, the development of anticancer agents was hampered by a scarcity of metabolic "targets" against which to test activity of drug candidates. Now, each protein implicated in errant signal transduction in cancer cells has become a validated target for therapeutic development. Robotic, high-throughput methods are then employed to screen the target against thousands of small molecules in chemical libraries. In vitro assays, typically performed in a 96-well microtiter plate, search for molecules that alter transcription of the target gene or which interact directly with the target protein to stabilize its activity, prevent its activation by phosphorylation, or block binding sites for interaction with other molecules in the pathway.

```
c-kit   LSFGKTLGAGAFGKVVEATAYGLIKSDAAMTVAVKMLKPSAHLTEREALMSELKVLSYLG
PDGFR   LVLGRTLGSGAFGQVVEATAHGLSHSQATMKVAVKMLKSTARSSEKQALMSELKIMSHLG
c-Abl   ITMKHKLGGGQYGEVYEG-----VWKKYSLTVAVKTLKEDT--MEVEEFLKEAAVMKEIK
              *    *  *                 * *                *   *
c-kit   NHMNIVNLLGACTIGGPTLVITEYCCYGDLLNFLRRKRDSFICSKQED--HAEAALYKN-
PDGFR   PHLNVVNLLGACTKGGPIYIITEYCRYGDLVDYLHRNKHTFLQHHSDKRRPPSAELYSNA
c-Abl   -HPNLVQLLGVCTREPPFYIITEFMTYGNLLDYLR-------------------------
                           * * **  **
c-kit   -----LLHSKESSCSDSTNEYMDMKPG--VSYVVPTKAD---KRRSVRIGSYIER--DVT
PDGFR   LPVGLPLPSHVSLTGESDGGYMDMSKDESVDYVPMLDMKGDVKYADIESSNYMAPYDNYV
c-Abl   ------------------------------------------------------------

c-kit   P--------AIMEDDELALDLEDLLSFSYQVAKGMAFLASKNCIHRDLAARNILLTHGRI
PDGFR   PSAPERTCRATLINESPVLSYMDLVGFSYQVANGMEFLASKNCVHRDLAARNVLICEGKL
c-Abl   ------------ECNRQEVNAVVLLYMATQISSAMEYLEKKNFIHRDLAARNCLVGENHL
                                                             *
c-kit   TKICDFGLARDIKNDSNYVVKGNARLPVKWMAPESIFNCVYTFESDVWSYGIFLWELFSL
PDGFR   VKICDFGLARDIMRDSNYISKGSTFLPLKWMAPESIFNSLYTTLSDVWSFGILLWEIFTL
c-Abl   VKVADFGLSRLMTGDT-YTAHAGAKFPIKWTAPESLAYNKFSIKSDVWAFGVLLWEIATY
            **
c-kit   GSSPYPGMPVDSKFYKMIKEGFRMLSPEHAPAEMYDIMKTCWDADPLKRPTFKQIVQLI
PDGFR   GGTPYPELPMNEQFYNAIKRGYRMAQPAHASDEIYEIMQKCWEEKFEIRPPFSQLVLLL
c-Abl   GMSPYPGIDLS-QVYELLEKDYRMERPEGCPEKVYELMRACWQWNPSDRPSFAEIHQAF
```

Amino Acid Alignment of the Tyrosine Kinase Domains of c-Kit, PDGFR, and c-Abl

Amino acids in the c-Abl protein that interact with the kinase inhibitor Gleevec™ are indicated with an asterisk. The first marked residue (L) represents amino acid 248 of c-Abl. Identical residues are indicated by a blue background. Conservative changes, where substituted amino acids have similar properties, are indicated by a gray background.

Judah Folkman
(Courtesy of Cold Spring Harbor Laboratory Archives.)

Gleevec, approved in 2001 for treatment of chronic myeloid leukemia (CML), became the first cancer therapeutic developed using this method, Gleevec interacts specifically with the BCR-Abl protein, a mutant form of the EGF receptor. The drug occupies a binding site for a protein tyrosine kinase, preventing phosphorylation of the receptor and halting signal transduction through the BCR-Abl pathway. Gleevec's pinpoint activity within a known biochemical pathway makes it similar to modern cholesterol-lowering drugs, such as Lipitor and Lovastatin. Although initially approved for CML, Gleevec also inhibits the related PDGF receptor and c-Kit, the receptor for stem cell factor (SCF). Thus, it should prove useful in treating a range of tumors in which these receptors are mutated.

A solid tumor that remains in its place of origin is rarely life-threatening. The most dangerous tumors are those that acquire their own blood supply and then use the circulatory system to spread to other sites around the body. When a tumor grows beyond approximately 1 million cells—about the size of a grain of rice—it can no longer rely on the existing local blood supply to provide enough nutrients. Thus, growing tumors secrete angiogenesis factors, including vascular endothelial growth factor (VEGF), a class of cytokines that stimulate formation of blood vessels. The promise of anti-angiogenesis drugs, or statins, in cancer treatment was brought to the fore in the late 1990s by Judah Folkman, of Harvard Medical School. Nobel Laureate James Watson's praise of Folkman's work caused a rash of speculation in a company commercializing his natural anti-angiogenesis compound, endostatin. Although early clinical trials proved disappointing, few doubt that statins will one day take their place among the new generation of molecular genetic treatments for cancer.

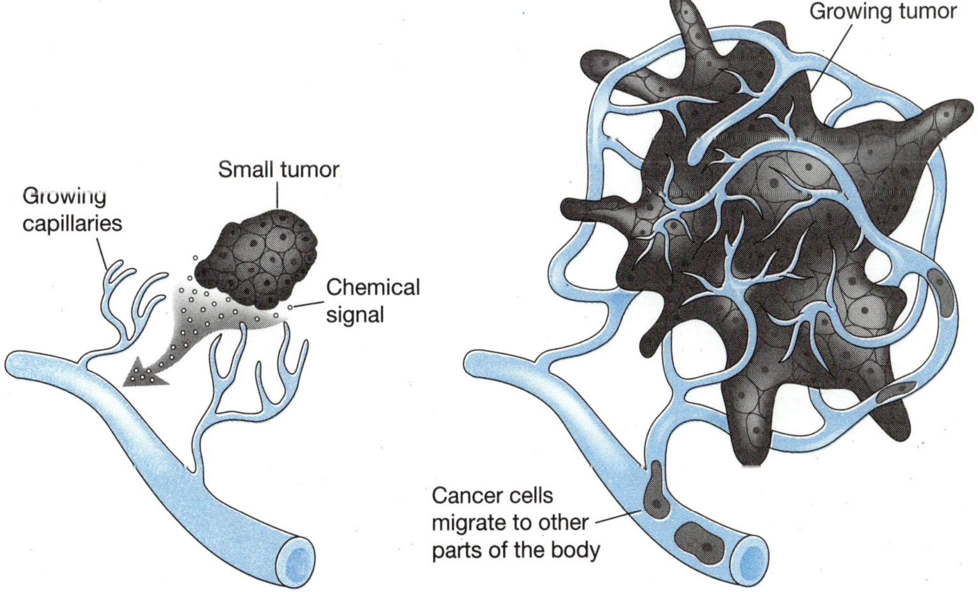

Angiogenesis in Tumor Growth
The small tumor sends out chemical signals to blood vessels, causing capillaries to grow out. The tumor grows and becomes vascularized.

REFERENCES

Angier N. 1999. *Natural obsessions: Striving to unlock the deepest secrets of the cancer cell.* Houghton Mifflin, Boston.

Cooke R. 2001. *Dr. Folkman's War: Angiogenesis and the struggle to defeat cancer.* Random House.

Evan G.I. and Vousden K.H. 2001. Proliferation, cell cycle, and apoptosis in cancer. *Nature* **411:** 342–348.

Gibbs J.B. 2000. Mechanism-based target identification and drug discovery in cancer research. *Science* **287:** 1969–1973.

Goldberg A.L., Elledge S.J., and Harper J.W. 2001. The cellular chamber of doom. *Sci. Am.* **284:** 68–73.

Hanahan D. and Weinberg R.A. 2000. The hallmarks of cancer. *Cell* **100:** 57–70.

Knudson A.G. 2001. Two genetic hits (more or less) to cancer. *Nature Rev. Cancer* **1:** 157–162.

Lengauer C., Kinzler K.W., and Vogelstein B. 1998. Genetic instabilities in human cancer. *Nature* **396:** 643–649.

Malumbres M. and Barbacid M. 2001. To cycle or not to cycle: A critical decision in cancer. *Nature Rev. Cancer* **1:** 222–231.

McKinnell R.G., Parchment R.E., Perantoni A.O., and Pierce G.B., eds. 1998. *The biological basis of cancer.* Cambridge University Press, United Kingdom.

Micklos D., ed. 1999. DNA from the beginning (http://www.dnaftb.org). Dolan DNA Learning Center, Cold Spring Harbor, New York.

Peto J. 2001. Cancer epidemiology in the last century and the next decade. *Nature* **411:** 390–395.

Rowley J.D. 2001. Chromosome translocations: Dangerous liasons revisited. *Nature Rev. Cancer* **1:** 245–250

Scientific American Special Issue. *What you need to know about cancer.* 1996. *Sci. Am.* **275(3).**

Vogelstein B., Lane D., and Levine A.J. 2000. Surfing the p53 network. *Nature* **408:** 307–310.

Weinberg R.A. 1998. *Racing to the beginning of the road: The search for the origin of cancer.* W.H. Freeman, New York.

——. 1999. *One renegade cell: How cancer begins.* Basic Books, New York.

CHAPTER 8

Applying DNA Science to Human Genetics and Evolution

WE STAND AT THE THRESHOLD OF A NEW CENTURY with the whole human genome stretched out before us. It is, at once, a public and intensely private record. Written in each person's DNA is a shared history of the evolution of our species and a personal portent of both the health and disabilities we may encounter as individuals.

High-throughput analyses using bioinformatic tools and DNA arrays are uncovering an exponentially increasing number of genes implicated in human diseases. Each gene identified in a known disease pathway automatically becomes a validated target for therapeutic development. Pharmaceutical companies are racing to convert this new knowledge into blockbuster drugs while the medical profession ponders how it will keep up with the flood of new diagnostic and treatment options. The implications extend far beyond medicine, as scientists seek genes behind the behaviors—and misbehaviors—that make us uniquely human.

The border between science fact and fiction will blur as we move further into the genome age. A tenfold increase in the information capacity of photolithographic DNA arrays could condense the entire human genome onto a chip the size of a postage stamp. Perhaps not everyone will be able to afford to carry a complete copy of their genome in a medical alert locket, but rapid scans of hundreds of medically and behaviorally important genes will almost certainly become part of standard medical care in developed countries.

What will it be like when we have a precise catalog of all the good, bad, and middling genes—and the wherewithal to determine who has which? On a personal level, will a genome-wide scan take on the aspect of genetic tarot, predicting the future course of our lives? In the face of such knowledge, will society continue to acquiesce to those who prefer to let nature take its course or will we gravitate toward a prescribed definition of the "right" genetic stuff?

This is not the first time that we have asked such questions. At the turn of the last century, science and society faced a similar rush to understand and exploit human genes. Eugenics was the name of the effort to apply principles of Mendelian genetics to improve the human species.

The eugenics movement began benignly in England with positive efforts by families to improve their own heredity. It took a negative turn in the United States, as well as in Scandinavia, where flawed data became the basis for laws to sterilize individuals and restrict immigration by ethnic groups deemed "unfit." These misguided attempts at eugenic social engineering formed part of the basis

Harry Laughlin and Charles Davenport Outside the Eugenics Record Office, 1912
(Courtesy of the Harry H. Laughlin Archives, Truman State University; http://www.eugenicsarchive.org.)

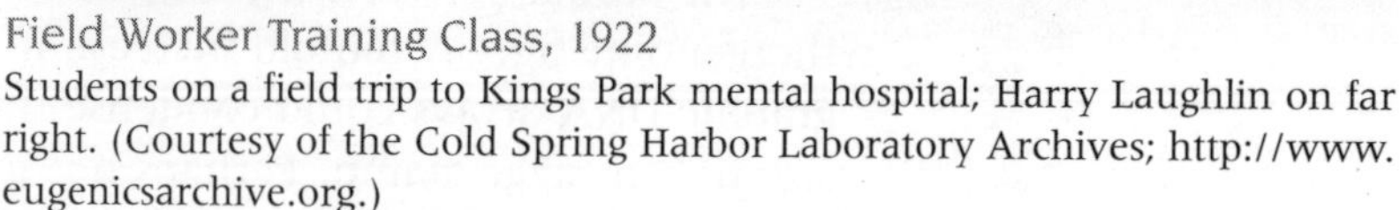

Field Worker Training Class, 1922
Students on a field trip to Kings Park mental hospital; Harry Laughlin on far right. (Courtesy of the Cold Spring Harbor Laboratory Archives; http://www.eugenicsarchive.org.)

Eugenic and Health Exhibit, 1922
Fitter Families exhibit and examination building at the Kansas State Free Fair. (Courtesy of American Philosophical Society; http://www.eugenicsarchive.org.)

of the Nazi "final solution" to achieve racial purity, which resulted in the murder of more than 10 million Jews, gypsies, and other groups considered unfit.

We will thus begin the story of human genetics with the cautionary tale of its conjoined birth with eugenics, focusing on its development in the United States. Partly due to the stigma of association with the Holocaust and partly due the difficulty of conducting rigorous experimental studies, human genetics languished as a sort of scientific backwater after World War II, until molecular genetics provided new methods to track gene inheritance. After discussing the applications of genetics to the practical problems of human disease and identity, we will link these to the emerging story of how humans evolved and populated the earth.

CHARLES DAVENPORT AND THE EUGENICS RECORD OFFICE

Francis Galton
(Copyright The Galton Collection, University College London; http://www.eugenicsarchive.org.)

The term eugenics—meaning well born—was coined in 1883 by Francis Galton, a scientist at University College in London. Galton's conception of eugenics arose from his earlier study, *Hereditary Genius* (1869), in which he concluded that superior intelligence and abilities were inherited with an efficiency of about 20% among primary relatives in noteworthy British families. Galton's emphasis on the voluntary improvement of a family's genetic endowment became known as "positive eugenics" and remained the focus of the British movement.

In Chapter 1, we saw how evolutionary thinking flowed into genetics by way of the ephemeral discipline of experimental evolution. In a similar way, evolutionary theory flowed into human genetics by way of eugenics. During the first decade of the 20th century, eugenics was organized as a scientific field by the confluence of ideas from evolutionary biology, Mendelian genetics, and experimental breeding. This synthesis was embodied by Charles Davenport.

Recall that, in 1898, Davenport became director of the Biological Laboratory at Cold Spring Harbor, New York, a field station to study evolution in the natural world. In 1904, on an adjacent property, he founded the Station for Experimental Evolution, whose researchers were among the very first adherents of Mendelian genetics. Davenport was among the first scientists to contribute to the genetic description of *Homo sapiens*. In 1907, he published the classical (although still incomplete) description of the inheritance of human eye color. He continued on to do early studies on the genetics of skin pigmentation, epilepsy, Huntington's disease, and neurofibromatosis. Other researchers, including Archibald Garrod, described Mendelian inheritance in alkaptonuria, brachydactyly, hemophilia, and color blindness.

Davenport became interested in eugenics through his association with the American Breeders Association (ABA), the first scientific body in the United States to actively support eugenics research. The ABA members—including Luther Burbank, a pioneer of the American seed business—were literal in their aim to directly apply principles of agricultural breeding to human beings. This is aptly illustrated by Davenport's book, *Eugenics: The Science of Human Improvement by Better Breeding*, as well as by the "Fitter Family" contests held at state fairs throughout the United States during the 1920s. These competitions judged families in the same context as the fastest racehorses, the fattest pigs, and the largest pumpkins.

In 1910, Davenport obtained funding to establish a Eugenics Record Office (ERO) on property adjacent to the Station for Experimental Evolution. A series of ERO bulletins, including Davenport's *Trait Book* and *How to Make a Eugenical Family Study*, helped to standardize methods and nomenclature for constructing pedigrees to track traits through successive generations. Constructing a pedigree entailed three important elements: (1) finding extended families that express the trait under study, (2) "scoring" each family member for the presence or absence of the trait, and (3) then attempting to discern one of three basic modes of Mendelian inheritance: dominant, recessive, or sex-limited (X-linked).

Eugenicists fared well on the first element, because large families were common in the first decades of the 20th century. However, scoring traits was a difficult problem, especially when eugenicists attempted to measure complex traits (such as intelligence or musical ability) and mental illnesses (such as schizophrenia or manic depression). In general, eugenicists were lax in defining the criteria for measuring many of the "traits" they studied. This led them to conclude that many real and imagined traits—including alcoholism, feeblemindedness, pauperism, social dependency, shiftlessness, nomadism, and lack of moral control—were single-gene defects inherited in a simple Mendelian fashion.

Much eugenical information was submitted voluntarily on questionnaires. Some families were proud to make known their pedigrees of intellectual or artistic achievement, whereas others sought advice on the eugenical fitness of proposed marriages. The circus performers on midways of nearby Coney Island offered eugenics researchers a trove of unusual physical trait differences, including giantism, dwarfism, polydactyly, and hypertrichosis. Notably, Davenport's correspondence with an albino circus family resulted in the first Mendelian study of albinism, published in the *Journal of Human Heredity*.

In addition to interviewing living family members, eugenics workers also used data from insane asylums, prisons, orphanages, and homes for the blind. Surveys filled out by superintendents were used to calculate the ethnic makeup of societal "dependents" and the costs of maintaining them in public institutions. With the mobilization for World War I, tens of thousands of men inducted for the draft provided a ready source of anthropometric and intelligence data. Notably, the Army Alpha and Beta Intelligence Tests, developed by Robert Yerkes of Harvard University, supposedly measured the innate intelligence of army recruits. African-American and foreign-born recruits were much more likely to do poorly on the Yerkes tests, because they mostly measured knowledge of white American culture and language.

THE CONSTRUCTION OF GENETIC BLAME

Whereas Francis Galton had focused on the positive aspects of human inheritance, the American movement increasingly focused on a "negative eugenics" program to prevent the contamination of the American germ plasm with supposedly unfit traits. The concept that some groups of people are genetically unfit dates back to Biblical references to the Amalekites. By about 1700, degeneracy theory supplied the "scientific" explanation that unfit people arose from bad environments which damaged heredity and perpetuated degenerate offspring.

Richard Dugdale, of the Executive Committee of the New York Prison Association, brought the concept of degenerate inheritance to eugenics in *The Jukes* (1877), a pedigree study of a clan of 700 petty criminals, prostitutes, and paupers living in the Hudson River Valley north of New York City.

Dugdale held the Lamarckian view that the environment induces heritable changes in human traits. He compassionately concluded the Jukes' situation could be corrected by providing them improved living conditions, schools, and job opportunities. However, this interpretation was discredited by American eugenicists, who embraced Mendel's genetics and Weismann's theory of the germ plasm. Together, these formed an interpretation that human traits are determined by genes which are passed from generation to generation without any interaction with the environment. Thus, when the ERO's field worker Arthur Estabrook reevaluated the Jukes in 1915, he found continued degeneration and placed the blame squarely on bad genes and the people who carried them.

Davenport's study of naval officers amusingly illustrates the extent to which eugenicists sought genetic explanations of human behavior to the exclusion of environmental influences. After analyzing the pedigrees of notable seamen—including Admiral Lord Nelson, John Paul Jones, and David Farragut—Davenport concluded that they shared several heritable traits. Among these was thalassophilia, "love of the sea," which he determined was a "sex limited" trait, because it was found only in men. Davenport failed to consider the equally likely explanations that sons of naval officers often grew up in environments dominated by boats and tales of the sea or that women were prohibited from seafaring occupations throughout the 19th and early 20th centuries.

Eugenic Social Engineering

Eugenics arose in the wake of the Industrial Revolution, when the fruits of science were improving public and private life. A growing middle class of professional managers believed that scientific progress offered the possibility of rational cures for social problems. Placing the blame for social ills on bad genes—and the people who carried them—raised the question: Why bother to build more insane asylums, poorhouses, and prisons when the problems that necessitated them could be eliminated at their source? Thus, negative eugenics seemed to offer a rational solution to age-old social problems. American eugenicists were largely successful in lobbying for social legislation on three fronts: to restrict European immigration, to prevent race mixing (miscegenation), and to sterilize the "genetically unfit."

As the eugenics movement was gathering strength, a phenomenal tide of immigration was rolling into the United States. During the first two decades of the 20th century, 600,000 to 1,250,000 immigrants per year entered the country through the facility on Ellis Island in New York Harbor (except during World War I). It is estimated that 100 million Americans alive today can trace their ancestry to an immigrant who arrived at Ellis Island. Also during this period, the nativity of the majority of immigrants shifted away from the northern and western European countries that had contributed most immigrants during the Colonial, Federal, and Victorian eras. Increasingly, the immigrant stream was dominated by southern and eastern Europeans, including large numbers of displaced Jews.

Doctor Examining Immigrants at Ellis Island, 1904
(Courtesy of National Park Service: Statue of Liberty Monument; http://www.eugenicsarchive.org.)

Immigrants Bound for New York on an Atlantic Liner

American eugenicists were overwhelmingly white, of northern and western European extraction, and members of the educated middle and upper classes. They looked with disdain on the new immigrants, many of whom settled in lower Manhattan. The plight of many of these immigrants—packed into tenements, plagued by tuberculosis and crime, and reduced to virtual serfdom in sweat shops—was sympathetically chronicled by Jacob Riis and the "muckraking" journalists. But to eugenicists, the immigrants' lot had little to do with poverty or lack of opportunity and had everything to do with their bad genes, which eugenicists feared would quickly "pollute" the national germ plasm. The eugenics movement provided a scientific rationale for growing anti-immigration sentiments in American society. Labor organizations fed on fears that working class Americans would be displaced from their jobs by an oversupply of cheap immigrant labor, while anti-Communist factions stirred up fears of the "red tide" entering the United States from Russia and eastern Europe.

As "expert agent" for the Committee on Immigration and Naturalization of the U.S. House of Representatives, ERO Superintendent Harry Laughlin became the anti-immigration movement's most persuasive lobbyist in the early 1920s. During three separate testimonies, he presented data that purported to show that southern and eastern European countries were "exporting" genetic defectives to the United States who had disproportionately high rates of mental illness, crime, and social dependency. The resulting Immigration Restriction Act of 1924 cut immigration to 165,000 per year and restricted immigrants from each country according to their proportion in the U.S. population in 1890—a time prior to the major waves of immigration from southern and eastern Europe. This had the desired effect of reducing southern and eastern European immigrants to less than 15,000 per year. Immigration did not regain prerestriction levels again until the late 1980s.

Of all the legislation enacted during the first four decades of the 20th century, sterilization laws adopted by 30 states most clearly bear the stamp of the eugenics lobby. Although the earliest sterilization law, passed in Indiana in 1907, was aimed at convicts and sex offenders, "feebleminded" persons became the major targets for eugenic sterilization. This owed much to Henry Goddard's influential book, *The Kallikaks* (1912). This effectively related study of the descendents of Martin Kallikak (the name is fictitious) was, in effect, a controlled experiment in positive and negative eugenics. Martin's marriage to a normal woman produced a normal lineage (from the Latin *kallos* for "goodness and beauty"). However, as a young militiaman in the Revolutionary War, he had an elicit union with an attractive but feebleminded barmaid, producing a second, "bad" lineage (*kakos*, for "bad"). Thus, the primary intent of eugenic sterilization was to curb the supposed promiscuous tendencies of the feebleminded, who threatened to perpetuate their kind and to contaminate good lineages, as surely as the case of Martin Kallikak.

Many of the early sterilization laws were legally flawed and did not meet the challenge of state court tests. To address this problem, Laughlin designed a model eugenics law that was reviewed by legal experts. Virginia's use of the model law was tested in *Buck v. Bell*, heard before the Supreme Court in 1927. Oliver Wendell Holmes, Jr., delivered the Court's decision upholding the legality of eugenic sterilization, which included the infamous phrase, "Three generations of imbeciles are enough!"

Carrie Buck, the subject of the case, had given birth to an illegitimate daughter and been institutionalized in the Virginia Colony for the Epileptic and the Feebleminded. Carrie was judged to be "feebleminded" and promiscuous. Arthur Estabrook examined Carrie's infant daughter Vivian and found her "not

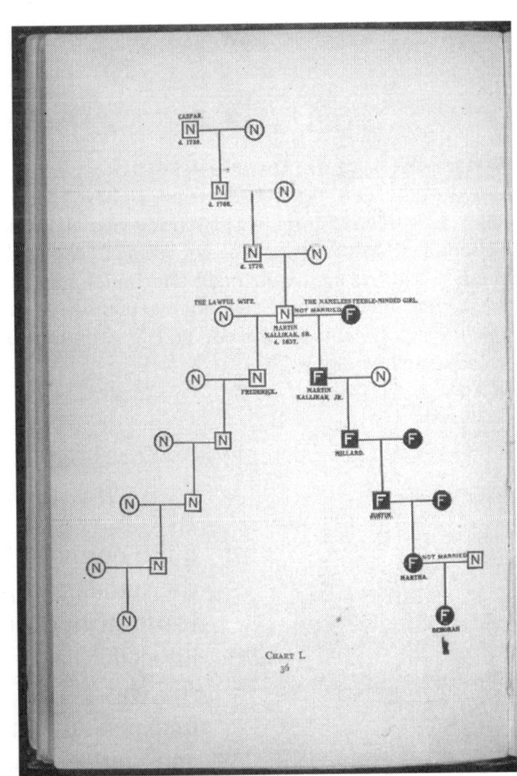

Henry Goddard's Pedigree of the Kallikak Family, 1912

The pedigree shows normal (N) and feebleminded (F) lines. (Courtesy of the University of Albany, SUNY; http://www.eugenicsarchive.org.)

Carrie and Emma Buck, 1924
This evocative photo of daughter and mother on a bench at the Viriginia Colony for the Epileptic and the Feebleminded, in Lynchburg, was taken the day before the start of the Virginia trial that would lead all the way to the U.S. Supreme Court. (Courtesy of the University of Albany, State University of New York; http://www.eugenicsarchive.org.)

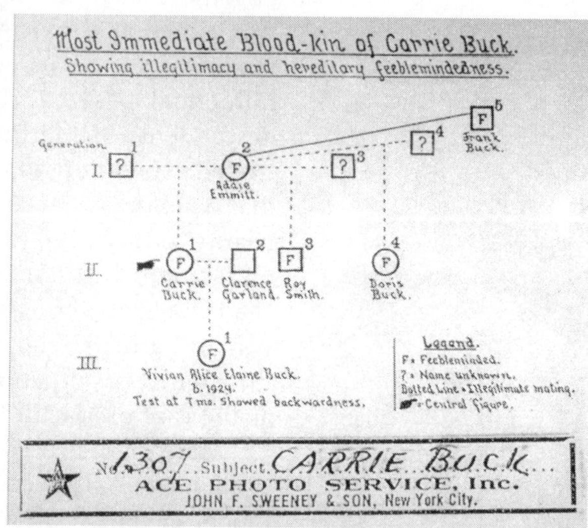

Pedigree of the Buck Family, 1924
This exhibit from the Virginia trial clearly illustrates three supposed generations of feebleminded females in the Buck family: Emma (Addie Emmitt), Carrie, and Vivian. Vivian's father, Clarence Garland, was the nephew of Carrie's foster parents, Mr. and Mrs. John Dobbs. Clarence had promised to marry Carrie, but disappeared by the time of her trial in 1924. (Courtesy of Paul Lombardo, Ph.D., J.D., and the American Philosophical Society; http://www.eugenicsarchive.org.)

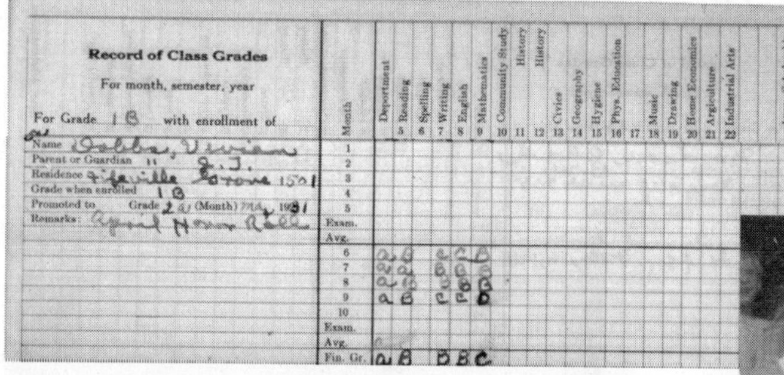

Record of Class Grades

For month, semester, year

For Grade 1 B with enrollment of

Name Dobbs, Vivian
Parent or Guardian " J. J.
Residence [illegible] 1501
Grade when enrolled 1 B
Promoted to Grade 2 [illegible] (Month) [illegible] 1931
Remarks: [illegible]

Month: 1, 2, 3, 4, 5, Exam., Avg., 6, 7, 8, 9, 10, Exam., Avg., Fin. Gr.

Deportment, 5 Reading, 6 Spelling, 7 Writing, 8 English, 9 Mathematics, 10 Community Study, 11 History, 12 History, 13 Civics, 14 Geography, 15 Hygiene, 16 Phys. Education, 17, 18 Music, 19 Drawing, 20 Home Economics, 21 Agriculture, 22 Industrial Arts, Aver. Scholarship

Vivian's First-grade Report, 1931
Listed under her adoptive surname, Dobbs, Vivian was a solid "B" student during her first-grade year at the Venable School in Charlottesville. She got straight "As" in deportment (conduct) and even made the honor role in April, 1931. She died a year later after a bout with measles. (Courtesy of Paul Lombardo, Ph.D., J.D.; http://www.eugenicsarchive.org.)

Carrie's Baby, Vivian, 1924.
This photo, taken the day before the Virginia trial, is believed to capture the "standard mental test" used by Arthur Estabrook to determine that Vivian Buck was feebleminded. Six-month old Vivian appears uninterested as foster mother Mrs. Dobbs attempts to catch her attention with a coin. (Courtesy of the University of Albany, State University of New York; http://www.eugenicsarchive.org.)

quite normal." It is impossible to judge whether Carrie was "feebleminded" by the standards of her time, but the child that Carrie bore out of wedlock was the result of her rape by the nephew of her foster parents. Clearly, Vivian was no imbecile. Later scholarship turned up Vivian's first-grade report, showing that she was a solid "B" student and received an "A" in deportment. Carrie was the first person sterilized under Virginia's law. *Buck v. Bell* was never overturned, and sterilization of the mentally ill continued into the 1970s, by which time about 60,000 Americans had been sterilized—most without their consent or the consent of a legal guardian.

Opposition and the End of Eugenics

Scientific opposition to eugenics came on many fronts and began even as it was being organized as a scientific discipline. In 1909, George Shull, at the Carnegie Station for Experimental Evolution, showed that the hybrid offspring of two inbred strains of corn are more vigorous than their inbred parents. The phenomenon of hybrid vigor also held true in mongrel animals, refuting eugenicists' notion that racial purity offers any biological advantage or that race mixing destroys "good" racial types.

Wilhelm Weinberg
(Reprinted, with permission, from Stern C. 1962. Wilhelm Weinberg, 1862–1937. *Genetics* 47: 1–5; (©Genetics Society of America.)

Work by a number of scientists countered the eugenicists' simplistic assertions that complex behavioral traits are governed by single genes. Hermann Muller's survey of mutations in *Drosophila* and other organisms from 1914 to 1923 showed variation in the "gene-to-character" relation that defied simple Mendelian analysis. Many genes are highly variable in their expression, and a single gene may affect several characteristics (traits) at one time. Conversely, mutations in many different genes can affect the same trait in similar ways. Moreover, the expression of a gene can be altered significantly by the environment. Twin studies conducted by Horatio Hackett Newman also showed that identical twins raised apart after birth averaged a 15-point difference in I.Q. Lionel Penrose found that most cases at a state-run institution in Colchester, England, resulted from a combination of genetic, environmental, and pathological causes.

Mathematical models of population genetics provided evidence against the simplistic claim that degenerate families were increasing the societal load of dysgenic genes. The equilibrium model of Godfrey Hardy and Wilhelm Weinberg showed that, although the absolute number of dysgenic family members might increase over time, the frequency of any "negative" trait does not increase relative to the normal population. Feeblemindedness, thought to be a recessive dis-

Godfrey Hardy
(Courtesy of Trinity College, Cambridge, England.)

order, presented a particular quandary. Although geneticists almost universally agreed that the feebleminded should be prevented from breeding, the Hardy-Weinberg equation showed that sterilization of affected individuals would never appreciably reduce the incidence of the disorder. Only a hideously massive program of sterilizing the vast reservoir of heterozygous carriers predicted by the equation would have any hope of significantly reducing the incidence of mental illness. Despite this, feeblemindedness was thought to be so rampant that many geneticists believed reproductive control could still prevent the birth of tens of thousands of affected individuals per generation.

Although he was a founding member of the board of the ERO, Thomas Hunt Morgan resigned after several years. He criticized the movement in the 1925 edition of his popular textbook, *Genetics and Evolution,* warning against the wholesale application of genetics to mental traits, and against comparing whole races as superior or inferior. He offered this advice: "...until we know how much the environment is responsible for, I am inclined to think that the student of human heredity will do well to recommend more enlightenment on the social causes of deficiencies...in the present deplorable state of our ignorance as to the causes of mental differences."

In 1928, Johns Hopkins geneticist Raymond Pearl charged that most eugenics preaching was "contrary to the best established facts of genetical science." A visiting committee of the Carnegie Institution in 1935 concluded that the body of work collected at the ERO was without scientific merit and recommended that it end its sponsorship of programs in sterilization, race betterment, and immigration restriction. Thus, the negative emphasis of American eugenics was completely discredited among scientists by the mid 1930s. Growing public knowledge of Germany's radical program of race hygiene led to a wholesale abandonment of popular eugenics. The ERO was closed in December 1939.

In the meantime, eugenics was gathering steam in Germany. Laughlin's model sterilization law was the basis for Nazis' own law in 1933, and his contributions to German eugenics were recognized by an honorary degree from the University of Heidelberg in 1936. Over the next several years, some 400,000 people—mainly in mental institutions—were sterilized. In 1939, euthanasia replaced sterilization as a solution for mental illness, and the lives of nearly 100,000 patients were ended "mercifully" with lethal gas. Overt euthanasia of mental patients ceased in 1941, when physicians with experience in euthanasia were reassigned to concentration camps in Poland, where they were needed to apply the "final solution" for Nazi racial purity.

PROBLEMS ON THE ROAD TO MODERN HUMAN GENETICS

Following the shocking revelations of euthanasia and human experimentation that took place in the Nazi concentration camps, it is not difficult to understand why human genetics research was largely avoided during the years following World War II. The fact also remained that prior to the advent of restriction enzymes and recombinant DNA, researchers simply did not have the tools to identify genes or to precisely locate them on chromosome maps.

Genetics had succeeded in *Drosophila,* for example, because a fly with a new mutation can be identified by visual inspection. The mutant fly can then be selectively mated with other flies of known characteristics to determine the mode of inheritance and chromosome position of the mutated gene. These analyses are simplified by the fly's rapid generation time and many offspring per generation. Furthermore, the lineage of each individual is known at the outset of a breeding experiment. Members of the experimental pool are most often physically and genetically identical, differing from one another only by one or, at most, several traits. This genetic homogeneity allows a specific trait or mutation to be observed against an essentially neutral background.

Human genetics differs from classical genetics in that the system under study cannot be easily manipulated. Although arranged marriages still take place in some cultures, people, for the most part, choose their own spouses and are generally opposed to being selectively mated. Thus, human geneticists must be content to work with the existing genetic makeup of related family members. In addition, they seldom have the luxury of following a single well-defined trait through successive generations; rather, they often must deal with a perplexing syndrome of variable traits. Most human populations tend to be outbred, meaning that they are physically and genetically heterogeneous. Thus, it is more difficult to identify genes—especially those with variable phenotypes—against this heterogeneous background.

Certain human populations, however, whose members are genetically isolated by geography or customs have a degree of genetic homogeneity. The relatively closed gene pools of the Icelandic people, the Old Order Amish, and the Mormons, combined with their habit of keeping meticulous genealogical records and having relatively large families, have made them amenable to genetic analysis. Customs prohibiting alcohol consumption among the Amish and Mormons make easier the analysis of mental and behavioral disorders, such as manic depression and schizophrenia, whose symptoms may be masked by alcohol or drug abuse.

The problems of human genetics were only solved as time eroded memories of Nazi eugenics, and when restriction enzymes and polymerase chain reaction (PCR) provided markers whose presence or absence can be scored with great certainty. It is worth remembering that during the entire reign of eugenics, DNA had not yet been shown to be the molecule of heredity, and nothing was known about the physical basis of mutation and gene variation.

With an understanding of gene variation has come a deeper understanding of disease complexity. Twin studies strongly indicate that genes have a dominant role in all aspects of human health and behavior. However, just as Hermann Muller observed in fruit flies, human diseases do not always exhibit a simple gene-to-character relation. An identical mutation may produce different physical symptoms (phenotypes) in different people. Conversely, different mutations may produce similar phenotypes in different people. As George W. Beadle and Edward Tatum found in *Neurospora*, mutations in any of several enzymes can have the same end effect—of altering or knocking out a biochemical pathway.

At one end of the spectrum are "simple," genetically homogeneous diseases, such as sickle cell anemia and cystic fibrosis, for which affected individuals share common mutations and highly similar symptoms. In the middle are diseases, such as β-thalassemia and neurofibromatosis 1, in which a variety of types of mutations in a single gene produce variable symptoms. At the other end of spectrum are "complex," genetically heterogeneous diseases, such as asthma and bipolar disorder, in which mutations in a number of genes—in combination with environmental factors—likely account for extremely variable symptoms.

DETERMINING THE CHROMOSOMAL BASIS OF HUMAN DISEASE

The development of new cytological methods, beginning in the 1950s, helped to bring human genetics out of its "dark ages." T.C. Hsu's treatment with hypotonic (low-salt) solution caused cells to swell, separating the chromosomes and making them easier to count. Wright stain, and later Giemsa and quinacrine stains, made it possible to identify each chromosome by size and distinctive staining patterns.

The state of human cytogenetics in the post-World War II years is best summed up by the fact that until 1956, humans were thought to have 48 chromosomes. This number was prejudiced by an earlier, and accurate, determination of 48 chromosomes in chimps. J.H. Tjio and A. Leven, of the National Institutes of Health, cleared up the matter when they showed conclusively that humans have 46 chromosomes. Within several years, cytologists established a direct relationship between human genetic disorders and abnormal chromosome number, or aneuploidy. Trisomy—an extra chromosome copy—was found in Down's syndrome (chromosome 21), Patau's syndrome (13), and Edward's syndrome (18). Abnormalities in sex chromosome number were also described for Turner's syndrome (X) and Klinefelter's syndrome (XXY). In the early 1960s, translocations were identified in some cases of Down's syndrome. Studies in the 1970s showed that chronic myeloid leukemia, Burkitt's lymphoma, and several other blood cancers are characterized by specific chromosome translocations (see Chapter 7).

Sickle Cell Brings Human Genetics into the Molecular Era

Recall Archibald Garrod's prophetic hypothesis, in 1908, that alkaptonuria and, by extension, other inherited disorders are caused by "inborn errors in metabolism." Beadle and Tatum proved that this is exactly the case in *Neurospora*. They showed that mutations in specific genes produce corresponding changes to enzymes, evidenced by heritable metabolic deficiencies. The elucidation of the molecular mechanism of sickle cell disease provided the first proof of this concept in humans and illustrates the accumulation of genetic knowledge over time.

Sickle cell disease was first described in 1910 by Chicago physician James Herrick, whose patient had anemia characterized by unusual sickle-shaped red cells. Over the years, evidence accumulated that it is a recessive disorder. In the mid 1940s, Irving Sherman, a medical student at Johns Hopkins School of Medicine, found that sickled blood transmits light differently than normal blood, suggesting structural differences in the hemoglobin molecule. William Castle, of Harvard Medical School, relayed this information to Linus Pauling at the California Institute of Technology, who had become interested in the molecular structure of hemoglobin. Castle supplied blood samples from sickle cell patients and healthy controls, from which Pauling and Harvey Itano isolated hemoglobin. When separated by electrophoresis, sickle-cell hemoglobin (Hb_s) migrates more slowly, showing it is less negatively charged than normal hemoglobin (Hb).

This was consistent with the work by Vernon Ingram, of the Cavendish Laboratory in Cambridge, England, showing that Hb contains more glutamic acid (a negatively charged amino acid) and Hb_s contains more valine (a neutral

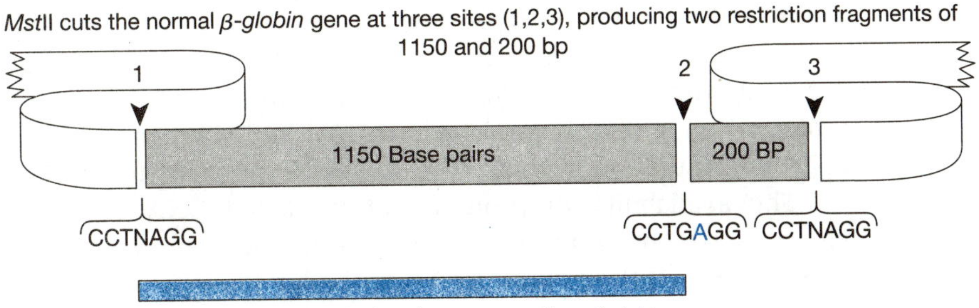

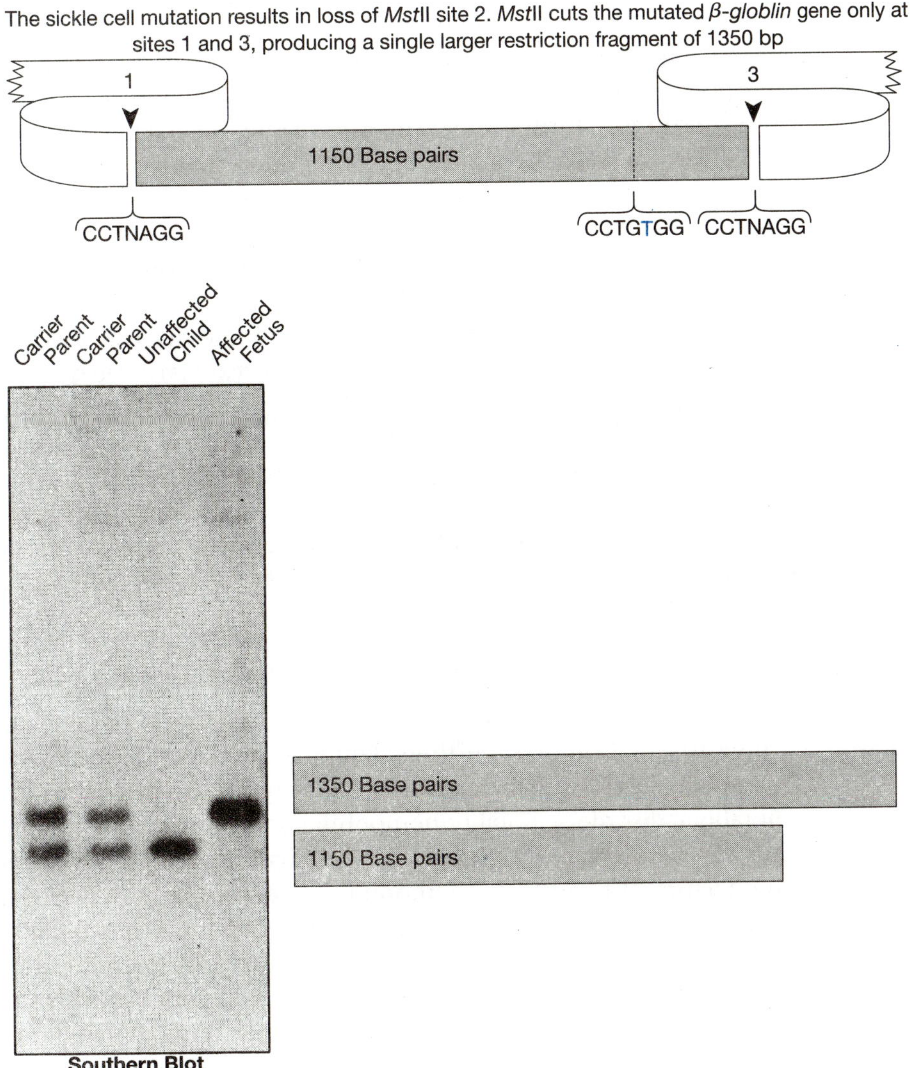

RFLP Diagnosis of Sickle Cell Anemia, 1982

The Southern blot shows the RFLP patterns of two carrier parents, an unaffected offspring, and amniotic fluid from an affected fetus. The carrier parents show a single copy of each RFLP: the 1350-bp fragment associated with the disease allele and the 1150-bp fragment associated with the normal allele (the 200-bp fragment is not detected by the probe). The unaffected child shows the 1150-bp fragment, and the affected fetus shows the 1350-bp fragment. Both offspring are homozygous and thus show a single relatively thick band, denoting two chromosomal copies of the normal (Hb) or mutated (Hb_s) gene. (Reprinted, with permission, from Chang J.C. and Kan Y.W. 1982. A sensitive new prenatal test for sickle-cell anemia. *N. Engl. J. Med. 307:* 30–32.)

amino acid). In 1956, Vernon Ingram and John Hunt independently sequenced the Hb and Hb_s proteins, finding that a glutamic acid at position 6 in Hb is replaced with valine in Hb_s. From this information, they used a genetic code table (showing that glutamic acid = GAG and valine = GTG) to predict that the A-T point mutation in the sixth codon is responsible for sickle cell disease.

The availability of protein and predicted DNA sequence facilitated the cloning of the α- and β-*globin* genes from a human genomic library in the early 1980s. (The methods used by Philip Leder to clone the β-*globin* gene are discussed in Chapter 5.) The combination of Southern blot and restriction enzyme analysis made DNA diagnoses possible for the causative lesions of many hemoglobinopathies.

In constructing early restriction maps of cloned human DNA, it became obvious that a point mutation can change a restriction enzyme recognition site, producing different-sized fragments, termed a restriction-fragment-length polymorphism (RFLP). These were the first DNA polymorphisms (poly for "many" and morph for "form") that could be readily detected. As discussed in Chapter 6, RFLPs also were the major type of marker employed in the early physical and linkage maps of the human chromosomes. Used in a local region of a chromosome, an RFLP also might detect the causative mutations of disease. Initially, RFLPs were detected by Southern blot analysis, using a radioactive probe that hybridizes to the polymorphic region.

The mutation responsible for sickle cell anemia was first detected by RFLP analysis in 1978 by Yuet Wai Kan and Andrea-Marie Dozy at the University of California, San Francisco. They used the restriction enzyme *Mst*II, which recognizes the sequence CCTNAGG (where N equals any nucleotide). The A-T mutation results in the loss of an *Mst*II recognition site that spans the region of sixth codon of the β-*globin* gene. Thus, the DNA from normal homozygous individuals, heterozygous carriers of the sickle cell trait, and homozygous sickle cell patients produces different restriction fragments when cut with *Mst*II.

Making Therapeutics from Cloned Genes

The isolation of insulin in 1921 by Frederick Banting and Charles Best of the University of Toronto, and their demonstration that it successfully corrects the metabolic defect of diabetes, paved the way for the treatment of other common metabolic disorders, notably hemophilia and pituitary dwarfism. Such deficiencies can be corrected by supplying the missing or underproduced protein: clotting factors VIII and IX for hemophilia, and human growth hormone (HGH) for dwarfism. However, ensuring adequate, contagion-free supplies of these therapeutic proteins proved difficult. Hormones are often produced in minute quantities in the body and, hence, are laborious and very expensive to isolate. In the case of human growth hormone, the number of patients who could be treated was limited by availability.

Approximately 8000 pints of blood were processed to yield enough clotting factor to treat a single hemophiliac for 1 year, and 7–10 pounds of pancreas from approximately 70 pigs or 14 cows were needed to purify enough insulin for 1 year's treatment of a single diabetic. The extraction of HGH was most onerous, requiring the pituitary glands from approximately 80 human cadavers to produce enough for a single year's therapy. The magnitude of the supply problem becomes obvious when one considers that patients suffering from these diseases require long-term treatment lasting a *minimum* of 5–10 years.

The risk of virus contamination is a most important consideration in any therapeutic product purified from mammalian cells. Simian virus 40 (SV40), which has proved so important in cancer research, was first isolated as a contaminant in poliovirus vaccine produced in monkey cells. Although there is no evidence of illness as a result of SV40-contaminated poliovirus vaccines, supplies of both human growth hormone and clotting factors have at one time or another been infected with life-threatening pathogens.

Prior to the identification of human immunodeficiency virus type 1 (HIV-1) and the development of virtually foolproof screening procedures, patients with hemophilia had a significant risk of contracting AIDS (acquired immune deficiency syndrome) from transfusion of contaminated clotting factors, as well as whole blood. During the window of time between the onset of the AIDS pandemic and the development of effective methods to screen for HIV and disable it in blood products, in 1983–1984, is it estimated that half of all hemophiliacs developed AIDS. According to 2001 statistics from the Centers for Disease Control, a total of 5234 American hemophiliacs have died of AIDS.

Therapeutic proteins isolated from animals, including porcine or bovine insulin, differ in amino acid makeup from the human protein they replace. The biological activity of an animal substitute may differ slightly from the native human protein, or it may elicit an immune response. Some diabetics had allergic reactions to porcine or bovine insulin, although this may have been due to impurities in the preparations and not necessarily differences in the amino acid sequence.

The development of new genetic tools made possible the cloning and production of a number of genes for medically important proteins, including insulin, clotting factors, tissue plasminogen activator, interleukin, interferon, erythropoietin, and colony stimulating factors. The Boyer-Cohen experiment (Chapter 4) showed that recombinant DNA methods can be used to transfer essentially any gene into *E. coli*, where the encoded protein may be expressed. This established a new paradigm of using cultured cells to produce therapeutic proteins to treat human metabolic disorders.

Producing therapeutic proteins from cloned human genes inside *Escherichia coli* hosts eliminates the risk of virus contamination and allergic sensitivity. Mammalian viruses cannot reproduce inside *E. coli* and hence cannot be co-isolated with the protein from a bacterial culture. The protein harvested from the bacterial culture has been expressed from a human coding region and is identical (or very nearly so) to the native protein. Thus, diabetics sensitive to bovine or porcine insulin do not have an adverse reaction to human insulin of recombinant origin.

Expressing Insulin and Growth Hormone in *E. coli*

The production of human insulin and growth hormone illustrate that expression of a human protein in *E. coli* typically requires detailed understanding of its biological synthesis and of the biochemical limitations of the bacterial cell. *E. coli* is incapable of processing eukaryotic pre-mRNAs or of performing the several posttranslational modifications needed to produce a biologically active form of insulin from its protein precursors.

Some Approved Drugs Produced from Cloned Genes

Product	Generic name/Company	Year of first U.S. Approval	Approved for
Recombinant human insulin	Humulin/Eli Lilly & Co.	1982	Diabetes mellitus
Recombinant human growth hormone	Protropin/Genentech, Inc.	1985	Growth hormone deficiency in children
Recombinant interferon-α	Intron A/Scherin-Plough	1986	Hairy cell leukemia
		1988	Genital warts
		1988	Kaposi's sarcoma
		1991	Hepatitis C
		1992	Hepatitis B
Recombinant hepatitis B vaccine	Recombivax HB/Merck & Co.	1986	Hepatitis B prevention
Epoetin alfa	EPOGEN/Amgen Ltd.	1989	Anemia of chronic renal failure
Recombinant interferon-γ	Acctimune/Genentech, Inc.	1990	Chronic granulomatous disease
Colony-stimulating factor (CSF)	Leukine/Immunex Corp.	1991	Bone marrow transplantation
Recombinant anti-hemophiliac factor	Recombinate rAHF/Baxter Healthcare	1992	Hemophilia A
Recombinant DNase I	Pulmozyme/Genentech, Inc.	1993	Cystic fibrosis
Recombinant coagulation factor IX	AlphaNine SD/Alpha Therapeutic Corp.	1996	Christmas Disease Hemophilia B

The mature insulin molecule consists of two polypeptide chains—an A chain of 30 amino acids and a B chain of 21 amino acids—which are held together by disulfide linkages. However, this active insulin results from sequential modifications in two precursor molecules: preproinsulin and proinsulin. The gene for insulin consists of two coding exons separated by a single intron. Following splicing of the pre-mRNA, a functional transcript is translated into a large polypeptide called preproinsulin. The molecule includes a 24-amino-acid signal peptide at its amino terminus, a feature of many secreted proteins needed for their proper transport through the cytoplasm. The signal peptide, which is the first part of the preproinsulin molecule produced, anchors the free-floating ribosome to the endoplasmic reticulum (ER) and is subsequently clipped off as the molecule passes through the ER membrane.

The result is a molecule of 84 amino acids called proinsulin, whose looped shape is maintained by cross-linking disulfide bonds. Proinsulin makes its way to the Golgi apparatus, where a converting enzyme removes 33 amino acids from the middle of the connecting loop (the C chain), leaving the remaining A and B chains held together by the disulfide linkages. This yields active insulin, which is stored in a secretory granule for eventual release into the bloodstream.

The human genomic sequence could not be used directly to produce active insulin in *E. coli*, because the bacterium lacks the enzyme systems needed (1) to splice out the intron sequence to produce a mature mRNA, (2) to remove the signal sequence from preproinsulin, and (3) to remove the C chain from proinsulin. Although other methods have been used, the strategy first employed in 1979 by Eli Lilly & Co. to produce recombinant human insulin neatly sidesteps these constraints by simply omitting the above sequences. The nucleotide sequences coding for the A and B chains of active insulin were chemically synthesized and cloned into separate expression plasmids. A bacterial strain containing each expression vector produces a fusion bacterial/human polypeptide, which is harvested and subsequently treated with cyanogen bromide to remove the bacterial amino acids. Cyanogen bromide cleaves the fusion polypeptide at the methionine residue that begins the human sequence. This treatment, which also alters tryptophan, is only useful because, by happenstance, neither the A nor B insulin chain includes tryptophan or additional methionine residues.

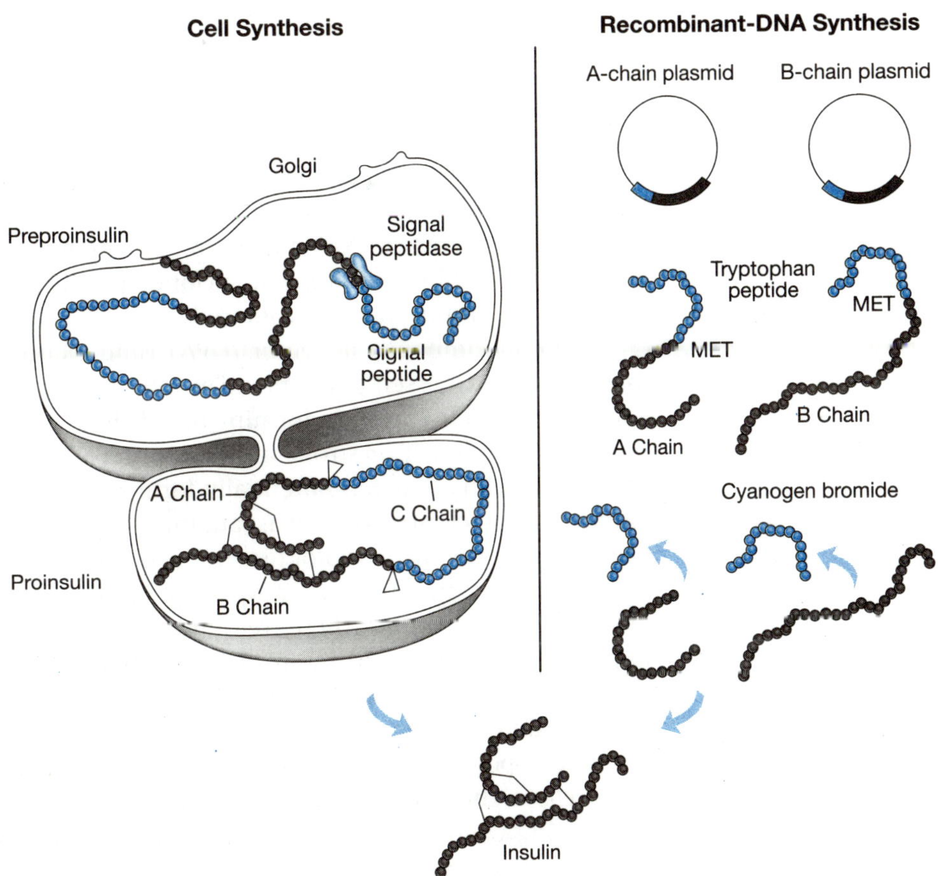

Cellular vs. Recombinant DNA Synthesis of Insulin

In the pancreas, insulin is synthesized as preproinsulin. Within the Golgi apparatus, the signal peptide is removed to yield inactive proinsulin. The C-chain peptide is subsequently removed to produce active insulin; A and B chains are linked by two disulfide bonds. Insulin of recombinant DNA origin (Humulin) is derived from plasmids in which the coding information for the A or B chain is fused to the promoter and the first few codons of the *E. coli* tryptophan gene (*trp*). Separate cultures of *E. coli* are transformed with the A or B constructs and produce large amounts of the fusion peptides: either *trp*/A or *trp*/B. The tryptophan sequences are removed by treatment with cyanogen bromide (CNBr), which cleaves at a methionine (MET) residue at the junction of the insulin gene. The A and B chains are mixed together, and disulfide bonds are formed by a chemical process.

Purified A and B chains are then mixed in equal portions and incubated under conditions that form the disulfide linkages.

Human growth hormone, a polypeptide of 191 amino acids, is also produced in recombinant *E. coli*. The coding sequence for the first 24 amino acids of the expressed gene is synthesized chemically, whereas amino acids 25 through 191 are derived from a cDNA copy of *HGH* mRNA isolated from pituitary cells. The recombinant HGH differs by one amino acid from normal HGH due to the fact that *E. coli* is unable to remove the initiator methionine residue that is removed posttranslationally in human cells.

Expressing t-PA, Erythropoietin, and Interferons in Mammalian Systems

Both insulin and HGH are relatively simple proteins that do not undergo glycosylation or other posttranslational modifications. Bacteria do not possess the enzymatic machinery for making posttranslational modifications to proteins, so genes encoding extensively modified proteins must be cloned into eukaryotic expression systems. Chinese hamster ovary (CHO) cells have proven to be the most popular mammalian system for expressing human therapeutic proteins. However, expressing proteins in most eukaryotic cells is much more costly than in bacteria. Whereas a cloned gene can be engineered to produce up to 40% of total protein production of *E. coli*, a cloned product may account for only 8% of total protein output in eukaryotic cells.

Tissue plasminogen activator (t-PA) is an example of a therapeutic protein that must be expressed in mammalian cells. t-PA is a protease that attacks fibrin, a major protein involved in forming blood clots. Patients that demonstrate early signs of heart attack or stroke are administered t-PA, which acts by destroying small blood clots that can potentially form blockages in arteries. Until the over-expression of cloned t-PA was attained in eukaryotic cells, the major clot destroyer in use was streptokinase, a protease isolated from *Streptococcus*. As a foreign protein, streptokinase may cause immune reactions, hemorrhaging, and other side effects. For these reasons, t-PA was initially hailed as a major improvement in the treatment of heart attack. However, retrospective studies have shown little difference in the recovery of patients treated with t-PA versus those treated with streptokinase.

Erythropoietin (EPO), which stimulates production of red blood cells from stem cells in the bone marrow, is useful for treating anemia, AIDS, and patients undergoing chemotherapy and bone marrow transplants. Both EPO and HGH have less reputably been used as performance boosters among athletes. Whereas HGH increases strength by increasing muscle mass, EPO increases endurance by increasing the oxygen-carrying capacity of the blood.

Discovered in 1957, interferon gained a reputation as having wondrous antiviral, anticancer, and immune modulatory effects. However, quantities were too limited to prove its usefulness in laboratory and clinical trials. When cloned interferon made significant quantities available, it was found that interferon modulates the function of several types of immune cells—macrophages, cytotoxic lymphoctyes, and B cells—increasing expression of immunoglobulins and human leukocyte antigens (HLAs). In 1992, James Darnell, at The Rockefeller

University, elucidated how interferon initiates a signal transduction pathway through which immune cells are primed to recognize and degrade the RNA of infecting viruses. However, interferon proved to be relatively toxic at pharmacological doses and generally failed to live up to its wonder drug hype. Although interferon has not proven to be the "magic bullet" that some had hoped for, its several forms have proven broadly useful in treating a number of diseases. Interferon-α is the most effective treatment available for hepatitis B and C, which infect hundreds of millions people worldwide, and is also used in treating leukemias. Interferon-β is the most effective treatment for multiple sclerosis, although it is still uncertain how it functions in controlling this disease. Interferon-γ is used to treat osteopetrosis and chronic granulomatous disease.

THE IMPORTANCE OF DNA POLYMORPHISMS

The obvious candidates for medications from cloned genes were soon virtually exhausted. Biologists had to come to grips with the harder work of cloning a disease gene in the absence of knowledge about its protein product. To do this, a disease gene first must be mapped to a chromosome position by linkage to known loci. After a precise location is identified, the region containing the gene is identified in a genomic library and then subcloned and examined bit by bit. Comparing DNA from affected versus unaffected individuals in a family can then potentially turn up obvious mutations that confirm the identity of the disease gene among a number of nearby candidates. This method would come to be known as positional cloning.

Unfortunately, there simply were not enough genes on the human chromosome maps to support the sort of linkage studies needed to find diseases. In 1911, Edmund B. Wilson of Columbia University had, by default, mapped the first human gene to the X chromosome, when he discovered that the inheritance of color blindness is "sex-limited." However, human chromosome mapping progressed slowly over the ensuing decades. The presence of a testis determining factor was inferred in 1959, making it the first Y-linked "gene."

By 1980, only 120 human genes had been assigned chromosome locations. A similar number of genes (135) had been mapped in *Drosophila* chromosomes. However, the lag in human gene mapping becomes clear when one compares the average number of genes mapped per chromosome (4 pairs for *Drosophila*, 23 pairs for humans). Excluding the gene-poor chromosome that determines maleness in each species, that represented an average of 34 genes per chromosome for *Drosophila*—a gene map almost seven times denser than the human map, which had only five genes per chromosome. Moreover, the majority of *Drosophila* genes had been given precise locations by linkage analysis, as well as in situ hybridization, to banded chromosomes. Few human genes had been precisely located.

This impasse was solved by a simple proposition made by David Botstein, Ronald Davis, and Mark Skolnick at a scientific meeting at Alta, Utah in 1978. They proposed that the human chromosome map could be populated with physical variations in the DNA molecule itself. These DNA polymorphisms, which are assayed by gel or capillary electrophoresis, would substitute for phenotypic or biochemical variants used in classical linkage analysis.

DNA Polymorphisms and Human Identity

At this point, it makes sense to leave our story of finding human disease genes to discuss the parallel development of DNA polymorphisms in establishing human identity. We will return to gene cloning a little later, armed with a more detailed understanding of the evolution of DNA markers. During the 1980s and 1990s, molecular analysis would become increasingly powerful as the RFLP polymorphisms, usually having only two alleles, were supplanted by repeat polymorphisms—VNTRs (variable number of tandem repeats) and STRs (short tandem repeats)—with increasing numbers of alleles.

Although fingerprints and thumbprints have been used as personal identifiers since ancient times, only in the 20th century did they come into use in criminal cases. Francis Galton, the father of the eugenics movement, studied the fingerprints of thousands of schoolchildren from around the British Isles. His 1892 treatise, *Finger Prints,* showed how to analyze the various patterns of whorls and loops, noted their relative frequencies, and suggested how fingerprints could be rigorously used in criminal cases. This book formed the basis of forensic DNA fingerprinting still in use today.

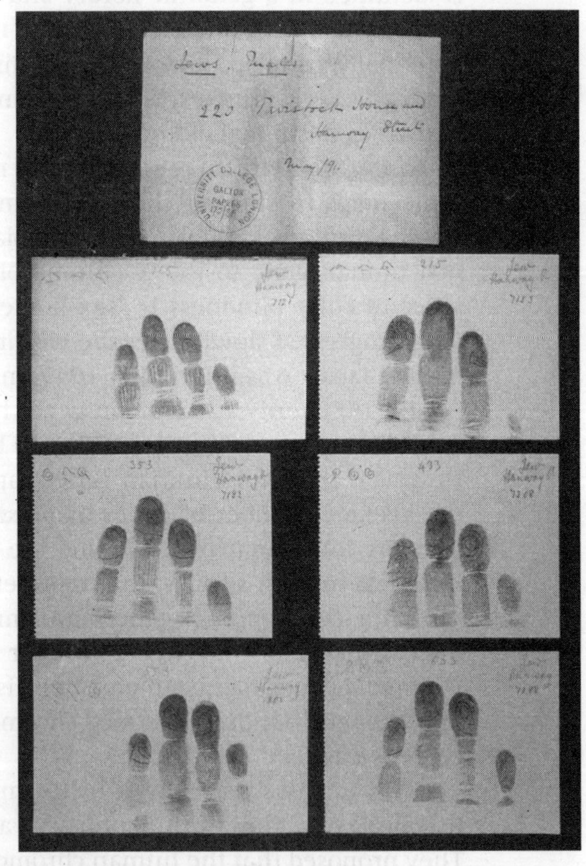

Francis Galton's Analysis of Fingerprints, 1891

Many of Galton's examples came from schoolboys. These fingerprints are from the Hanway Street School, London. (Copyright The Galton Collection, University College London.)

The term "DNA fingerprinting" was coined to allude to the traditional use of fingerprints as a unique means of human identification. Whereas classic fingerprinting analyzes a phenotypic trait, DNA typing directly analyzes genotypic information. When properly conducted, DNA-based testing can provide positive evidence of a person's identity. In contrast, the phenotypes detected by blood grouping and leukocyte antigen testing are shared by sufficiently large numbers of individuals that they are not, in the strictest sense, tests of identity. Rather, they are exclusionary tests that can only prove that forensic evidence does not match a suspect or that persons are not related.

All that is required for DNA fingerprinting is a small tissue sample from which DNA can be extracted. This can be blood or cheek cell samples in a paternity case, a semen sample from a rape victim, dried blood from fabric, skin fragments from under the fingernails of a victim after a struggle, or even several hairs (with the attached roots) combed from a crime scene. Ted Kaczynski, the "Unabomber," was definitively linked to the case when his DNA type matched the one obtained from cells left when he licked a stamp used on a letter. Using the best available techniques, a DNA type can be obtained from cells in *fingerprints* on a glass or other hard surface. The time is approaching when a criminal will not be able to afford to leave even a single cell at a crime scene.

Variable Number of Tandem Repeats

British researcher Alec Jeffreys was the first to realize that DNA polymorphisms can be used to establish human identity. He coined the term DNA fingerprinting and was the first to use DNA polymorphisms in paternity, immigration, and murder cases. The discovery, in 1984, of the so-called "Jeffreys' probes" arose from the investigation of the "minisatellite" fraction of highly repetitive DNA in the human genome. (Recall Roy Britten's experiments from Chapter 6.) Minisatellites, composed of short repeated DNA sequences that "hover" in a chromosome region, were first described in 1980 by Arlene Wyman and Ray White at the University of Utah. Each minisatellite proved to be composed of

Alec Jeffreys, 1989
(Courtesy of A. Jeffreys, University of Leicester.)

tandem repeated units ranging in size from 9 to 80 bp. The number of repeats at a particular locus was variable between homologous chromosomes, hence the acronym VNTR (variable number of tandem repeats).

Working at the University of Leicester, Jeffreys found two "core" sequences that are common to a set of VNTRs associated with the myoglobin gene locus. Assaying for them by Southern blotting produced a DNA fingerprint that was a composite of VNTRs at multiple loci, leading to the term "multilocus probes." In his analysis, radioactive probes hybridize to restriction fragments that have partial homology with the core sequence, typically detecting 20–30 interpretable bands. These distinctive banding patterns are inherited in a Mendelian fashion, with half of the bands derived from the mother and half from the father.

The "Ghana Immigration Case" (1985) provided the first practical test of DNA fingerprinting. The case involved Christiana Sarbah and her teenage son Andrew,

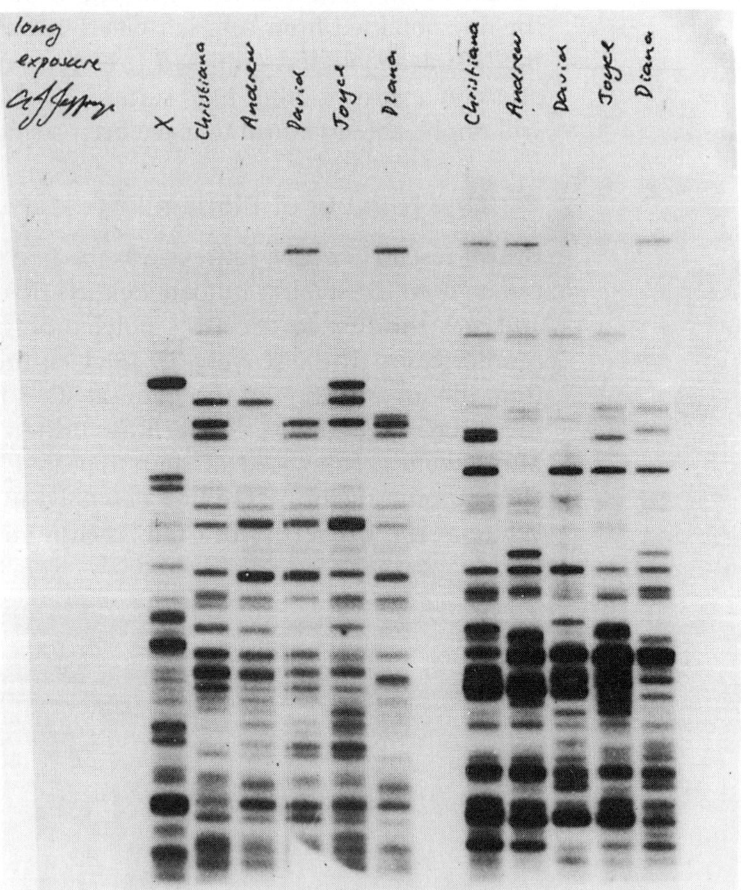

Use of Multilocus Probes in the Ghana Immigration Case, 1985

This Southern blot shows the first use of DNA fingerprints as evidence in a court of law. The match between many bands in their DNA fingerprints proved a family relationship between a Ghanian boy, Andrew, who wished to remain in England with his mother (Christiana) and his siblings (David, Joyce, and Diana). Bands not shared by a child and the mother were inherited from the father. The DNA fingerprint of an unrelated person is shown in lane X. The evidence was prepared by Alec Jeffreys, who originated the use of multilocus probes and coined the term "DNA fingerprint." (Courtesy of A. Jeffreys, University of Leicester.)

who immigrated to England after living for some time with his father in Ghana. Although depositions and other information showed that Christiana and Andrew were almost certainly related, the British Home Office ordered that Andrew be deported in the absence definitive evidence to prove Christiana's parentage. Jeffreys agreed to assist with the appeal case, believing it would be an ideal test of the DNA fingerprint technology he had recently developed. He used his myoglobin VNTR probes to produce DNA profiles from blood samples from Christiana, Andrew, and three siblings—David, Joyce, and Diana. Because of the lack of the father's blood sample, Jeffreys reconstructed the father's fingerprint from bands present in the three undisputed children, but absent in Christiana. About half of Andrew's bands matched bands in the father's compilation and the remaining bands were all present in Christiana's fingerprint. The possibility of this happening by chance is greater than one in a trillion. The Home Office accepted the DNA fingerprint evidence and allowed Andrew to stay in England.

Jeffreys' probes essentially analyzed a number of VNTR polymorphisms simultaneously. The multiple bands created by the multilocus system proved difficult to analyze and standardize. The system was prone to produce artifact bands whenever a restriction enzyme failed to cut entirely, and it could be difficult to determine whether a sample had digested to completion. Furthermore, the number and frequency of alleles were never rigorously worked out, making it impossible to accurately determine the relative rarity of one fingerprint over another.

Jeffreys' multilocus probes were supplanted by single-locus probes which identify a polymorphism that occurs at a single location on one chromosome. The majority of RFLPs that had been discovered through the mid 1980s were point mutations that destroy or create a restriction enzyme recognition site. This type of RFLP has only two alleles and three genotypes (++, +–, and ––). Thus, gene mappers and forensic biologists alike sought out more variable polymorphisms as they became increasingly available in the late 1980s.

Beginning in 1987, Yusuke Nakamura, Ray White, and others at the University of Utah began a systematic search for single-locus VNTRs, ultimately providing more than 100 useful polymorphic loci scattered throughout the genome. Probes for these VNTRs became widely used in gene mapping and DNA fingerprinting. Each probe hybridizes to a unique hypervariable region of the genome and generates a pattern consisting of one or two bands from an individual's DNA, depending on whether they are homozygous or heterozygous at that locus. Used alone, a single-locus probe only detects one or two differences; however, "cocktails" of several probes came into use for forensic purposes.

Because each probe identifies a discrete locus, the frequency of each allele can be determined in population studies and the Hardy-Weinberg equation used to calculate the occurrence probability of each genotype. The VNTR loci that became most useful in forensics were those with 10 or more alleles and with a high degree of heterozygosity in many human populations, thus maximizing the ability to discriminate between two individuals. Consider the case of D1S80, a VNTR on chromosome 1 in which a 16-nucleotide unit is repeated from 14 to more than 41 times, creating 29 different alleles.

By studying the occurrence of alleles in human populations, one can calculate the probability of an individual's DNA fingerprint having this or that combi-

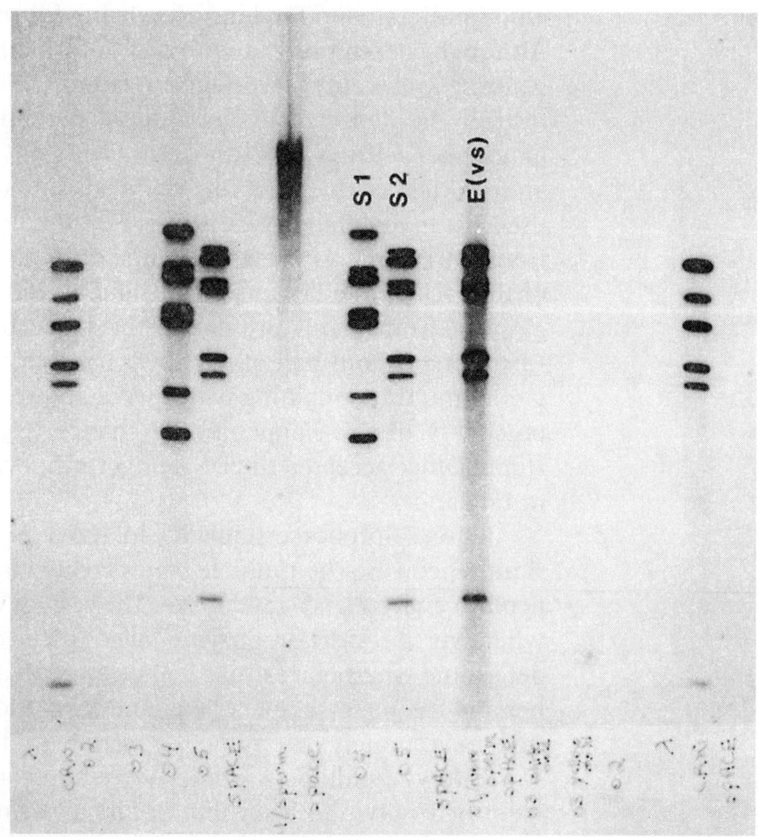

Use of Single-locus Probes in a Criminal Case, Palatka, Florida, 1988

This Southern blot is from a case in which two friends, Randall Jones (S2) and Chris Reesh (S1), were accused in the double rape-murder of a Florida woman and her boyfriend. A cocktail of single-locus probes showed an exact match between the DNA fingerprint of semen obtained from the female victim, E(vs), and the DNA fingerprint from Jones' blood sample. Jones received the death penalty—the first time this sentence was handed down in the United States on the strength of DNA fingerprint evidence. Jones is currently on Death Row at the Union Correctional Institute in Raiford, Florida. Reesh served 9 months of an 8-year sentence as accessory. (Courtesy of Cellmark Diagnostics.)

nation of DNA bands, or the probability of two DNA samples matching each other. Adding a second VNTR increases the ability to distinguish between individuals. Different VNTRs used in identity testing have been chosen on different chromosomes. This way, one can be assured that each VNTR is unlinked from the others—that each VNTR is inherited independently. If they are unlinked, then the probability of any two bands being inherited together is the product of their individual occurrences. By the mid 1990s, most forensic laboratories were producing types with five to eight unlinked markers.

Although it was initially challenged in the courts, due to its extreme sensitivity and potential for contamination, PCR eventually supplanted Southern blotting in forensic analysis. PCR made possible extremely rapid protocols that required very small amounts of template and obviated the use of radioactivity. D1S80 was among the first to be adapted for use in a forensic PCR kit. Briefly, a small sample of blood or other cells is lysed by boiling, the cell debris is removed by centrifugation, and the PCR reagents are added directly to the crude extract.

1	2 Type	3 Allele	4 Frequency	5 Hardy-Weinberg	6 Calculation	7 D1S80 Probability	8 Locus 2 Probability	9 Combined Probability
C	18/31	18	0.263	2pq	2 (0.263 x 0.058)	0.0305	0.0050	0.000153
		31	0.058					
1	24/37	24	0.318	2pq	2 (0.318 x 0.003)	0.0002	0.0035	0.0000007
		37	0.003					
2	18/18	18	0.263	p^2	(0.263 x 0.263)	0.0692	0.0075	0.000519
		18	0.263					
3	28/31	28	0.050	2pq	2 (0.050 x 0.058)	0.0058	0.0025	0.0000145
		31	0.058					
4	18/25	18	0.263	2pq	2 (0.263 x 0.055)	0.0289	0.0045	0.000013
		25	0.055					
5	17/24	17	0.013	2pq	2 (0.013 x 0.318)	0.0008	0.0065	0.0000052
		24	0.318					

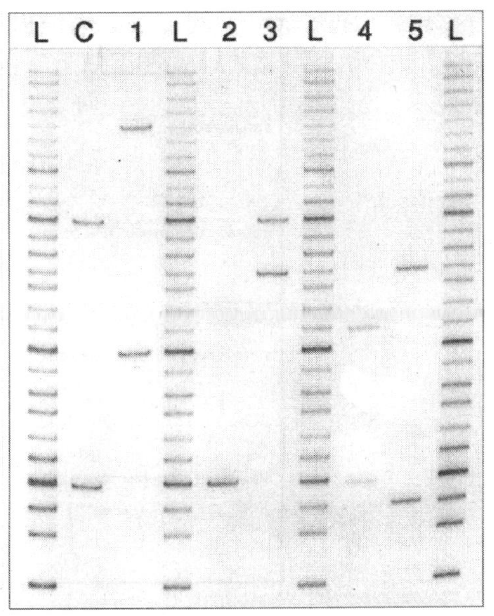

Use of PCR to Amplify the D1S80 Locus, 1991

Shown are a control (C) and five types (1–5). Type lanes show the major alleles of the system, composed of 14–41 repeats of a 16-nucleotide unit (L lanes show DNA size marker ladder). The table shows how to "score" the alleles (Columns 1–3), with allele frequencies for a Hispanic-American population (Column 4). Assuming Hardy-Weinberg equilibrium, the frequency of each DNA type is calculated in Columns 5–7. Hypothetical frequencies for DNA types at a second locus are given in Column 8. Provided the two loci are unlinked, then the combined frequency of their coinheritance is the product of their individual occurrences (Column 9). With each additional marker, the DNA types become increasingly diversified, with some types being orders of magnitude rarer than others. (Courtesy of Applied Biosystems/Perkin Elmer.)

Following the appropriate number of synthesis cycles, the amplified DNA is separated by electrophoresis in a polyacrylamide gel, stained with ethidium bromide or silver, and visualized directly.

Short Tandem Repeats

The most recent stage in the evolution of DNA polymorphisms has been the employment of "microsatellites," with repeat units of two to five nucleotides. Their potential use in forensic DNA science was first suggested in 1992 by Thomas Caskey of the Baylor College of Medicine. The short repeat unit of STRs (short tandem repeats) creates smaller alleles, providing a greater chance of "rescuing" an STR polymorphism from degraded DNA samples than a longer VNTR. Because they came into popular use later than other polymorphic systems, STR analysis was developed primarily using PCR technology. Although STR alleles can be separated in various electrophoresis systems, their small size allows automated analysis using DNA sequencers.

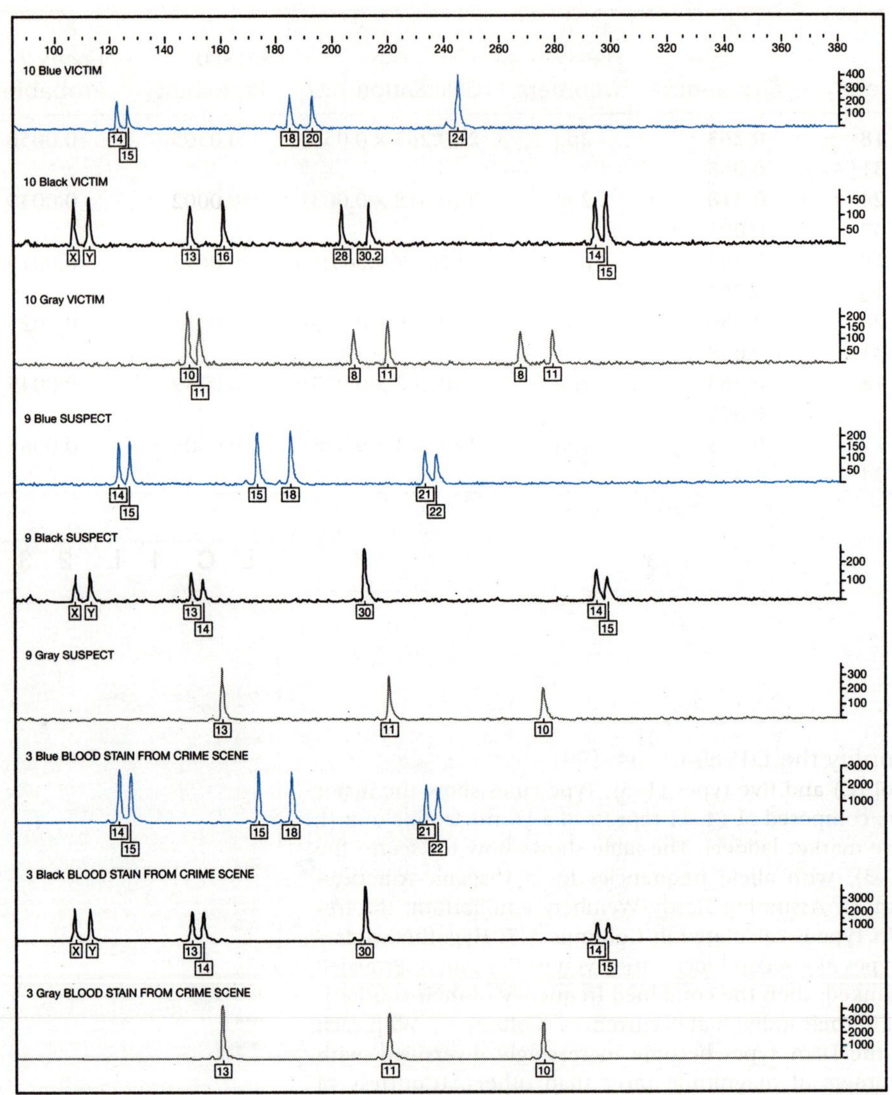

Alleles Detected

Sample	Amelogenin	D3S1358	vWA	FGA	D8S1179	D21S11	D18S51
Victim	XY	14, 15	18, 20	24	13, 16	28, 30.2	14, 15
Suspect	XY	14, 15	15, 18	21, 22	13, 14	30	14, 15
Blood Stain From Crime Scene	XY	14, 15	15, 18	21, 22	13, 14	30	14, 15

Sample	D5S818	D13S317	D7S820	D16S539	THO1	TPOX	CSF1PO
Victim	10,11	8, 11	8, 11	9, 11	7, 9	9, 11	10, 12
Suspect	13	11	10	9, 12	6, 9	8, 11	9, 12
Blood Stain From Crime Scene	13	11	10	9, 12	6, 9	8, 11	9, 12

Use of STRs in a Criminal Case, Suffolk County, New York, 2000

A typical criminal case from a 1997 homicide in which a blood stain from the crime scene was tested against blood from the victim and from the suspect, who was wounded during a struggle. Shown are sequencing results for nine STR loci, plus an XY marker, using three color channels. Four additional loci were run in another channel, which is not shown. The frequency of the suspect/blood stain type in different populations is: Caucasian 0.000000000000000372 (3.73×10^{-16}); African-American 0.000000000000000103 (1.03×10^{-16}); Hispanic 0.0000000000000000267 (2.67×10^{-17}). (Courtesy of J. Galdi, Suffolk County Crime Laboratory.)

STR alleles are perfectly suited to fluorescent detection on a DNA sequencer, allowing forensic scientists to make use of the four dye labels (red, green, blue, and yellow) as a separate "channel." Since each STR polymorphism typically produces a tight range of common alleles, three to four STRs with differing ranges in allele size can be labeled with the same dye and detected in a single channel. With this step, "multiplex" and "megaplex" polymorphism analyses gained a mechanization and reproducibility akin to hospital metabolite testing.

In 1997, the Federal Bureau of Investigation (FBI) recommended that a 13-marker panel of STRs, plus an XY marker, become the standard in criminal investigations. With this number of independently inherited polymorphisms, the probability of even the most common combination is in the tens of billions. Thus, modern DNA testing has the capability of uniquely identifying each and every person alive today. As of June 2002, the FBI's Combined DNA Index System (CODIS) contained 1,013,746 DNA profiles, including 977,895 profiles of convicted offenders.

GENE CLONING: FROM LINKAGE TO DNA DIAGNOSIS

During meiosis, paired chromosomes align and homologous regions are exchanged when chromatids "cross over" with one another. Usually, large DNA fragments, on the order of tens of millions of nucleotides, are moved between chromatids. The further apart two chromosome loci (locations), the greater the possibility that they will become separated during a crossover event. As discussed in Chapter 1, the frequency of recombination is a measure of the genetic distance between two sites on a chromosome. If two loci have a recombination frequency of 1%, they will become separated once in 100 meiotic recombinations. A recombination frequency of 1% is referred to as 1 centiMorgan (cM).

For any two chromosome loci, there is an equilibrium between two states: crossover and linkage. For distant loci, the equilibrium shifts toward a high recombination frequency. At a distance of 50 cM, recombination reaches a maximum of 50%. Loci separated by 50 cM or more are said to be unlinked, with an equal chance that they stay together or become separated during meiosis. Linkage increases as distance and recombination frequency decrease, reaching 0% for loci very close to one another.

At 0 cM distance, there is only one possible state—linkage—so equilibrium is formally impossible. Thus, when crossover never occurs between loci, they are said to be in linkage disequilibrium. The loci, and the entire region between them, are inherited as a single unit. Since the extent of linkage disequilibrium varies from region to region on the chromosome, defining blocks of linkage disequilibrium on human chromosomes is an important ongoing effort.

To pass the threshold for linkage, a marker must lie within 20 cM of a disease gene locus. This means that the two loci will be separated in 20 of 100 meiotic recombinations or, conversely, that the marker will be present in 80% of patients screened. Assuming that the probes are generated randomly and that they are evenly distributed throughout the genome, there is a 1 in 135 probability of any probe being linked to the disease (2700 cM per genome/20 cM linkage distance).

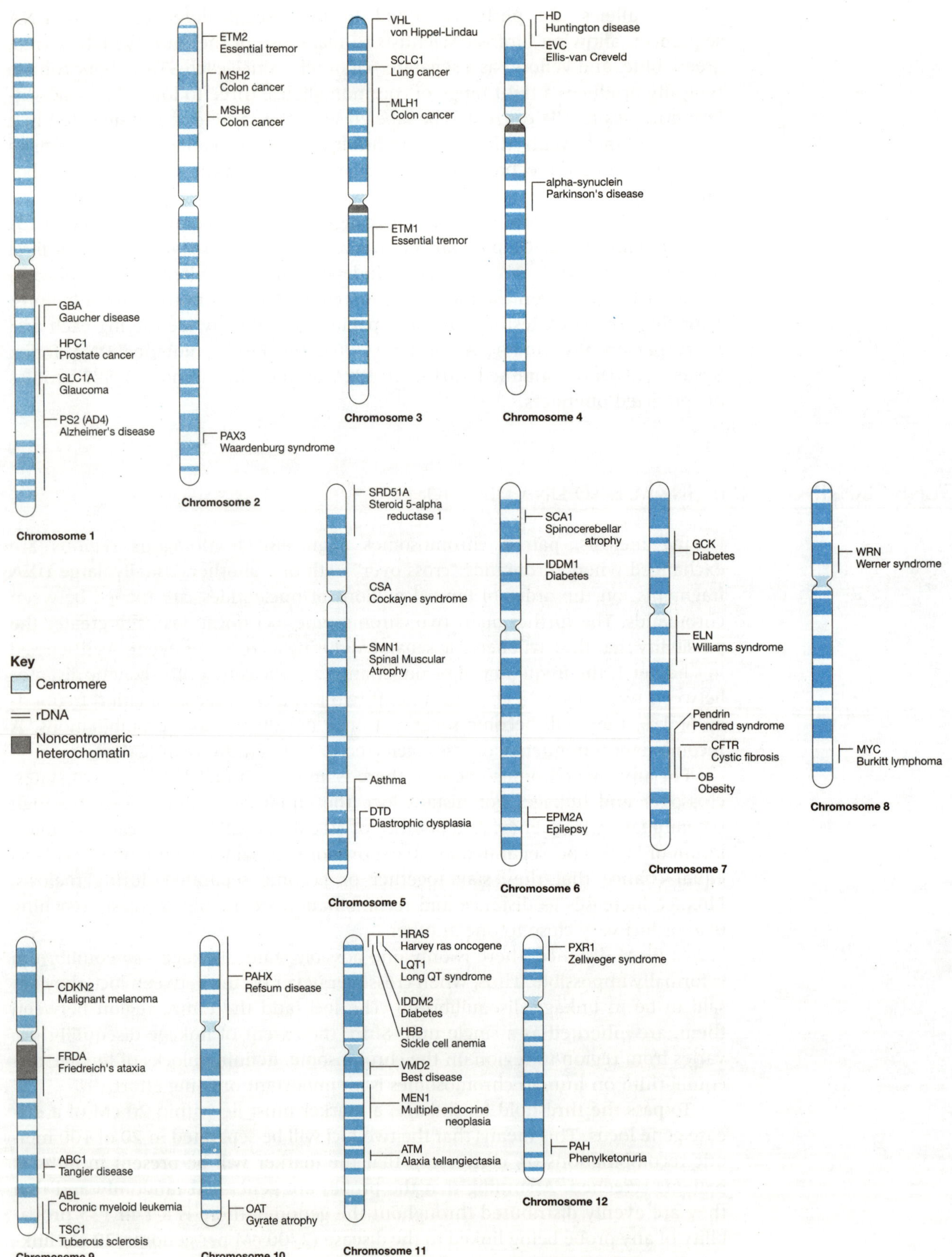

Cloned Genes Involved in Genetic Diseases and Cancer

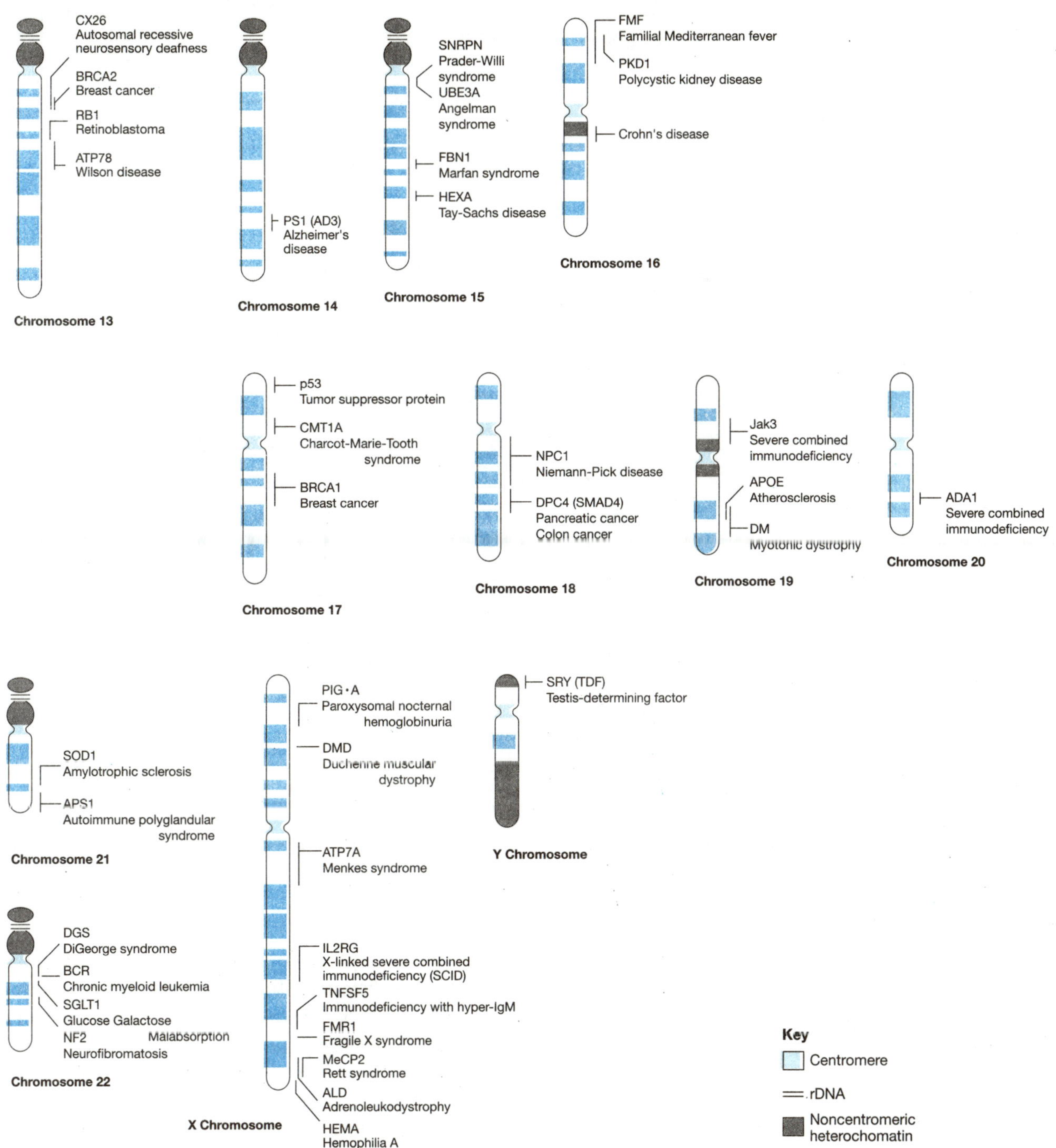

Cloned Genes Involved in Genetic Diseases and Cancer (continued)
(Adapted, with permission, from the National Center for Biotechnology Information, National Library of Medicine, National Institutes of Health, Bethesda, Maryland.)

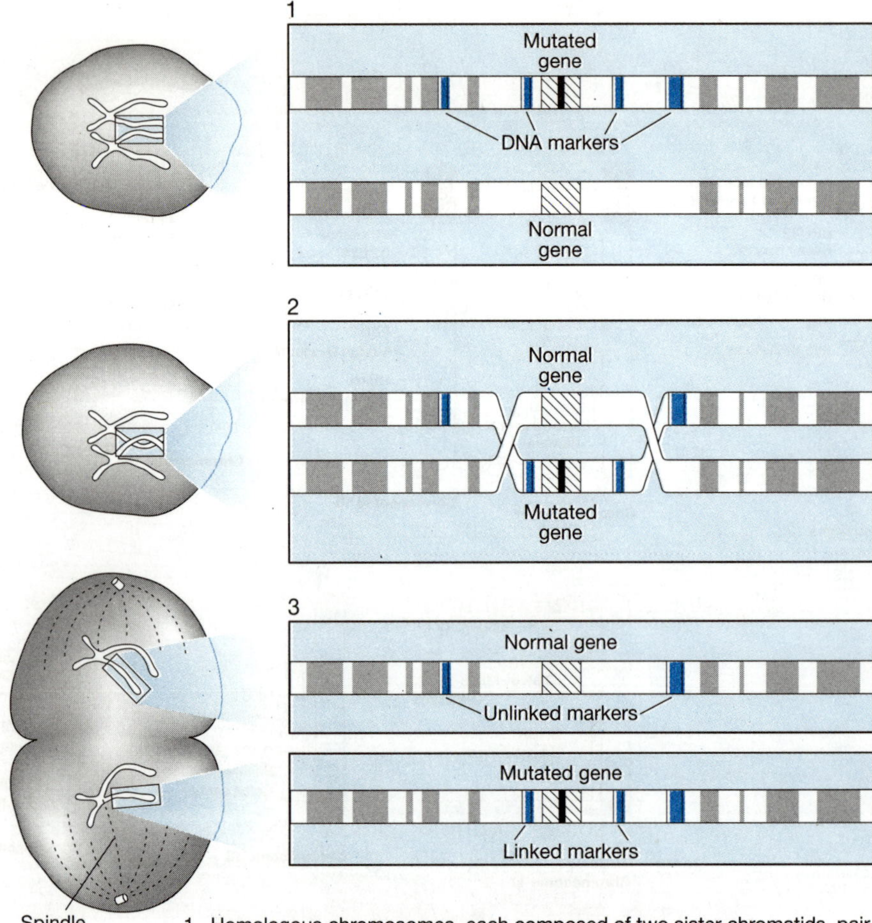

1 Homologous chromosomes, each composed of two sister chromatids, pair (synapse) during prophase of the first meiotic division.

2 Recombination occurs when chromatids cross over, exchanging DNA fragments. Linked markers remain with the original chromatid, whereas unlinked markers become separated from it.

3 During anaphase, the recombined chromatids separate into two different daughter cells (in a subsequent meiotic division, the sister chromatids will segregate into separate haploid sex cells).

Fate of Linked and Unlinked Markers during Meiotic Recombination

Accurate DNA diagnosis becomes feasible once a marker has been located within 5 cM of the disease gene. At this distance, the recombination frequency (and probability that the marker and gene become unlinked) is 5%. Conversely, the coinheritance of the marker and the disease gene, as well as the accuracy of diagnosis, is 95%. The addition of a "flanking" marker within 5 cM on the other side of the gene theoretically increases the accuracy of predictions to 99.75%. (The chance of both markers becoming unlinked is $0.05 \times 0.05 = 0.0025$.)

DNA diagnosis relies on linking one allele of a polymorphic marker to the inheritance of a disease phenotype. Because the alleles present at the polymorphic locus may differ from family to family, it is often necessary to follow a linked polymorphism through the pedigree of the family under study. This establishes which particular polymorphic allele is associated with the disease state in that particular family. It is also necessary to identify heterozygous carriers of the disease gene in whom one polymorphic allele segregates with the disease gene and a different polymorphism segregates with the normal gene.

In general, the closer the marker to the disease locus, the more accurate the diagnosis. A marker and a gene in extremely close proximity may be in linkage disequilibrium, in which case, they are always coinherited. Such markers, including those actually located within the disease gene, can provide accurate diagnosis *without* a family history. (The sickle cell polymorphism discussed earlier falls into this category.) However, markers located within a very large disease locus may not even be in linkage disequilibrium with the gene itself. For example, markers located at the 5′ end of the *dystrophin* gene have a recombination frequency of 5%—an apparent distance of 5 cM, or about 6 million nucleotides!

Although many DNA diagnoses originally relied on Southern blotting, most new diagnostic tests rely almost exclusively on PCR. Southern analysis typically takes 24 hours or more to complete, whereas a PCR analysis can be completed in several hours.

The Triumph and Frustration of Cloning the Huntington's Disease Gene

In 1983, Huntington's disease (HD) was the first major disease locus mapped by RFLP/linkage analysis and illustrates the method's power—and difficulties. HD is a degenerative nervous system disorder that invariably leads to loss of motor function, mental incapacitation, and early death. It is a rare example of an autosomal dominant lethal. HD has been perpetuated in the human gene pool because the onset of symptoms usually occurs well after the affected individual is capable of reproducing.

The linkage analysis performed by James Gusella's group, at the Massachusetts General Hospital, was based on studies carried out simultaneously by Nancy Wexler, of Columbia University and the Hereditary Disease Foundation, on the inheritance of HD in a group of patients living at Lake Maracaibo, Venezuela. This population made an ideal case study, because it is an extended family composed of 9000 members. In addition to establishing a pedigree showing the inheritance of the disease, Wexler and her co-workers collected blood samples from more than 2000 affected persons and family members. These samples were returned to the laboratory and used to establish lymphoblastoid cell lines that can be contin-

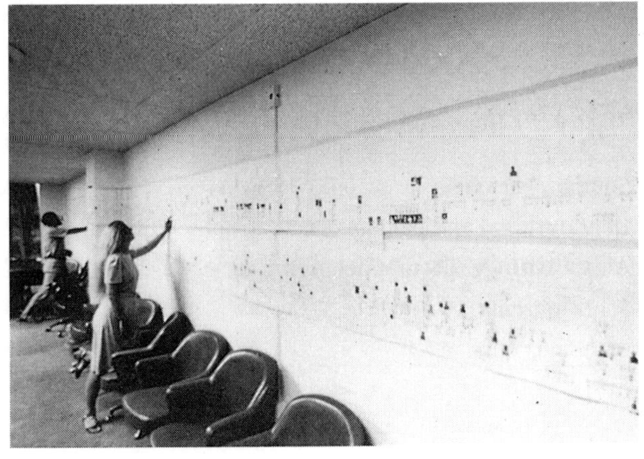

Nancy Wexler and the "Pedigree Wall" at Columbia University, 1987

The pedigree traces the inheritance of Huntington's disease through extended families living at Lake Maracaibo, Venezuela. (Courtesy of S. Uzzell.)

uously cultured. This is accomplished by fusing white cells in the blood with immortal cancer cells or by transforming them with an immortalizing oncogene from a tumor virus. In either case, the cultured white blood cells provide an easy source of DNA needed for the next, and most laborious, step.

James Gusella radioactively labeled at random a number of cloned DNA fragments from a human genomic library. These were then used to probe Southern blots of DNA from HD patients and unaffected family members. Gusella was looking for a probe that identifies a polymorphism whose appearance parallels the pattern of inheritance of the disease. The assumption was that all members of the extended Lake Maracaibo family inherited the disease from a single common ancestor. Thus, if a polymorphic marker is tightly linked to the disease gene, it should be coinherited by all of the affected individuals. Luck was with Gusella: A linked marker was identified with the 12th probe tested. Using this probe, he demonstrated that the HD gene is located near the telomere of the short arm of chromosome 4. Gusella's initial good fortune did not continue, and it took almost 4 years to identify a tightly linked marker, this one also located on the centromere side of the disease locus. Theoretically, he should have been able to find other linked markers located some distance from these original markers. The intervening distance would be occupied by the disease gene, thus providing flanking markers on either side. This ultimately proved impossible, because the HD gene lies very near the telomere.

HD patients turned out to have different profiles of genetic markers, indicating independent origins of the disease. Using a complex analysis of haplotype linkage disequilibrium between markers in different pedigrees, the Huntington's Disease Collaborative Research Group finally cloned the HD gene in 1993, 10 years after its initial mapping. Affected members in all 75 HD families used in the study showed a length polymorphism in the coding region of the *huntingtin*

Triplet Repeat Disorders

Disease	Protein	Number of triplet repeats (normal/disease)
Huntington's disease	huntingtin	6–35/36–180
Fragile X mental retardation	FMR1	30/60–200
Spinocerebellar ataxia		
SCA1	Ataxin-1	6–39/40–88
SCA2	Ataxin-2	14–32/33–77
SCA3	Ataxin-3	12–40/55–86
SCA6	Ataxin-P/Q Ca^{++} channel	4–18/21–31
SCA7	Ataxin-7	7–17/34–200
SCA12	PPP2R2B	7–32/55–93
Spinobulbar muscular atrophy (SBMA)	Androgen receptor	9–36/38–65
Dentatorubral and pallidolyusian atrophy (DRPLA)	Atrophin-1	3–36/49–88
Ataxia with intellectual deterioration	TATA-binding protein	25–42/45–63
Schizophrenia	KCNN3	12–28/Long alleles over-represented
Male infertility	POLG1	10/0

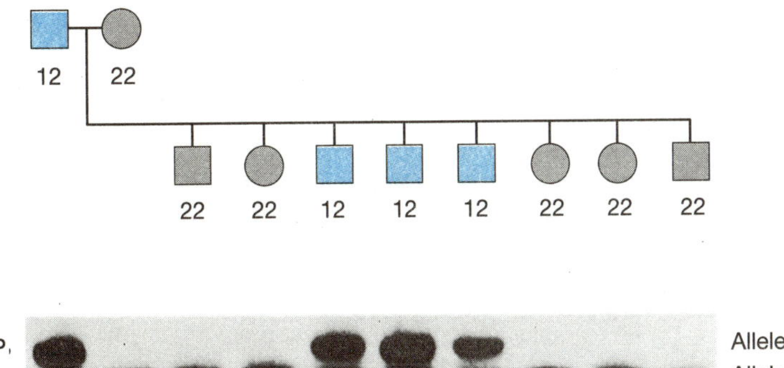

DNA Diagnosis of Huntington's Disease, 1987

This pedigree shows the coinheritance of a polymorphic allele and Huntington's disease. Allele 1 is linked to the Huntington's disease gene, and allele 2 is linked to the normal gene. The affected father and three affected sons (*blue*) all carry one copy of each allele. The unaffected mother and offspring (*gray*) have copies of the normal allele. (Courtesy of T.C. Gilliam, Columbia University.)

gene caused by a CAG repeat. The expansion of nucleotide triplets had been previously identified in Fragile X mental retardation and has a role in several other disorders.

The normal *huntingtin* gene has 6–35 CAG repeats; the mutated version in HD patients has 36–180 repeats. The number of repeats correlates with age when symptoms appear: Individuals with 36–41 repeats may never have symptoms, whereas those with more than 50 repeats develop symptoms before age 20. Since the repeat occurs in a coding exon, each additional repeat adds another unit of the amino acid glutamine to the expressed huntingtin protein. This alters the three-dimensional structure of the huntingtin protein, changing its interactions with other cell proteins.

Cloning the Duchenne Muscular Dystrophy Gene

Several important genes were cloned during the interval between the mapping of the HD locus and the eventual cloning of *huntingtin*. The X-linked disease Duchenne muscular dystrophy (DMD) was among the first disease loci to actually be cloned in the absence of knowledge of its protein product. The early success in identifying this gene relied on evidence showing that a number of DMD patients have large deletions clustered in a region of the X chromosome known as Xp21. In addition, females having the disease were found to have a break in their active X chromosome at position Xp21. This suggested that the deletions are associated with pathology, causing a loss of part of the normal gene at this locus. A combination of strategies, including RFLP analysis, was used to isolate the disease gene.

The *dystrophin* gene, as it has become known, is one of the largest and most complex genes yet discovered. Encompassing more than 2,000,000 bp and possessing more than 60 exons, it produces a 14,000-bp mRNA that codes for a protein containing 4000 amino acids. The *dystrophin* gene appears to be prone to

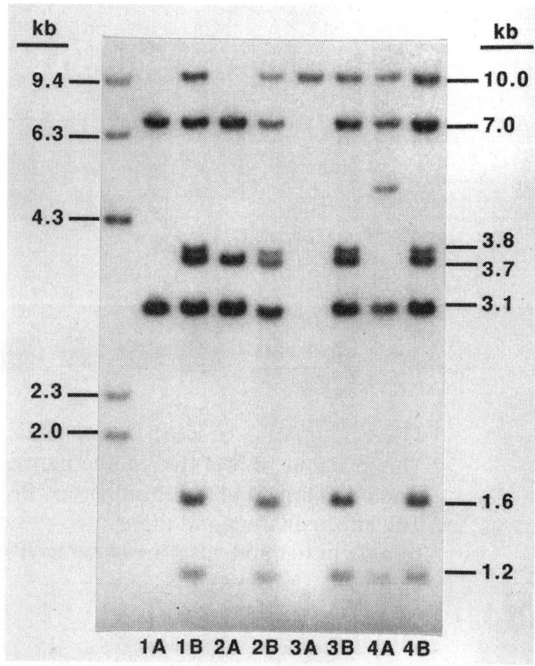

DNA Diagnosis of Duchenne Muscular Dystrophy, 1986
This early Southern blot analysis shows the DNA-banding patterns of four sets of brothers (1–4). Unaffected boys (in *B* lanes) show seven DNA bands, which are protein-coding exons of the *dystrophin* gene on the X chromosome. Brothers with muscular dystrophy (in *A* lanes) have deletions that eliminate one or more exons. (Courtesy of Jan Witkowski, Banbury Center, Cold Spring Harbor Laboratory.)

damage. DMD patients show many different deletions of exons of the gene, which effectively knock out production of any functional dystrophin. In the milder form of Becker muscular dystrophy, deletions in the *dystrophin* gene produce a semifunctional dystrophin protein. Diagnosis by Southern blot analysis, using a cDNA of the *dystrophin* mRNA, could detect various exon deletions. This was replaced by multiplex PCR, where multiple sets of primers are used to amplify ten of the commonly deleted exons in a single PCR experiment.

Cloning the Cystic Fibrosis Gene

The 1989 isolation and analysis of the causative gene for cystic fibrosis (CF) on chromosome 7 was a case study of the practice and power of modern molecular genetics. First, it resulted from an interdisciplinary collaboration among more than 25 scientists at the Hospital for Sick Children and University of Toronto, the University of Michigan, and the University of Pittsburgh. Second, it was the first disease gene identified *entirely* using the methods of positional cloning. Unlike DMD, CF is not characterized by large-scale deletions or rearrangements that could be used to map the gene to its chromosomal location.

Once a linked marker was identified within 1 cM of the disease locus, the researchers used the strategy of "chromosome walking" to clone the gene responsible for CF. Generally, chromosome walking works as follows: First, a genomic library of large DNA fragments (20,000–40,000 bp) is constructed that encompasses the disease locus. The closest linked marker is used as a probe to isolate its corresponding genomic clone. Following restriction mapping of the clone, a restriction fragment is isolated from the end of the clone closest to the disease locus. This fragment is used to reprobe the library to identify an overlapping clone. The endmost fragment of this clone is then used to reprobe the library, and another overlapping clone is isolated. Through such a succession of overlapping clones, one "walks" along the chromosome region spanning the disease locus, eventually reaching the flanking marker on the other side. The overlapping fragments are then assembled to produce a map of the disease locus.

Researchers screened ten genomic libraries and isolated the *cystic fibrosis transmembrane conductance regulator* (*CFTR*) gene, which spans approximately 250,000 nucleotides. The coding exons of *CFTR* predict a protein of 1480 amino acids. The CFTR protein is involved in the transport of sodium chloride and water in and out of the epithelial cells that line the lungs and the digestive system. As in sickle cell anemia, the primary genetic lesion in CF is a specific mutation affecting a single amino acid. Approximately 70% of CF patients show a 3-bp deletion, named deltaF508, that results in loss of a single phenylalanine residue at amino acid position 508 of the CFTR polypeptide. With this mutation, the cell excretes out too much salt, and too little water, resulting in a sticky mucus that clogs the lungs and extra salt in the patient's sweat.

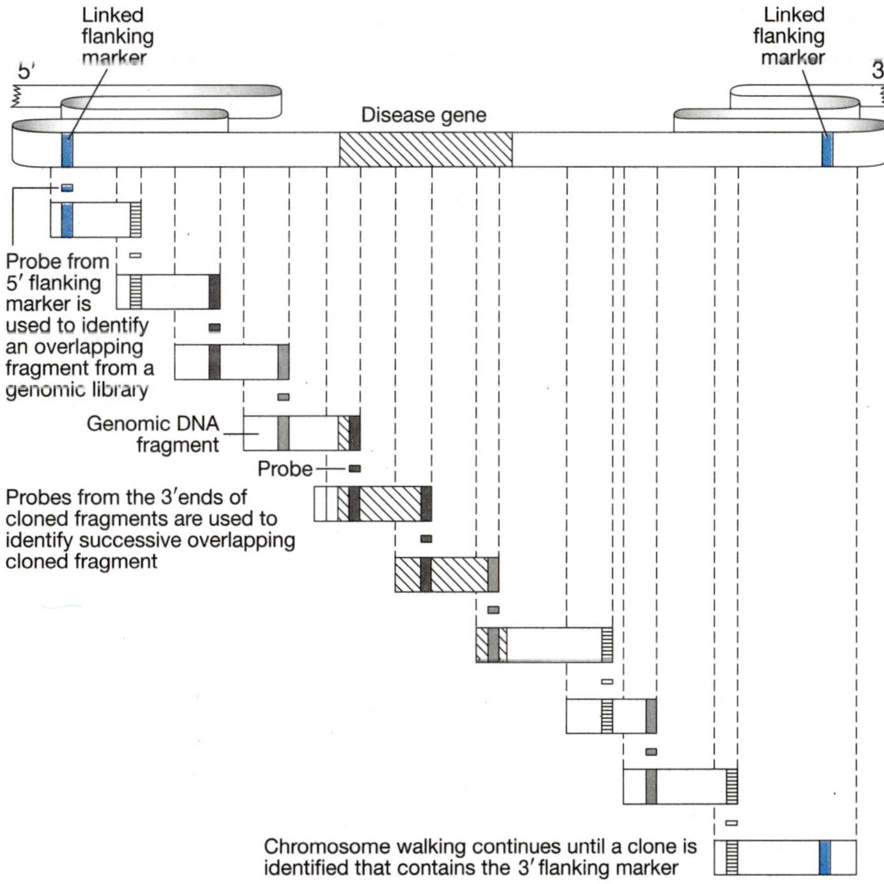

Using Chromosome Walking to Clone a Disease Gene

PHARMACOGENOMICS

Throughout the second half of the 20th century, major pharmaceutical companies amassed "libraries" containing hundreds of thousands of chemical compounds. These numbers have increased dramatically with the advent of combinatorial chemistry, which builds up compounds from simple chemical components—analogous to the way DNA probes are built up on a photolithographic DNA chip (Chapter 6). Each of these compounds is a potential pharmaceutical that can fight disease by altering the activity of a gene or its corresponding protein. For example, a number of compounds have been developed against histamines and other molecules involved in allergic reactions.

Pharmaceutical development, however, has been hampered by a relative lack of metabolic "targets" against which companies can test their huge compound libraries. The availability of the human genome sequence promises to solve this problem by presenting drug developers with a trove of new targets. Using the human genome sequence to inform drug discovery is termed pharmacogenomics. Each gene that is definitively linked to a disease becomes a validated target for drug discovery. Knowledge of mutations in that gene, and the corresponding changes in the three-dimensional structures of the encoded protein, allows one to develop strategies for screening compound libraries. Rational drug design carries this concept a step further by using the target protein's own structure to predict the properties of small molecules that can bind to an active site or otherwise modulate the protein's activity. This was the triumph of Gleevec, the first anticancer drug developed using detailed knowledge of protein kinase receptors (Chapter 7).

As more genes, and therefore proteins, are identified in a disease pathway, treatments increasingly can be tailored to specific defects in a metabolic or signal transduction pathway. Defects in different proteins in the same pathway may cause the same disease or symptoms. (Recall Beadle and Tatum's experiment, where mutations in different genes produced the same metabolic phenotype.) Thus, the same apparent disease may present different drug targets, depending on which gene in a pathway is mutated. For example, a mutation in the cytoplasmic domain of the EGF (epidermal growth factor) receptor is blocked by Gleevec; however, a different drug would be needed to counter a mutation in the EGF extracellular domain.

FINDING GENES BEHIND COMPLEX DISORDERS

Recall that most products made from cloned genes treat uncommon or tightly defined disorders. With the notable exception of statins used to treat hypercholesterolemia, the DNA revolution has not been successful in offering new or improved treatments for common disorders, such as asthma and noninsulin-dependent diabetes. Furthermore, molecular genetics and genomic biology have not yet produced a rational drug for the treatment of major behavioral disorders, notably schizophrenia and bipolar disorder (manic depression).

Asthma, diabetes, schizophrenia, and bipolar disorder are all examples of complex, or heterogeneous, disorders. Each appears to involve multiple genes whose expression is further modified by environmental factors, for example, air quality in asthma, diet in diabetes, and drug or alcohol abuse in schizophrenia and

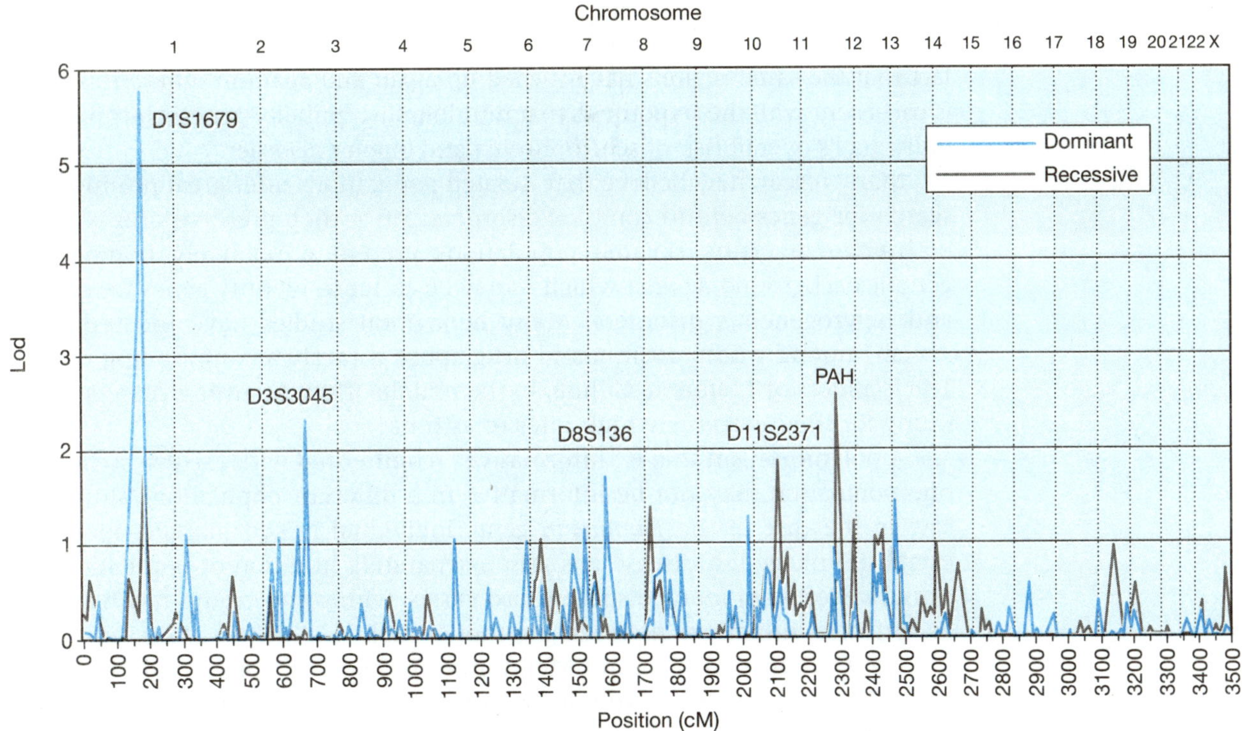

LOD Scores for a Genome-wide Scan for Schizophrenia, 2000

Affected and unaffected individuals in 22 families with schizophrenia were genotyped at 381 marker loci throughout the genome producing significant linkage on one long arm of chromosome 1. (Reprinted, with permission, from Brzustowicz L.M., Hodgkinson K.A., Chow E.W.C., Honer W.G., and Bassett A.S. 2000. Location of a major susceptibility locus for familial schizophrenia on chromosome 1q21-q22. *Science 288:* 678–682.)

bipolar disorder. To further complicate matters, each disorder has a range of severity and expression that may make it difficult to standardize diagnosis to the point that all researchers would evaluate the same pedigree in an identical manner.

The goal of population studies, like family studies, is to link particular DNA polymorphisms to a disease phenotype. The LOD (logarithm of the odds) score is the key statistical method used to establish linkage in family and population studies. On the basis of an observed recombination frequency between a marker and a putative disease locus, the LOD score is a ratio of the probability (odds) of a pedigree occurring at that linkage value divided by the probability (odds) of no linkage. A LOD score of 3, which is generally considered the threshold for possible linkage in a complex disorder, means that linkage is 1000 times more likely than no linkage. As a logarithmic function, like the Richter earthquake scale, each LOD score increases by a factor of 10. Thus, a LOD score of 3 represents a 10 times closer association between a marker and a locus than does a LOD score of 2.

The LOD score is extremely sensitive to changes in data analysis and laboratory errors. The change in diagnosis of a single person in an extended pedigree may be enough to lessen the association between the marker and a phenotype and, thus, weaken statistical linkage. Despite these problems, a number of genome scans and family studies during the past 10 years have reported linkage for schizophrenia with loci on chromosomes 1, 6, 8, 10, 13, 15, and 22. The case is similar for bipolar disorder, where linkage has been reported on chromosomes 4, 12, 13, 18, 21, and 22. Although the linkage reported in any single study is modest, the fact that the same regions have turned up again and again in different pedigrees is consistent with the hypothesis that multiple susceptibility genes contribute to an individual's overall risk of schizophrenia and bipolar disorder.

Many researchers believe that isolated populations offer great promise in the search for genes behind complex disorders. Since they preserve only a fraction of human diversity, isolated populations present a relatively homogeneous genetic background against which it may be easier to identify genes for common and heterogeneous disorders. Many behavioral studies have focused on the Amish, among whom alcohol and drug abuse is rarely a confounding problem. The islanders of Tristan de Cuhna, in the middle of the Atlantic Ocean, are interesting for their extremely high rates of asthma.

A polymorphism that is "informative" (coinherited with the disease locus) in one population may not be informative in a different population. Thus, as we saw in the case of the *Huntingtin* gene, individual linked markers may fail to establish linkage if a disease has arisen separately in different populations. The causative lesion of many disorders also varies, with some groups having unique, or "private," mutations not seen in other groups.

The Old Order Amish and Mennonites provide an object study in founder effect. Members of these two religious sects, known as the "plain people," live in agricultural communities where they eschew most modern technology and dress in simple clothing without adornment or buttons. To this day, they disdain motor vehicles in favor of horse and buggy. Lancaster County, Pennsylvania, remains a homeland for both groups, each of which is predominately derived from fewer than 100 individuals who settled there in the 1700s. Like many groups with small founding populations, they have concentrated mutations for some otherwise rare metabolic disorders. For example, maple syrup urine disease (MSUD) affects about 1/250,000 children in the general population but strikes about 1/400 Amish and Mennonite children.

Methylcrotonyl-CoA carboxylase (MCC) deficiency, a related but usually less severe disorder of the breakdown of the amino acid leucine, provides an example of a "private mutation." This disease has an overall frequency of about 1/50,000 in Caucasian populations, but it reaches a frequency of about 1/1500 among the Old Order Amish and Mennonites of Lancaster County. Affected Amish children have a G-to-C missense mutation at position 295 of the β-subunit of the *MCC* gene, whereas Mennonites have a frameshift mutation caused by a T insertion at position 518. Thus, although they live nearby and share a closely related religion, lifestyle, and ethnic background, each of these groups has inherited a different point mutation responsible for a rare disorder.

Single-nucleotide Polymorphisms

Imagine the complexity of searching for any of several potential genes involved in a heterogeneous disorder in which different genes or gene combinations may produce similar phenotypes in different populations. The various genes are likely to be associated with different markers in different population groups. Association studies using single-nucleotide polymorphisms (SNPs) hold potential in solving the problem of linking markers to the genes involved complex disorders.

Although the term SNP burst on the scene in the late 1990s, they are nothing but point mutations. To put this into perspective, there is about 1 nucleotide difference per 1200 nucleotides in two comparable chromosomes. This translates into about 3 million single-base differences and 100,000 amino acids differences between any two people. However, most single-nucleotide mutations are rare in a population; to be generally useful in gene scans, an SNP must have a population frequency of at least 1%.

Because SNPs are the most frequent type of polymorphism, there are potentially hundreds of useful SNP markers in a region of linkage disequilibrium that is associated with a disease gene. A region of linkage disequilibrium is termed a haploblock, because it is inherited, without recombination, like the haploid mitochondrial DNA (mtDNA) or the Y chromosome. A set of SNPs, or other markers, within the haploblock are inherited together as a haplotype.

Different populations have accumulated different SNPs within the haploblock. Thus, affected individuals from different populations may share certain markers within the haploblock, whereas other markers will be unique in certain populations. Just as one may find a consensus sequence for promoter regions and intron/exon splice junctions, haplotypes can be identified that represent a consensus of SNPs that are coinherited with the disease gene across many populations. Although no individual SNP is likely to have great predictive value, the combinatorial effect of an SNP haplotype can be a powerful tool in linkage studies. Thus, in 2002, the National Human Genome Research Institute announced a 100-million-dollar project to establish a haplotype map of the human genome, potentially containing 200,000 haploblocks.

Many researchers are confident that once the human genome map is heavily populated with SNPs, disease genes can be identified in heterogeneous populations of unrelated individuals. For example, a sample could be drawn from a database of all individuals who suffer from severe asthma, irrespective of their population group. An equivalent control sample is then drawn of healthy people. Each patient and control are SNP typed across his or her whole genome or across a specific candidate region. A haplotype is constructed for each person, using SNPs that occur in haploblock regions of the genome. Then, computer algorithms search for a consensus haplotype that is associated with the disease locus. Since haplotypes may encompass tens or hundreds of SNPs, this type of association analysis is much more complex for determining linkage with one marker at a time. It is not clear how saturated with SNPs the genome map must be before pure association analysis of this type will become possible, but it may be as few as 1 million SNPs.

Pharmacogenetics

Everyone at one time must have taken pause at the paradox of a physician asking us if we are allergic to a particular drug. After all, the doctor should be the one to inform us of a potential problem. Unfortunately, trial and error is the only way to determine a response to most drugs—it takes an allergic reaction to know if we are allergic! SNPs offer the potential of predicting a negative response *before* a drug is taken. Thus, the endgame of genetic medicine is pharmacogenetics, predicting drug response and tailoring treatment to each person's genetic makeup. However, before we enter this era of personalized medicine, experts today believe we must pass through a period of "population medicine," where drugs are targeted according to a generalized profile of the population group that most closely matches the patient.

Although it is very much in vogue today, the term "pharmacogenetics" was first coined in 1959 by Freidrich Vogel. This was based on earlier evidence that drug responses are inherited and vary between population groups. Notably, African American soldiers serving in Italy during World War II suffered adverse effects, including hemolysis, from the antimalaria drug primiquine. This was correlated with glucose-6-phosphate dehydrogenase (G6PD) deficiency, which, ironically, provides some protection against malaria.

Drug response is largely mediated by so-called metabolic enzymes in the liver—the cytochrome P450 monooxidases (CPY450s)—which detoxify compounds and metabolize many drugs into their bioactive forms. People who are "extensive metabolizers" efficiently convert a given drug to its active form and/or metabolize it at a rate that provides the desired therapeutic effect. "Poor metabolizers" fail to convert enough of the drug to its active form or metabolize it at a rate that fails to produce a therapeutic effect. "Toxic metabolizers" convert the drug into a toxic product or metabolize it so slowly that it accumulates to toxic levels.

In the late 1970s, Robert Smith of St. Mary's Hospital, London, noticed an unusually high incidence of side effects, including an unusual fainting response, among patients prescribed the antihypertension drug debrisoquine. He found that about 8% of Caucasians (but less than 2% of Black and Asian populations) are poor metabolizers, handling debrisoquine 10–200 times less efficiently than extensive metabolizers. Michel Eichelbaum, of the University of Bonn, found similar disparities in the metabolism of sparteine, an anti-arrythymic. This led to the realization that both drugs are metabolized by the CPY2D6 enzyme and that poor metabolizers inherit a defective CPY2D6 enzyme. Subsequent work revealed that CPY2D6 is involved in deficient responses to at least 40 common drugs, including codeine, dextromethorphan, beta-blockers, monoamine oxidase inhibitors, tricyclic antidepressants, antipsychotics, neuroleptics, and fluoxetine (Prozac). Cloning and sequencing of the *CPY2D6* gene in 1988 showed that poor metabolizers have polymorphisms that produce splicing errors or amino acid substitutions. Recent research showed that several SNP haplotype pairs in *CPY2D6* predict response to the anti-asthma drug albuterol, with striking differences in haplotype distribution between population groups.

Screening for relevant *CPY450* polymorphisms would be a perfect application for gene chip and would be a logical first step in the development of pharmacogenetics.

THINKING ABOUT HUMAN HISTORY AND POPULATIONS

Each person's unique disease susceptibilities and responses to drugs are, in large part, the balance between our uniqueness as individuals and the similarities we share with others in historical population groups. Written in each person's DNA is a record of our shared ancestry and our species' struggle to populate the earth. Our ancient ancestors moved around and eventually out of Africa. They moved in small groups, following river valleys and coastlines, reaching Asia and Europe. Land bridges that appeared during recurring Ice Ages allowed them to reach Australia and the Americas.

As these early people wandered, their DNA accumulated mutations. Some provided advantages that allowed these pioneers to adapt to new homes and ways of living. Most were nonessential. Mutations are the grist of evolution, producing gene and protein variations that have allowed humans to adapt to a variety of environments—and to become the most far-ranging mammal on the planet. The same mutational processes that generated human diversity—point mutations, insertions/deletions, transpositions, and chromosome rearrangements—also generated disease.

It may be hard to see from our current vantage point, but the entire industrial revolution has occupied only about 0.1% of our 150,000-year history as a species. The cradles of western civilization—classical Greece and Rome—take us back into only 2% of our history. The earliest city-states of Mesopotamia, Babylonia, Assyria, and China take us back only 4% of way into our past. At 6%, we reach the watershed of agriculture, which changed forever the way humans live and work. After language, the domestication of plants and animals is the single greatest civilizing factor in human history. Increased production and performance of domesticated organisms made possible urbanization and task specialization in human society. Thus, the labor of fewer and fewer farmers produced enough food and clothing materials to satisfy the needs of growing numbers of nonfarmers—artisans, engineers, scribes, and merchants—freeing them to develop other elements of culture. Reaching back the remaining 93% of our history, to the dawn of the human species, we lived only as hunter-gatherers.

The fastest evolving part of our genome, the mitochondrial control region, accumulates about one new mutation every 20,000 years. Mutations are five- to tenfold less frequent in most regions of the nuclear chromosomes. Thus, virtually every gene in our genome is, on average, only one event away from our hunter-gatherer heritage. This leads to two far-reaching conclusions that substantially broaden our understanding of evolutionary processes and the origin of human disease:

- Throughout most of human history, the hunter-gatherer group was a basic population unit upon which evolution acted.
- Our basic anatomy, physiology, and many aspects of behavior are essentially identical to the hunter-gatherers who ranged through the ancient landscapes of Africa, Europe, Asia, Australia, and the Americas.

It may be difficult for many people today to conceive of what is meant, in a genetic sense, by a human population. This is because, over the past quarter century, people have become extremely mobile. Airplanes and four-wheel vehicles have made it possible to travel virtually anywhere in the inhabited world within a day or two. Major urban centers have become cosmopolitan, with mixes of people representing many races and cultures. Even so, today there are still regions of the world where people are born, reproduce, and die all in the same village. This essentially defines the "classical" definition of a human population: a group of people who, by reason of geography, language, or culture, preferentially mate with another.

Unique human populations—for example, the Saami of Finland, the Ainu of Japan, the Nanuit of Alaska, the Yanomami of Brazil, the Pygmies of Central Africa, and the Bushmen of Southern Africa—have preserved unique cultures and languages. Their genomes preserve the genetic residue of a time when all human beings lived in smaller and more cohesive groups. Small populations are subject to the founder effect, "inbreeding," and genetic drift (a random fluctuation of nonessential alleles). Over millennia, these effects join with selection to concentrate particular gene variations within different population groups. Gene variations come into equilibrium when a population grows to several thousand individuals.

THE BIOLOGICAL CONCEPT OF RACE

Most people can readily define characteristics that make them different from others. The most obvious difference between people is the color of their skin, followed by hair and eye color, hair texture, and shapes of body and facial features. These physical characteristics, in combination with cultural and religious practices, have been generalized into the related concepts of race and ethnicity. Unfortunately, racial and ethnic prejudices have fueled many of the worst events in human history.

The physical characteristics we associate with race and ethnicity likely are controlled by a mere handful of the 30,000 to 50,000 genes in the human genome. Variation in human skin color is determined by levels of two different forms of a pigment produced by melanocytes in the dermis layer of the skin. Eumelanin is brown-black and pheomelanin is red-yellow. However, the genetic basis of pigmentation is not well understood, and only one gene involved in human pigment variation has been located. But why did different population groups develop different skin colors in the first place?

Biologists assume that early human ancestors had light skin covered by dense hair—like our near primate relative, the chimpanzee. Australopithecines and other early human ancestors probably looked and acted like tall chimpanzees, but with the ability to walk upright for longer periods of time. *Homo erectus*, with its striding gate, spent more time tracking prey and foraging in the open African savannas. Increased activity in the open sun, and a larger brain to be protected from overheating, necessitated efficient evaporative cooling. This would have selected for individuals with larger numbers of sweat glands. (Chimps have very few sweat glands.) However, wet hair hinders evaporation, so a trend toward evaporative cooling also favored a reduction in body hair. This is a plausible explanation of how humans came to have nearly hairless bodies.

In the absence of protective hair, it is generally assumed that dark-pigmented skin developed in hominid populations in Africa as protection against the damaging effects of UV radiation. Most skin cancers develop later in life, well after reproductive age. Thus, it seems unlikely that melanin's anti-cancer effect, alone, could have provided enough selective advantage for dark skin.

In 1967, W. Farnsworth Loomis, of Brandeis University, offered an explanation of why lighter skin evolved among populations living at higher latitudes. Short-wavelength UV (UVB) radiation in sunlight triggers a reaction in the skin to produce vitamin D, which is important in skeletal formation and immune function. Vitamin D synthesis by the skin is especially important for people without diets rich in this vitamin, as would have been the case for most early hominids. Thus, Loomis hypothesized that lighter skin offered a selective advantage as people migrated out of Africa, allowing them to absorb more of the reduced UV light that penetrates the atmosphere in the higher latitudes.

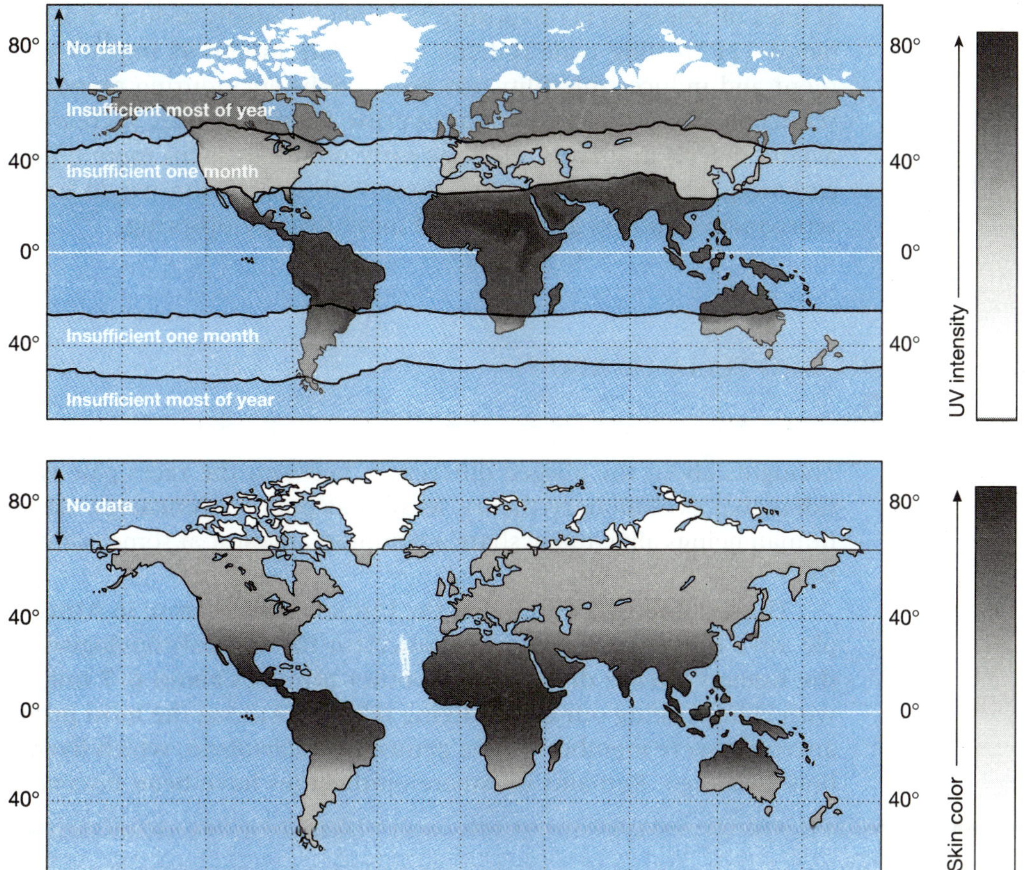

Ultraviolet Light and Vitamin D Production, and Skin Color

(*Top map*) Populations that live in the tropics near the equator receive enough UV light from the sun to synthesize vitamin D all year long. In temperate zones, people lack sufficient UV to make vitamin D at least one month of the year. Those nearer the poles do not get enough UV light most months for vitamin D synthesis. (*Bottom map*) Shown are predicted skin colors for humans based on UV light levels. In the Old World, the skin color of indigenous peoples closely matches predictions. In the New World, however, the skin color of long-term residents is generally lighter than expected, probably because of their recent migration and factors such as diet. (Adapted, with permission, from Jablonski N.G. and Chaplin G. 2002. Skin deep. *Sci. Am. 287:* 74–81.)

Recent modeling of worldwide UVB radiation, based on satellite mapping of the earth's ozone layer, shows a correlation between levels of UVB that are sufficient for vitamin synthesis and skin pigmentation. Thus, dark-pigmented skin is found in the tropics where there is sufficient UVB to synthesize vitamin D year-round. Lighter skin, but with the ability to tan, is found in the subtropical and temperate regions, which have at least 1 month of insufficient UVB radiation. Very light skin that burns easily is found north of 45 degrees, where there is insufficient UVB year-round.

In 2000, Nina Jablonski and George Chaplin, of the California Academy of Sciences, offered a more complete explanation for the evolution of dark-pigmented skin among early hominids in Africa. They proposed that melanin protects the body's stores of the B vitamin folate, which is essential for reproduction and embryonic development. This conclusion came from the synthesis of several lines of research: (1) Exposure to sunlight rapidly reduces folate levels in the blood. (2) Treating male rodents with folate inhibitors impairs sperm development and induces infertility. (3) Folate deficiency during pregnancy, including reduction apparently induced by overuse of tanning beds, increases risk of neural cord defects in infants. Thus, as early hominids spent more time hunting and gathering on the open savanna, those with darker skin would have had greater reproductive success and produced more healthy offspring.

WHAT THE FOSSIL RECORD TELLS US ABOUT HUMAN EVOLUTION

Thoughts about the alleged differences between the races pale when one considers that evolutionary theory, as well as popular genealogy, demands that all human beings alive today share a common ancestor at some point in the distant past.

The fossil record shows that the human species arose in Africa, and all people alive today share a common ancestor there. Anthropologists estimate that the human lineage diverged from other primates about 6–7 million years ago, with chimps being our closest living relative. Among the most primitive human ancestors were members of the genus *Australopithecus*, which lived about 3 million years ago. Remains of Australopithecines have been discovered primarily

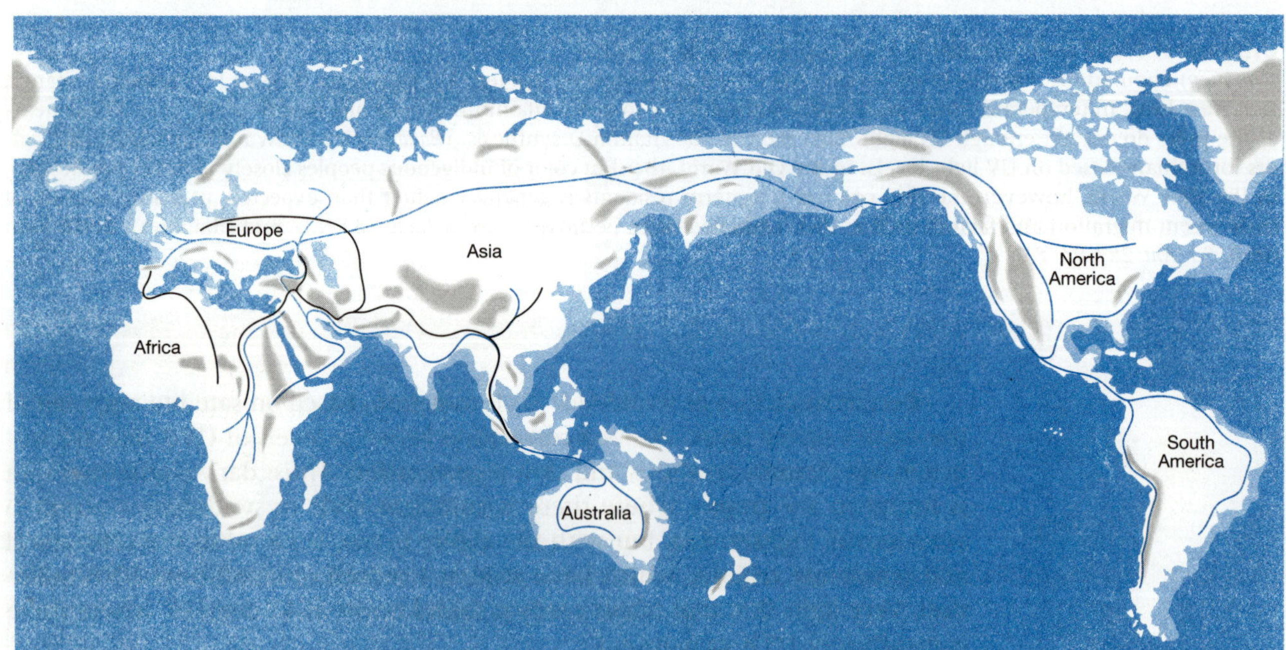

"Out of Africa" Theory of Human Evolution

Ancient humans of the species *Homo erectus* left Africa 1.8 million years ago, reaching Europe and Asia (*black lines*). Groups of *Homo sapiens,* from whom all modern humans are descended left Africa about 70,000 years ago (*blue lines*). These groups replaced any remaining ancient populations, reaching Asia and Australia about 60,000 years ago and entering Europe about 45,000 years ago.

in the Rift Valley of Africa. Early members of our own genus, *Homo erectus*, arose in the same region about 2.5 million years ago. These "archaic" hominids migrated out of Africa approximately 1.8 million years ago to found populations in Europe, the Middle East, and southern Asia.

The earliest fossils of modern humans, or artifacts made by them, have been found in southern and eastern Africa, dating to about 140,000 years ago. Remains of modern humans dating to 100,000 years ago have been found in the Middle East, to 60,000 years in Asia and Australia, and to 45,000 years in Europe. By modern humans, we mean members of our own species *Homo sapiens*, who share with us important anatomical features (skull shape and size) and behavioral attributes (use of blades, bone tools, pigments, burial goods, representational art, long-distance trade, and varied environmental resources). These humans subsequently spread to Micronesia, Polynesia, and the "New World" (North and South America).

How modern humans emerged is a matter of debate between proponents of two opposing theories. Supporters of the *multiregional theory* contend that modern human populations developed independently from archaic hominid (*Homo erectus*) populations in Africa, Europe, and Asia. Early modern groups evolved in parallel with one another and exchanged members to give rise to modern population groups.

Supporters of the *displacement theory*, commonly known as "out of Africa," contend that all modern human populations are derived from one or several modern population groups that left Africa beginning about 70,000 years ago. These founding groups migrated throughout the Old World, displacing any surviving archaic hominids. Thus, scientists all agree that our earliest hominid relatives arose in Africa, but disagree on when the direct ancestors of living humans left Africa to populate the globe.

THE DNA MOLECULAR CLOCK

At first thought, there does not seem to be any way in which DNA could provide us information about the origin of modern humans. The oldest hominid DNA isolated thus far dates back only about 60,000 years, well after our emergence as a species. In fact, we can study our evolutionary past by looking at the DNA variation of humans and primates alive today. Although this might seem to be a contradiction, remember that the DNA of any individual bears the accumulated genetic history of its species.

When two groups split off from a common ancestor, each accumulates a unique set of random DNA mutations. Provided mutations accumulate at a constant rate, and occur sequentially (one at a time), then the number of mutations is proportional to the length of time that two groups have been separated. This relatively constant accumulation of mutations in the DNA molecule over time is called the "molecular clock." An event that has been independently established by anatomical, anthropological, or geochronological data is used to attach a time scale to the clock. For example, the human molecular clock is typically set using fossil and anatomical evidence suggesting that humans and chimps diverged 6–7 million years ago.

Because of its high mutation rate, the mitochondrial control region evolves more quickly than other chromosome regions—it has a faster molecular clock. The fast mutation rate means that lineages diversify rapidly, amplifying differences between populations. However, rapid mutation also introduces the confounding problem of "back mutation" where the same nucleotide mutates more than once, returning it to its original state. Multiple mutations at the same position also cause an underestimation of the total number of mutation events. Thus, the number of observed differences between human and chimp sequences are less than one would expect to have occurred in the 6–7 million years since the lineages diverged. The chance of back or multiple mutations is much smaller over the period during which modern humans have arisen. So, the number of observed mutations among living humans is very close to the actual number that has accumulated since we arose as a species.

Mitochondrial DNA (mtDNA) offers another important advantage in reconstructing human evolution: With very few documented exceptions, the mitochondrial chromosome is inherited exclusively from the mother. This is because mitochondria are inherited from the cytoplasm of the mother's large egg cell. Any paternal mitochondria that may enter the ovum at the moment of conception are identified by different ubiquitin proteins expressed on their surface and destroyed. The lack of paternal chromosomes with which to recombine greatly simplifies the analysis of mitochondrial inheritance. The mitochondrial genome is inherited intact over thousands of generations, without the confounding effect of crossover with a paternal chromosome. Because the mitochondrial genome is haploid, having only a contribution from the mother, mtDNA types are termed haplotypes ("half-types").

WHAT DNA TELLS US ABOUT HUMAN EVOLUTION

Allan Wilson
(Courtesy of Cold Spring Harbor Laboratory Archives.)

Throughout the 20th century, fossils provided the only tools for reconstructing human origins and the field remained virtually the sole province of anthropologists. In 1987, Allan Wilson and co-workers at the University of California at Berkeley moved anthropology into the molecular age when they made mtDNA haplotypes for 145 living humans. Using a molecular clock like that described above, they constructed a tree that extrapolated back to a common ancestor who lived about 200,000 years ago.

It is important to note that the tips of the branches of Wilson's evolutionary tree were the 145 individual humans whose mtDNA types he had determined. Although most individuals generally came out on branches with others from their regional population group, some individuals fell on branches with other groups. This illustrated a high degree of mixing between human population groups. Importantly, Africans turned up on several non-African branches, but only African individuals were found on the branch closest to the root of the tree. Thus, Wilson concluded that the so-called mitochondrial "Eve" most likely lived in Africa, providing DNA evidence for the recent dispersion of modern humans "out of Africa."

During the next decade, molecular reconstructions of human lineages were conducted with autosomal and Y chromosome polymorphisms. Unlike mitchondrial SNPs, which have a high rate of back mutation, each Y chromosome SNP is believed to represent a unique mutation event that occurred once in evolu-

A. Six allelic sequences

```
Sequence 1) agctggctgaatgctatctgcgtcgcgcgaaataacgtcagcaattcgttacatctctctagggc
Sequence 2) agctggctgaatgctatctgcctcgcgcgaaataacgtcagcaattcgttacatttctctagggc
Sequence 3) agctggctgaatgctatctgcgtcgcgcgaaacaacgtcagcaattcgttacatttctctagggc
Sequence 4) agctggctgaatgctatctgcgtcgcgcgaaataacgtcagcaattcgttacatttctctagggc
Sequence 5) agctggctgagtgctatctgcctcgcgcgaaataacgtcagcaattcgttacatttctctagggc
Sequence 6) agctggctgaatgctatctgcgtcgcgcgaaataacgtcagcagttcgttacatctctctagggc
```

B. The six haplotypes for the five SNPs in the allelic sequences above.

```
Position            11          22          33          44          55
Sequence 1) ----------a----------g----------t----------a----------c----------
Sequence 2) ----------a----------c----------t----------a----------t----------
Sequence 3) ----------a----------g----------c----------a----------t----------
Sequence 4) ----------a----------g----------t----------a----------t----------
Sequence 5) ----------g----------c----------t----------a----------t----------
Sequence 6) ----------a----------g----------t----------g----------c----------
```

C. Two possible trees for the evolutionary relationships among the haplotypes above, assuming that mutations occur sequentially.

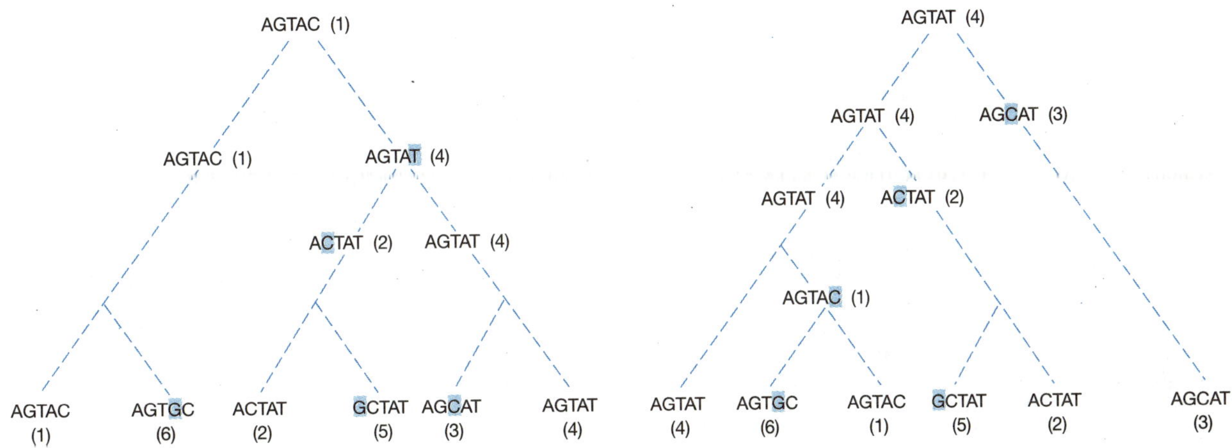

Using Haplotypes to Create Trees Showing Evolutionary Relationships

tionary history. The accumulated DNA evidence has confirmed Wilson's original story. Several lines of evidence point to the emergence of modern humans in Africa about 150,000 years ago.

- The greatest amount of DNA variation occurs in Africa, suggesting that African populations have been accumulating mutations for the longest period of time. Europeans have only about half the variation of African groups, suggesting that they are only about half as old.
- Most Asian and European variations are a subset of variations found in African populations, suggesting that Asian and European populations are derived from an African source.
- The deepest roots of a tree diagram of human variation contain only Africans. Ancient alleles have not been found in non-African populations.

Interestingly, comparisons of mitochondrial and Y chromosome polymorphisms suggest that men and women have had different roles in the peopling of

A

Greece	GGTACCACCCAAGTATTGACTCACC	
Japan	GGTACCACCCAAGTATTGACTCACC	25
Greece	CATCAACAACCGCTATGTATTTCGT	
Japan	CATCAACAACCGCTATGTATTTCGT	50
Greece	ACATTACTGCCAGCCACCATGAATA	
Japan	ACATTACTGCCAGCCACCATGAATA	75
Greece	TTGTACGGTACCATAAATACTTGAC	
Japan	TTGTACGGTACCATAAATACTTGAC	100
Greece	CACCTGTAGTACATAAAAACCCAAT	
Japan	CACCTGTAGTACATAAAAACCCAAT	125
Greece	CCACATCAAAACCCCCTCCCCATGC	
Japan	CCACATCAAAACCCCCCCCCCGCGC	150
Greece	TTACAAGCAAGTACAGCAATCAACC	
Japan	TTACAAGCAAGTACAGCAATCAACC	175
Greece	CTCAACTATCACACATCAACTGCAA	
Japan	TTCAGCTATCACACATCAACTGCAA	200
Greece	CTCCAAAGCCACCCCTCACCCACTA	
Japan	CTCCAAAGCCACCCCTCACCCACTA	225
Greece	GGATACCAACAAACCTACCCACCCT	
Japan	GGATATCAACAAACCTACCCACCCT	250
Greece	TAACAGTACATAGTACATAAAGCCA	
Japan	TAACAGTACATAGTACATAAAGCCA	275

C

Chimpanzee	TTCTTTCATGGGGAAGCAAATTTAA	
Greece	TTCTTTCATGGGGAAGCAGATTTGG	25
Chimpanzee	GTACCACCTAAGTACTGGCTCATTC	
Greece	GTACCACCCAAGTATTGACTCACCC	50
Chimpanzee	ATTA-CAACCGCTATGTATTTCGTA	
Greece	ATCAACAACCGCTATGTATTTCGTA	75
Chimpanzee	CATTACTGCCAGCCACCATGAATAT	
Greece	CATTACTGCCAGCCACCATGAATAT	100
Chimpanzee	TGTACAGTACCATAATCACCCAACC	
Greece	TGTACGGTACCATAAATACTTGACC	125
Chimpanzee	ACCTATAGCACATAAAATCCACCTC	
Greece	ACCTGTAGTACATAAAAACCCAATC	150
Chimpanzee	-ACATTAAAACCTTCACCCCATGCT	
Greece	CACATCAAAACCCCCTCCCCATGCT	175
Chimpanzee	TACAAGCACGCACAACAATCAACCC	
Greece	TACAAGCAAGTACAGCAATCAACCC	200
Chimpanzee	CCAACTATCGAACATAAAACACAAC	
Greece	TCAACTATCACACATCAACTGCAAC	225
Chimpanzee	TCCAACGACACTTCTCCCCCACCCT	
Greece	TCCAAAGCCACCCCTCACCCACTAG	250
Chimpanzee	AATACCAACAAACCTACCCTCCCTT	
Greece	GATACCAACAAACCTACCCACCCTT	275

B

Greece	CCAAGTATTGACTCACCCATCAACA	
Neandertal	CCAAGTATTGACTCACCCATCAGCA	25
Greece	ACCGCTATGTATTTCGTACATTACT	
Neandertal	ACCGCTATGTATCTCGTACATTACT	50
Greece	GCCAGCCACCATGAATATTGTACGG	
Neandertal	GTTAGTTACCATGAATATTGTACAG	75
Greece	TACCATAAATACTTGACCACCTGTA	
Neandertal	TACCATAATTACTTGACTACCTGCA	100
Greece	GTACATAAAAACCCAATCCACATCA	
Neandertal	GTACATAAAAACCTAATCCACATCA	125
Greece	AAACCCCCTCCCCATGCTTACAAGC	
Neandertal	AACCCCCCCCCCCATGCTTACAAGC	150
Greece	AAGTACAGCAATCAACCCTCAACTA	
Neandertal	AAGCACAGCAATCAACCTTCAACTG	175
Greece	TCACACATCAACTGCAACTCCAAAG	
Neandertal	TCATACATCAACTACAACTCCAAAG	200
Greece	CCACCCCT-CACCCACTAGGATACC	
Neandertal	ACGCCCTTACACCCACTAGGATATC	225
Greece	AACAAACCTACCCACCCTTAACAGT	
Neandertal	AACAAACCTACCCACCCTTGACAGT	250
Greece	ACATAGTACATAAAGCCATTTACCG	
Neandertal	ACATAGCACATAAAGTCATTTACCG	275

D

Neandertal	CCAAGTATTGACTCACCCATCAGCA	
Neandertal	CCAAGTATTGACTCACCCATCAGCA	25
Neandertal	ACCGCTATGTATCTCGTACATTACT	
Neandertal	ACCGCTATGTATTTCGTACATTACT	50
Neandertal	GTTAGTTACCATGAATATTGTACAG	
Neandertal	GCCAGCCACCATGAATATTGTACAG	75
Neandertal	TACCATAATTACTTGACTACCTGCA	
Neandertal	TACCATAATTACTTGACTACCTGCA	100
Neandertal	GTACATAAAAACCTAATCCACATCA	
Neandertal	GTACATAAAAACCTAATCCACATCA	125
Neandertal	AACCCCCCCCCCCATGCTTACAAGC	
Neandertal	ACCCCCCCCCCCCATGCTTACAAGC	150
Neandertal	AAGCACAGCAATCAACCTTCAACTG	
Neandertal	AAGCACAGCAATCAACCTTCAACTG	175
Neandertal	TCATACATCAACTACAACTCCAAAG	
Neandertal	TCATACATCAACTACAACTCCAAAG	200
Neandertal	ACGCCCTTACACCCACTAGGATATC	
Neandertal	ACGCCCTTACACCCACTAGGATATC	225
Neandertal	AACAAACCTACCCACCCTTGACAGT	
Neandertal	AACAAACCTACCCACCCTTGACAGT	250
Neandertal	ACATAGCACATAAAGTCATTTACCG	
Neandertal	ACATAGCACATAAAGTCATTTACCG	275

Was Neandertal Our Direct Ancestor?

Representative two-way comparisons of mitochondrial control region sequences used to determine whether Neandertal was the direct ancestor of modern humans. Sequence differences are highlighted in blue. (*A*) Comparison of modern Greek and Japanese humans shows 6 differences over 275 nucleotides (less than the 379 nucleotides analyzed by Svante Pääbo). (*B*) Modern Greek and Neandertal comparison shows 26 differences. (*C*) Modern Greek and a chimpanzee comparison shows 48 differences. (*D*) Comparison between two Neandertals shows 6 differences.

the planet and in the mixing of genes among population groups. Generally, mtDNA types show a gradation (or cline) of allele frequencies from one geographic region to another. This is the signature of "gene flow," the slow and steady exchange of genes between adjacent populations, which occurs in many cultures when women leave their families to live in their husbands' villages. Y polymorphisms, on the other hand, show discontinuities between adjacent regions, suggesting that men have not moved freely between local groups. However, related Y chromosome types do leave the signature of migrations and war campaigns that abruptly transplant genes over long distances. Thus, members of the Black South African tribe, the Lemba, have the telltale signature of Cohanin Jewry displaced from the Middle East. The most common Y chromosomes in cosmopolitan southern Japan clearly were transplanted from Korea in the last several hundred years, but the Ainu of the northern islands of Japan have an ancient affinity with Tibetans.

Was Neandertal Our Direct Ancestor?

Since their initial discovery in the Neander Valley of Germany in 1856, the heavy-set bones of Neandertal have fascinated scientists, as well as the general public. Neandertal had a brain capacity within the range of modern humans and was certainly an archaic member of the genus *Homo*. Neandertal ranged through Europe, the Middle East, and Western Russia beginning about 300,000 years ago and became extinct about 30,000 years ago. Clearly, during part of its span on earth, Neandertal shared its habitat with modern humans. Thus, there was a longstanding controversy about whether Neandertal was the direct ancestor of modern humans.

According to the multiregional model, modern humans developed concurrently from several distinct archaic populations living in different parts of the world. Under this model, Neandertal must be the intermediate ancestor of modern Europeans. Other archaic hominid fossils, Java and Peking man (*Homo erectus*), were the ancestors of modern Asians. According to the "Out of Africa" model, Neandertal was displaced by modern *Homo sapiens* who arrived in Europe about 40,000 years ago.

Svante Pääbo, 1997
(Photo by Margot Bennett, Cold Spring Harbor Laboratory.)

In 1997, at the University of Munich, Svante Pääbo, a student of Allan Wilson, further revolutionized human molecular anthropology when he added a 40,000-year-old DNA sample to the reconstructions of hominid evolution. Pääbo extracted DNA from the humerus of the original Neandertal-type specimen, amplified the sample by PCR, and cloned the resulting products in *E. coli*. The cloned fragments were then used to reconstruct a 379-bp stretch of the mitochondrial control region. Pääbo drew about 1000 human mitochondrial control region sequences from the Genbank database and compared them in pairs. He found an average of 7 mutations between these pairs of modern humans, representing the average variation accumulated since the divergence from a common ancestor. However, he found an average of 27 mutations when he compared each of the 1000 modern human sequences against the reconstructed Neandertal sequence. This put Neandertal outside the range of variation of modern humans.

If one takes the mutation rate of the mitochondrial control region to be about 1/20,000 years, then 7 mutations equals 140,000 years to a convergence to a common ancestor, in accordance with the earliest fossil record of modern humans. The 27-mutation difference between living humans and Neandertal suggests that our lineage converges on (or diverges from) a common ancestor about 550,000 years ago. This weighed still more DNA data in favor of the "Out of Africa" model.

"Bushy" Evolution

Until about 30 years ago, the "single-species hypothesis" dictated that only a single human ancestor lived on the earth at any point in prehistory. Each new Australopithecine or hominid-like fossil was immediately considered one of a direct succession of ancestors of modern humans. This created the image of an evolutionary "tree" with a long, straight trunk and essentially no branches until one reached racially and ethnically diverse modern humans. Thus, when racial segregation was still a way of life in the United States and elsewhere around the world, "straight-line" evolution provided scientifically minded people the comfort of believing in a shared evolutionary past, but put some distance between modern human groups who considered themselves different from each other.

This followed from a concept of gradual evolution, in which features that define a new stage are essentially properties of the preceding stage. In other words, there was a certain preordained "sense" in the way evolution proceeded. However, as more and more distinctive fossils emerged, it became impossible to fit them all into one straight path toward modern humans. Some had to be evolutionary dead ends and not a part of our direct ancestry. The concept of a "bushy" human pedigree, with short branches due to frequent extinctions, was

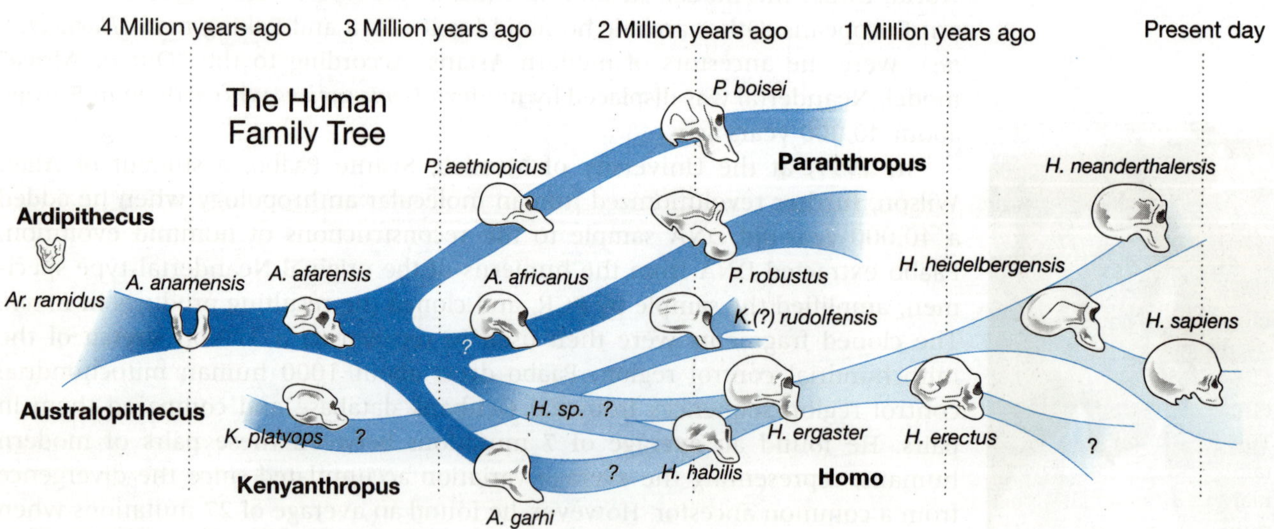

Bushy Evolution
Current theories of hominid evolution show a bushy lineage, rather than a straight line. This evolutionary scheme is based on the work of Donald Johanson. (Redrawn, with permission, from the Institute of Human Origins, Arizona State University.)

predicted by the model of punctuated evolution articulated in the early 1970s by Niles Eldridge of the American Museum of Natural History and Stephen Jay Gould of Harvard University. According to this model, evolution proceeds in abrupt fits and starts, as adaptive changes arise in small, dispersed populations.

Svante Pääbo's demonstration that the mtDNA variation of Neandertal lies outside the variation of modern humans provided DNA evidence to support the bushy tree concept. It makes better sense of the fossil evidence, but also makes better sense of the fact that most genetic variation accumulated during a time in human history when hunter-gatherer populations averaged perhaps 50 persons. A bushy evolutionary tree is exactly what one would expect from selection acting upon "private" sets of mutations that arise in small, relatively segregated groups of hunter-gatherers. It is most useful to consider that mutation and selection are separate events, often widely separated by time. Thus, neutral mutations anticipate future circumstances—environmental or behavioral—when they may provide a selective advantage.

Working from the premise that most mutations did not confer an immediate selective advantage to the hunter-gatherer, then most new mutations would survive or be extinguished in the group according to the whims of genetic drift. In this way, different hunter-gatherer groups accumulated different sets of mutations. On occasions, one or more of the accumulated mutations would provide a selective advantage to one population or another—and groups exchanged alleles through intermarriage. However, as climate, circumstances, or the luck of genetic drift changed, different lineages would become extinct, leaving behind a fossil record of its distinctive features.

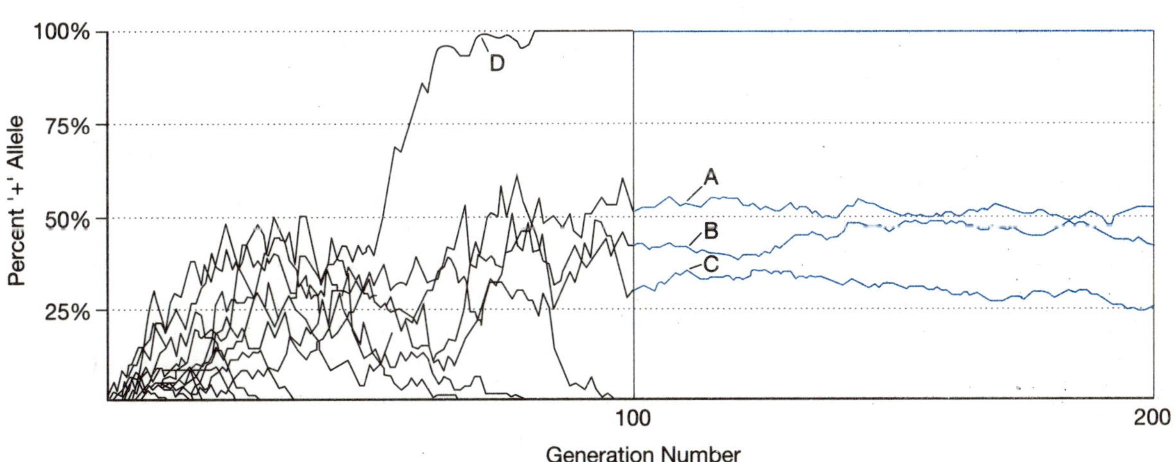

Genetic Drift in Hunter-Gatherer Groups

This simulation illustrates the fate of a new, neutral mutation that occurs in 100 hunter-gatherer groups. Each group maintains a population of 50 individuals who mate randomly, with respect to the new allele (+). The + allele is lost from the majority of populations within 10 generations and from 96% of populations after 100 generations. However, the + allele rises to frequencies of 30–50% in three populations (*A,B,C*) and reaches 100% in a single population (*D*). At the end of 100 generations, each population expands to 2000 individuals, and the + allele frequency stabilizes in populations *A*, *B*, and *C* over the next 100 generations. This represents the sort of population growth enabled by the advent of agriculture (although it would not have occurred over a single generation). Populations *A*, *B*, *C*, or *D* would thrive under a new environmental challenge to which the mutation confers a selective advantage. Thus, the drift of a neutral allele may favor certain populations in the future.

Climate Changes and Population Bottlenecks

The bushy nature of the tree of human evolution tells us that many hominid lineages have become extinct in the past. Several lines of evidence suggest that the lineage shared by all humans alive today also came perilously close to extinction at one or more points in the last 100,000 years.

On the surface, humans appear very diverse. Different populations have acquired distinctive morphological adaptations, including skin color, body shape, and pulmonary capacity, that allowed humans to inhabit virtually every biogeographical region of the earth. Despite these morphological differences, the human species as a whole has surprisingly little genetic diversity. Differences between populations account for only about 10% of human genetic variance, and there is no evidence for separate human subspecies.

This contrasts sharply with chimpanzees, which are restricted to similar habitats in equatorial Africa and have few morphological differences. Despite this seeming homogeneity, scientists recognize as many as four distinct subspecies of chimps living in eastern, central, and western Africa and Nigeria. There is substantial genetic diversity between, and within, the chimp subspecies. Notably, there is a greater diversity of mtDNA types among members of a single troop of western chimps than among all humans alive today. This striking lack of genetic diversity in the human species supports the contention that we are a young species that has gone through several "bottlenecks" that drastically reduced the human population—and genetic diversity. During these periods, the entire human population may have shrunk to as few as 1000 individuals, clinging to life in scattered refuges.

The fossil record shows that modern humans first left Africa about 100,000 years ago, traveling via the Sinai Peninsula into the Middle East. However, this group seems to have stalled, never reaching Asia or Europe. The track of modern human migrations grows cold worldwide until about 60,000 years ago, when we find evidence of a second movement out of Africa, hugging the coast of the Sinai Peninsula and across the narrow neck of the Red Sea into Asia and Australia. There is no sign of *Homo sapiens* in Europe until about 45,000 years ago, when they appear to burst on the scene at a number of sites almost simultaneously.

Stanley Ambrose, of the University of Illinois, has provided a plausible explanation for the aborted first venture of *Homo sapiens* out of Africa. This was the volcanic eruption of Mt. Toba in Sumatra, about 71,000 years ago. Toba's 30 x 100-km caldera (more than 30 times the size of Crater Lake in Oregon) was produced by the largest known eruption of the Quaternary Period, the most recent 1.8 million years of the earth's history. The eruption produced as much as 2000 km^3 of ash, leaving ash beds across the Indian Ocean and into mainland India. By comparison, the 1984 eruption of Mt. St. Helens in Oregon produced 0.2 km^3 of ash.

Although the airborne ash would have settled after several months, Mt. Toba also injected huge amounts of sulfur into the atmosphere, where it combined with water vapor to form sulfuric acid. Greenland ice cores show heavy sulfur deposition for six years following the eruption, indicating a lingering, sun-obscuring haze worldwide. The reduced solar radiation reaching the earth's surface would have lowered sea surface temperatures by about 3°C for several

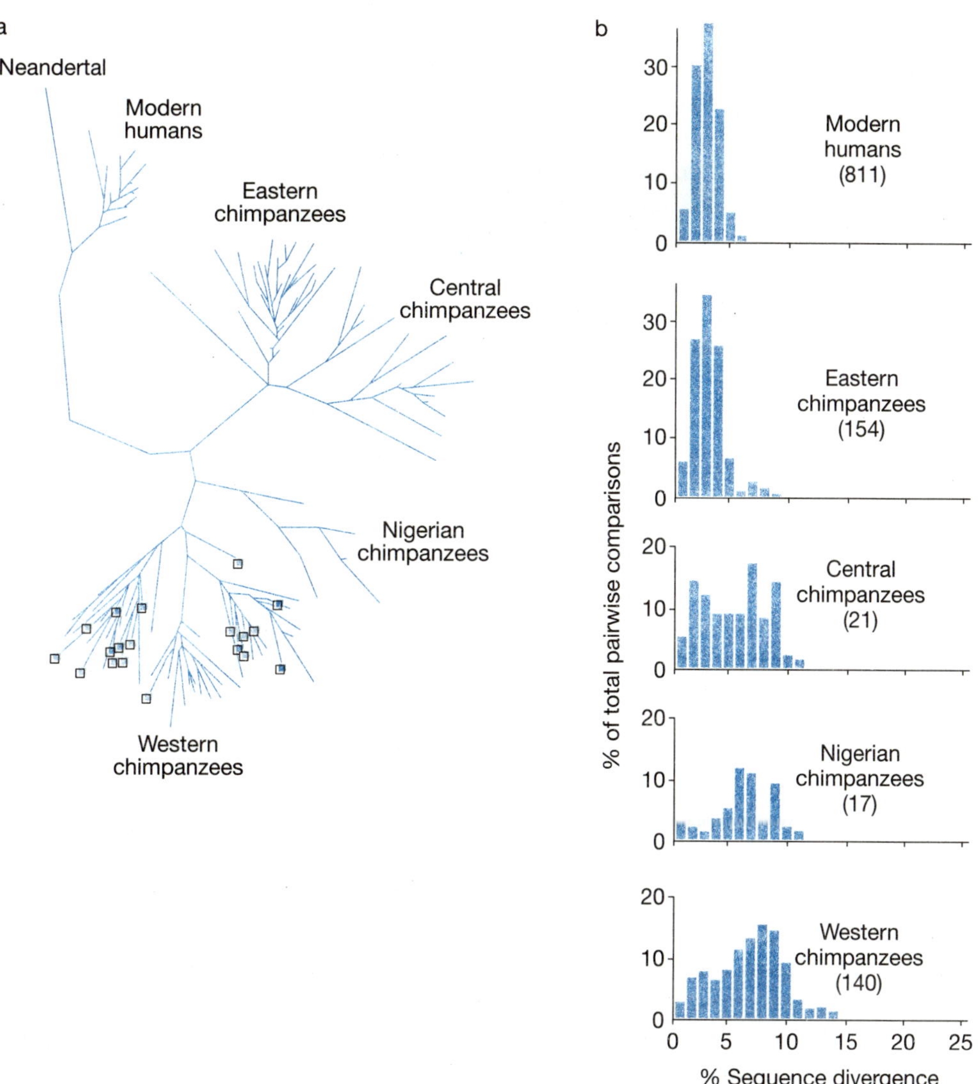

Comparison of Genetic Diversity in Humans and Chimpanzees

(*a*) The family tree compares mitochondrial control region sequences from 811 humans and 332 chimps. Note the extensive branching of chimp groups, especially the western subspecies, indicating a high degree of diversity. Blue boxes indicate chimps from two social groups, each of which exhibits greater diversity than the entire human population. (*b*) The bar graphs show pairwise sequence differences. Note the tight clustering of humans with less than 5% sequence divergence, compared to the broader distribution and greater sequence divergence among central, Nigerian, and western chimps. (Reprinted, with permission, from Robinson R., ed. 2003. *Genetics*. MacMillan Reference USA, New York; Figure created by Dr. Stanley H. Ambrose, Department of Anthropology, University of Illinois, Urbana.)

years. Pollen records suggest that much of Southeast Asia was deforested following the eruption, and significant changes are also recorded in the pollen profile of Grand Pile, in France. Greenland ice cores show that the eruption of Toba was followed by 1000 years of the lowest oxygen isotope ratios of the last glacial period, indicating the lowest temperatures of the last 100,000 years. Thus, it is not difficult to believe that the eruption of Toba produced several years of volcanic winter, followed by 1000 years of unrelenting cold. This surely would have decimated human populations outside of the scattered refuges in tropical Africa.

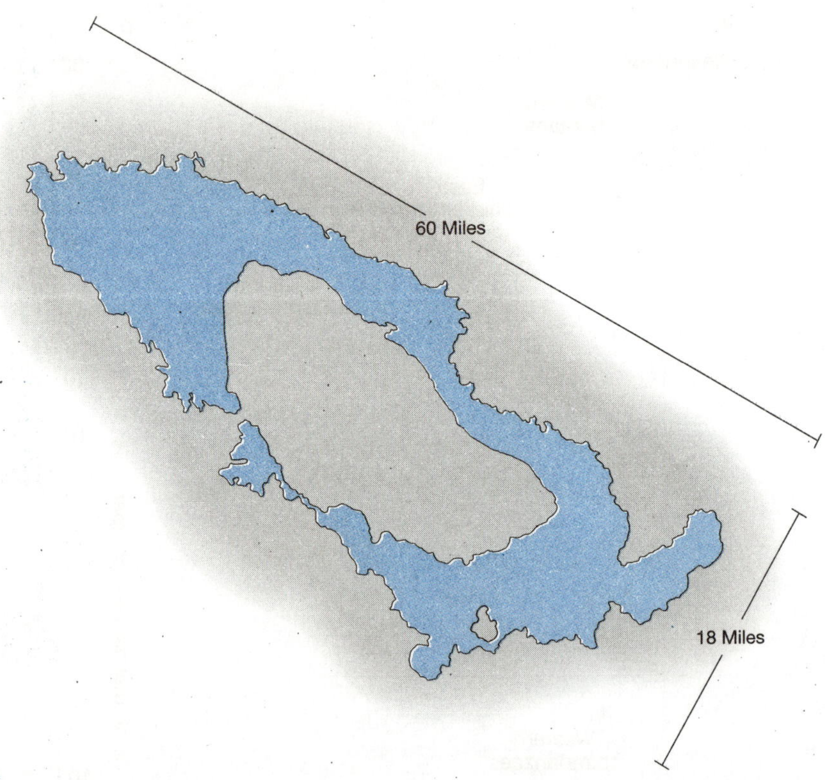

Toba Caldera, Sumatra, Indonesia
(Adapted from the Landsat Pathfinder Project at: http://edcdaac.usgs.gov/pathfinder/pathpage.html.)

The fossil record shows that the modern humans who first reached the Middle East were replaced after the Toba eruption by cold-tolerant Neandertals, illustrating that adaptations are relative to environmental factors. Interestingly, Neandertal seem to have gone through similar population bottlenecks during its 250,000 or so years on earth. Mitochondrial control region sequences have been obtained from two additional Neandertal specimens from Croatia and the Caucasus. Added to the German sample, these represent about half the range of Neandertal. Comparisons of these samples suggest that Neandertal had only about the same (limited) level of diversity as modern human populations, despite the fact that this species existed nearly twice as long as currently has *Homo sapiens*.

The Hunter-Gatherer Remains

Before the advent of RFLP and SNP data, disease and protein polymorphisms of the blood system, including ABO and Rh groups, human leukocyte antigens, and globin variants, provided the only means to study human population variation quantitatively. Using these data, Luigi Luca Cavalli-Sforza, Paolo Menozzi, and Alberto Piazza were among the first to attempt to reconstruct the genetic history of Europe. They identified gradients, or clines, in gene frequencies across Europe using principal component analysis. The first principal component identified a gradient emanating from the Middle East and diminishing through northwestern Europe. They interpreted this east-west cline as genes that origi-

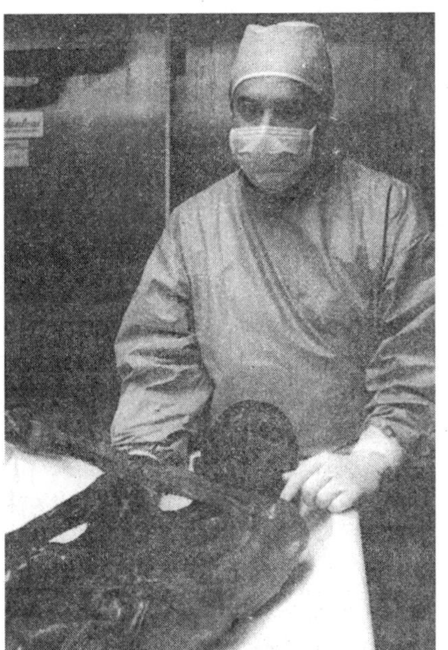

Ötzi the Iceman
(Reprinted, with permission from Schiermeir Q. and Stehle K. 2000. Frozen body offers chance to travel back in time. *Nature 407:* 550.)

nated in the Fertile Crescent and spread westward with agriculture, at a rate of about 1 km per year. Farming arose about 10,000 years ago and roughly marks the boundary between the Paleolithic (Old Stone Age) and Neolithic (New Stone Age). Because the east-west component accounted for the greatest variance across many genes, they concluded that Neolithic genes had essentially replaced Paleolithic genes. In their quest for new agricultural lands, farmers moved inexorably northwest through Europe, mixing with and eventually displacing the hunter-gatherer populations they encountered.

The advent of mtDNA typing of ancient remains and living Europeans challenged Cavalli-Sforza, Menozzi, and Piazza's "demic diffusion" model of the wave-like expansion of Neolithic farmers. The first data came from "Ötzi the Iceman," a 5000-year-old mummy found frozen in the Tyrolean Alps in 1991. An international team obtained mitochondrial control region sequences from tissue samples in 1994. Ötzi's mtDNA type, and others differing from his by only a single nucleotide, proved common among people alive today of European ancestry. Ötzi's DNA looks modern, because he is fully modern. Five thousand years is less than a single mutation from the present, even by the fast mitochondrial clock.

Bryan Sykes of Oxford University extended the human mitochondrial lineage directly back to the Paleolithic, when he analyzed DNA from a human tooth from the Cheddar Gorge in the south of England. Excavated from a limestone cave, the tooth was carbon dated to 12,000 years ago, approximately 6,000 years before farming reached England. Even so, like Ötzi the Iceman, Cheddar Man's identical DNA type and many others differing from his by only a single nucleotide are common among Europeans alive today. Clearly, Cheddar Man was a hunter-gatherer, yet his DNA has survived into the present time. So there could not have been a virtually complete replacement of Paleolithic hunter-gatherers by Neolithic farmers.

Analysis of mtDNA from living Europeans by Sykes and Antonio Torroni (of the University of Rome) identified seven major mitochondrial haplogroups.

"Seven Daughters of Eve"

"Daughters"	Age	Origin	% of Modern Europeans
Ursula	45,000	Greece	11
Xenia	25,000	Southern Russia	6
Helena	20,000	Southern France	46
Velda	17,000	Northern Spain	5
Tara	17,000	Central Italy	9
Katrine	15,000	Northeastern Italy	6
Jasmine	10,000	Middle East	17

The fictitious names given by Bryan Sykes to the founders of the seven major European mitochondrial haplogroups are based on the alphabetic classification system of Antonio Torroni.

Sykes popularized the founders of these lineages as "the seven daughters of Eve." Six of these lineages, representing about 80% of Europeans alive today, are derived from Paleolithic stocks dating back before the advent of agriculture. Only about 20% of European haplotypes are young enough to represent the new genes of Neolithic farmers. This is not far different from the 28% of variance described by Cavalli-Sforza's first principal component.

mtDNA has thus shown an unbroken continuity of inheritance from the Paleolithic hunter-gatherers clear through to suit-clad urban humans alive today. Although agriculture did eventually reach England and the rest of Europe, it came largely on its own. The culture of farming diffused, rather than the farmers themselves. This will almost certainly hold true for the diffusion of agriculture from other ancient centers in Africa, Asia, and the Americas. "Modern" Neolithic farmers did not outpace the Paleolithic hunters, as they had replaced the Neandertals before them. The Paleolithic hunter-gatherers lacked nothing—genetically, physiologically, or behaviorally—that they needed to move into the modern age. These hunter-gatherers became farmers—and they became us.

FOR FURTHER READING

Carlson E.A. 2001. *The unfit: The history of a bad idea.* Cold Spring Harbor Laboratory Press, Cold Spring Harbor, New York.

Cavalli-Sforza L.L., Menozzi P., and Piazza A. 1994. *The history and geography of human genes.* Princeton University Press, Princeton, New Jersey.

Gould S.J. 1985. Carrie Buck's daughter. In *The flamingo's smile: Reflections in natural history,* pp. 306–318. W.W. Norton, New York.

Jablonski N.G. and Chaplin G. 2002. Skin deep. *Sci. Am.* **287:** 74–81.

Kevles D.J. 1995. *In the name of eugenics: Genetics and the uses of human heredity.* Harvard University Press, Cambridge, Massachusetts.

Micklos D., ed. 1999. DNA from the beginning (http://www.dnaftb.org). Dolan DNA Learning Center, Cold Spring Harbor, New York.

Micklos D., ed. 2000. Genetic Origins (http://www.geneticorigins.org). Dolan DNA Learning Center, Cold Spring Harbor, New York.

Micklos D., ed. 2000. *Image archive on the American eugenics movement.* Dolan DNA Learning Center, Cold Spring Harbor, New York. At: http://www.eugenicsarchive.org

Micklos D., ed. 2002. Your Genes/Your Health (http://www.ygyh.org). Dolan DNA Learning Center, Cold Spring Harbor, New York.

Stanley S.M. 1996. *Children of the ice age: How a global catastrophe allowed humans to evolve.* Harmony Books, New York.

Sykes B. 2001. *The seven daughters of Eve: The science that reveals our genetic ancestry.* W.W. Norton, New York.

LABORATORIES

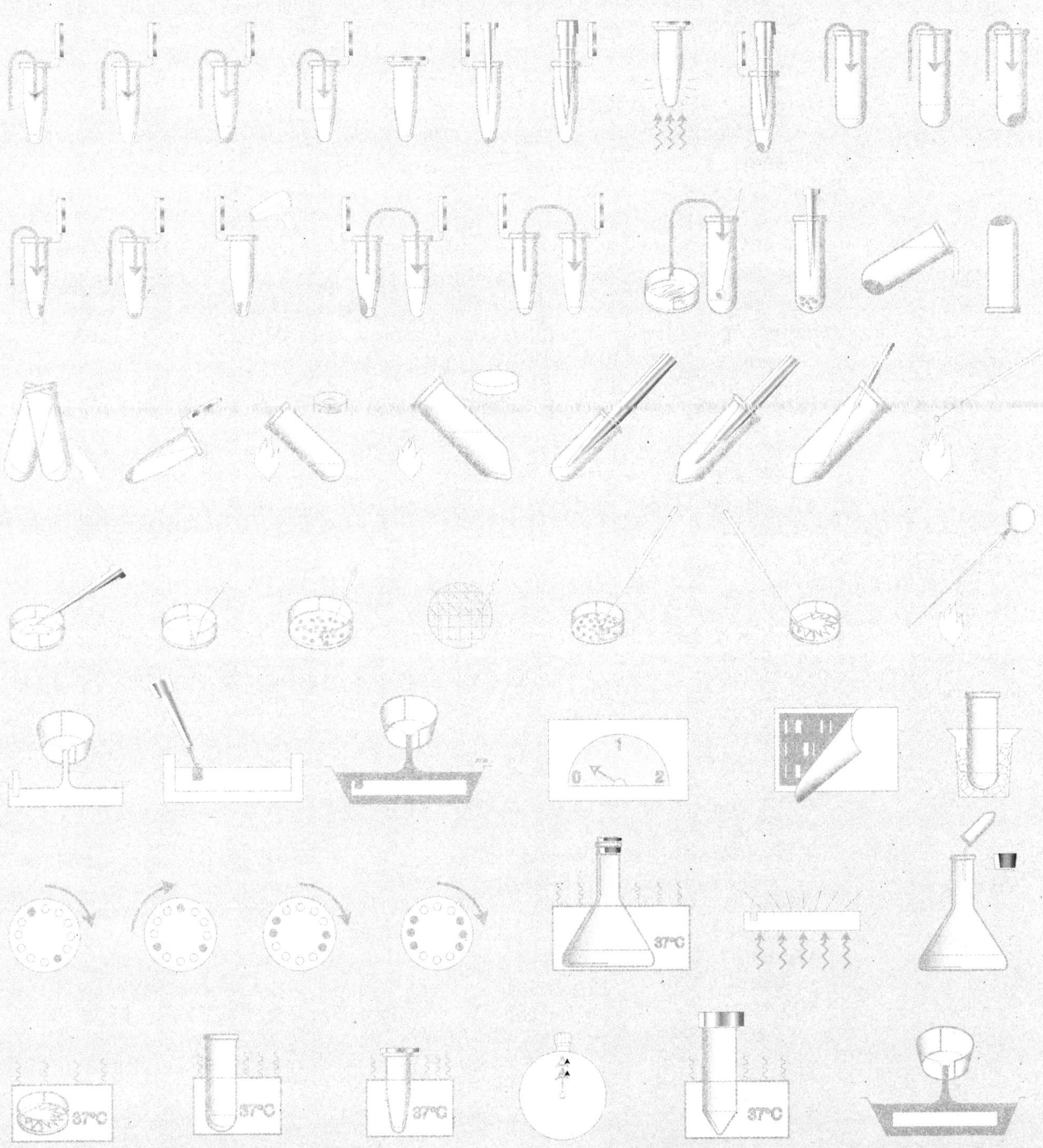

Lab Safety and Adherence to National Institutes of Health Guidelines

WARNING

Individuals should use this manual only in accordance with prudent laboratory safety precautions and under the supervision of a person familiar with such precautions. Use of this manual by unsupervised or improperly supervised individuals could result in serious injury.

Instructors and students are urged to pay particular attention to specific institution guidelines for safety and waste management as well as cautions placed throughout this manual, particularly the instructions contained within caution boxes and in Appendix 4.

LABORATORY WASTE DISPOSAL

Alternative procedures for the handling and disposal of laboratory wastes can and should be used instead of those suggested in this manual when local requirements, conditions, or practices dictate. *Instructors should be familiar with and follow all national, state, local, or institutional regulations or practices pertaining to the use and disposal of materials utilized in this manual.*

WORKING WITH *ESCHERICHIA COLI*

The experiments in this manual make extensive use of the bacterium *Escherichia coli*. Working with *E. coli* should not be a threatening experience, because it is a rather safe organism with which to work. *E. coli* is part of the normal bacterial fauna of the large intestines, and it is rarely associated with serious illness in otherwise healthy individuals. Although one might be concerned about the consequences of accidentally ingesting some *E. coli*, research has shown that strain MM294 used in this laboratory sequence is unlikely to grow inside the human intestines.

MM294 and other *E. coli* strains commonly used in research today are derivatives of wild-type *E. coli* strain K-12, which was originally isolated from the feces of a diphtheria patient at Stanford University in 1922. The work of Edward Tatum in the 1940s popularized its use in biochemical and genetic studies. Research in the 1970s showed that strain K-12 cannot effectively colonize the human gut, probably because of genetic changes accumulated during countless generations over decades of in vitro culture (Bachman 1987). In evaluating biological agents that pose theoretical health risk to humans, the National Institutes of Health (NIH) assigns K-12 *E. coli* to Risk Group 1 (RG1): Agents that are not associated with disease in healthy adult humans (National Insitutes of Health 2001).

The Prelab Notes of Laboratory 2 contain detailed instructions for the responsible handling and disposal of E. coli *which should be followed throughout this laboratory course.*

NIH GUIDELINES

Instructors and students should be aware that the NIH oversees all research involving interspecific DNA transfer. Experiments that entail moving a gene or DNA sequence from one species to another are regulated by the *NIH Guidelines for Research Involving Recombinant DNA Molecules*. However, having overseen recombinant DNA research for more than 20 years, the NIH recognizes that many of these experiments pose such limited risk that they need not be regulated.

The laboratories in this manual, including those suggested in the *For Further Research* section, have been designed so that they fall below the threshold for NIH regulation. Provision III-F-6 of the *Guidelines* exempts experiments from regulation "those that do not present a significant risk to health or the environment." This includes "Experiments which use *Escherichia coli* K-12 host-vector systems...provided that...non-conjugative plasmids shall be used as vectors" (National Insitutes of Health 2001).

Most "modern" plasmids—including pAMP, pKAN, pBLU, and pGREEN—are termed nonconjugative, meaning that they are poorly transferred from one *E. coli* cell to another during conjugation. Efficient transport during conjugation requires a mobility protein (encoded by a plasmid-borne *mob* gene) and a specific site (*nic*) on the plasmid. The Mob protein nicks the plasmid at *nic* and then attaches to the nicked strand to conduct the plasmid through a mating channel into a recipient cell. Modern cloning vectors have been designed that lack both the *mob* gene and the *nic* site and so cannot be mobilized for transport (Lauria and Suit 1987; Sambrook and Russell 2001). This means that, for all intents and purposes, transformation experiments end in the culture dish. It is unlikely that a transformed K-12 *E. coli* would survive very long outside a culture dish. It is equally unlikely that, if accidentally ingested, the transformed *E. coli* would transfer a resistance gene by conjugation to the *E. coli* naturally living in the gut.

STANDARD MICROBIOLOGICAL PRACTICES

The *NIH Guidelines* also specify four levels of physical containment needed "to confine organisms containing recombinant DNA molecules and to reduce the potential for exposure of the laboratory worker, persons outside the laboratory, and the environment to organisms containing recombinant DNA." All of the experiments in this course fall under the lowest level of containment, Biosafety Level 1 (BL1), which includes the following standard microbiological practices:

- Access to the laboratory is limited or restricted at the discretion of the Principal Investigator when experiments are in progress.

- Work surfaces are decontaminated once a day and after any spill of viable material.
- All contaminated liquid or solid wastes are decontaminated before disposal.
- Mechanical pipetting devices are used; mouth pipetting is prohibited.
- Eating, drinking, smoking, and applying cosmetics are not permitted in the work area. Food may be stored in cabinets or refrigerators designated and used for this purpose only.
- Persons wash their hands: (i) after they handle materials involving organisms containing recombinant DNA molecules and animals, and (ii) before exiting the laboratory.
- All procedures are performed carefully to minimize the creation of aerosols.
- In the interest of good personal hygiene, facilities (e.g., hand washing sink, shower, and changing room) and protective clothing (e.g., uniforms and laboratory coats) shall be provided that are appropriate for the risk of exposure to viable organisms containing recombinant DNA molecules.
- Special containment equipment is generally not required for manipulations of agents assigned to BL1.
- The laboratory is designed so that it can be easily cleaned.
- Bench tops are impervious to water and resistant to acids, alkalis, organic solvents, and moderate heat.
- Laboratory furniture is sturdy. Spaces between benches, cabinets, and equipment are accessible for cleaning.
- Each laboratory contains a sink for hand washing.
- If the laboratory has windows that open, they are fitted with fly screens.

USE OF ETHIDIUM BROMIDE

Ethidium bromide is the most rapid and sensitive means to stain DNA. However, it is a mutagen by the Ames microsome assay and a suspected carcinogen. For this reason, instructors may elect (or be required) to use methylene blue—a safe but less-sensitive stain that can be used for all staining procedures in this manual. We have provided directions for both staining methods, and the pros and cons of each are discussed in the *Prelab Notes* of Laboratory 3. *These Notes also contain instructions for the responsible handling and decontamination of ethidium bromide solutions, which should be reviewed prior to performing any staining procedure using ethidium bromide.*

The greatest risk is to inhale ethidium bromide powder when mixing a 5 mg/ml stock solution, and therefore we strongly recommend purchasing a *premixed* 5 mg/ml solution from a supplier. The stock solution should be handled according to the instructions contained in the Material Safety Data Sheet provided by the supplier. The 5 mg/ml stock solution is diluted to make a staining solution with a final concentration of 1 μg/ml according to the procedure described in Appendix 2. With responsible handling, the dilute solution (1 μg/ml) used for gel staining poses minimal risk.

REFERENCES

Bachman B. 1987. Derivations and genotypes of some mutant derivatives of *Escherichia coli* K-12. In Escherichia coli *and* Salmonella typhimurium: *Cellular and molecular biology* (ed. F.C. Neidhardt), vol. 2, p. 1190. American Society for Microbiology, Washington, D.C.

Luria S.E. and Suit J.L. 1987. Colicins and Col plasmids. In Escherichia coli *and* Salmonella typhimurium: *Cellular and molecular biology* (ed. F.C. Neidhardt), vol. 2, pp. 1620–1621. American Society for Microbiology, Washington, D.C.

National Institutes of Health. 2001. Guidelines for research involving recombinant DNA molecules. *Federal Register:* January 5, 2001 (at: http://www4.od.nih.gov/oba/rac/guidelines/guidelines.html).

Sambrook J. and Russell D. 1989. *Molecular cloning: A laboratory manual,* 3rd edition, pp. A8.27–A8.28. Cold Spring Harbor Laboratory Press, Cold Spring Harbor, New York.

Measurements, Micropipetting, and Sterile Techniques

LABORATORY 1 INTRODUCES MICROPIPETTING AND STERILE PIPETTING techniques used throughout this course. Mastery of these techniques is important for good results in all of the experiments that follow. Most of the laboratories are based on *microchemical* protocols that use very small volumes of DNA and reagents. These require use of an adjustable micropipettor (or microcapillary pipette) that measures as little as one microliter (μl)—a millionth of a liter.

Many experiments require growing *Escherichia coli* in a culture medium that provides an ideal environment for other microorganisms as well. Therefore, it is important to maintain sterile conditions to minimize the chance of contaminating an experiment with foreign bacteria or fungi. *Sterile conditions* must be maintained whenever living bacterial cells are to be used in further cultures. Use sterilized materials for everything that comes in contact with a bacterial culture: nutrient media, solutions, pipettes, micropipettor tips, inoculating and spreading loops, flasks, culture tubes, and plates.

Remember this rule of thumb: Use sterile technique if live bacteria are needed at the end of a manipulation (general culturing and transformations). Sterile technique is not necessary when the bacteria are destroyed by the manipulations in the experiment or when working with solutions for DNA analysis (plasmid isolation, DNA restriction, and DNA ligation).

Equipment and materials for this laboratory are available from the Carolina Biological Supply Company (see Appendix 1).

I. Small-volume Micropipettor Exercise

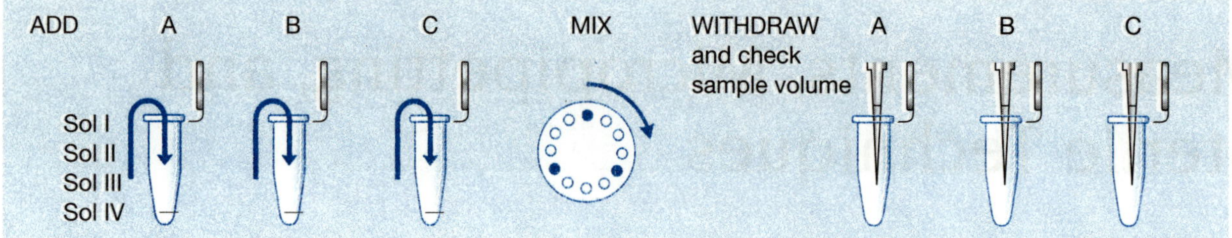

II. Large-volume Micropipettor Exercise

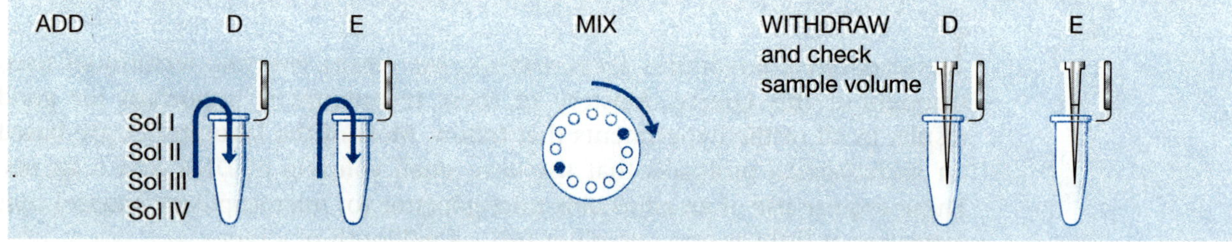

III. Sterile Use of 10-ml Standard Pipette

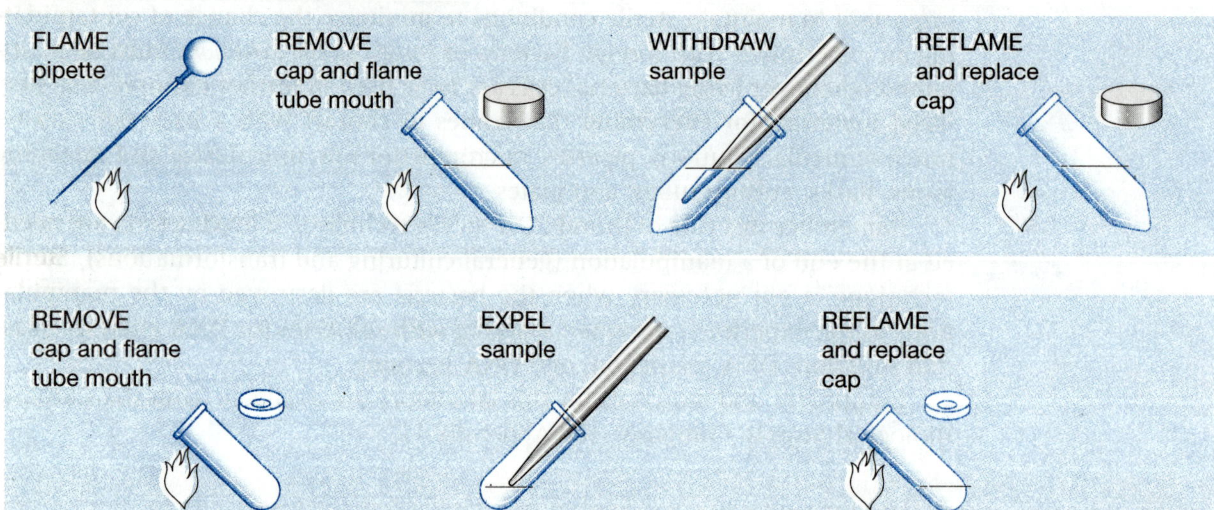

PRELAB NOTES

Digital Micropipettors

The volume range of digital micropipettors varies from manufacturer to manufacturer. Select both a small-volume micropipettor with a range of 0.5–10 μl or 1–20 μl and a large-volume micropipettor with a range of 100–1000 μl.

Microcapillary Pipettes

Microcapillary pipettes are an inexpensive alternative to adjustable micropipettors. These disposable glass capillary tubes come in sizes that cover the range of volumes used in this course. Several types of inexpensive micropipette aids are available. A thumbscrew micrometer may be easier to use than a pipettor bulb. A no-cost and easily controllable pipette bulb can be made by tying a knot in a length of latex tubing, which is usually provided with the capillary pipettes.

Under conditions of high static electricity, capillary pipetting can be very difficult and, at times, impossible. The reagent droplet adheres stubbornly to the side of the pipette and cannot be transferred to the side of a polypropylene reaction tube. Even under the best of circumstances, microcapillary pipettes are more difficult to master. Allow students sufficient time to become competent with them before attempting any experiments.

Transfer Pipettes

Small polypropylene transfer pipettes are handy because they have an integrated bulb. The smallest size, which holds a *total* volume of approximately 1 ml, has a thin tip that can be used to measure microliter amounts. Before use, calibrate the transfer pipette using a digital micropipettor or microcapillary pipette. Pressing on the pipette barrel, rather than the bulb, creates less air displacement and makes measuring small volumes easier.

10-ml Pipettes

Presterilized, disposable 10-ml plastic pipettes are most convenient and are supplied in bulk pack or individually wrapped. Bulk-packed pipettes should be opened immediately before use. To dispense, cut one corner of the plastic wrapper at the end opposite the pipette tips. Avoid touching and contaminating the wrapper opening; tap bag to push the pipette end through the cut opening. Reclose with tape to keep sterile for future use. To use individually wrapped pipettes properly, peel back only enough of the wrapper to expose the wide end of the pipette and affix the end into the pipette aid or bulb. Completely peel back the wrapper immediately before use.

To Flame or Not to Flame?

There is general disagreement about whether it is necessary to flame pipettes and mouths of tubes as part of the sterile technique. Flaming warms the air at the mouth of the container, creating an outward convection current that prevents microorganisms from falling in. Even so, the effect of flaming may be primarily psychological when fresh sterile supplies are used and manipulations are done quickly. Especially when using individually wrapped supplies, flaming can be omitted without compromise to sterility. When flaming plasticware, do so briefly to avoid melting the plastic.

Microfuge

Although not essential, a microfuge is very useful for pooling and mixing droplets of pipetted solutions in the bottom of a 1.5-ml reaction tube.

PRELAB PREPARATION

1. To simplify initial practice with a micropipettor, use colored solutions that are easily visible. Prepare five colored solutions using food coloring or other dyes mixed with water.
2. Prepare for each experiment:
   - Four 1.5-ml tubes, each containing 1 ml of a different colored solution, marked I, II, III, and IV.
   - One 50-ml conical tube containing 25 ml of colored solution, marked V.

MATERIALS

REAGENTS	SUPPLIES AND EQUIPMENT
Solution I (1 ml), colored	Beaker for waste/used tips
Solution II (1 ml), colored	Bunsen burner (optional)
Solution III (1 ml), colored	Conical tube (50-ml)
Solution IV (1 ml), colored	Culture tube (15-ml)
Solution V (25 ml), colored	Microfuge (optional)
	Micropipettor (0.5–10-µl) + tips
	Micropipettor (100–1000-µl) + tips
	Permanent marker
	Pipette (10-ml)
	Pipette aid or bulb
	Test tube rack
	Tubes (1.5-ml)

METHODS

Metric Conversions

Become familiar with metric units of measurement and their conversions. We concentrate on liquid measurements based on the liter, but the same prefixes also apply to dry measurements based on the gram. The two most useful units of liquid measurement in molecular biology are the milliliter (ml) and microliter (µl).

1 ml	=	0.001 liter	1,000 ml	=	1 liter
1 µl	=	0.000001 liter	1,000,000 µl	=	1 liter

Complete the following conversions:

1 µl	=	_______	ml	_______ µl	=	1 ml
10 µl	=	_______	ml			
100 µl	=	_______	ml			

Use of Digital Micropipettors (10 minutes)

"Nevers"

- Never rotate volume adjustor beyond the upper or lower range of the micropipettor, as stated by the manufacturer.
- Never use a micropipettor without the tip in place; this could ruin the precision piston that measures the volume of fluid.
- Never lay down a micropipettor with a filled tip; fluid could run back into the piston.
- Never let the plunger snap back after withdrawing or ejecting fluid; this could damage the piston.
- Never immerse the barrel of the micropipettor in fluid.
- Never flame the micropipettor tip.

Micropipetting Directions

1. Rotate the volume adjustor to the desired setting. Note the change in the plunger length as the volume is changed. Be sure to locate the decimal point properly when reading volume setting.
2. *Firmly* seat proper-sized tip on the end of the micropipettor.
3. When withdrawing or expelling fluid, always hold the tube firmly between thumb and forefinger. Hold the tube at nearly eye level to observe the change in the fluid level in the pipette tip. Do not pipette with the tube in the test tube rack or have another person hold the tube while pipetting.
4. Each tube must be held in the hand during each manipulation. Grasping the tube body, rather than the lid, provides more control and avoids contamination from the hands.
5. Hold micropipettor almost vertical when filling.

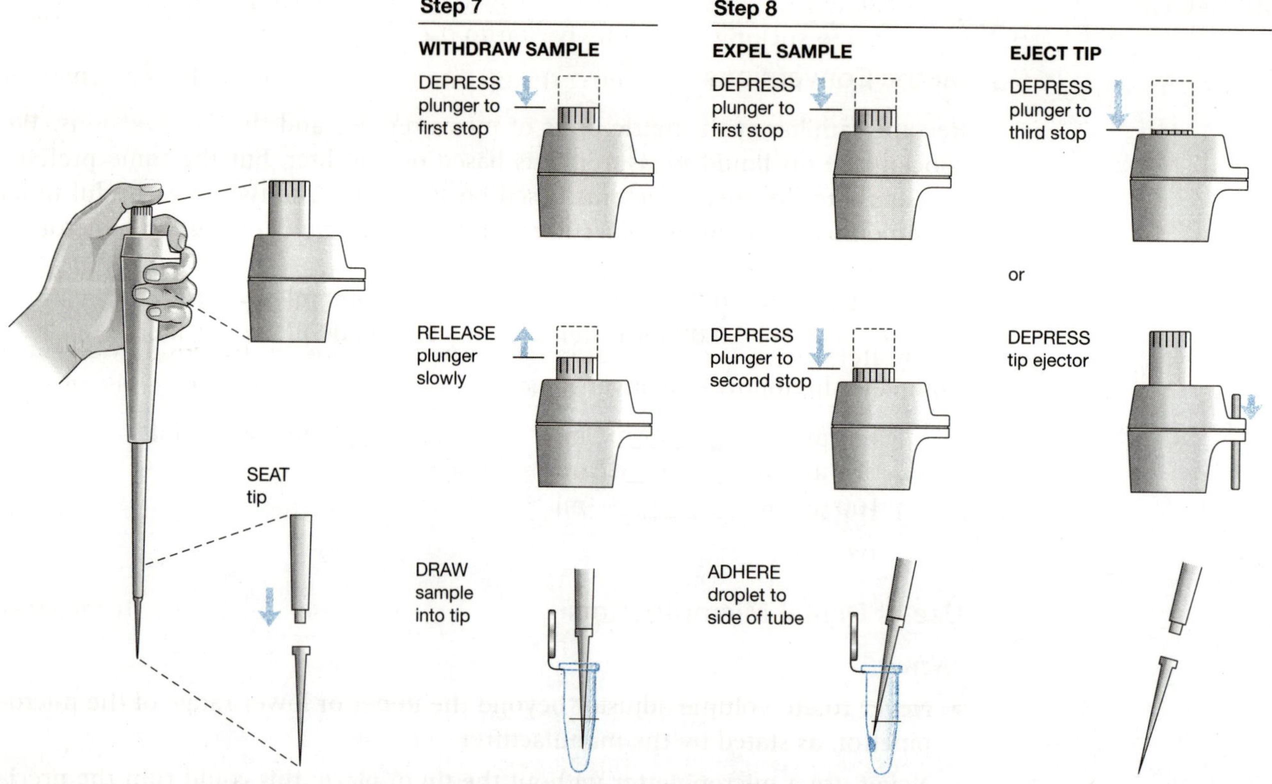

Use of Digital Micropipettor (Steps 7 and 8)

6. Most digital micropipettors have a two-position plunger with friction "stops." Depressing to the first stop measures the desired volume. Depressing to the second stop introduces an additional volume of air to blow out any solution remaining in the tip. Pay attention to these friction stops, which can be felt with the thumb.
7. To withdraw sample from reagent tube:
   a. Depress the plunger *to the first stop* and hold in this position. Dip the tip into the solution to be pipetted, and draw the fluid into the tip by *gradually* releasing the plunger.
   b. Slide the tip out along the inside wall of the reagent tube to dislodge excess droplets adhering to the outside of the tip.
   c. Check that there is no air space at the very end of the tip. To avoid future pipetting errors, learn to recognize the approximate level that particular volumes reach in the tip.
8. To expel sample into reaction tube:
   a. Touch the tip to the inside wall of the reaction tube into which the sample will be emptied. This creates a capillary effect that helps draw fluid out of the tip.
   b. *Slowly* depress the plunger to the first stop to expel sample. Depress to the second stop to blow out the last bit of fluid. Hold plunger in depressed position.

c. Slide the pipette out of the reagent tube with the plunger depressed to avoid sucking any liquid back into the tip.

d. Manually remove or eject the tip into a beaker kept on the lab bench for this purpose. The tip is ejected by depressing the measurement plunger beyond the second stop or by depressing a separate tip-ejection button, depending on the particular micropipettor being used.

9. To prevent cross-contamination of reagents:

   a. Always add appropriate amounts of a single reagent sequentially to all reaction tubes.

   b. Release each reagent drop onto a new location on the inside wall, near the bottom of the reaction tube. In this way, the same tip can be used to pipette the reagent into each reaction tube.

   c. Use a *fresh tip* for each new reagent to be pipetted.

   d. If the tip touches one of the other reagents in the tube, switch to a new tip.

10. Eject used tips into a beaker kept on the lab bench for this purpose.

I. Small-volume Micropipettor Exercise (15 minutes)

This exercise simulates setting up a reaction, using a micropipettor with a range of 0.5–10 μl or 1–20 μl.

1. Use a permanent marker to label three 1.5-ml tubes A, B, and C.
2. Use the matrix below as a checklist while adding solutions to each reaction tube.

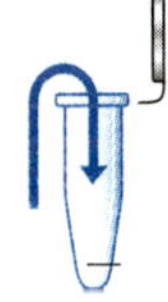

Tube	Sol. I	Sol. II	Sol. III	Sol. IV
A	4 μl	5 μl	1 μl	–
B	4 μl	5 μl	–	1 μl
C	4 μl	4 μl	1 μl	1 μl

3. Set the micropipettor to 4 μl and add Solution I to each reaction tube.
4. Use a *fresh tip* to add appropriate volume of Solution II to a clean spot on reaction Tubes A, B, and C.
5. Use a *fresh tip* to add 1 μl of Solution III to Tubes A and C.
6. Use a *fresh tip* to add 1 μl of Solution IV to Tubes B and C.
7. Close tops. Pool and mix reagents by using one of the following methods:

   a. Sharply tap the tube bottom on the bench top. Make sure that the drops have pooled into one drop at the bottom of the tube.

   or

   b. Place the tubes in a microfuge and apply a short, few-second pulse. Make sure that the reaction tubes are placed in a *balanced* configuration in the microfuge rotor. Spinning tubes in an unbalanced position will damage the microfuge motor.

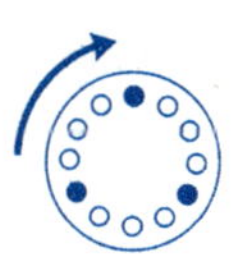

An empty 1.5-ml tube can be used to balance a sample with a volume of 20 μl or less.

8. A total of 10 μl of reagents was added to each reaction tube. To check that the measurements were accurate, set the pipette to 10 μl and very carefully withdraw solution from each tube.
   a. Is the tip just filled?

   or

   b. Is a small volume of fluid left in tube?

   or

   c. After extracting all fluid, is an air space left in the tip end? (The air can be displaced and actual volume determined simply by rotating volume adjustment to push fluid to very end of tip. Then, read the volume directly.)
9. If several measurements were inaccurate, repeat the exercise to obtain a near-perfect result.

II. Large-volume Micropipettor Exercise (10 minutes)

This exercise simulates a bacterial transformation or plasmid preparation, for which a 100–1000-μl micropipettor is used. It is far easier to mismeasure when using a large-volume micropipettor. If the plunger is not released slowly, an air bubble may form or solution may be drawn into piston.

1. Use a permanent marker to label two 1.5-ml reaction tubes D and E.

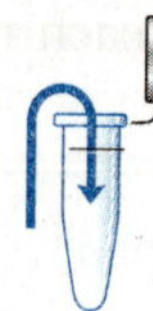

2. Use the matrix below as a checklist while adding solutions to each reaction tube.

Tube	Sol. I	Sol. II	Sol. III	Sol. IV
D	100 μl	200 μl	150 μl	550 μl
E	150 μl	250 μl	350 μl	250 μl

3. Set the micropipettor to add appropriate volumes of Solutions I–IV to reaction tubes D and E. Follow the same procedure as for the small-volume micropipettor.

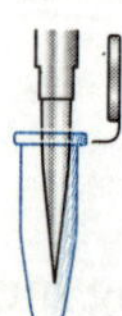

4. A total of 1000 μl of reactants was added to each tube. To check that the measurements were accurate, set the micropipettor to 1000 μl and carefully withdraw solution from each tube.
   a. Is the tip just filled?

   or

   b. Is a small volume of fluid left in tube?

   or

   c. After extracting all fluid, is an air space left in the tip end? (The air can be displaced and actual volume determined simply by rotating volume adjustment to push fluid to very end of tip. Then, read the volume directly.)
5. If the measurements were inaccurate, repeat the exercise to obtain a near-perfect result.

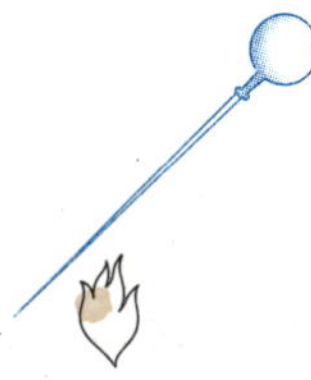

III. Sterile Use of 10-ml Standard Pipette (10 minutes)

The following directions include flaming the pipette and tube mouth. It is probably best to learn to flame, then omit flaming when safety or situation dictates. The directions also assume one-person pipetting, which is rather difficult. The process is much easier when working as a team: One person handles the pipette, while the other removes and replaces the caps of the tubes.

The key to successful sterile technique is to work quickly and efficiently. Before beginning, clear off the lab bench and arrange tubes, pipettes, and culture medium within easy reach. Locate Bunsen burner in a central position on the lab bench to avoid reaching over the flame.

Loosen caps so that they are ready for easy removal. Remember, the longer the top is off the tube, the greater the chance of microbe contamination. Do not place a sterile cap on a nonsterile lab bench.

> CAUTION
> Always use a pipette aid or bulb to draw solutions up the pipette. Never pipette solutions using mouth suction: This method is not sterile and can be dangerous.

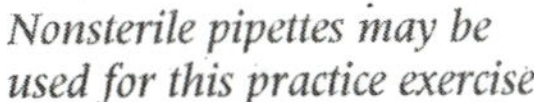

Nonsterile pipettes may be used for this practice exercise.

1. Light Bunsen burner.

This expels contaminated air and prepares vacuum to withdraw fluid.

2. Set pipette aid to 5 ml or depress pipette bulb if using a pipette.
3. Select a sterile 10-ml pipette and insert into pipette aid or bulb. *Remember to handle only the large end of the pipette; avoid touching the lower two thirds.*

When using an individually wrapped pipette, be careful to open wrapper end opposite the pipette tip. Unwrap only enough of the pipette to attach end into pipette aid or bulb.

4. Quickly pass the lower two thirds of the pipette cylinder through the Bunsen flame several times. *Be sure to flame any portion of the pipette that will enter the sterile container.* Pipette should become warm, but not hot enough to melt the plastic pipette or to cause the glass pipette to crack when immersed in solution to be pipetted.
5. Hold a 50-ml conical tube containing Solution V in free hand and remove cap using little finger of hand holding pipette aid or bulb. *Do not place cap on lab bench.*

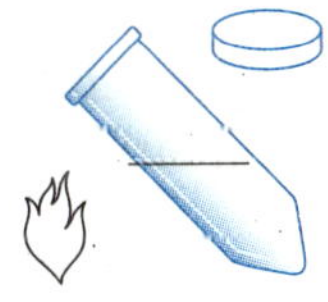

6. Quickly pass mouth of conical tube through the Bunsen flame several times, being careful not to melt the plastic.
7. Withdraw 5 ml of Solution V from the conical tube.
8. Reflame the mouth of the tube and replace top.
9. Remove the top of a sterile 15-ml culture tube with little finger of hand holding the pipette. Quickly flame the mouth of the tube.

10. Expel fluid into the culture tube. Reflame the mouth of the tube and replace top.

RESULTS AND DISCUSSION

Inaccurate pipetting and improper sterile technique are the chief contributors to poor laboratory results. If the handling of micropipettors and/or the use of sterile technique are still uncomfortable or difficult, take time now for additional practice. These techiques will soon become second nature.

1. Why must tubes be balanced in a microfuge rotor?
2. Use the rotor diagrams below to show how to balance 3–11 tubes (for the 12-place rotor) or 3–15 tubes (for the 16-place rotor). When balancing an odd numbers of tubes, begin with a balanced triangle of three tubes, and then add balanced pairs. Which number of tubes cannot be balanced in the 12- or 16-place rotors?

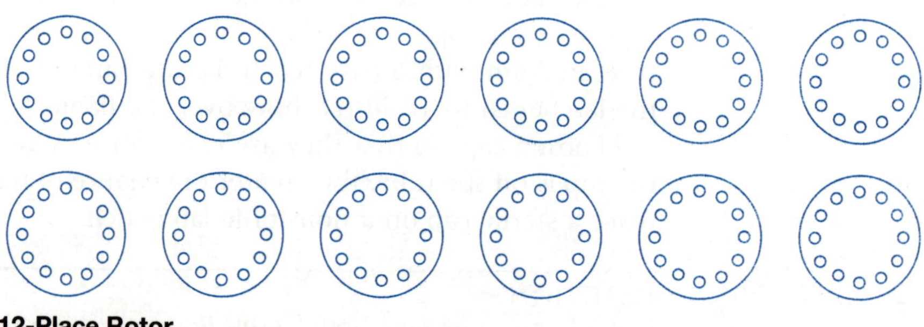

12-Place Rotor

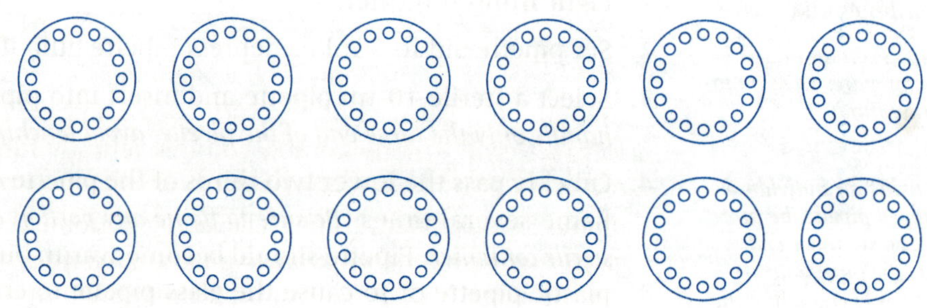

16-Place Rotor

3. What common error in handling a micropipettor can account for pipetting too much reagent into a tube? What errors account for underpipetting?
4. When is it necessary to use sterile technique?
5. What does flaming accomplish?
6. Convert the following into microliters (μl):

   0.130 ml 0.025 ml
   0.002 ml 1.034 ml

   Convert the following into milliliters (ml):

   0.036 liter 0.803 liter
   345 μl 1345 μl

FOR FURTHER RESEARCH

Devise a method to determine the percentage of error in micropipetting. Then determine the percentage of error when purposely make pipetting mistakes with small- and large-volume micropipettors.

LABORATORY 2

Bacterial Culture Techniques

LABORATORY 2 CONTAINS MOST OF THE CULTURE TECHNIQUES used throughout the course. We suggest that Part A (Isolation of Individual Colonies) be done in sequence between Laboratory 1 and Laboratory 3. Part B (Overnight Suspension Culture) need be done only in conjunction with plasmid purification in Laboratories 8 and 12. Part C (Mid-log Suspension Culture) is done in preparation for making competent cells by the standard calcium chloride procedure in Laboratory 10.

- Part A provides a technique for streaking *E. coli* cells onto LB agar plates such that single cells are isolated from one another. Each cell then reproduces to form a visible colony composed of genetically identical clones. Streaking cells to obtain individual colonies is usually the first step in genetic manipulations of microorganisms. Using cells derived from a single colony minimizes the chance of using a cell mass contaminated with a foreign microorganism. To demonstrate antibiotic resistance, the growth of wild-type *E. coli* and that of an *E. coli* containing an ampicillin resistance gene are compared, using LB medium containing ampicillin. The resistant strain contains the plasmid pAMP, which produces an enzyme that destroys the ampicillin in the medium, thus allowing these cells to grow.
- Part B provides a protocol for growing small-scale suspension cultures of *E. coli* that reach stationary phase with overnight incubation. Overnight cultures are used for purification of plasmid DNA and for inoculating mid-log cultures. When growing *E. coli* strains that contain a plasmid, it is best to maintain selection for antibiotic resistance by growing in LB broth containing the appropriate antibiotic. Strains containing an ampicillin resistance gene (such as pAMP) should be grown in LB broth plus ampicillin.
- Part C provides a protocol for preparing a mid-log culture of *E. coli*. Cells in mid-log growth can generally be rendered more competent to uptake plasmid DNA than can cells at stationary phase. Mid-log cells are used in the classic transformation protocol described in Laboratory 10. The protocol begins with an overnight suspension culture of *E. coli*. Incubation with agitation has brought the culture to stationary phase and ensures a large number of healthy cells capable of further reproduction. The object is to subculture (reculture) a small volume of the overnight culture in a large volume of fresh nutrient broth. This "re-sets" the culture to zero growth, where after a short

lag phase, the cells enter the log-growth phase. As a general rule, 1 volume of overnight culture (the *inoculum*) is added to 100 volumes of fresh LB broth in an Erlenmeyer flask. To provide good aeration for bacterial growth, the flask volume should be at least four times the total culture volume.

Equipment and materials for this laboratory are available from the Carolina Biological Supply Company (see Appendix 1).

PART A

Isolation of Individual Colonies

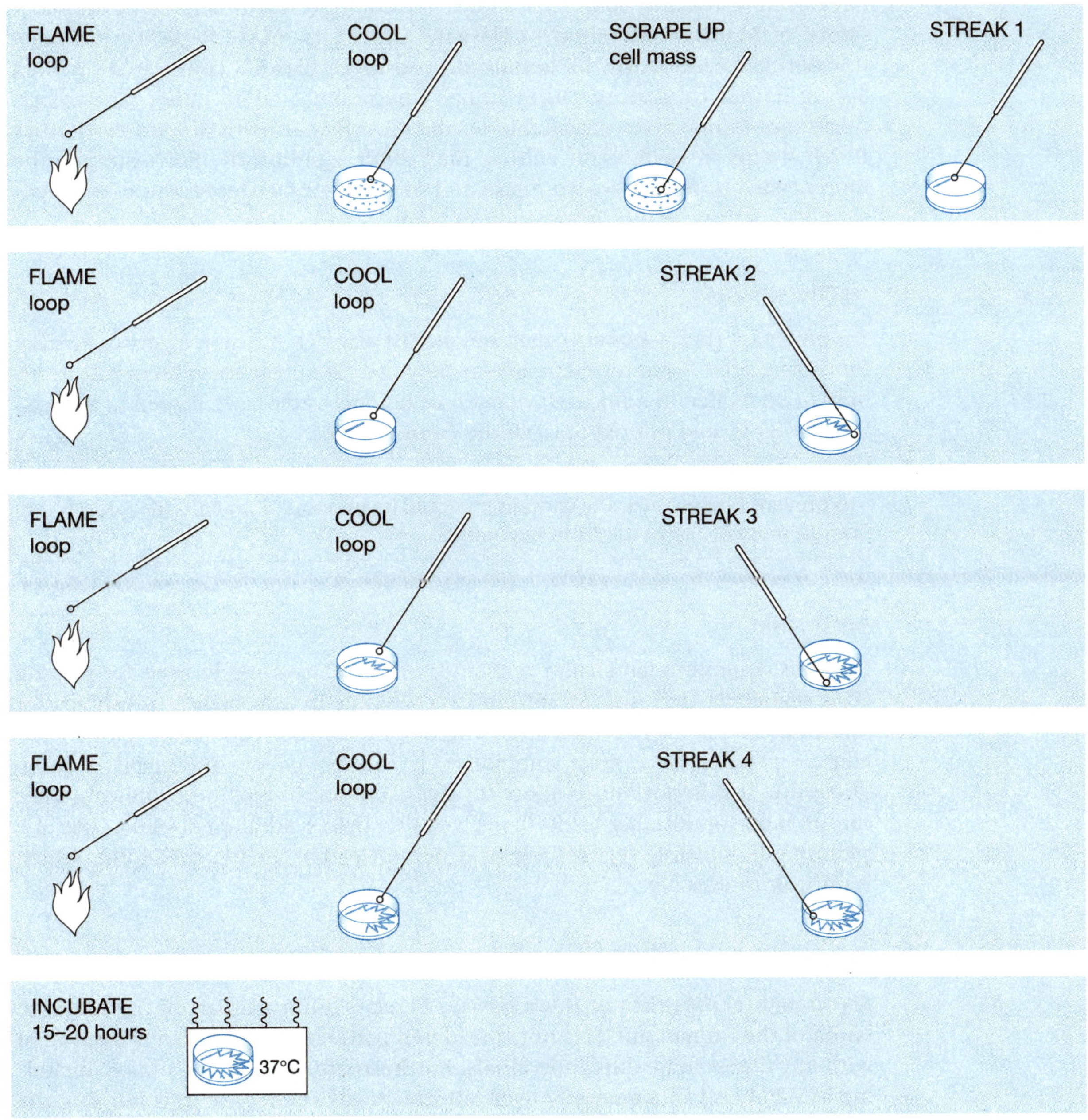

PRELAB NOTES

E. coli Strains

All protocols involving bacterial growth, transformation, and plasmid isolation have been tested and optimized with *E. coli* strain MM294, derived in the laboratory of Matthew Meselson at Harvard University. MM294/pAMP has been transformed with pAMP, an amplicillin resistance plasmid constructed at Cold Spring Harbor Laboratory. Other strains commonly used for molecular biological studies should give comparable results. However, growth properties of other *E. coli* strains in suspension culture may differ significantly. For example, the time needed to reach mid-log phase and the cell number represented by specific optical densities differ from strain to strain.

Nutrient Agar

We prefer LB (Luria-Bertani) agar, but almost any rich nutrient agar can be used for plating cells. Presterilized, ready-to-pour agar is a great convenience. It needs only to be melted in a microwave oven or boiling water bath, cooled to approximately 60°C, and poured onto sterile culture plates.

> CAUTION
> To prevent boiling over, the container should be no more than half full. Loosen the cap to prevent the bottle from exploding.

Ampicillin

Plasmids containing ampicillin resistance are most commonly used for cloning DNA sequences in *E. coli.* Ampicillin is very stable in agar plates, thresholds for selection are relatively broad, and contaminants are infrequent. Despite its stability, ampicillin, like most antibiotics, is inactivated by prolonged heating. Therefore, it is important to allow the agar solution to cool until the container can be held comfortably in the hand (~60°C) before adding antibiotic. Use the sodium salt, which is very soluble in water, instead of the free acid form, which is difficult to dissolve.

Responsible Handling and Disposal of *E. coli*

A commensal organism of *Homo sapiens, E. coli* is a normal part of the bacterial fauna of the human gut. It is not considered pathogenic and is rarely associated with any illness in healthy individuals. Furthermore, K-12 *E. coli* strains, including MM294 and all commonly used lab strains, are ineffective in colonizing the human gut. Adherence to simple guidelines for handling and disposal makes work with *E. coli* a nonthreatening experience.

1. To avoid contamination, always reflame inoculating loop or cell spreader one final time before placing it on the lab bench.
2. Keep nose and mouth away from tip end when pipetting suspension culture to avoid inhaling any aerosol that might be created.

3. Do not over-incubate plates. Because a large number of cells are inoculated, *E. coli* is generally the only organism that will appear on plates incubated for 15–20 hours. However, with longer incubation, contaminating bacteria and slower-growing fungi can arise. If plates cannot be observed following initial incubation, refrigerate them to retard growth of contaminants.
4. Collect for treatment bacterial cultures *and* tubes, pipettes, and micropipettor tips that have come into contact with the cultures. Disinfect these materials as soon as possible after use. Contaminants, often odorous and sometimes potentially pathogenic, are readily cultured over a period of several days at room temperature. Disinfect bacteria-contaminated materials in one of two ways:
   a. Autoclave materials for 15 minutes at 121°C. Tape three to four culture plates together and loosen tube caps before autoclaving. Collect contaminated materials in a "bio bag" or heavy-gauge trash bag; seal bag before autoclaving. Dispose of autoclaved materials in accordance with local regulations.

   or

   b. Treat with solution containing 5000 parts per million (ppm) available chlorine (10% bleach solution). Immerse contaminated pipettes, tips, and tubes (open) directly into sink or tub containing bleach solution. Plates should be placed, with lids open, in sink or tub, and flooded with bleach solution. Allow materials to stand in bleach solution for 15 minutes or more. Then drain excess bleach solution, seal materials in plastic bag, and dispose in accordance with local regulations.
5. Wipe down lab bench with soapy water, 10% bleach solution, or disinfectant (such as Lysol) at the end of lab.
6. Wash hands before leaving lab.

PRELAB PREPARATION

Before performing this Prelab Preparation, please refer to the cautions indicated on the Laboratory Materials list.

1. About 1 day to 1 week before class, MM294 must be streaked onto several LB agar plates and MM294/pAMP must be streaked onto several LB+ampicillin (LB/amp) plates. Following overnight incubation at 37°C, wrap the plates in Parafilm or plastic wrap to prevent drying and store at 4°C (in a refrigerator) until they are needed. Alternately, streak directly from stab or slant cultures.
2. Prepare for each experiment:

   2 LB agar plates
   2 LB/amp plates
3. Make sure that the plate type is clearly marked on the bottom of the plate (not on the lid).
4. Prewarm incubator to 37°C.

MATERIALS

CULTURE AND PLATES

LB agar (LB) plates (2)
LB+ampicillin▼ (LB/amp) plates (2)
MM294 culture
MM294/pAMP culture

SUPPLIES AND EQUIPMENT

"Bio bag" or heavy-duty trash bag
Bleach (10%)▼ or disinfectant
Bunsen burner
Incubator (37°C)
Inoculating loop
Permanent marker

▼ See Appendix 4 for Cautions list.

METHODS

Plate-streaking Technique

(15 minutes)

As with sterile pipetting, plan out manipulations before beginning to streak plates. Organize the lab bench to allow plenty of room and work quickly. If working from a stab or slant culture, loosen the cap before starting.

1. Use a permanent marker to label the *bottom* of each agar plate with your name and the date. Each plate will have been previously marked to indicate whether it is plain LB agar (LB) or LB agar+ampicillin (LB/amp).
2. Select the two LB plates. Mark one plate –pAMP for cells without plasmid and the other plate +pAMP for cells with plasmid.
3. Select the two LB/amp plates. Mark one plate –pAMP for cells without plasmid and the other plate +pAMP for cells with plasmid.
4. Hold the inoculating loop like a pencil, and sterilize the loop in the Bunsen burner flame until it glows red hot. Then continue to pass lower half of shaft through flame.
5. Cool for 5 seconds. *To avoid contamination, do not place inoculating loop on lab bench.*
6. Use one of the techniques below to scrape up *E. coli*.

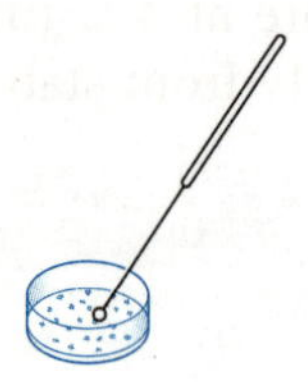

If working from culture plate:

a. Remove the lid from the *E. coli* culture plate with free hand. *Do not place lid on lab bench.* Hold the lid face down just above the culture plate to help prevent contaminants from falling on the plate or lid.
b. Stab inoculating loop into a clear area of the agar several times to cool.
c. Use the loop tip to scrape up a visible cell mass from a colony. Do not gouge agar. Replace culture plate lid, and proceed to Step 7.

or

If working from stab culture:

a. Grasp the bottom of the *E. coli* culture vial between thumb and two fingers of free hand. Remove vial cap using little finger of same hand that holds inoculating loop. *Avoid touching rim of cap.*

b. Quickly pass mouth of vial several times through burner flame.

c. Stab inoculating loop into side of agar several times to cool.

d. Scrape the loop several times across area of culture where bacterial growth is apparent. Remove loop, flame vial mouth, and replace cap. Proceed to Step 7.

7. Select LB –pAMP plate and lift lid only enough to perform streaking as shown below. *Do not place top on lab bench.*

   - *Streak 1:* Glide inoculating loop tip back and forth across the agar surface to make a streak across the top of the plate. Avoid gouging agar. Replace lid of plate between streaks.

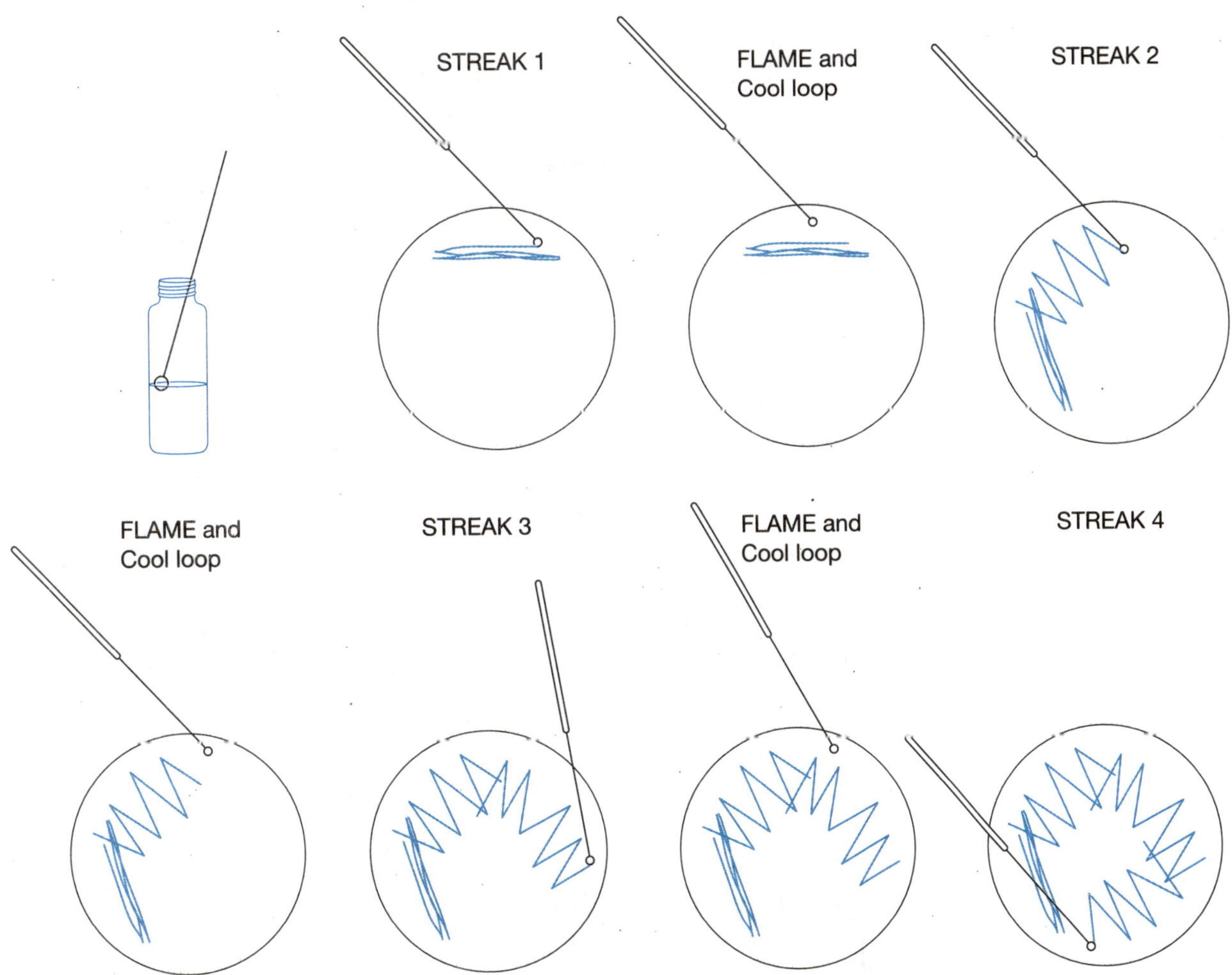

Streaking to Isolate Individual Colonies (Step 7)

- *Streak 2:* Reflame inoculating loop and *cool* by stabbing it into the agar away from the first (primary) streak. Draw loop tip through the end of the primary streak and, without lifting loop, make a zigzag streak across one quarter of the agar surface. *Replace plate lid.*
- *Streak 3:* Reflame loop and *cool* in the agar as above. Draw loop tip through the end of the secondary streak, and make another zigzag streak in the adjacent quarter of the plate *without touching the previous streak.*
- *Streak 4:* Reflame loop and *cool* it as above. Draw tip through the end of the tertiary streak, and make a final zigzag streak in remaining quarter of plate.

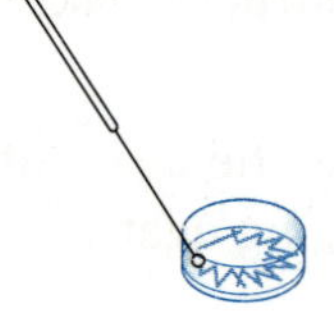

8. Repeat Steps 4–7 to streak *E. coli* onto LB/amp −pAMP plate.
9. Repeat Steps 4–7 to streak *E. coli*/pAMP onto LB +pAMP plate.
10. Repeat Steps 4–7 to streak *E. coli*/pAMP onto LB/amp +pAMP plate.
11. Reflame the loop, and allow it to cool, before placing it on the lab bench. Make it a habit to always flame loop one last time.

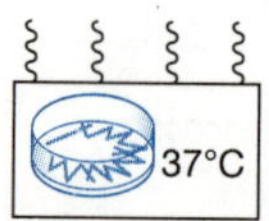

If cells are to be used for colony transformation, continue incubating plates at room temperature for 1 day. During this time, colonies grow large and become sticky, making them easier to pick up with an inoculating loop.

12. Place plates upside down in a 37°C incubator and incubate for 15–20 hours. (Plates are inverted to prevent condensation that might collect on the lids from falling back on the agar and causing colonies to run together.)
13. Optimal growth of well-formed, individual colonies is achieved in 15–20 hours. At this point, colonies should range in diameter from 0.5 mm to 3 mm.
14. Take time for responsible cleanup.
    a. Segregate bacterial cultures for proper disposal.
    b. Wipe down lab bench with soapy water, 10% bleach solution, or disinfectant (such as Lysol) at end of lab.
    c. Wash hands before leaving lab.

RESULTS AND DISCUSSION

This laboratory demonstrates a method to streak bacteria to single colonies. It also introduces antibiotics and plasmid-borne resistance to antibiotics—topics that will be important in several laboratories that follow.

There are two classes of antibiotics: *bacteriostats,* which prevent cell growth, and *bacteriocides,* which kill cells outright. Two antibiotics are used in this course. Ampicillin is a bacteriostatic agent (inhibits the growth of bacteria) and kanamycin is a bacteriocide (kills bacteria). Ampicillin, a derivative of penicillin, blocks synthesis of the peptidoglycan layer (sometimes referred to as the "cell wall") that lies between the *E. coli* inner and outer cell membrane. Thus, it does not affect existing cells with intact cell walls, but kills dividing cells as they synthesize new peptidoglycan. Kanamycin (introduced in later laboratories) is a member of the aminoglycoside family of antibiotics, which block protein synthesis by covalently modifying the bacterial ribosome. Thus, kanamycin quickly kills both dividing and quiescent cells.

The ampicillin resistance gene carried by the plasmid pAMP produces a protein, β-lactamase, that disables the ampicillin molecule. β-lactamase cleaves a specific bond in the β-lactam ring, a four-membered ring in the ampicillin molecule that is essential to its antibiotic action. β-lactamase not only disables ampicillin within the bacterial cell, but because it leaks through the cell envelope, it also disables ampicillin in the surrounding medium. The enzyme kanamycin phosphotransferase prevents kanamycin from interacting with the ribosome.

Antibiotic-resistant Growth

If the plates cannot be observed on the day after streaking, store them at 4°C to arrest *E. coli* growth and to slow the growth of any contaminating microbes. Wrap in Parafilm or plastic wrap to retard drying.

Observe plates and use the matrix below to record which plates have bacterial *growth* and which have *no growth*. On plates with growth, distinct, individual colonies should be observed within one of the streaks.

On the LB/amp plate, growth must be observed in the secondary streak to count as antibiotic-resistant growth. In a heavy inoculum, nonresistant cells in the primary streak may be isolated from the antibiotic on a bed of other nonresistant cells.

	Transformed cells +pAMP	Wild-type cells −pAMP
LB/amp	experiment	negative control
LB	positive control	positive control

On the LB/amp +pAMP plate, tiny "satellite" colonies may be observed radiating from the edges of large well-established colonies. These satellite colonies are not ampicillin-resistant, but grow in an "antibiotic shadow," where ampicillin in the media has been broken down by the large resistant colony. Satellite colonies are generally a sign of antibiotic weakened by not cooling the medium enough before adding the antibiotic, long-term storage of more than 30 days, or overincubation.

1. Were results as expected? Explain possible causes for variations from expected results.
2. In Step 7:
   a. What is the reason for the zigzag streaking pattern?
   b. Why is the inoculating loop resterilized between each new streak?
   c. Why should a new streak intersect only the end of the previous one only at a single point?
3. Describe the appearance of a single *E. coli* colony. Why can it be considered genetically homogeneous?
4. Upcoming laboratories use cultures of *E. coli* cells derived from a single colony or from several discrete parental colonies isolated as described in this experiment. Why is it important to use this type of culture in genetic experiments?
5. *E. coli* strains containing the plasmid pAMP are resistant to ampicillin. Describe how this plasmid functions to bring about resistance.
6. A major medical problem is the ever-increasing number of bacterial strains that are resistant to specific antibiotics. Antibiotic resistance is carried on circular DNA molecules, called plasmids, which are generated separately from the cell's chromosome. With this in mind, suggest a mechanism through which new antibiotic-resistant strains of bacteria arise.

PART B
Overnight Suspension Culture

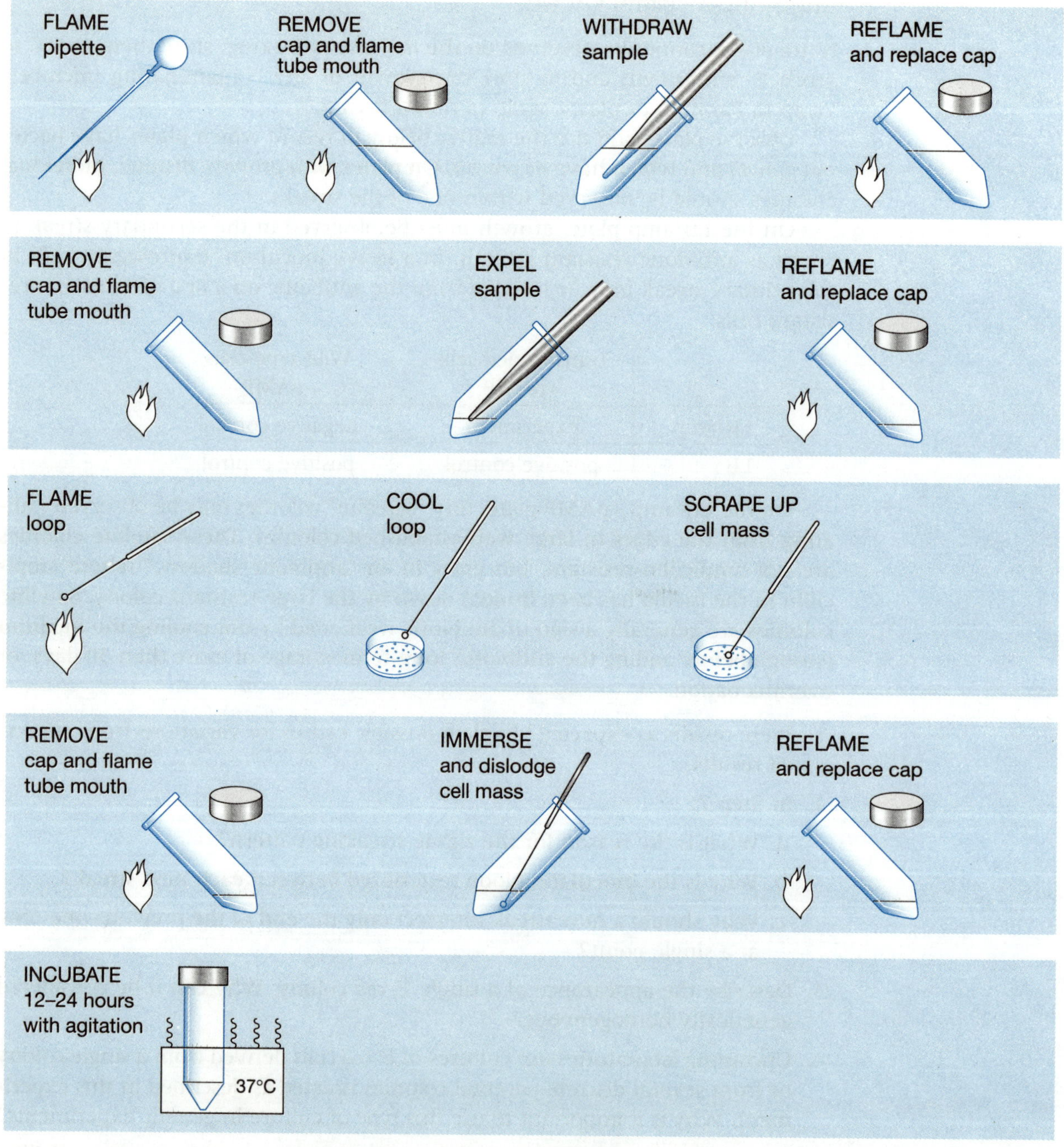

PRELAB NOTES

Review Prelab Notes in Laboratory 2A, Isolation of Individual Colonies.

Although it is always best to grow any overnight culture with shaking, it is not absolutely essential when growing cells for purifying plasmid or for inoculating a larger culture. For these purposes, suspensions can be incubated, without shaking, in a rack within a 37°C incubator. However, the cultures will need to incubate for 1 day or more to obtain an adequate number of cells.

A 50-ml conical tube is preferable for growing overnight cultures. It provides greater surface area for aeration than does a 15-ml culture tube, although a 15-ml tube can be also be used. In overnight cultures for plasmid preparations, it is best, but not essential, to maintain antibiotic selection of the transformed strain by growing in medium containing the appropriate antibiotic. It is prudent to inoculate a "back up" overnight culture, in case the first was not inoculated properly.

PRELAB PREPARATION

Before performing this Prelab Preparation, please refer to the cautions indicated on the Laboratory Materials list.

1. For plasmid purification: Make sure a freshly streaked plate (less than 1 week old) of a transformed or plasmid-bearing *E. coli* strain is available.
2. For mid-log culture: Make sure a freshly streaked plate of wild-type MM294 or other *E. coli* strain is available. It is convenient to make the overnight culture exactly the size of the inoculum needed to start mid-log culture: 1-ml overnight per 100-ml mid-log culture. The entire overnight culture is then simply poured into a flask of fresh, sterile LB broth.
3. Prewarm shaking water bath or incubator to 37°C.

MATERIALS

CULTURE AND MEDIA	SUPPLIES AND EQUIPMENT
E. coli plate	Bleach (10%)▼ or disinfectant
LB broth with appropriate antibiotic	Bunsen burner
	Conical tube (50-ml), sterile
	Inoculating loop
	Permanent marker
	Pipette (10-ml standard)
	Pipette aid or bulb
	Shaking water bath (37°C) (or 37°C dry shaker or dry shaker + 37°C incubator)

▼ See Appendix 4 for Cautions list.

METHODS

Prepare Overnight Culture (10 minutes)

Review Sterile Use of 10-ml Standard Pipette, section III in Laboratory 1, Measurements, Micropipetting, and Sterile Techniques. Think sterile! A pipette should be considered contaminated whenever the tip end comes into contact with anything in the environment—lab bench, hands, or clothing. When contamination is suspected, discard pipette and start with a fresh one. Plan out steps to perform, organize lab bench, and work quickly.

If working as a team, one partner should handle the pipette and the other should handle the tubes and caps.

Pipette flaming can be eliminated if individually wrapped pipettes are used.

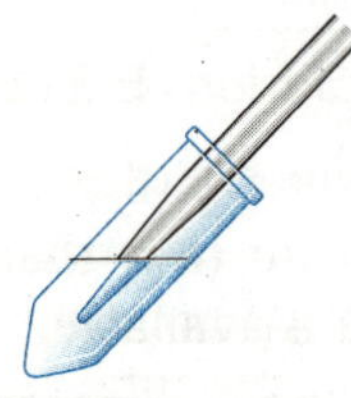

Loop flaming is eliminated if an individually wrapped, sterile plastic loop is used.

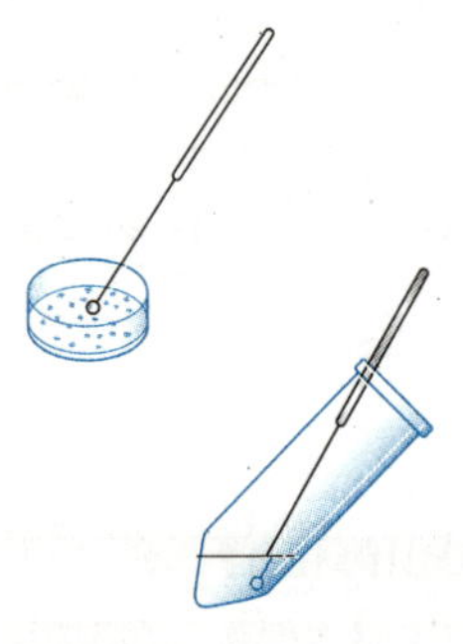

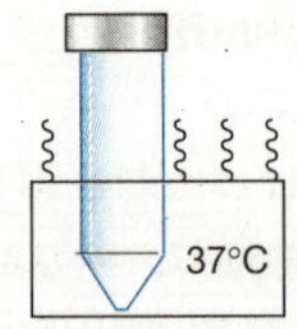

1. Label a sterile 50-ml tube with your name and the date. The large tube provides a greater surface area for good aeration of culture.
2. Use a 10-ml pipette to sterilely transfer 5 ml of LB broth into the tube.
   a. Attach pipette aid or bulb to pipette. Briefly flame pipette cylinder.
   b. Remove cap of LB bottle using little finger of hand holding pipette bulb. Flame mouth of LB bottle.
   c. Withdraw 5 ml of LB. Reflame mouth of bottle, and replace cap.
   d. Remove cap of sterile 50-ml culture tube. Briefly flame mouth of culture tube.
   e. Expel sample into tube. Briefly reflame mouth of tube, and replace cap.
3. Locate well-defined colony 1–4 mm in diameter on a freshly streaked plate.
4. Sterilize inoculating loop in the Bunsen burner flame until it glows red hot. Then, continue to pass lower half of its handle through the flame.
5. Cool loop tip by stabbing it several times into agar near the edge of plate.
6. Use loop to scrape up a visible cell mass from selected colony.
7. Sterilely transfer colony into culture tube:
   a. Remove cap of the culture tube using little finger of hand holding loop.
   b. Briefly flame mouth of culture tube.
   c. Immerse loop tip in broth, and agitate to dislodge cell mass.
   d. Briefly reflame mouth of culture tube, and replace cap.
8. Reflame loop before placing it on lab bench.
9. Loosely replace cap to allow air to flow into culture. Affix a loop of tape over the cap to prevent it from becoming dislodged during shaking.
10. Incubate for 12–24 hours at 37°C, preferably with continuous agitation. Shaking is not essential for a culture to be used for plasmid purification. The culture can be incubated at 37°C, without shaking, for 1 or more days.
11. Take time for responsible cleanup.
    a. Segregate for proper disposal bacterial cultures *and* tubes, pipettes, and micropipettor tips that have come into contact with the cultures.
    b. Disinfect overnight culture and pipettes and tips with 10% bleach or disinfectant.
    c. Wipe down lab bench with soapy water, 10% bleach solution, or disinfectant (such as Lysol).
    d. Wash hands before leaving lab.

RESULTS AND DISCUSSION

E. coli has simple nutritional requirements and grows slowly on "minimal" medium containing (1) an energy source such as glucose, (2) salts such as NaCl and $MgCl_2$, (3) the vitamin biotin, and (4) the vitamin thiamine (B1). *E. coli* sythesizes all necessary vitamins and amino acids from these precursors and grows rapidly in a complete medium, such as LB. Yeast extract and hydrolized milk protein (casein) provide a ready supply of vitamins and amino acids.

A liquid bacterial culture goes through a series of growth phases. For approximately 30–60 minutes following inoculation, there is a *lag phase* during which there is limited cell growth. The bacteria begin dividing rapidly during *log phase,* when the cell number doubles every 20–25 minutes. As nutrients in the media are depleted, the cells nearly stop dividing and the culture enters *stationary phase*. During *death phase,* waste products accumulate and cells begin to die.

Optimum growth in liquid culture is achieved with continuous agitation, which aerates the cells, facilitates the exchange of nutrients, and flushes away waste products of metabolism. It can be safely assumed that a culture in complete medium has reached stationary phase following overnight incubation with continuous shaking.

A stationary-phase culture will look very cloudy and turbid. Discard any overnight culture where vigorous growth is not evident. Expect less growth in cultures incubated for 1–2 days *without continuous shaking.* To gauge growth, shake the tube to suspend cells that have settled at the bottom of the tube.

1. Why is 37°C the optimum temperature for *E. coli* growth?
2. Give two reasons why it is ideal to provide continuous shaking for a suspension culture.
3. What growth phase is reached by a suspension of *E. coli* following overnight shaking at 37°C?
4. Approximately how many *E. coli* cells are in a 5-ml suspension culture at stationary phase?

FOR FURTHER RESEARCH

The number of cells in an overnight culture can be determined with a simple experiment.

1. Set up five tubes, marked -2, -4, -6, -8, -10, with 1 ml of sterile water or LB in each of them.
2. Use a micropipettor with a sterile tip to add 10 µl of overnight culture to the first tube (labeled "-2"). Mix the tube thoroughly and use a fresh tip to transfer 10 µl of diluted cells in the -2 tube to the -4 tube. Mix thoroughly and transfer 10 µl from the -4 tube to the -6 tube, to the -8 tube, and finally to the -10 tube.
3. Use a sterile tip to transfer 100 µl of diluted cells from the -10 tube onto an LB plate and spread the cells across the plate. See instructions in Laboratory 5 for spreading bacteria onto plates.

4. Repeat with fresh plates for the -8 and -6 tubes. Incubate the plates overnight and count colonies. Each colony represents a single original cell.
5. Based on the number of colonies/original cells counted and the dilution factor for each tube, calculate the concentration of the cells in the original culture. For example, if there were 15 colonies on the plate spread from the -8 tube, then there were 150 bacteria in the -8 tube (you removed 100 μl out of 1 ml, which is a tenth of the total, so 15 colonies x 10 = 150 bacteria in the -8 tube). The -8 tube was a 10^8-fold dilution of the original culture, so 150 bacteria x 10^8 = 150 x 10^8 or 1.5 x 10^{10} bacteria per milliliter in the original overnight culture.

PART C

Mid-log Suspension Culture

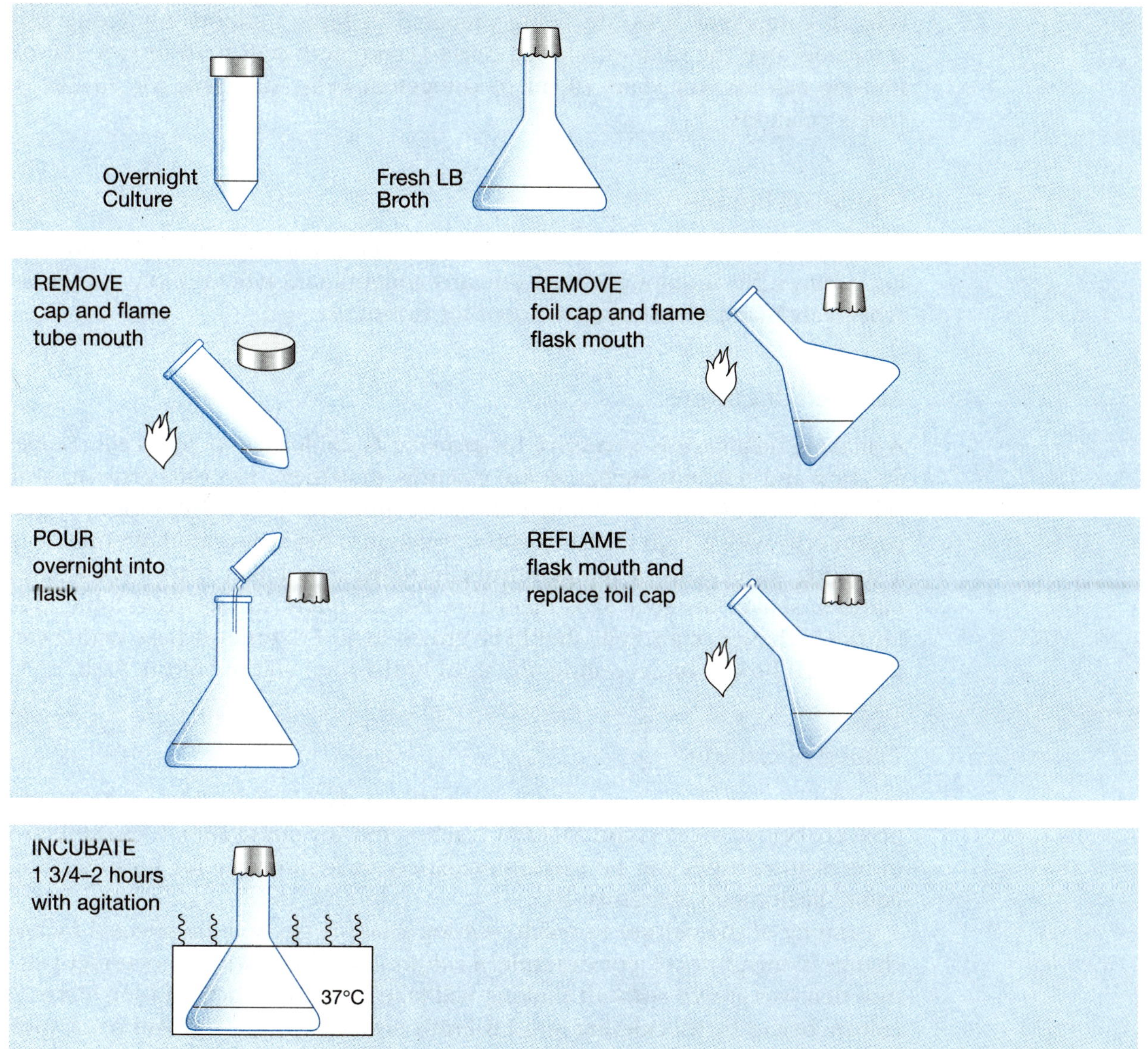

PRELAB NOTES

Competent Cell Yield

In Laboratory 10A—Classic Procedure for Preparing Competent Cells—each experiment will require 20 ml of mid-log cells, which yields 2 ml of competent cells. If competent cells are being prepared in large quantity for group use, remember that the ratio of mid-log cells to competent cells is 10 to 1. A 100-ml mid-log culture will yield 10 ml of competent cells, sufficient for 50–200 μl transformations.

Sterile Technique

Scrupulous sterile technique must be used when preparing overnight and mid-log cultures. No antibiotic is used, and any contaminant will multiply as cells are repeatedly manipulated and/or stored for future use.

Aeration of Culture

A shaking incubator is necessary for growing *E. coli* for competent cells. Proper aeration and nutrient exchange are essential to achieve vigorous growth; only cells collected during the middle part of log (mid-log) phase will produce competent cells with a high transformation frequency. An economical alternative to a shaking water bath or temperature-controlled dry shaker is to place a small platform shaker inside a 37°C incubator. For adequate surface-to-volume ratio for the exchange of air, cells should be grown in an Erlenmeyer flask with a volume of LB broth not exceeding one third of the total volume of the flask.

Timing of Culture

Inoculate a mid-log culture 2–4 hours before Laboratory 10 begins. Using the protocol below, *E. coli* strain MM294 reaches mid-log phase after 1.75–2.0 hours of incubation. Cells can be used immediately or held on ice for up to 2 hours before beginning Laboratory 10.

Timing of the culture to reach mid-log phase is likely to be affected by any change in the protocol. For example, a culture inoculated with an overnight culture that was grown without shaking will take longer to reach mid-log phase. A culture begun by inoculating into LB broth prewarmed to 37°C will reach mid-log phase more quickly than one begun at room temperature. Different strains of *E. coli* display different growth properties. Strain MM294 may exhibit different growth properties in a nutrient broth other than LB broth.

PRELAB PREPARATION

Before performing this Prelab Preparation, please refer to the cautions indicated on the Laboratory Materials list.

1. The day before performing this protocol, start a culture of MM294 or other *E. coli* strain to be transformed with plasmid DNA according to instructions in Laboratory 2B, Overnight Suspension Culture.

2. It is convenient to make the overnight culture exactly the size of the inoculum needed to start a mid-log culture, for example, 1 ml of overnight culture for 100 ml of mid-log culture. Then, simply pour the entire overnight culture into a flask of fresh LB broth. Pipetting is not necessary.
3. Sterile flasks of LB broth can be prepared weeks ahead of time. Add 100 ml of LB broth to a 500-ml Erlenmeyer flask and cap with aluminum foil as a microbe barrier. Autoclave for 15 minutes at 121°C (250°F). Cool and store at room temperature, until ready for use.

MATERIALS

CULTURES AND MEDIA	SUPPLIES AND EQUIPMENT
LB broth (sterile) MM294 overnight culture	Bleach (10%)▼ or disinfectant Bunsen burner Erlenmeyer flask (500-ml), sterile Pipettes (10-ml), sterile (optional) Shaking water bath (37°C) (or 37°C dry shaker or dry shaker + 37°C incubator) Spectrophotometer (optional)

▼ See Appendix 4 for Cautions list.

METHODS

Prepare Culture

(2 hours, including incubation)

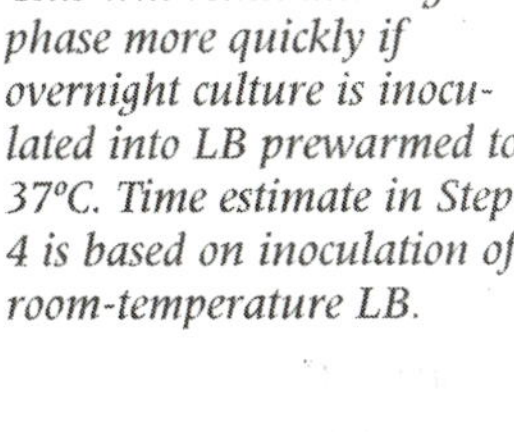

Cells will reach mid-log phase more quickly if overnight culture is inoculated into LB prewarmed to 37°C. Time estimate in Step 4 is based on inoculation of room-temperature LB.

1. Sterilely transfer 1 ml of overnight culture into 100 ml of LB broth at *room temperature.*
2. *If using a 1-ml overnight culture:*
   a. Remove cap from overnight culture tube, and flame mouth. *Do not place cap on lab bench.*
   b. Remove foil cap from flask, and flame mouth. Do not place cap on lab bench.
   c. Pour entire overnight culture into flask. Reflame mouth of flask, and replace foil cap.

   If transferring only a portion of larger overnight culture:
   a. Flame pipette cylinder.
   b. Remove cap from overnight culture tube, and flame mouth of tube. Do not place cap on lab bench.
   c. Withdraw 1 ml of overnight suspension. Reflame mouth of overnight culture tube, and replace cap.

d. Remove foil cap from flask, and flame mouth. Do not place cap on lab bench.

e. Expel overnight sample into flask. Reflame mouth of flask, and replace foil cap.

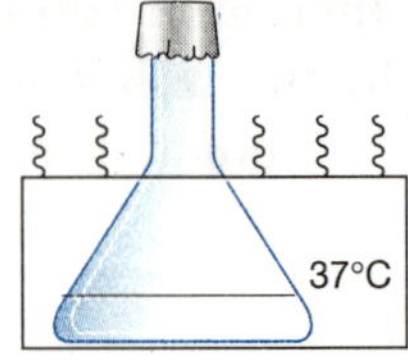

3. Incubate at 37°C with continuous shaking.

4. *If a spectrophotometer is available:* Approximately 1 hour after inoculation, sterilely withdraw a 1-ml sample of the culture, and measure absorbance (optical density at 550 nm). Repeat procedure at approximately 20-minute intervals. An MM294 culture should be grown to an OD_{550} of 0.3–0.5.

   If spectrophotometer is not available: It can be safely assumed that an MM294 culture has reached OD_{550} 0.3–0.5 after 2 hours, 15 minutes of incubation with continuous shaking. Note that under ideal conditions, as represented below, an MM294 culture reaches an OD_{550} of 0.3–0.5 in 1 hour, 30 minutes. However, less ideal conditions often result in slower growth.

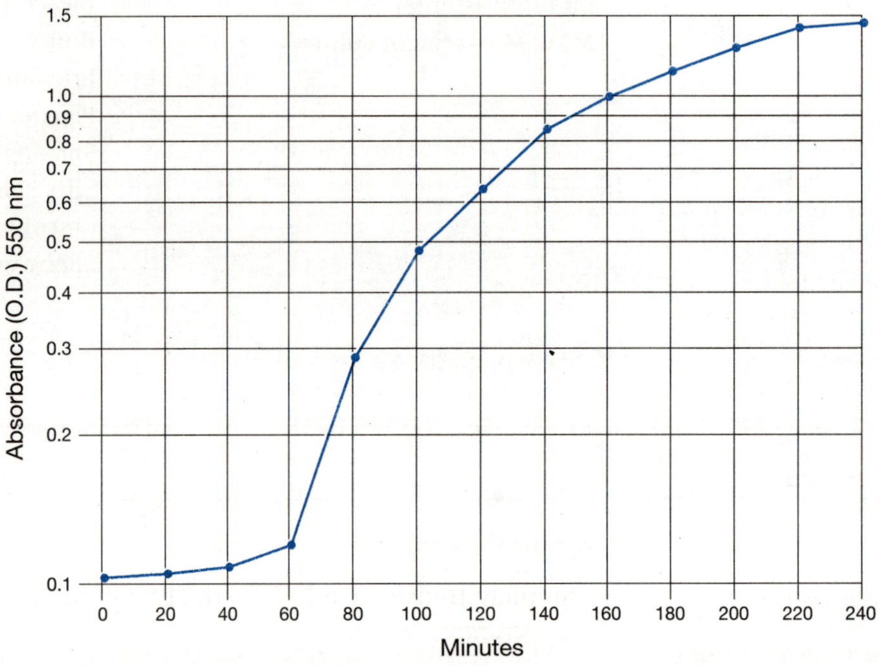

Growth Curve for *E. coli* Strain MM294

5. Store mid-log culture on ice until ready to begin Laboratory 10. This arrests cell growth. Cells can be stored on ice for up to 2 hours prior to use.

6. Take time for responsible cleanup.

   a. Segregate for proper disposal bacterial cultures *and* tubes and pipettes that have come into contact with the cultures.

   b. Disinfect overnight culture, mid-log culture, tubes, and pipettes with 10% bleach or disinfectant.

   c. Wipe down lab bench with soapy water, 10% bleach solution, or disinfectant (such as Lysol).

   d. Wash hands before leaving lab.

RESULTS AND DISCUSSION

1. What variables influence the length of time for an *E. coli* culture to reach mid-log phase?
2. What are the disadvantages of beginning a mid-log culture from a colony scraped off a plate, as opposed to an inoculum of overnight culture?

FOR FURTHER RESEARCH

This experiment can be started in the morning by one experimenter and continued by others throughout the day, until late afternoon.

1. Start a 500-ml *E. coli* culture as described in the above protocol. Determine the optical density of samples sterilely withdrawn at 20-minute intervals, from time zero for as many hours as possible. Plot a graph of time *versus* OD_{550}.
   a. What is the slope of the curve at a point that corresponds to an OD_{550} of 0.3?
   b. Describe the growth of the culture at this point.
2. Perform the following experiment to correlate the optical density of culture with actual number of viable *E. coli* cells. Observe sterile technique.
   a. Inoculate 500 ml of LB with 5 ml of *E. coli* overnight culture. Swirl to mix.
   b. Immediately remove a 10-ml aliquot (time = 0) of the culture, and place on ice to arrest growth. Then incubate the remaining culture at 37°C with vigorous shaking.
   c. Remove additional aliquots from shaking culture every 20 minutes for a total of 4 hours. Hold each aliquot on ice until ready to perform Steps d–f.
   d. Determine the OD_{550} of each aliquot.
   e. Make a 10^2 dilution by mixing 10 µl of the aliquot with 990 µl of fresh LB broth. Prepare three serial dilutions of each aliquot for plating in Step f:

      10^4 = 10 µl of 10^2 culture + 990 µl of LB
      10^5 = 100 µl of 10^4 culture + 900 µl of LB
      10^6 = 100 µl of 10^5 culture + 900 µl of LB

   f. Spread 100 µl of each dilution onto an LB agar plate, for a total of three plates for each time point (aliquot). *Label each plate bottom with time point and dilution.* Invert plates, and incubate for 15–20 hours at 37°C.
   g. For each time point, select a dilution plate that has between 30 and 300 colonies. Multiply the number of colonies by the appropriate dilution factor to give cell number per milliliter in the original aliquot.
   h. Plot two graphs:

      time (*x* axis) *versus* OD_{550} and cell number (*y* axis)
      cell number (*x* axis) *versus* OD_{550} (*y* axis)

   i. An OD_{550} 0.3–0.4 corresponds to what number of cells?

j. What is the average cell number at each of the following points:

lag phase
first third of log phase (early log)
second third of log phase (mid log)
final third of log phase (late log)
stationary phase

k. Do OD_{550} measurements distinguish between living and dead cells?

DNA Restriction Analysis

LABORATORY 3 INTRODUCES THE ANALYSIS OF DNA using restriction enzymes and gel electrophoresis. Three samples of purified DNA from bacteriophage λ (48,502 bp in length) are incubated at 37°C, each with one of three restriction endonucleases: *Eco*RI, *Bam*HI, and *Hin*dIII. Each enzyme has five or more restriction sites in λ DNA and therefore produces six or more restriction fragments of varying lengths. A fourth sample of λ DNA, the negative control, is incubated without an endonuclease and remains intact.

The digested DNA samples are then loaded into wells of a 0.8% agarose gel. An electrical field applied across the gel causes the DNA fragments to move from their origin (the sample well) through the gel matrix toward the positive electrode. The gel matrix acts as a sieve through which smaller DNA molecules migrate faster than larger molecules; restriction fragments of differing sizes separate into distinct bands during electrophoresis. The characteristic pattern of bands produced by each restriction enzyme is made visible by staining with a dye that binds to the DNA molecule.

Kits based on this laboratory are available from the Carolina Biological Supply Company.

- Catalog no. 21-1103 (with ethidium bromide stain)
- Catalog no. 21-1104 (with *Carolina*Blu™ stain)

I. Set Up Restriction Digest

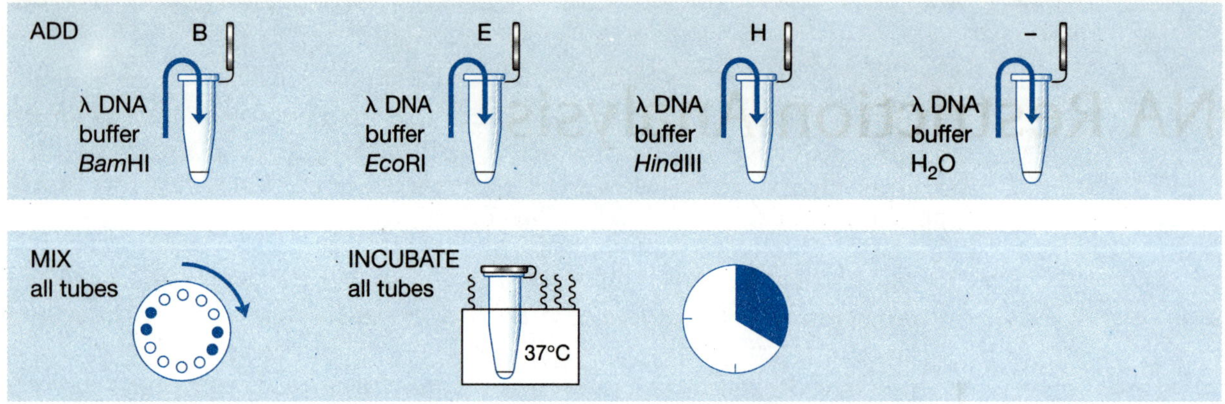

II. Cast 0.8% Agarose Gel

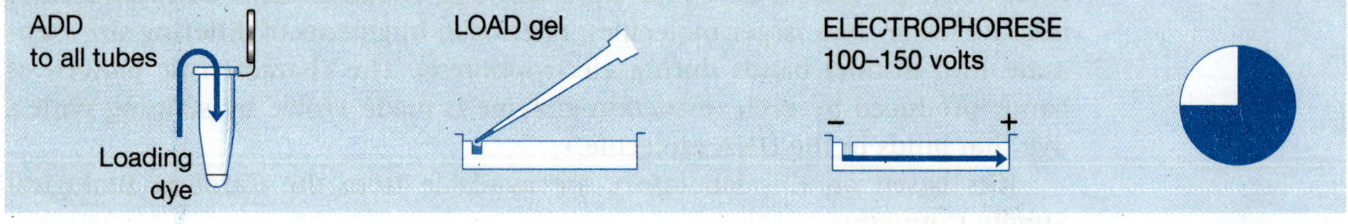

III. Load Gel and Separate by Electrophoresis

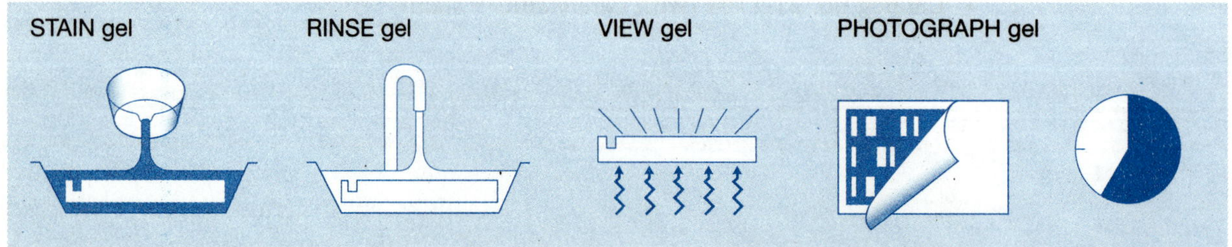

IV. Stain Gel and View (Photograph)

STAIN gel

RINSE gel

VIEW gel

PHOTOGRAPH gel

PRELAB NOTES

Storing and Handling Restriction Enzymes

Restriction enzymes, like many enzymes, are most stable at cold temperatures and lose activity if warmed for any length of time. Since maintaining these enzymes in good condition is critical to the success of experiments in this course, follow the guidelines listed below for handling.

1. Always store enzymes in a NON-frost-free freezer that maintains a constant temperature between −10°C and −20°C. NON-frost-free freezers develop a layer of frost around the chamber that acts as an efficient insulator. Frost-free freezers go through freeze-thaw cycles that subject enzymes to repeated warming and subsequent loss of enzymatic activity. However, there are commercially available storage containers that are essentially plastic freezer packs with holes for tubes. Restriction enzymes stored in these containers (even in a frost-free freezer) are much more stable because the freezer packs hold the temperature for several hours. Even in a NON-frost-free freezer, enzymes can warm up when the freezer is opened or when enzymes are used. Storage in these containers maintains constant temperature regardless of freezer type.
2. When a large shipment of an enzyme is received, split it into several smaller aliquots of 50–100 µl in 1.5-ml tubes. Use a permanent marker on tape to clearly identify aliquots with enzyme type, concentration in units/µl, and date received. Use up one aliquot before starting another.
3. Remove restriction enzymes from the freezer directly onto crushed or cracked ice in an insulated ice bucket or cooler. Make sure that the tubes are pushed down into the ice and not just sitting on top. Keep enzymes on ice at all times during preparation, and return to the freezer immediately after use.
4. Keep aliquots of enzymes, buffer, and DNA in a cooler filled with ice while in use to ensure that the unused aliquots remain fresh.
5. Although it is good technique to set up restriction digests on ice, it is much simpler to set up reactions in a test tube rack at room temperature. Little loss of enzyme activity occurs during the brief time it takes to set up the reaction.

But take heart—the enzymes used in this course are all remarkably stable. Those used in training workshops presented by us have survived multiple day-long shipments on ice, freezer power failures, and various abuse by student interns. In 15 years and more than 100 training workshops, we have yet to experience enzyme failure. Most restriction enzymes can survive several hours if left out on the lab bench, but do not take any chances!

Storing DNA and Restriction Buffer

Purified DNA is generally stored in the refrigerator (~4°C). DNA can be kept at −10°C to −20°C for long-term storage of several months or more. However, repeated freezing and thawing damages DNA. Restriction buffer is best kept frozen; freezing-thawing does not affect restriction buffer.

Buffers

Several types of buffers are used in this course: restriction buffer, electrophoresis buffer, and ligation buffer. Each has a different chemical composition and use. Always double-check to ensure that the proper buffer is being used.

Tris/Borate/EDTA (TBE) electrophoresis buffer can be reused several times. Collect used buffer and store in large carboy. If different gels are to be run over a period of several days, store the buffer in the electrophoresis box with the cover in place to retard evaporation. Prior to reusing buffer stored in the electrophoresis box, rock the box back and forth to mix the buffer at either end. This reequilibrates ions that differentially accumulate at either end during electrophoresis.

Groups of restriction enzymes operate under different conditions of salt and pH. For optimal activity, several different buffers are needed for the enzymes used in this course. To simplify procedures, we use a "compromise" restriction buffer—a universal buffer that is a compromise between the conditions preferred by various enzymes.

All buffers are used at a final concentration of 1x. Rely on the standard $C_1V_1 = C_2V_2$ formula to determine how much buffer to add to obtain a 1x solution:

(vol. buffer)	(conc. of buffer)	=	(total vol. of reaction)	(1x buffer)
(5 µl)	(2x)	=	10 µl	(1x)
(1 µl)	(10x)	=	10 µl	(1x)

For convenience, use 2x restriction buffer whenever possible—it can save a pipetting step to add water to bring a reaction up to 10 µl total volume. It is also easier and more accurate to pipette 5 µl than to pipette 1 µl. Compare a typical restriction reaction using 2x versus 10x restriction buffer:

	2x Buffer	10x Buffer
DNA	4 µl	4 µl
Enzyme	1 µl	1 µl
Buffer	5 µl	1 µl
Water	–	4 µl
Total Solution	10 µl	10 µl

Bacteriophage λ DNA

Because it is inexpensive and readily available, purified DNA from bacteriophage λ is most suitable for demonstrating the concept of DNA restriction. λ DNA costs approximately $0.10 per microgram, compared to plasmid DNA which ranges typically from $2.00 to $3.00 per microgram. Most commercially available λ is derived from a temperature-sensitive lysogen of *E. coli* called *c*I857 and is 48,502 bp in length. Restriction of chromosomal DNA, even from a simple organism such as *E. coli*, will generate thousands of DNA fragments that appear as a smear in an agarose gel.

Diluting DNA

DNA for near-term use can be diluted with distilled or deionized water. However, dilute the DNA with Tris-EDTA (TE) buffer for long-term storage.

EDTA in the buffer binds divalent cations, such as Mg^{++}, that are necessary cofactors for DNA-degrading nucleases. Always dilute DNA to the concentration specified by the protocol.

1. Determine the total volume of DNA required by multiplying the number of experiments times the total volume of DNA per experiment, including overage.

   (10 experiments) (20 μl DNA) = 200 μl DNA

2. Use this number in the $C_1V_1 = C_2V_2$ formula, along with the desired final DNA concentration and the concentration of the stock DNA. Solve for V_1, the volume of stock DNA needed in the dilution.

   (C_1 stock DNA) (V_1) = (C_2 final DNA) (V_2 total volume)
   (0.5 μg/μl) (V_1) = (0.1 μg/μl) (200 μl)
   (V_1) = (0.1 μg/μl) (200 μl)/(0.5 μg/μl) = 40 μl stock DNA

3. Add water or TE to make total volume of final solution.

   40 μl stock DNA + 160 μl H_2O or TE = 200 μl final solution

Making Aliquots of Reagents

1. We find that it is safest to prepare separate aliquots of enzyme, DNA, and buffer in 1.5-ml tubes for each experiment. Each aliquot should contain slightly more than is required for the lab. Following the experiment, discard the tubes, and make new aliquots for the next experiment. Although this procedure appears to be wasteful, it avoids cross-contamination that invariably occurs if aliquots are reused or shared between groups. It is a small price to pay for consistent results.
2. For aliquots of restriction enzymes, add 1 μl extra when 1–3 μl is called for and 2 μl extra when 4–6 μl is actually needed. The overage aids in visualizing the reagent in the tube and allows for small pipetting errors.
3. It is probably unwise to make small aliquots of enzymes more than 1 or 2 days in advance. A several-microliter droplet, clinging to the side of a 1.5-ml tube, has a large surface-to-volume ratio. For this reason, it may be affected by temperature fluctuations. If aliquots are made in advance, store them in the freezer until needed. When setting up, remove aliquots from the freezer and place on ice, making sure that the tubes are fully submerged in ice while awaiting use.
4. Aliquots of DNA and buffer should be approximately 20% larger than the volume actually needed in the experiment. This allows for overpipetting and other mishaps. Considering that DNA is generally the most expensive component of an experiment, making aliquots of the exact amount and adding other reagents directly to the DNA tube may be more cost-effective.
5. Reagent volumes listed in the Prelab Preparation section of each laboratory have been scaled to include the overage suggested here.
6. Large aliquots of distilled water and loading dye can be used for several experiments.
7. Colored 1.5-ml tubes are very handy for color-coding each reagent aliquot.

Pooling Reagents

During aliquot preparation and movement to and from the freezer or refrigerator and ice bucket, reagent aliquots often become spread in a film around the sides or caps of the 1.5-ml tubes. Use one of the following methods to pool reagent droplets to make them easier to find in the tube.

- Spin tubes briefly in a microfuge.

  or

- Spin tubes briefly in a preparatory centrifuge, using adaptor collars for 1.5-ml tubes. Alternately, spin tubes within 15-ml tube, and remove carefully.

  or

- Tap tubes sharply on bench top.

Restriction Enzyme Activity

The "unit" is the standard measure of restriction enzyme activity and is defined as the amount of enzyme needed to digest to completion 1 μg of λ DNA in a 50-μl reaction in 1 hour. The unit concentration of various restriction enzymes varies from batch to batch and from manufacturer to manufacturer. Typical batches of commercially available enzymes have activities in the range of 5–20 units/μl.

We suggest using enzymes at full strength as supplied—a working concentration of approximately 1 unit per microliter of reaction mix. Although this is technically far more enzyme than is required, such "overkill" assures complete digestion by compensating for the following experimental conditions:

1. As a time saver, reaction times for restriction digests have been shortened to 20 minutes. Complete digestion of the DNA would not occur during an abbreviated incubation if the restriction enzyme was used at the standard condition of 1 unit/μg of DNA.
2. Many enzymes do not exhibit 100% activity in a compromise buffer.
3. Enzymes lose activity over time, due to imperfect handling.
4. It is easy enough to underpipette when measuring 1 μl of enzyme, especially considering that the micropipettor's mechanical error is greatest at the low end of its volume range.

Incubating Restriction Reactions

A constant-temperature water bath for incubating reactions can be made by maintaining a trickle flow of tap water into a Styrofoam box. Monitor temperature with a thermometer. An aquarium heater can be used to maintain constant temperature.

Twenty minutes is the bare minimum incubation time for the restriction reaction to go to completion. If electrophoresis is to be done the following day, incubate the reactions for 1–24 hours. After several hours, enzymes lose their activity, and the reaction simply stops. Stop incubation whenever it is convenient; reactions may be stored in a freezer (–20°C) until ready to continue. Thaw reactions before adding loading dye.

Casting Agarose Gels

Remarkably, the enzyme digest of DNA is not the main determining factor of good results in restriction analysis. Enzymes rarely fail, and λ DNA is inexpensive enough that there is no need to scrimp. Measurements are not extremely critical—restriction reactions come out fine as long as some DNA, some enzyme, and some buffer make it into the test tube. Most unsatisfactory results in restriction analysis can be traced to the problems in casting agarose gels.

- Thin gels yield dramatically better results than thick gels, so cast gels only thick enough to contain the volume of DNA that will be loaded. A thin gel stains quickly and concentrates the DNA in a shorter vertical distance. This improves contrast, and the stained DNA appears brighter. The reason is that, in addition to binding to DNA, stain also accumulates in the gel itself. This unbound stain creates a background that decreases contrast, and relative brightness, of the stained DNA. Thick gels inherently have less contrast, because the DNA is viewed through a thicker background of unbound stain. In addition, thick gels distribute the DNA over a larger vertical distance, decreasing the width of the band. Considering that thick gels consume more agarose and take longer to set and stain, there is no reason to cast them.

  Although casting thin gels is the desirable way to go, remember that they are relatively fragile! For all intents and purposes, the least concentrated gel that can be handled without great fear of breaking is 0.8%. If problems are encountered, increase the concentration to 1.0%.

- Clean, properly shaped wells are key to producing technically excellent gels—those with straight, well-focused bands. The front edge of the well determines how the DNA enters the gel and what kinds of bands are formed as the DNA sorts by size. Ideally, the front edge of the well should be smooth and perpendicular to the gel. This forces the DNA to enter the gel as a single vertical front. Given relatively constant current, the DNA will retain this focused configuration as it moves down the gel and resolves into bands. However, a loose or bent comb may not rest perpendicular to the casting tray, which produces a slanted well, allowing the DNA to enter the gel over a broad front. Viewed from above, the DNA is spread out along the slanted front. This effectively dilutes the DNA, producing a diffuse, or "fuzzy," band. This effect can be seen by moving your head in an arc when viewing a stained gel. The fuzzy bands produced by an angled well will become sharply focused when your viewing angle aligns with the angle of the slanted wells.

  Even more insidious is the effect of removing the comb when the gel is not completely set. In this case, the weight of the gel compresses the partially set gel into the well, causing the edges to bow inward. This destroys the perpendicular edge, and the DNA enters the well over a broader front, which produces fuzzy bands that cannot be resolved by changing the viewing angle of the stained gel. Changes in the shape of the well, due to partially set gel, are almost impossible to detect ahead of time. However, they are likely the major cause of chronically fuzzy bands—especially when short lab periods impose a need for speed. So it pays to cast gels well in advance of when they will be used.

  The problems described above offer good arguments for immersing a gel in buffer prior to removing the comb. The buffer aids in cooling a recently cast gel, helping to ensure that it has solidified throughout. In addition, the buffer

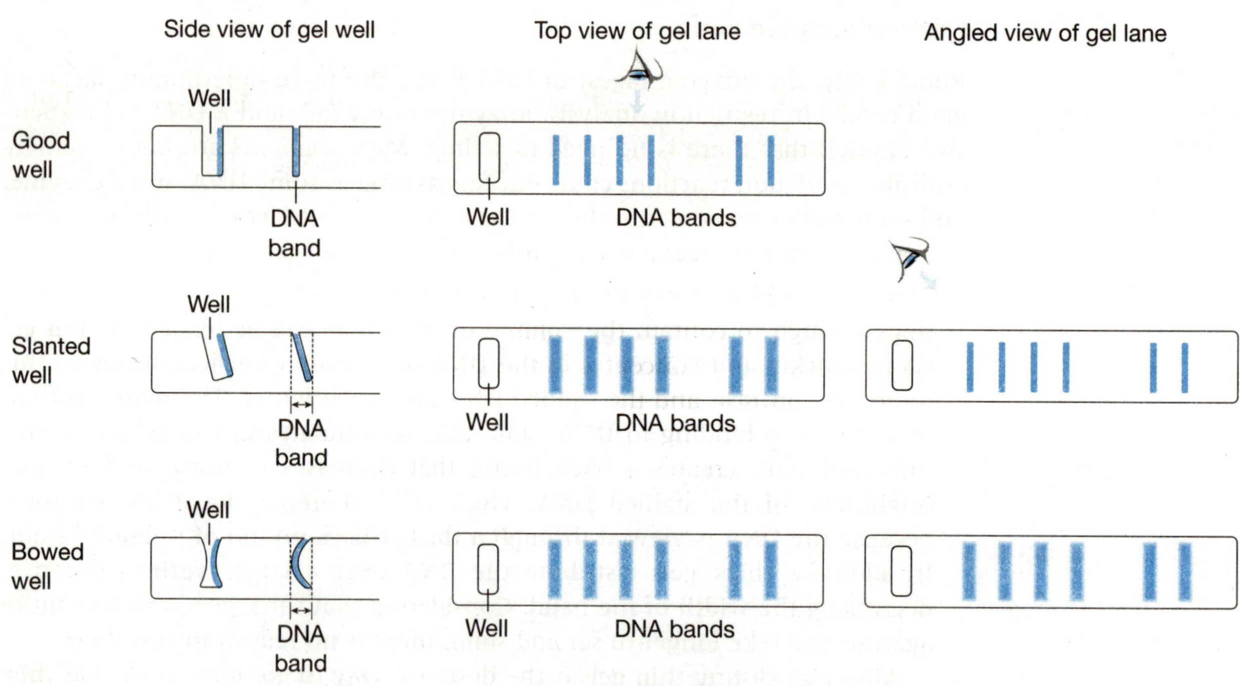

Problems with Casting Gels

helps to lubricates the comb, reducing the chance of damaging the front edge of the well when the comb is withdrawn.

- It is a sad fact that the design of some commercially available combs and casting trays makes it difficult to cast an excellent gel. Poorly molded combs may have burrs that nick or scrape the well edge; combs molded of nonrigid materials may bend; and ill-fitting combs may "wobble" in the casting case. At least these problems are relatively easy to fix. A difficult problem to fix is a comb whose teeth are set too far above the bed of the casting tray. The distance between the comb and the tray surface should be about 1 mm. Any more distance produces an overly thick gel, with the attendant problems of staining and contrast. The only remedy is to file the comb or casting tray to allow the comb to come to rest closer to the tray surface.

Storing Cast Agarose Gels

Gels can be cast 1 or 2 days before use. Keep gels covered with TBE electrophoresis buffer or wrapped in plastic wrap to prevent drying.

Separating by Electrophoresis

Shortly after the current is applied to the electrophoresis system, the loading dye should be seen moving through the gel toward the positive side of the gel apparatus. It will appear as a blue band, eventually resolving into two bands of color. The faster-moving, purplish band is bromophenol blue. The slower-moving, aqua band is xylene cyanol. In a 0.8% gel, bromophenol blue migrates through the gel at the same rate as a DNA fragment of approximately 300 bp. Xylene cyanol migrates at a rate equivalent to approximately 9000 bp. Best separation for analysis of plasmid DNA is achieved when the bromophenol blue migrates 4–7 cm or more from the origin. However, be careful not to let the bromophenol blue band run off the end of the gel.

The migration of DNA through the agarose gel is dependent on voltage—the higher the voltage, the faster the rate of migration. Refer to the chart below for approximate running times at various voltages. The times below are for a "minigel" system with a 84 x 96-mm gel; times will vary according to apparatus.

Voltage	150	125	100	75	50	25	12.5
Time	0:40	0:50	1:20	1:40	3:20	6:40	13:00

Responsible Handling of Ethidium Bromide

The protocols in this manual limit the use of ethidium bromide to a single procedure that can be performed by the instructor in a controlled area. With responsible handling, the dilute solution (1 μg/ml) used for gel staining poses minimal risk. Ethidium bromide, like many natural and man-made substances, is a mutagen by the Ames microsome assay and a suspected carcinogen.

CAUTION/HANDLING AND DECONTAMINATION OF ETHIDIUM BROMIDE

1. Always wear gloves while working with ethidium bromide solutions or stained gels.
2. Limit ethidium bromide use to areas covered with bench paper.
3. Following gel staining, use a funnel to decant as much of the ethidium bromide solution as possible into a storage container for reuse or decontamination and final disposal.
4. Dispose of stained gels and used staining solution according to current chemical waste disposal regulations. Consult your local safety office for these regulations. For further information, see Appendix 4.

The greatest risk is to inhale ethidium bromide powder when mixing a 5 mg/ml stock solution. Therefore, we suggest purchasing a ready-mixed stock solution from a supplier. The stock solution is diluted to make a staining solution with a final concentration of 1 μg/ml.

DNA Staining with Methylene Blue

The volumes and concentrations of DNA used in these experiments have been optimized for staining with ethidium bromide, which is the most rapid and sensitive method. If methylene blue staining (or other proprietory staining procedure) is preferred, increase stated concentrations 4–5 times for λ DNA and 2 times for plasmid DNA. If DNA *concentration* is increased, volumes used in laboratories remain as stated.

Viewing Stained Gels

Transillumination, where light passes up through gel, gives superior viewing of gels stained with either ethidium bromide or methylene blue.

A mid-wavelength (260–360 nm) ultraviolet (UV) lamp emits in the optimum range for illuminating ethidium-bromide-stained gels. Commercially available gel illumination systems are designed for optimal illumination of ethidium bromide. These systems use UV light at a specific wavelength that does not contain the most harmful rays. However, use great caution to ensure that skin is not exposed and that eye protection is used.

CAUTION
Ultraviolet light can damage the retina of the eye. Never look at an unshielded UV light source with naked eyes. View only through a filter or safety glasses that absorb harmful wavelengths. For further information, see Appendix 4.

A fluorescent light box for viewing slides and negatives provides ideal illumination for methylene-blue-stained gels. An overhead projector may also be used. Cover the surface of the light box or projector with plastic wrap to protect the apparatus from liquid spills.

Photographing Gels

Photographs of DNA gels provide a permanent record of the experiment, allowing time to analyze results critically, to discover subtleties of gel interpretation, and to correct mistakes. Furthermore, time exposure can record bands that are faint or invisible to the unaided eye.

A Polaroid "gun" camera, equipped with a close-up diopter lens, is used to photograph gels on either a UV or white-light transilluminator. A plastic hood extending from the front of the camera forms a mini-darkroom and provides correct lens-to-subject distance. Alternatively, a close-focusing 35-mm camera can be used. For UV photography, two filters are placed in front of the lens: a 23A orange filter is closest to the camera and a 2B UV-blocking filter (clear) is closest to the subject. Any yellow or orange filter will intensify contrast in gels stained with methylene blue. The UV filter set described above works well and can be left in place for both ethidium bromide and methylene blue photography.

Exposure times vary according to the mass of DNA in the lanes, level of staining, degree of background staining, thickness of gel, and density of filter. Experiment to determine the best exposure. When possible, stop the lens down (to higher f/number) to increase the depth of field and the sharpness of bands.

For UV photography of ethidium-bromide-stained gels, use Polaroid high-speed film Type 667 (ASA 3000), black and white. Set camera aperture to f/8 and shutter speed to B. Depress shutter for a 2–3-second time exposure. For white-light photography of methylene-blue-stained gels, use Polaroid film Type 667, with an aperture of f/8 and shutter speed of 1/125 second.

Digital photography is becoming increasingly popular for gel documentation. Digital cameras, especially those with close focus capability, can provide high-quality images. The addition of a hood (extending from the lens) will block ambient light and increase contrast. Image files can be opened and edited in any photo-editing program, such as Microsoft Photo Editor and Adobe Photoshop. A hard copy is printed on regular or glossy printer paper, saving the cost of Polaroid or photographic film/printing. The digital images can also be distributed via e-mail or an Internet site.

For Further Information

The protocol presented here is based on the following published methods:

Aaij C. and Borst P. 1972. The gel electrophoresis of DNA. *Biochim. Biophys. Acta* **269:** 192–200.

Helling R.B., Goodman H.M., and Boyer H.W. 1974. Analysis of R-*Eco*RI fragments of DNA from lamboid bacteriophages and other viruses by agarose-gel electrophoresis. *J. Virol.* **14:** 1235–1244.

Sharp P.A., Sugden B., and Sambrook J. 1973. Detection of two restriction endonuclease activities in *Haemophilus parainfluenzae* using analytical agarose-ethidium bromide electrophoresis. *Biochemistry* **12:** 3055–3063.

PRELAB PREPARATION

Before performing this Prelab Preparation, please refer to the cautions indicated on the Laboratory Materials list.

The volumes of agarose solution and Tris/Borate/EDTA (TBE) buffer needed vary according to the electrophoresis apparatus used. The volumes quoted here are based on typical "minigel" systems.

1. Prepare aliquots for each experiment:

   20 μl of 0.1 μg/μl λ DNA (store on ice)
   25 μl of 2x restriction buffer (store on ice)
   2 μl each of *Bam*HI, *Eco*RI, and *Hin*dIII (store on ice)
   500 μl of distilled water
   500 μl of loading dye

2. Prepare 0.8% agarose solution (40–50 ml per experiment). Keep agarose liquid in a hot-water bath (at ~60°C) throughout the lab. Cover the solution with aluminum foil to retard evaporation.
3. Prepare 1x TBE buffer for electrophoresis (400–500 ml per experiment).
4. Prepare ethidium bromide or methylene blue staining solution (100 ml per experiment).
5. Adjust water bath to 37°C.

MATERIALS

REAGENTS	SUPPLIES AND EQUIPMENT
Agarose (0.8%)	Aluminum foil
Distilled water	Beakers for agarose, for waste/used tips, and for TBE buffer
Ethidium bromide▼(1 µg/ml) (or 0.025% methylene blue▼)	Camera and film (optional)
λ DNA (0.1 µg/µl)	Electrophoresis box
Loading dye	Latex gloves
2x Restriction buffer	Masking tape
Restriction enzymes	Microfuge (optional)
*Eco*RI	Micropipettor (0.5–10 µl) + tips
*Bam*HI	Parafilm or wax paper (optional)
*Hin*dIII	Permanent marker
1x Tris▼/Borate/EDTA (TBE) buffer	Plastic wrap (optional)
	Power supply
	Test tube rack
	Transilluminator (optional)▼
	Tubes (1.5-ml)
	Water bath (37°C)

▼ See Appendix 4 for Cautions list.

METHODS

I. Set Up Restriction Digest

(30 minutes, including incubation)

1. Use a permanent marker to label four 1.5-ml tubes, in which restriction reactions will be performed:

   B = *Bam*HI
   E = *Eco*RI
   H = *Hin*dIII
   – = no enzyme

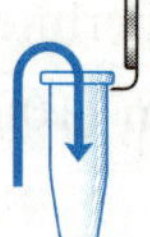

2. Use the matrix below as a checklist while adding reagents to each reaction. Read down each column, adding the same reagent to all appropriate tubes. *Use a fresh tip for each reagent.* Refer to detailed directions that follow.

Tube	λ DNA	Buffer	*Bam*HI	*Eco*RI	*Hin*dIII	H_2O
B	4 µl	5 µl	1 µl	—	—	—
E	4 µl	5 µl	—	1 µl	—	—
H	4 µl	5 µl	—	—	1 µl	—
–	4 µl	5 µl	—	—	—	1 µl

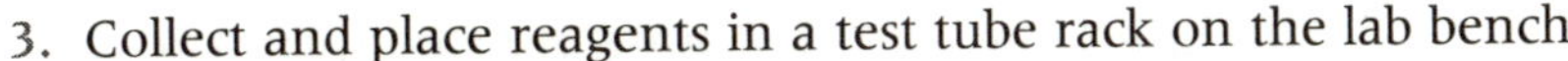

3. Collect and place reagents in a test tube rack on the lab bench.

It is not necessary to change tips when adding the same reagent. The same tip may be used for all tubes, provided the tip has not touched solution already in tubes.

4. Add 4 µl of DNA to each reaction tube. Touch the pipette tip to the side of the reaction tube, as near to the bottom as possible, to create capillary action to pull the solution out of the tip.
5. Always add buffer to the reaction tubes *before* adding enzymes. Use a *fresh tip* to add 5 µl of restriction buffer to a clean spot on each reaction tube.
6. Use *fresh tips* to add 1 µl of *Eco*RI, *Bam*HI, and *Hin*dIII to the appropriate tubes.
7. Use a *fresh tip* to add 1 µl of deionized water to tube labeled "–."

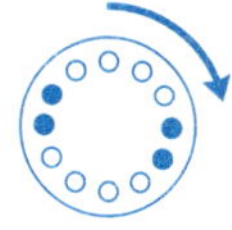

8. Close tube tops. Pool and mix reagents by pulsing in a microfuge or by sharply tapping the tube bottom on the lab bench.

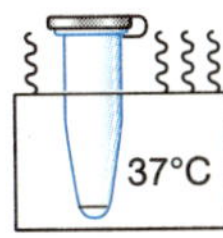

9. Place the reaction tubes in a 37°C water bath, and incubate them for a minimum of 20 minutes. Reactions can be incubated for a longer period of time.

After several hours, enzymes lose activity and reaction stops.

Following incubation, freeze reactions at –20°C until ready to continue. Thaw reactions before continuing to Section III, Step 1.

II. Cast 0.8% Agarose Gel (15 minutes)

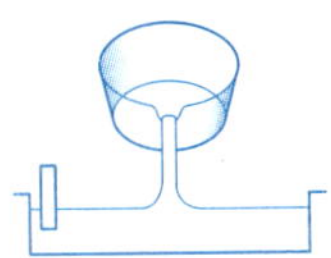

Gel is cast directly in box in some electrophoresis apparatuses.

1. Seal the ends of the gel-casting tray with tape, and insert well-forming comb. Place the gel-casting tray out of the way on the lab bench so that agarose poured in next step can set undisturbed.
2. Carefully pour enough agarose solution into the casting tray to fill to a depth of about 5 mm. Gel should cover only about one-third the height of comb teeth. Use a pipette tip to move large bubbles or solid debris to the sides or end of tray while gel is still liquid.
3. Gel will become cloudy as it solidifies (~10 minutes). *Do not move or jar casting tray while agarose is solidifying*. Touch corner of agarose away from comb to test whether gel has solidified.
4. When agarose has set, unseal ends of casting tray. Place tray on the platform of the gel box so that comb is at negative black electrode (cathode).

Too much buffer will channel current over top of gel rather than through gel, increasing the time required to separate DNA. TBE buffer can be used several times; do not discard. If using buffer remaining in electrophoresis box from a previous experiment, rock chamber back and forth to remix ions that have accumulated at either end.

5. Fill box with TBE buffer, to a level that just covers entire surface of gel.
6. Gently remove comb, taking care not to rip the wells.

Buffer solution helps to lubricate the comb. Some gel boxes are designed such that comb must be removed prior to inserting the casting tray into the box. In this case, flood casting tray and gel surface with running buffer before removing the comb. Combs removed from a dry gel can cause tearing of the wells.

7. Make sure that the sample wells left by the comb are completely submerged. If "dimples" appear around the wells, slowly add buffer until they disappear.

Cover electrophoresis tank and save gel until ready to continue. Gel will remain in good condition for at least several days if it is completely submerged in buffer.

III. Load Gel and Separate by Electrophoresis (50–70 minutes)

1. Add loading dye to each reaction. Either

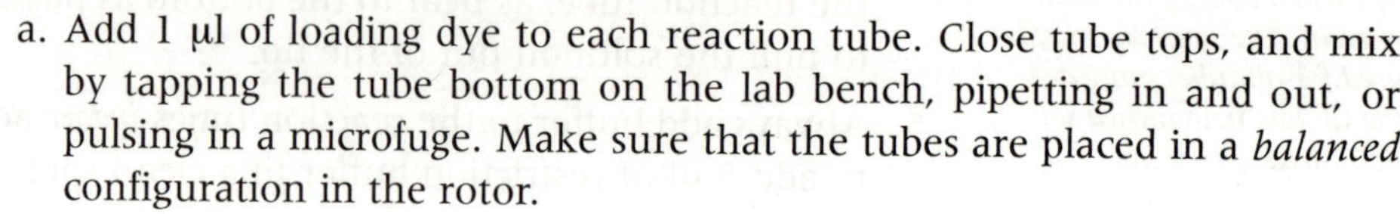

   a. Add 1 μl of loading dye to each reaction tube. Close tube tops, and mix by tapping the tube bottom on the lab bench, pipetting in and out, or pulsing in a microfuge. Make sure that the tubes are placed in a *balanced* configuration in the rotor.

   or

   b. Place four individual droplets of loading dye (1 μl each) on a small square of Parafilm or wax paper. Withdraw contents from the reaction tube and mix with a loading dye droplet by pipetting in and out. Immediately load dye mixture according to Step 2. Repeat successively, *with a clean tip*, for each reaction.

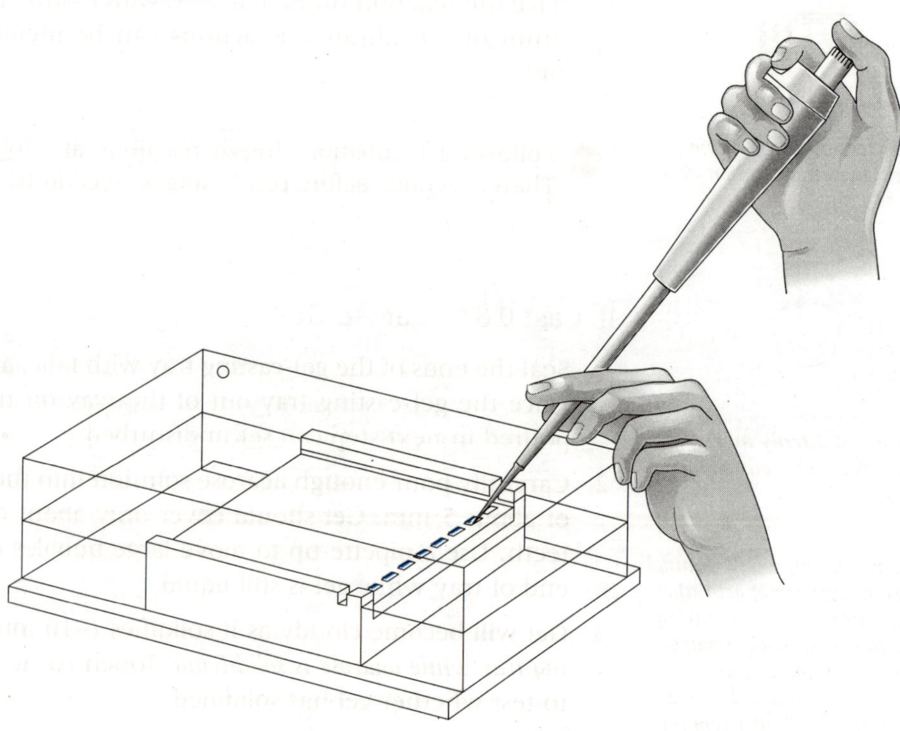

Hand Positions for Loading an Agarose Gel (Step 2)

A piece of dark construction paper beneath the gel box will make the wells more visible.

2. Use a micropipettor to load 10 μl of each reaction tube into a separate well in the gel, as shown in the diagrams. *Use a fresh tip for each reaction.*

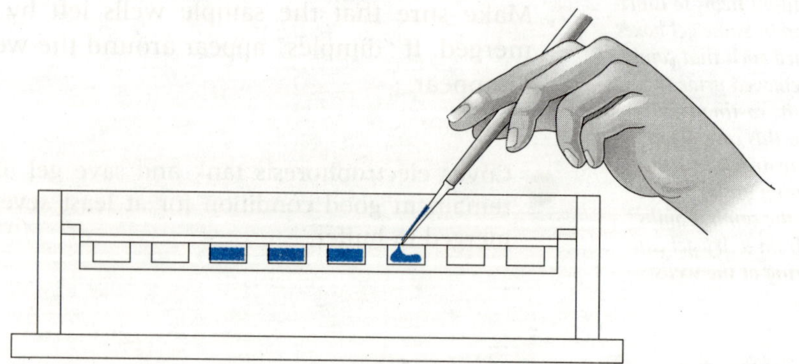

a. Use two hands to steady the micropipettor over the well.

b. Before loading the sample, make sure that there are no bubbles in the wells. If bubbles exist, move them with a micropipettor tip.

c. If there is air in the end of the tip, carefully depress plunger to push the sample to the end of the tip. (If an air bubble forms a "cap" over the well, DNA/loading dye will flow into buffer around edges of well.)

d. Dip micropipettor tip through surface of buffer, center it over the well, and gently depress micropipettor plunger to slowly expel sample. Sucrose in the loading dye weighs down the sample, causing it to sink to the bottom of the well. *Be careful not to punch the tip of the micropipettor through the bottom of the gel.*

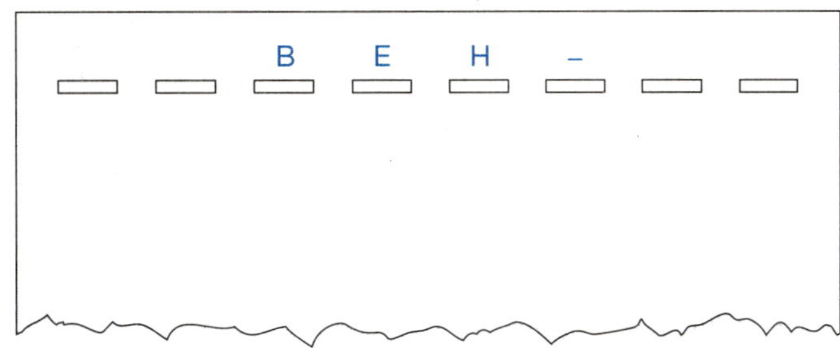

Alternately, set power supply on lower voltage, and run gel for several hours. When running two gels from the same power supply, the current is double that for a single gel at the same voltage.

3. Close the top of the electrophoresis box, and connect electrical leads to a power supply, anode to anode (red-red) and cathode to cathode (black-black). Make sure that both electrodes are connected to the same channel of the power supply.

4. Turn the power supply on, and set to 100–150 volts. The ammeter should register approximately 50–100 milliamperes. If current is not detected, check connections and try again.

5. Separate by electrophoresis for 40–60 minutes. Good separation will have occurred when the bromophenol blue band has moved 4–7 cm from the wells. If time allows, carry out electrophoresis until the bromophenol blue band nears the end of the gel. *Stop* electrophoresis before the bromophenol blue band runs off the end of the gel.

6. Turn off power supply, disconnect leads from the inputs, and remove top of electrophoresis box.

7. Carefully remove the casting tray from the electrophoresis box, and slide the gel into a disposable weigh boat or other shallow tray. Label staining tray with your name.

STOP Cover electrophoresis tank and save gel until ready to continue. Gel can be stored in a zip-lock plastic bag and refrigerated overnight for viewing/photographing the next day. However, over longer periods of time, the DNA will diffuse through the gel, and the bands will become indistinct or disappear entirely.

Staining may be performed by an instructor in a controlled area when students are not present.

8. Stain and view gel using one of the methods described in Sections IVA and IVB.

IVA. Stain Gel with Ethidium Bromide and View (Photograph)

(10–15 minutes)

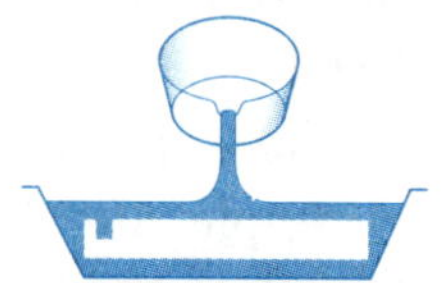

CAUTION
Review Responsible Handling of Ethidium Bromide in the Prelab Notes. Wear latex gloves when staining, viewing, and photographing gel and during clean up. Confine all staining to a restricted sink area. For further information, see Appendix 4.

Staining time increases markedly for thicker gels. Do not be tempted to use a higher concentration of ethidium bromide in the staining solution. This will not enhance the DNA bands; it only increases the background staining of the agarose gel itself.

Ethidium bromide solution may be reused to stain 15 or more gels. When staining time increases markedly, dispose of ethidium bromide solution as explained in the Prelab Notes.

1. Flood gel with ethidium bromide solution (1 μg/ml), and allow to stain for 5–10 minutes.
2. Following staining, use a funnel to decant as much ethidium bromide solution as possible from the staining tray back into the storage container.
3. Rinse gel and tray under running tap water.
4. If desired, the gel can be destained in tap water or distilled water for 5 minutes or more to help remove background ethidium bromide from the gel.

Staining intensifies dramatically if rinsed gels set overnight at room temperature. Stack staining trays, and cover top gel with plastic wrap to prevent desiccation.

5. View under UV transilluminator or other UV source.

CAUTION
UV light can damage eyes. Never look at unshielded UV light source with naked eyes. View only through a filter or safety glasses that absorb harmful wavelengths. For further information, see Appendix 4.

6. Photograph with a Polaroid or digital camera.
7. Take time for responsible cleanup.
   a. Wipe down camera, transilluminator, and staining area.
   b. Decontaminate gels and any staining solution not to be reused.
   c. Wash hands before leaving lab.

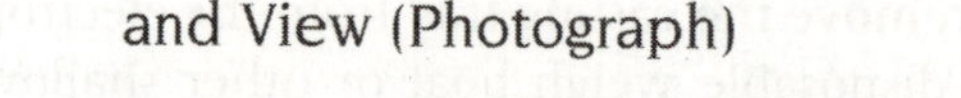

IVB. Stain Gel with Methylene Blue and View (Photograph)

(30+ minutes)

1. Wear latex gloves during staining and cleanup.
2. Flood gels with 0.025% methylene blue, and allow to stain for 20–30 minutes.
3. Following staining, use a funnel to decant as much methylene blue solution as possible from the staining tray back into the storage container.
4. Rinse the gel in running tap water. Let the gel soak for several minutes in several changes of fresh water. DNA bands will become increasingly distinct as the gel destains.

Destaining time is decreased by agitating and rinsing in warm water.

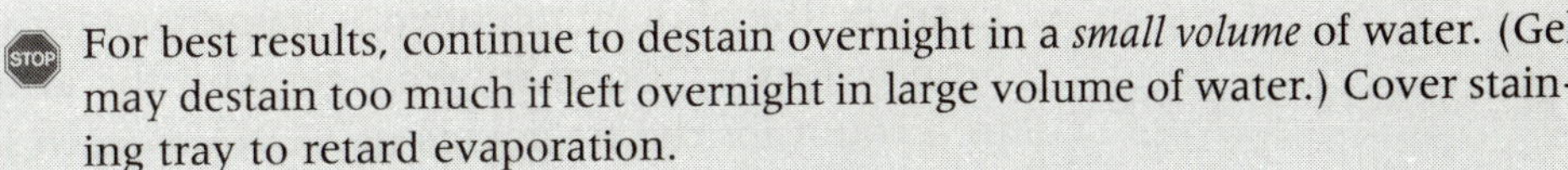

For best results, continue to destain overnight in a *small volume* of water. (Gel may destain too much if left overnight in large volume of water.) Cover staining tray to retard evaporation.

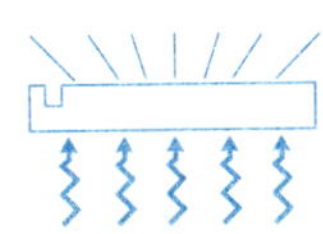

5. View gel over light box; cover the surface with plastic wrap to prevent staining.

6. Photograph with a Polaroid or digital camera.

RESULTS AND DISCUSSION

Agarose gel electrophoresis combined with ethidium bromide staining allows the rapid analysis of DNA fragments. However, prior to the introduction of this method in 1973, analysis of DNA molecules was a laborious task. The original separation method, involving ultracentifugation of DNA in a sucrose gradient, gave only crude size approximations and took more than 24 hours to complete.

Electrophoresis using a polyacrylamide gel in a glass tube was an improvement, but it could only be used to separate small DNA molecules of up to 2000 bp. Another drawback was that the DNA had to be radioactively labeled prior to electrophoresis. Following electrophoresis, the polyacrylamide gel was cut into thin slices, and the radioactivity in each slice was determined. The amount of radioactivity detected in each slice was plotted versus distance migrated, producing a series of radioactive peaks representing each DNA fragment.

DNA restriction analysis is at the heart of recombinant DNA technology and of the laboratories in this course. The ability to cut DNA predictably and precisely enables DNA molecules to be manipulated and recombined at will. The fact that discrete bands of like-sized DNA fragments are seen in one lane of an agarose gel shows that each of the more than 1 billion λ DNA molecules present in each restriction reaction was cut in precisely the same place.

By convention, DNA gels are "read" from left to right, with the sample wells oriented at the top. The area extending from the well down the gel is termed a "lane." Thus, reading down a lane identifies fragments generated by a particular restriction reaction. Scanning across lanes identifies fragments that have comigrated the same distance down the gel and are thus of like size.

1. Why is water added to tube labeled "–" in Part I, Step 7?
2. What is the function of compromise restriction buffer?
3. What are the two functions of loading dye?
4. How does ethidium bromide stain DNA? How does this relate to the need to minimize exposure to humans?
5. Troubleshooting electrophoresis. What would occur
    a. if the gel box is filled with water instead of TBE buffer?
    b. if water is used to prepare the gel instead of TBE buffer?
    c. if the electrodes are reversed?

6. Examine the photograph of your stained gel (or view on a light box or overhead projector). Compare your gel with the ideal gel shown below and try to account for the fragments of λ DNA in each lane. How can you account for differences in separation and band intensity between your gel and the ideal gel?
7. Troubleshooting gels. What effect will be observed in the stained bands of DNA in an agarose gel
   a. if the casting tray is moved or jarred while agarose is solidifying in Part II, Step 3?
   b. if the gel is run at very high voltage?
   c. if a large air bubble or clump is allowed to set in agarose?
   d. if too much DNA is loaded in a lane?
8. Linear DNA fragments migrate at rates inversely proportional to the $\log_{10}$ of their molecular weights. For simplicity's sake, base-pair length is substituted for molecular weight.
   a. The matrix on the facing page gives the base-pair size of λ DNA fragments generated by a *Hin*dIII digest.

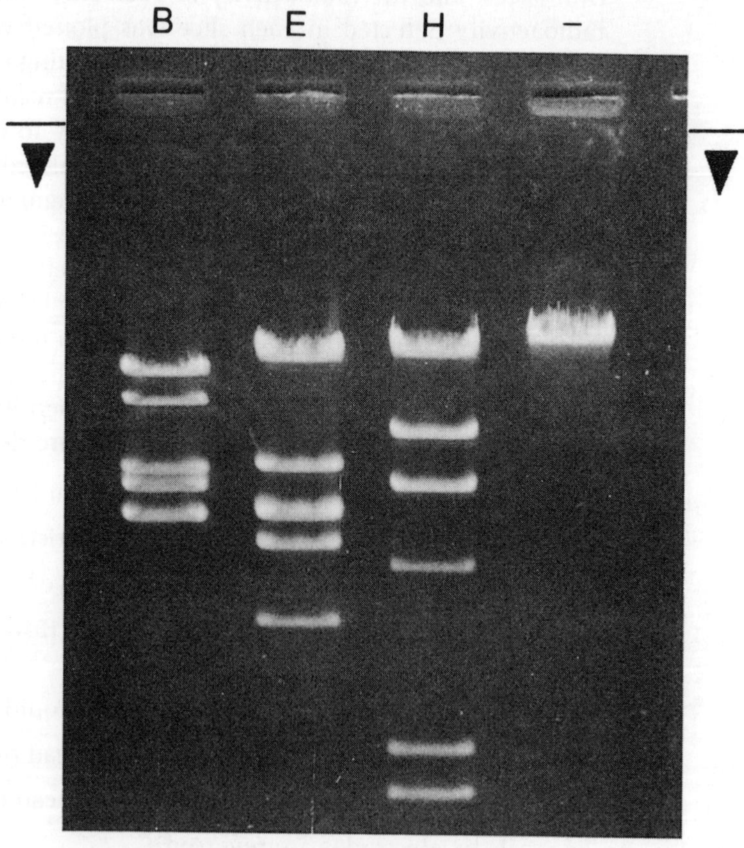

Ideal Gel

*Hin*dIII		*Eco*RI			*Bam*HI		
Dis.	Act. bp	Dis.	Cal. bp	Act. bp	Dis.	Cal. bp	Act. bp
	27,491[a]						
	23,130[a]						
	9,416						
	6,557						
	4,361						
	2,322						
	2,027						
	564[b]						
	125[c]						

[a]Pair appears as a single band on the gel.
[b]Band may not be visible in methylene-blue-stained gel.
[c]Band runs off the end of the gel when bromophenol blue is approximately 2 cm from the end of the gel. When present on the gel, the band is not detected by methylene blue and is usually difficult to detect with ethidium bromide staining.

b. Using the ideal gel shown on the facing page, carefully measure the distance (in millimeters) each *Hin*dIII, *Eco*RI, and *Bam*HI fragment migrated from the origin. Measure from the front edge of the well to the front edge of each band. Enter distances into the matrix. Alternatively, measure the distances on the overhead-projected image of the methylene-blue-stained gel.

c. Match base-pair sizes of *Hin*dIII fragments with bands that appear in the ideal digest. Label each band with kilobase pair (kbp) size. For example, 27,491 bp equals 27.5 kbp.

d. Set up semilog graph paper with distance migrated as the *x* (arithmetic) axis and log of base-pair length as the *y* (logarithmic) axis. Then, plot the distance migrated versus the base-pair length for each *Hin*dIII fragment.

e. Connect data points with a line.

f. Locate on the *x* axis the distance migrated by the first *Eco*RI fragment. Use a ruler to draw a vertical line from this point to its intersection with the best-fit data line.

g. Now extend a horizontal line from this point to the *y* axis. This gives the base-pair size of this *Eco*RI fragment.

h. Repeat Steps f and g for each *Eco*RI and *Bam*HI fragment. Enter the results in the calculated base-pair (Cal. bp) columns for each digest.

i. Enter the actual base-pair size of *Eco*RI and *Bam*HI fragments (as provided by your instructor) into Act. bp column.

j. For which fragment sizes was your graph most accurate? For which fragment sizes was it least accurate? What does this tell you about the resolving ability of agarose gel electrophoresis?

9. DNA fragments of similar size will not always resolve on a gel. This is seen in lane E in the Ideal Gel, where *Eco*RI fragments of 5804 bp and 5643 bp migrate as a single heavy band. These are referred to as a doublet and can be recognized because they are brighter and thicker than similarly sized singlets. What could be done to resolve the doublet fragments?

10. Determine a range of sensitivity of DNA detection by ethidium bromide by comparing the mass of DNA in the bands of the largest and smallest detectable fragments on the gel. To determine the mass of DNA in a given band:

$$\frac{\text{number of bp in fragment} \times (\text{conc. of DNA}) \times (\text{vol. of DNA})}{\text{number of bp in } \lambda \text{ DNA}}$$

For example:

$$\frac{24{,}251 \text{ bp } (0.1\ \mu\text{g}/\mu\text{l})\ (4\ \mu\text{l})}{48{,}502 \text{ bp}} = 0.2\ \mu\text{g}$$

Now, compute the mass of DNA in the largest and smallest *singlet* fragments on the gel.

11. λ DNA can exist both as a circular molecule and as a linear molecule. At each end of the linear molecule is a single-stranded sequence of 12 nucleotides, called a COS site. The COS sites at each end are complementary to each other and thus can base pair to form a circular molecule. These complementary ends are analogous to the "sticky ends" created by some restriction enzymes. Commercially available λ DNA is likely to be a mixture of linear and circular molecules. This leads to the appearance of more bands on the

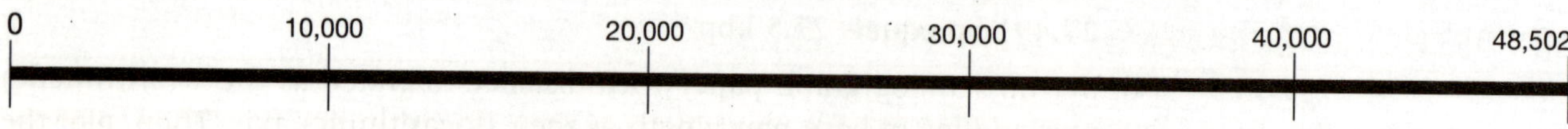

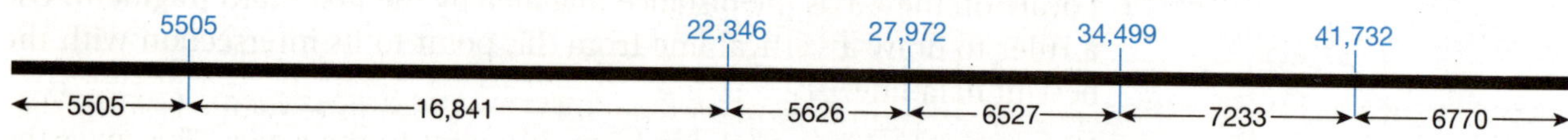

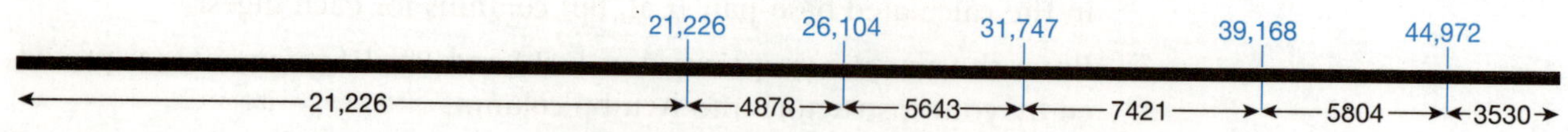

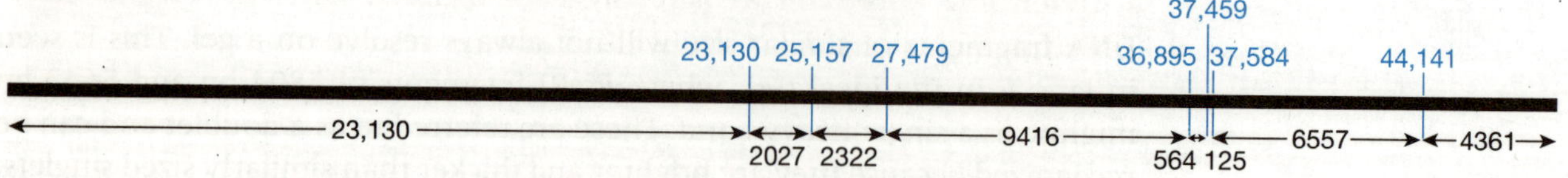

Restriction Maps of the Linear λ Genome

gel than would be predicted from a homogeneous population of linear DNA molecules. This also causes the partial loss of other fragments. For example, the left-most *Hin*dIII site is 23,130 bp from the left end of the linear λ genome, and the right-most site is 4361 bp from the right end. The 4361-bp band is faint in comparison to other bands on the gel of similar size. This indicates that a percentage of the DNA molecules are circular—combining the 4361-bp terminal fragment with the 23,130-bp terminal fragment to produce a 27,491-bp fragment. However, the combined 27,491-bp fragment usually runs as a doublet along with the 23,130-bp fragment from the linear molecule.

a. Use a protractor to draw three circles about 3 inches in diameter. These represent λ DNA molecules with base-paired COS sites.

b. Label a point at 12:00 on each circlet 48/0. This marks the point where the COS sites are joined.

c. Use data from the restriction maps of the linear λ genome to make a rough map of restriction sites for *Hin*dIII on one of the circles. Note the situation described above.

d. Next make rough restriction maps of *Bam*HI and *Eco*RI sites on the remaining two circles.

e. What *Bam*HI and *Eco*RI fragments are created in the circular molecules? Why (or why not) can you locate each of these fragments on your gel or the ideal gel above?

FOR FURTHER RESEARCH

1. Some of the circular λ molecules are covalently linked at the COS sites. Other circles are only hydrogen-bonded and can dissociate to form linear molecules. Heating λ DNA to 65°C for 10 minutes linearizes any noncovalent COS circles in the preparation by breaking hydrogen bonds that hold the complementary COS sites together.

   a. Set up duplicate restriction digests of λ DNA with several enzymes. Then heat one reaction from each set at 65°C for 10 minutes, while holding the duplicates on ice. After 10 minutes, immediately place the heated tubes on ice. Relate changes in restriction patterns of heated versus unheated DNA to a restriction map of the circular λ genome as in Question 11 in Results and Discussion.

   b. How can the data generated by this experiment be used to quantify the approximate percentage of circular DNA in your preparation?

2. Design and carry out a series of experiments to study the kinetics of a restriction reaction.

   a. Determine approximate percentage of digested DNA at various time points.

   b. Repeat experiments with several enzyme dilutions and several DNA dilutions.

   c. In each case, at what time point does the reaction appear to be complete?

3. Design and test an assay to determine the relative stabilities of *Bam*HI, *Eco*RI, and *Hin*dIII at room temperature.
4. Determine the identity of an unknown restriction enzyme.
   a. Perform single digests of λ DNA with the unknown enzyme, as well as with several known restriction enzymes. Run the restriction fragments in an agarose gel at 50 volts to produce well-spread and well-focused bands.
   b. For each fragment, plot distance migrated *versus* base-pair size, as in Question 8 in Results and Discussion. Use the graph to determine the base-pair lengths of the unknown fragments and compare with restriction maps of commercially available enzymes.
5. Research the steps needed to purify a restriction enzyme from *E. coli* and characterize its recognition sequence.

Field Guide to Electrophoresis Effects

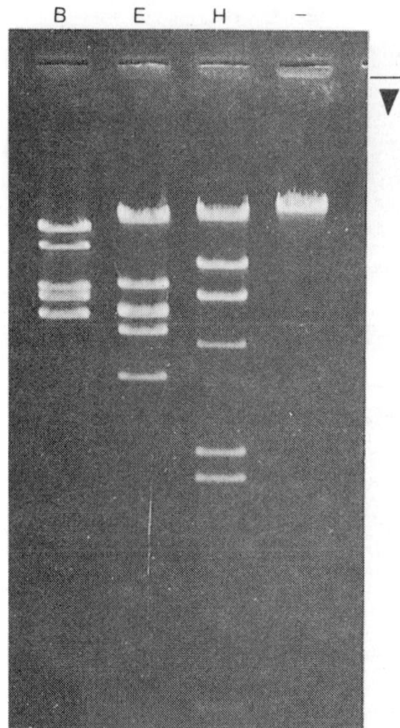

Ideal Gel

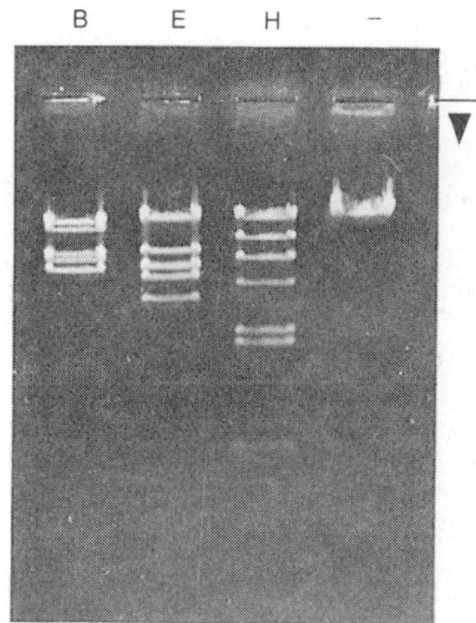

Short Run
Bands compressed. Short time electrophoresing.

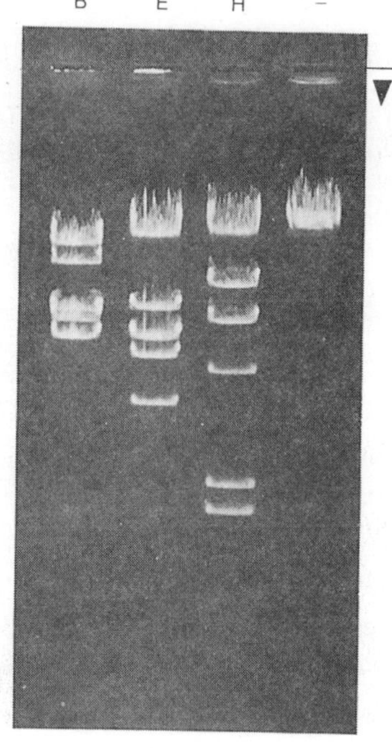

Overloaded
Bands smeared in all lanes. too much DNA in digests.

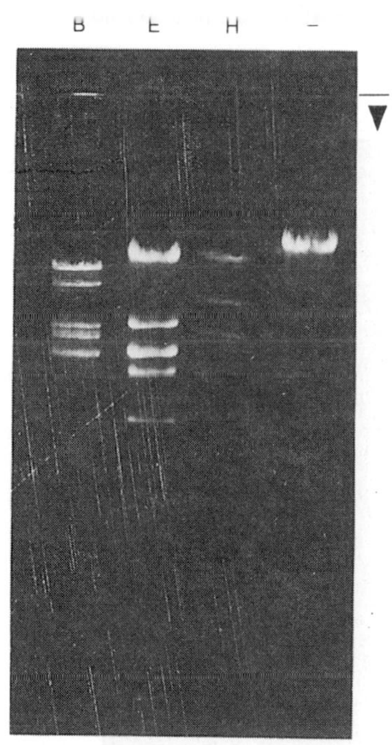

Punctured Wells
Bands faint in Lanes B and H. DNA lost through hole punched in bottom of well with pipette tip.

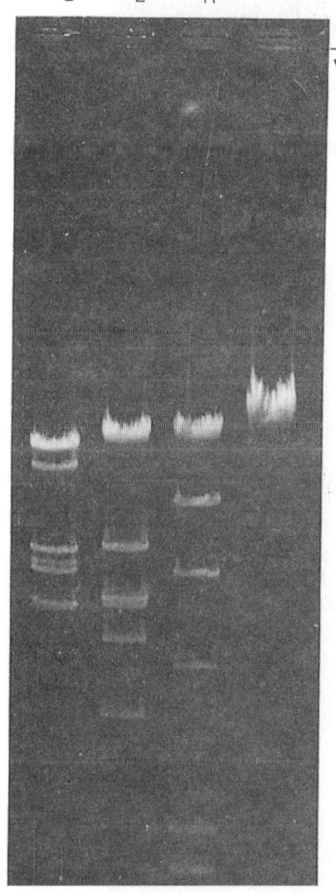

Long Run
Bands spread. Long time electrophoresing.

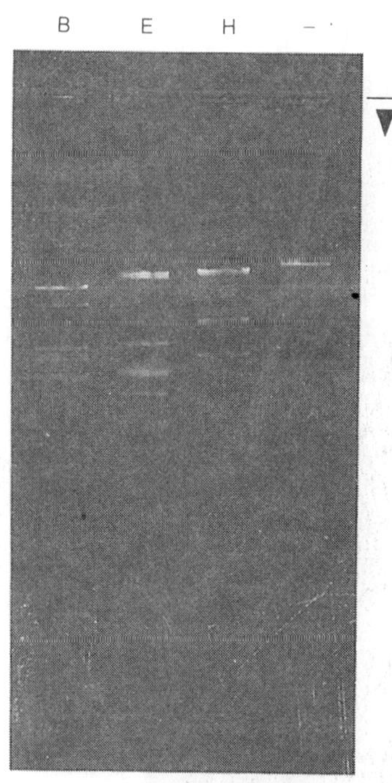

Underloaded
Bands faint in all lanes. Too little DNA in digests.

(Field Guide is continued on next page.)

Field Guide to Electrophoresis Effects (*continued*)

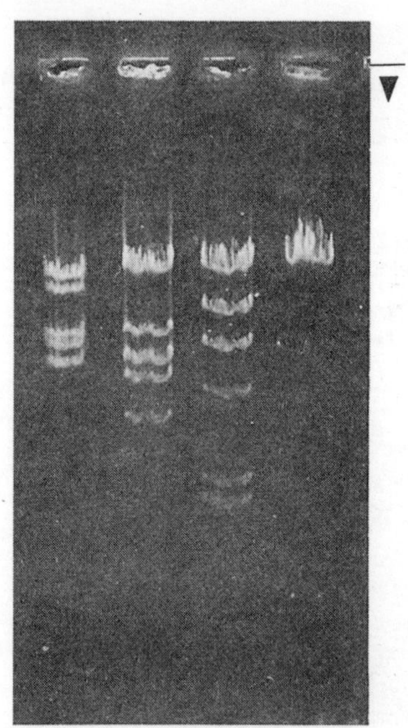

Poorly Formed Wells

Wavy bands in all lanes. Comb removed before gel was completely set.

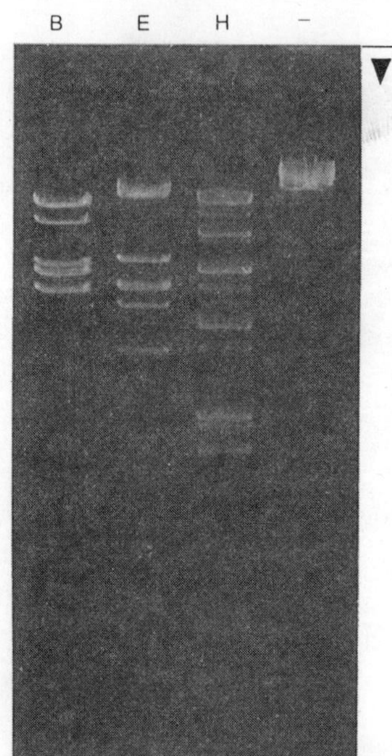

Enzymes Mixed

Extra bands in Lane H. *Bam*HI and *Hin*dIII mixed in digest.

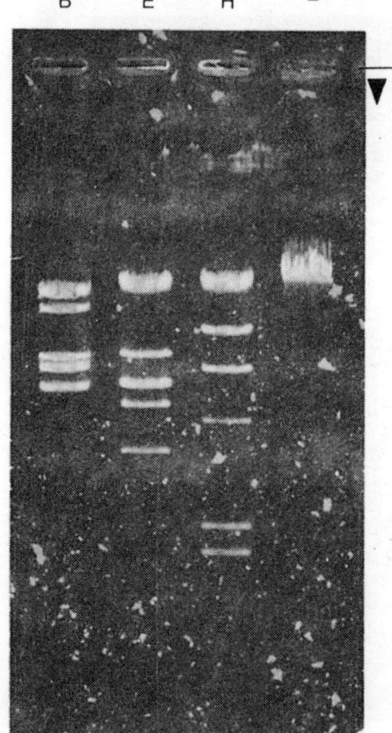

Precipitate

Precipitate in TBE buffer used to make gel.

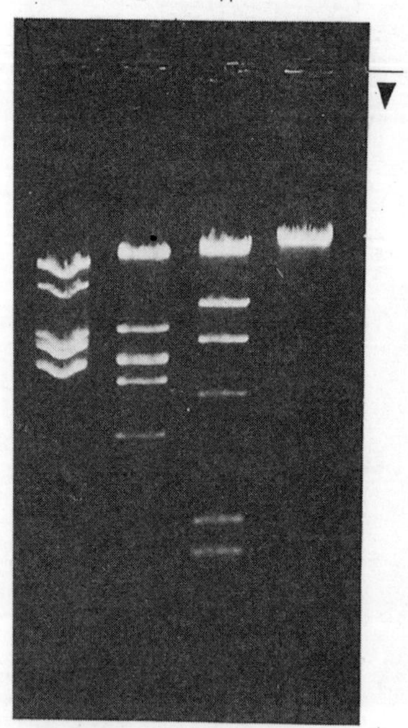

Bubble in Lane

Bump in band in Lane B. Bubble in lane.

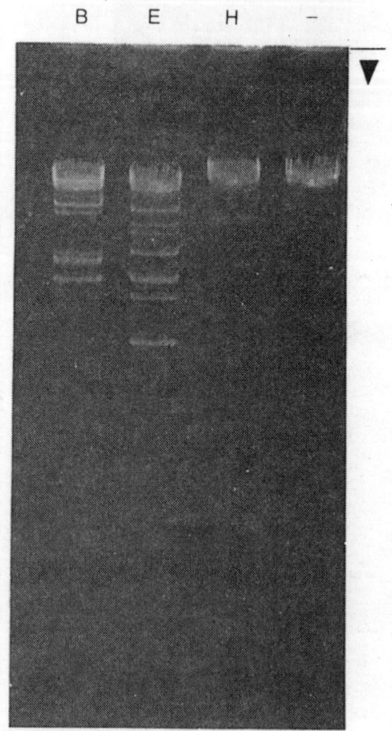

Incomplete Digest

Bands faint in Lane H. Very little *Hin*dIII in digest. Also, extra bands are present in Lanes B and E.

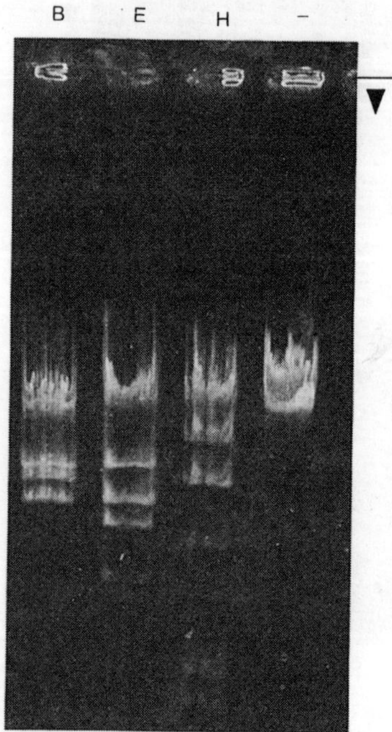

Gel Made with Water

Bands smeared in all lanes. Gel made with water or wrong concentration of TBE buffer.

Effects of DNA Methylation on Restriction

IN LABORATORY 4, THE *Eco*RI METHYLATION SYSTEM is used to illustrate the sequence specificity of a modifying enzyme that protects DNA from restriction enzyme digestion. *Eco*RI methylase adds a methyl group to the second adenine residue in the *Eco*RI recognition site, thus preventing the endonuclease from binding and cutting the DNA. *S*-Adenosyl methionine (SAM), included in the methylation reaction, donates the methyl group that is attached to the DNA molecule by the methylase.

Three samples of bacteriophage λ DNA are incubated at 37°C with *Eco*RI methylase, one of which is subsequently incubated with *Eco*RI and another of which is incubated with *Hin*dIII. The third sample, a control, is incubated without a restriction enzyme. Three control samples of nonmethylated DNA are also incubated with *Eco*RI, *Hin*dIII, and no enzyme. All of the samples are separated by electrophoresis in an agarose gel and stained. Comparison of the band patterns reveals that the methylated DNA is protected from digestion by *Eco*RI. However, methylation at the *Eco*RI site has no effect on the activity of the restriction enzyme *Hin*dIII.

Equipment and materials for this laboratory are available from the Carolina Biological Supply Company (see Appendix 1).

I. Set Up Methylase Reaction

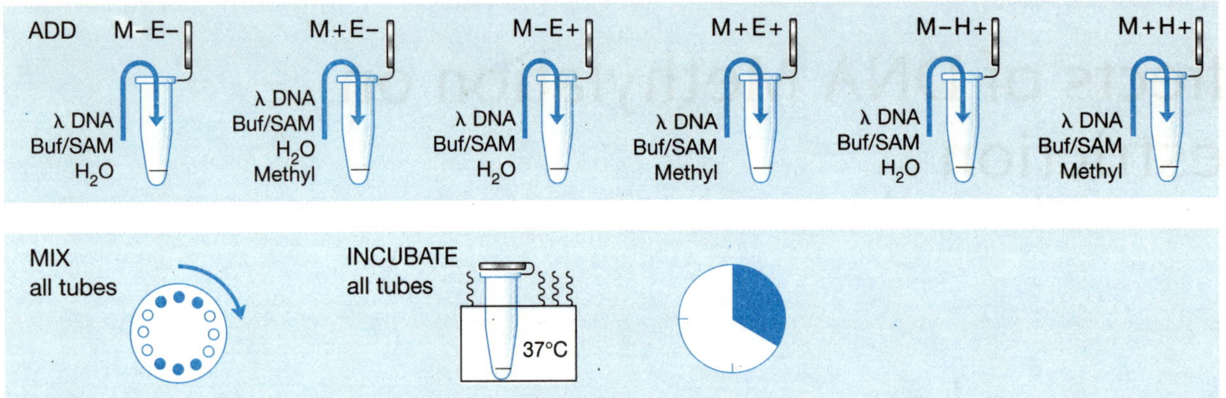

II. Cast 0.8% Agarose Gel

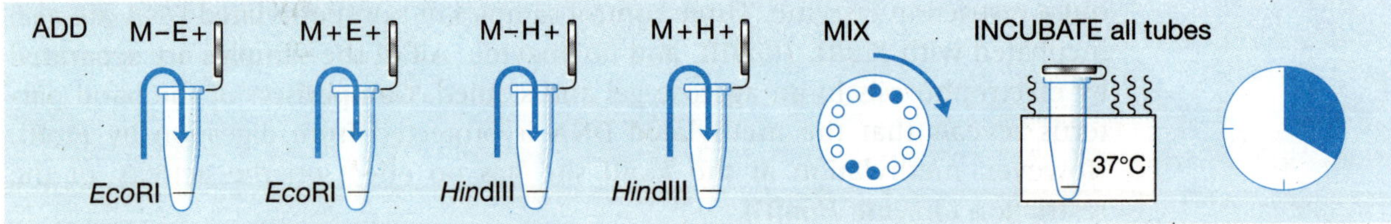

III. Set Up Restriction Reaction

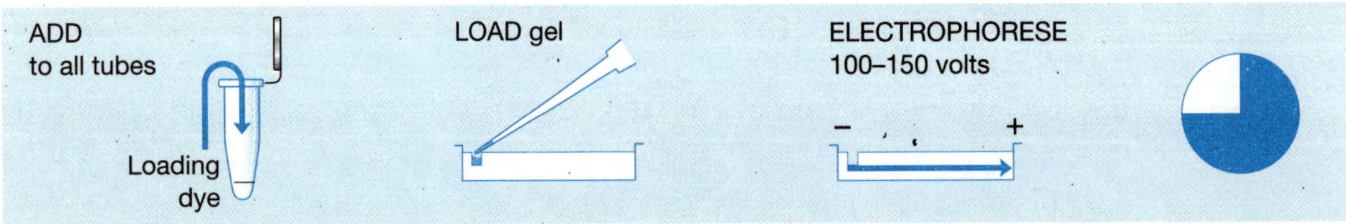

IV. Load Gel and Separate by Electrophoresis

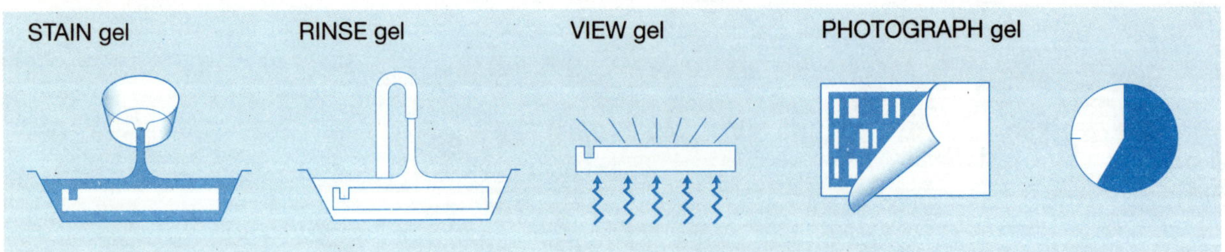

V. Stain Gel and View (Photograph)

STAIN gel

RINSE gel

VIEW gel

PHOTOGRAPH gel

PRELAB NOTES

Review Prelab Notes in Laboratory 3, DNA Restriction Analysis.

S-Adenosyl Methionine

S-Adenosyl methionine (SAM) is incorporated into a 2x restriction buffer, so that the same buffer is used for *both* methylation and restriction reactions. Because SAM is not very stable, mix the buffer/SAM solution just prior to the lab and discard after use. In addition, make sure to use a fresh stock of SAM not more than several months old.

To Avoid Confusion

This laboratory has two distinct steps involving two similar-sounding reagents. In the first step, DNA is preincubated with *Eco*RI methylase. In the second step, the methylated DNA is incubated with *Eco*RI restriction enzyme. To avoid mishaps, do not set out the endonuclease *Eco*RI restriction enzyme until *after* the methylation reactions are set up.

For Further Information

The protocol presented here is based on the following published methods:

Aaij C. and Borst P. 1972. The gel electrophoresis of DNA. *Biochim. Biophys. Acta* **269:** 192–200.

Greene P.H., Poonian M.S., Nussbaum A.L., Tobias L., Garfin D.E., Boyer H.W., and Goodman H.M. 1975. Restriction and modification of a self-complementary octanucleotide containing the *Eco*RI substrate. *J. Mol. Biol.* **99:** 237–261.

Helling R.B., Goodman H.M., and Boyer H.W. 1974. Analysis of R-*Eco*RI fragments of DNA from lambdoid bacteriophages and other viruses by agarose-gel electrophoresis. *J. Virol.* **14:** 1235–1244.

Sharp P.A., Sugden B., and Sambrook J. 1973. Detection of two restriction endonuclease activities in *Haemophilus parainfluenzae* using analytical agarose–ethidium bromide electrophoresis. *Biochemistry* **12:** 3055–3063.

PRELAB PREPARATION

Before performing this Prelab Preparation, please refer to the cautions indicated on the Laboratory Materials list.

1. Prepare 2x restriction buffer plus SAM within 1–2 days before the experiment.
2. Prepare aliquots for each experiment:

   30 μl of 0.1 μg/μl λ DNA (store on ice)
   40 μl of 2x restriction buffer/SAM (store on ice)
   5 μl of *Eco*RI methylase (store on ice)
   3 μl each of *Eco*RI and *Hin*dIII (store on ice)
   500 μl of distilled water
   500 μl of loading dye

3. Prepare 0.8% agarose solution (40–50 ml per experiment). Keep agarose liquid in a hot-water bath (~60°C) throughout lab. Cover with aluminum foil to retard evaporation.
4. Prepare 1x Tris/Borate/EDTA (TBE) buffer for electrophoresis (400–500 ml per experiment).
5. Prepare ethidium bromide or methylene-blue-staining solution (100 ml per experiment).
6. Adjust water bath to 37°C.

MATERIALS

REAGENTS	SUPPLIES AND EQUIPMENT
Agarose (0.8%)	Aluminum foil
Distilled water	Beakers for agarose and for waste/used tips
Ethidium bromide▼(1 μg/ml) (or 0.025% methylene blue▼)	Camera and film (optional)
λ DNA (0.1 μg/μl)	Electrophoresis box
Loading dye	Latex gloves
2x Restriction buffer/SAM▼	Masking tape
Enzymes:	Microfuge (optional)
*Eco*RI	Micropipettor (0.5–10-μl) + tips
*Hin*dIII	Parafilm or wax paper (optional)
*Eco*RI methylase	Permanent marker
1x Tris▼/Borate/EDTA (TBE) buffer	Plastic wrap (optional)
	Power supply
	Test tube rack
	Transilluminator (optional)▼
	Tubes (1.5-ml)
	Water baths (37°C and 60°C)

▼ See Appendix 4 for Cautions list.

METHODS

I. Set Up Methylase Reaction

(30 minutes, including incubation)

1. Use a permanent marker to label six 1.5-ml tubes, in which methylation and restriction reactions will be performed:

M– E– = no methylase, no *Eco*RI
M+ E– = methylase, no *Eco*RI
M– E+ = no methylase, *Eco*RI
M+ E+ = methylase, *Eco*RI
M– H+ = no methylase, *Hin*dIII
M+ H+ = methylase, *Hin*dIII

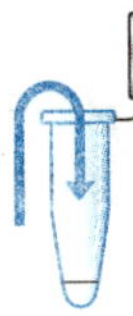

2. Use the matrix below as a checklist while adding reagents to each reaction. Read down each column, adding the same reagent to all appropriate tubes. *Use a fresh tip for each reagent.* Refer to detailed instructions that follow.

Tube	λ DNA	Buffer/SAM	H_2O	*Eco*RI methylase
M–E–	4 µl	5 µl	2 µl	—
M+E–	4 µl	5 µl	1 µl	1 µl
M–E+	4 µl	5 µl	1 µl	—
M+E+	4 µl	5 µl	—	1 µl
M–H+	4 µl	5 µl	1 µl	—
M+H+	4 µl	5 µl	—	1 µl

It is not necessary to change tips when adding the same reagent. The same tip may be used for all tubes, provided the tip has not touched the solution already in the tubes.

3. Collect and place reagents in the test tube rack on the lab bench.
4. Add 4 µl of DNA to each reaction tube. Touch the micropipettor tip to the side of the reaction tube, as near to the bottom as possible, to create capillary action to pull the solution out of the tip.
5. Use a *fresh tip* to add 5 µl of restriction buffer/SAM to a clean spot on each reaction tube.

To avoid confusing methylase with reagents in Part III, discard after completing Step 7.

6. Use a *fresh tip* to add specified volume of distilled water to appropriate tubes.
7. Use a *fresh tip* to add 1 µl of *Eco*RI methylase to Tubes M+E–, M+E+, and M+H+.

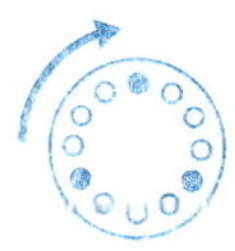

8. Close tube tops. Pool and mix reagents by pulsing in a microfuge or sharply tapping the tube bottom on the lab bench.

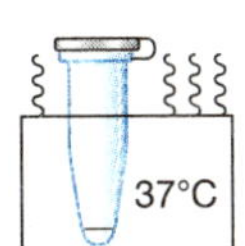

9. Place the reaction tubes in a 37°C water bath, and incubate for a minimum of 20 minutes. Reactions can be incubated for longer periods of time.

After several hours, methylase loses activity and the reaction stops.

STOP Following incubation, freeze reactions at –20°C until ready to continue. Thaw reactions before continuing to Section III, Step 1.

II. Cast 0.8% Agarose Gel (5 minutes)

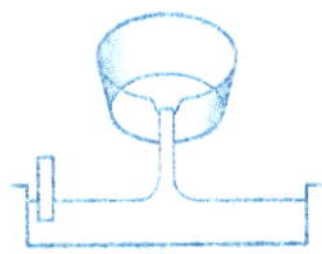

Gel is cast directly in the box in some electrophoresis apparatuses.

1. Seal the ends of the gel-casting tray with tape, and insert a well-forming comb. Place gel-casting tray out of the way on the lab bench, so that agarose poured in next step can set undisturbed.
2. Carefully pour enough agarose solution into the casting tray to fill to a depth of about 5 mm. Gel should cover only about one-third the height of comb teeth. Use a pipette tip to move large bubbles or solid debris to the sides or end of the tray while the gel is still liquid.
3. Gel will become cloudy as it solidifies (~10 minutes). *Do not move or jar the casting tray while the agarose is solidifying.* Touch corner of agarose *away* from the comb to test whether the gel has solidified.
4. When agarose has set, unseal ends of casting tray. Place the tray on the platform of the gel box, so that comb is at negative black electrode (cathode).
5. Fill box with TBE buffer, to a level that just covers the entire surface of the gel.

Too much buffer will channel the current over the top of the gel rather than through the gel, increasing the time required to separate DNA. TBE buffer can be used several times; do not discard. If using buffer in electrophoresis box from a previous experiment, rock the chamber back and forth to remix ions that have accumulated at either end.

Buffer solution helps to lubricate the comb. Some gel boxes are designed such that the comb must be removed prior to inserting the casting tray into the box. In this case, flood the casting tray and gel surface with running buffer before removing the comb. Combs removed from a dry gel can tear the wells.

6. Gently remove comb, taking care not to rip the wells.
7. Make sure that the sample wells left by the comb are completely submerged. If "dimples" appear around the wells, slowly add buffer until they disappear.

Cover electrophoresis tank and save gel until ready to continue. Gel will remain in good condition for at least several days if it is completely submerged in buffer.

III. Set Up Restriction Reaction (30 minutes, including incubation)

1. Remove methylation reactions from the water bath or thaw the tube stored in the freezer.
2. Use the matrix below as a checklist while adding reagents to each reaction. Read down each column, adding the same reagent to all appropriate tubes. *Use a fresh tip for each reagent.* Refer to detailed instructions that follow.

Tube	*Eco*RI	*Hin*dIII
M–E–	—	—
M+E–	—	—
M– E+	1 μl	—
M+E+	1 μl	—
M–H+	—	1 μl
M+H+	—	1 μl

3. Collect *Eco*RI and *Hin*dIII, and place them on ice on the lab bench.
4. Add 1 μl of *Eco*RI to Tubes M–E+ and M+E+.
5. Use a *fresh tip* to add 1 μl of *Hin*dIII to tubes labeled M–H+ and M+H+.

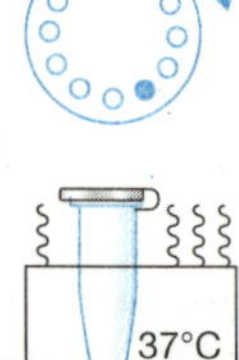

6. Close tube tops. Pool and mix reagents by pulsing in a microfuge or sharply tapping the tube bottom on the lab bench.
7. Place the reaction tubes in a 37°C water bath, and incubate restriction reactions for a minimum of 20 minutes. Reactions can be incubated for longer periods of time.

After several hours, enzymes lose activity and the reaction stops.

Following incubation, freeze reactions at –20°C until ready to continue. Thaw reactions before continuing to Section IV, Step 1.

IV. Load Gel and Separate by Electrophoresis (50–70 minutes)

1. Add loading dye to each reaction. Either
   a. Add 1–2 μl of loading dye to each reaction tube. Close tube tops, and mix by tapping the tube bottom on the lab bench, pipetting in and out, or pulsing in a microfuge. Make sure that the tubes are placed in a *balanced* configuration in the rotor.

   or

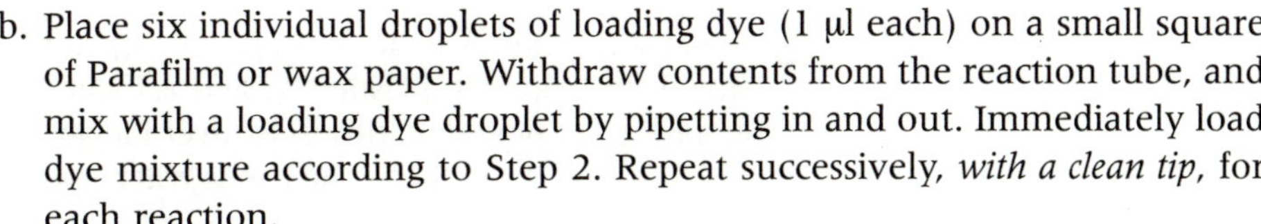

b. Place six individual droplets of loading dye (1 µl each) on a small square of Parafilm or wax paper. Withdraw contents from the reaction tube, and mix with a loading dye droplet by pipetting in and out. Immediately load dye mixture according to Step 2. Repeat successively, *with a clean tip*, for each reaction.

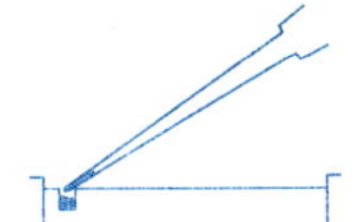

A piece of dark construction paper beneath the gel box will give the wells more visibility.

2. Use a micropipettor to load 10 µl of each reaction tube into a separate well in the gel, as shown in diagram below. *Use a fresh tip for each reaction.*
   a. Use two hands to steady the micropipettor over the well.
   b. Before loading sample, make sure that there are no bubbles in the wells. If bubbles exist, remove them with a micropipettor tip.
   c. If there is air in the end of the tip, carefully depress the plunger to push the sample to the end of the tip. (If an air bubble forms a "cap" over the well, DNA/loading dye will flow into buffer around edges of well.)
   d. Dip the micropipettor tip through the surface of the buffer, center it over the well, and gently depress the micropipettor plunger to slowly expel the sample. Sucrose in the loading dye weighs down the sample, causing it to sink to the bottom of the well. *Be careful not to punch the tip of the micropipettor through the bottom the gel.*

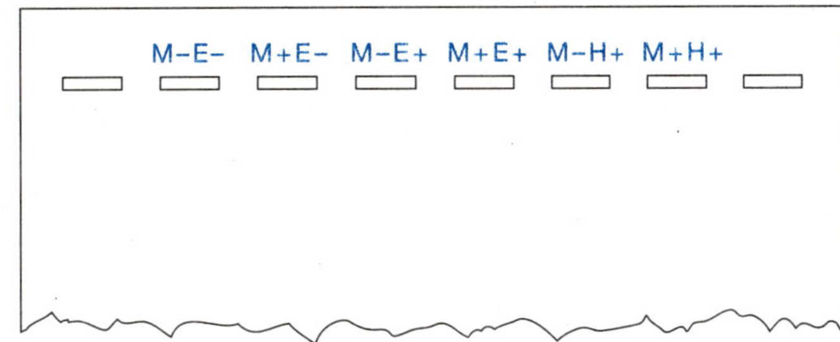

3. Close the top of the electrophoresis box, and connect the electrical leads to a power supply, anode to anode (red-red) and cathode to cathode (black black). Make sure that both electrodes are connected to the same channel of power supply.
4. Turn power supply on, and set to 100–150 volts. The ammeter should register approximately 50–100 milliamperes. If current is not detected, check connections and try again.

Alternately, set power supply on lower voltage, and run the gel for several hours. When running two gels from the same power supply, the current is double that for a single gel at the same voltage.

5. Separate by electrophoresis for 40–60 minutes. Good separation will have occurred when the bromophenol blue band has moved 4–7 cm from the wells. If time allows, carry out electrophoresis until the bromophenol blue band nears the end of the gel. *Stop* electrophoresis before bromophenol blue band runs off end of gel.
6. Turn off power supply, disconnect leads from the inputs, and remove top of electrophoresis box.
7. Carefully remove casting tray from electrophoresis box, and slide the gel into a disposable weigh boat or other shallow tray. Label staining tray with your name.

Staining may be performed by an instructor in a controlled area when students are not present.

8. Stain and view gel using one of the methods described in Sections VA and VB.

 Cover electrophoresis tank and save gel until ready to continue. Gel can be stored in a zip-lock plastic bag and refrigerated overnight for viewing/photographing the next day. However, over longer periods of time, the DNA will diffuse through the gel, and the bands will become indistinct or disappear entirely.

VA. Stain Gel with Ethidium Bromide and View (Photograph)

(10–15 minutes)

CAUTION
Review Responsible Handling of Ethidium Bromide in Laboratory 3. Wear latex gloves when staining, viewing, and photographing gels and during cleanup. Confine all staining to a restricted sink area. For further information, see Appendix 4.

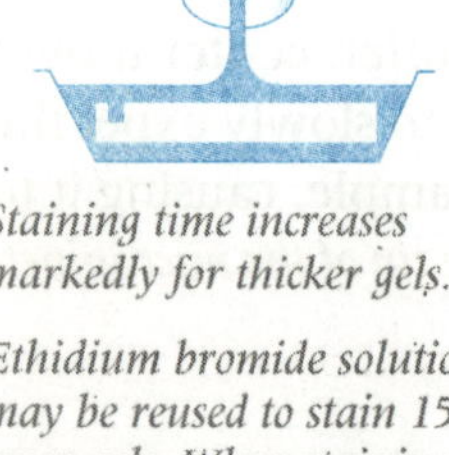

Staining time increases markedly for thicker gels.

Ethidium bromide solution may be reused to stain 15 or more gels. When staining time increases markedly, dispose of ethidium bromide solution as explained in Laboratory 3.

1. Flood gel with ethidium bromide solution (1 μg/ml), and allow to stain for 5–10 minutes.
2. Following staining, use a funnel to decant as much ethidium bromide solution as possible from the staining tray back into the storage container.
3. Rinse the gel and tray under running tap water.
4. If desired, gels can be destained in tap water or distilled water for 5 minutes or more to remove background ethidium bromide.

 Staining intensifies dramatically if rinsed gels set overnight at room temperature. Stack staining trays, and cover top gel with plastic wrap to prevent desiccation.

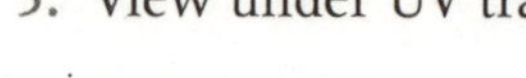

5. View under UV transilluminator or other UV source.

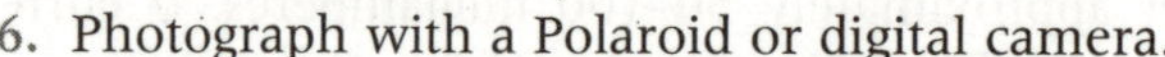

CAUTION
Ultraviolet light can damage eyes. Never look at unshielded UV light source with naked eyes. View only through a filter or safety glasses that absorb harmful wavelengths. For further information, see Appendix 4.

6. Photograph with a Polaroid or digital camera.
7. Take time for responsible cleanup.
   a. Wipe down camera, transilluminator, and staining area.
   b. Wash hands before leaving lab.

VB. Stain Gel with Methylene Blue and View (Photograph)

(30+ minutes)

1. Wear latex gloves during staining and cleanup.
2. Flood gels with 0.025% methylene blue, and allow to stain for 20–30 minutes.
3. Following staining, use a funnel to decant as much methylene blue solution as possible from the staining tray back into the storage container.

Destaining time is decreased by rinsing the gel in warm water, with agitation.

4. Rinse the gel in running tap water. Allow the gel to soak for several minutes in several changes of fresh water. DNA bands will become increasingly distinct as the gel destains.

> STOP For best results, continue to destain overnight in a *small volume* of water. (Gel may destain too much if left overnight in large volume of water.) Cover staining tray to retard evaporation.

5. View gel over light box; cover surface with plastic wrap to prevent staining.
6. Photograph with a Polaroid or digital camera.

RESULTS AND DISCUSSION

Each type II restriction enzyme has a corresponding methylase that recognizes the same nucleotide sequence. However, rather than cutting the DNA at this point, the methylase adds a methyl group within the recognition sequence. This "modification" blocks the restriction endonuclease from recognizing and binding to the restriction site. Within the bacterium, methylation protects the host DNA from cleavage by its own restriction enzyme. Unmethylated foreign DNA is not protected from cleavage.

The methylation reaction requires *S*-adenosyl methionine (SAM), the universal methyl-donating molecule in both prokaryotes and eukaryotes. As its name implies, SAM is composed of the nucleoside adenosine and the amino acid methionine. The donation of a methyl from the methionine portion of the molecule converts it into *S*-adenosyl homoserine.

One common occurrence in this laboratory is partial methylation, where methyl groups are added to only a fraction of the *Eco*RI sites within the λ DNA molecules. Cleavage at the unprotected sites produces a partial digest, yielding restriction fragments of varying lengths. These fragments are evidenced as lower-molecular-weight bands in an agarose gel. The intensity of the bands is inversely proportional to the level of DNA methylation.

1. Examine the photograph of your stained gel (or view on a light box or overhead projector). Compare your gel to the ideal gel shown on the next page. How can you account for differences in separation and band intensity?
2. What does the M+H+ control tell you about *Eco*RI methylation?
3. What does the M+E– control tell you about methylation?
4. What biological function do methylases perform in bacteria? What adaptive value do they have for a bacterium?
5. Experimental constraints demand that a plasmid be constructed in two major steps, whose order cannot be reversed. In Step 1, a *Bam*HI fragment is inserted into the *Bam*HI site of plasmid pAMP. In Step 2, an *Eco*RI fragment must be cloned into an *Eco*RI site *within* the *Bam*HI insert. Experimental constraints demand that the order of the steps not be reversed. Unfortunately,

M−E− M+E− M−E+ M+E+ M−H+ M+H+

Ideal Gel

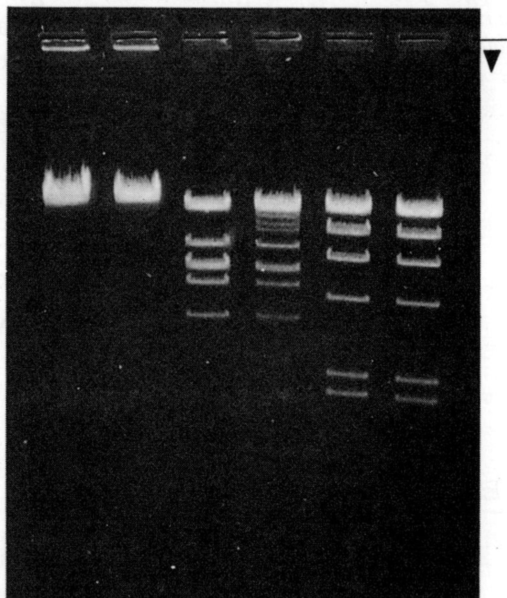

Incomplete Methylation
Faint bands in Lane M+E+ DNA partially cut by *Eco*RI

the pAMP "backbone" also contains an *Eco*RI site, which is not the intended cloning site for the *Eco*RI fragment in Step 2.

a. Draw a diagram of this cloning experiment.

b. Explain how *Eco*RI methylase could be used to solve this experimental problem.

6. Which nucleotide(s) is (are) methylated by *Eco*RI methylase? Draw the structure of the newly methylated base.

FOR FURTHER RESEARCH

1. Design and execute a series of experiments to study the kinetics of a methylation reaction.
   a. Determine the approximate percentage of sites protected at various time points.
   b. Repeat the experiments with several methylase dilutions and several DNA dilutions.
   c. In each case, at what time point does protection appear to be complete?
2. Design and execute experiments to use *Eco*RI methylase to map the locations of *Eco*RI restriction sites in the λ genome.
3. Research the use of methylases in constructing a genomic library.
4. Research the role of DNA methylation in gene regulation in higher organisms.
5. Research the role of methylation in controlling the movement of transposable elements in maize (corn).

LABORATORY 5

Rapid Colony Transformation of *E. coli* with Plasmid DNA

LABORATORY 5 DEMONSTRATES A RAPID METHOD to transform *E. coli* with a foreign gene. The bacterial cells are rendered "competent" to uptake plasmid DNA containing a gene for resistance to the antibiotic ampicillin. A bacteria that successfully takes up plasmid and expresses the gene for antibiotic resistance can be detected by its ability to grow in the presence of ampicillin.

Samples of *E. coli* cells are scraped off a nutrient agar plate (LB agar) and suspended in two tubes containing a solution of calcium chloride. One of three plasmids (pAMP, pBLU, or pGREEN) is added to one cell suspension, and both tubes are incubated for 15 minutes at 0°C. Following a brief heat shock at 42°C, cooling, and addition of LB broth, samples of the cell suspensions are plated on two types of media: LB agar and LB agar plus ampicillin (LB/amp).

The plates are incubated for 15–20 hours at 37°C and then checked for bacterial growth. Only cells that have been transformed by taking up the plasmid DNA with the ampicillin resistance gene will grow on the LB/amp plate. Subsequent division of a single antibiotic-resistant cell produces a colony of resistant clones. Thus, each colony seen on an ampicillin plate represents a single transformation event. In addition, cells transformed with pBLU will have a blue color (on specially prepared plates), and cells transformed with pGREEN fluoresce under ultraviolet (UV) light.

Kits based on this laboratory are available from the Carolina Biological Supply Company.

- Catalog no. 21-1142: Colony Transformation Kit
- Catalog no. 21-1082: Green Gene Colony Transformation Kit
- Catalog no. 21-1088: Glow-in-the-Dark Transformation Kit
- Catalog no. 21-1146: pBLU® Colony Transformation Kit

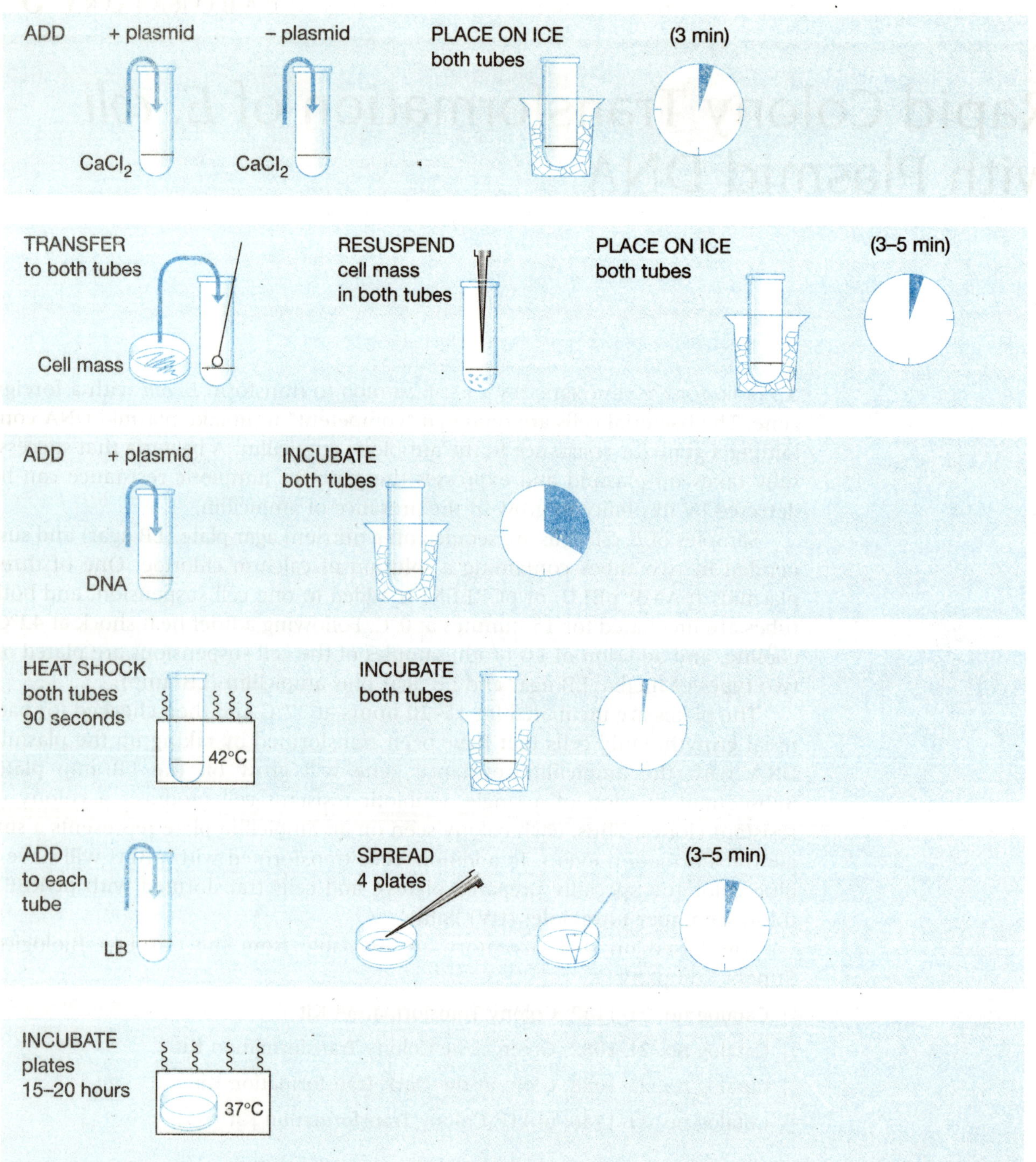
ADD
+ plasmid
– plasmid
$CaCl_2$
$CaCl_2$
PLACE ON ICE
both tubes
(3 min)
TRANSFER
to both tubes
Cell mass
RESUSPEND
cell mass
in both tubes
PLACE ON ICE
both tubes
(3–5 min)
ADD
+ plasmid
DNA
INCUBATE
both tubes
HEAT SHOCK
both tubes
90 seconds
42°C
INCUBATE
both tubes
ADD
to each
tube
LB
SPREAD
4 plates
(3–5 min)
INCUBATE
plates
15–20 hours
37°C

PRELAB NOTES

Review Prelab Notes in Laboratories 1 and 2 regarding sterile technique and *E. coli* culture.

Colony transformation is a simplification of the classic transformation protocol used in Laboratory 10 which requires mid-log phase cells grown in liquid culture. This abbreviated protocol begins with *E. coli* colonies scraped from an agar plate. Since liquid culturing is not used, equipment for shaking incubation is not required. The procedure entails minimal preparation time and is virtually foolproof. However, what is gained in simplicity and time is lost in efficiency. This protocol, although fine for transforming intact plasmids, is not efficient enough to use when transforming ligated DNA.

Transformation Scheme

Most transformation protocols can be conceptualized as four major steps.

1. *Preincubation:* Cells are suspended in a solution of cations and incubated at 0°C. The cations are thought to complex with exposed phosphates of lipids in the *E. coli* cell membrane. The low temperature freezes the cell membrane, stabilizing the distribution of charged phosphates.
2. *Incubation:* DNA is added, and the cell suspension is further incubated at 0°C. The cations are thought to neutralize negatively charged phosphates in the DNA and cell membrane. With these charges neutralized, the DNA molecule is free to pass through the cell membrane.
3. *Heat shock:* The cell/DNA suspension is briefly incubated at 42°C and then returned to 0°C. The rapid temperature change creates a thermal imbalance on either side of the *E. coli* membrane, which is thought to create a draft that sweeps plasmids into the cell.
4. *Recovery:* LB broth is added to the DNA/cell suspension and incubated at 37°C (ideally with shaking) prior to plating on selective media. Transformed cells recover from the treatment, amplify the transformed plasmid, and begin to express the antibiotic resistance protein.

The incubation and heat shock steps are critical. Since preincubation and recovery steps do not consistently improve the efficiency of colony transformation, they have been omitted from this protocol. If time permits, a preincubation of 5–15 minutes and/or a recovery of 5–30 minutes may be included.

Relative Inefficiency of Colony Transformation

The transformation efficiencies achieved with the colony protocol (5×10^3 to 5×10^4 colonies per microgram of plasmid) are 2–200 times less than those of the classic protocol (5×10^4 to 5×10^6 colonies per microgram). Colony transformation is perfectly suitable for transforming *E. coli* with purified intact plasmid DNA. However, it will give poor results with ligated DNA, which is composed of relaxed circular plasmid and linear plasmid DNA. These forms yield 5–100 times fewer transformants than an equivalent mass of intact supercoiled plasmid.

Maintenance of *E. coli* Strains for Colony Transformation

Prolonged reculturing (passaging) of *E. coli* can result in a loss of competence that makes the bacterium virtually impossible to transform using the colony method. There is some evidence that loss of transforming ability in MM294 may result from exposure of cells to temperatures below 4°C. Therefore, take care to store stab/slant cultures and streaked plates at room temperature. If there is a severe drop in number of transformants—from the expected 50–500 colonies per plate to essentially zero—discard the culture and obtain a fresh one.

Plasmids Used in This Experiment

Almost any plasmid containing a selectable antibiotic resistance marker can be substituted for pAMP for the purpose of demonstrating transformation of *E. coli* to an antibiotic-resistant phenotype. However, pAMP, pBLU, and pGREEN were constructed specifically as teaching molecules and offer advantages in other contexts:

1. All are derived from a pUC expression vector that replicates to a high number of copies per cell. Therefore, yields from plasmid preparations are significantly greater than those obtained with pBR322 and other less highly amplified plasmids.
2. pAMP was designed for use with another teaching plasmid, pKAN. Each produces unique and readily recognizable restriction fragments when separated on an agarose gel. Thus, recombinant molecules formed by ligating these fragments can be easily characterized.
3. pBLU carries ampicillin resistance and the full-length gene coding for β-galactosidase (*lacZ*). Other plasmids for expressing β-galactosidase contain only a small part of the *lacZ* gene and thus only make a small part of the protein. These vectors depend on using specific host cells that contain the remaining part of the *lacZ* gene. The pieces "complement" each other to make a complete protein. This is called α complementation.

   Note that wild-type *E. coli* does possess an endogenous *lacZ* gene. However, in the absence of lactose, this gene is suppressed. The *lacZ* gene in pBLU is expressed from the *lac* promoter but does not have the normal regulatory sequences (known as the operator), nor does it express the Lac repressor protein made from the *lacI* gene. This means that cells transformed with pBLU will constitutively express β-galactosidase, presumably at high levels. To detect the expression of β-galactosidase, prepare plates with X-gal (see Prelab Preparation).
4. pGREEN carries ampicillin resistance and the gene coding for GFP (green fluorescent protein). pGREEN contains an enhanced GFP mutant, which allows the expression of GFP to be visualized with ambient light alone. In addition, pGREEN does not use an inducible expression system. Therefore, no inducer (IPTG, arabinose, etc.) must be added to the media for the expression of recombinant GFP.

NOTE When viewing GFP expression, use a long-wavelength UV light source ("black light"). Do not use a short-wavelength light source. Use a Plexiglas shield or UV-blocking glasses when viewing with a *mid*-wavelength DNA transilluminator.

Antibiotic Selection

Ampicillin is the most practical antibiotic resistance marker for demonstration purposes, especially in the rapid transformation protocol described here. Ampicillin interferes with construction of the peptidoglycan layer and only kills dividing cells that are assembling new cell walls. It does not kill outright preexisting *E. coli* with intact cell walls. Thus, cells can be plated onto ampicillin-containing medium directly following heat shock, omitting the recovery step. Kanamycin selection, on the other hand, is less amenable to rapid transformation. A recovery step prior to plating is essential, because the antibiotic acts quickly to block protein synthesis and to kill *any* preexisting cells that are not actively expressing the resistance protein.

Test Tube Selection

The type of test tube used is a critical factor in achieving high-efficiency transformation and may also be important in the colony protocol. Therefore, we recommend using a presterilized 15-ml (17 x 100 mm) polypropylene culture tube. The critical heat-shock step has been optimized for the thermal properties of a 15-ml polypropylene tube. Tubes of a different material (such as polycarbonate) or thickness conduct heat differently. In addition, the small volume of cell suspension forms a thin layer across the bottom of a 15-ml tube, allowing heat to be quickly transferred to all cells. A smaller tube (such as 1.5-ml) increases the depth of the cell suspension through which heat must be conducted. Thus, *any* change in the tube specifications requires recalibrating the duration of the heat shock. The Becton Dickinson Falcon 2059 is the standard for transformation experiments.

Purified Water

Extraneous salts and minerals in the transformation buffer can also affect results. Use the most highly purified water available; pharmacy-grade distilled water is recommended. It might pay to obtain from a local research center or hospital several liters of water purified through a multistage ion-exchange system, such as Milli-Q.

Presterilized Supplies

Presterilized supplies can be used to good effect in transformations; 15-ml culture tubes and individually packaged 100–1000-μl micropipettor tips are handy. A 3-ml transfer pipette, marked in 250-μl gradations, can be substituted for a 100–1000-μl micropipettor with no loss of speed or accuracy.

Technically, everything used in this experiment should be sterilized. However, it is acceptable to use clean, but nonsterile, 1.5-ml tubes for aliquots of calcium chloride, LB broth, and plasmid DNA, *provided they will be used within 1 or 2 days*. Clean, nonsterile 1–10-μl micropipettor tips can be used for adding DNA to cells in Step 9. Plastic supplies, if not handled before use, are rarely contaminated. Antibiotic selection covers such minor lapses of sterile technique.

Plating Cell Suspensions

An alternative to using a traditional cell spreader is to use sterile glass spreading beads. Five to seven glass beads are placed on each agar plate after adding the cell suspension. The beads are swirled around the plate until the cells have been evenly spread. No flame/ethanol is required for this method, thus lowering potential fire hazards. Use 3-mm silica beads. Beads can be used directly from the package or autoclaved prior to use.

For Further Information

The protocol presented here is based on the following published methods:

Cohen S.N., Chang A.C., and Hsu L. 1972. Nonchromosomal antibiotic resistance in bacteria: Genetic transformation of *Escherichia coli* by R-factor DNA. *Proc. Natl. Acad. Sci.* **69:** 2110–2114.

Hanahan D. 1983. Studies on transformation of *Escherichia coli* with plasmids. *J. Mol. Biol.* **166:** 557–580.

———. 1987. Techniques for transformation of *E. coli*. In *DNA cloning: A practical approach* (ed. D.M. Glover), vol. 1. IRL Press, Oxford.

Mandel M. and Higa A. 1970. Calcium-dependent bacteriophage DNA infection. *J. Mol. Biol.* **53:** 159–162.

PRELAB PREPARATION

Before performing this Prelab Preparation, please refer to the cautions indicated on the Laboratory Materials list.

1. The day before the laboratory, streak out several fresh "starter plates" of MM294 or other *E. coli* host strain. Follow the procedure in Laboratory 2A, Isolation of Single Colonies. Following initial overnight incubation at 37°C, use cells.
2. PLAN AHEAD. Be sure to have a streaked plate or stab/slant culture of viable *E. coli* cells from which to streak starter plates. Also, streak the *E. coli* strain on an LB/amp plate to ensure that an ampicillin-resistant strain has not been used by mistake.
3. Sterilize 50 mM calcium chloride ($CaCl_2$) solution and LB broth by autoclaving or filtering through a 0.45-μm or 0.22-μm filter (Nalgene or Corning). To eliminate autoclaving completely, store filtered solutions in presterilized 50-ml conical tubes.
4. Prepare for each experiment:

   two LB agar plates
   two LB + ampicillin plates (LB/amp)

   When transforming with pBLU, X-gal (5 bromo-4-chloro-3-indolyl-β-D-galactopyranoside) must be included in the LB/amp plate in order to detect the expression of the *lacZ* gene. Buy premade plates or make up a 2% solution of X-gal in dimethyl formamide▼. When pouring plates, cool media to 60°C and add ampicillin to 100 μg/ml and 2 ml of the X-gal stock solution per liter of media. These LB/amp/X-gal plates can also be used as regular LB-ampicillin plates.
5. Prepare aliquots for each experiment:

   1 ml of sterile 50 mM $CaCl_2$ in a 1.5-ml tube (store on ice)
   1 ml of sterile LB broth in a 1.5-ml tube
   12 μl of 0.005 μg/μl pAMP, pBLU, or pGREEN in a 1.5-ml tube (store on ice)

6. Adjust the water bath to 42°C. A constant-temperature water bath can be made by maintaining a trickle flow of 42°C tap water into a Styrofoam box. Monitor temperature with a thermometer. An aquarium heater can be used to maintain temperature.
7. Prewarm incubator to 37°C.
8. To retard evaporation, keep ethanol in a beaker covered with Parafilm, plastic wrap, or, if using a small beaker, the lid from a Petri dish. Retrieve and reuse ethanol exclusively for flaming.
9. If using spreading beads, carefully place five to seven beads into a sterile 1.5-ml tube. Tube can be used as a scooper. Prepare four tubes per experiment.

MATERIALS

CULTURES, MEDIA, AND REAGENTS

$CaCl_2$ (50 mM)
LB/amp plates (2) (or 2 LB/amp/X-gal▼ plates, if using pBLU)
LB broth
LB plates (2)
MM294 starter culture
Plasmid (0.005 μg/μl) (pAMP, pBLU, or pGREEN)

SUPPLIES AND EQUIPMENT

Beakers for crushed or cracked ice and for waste/used tips
Beaker of 95% ethanol▼ and cell spreader (or spreading beads)
"Bio-bag" or heavy-duty trash bag
Bleach (10%)▼ or disinfectant
Bunsen burner
Culture tubes (two 15-ml)
Incubator (37°C)
Inoculating loop
Micropipettor (100–1000-μl) + tips (or 3-ml transfer pipettes)
Micropipettor (0.5–10-μl) + tips
Permanent marker
Test tube rack
Water bath (37°C) (optional)
Water bath (42°C)

▼ See Appendix 4 for Cautions list.

METHODS

Prepare *E. coli* Colony Transformation (40 minutes)

This entire experiment *must be performed under sterile conditions*. Review sterile techniques in Laboratory 1, Measurements, Micropipetting, and Sterile Techniques.

1. Use a permanent marker to label one sterile 15-ml tube +plasmid. Label another 15-ml tube –plasmid. Plasmid DNA will be added to the +plasmid tube; none will be added to –plasmid tube.

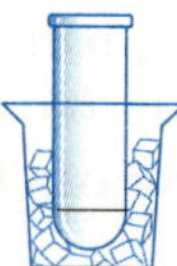

If there are no separate colonies on the starter plate, scrape up a small cell mass from a streak. Transformation efficiency decreases if too many cells are added to the calcium chloride.

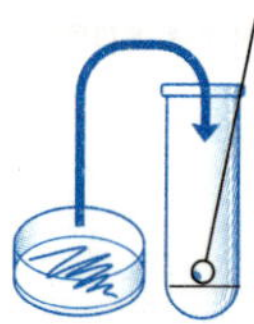

Optimally, flame the mouth of the 15-ml tube after removing and before replacing cap. Cells become difficult to resuspend if allowed to clump together in $CaCl_2$ solution for several minutes. Resuspending cells in the +plasmid tube first allows the cells to preincubate for several minutes at 0°C while –plasmid tube is being prepared. If time permits, both tubes can be preincubated on ice for 5–15 minutes.

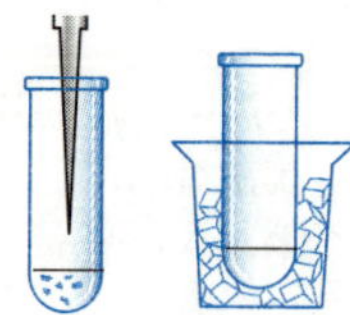

Double check both tubes for complete resuspension of cells, which is probably the most important variable in obtaining good results.

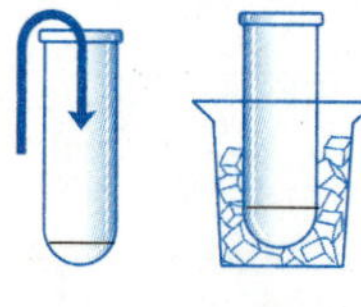

To save plates, different experimenters may omit either the +LB or the –LB plate.

2. Use a 100–1000-µl micropipettor and sterile tip (or sterile transfer pipette) to add 250 µl of $CaCl_2$ solution to each tube.
3. Place both tubes on ice.
4. Use a sterile inoculating loop to transfer one or two large (3-mm) colonies from the starter plate to the +plasmid tube:
   a. Sterilize the loop in a Bunsen burner flame until it glows red hot. Then pass the lower one half of the shaft through the flame.
   b. Stab the loop several times at the edge of the agar plate to cool.
   c. Pick a couple of large colonies and scrape up a visible cells mass, but be careful not to transfer any agar. (Impurities in the agar can inhibit transformation.)
   d. Immerse the loop tip in the $CaCl_2$ solution and *vigorously* tap it against the wall of the tube to dislodge the cell mass. Hold the tube up to the light to observe the cell mass drop off into the $CaCl_2$ solution. Make sure that the cell mass is not left on the loop or on the side of the tube.
   e. Reflame the loop before placing it on the lab bench.
5. Immediately resuspend the cells in the +plasmid tube by repeatedly pipetting in and out, using a 100–1000-µl micropipettor with a sterile tip (or sterile transfer pipette).

> CAUTION
> Keep nose and mouth away from the tip end when pipetting suspension culture to avoid inhaling any aerosol that might be created.

   a. Pipette carefully to avoid making bubbles in suspension or splashing suspension far up the sides of the tube.
   b. Hold the tube up to the light to check that the suspension is homogeneous. No visible clumps of cells should remain.
6. Return the +plasmid tube to ice.
7. Transfer a second mass of cells to the –plasmid tube as described in Steps 4 and 5 above.
8. Return the –plasmid tube to ice. Both tubes should be on ice.
9. Use a 1–10-µl micropipettor to add 10 µl of 0.005 µg/µl plasmid solution *directly into the cell suspension* in the +plasmid tube. Tap tube with a finger to mix. Avoid making bubbles in the suspension or splashing the suspension up the sides of tube.
10. Return the +plasmid tube to ice. Incubate both tubes on ice for an additional 15 minutes.
11. While the cells are incubating on ice, use a permanent marker to label two LB plates and two LB/amp plates with your name and the date. Remember, if transforming with pBLU, to use LB/amp/X-gal plates in place of regular LB/amp plates.

    Label one LB/amp plate +. This is the experimental plate.
    Label the other LB/amp plate –. This is the negative control.
    Label one LB plate +. This is a positive control.
    Label one LB plate –. This is a positive control.

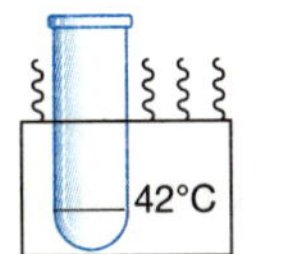

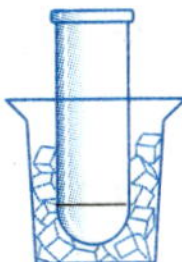

12. Following the 15-minute incubation, heat shock the cells in the +plasmid and –plasmid tubes. *It is critical that cells receive a sharp and distinct shock.*
    a. Carry the ice beaker to the water bath. Remove the tubes from ice, and *immediately* immerse them in the 42°C water bath for 90 seconds.
    b. Immediately return both tubes to ice for at least 1 additional minute.

If time permits, allow +plasmid and –plasmid cells to recover for 5–30 minutes at 37°C. Gentle shaking is also helpful.

STOP An extended period on ice following the heat shock will not affect the transformation. If necessary, store the +plasmid and –plasmid tubes on ice in the refrigerator (0°C) for up to 24 hours, until there is time to plate cells. Do not put cell suspensions in the freezer.

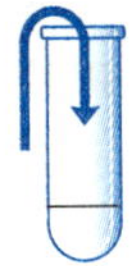

13. Place +plasmid and –plasmid tubes in the test tube rack at room temperature.
14. Use a 100–1000-μl micropipettor and sterile tip (or sterile transfer pipette) to add 250 μl of LB broth to each tube. Gently tap tubes with finger to mix.
15. Use the matrix below as a checklist as +plasmid and –plasmid cells are spread on each type of plate:

	Transformed cells +plasmid	Nontransformed cells –plasmid
LB/amp	100 μl	100 μl
LB	100 μl	100 μl

If too much liquid is absorbed by agar, cells will not be evenly distributed.

The object is to evenly distribute and separate cells on agar so that each gives rise to a distinct colony clones. It is essential not to overheat spreader in burner flame and to cool it before touching cell suspensions. A hot spreader will kill E. coli *cells on the plate.*

16. Use a micropipettor with a sterile tip (or transfer pipette) to add 100 μl of cell suspension from the –plasmid tube onto the –LB plate, and another 100 μl onto the –LB/amp plate. *Do not allow the suspensions to sit on the plates too long before proceeding to Step 17.* Spread cells using one of the methods described in Steps 17 and 18.

Sterile Spreading Technique (Steps 16 and 17)

17. Sterilize cell spreader, and spread cells over the surface of each –plate in succession.

    a. Dip the spreader into the ethanol beaker and *briefly* pass it through a Bunsen flame to ignite alcohol. Allow alcohol to burn off *away from* the Bunsen flame; spreading rod will become too hot if left in flame.

CAUTION
Be extremely careful not to ignite the ethanol in the beaker. Do not panic if the ethanol is accidentally ignited. Cover the beaker with a Petri lid or other cover to cut off oxygen and rapidly extinguish fire.

    b. Lift the lid of one –plate just enough to allow spreading; *do not place lid on lab bench.*

    c. Cool spreader by gently rubbing it on the surface of the agar *away* from the cell suspension or by touching it to condensation on the plate lid.

    d. Touch the spreader to the cell suspension, and gently drag it back and forth several times across the surface of the agar. Rotate plate one-quarter turn, and repeat spreading motion. Try to spread the suspension evenly across agar surface. *Be careful* not to gouge the agar.

    e. Replace plate lid. Return cell spreader to ethanol *without flaming.*

18. Use spreading beads to spread cells over the surface of each –plate in succession.

    a. Lift the lid of one –plate enough to allow adding beads; *do not place the lid on the lab bench.*

    b. Carefully pour five to seven glass spreading beads from a 1.5-ml tube onto the agar surface.

    c. Close plate lids and use a swirling motion to move glass beads around the entire surface of the plate. This evenly spreads the cell suspension on the agar surface. Continue swirling until the cell suspension is absorbed into the agar.

19. Use a micropipettor with a sterile tip (or transfer pipette) to add 100 μl of cell suspension from +plasmid tube onto +LB plate and to add another 100 μl of cell suspension onto +LB/amp plate. *Do not allow the suspensions to sit on the plate too long before proceeding to Step 20.*

20. Repeat Step 17a–e or Step 18a–c to spread cell suspension on +LB and +LB/amp plates.

21. If Step 17 was used, reflame the spreader one last time before placing it on the lab bench.

22. Allow the plates to set for several minutes so that the suspension absorbs into the agar. If Step 18 was used, invert plates and gently tap plate bottoms, so that the spreading beads fall into plate lids. Carefully pour beads from each lid into storage container for reuse.

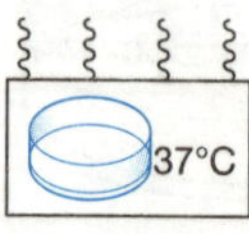

23. Stack plates and tape into a bundle to keep the experiment together. Place the plates upside down in a 37°C incubator, and incubate for 15–20 hours.

24. After initial incubation, store plates at 4°C to arrest *E. coli* growth and to slow the growth of any contaminating microbes.

25. If planning to do Laboratory 8, Purification and Identification of Plasmid DNA, save the +LB/amp plate as source of a colony to begin an overnight suspension culture.
26. Take time for responsible cleanup.
    a. Segregate for proper disposal culture plates and tubes, pipettes, and micropipettor tips that have come into contact with *E. coli.*
    b. Disinfect cell suspensions, tubes, and tips with 10% bleach or disinfectant.
    c. Wipe down lab bench with soapy water, 10% bleach solution, or disinfectant (such as Lysol).
    d. Wash hands before leaving lab.

RESULTS AND DISCUSSION

Count the number of individual colonies on the +LB/amp plate. Observe colonies through the bottom of the culture plate, and use a permanent marker to mark each colony as it is counted. If the transformation worked well, between 50 and 500 colonies should be observed on the +LB/amp plate; 100 colonies is equal to a transformation efficiency of 10^4 colonies per microgram of plasmid DNA. (Question 3 explains how to compute transformation efficiency.)

If plates have been overincubated or left at room temperature for several days, tiny "satellite" colonies may be observed that radiate from the edges of large, well-established colonies. Nonresistant satellite colonies grow in an "antibiotic shadow" where ampicillin has been broken down by the large resistant colony. Do not include satellite colonies in the count of transformants. Also examine the colonies carefully to detect any possible contamination. Contaminating organisms will usually look different in color, shape, or size of colony. Over time, you will improve at distinguishing *E. coli* colonies from other organisms. A "lawn" should be observed on positive controls, where the bacteria cover nearly the entire agar surface and individual colonies cannot be discerned.

If pBLU was used for transformation, you will observe blue colonies on the +LB/amp/X-gal plate because of the expression of β-galactosidase. The X-gal in the plates mimics the normal substrate for β-galactosidase, the disaccharide lactose. β-galactosidase cleaves the X-gal, removing the compound 5-bromo-4-chloro-3-indolyl from galactopyranoside, which is blue. Thus, the presence of a blue colony indicates the presence of β-galactosidase activity.

If pGREEN was used for transformation, you will observe green colonies under long-wavelength UV light (black light). Green colonies indicate the presence of GFP (green fluorescent protein).

1. Record your observation of each plate in matrix below. If cell growth is too dense to count individual colonies, record "lawn." Were the results as expected? Explain possible reasons for variations from expected results.

	Transformed cells +plasmid	Nontransformed cells −plasmid
LB/amp	experiment	negative control
LB	positive control	positive control

2. Compare and contrast the growth on each of the following pairs of plates. What does each pair of results tell you about the experiment?
   a. +LB and –LB
   b. –LB/amp and –LB
   c. +LB/amp and –LB/amp
   d. +LB/amp and +LB
3. Transformation efficiency is expressed as the number of antibiotic-resistant colonies per microgram of plasmid DNA. The object is to determine the mass of plasmid that was spread on the experimental plate and was therefore responsible for the transformants observed.
   a. Determine total mass (in micrograms) of plasmid used in Step 9.

      concentration x volume = mass
   b. Determine the fraction of the cell suspension spread onto +LB/amp plate (Step 19): volume suspension spread/*total* volume suspension (Steps 2 and 14) = fraction spread.
   c. Determine the mass of plasmid in the cell suspension spread onto +LB/amp plate: total mass plasmid (*a*) x fraction spread (*b*) = mass plasmid spread.
   d. Determine number of colonies per microgram of plasmid. Express answer in scientific notation: colonies observed/mass plasmid spread (*c*) = transformation efficiency.
4. What factors might influence transformation efficiency?
5. Your Favorite Gene (*YFG*) is cloned into pAMP, and 0.2 μg of pAMP/YFG is used to transform *E. coli* according to the protocol described in this laboratory. Using the information below, calculate the number of molecules of pAMP/YFG that are present in a culture 200 minutes after transformation.
   a. You achieve a transformation efficiency equal to 10^6 colonies per microgram of intact pAMP/YFG.
   b. pAMP/YFG grows at an average copy number of 100 molecules per transformed cell.
   c. Following heat shock (Step 12), the entire 250 μl of cell suspension is used to inoculate 25 ml of fresh LB broth. The culture is incubated, with shaking, at 37°C. Transformed cells enter log phase 60 minutes after inoculation and then begin to replicate an average of once every 20 minutes.
6. The transformation protocol above is used with 10 μl of intact plasmid DNA at different concentrations. The following numbers of colonies are obtained when 100 μl of transformed cells are plated on selective medium:

0.00001 μg/μl	4 colonies
0.00005 μg/μl	12 colonies
0.0001 μg/μl	32 colonies
0.0005 μg/μl	125 colonies
0.001 μg/μl	442 colonies
0.005 μg/μl	542 colonies
0.01 μg/μl	507 colonies
0.05 μg/μl	475 colonies
0.1 μg/μl	516 colonies

a. Calculate transformation efficiencies at each concentration.

b. Plot a graph of DNA mass *versus* colonies.

c. Plot a graph of DNA mass *versus* transformation efficiency.

d. What is the relationship between mass of DNA transformed and transformation efficiency?

e. At what point does the transformation reaction appear to be saturated?

f. What is the true transformation efficiency?

7. For cells transformed with pBLU or pGREEN, what color would you expect nontransformed satellite colonies to be?

FOR FURTHER RESEARCH

Interpretable experimental results can only be achieved when the colony transformation can be repeated with reproducible results. Attempt experiments below only when you are able to routinely achieve 100–500 colonies on the +LB/amp plate.

1. Design and execute an experiment to compare transformation efficiencies of linear *versus* circular plasmid DNAs. Keep molecular weight constant.

2. Design and execute a series of experiments to test the relative importance of each of the four major steps of most transformation protocols: (1) preincubation, (2) incubation, (3) heat shock, and (4) recovery. Which steps are absolutely necessary?

3. Design and execute a series of experiments to compare the transforming effectiveness of $CaCl_2$ *versus* the salts of other monovalent (+), divalent (++), and trivalent (+++) cations.

   a. Make up 50 mM solutions of each salt.

   b. Check pH of each solution, and buffer to approximately pH 7 when necessary.

   c. Is $CaCl_2$ unique in its ability to facilitate transformation?

   d. Is there any consistent difference in the transforming ability of monovalent *versus* divalent *versus* trivalent cations?

4. Carry out a series of experiments to determine the saturating conditions for transformation reactions.

   a. Transform *E. coli* using DNA concentrations list in Question 6 above.

   b. Plot a graph of DNA mass *versus* colonies per plate.

   c. Plot a graph of DNA mass *versus* transformation efficiency.

   d. At what mass does the reaction appear to become saturated?

   e. Repeat experiment with concentrations clustered on either side of the presumed saturation point to produce a fine saturation curve.

5. Repeat the experiment in Step 4 above, but transform with a 1:1 mixture of pAMP and pKAN at each concentration. Plate transformants on LB/amp, LB/kan, and LB/amp+kan plates. *Be sure to include a 40–60-minute recovery, with shaking.*

a. Calculate the percentage of double transformations at each mass.

$$\frac{\text{colonies LB/amp+kan plate}}{\text{colonies LB/amp plate + colonies LB/kan plate}}$$

b. Plot a graph of DNA mass *versus* colonies per plate.

c. Plot a graph of DNA mass *versus* percentage of double transformations. Under saturating conditions, what percentage of bacteria are doubly transformed?

6. Plot a recovery curve for *E. coli* transformed with pKAN. Allow cells to recover for 0–120 minutes at 20-minute intervals.

   a. Plot a graph of recovery time *versus* colonies per plate.

   b. At what time point is antibiotic expression maximized?

   c. Can you discern a point at which the cells began to replicate?

Assay for an Antibiotic Resistance Enzyme

LABORATORY 6 INTRODUCES WORKING WITH PROTEINS and performing an enzyme assay. In Laboratory 5, Rapid Colony Transformation of *E. coli* with Plasmid DNA, *E. coli* were transformed with a plasmid that rendered the cells resistant to the antibiotic ampicillin. The gene that confers ampicillin resistance codes for the enzyme β-lactamase, which breaks the β-lactam ring in ampicillin. In this laboratory, β-lactamase is isolated from ampicillin-resistant *E. coli*, and its enzyme activity is observed using an in vitro assay.

Ampicillin is a derivative of the antibiotic penicillin, and β-lactamase destroys penicillin even more efficiently than it destroys ampicillin. In this experiment, penicillin will therefore be used as the substrate in the β-lactamase assay. β-lactamase is isolated from overnight cultures of *E. coli* transformed with a plasmid that confers ampicillin resistance, such as pAMP. Because β-lactamase readily leaks out of the cells and into the culture medium, the media itself is used as a solution of β-lactamase. β-lactamase is then used to convert penicillin into penicilloic acid. The reaction is monitored using the pH indicator phenol red, which signals the formation of penicilloic acid by turning from red (basic pH) to yellow (acidic pH).

A kit based on this laboratory is available from the Carolina Biological Supply Company (Catalog no. 21-1137).

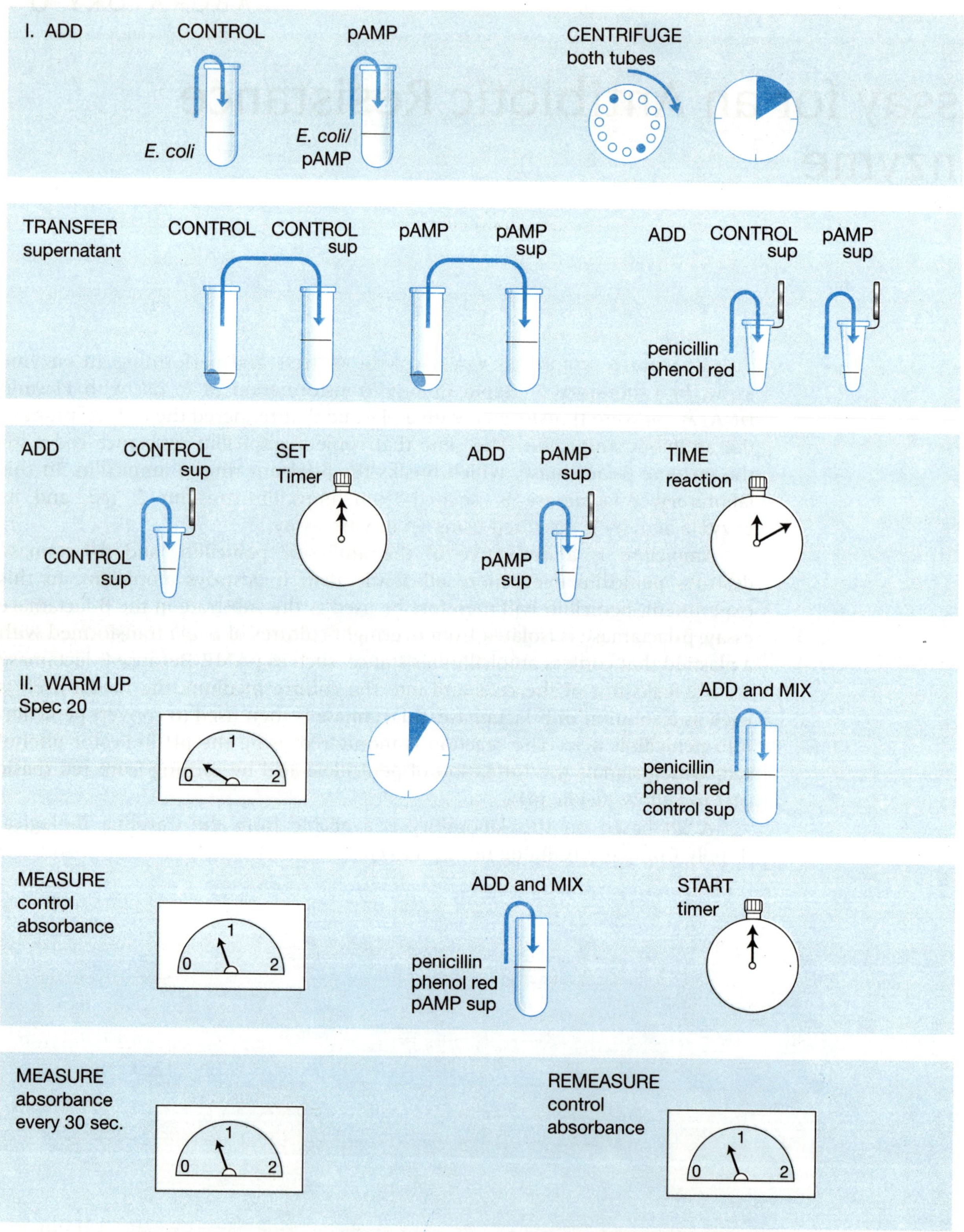
I. ADD
CONTROL
pAMP
E. coli
E. coli/
pAMP
CENTRIFUGE
both tubes
TRANSFER
supernatant
CONTROL
CONTROL
sup
pAMP
pAMP
sup
ADD
CONTROL
sup
pAMP
sup
penicillin
phenol red
ADD
CONTROL
sup
CONTROL
sup
SET
Timer
ADD
pAMP
sup
pAMP
sup
TIME
reaction
II. WARM UP
Spec 20
0
1
2
ADD and MIX
penicillin
phenol red
control sup
MEASURE
control
absorbance
ADD and MIX
penicillin
phenol red
pAMP sup
START
timer
MEASURE
absorbance
every 30 sec.
REMEASURE
control
absorbance

PRELAB NOTES

β-lactamase

β-lactamase is so-called because it acts to break a bond in a four-member ring in the ampicillin molecule called a β-lactam ring. The enzyme has a similar action on penicillin, which converts to penicilloic acid in the reaction. In fact, β-lactamase is commonly called penicillinase. Penicillin and penicilloic acid are shown below. The enzyme acts on the carbon marked with a circle.

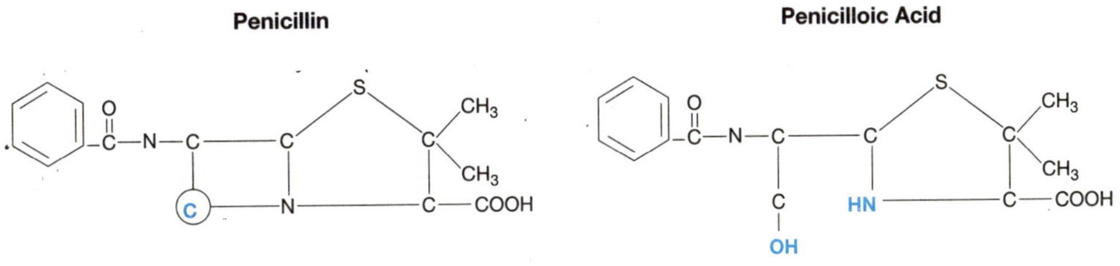

Penicillin **Penicilloic Acid**

β-lactamase is easy to isolate because it is stored in the *E. coli* periplasm, the space between the inner cell membrane and the cell wall, and leaks into the surrounding media. Thus, the growth media from an overnight culture of *E. coli* containing pAMP plasmid becomes a ready source of β-lactamase. Cells can be removed and the supernatant can be used as a β-lactamase solution. β-lactamase is fairly stable and can be frozen, thawed, and reused, but like any enzyme, it should not be left at room temperature for long periods of time. Keep β-lactamase on ice when not in use.

The β-lactamase solution may have a low pH, which can interfere with the assay. Test by mixing 100 μl of phenol red with 200 μl of β-lactamase solution. If the solution turns orange or yellow, the pH of the β-lactamase is too low. The pH of the β-lactamase can be adjusted to 7.0 using NaOH and a pH meter. Alternatively, increase the concentration of Tris (pH 7.9) in the phenol red from 10 mM to 20 mM or use Tris at a higher pH, such as 8.5 or 9.

Assay Conditions

The assay for β-lactamase is very simple. As the enzyme converts penicillin to penicilloic acid, the associated formation of hydrogen ions (H^+) can be monitored using a pH indicator. The pH indicator in this experiment is phenol red. At basic pH (>7.0), a solution of phenol red is red-pink. As the pH drops below 7.0 (i.e., becomes acidic), the solution changes first to orange and finally to yellow. Here, the reaction will be started at a pH of 7.9 (by adding 10 mM Tris [pH 7.9] to the phenol red solution). This colorimetric reaction can be quantitated using a spectrophotometer to measure the increase in absorbance at 430 nm, the absorption optimum for yellow. (Any spectrophotometer that can be used to read visible light, such as a Spectronic 20+, will suffice.) The rate of the reaction will vary, based on the quality of the β-lactamase. Generally, the reaction is complete within 10 minutes.

Plasmid Selection

Any plasmid carrying ampicillin resistance can be used in this experiment, although we recommend using pAMP. It is suggested that colonies obtained from Laboratory 5, Rapid Colony Transformation of *E. coli* with Plasmid DNA, be used as the source of ampicillin-resistant bacteria. Transformed colonies from Laboratory 5 can be stored at 4°C and used to make an overnight culture for up to 1 week. After 1 week, most or all of the cells on an LB-AMP plate will be dead.

Penicillin Solution

There are various forms of penicillin, but for this laboratory, we recommend using penicillin-G Procaine salt to make the penicillin solution. Do not worry if the salt does not completely dissolve; the solution is still usable. Note that penicillin-G Procaine salt is usually sold in units of activity. Generally, the number of units/gram is ~1000/mg. In addition, be aware that penicillin can break down into penicilloic acid spontaneously, especially if it is in solution for too long, if it is in solution at room temperature overnight or longer, or if the penicillin salt is old. Before use, check the penicillin solution by mixing 700 μl with 100 μl of phenol red solution. The phenol red may become pale red to pink, indicating the presence of a small amount of penicilloic acid. If the phenol red turns orange, we recommend that you make a fresh solution using a new stock of penicillin-G.

FOR FURTHER INFORMATION

The protocol presented here is based on the following published methods.

Skinner A. and Wise R. 1977. A comparison of three rapid methods for the detection of beta-lactamase activity in *Haemophilus influenzae*. *J. Clin. Pathol.* **30:** 1030–1032.

Wong K.W. and Soo-Hoo T.S. 1976. A rapid, simple agar-overlay method for the detection of pencillinase-producing *Staphylococcus aureus* in the clinical bacteriology laboratory. *Jpn. J. Microbiol.* **20:** 153–154.

PRELAB PREPARATION

1. Prepare a 0.04% solution of phenol red in 10 mM Tris (pH 7.9). This solution can be stored indefinitely at room temperature. Phenol red is typically sold as a 1% solution.
2. The day before the laboratory, prepare two *E. coli* cultures according to the protocol in Laboratory 2B, Overnight Suspension Culture. Inoculate the first culture with a cell mass scraped from one wild-type MM294 colony from an LB plate. You will need one 5-ml culture per experiment. Inoculate the second culture with colonies selected from the +LB/amp plate from Laboratory 5, Rapid Colony Transformation of *E. coli* with Plasmid DNA. You will need one 5-ml culture per experiment. Grow this culture in LB broth plus ampicillin. If possible, use a higher concentration of ampicillin (200 μg/ml) than previously suggested.

3. On the day of the laboratory, make up a 4 mg/ml solution of penicillin-G. Keep at 4°C until needed. If making the solution the day before, be aware that it will break down to penicilloic acid if left at room temperature. Store at 4°C overnight or at –20°C for long-term storage.
4. Test reagents on the day of the laboratory. Mix phenol red solution with the penicillin stock to check pH (see Prelab Notes).
5. Prepare aliquots for each experiment:

    1 *E. coli* overnight culture (5 ml) in 15-ml culture tube labeled "control"
    1 *E. coli*/pAMP overnight culture (5 ml) in 15-ml culture labeled "pAMP"
    6 ml of 4 mg/ml penicillin-G
    1 ml of 0.04% phenol red solution

MATERIALS

REAGENTS	SUPPLIES AND EQUIPMENT
E. coli overnight culture (1)	Beaker for waste/used tips
E. coli/pAMP overnight culture (1)	Bleach (10%)▼ or disinfectant
LB broth	Clinical centrifuge
LB broth/amp	Culture tubes (15 ml)
Penicillin-G▼(4 mg/ml)	Micropipettor (100–1000 µl) + tips
Phenol red▼solution (0.04%)	Permanent marker
	Spectrophotometer (optional)
	Spectrophotometer tubes (clear)
	Test tube rack
	Timer or stopwatch
	Tubes (1.5 ml)

▼ See Appendix 4 for Cautions list.

METHODS

I. Set Up β-lactamase Assay (30 minutes)

1. Collect the reagents and place them in a test tube rack on the lab bench.
2. Obtain two 15-ml tubes, one containing 5 ml of *E. coli*/pAMP overnight culture (labeled "pAMP") and a second containing 5 ml of wild-type *E. coli* culture (labeled "control"). Label the tubes with your name.

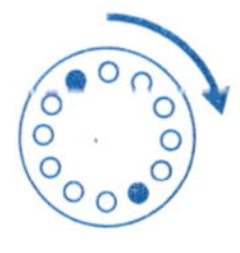

A tight pellet will form at the bottom of the tube after centrifugation and the supernatant should be clear. If the supernatant is cloudy, recentrifuge 5 minutes longer and if possible increase the speed.

3. Place two tubes in a balanced configuration in a rotor of a clinical centrifuge. Centrifuge cells at 2500–4000 rpm for 10–15 minutes to pellet cells on the bottom-side of the culture tube.
4. While the tubes are in the centrifuge, label one fresh 15-ml culture tube "control sup." Label a second fresh 15-ml culture tube "pAMP sup." Label both tubes with your name.

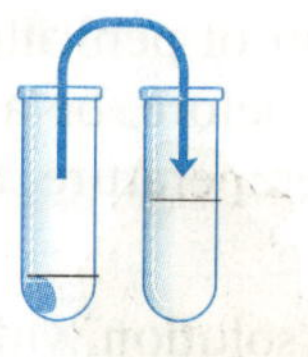

You do not want any of the cell pellet and have plenty of β-lactamase, so leave a little of the supernatant behind to avoid disturbing the pellet.

5. When the centrifuge stops, pour off the supernatant from each tube into the correspondingly labeled fresh 15-ml tube. *Do not to disturb cell pellet.* The "pAMP sup" tube contains β-lactamase that has leaked into the supernatant, and the "control sup" tube contains control supernatant. Bacterial pellets may be frozen for further analysis.
6. Obtain two 1.5-ml tubes. Label one tube "control sup" and the other "pAMP sup."
7. Use the matrix below as a checklist while adding reagents to each reaction. Read down each column, adding the same reagent to all appropriate tubes. *Use a fresh tip for each reagent.* Refer to detailed instructions that follow.

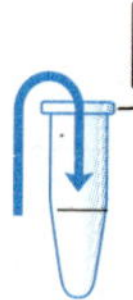

Tube	Penicillin-G	Phenol red	Control sup	pAMP sup
Control sup	700 μl	100 μl	200 μl	–
pAMP sup	700 μl	100 μl	–	200 μl

The phenol red solution may turn from dark red to a reddish orange in color. This is normal. If it turns orange or even yellow, the penicillin has already broken down to penicilloic acid. Get a fresh solution of penicillin.

8. Use a micropipettor to add 700 μl of penicillin solution to each 1.5-ml reaction tube.
9. Use a fresh tip to add 100 μl of phenol red to each reaction tube.
10. Use a fresh tip to add 200 μl of "control sup" solution to the 1.5-ml reaction tube labeled "control sup."
11. Note time, or set timer, and then use a fresh tip to add 200 μl of "pAMP sup" β-lactamase solution to the "pAMP sup" reaction tube. Invert tube to mix. Work quickly as the reaction will begin as soon as the β-lactamase is added. If working with a partner, one person adds the β-lactamase while the other sets the timer.

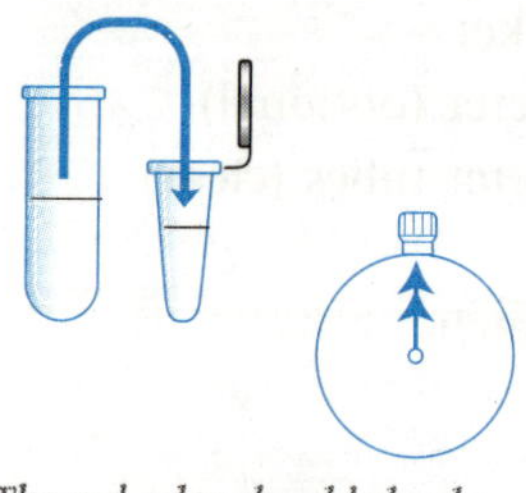

12. Begin timing the reaction. Place the tube in a rack to watch the color change. (Remember that temperature affects the rate of the reaction, so do not hold the tube in your hand as this will speed up the reaction.) Determine the time it takes for the reaction to turn yellow.

The red color should slowly fade to a rose color, then to orange, and finally, when the reaction is complete, completely to yellow. Try to be accurate in timing reaction. One variable may be your judgment of when the reaction is yellow. At this point, note the time again or stop the timer and record.

TIME FOR THE REACTION TO TURN FROM RED TO YELLOW ____________

II. Quantitative β-lactamase Reaction

(30 minutes)

Now follow the enzyme reaction quantitatively by monitoring in a spectrophotometer, such as the Spectronic 20+ (Spec 20). The increase in yellow absorbance is a measure of the formation of penicilloic acid.

1. Turn on the spectrophotometer to warm up and set wavelength to 430 nm.
   a. Zero the spectrophotometer following the manufacturer's instructions. A solution containing penicillin, phenol red, and fresh LB (mixed in the amounts shown in Step 2) can be used to "zero" the spectrophotometer. Peak absorbance of phenol red at basic pH (i.e., when red) is 560 nm, whereas peak absorbance at acidic pH (i.e., when yellow) is 430 nm.
   b. Test by measuring the absorbance of the phenol red diluted to 100 μl in 3 ml of water at several different wavelengths. Then add a little acid to the same phenol red solution so that it turns yellow and retake absorbance readings at different wavelengths.

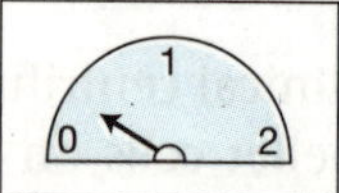

The control tube will be used to measure the background level of 430 nm absorbing material in the reaction. It should have a very low absorbance.

2. When the spectrophotometer has warmed up for 5 minutes, make up a control tube to determine the background in the reaction. (If using a spectrophotometer with a 1-ml capacity cuvette, add volumes indicated in parentheses.)
   a. Add 2 ml of penicillin solution to the spectrophotometer tube (600 μl for a 1-ml cuvette).
   b. Add 100 μl of phenol red solution (50 μl for a 1-ml cuvette).
   c. Add 600 μl of "control sup" solution to the tube (200 μl for a 1-ml cuvette).
   d. Cover the tube with Parafilm and with a thumb over the top, invert it twice to mix the reagents.
3. Place the control tube in the spectrophotometer. Record the absorbance at 430 nm. This is the "control sup" background absorbance. Save this tube and read the absorbance again in Step 10.
4. Set up a table in a notebook as shown below to record the data.

"CONTROL SUP" BACKGROUND INITIAL ABSORBANCE READING ______________

Time (seconds)	pAMP sup
0	
30	
60	
90	
120	
150	
180	
210	
240	
270	
300	
330	
360	
390	
420	
480	

"CONTROL SUP" BACKGROUND FINAL ABSORBANCE READING ______________

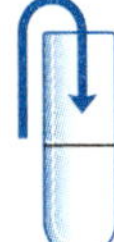

Using less pH indicator keeps spectrophotometer readings from going off scale. If reaction produces an absorbance value of 2, then repeat using half the amount of phenol red (50 μl for Spec 20 glass tube, or 25 μl if using 1-ml cuvette).

5. In a new spectrophotometer glass tube, add 2 ml of penicillin solution. (If using a regular spectrophotometer and a 1-ml capacity cuvette, add 0.6 ml of penicillin.)
6. To the same tube, add 100 μl of phenol red (50 μl for a 1-ml cuvette).
7. To the same tube, add 0.5 ml of "pAMP sup" β-lactamase solution (0.2 ml in a 1-ml cuvette).
8. Mix by pipetting up and down or by covering the top of the tube with Parafilm and inverting two or three times. Immediately transfer the tube to the Spec 20. The reaction starts as soon as the β-lactamase is added, so move quickly to establish the zero time reading.

If working with a partner, one person keeps track of the time and the other keeps track of the change in absorbance.

9. Simultaneously, start the timer and take an initial reading at 430 nm from the spectrophotometer. The initial reading of pAMP sup may not be as low as the control reading. Overnight *E. coli*/pAMP cultures contain cleaved ampicillin which, like penicilloic acid, is acidic and contributes to the color change from red to yellow. Record the readings every 30 seconds until no significant increase in absorbance is seen for 1 minute.
10. Place the control tube (from Step 3) in the spectrophotometer. Record this reading again to confirm that the absorbance of the control tube does not change over time.
11. Take time for responsible cleanup.
    a. Segregate for proper disposal culture tubes and micropipettor tips that have come in contact with *E. coli*.
    b. Disinfect cell suspensions, tubes, and tips with 10% bleach solution or disinfectant.
    c. Wipe down lab bench with soapy water, 10% bleach solution, or disinfectant (such as Lysol).
    d. Wash hands before leaving lab.

RESULTS AND DISCUSSION

In this laboratory, you have carried out an enzyme reaction using β-lactamase, also known as penicillinase. β-lactamase is the enzyme encoded by the gene *pAMP* that renders *E. coli* resistant to ampicillin, a form of penicillin. The enzyme converts penicillin into penicilloic acid. Although this reaction cannot be "seen," it can be monitored by following the associated accumulation of hydrogen ions. This experiment demonstrates the daily challenge of the molecular biologist: How to see what is not visible.

In this laboratory exercise, you should have observed that the "control sup" tube remained red throughout the incubation. On the other hand, you should have observed the "pAMP sup" tube change from pale red to orange to bright yellow. This color change represents the formation of hydrogen ions as a result of the formation of penicilloic acid. The color change in the enzyme reaction tubes should be fairly rapid. The rapidity of the change is a function of the activity and/or concentration of the enzyme.

1. Why is the phenol red solution made up in 10 mM Tris-HCl (pH 7.9)?
2. Write out the equation for the enzyme reaction that occurs in this experiment.
3. How would you predict the following factors to affect the rate of the reaction:
    a. Decreasing the temperature.
    b. Using a higher concentration penicillin-G.
    c. Heating the β-lactamase to 70°C for 10 minutes.
    d. Adding more β-lactamase.

e. Adding more phenol red.

f. Using 50 mM Tris (pH 7.9) to make up the phenol red.

g. Using 10 mM Tris (pH 9.0) to make up the phenol red.

4. Plot the spectrophotometer data for the "pAMP sup" β-lactamase reaction. Plot time in minutes on the *x* axis. Plot absorbance on the *y* axis.
5. Predict what the plots would look like if you took readings at 560 nm (the peak absorbance of phenol red at basic pH).
6. The rate of the reaction is defined as the amount of substrate formed over time. In this experiment, you indirectly measured substrate formation as an increase in [H^+], which you measured as the increase in absorbance at 430 nm. Look at your plot and describe the point at which the rate of the reaction is greatest. Write the rate of the reaction as change in absorbance per minute.

FOR FURTHER RESEARCH

1. Enzymes are generally sensitive to heat. Heat-inactivate 1 ml of the β-lactamase solution by incubating at 70°C for 10 minutes and then repeat the assay, comparing the reaction of the heat-inactivated enzyme with untreated enzyme.
2. Use β-lactamase activity to destroy ampicillin on an agar plate. Do the following:
   a. Sterilize β-lactamase solution by filtering through a syringe filter into a sterile 15-ml tube.

   b. Obtain sterile cotton applicators, an agar plate containing ampicillin, and a tube of *E. coli* culture. Soak a cotton applicator in the sterile β-lactamase.

   c. Moving quickly, lift the plate lid and write your initials with the applicator on the surface of the plate. (You can redip into the β-lactamase.)

   d. Allow the plate to set for a few minutes, so that the β-lactamase soaks into the agar.

   e. Soak a fresh sterile applicator in the *E. coli* culture. Open the plate again and use the applicator to coat the entire surface with bacteria.

   f. Incubate the plate at 37°C (take time for responsible cleanup).

   g. After 24 hours, remove the plate; store at 4°C until it can be analyzed.

   h. Examine the plate and explain what has happened.
3. β-lactamase resides in the periplasm between the cell wall and the cell membrane. Normally, β-lactamase leaks into the media through the cell wall, but some remains trapped in the periplasm. Lysozyme is an enzyme from egg whites that destroys bacterial cell walls and releases the trapped β-lactamase into the media. Treat bacteria with lysozyme to release β-lactamase.
   a. Make up a stock of 20 mg/ml solution of lysozyme (pH 7.9) (store at –20°C).

   b. Thaw the "pAMP" and "control" bacterial pellets frozen in Part I, Step 5.

c. Add 5 ml of fresh LB broth to each tube containing the pellets. Close caps securely and vortex or shake the tubes vigorously for 1 minute.

d. Add 500 μl of lysozyme solution to each tube. Close caps securely and mix by inverting tubes several times.

e. Incubate for 10 minutes at 37°C or 15 minutes at room temperature. Close caps securely and vortex or shake tubes vigorously for 1 minute.

f. Place tubes in a *balanced* configuration in a clinical centrifuge. Centrifuge at 2000–4000 rpm for 10 minutes to pellet cells.

g. Label two fresh 15-ml tubes "pAMP + lys" and "control + lys."

h. When the centrifuge stops, pour the supernatant from "pAMP" and "control" tubes into "pAMP + lys" and "control + lys" tubes, respectively. These tubes contain β-lactamase released from the periplasmic space (pAMP + lys) and control supernatant (control + lys).

i. Proceed exactly as in Steps 6 through 12 from Part I of the protocol, replacing "control sup" with "control + lys" and "pAMP sup" with "pAMP + lys."

j. Compare the times the "pAMP sup" and "pAMP + lys" enzyme reactions take to reach completion.

4. Overnight *E. coli*/pAMP cultures contain cleaved ampicillin which, like penicilloic acid, is acidic. Compare the absorbance of your "control sup" and "pAMP sup" solutions by setting up two reactions as in Part II of this lab, but using water in place of penicillin-G. To each reaction, add 600 μl of water and 50 μl of phenol red. To one tube, add 200 μl of "control sup" and to the other tube add 200 μl of "pAMP sup." Measure the absorbances at 430 nm using a spectrophotometer. Do the values differ?

5. You can purify β-lactamase from a 100-ml overnight culture of *E. coli*/pAMP.

   a. Add lysozyme to the culture to 2 mg/ml and incubate for 15 minutes at 37°C with shaking. Centrifuge to remove cells. Dissolve 15 g of ammonium sulfate in the supernatant. Place on ice for 10 minutes, and then centrifuge for 10 minutes. A precipitate will form at the bottom of the tube. Save this pellet and the supernatant. Resuspend the pellet in 2 ml of 10 mM Tris (pH 7.9) and set aside. Add another 25 g of ammonium sulfate to the supernatant and place on ice for 10 minutes. Centrifuge for 10 minutes and save the final supernatant. Resuspend the pellet in 2 ml of 10 mM Tris (pH 7.9).

   b. Now determine which fractions contain β-lactamase activity by assaying the original supernatant, the resuspended ammonium sulfate precipitates, and the final supernatant with penicillin-G and phenol red. Determine the protein concentration of the original supernatant using a Bradford assay or Bio-Rad solution and compare with the other fraction containing β-lactamase activity. Finally, use a spectrophotometer as described in this laboratory to determine the relative β-lactamase activity of the original supernatant and the active fraction. Determine activity versus protein concentration.

6. You can visualize your β-lactamase protein by SDS-polyacrylamide gel electrophoresis. For details, see Laboratory 7, Purification and Identification of Recombinant GFP. β-lactamase samples are mixed with an equal volume of 2x denaturing buffer (100 mM Tris [pH 6.8], 200 mM dithiothreitol, 4% SDS, 0.1% bromophenol blue, and 10% glycerol) and boiled for 10 minutes prior to loading onto the gel. In addition, obtain 5-ml cultures of wild-type *E. coli* and *E. coli*/pAMP. Centrifuge and resuspend the cells in 1x denaturing buffer, boil 10 minutes, and load on gel. After running gel, determine the predicted size of β-lactamase and look for the corresponding band on the gel. It should be present in *E. coli*/pAMP but not in wild-type cells.

Purification and Identification of Recombinant GFP

LABORATORY 7 INTRODUCES A RAPID METHOD to purify recombinant green fluorescent protein (GFP) using hydrophobic interaction chromatography (HIC). The purified protein is then identified using polyacrylamide gel electrophoresis (PAGE). This lab is divided into two parts: Purification of GFP by HIC and PAGE Analysis of Purified GFP.

- Part A provides a procedure for purifying GFP. Cells taken from a GFP/ampicillin-resistant colony are grown to stationary phase in suspension culture. The cells from 2 ml of culture are harvested and lysed on ice to liberate GFP and other cellular proteins. During the first step of the purification, the cell lysate is incubated in a high-salt binding buffer. The charged ions in the binding buffer repel ions on the exterior of the GFP molecule, essentially flipping the GFP molecule inside out to reveal the hydrophobic chromophore. The exposed chromophore then binds tightly to the HIC resin, and successive washes in mid-salt elute unbound and weakly interacting proteins. Finally, incubation with low-salt TE buffer restores the normal structure of GFP and releases the protein from the resin. The eluted protein is transferred to a separate tube and its characteristic fluorescence is detected by exposure to long-wavelength UV light ("black light").
- Part B provides a technique whereby samples of the cell lysate, mid-salt wash, purified GFP, and a protein molecular-weight ladder are coelectrophoresed in a polyacrylamide gel and stained with Coomassie Blue. A single band of GFP protein is present in the lane of purified protein and comigrates with a size marker of approximately 27 kD. Bands and smears representing numerous cellular proteins are visible in the lysate and wash lanes, but are absent from the lane containing purified GFP.

Kits based on this laboratory are available from the Carolina Biological Supply Company.

- Catalog no. 21-1082: Module 1 (Green Gene Colony Transformation Kit)
- Catalog no. 21-1070: Module 2 (GFP Purification)
- Catalog no. 21-1071: Module 3 (Electrophoretic Analysis of GFP Purification)

PART A

Purification of GFP by HIC

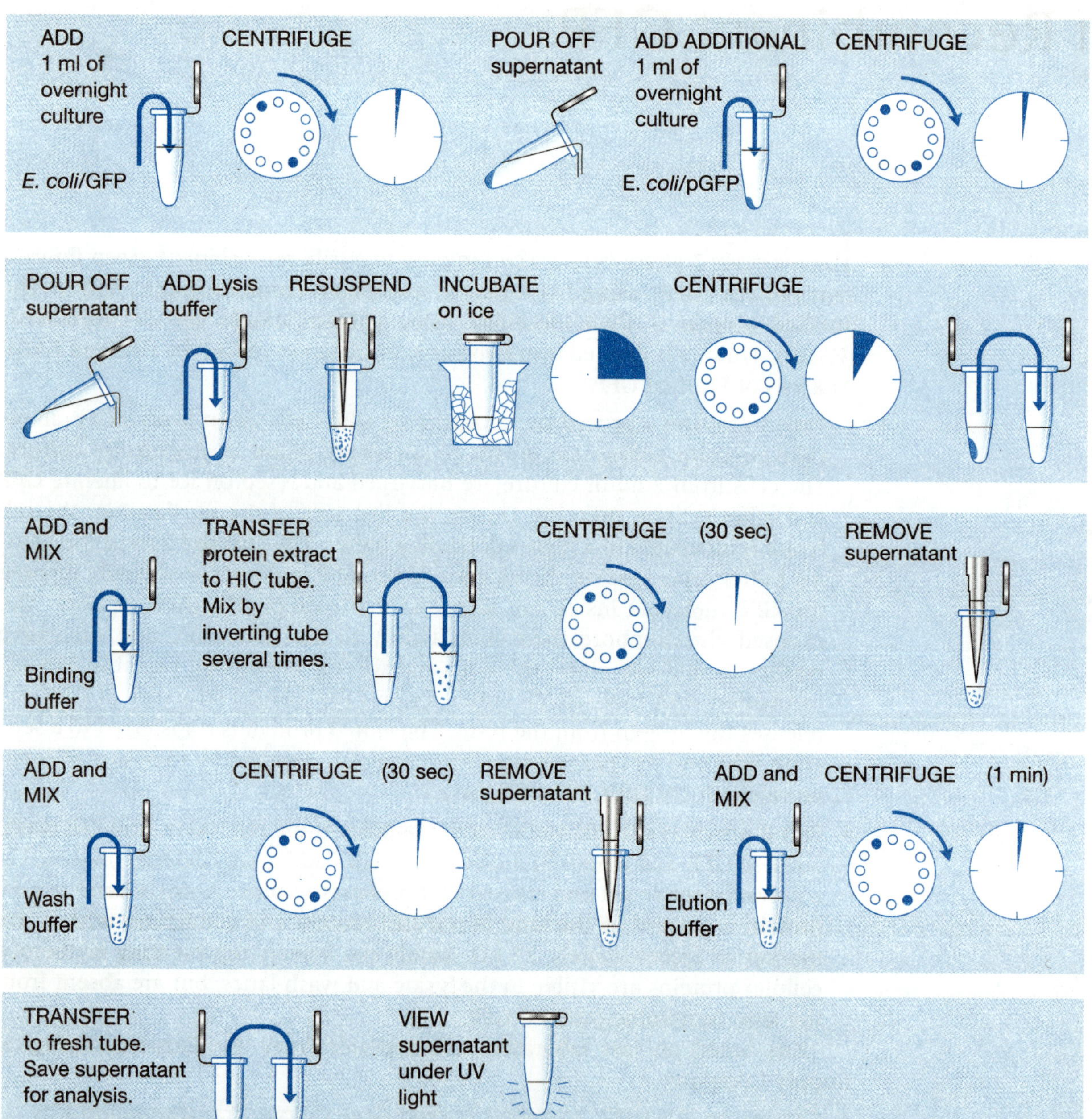

PRELAB NOTES

GFP Expression in *E. coli*

It is critical that overnight cultures of *E. coli*/GFP be shaken *vigorously* to provide good aeration. Shake the cultures hard enough so that some air bubbles are introduced at the surface of the culture medium. Of course, oxygen is required for *E. coli* reproduction, plasmid amplification, and transcription/translation needed for good expression of GFP. However, oxygen is also critical for the post-translational formation of the GFP chromophore, upon which fluorescence depends. Without proper aeration, GFP-transformed strains of *E. coli* will fail to fluoresce.

The best protein yield is obtained from *fresh* cells grown in overnight suspension culture. Use cells taken directly from overnight culturing of not more than 16 hours; GFP begins to degrade in stationary-phase cells.

Batch Technique for Protein Purification

The best known methods of HIC involve passing a cell lysate through a vertical column packed with resin. However, the passage of lysate and washes through a column is inherently time-consuming. Packing and handling a column can be problematic, since even small air spaces in the resin inhibit the capillary action on which flow-through depends. Fortunately, the strong binding of GFP to hydrophobic resin makes column chromatography unnecessary. Rather, this simplified method uses a "batch technique" in which the entire purification takes place in a 1.5-ml tube. Use only clear tubes, since it is important to remove as little of the HIC resin as possible during the purification process.

HIC Bead Resin

The HIC bead resin consists of 50-μm porous beads that contain methyl groups ($-CH_3$). The methyl groups provide a strong interaction with hydrophobic proteins under high-salt conditions.

SDS-PAGE Molecular-weight Standards

Molecular-weight markers are available from many suppliers and assist in identifying GFP protein. Since GFP protein is about 27 kD, protein markers should range from 20 to 150 kD.

For Further Information

The protocol presented here is based on the following published methods.

Lin F.Y., Chen W.Y., and Hearn M.T. 2001. Microcalorimetric studies on the interaction mechanism between proteins and hydrophobic solid surfaces in hydrophobic interaction chromatography: Effects of salts, hydrophobicity of the solvent, and structure of the protein. *Anal. Chem.* **73:** 3875–3883.

PRELAB PREPARATION

Before performing this Prelab Preparation, please refer to the cautions indicated on the Laboratory Materials list.

1. The day before starting this laboratory, prepare an *E. coli* culture according to the protocol in Laboratory 2B, Overnight Suspension Culture. Inoculate the culture with a cell mass scraped from one colony selected from a +pGREEN plate, from Laboratory 5, Rapid Colony Transformation of *E. coli* with Plasmid DNA. Maintain antibiotic selection with LB broth plus ampicillin. Before incubating the overnight culture, gently swirl the culture tube until the colony is dispersed. Vigorous aeration in a shaking water bath or environmental shaker is essential to GFP fluorescence. Although incubation at 33°C is recommended for best expression, we have found that 37°C also works well.
2. The day before the experiment, equilibrate the HIC bead resin. Add 1000 μl of equilibration buffer to 500 μl of HIC beads in a 1.5-ml tube. Invert the tube several times to mix. Centrifuge at high speed for 1 minute. Remove the buffer layer without disturbing the equilibrated HIC beads.
3. Prepare aliquots for each experiment:

   2 ml of *E. coli*/GFP overnight culture
   500 μl of preequilibrated HIC beads
   600 μl of lysis buffer
   300 μl of binding buffer
   600 μl of wash buffer
   300 μl of elution buffer (TE)

4. Review Part B, PAGE Analysis of Purified GFP.

MATERIALS

REAGENTS	SUPPLIES AND EQUIPMENT
Binding buffer	Beaker for waste/used tips
E. coli/GFP overnight culture	Microfuge
Equilibration buffer	Micropipettor (100–1000 μl) + tips
Glucose/Tris▼/EDTA (GTE)	Permanent marker
Hydrophobic (HIC) bead resin	Shaking water bath (37°C) or incubator
Lysis buffer	Test tube rack
Tris▼-EDTA (TE) buffer	Tubes (1.5-ml clear)
Wash buffer	

▼ See Appendix 4 for Cautions list.

METHODS

Purification of GFP by HIC

(40 minutes)

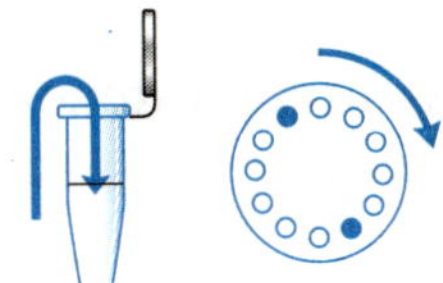

1. Shake the culture tube to resuspend *E. coli* cells.
2. Use a micropipettor to transfer 1 ml of the overnight *E. coli*/GFP culture into a 1.5-ml tube.
3. Close cap, and place the tube in a *balanced* configuration in the microfuge rotor. Spin for 1 minute to pellet cells.

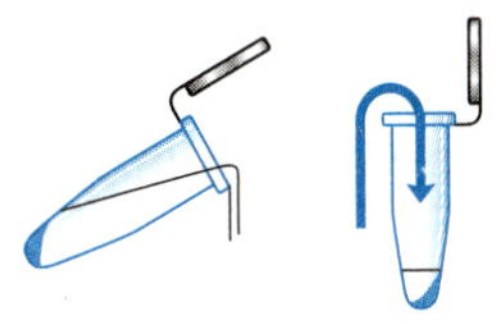

4. Carefully pour off the supernatant into a waste beaker for later disinfection. *Do not disturb the green cell pellet.*
5. Repeat Steps 1–4 *in the same 1.5-ml tube* to pellet cells from a second 1-ml sample on top of the first pellet. This will result in a large green cell pellet. Resuspend the pellet in 50 μl of GTE buffer.
6. Add 500 μl of lysis buffer to the tube. Resuspend the pellet by pipetting in and out several times.
7. Incubate the tube for 15 minutes on ice. Incubation on ice helps prevent protein degradation.
8. Place the tube in a *balanced* configuration in the microfuge rotor. Spin for 5 minutes to pellet the insoluble cellular debris.

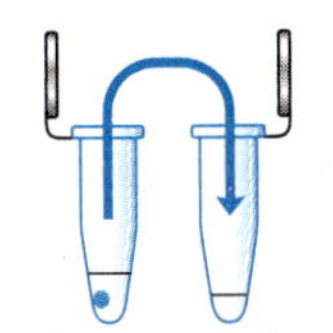

Transfer 100 μl of remaining lysate to a new tube and freeze at –20°C for PAGE analysis in Part B.

9. Transfer 250 μl of green cell extract (supernatant) into a clean 1.5-ml tube. *Do not disturb the pellet of cellular debris.*
10. Add 250 μl of binding buffer to tube containing 250 μl of green cell extract. Close cap, and mix solutions by rapidly inverting the tube several times.
11. Add 400 μl of the cell extract/binding buffer mixture to tube of hydrophobic bead resin. Close cap, and mix by inverting the tube several times.

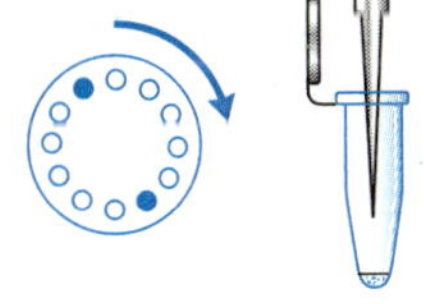

The hydrophobic chromophore of GFP binds to the resin beads.

12. Microfuge for 30 seconds. Gently remove the supernatant with a micropipettor. *Do not disturb the hydrophobic bead pellet.*
13. Add 400 μl of wash buffer to the hydrophobic bead pellet. Mix by inverting the tube several times. This step unbinds weakly interacting cellular proteins from the resin.
14. Microfuge for 30 seconds. Gently remove the supernatant with a micropipettor. *Do not disturb the hydrophobic bead pellet.*
15. Elute the recombinant GFP by adding 200 μl of TE buffer to the hydrophobic bead pellet. Mix by inverting the tube several times.

16. Microfuge for 1 minute. The supernatant now contains the purified GFP. Use a micropipettor to transfer the recombinant GFP to a new 1.5-ml Eppendorf tube.
17. After observing under UV light, freeze purified GFP at –20°C for PAGE analysis in Part B. GFP will retain fluorescence while frozen.

RESULTS AND DISCUSSION

HIC is a simple and efficient means of isolating proteins with a hydrophobic domain. You should be familiar with the biochemical interactions taking place during each step of the purification.

- Chloride ions in the binding buffer repel negative charges in the exterior β sheath of the GFP molecule. This repulsion causes the molecule to flip inside out, exposing the nonionic (hydrophobic) chromophore.
- The exposed chromophore binds tightly to nonpolar methyl groups attached to the HIC resin.
- The mid-salt wash maintains GFP in its hydrophobic state, but it elutes unbound or weakly bound proteins in the cell lysate. These remain in solution.
- The low-salt buffer (TE) allows the chromophore to return to its normal position on the inside of the GFP molecule, releasing GFP from the resin.

1. What class of molecules does the lysis buffer interact with to release GFP from *E. coli* cells?
2. What aspect of GFP structure allows it to interact so strongly with the HIC beads?
3. How does the TE buffer release the GFP molecules from the HIC beads in Step 15?
4. HIC chromatography does not yield 100% pure GFP. What other types of cellular proteins would most likely be found in the GFP preparation?

FOR FURTHER RESEARCH

Use a second method of protein chromatography to purify your HIC-purified GFP. Run protein samples from single and double chromatography purifications on SDS-polyacrylamide gel. Analyze the level of background native proteins between single and double purification schemes.

PART B

PAGE Analysis of Purified GFP

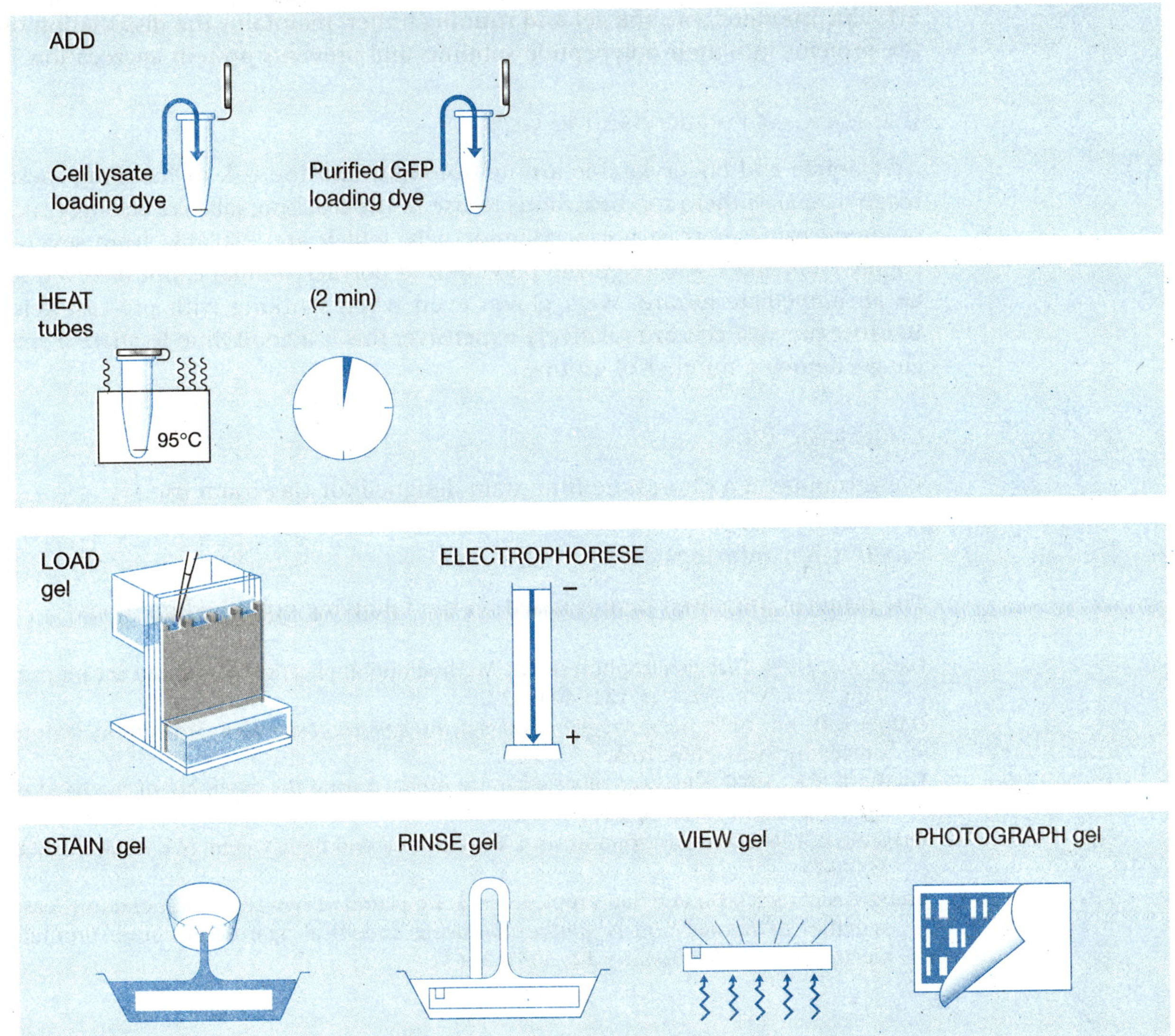

PRELAB NOTES

Analytical electrophoresis of proteins is usually carried out under denaturing conditions using a polyacrylamide gel. Before loading onto the gel, samples are incubated at 95°C to denature the proteins. The negatively charged detergent SDS, incorporated into the gel and running buffer, maintains the dissociation of the proteins into their polypeptide subunits and prevents protein aggregation.

Use Pre-Cast Polyacrylamide Gels

Acrylamide and bisacrylamide are neurotoxins, and the risk of handling these reagents makes them too hazardous to use in the teaching lab. We strongly recommend using pre-cast polyacrylamide gels, which are available from several supply companies. Although the polymerized polyacrylamide is not believed to be an immediate hazard, wear gloves even when working with pre-cast gels. Because pre-cast gels are relatively expensive, this is a good time to share a single gel among a number of groups.

Coomassie Blue

We recommend a Coomassie Blue stain designed for classroom use.

For Further Information

The protocol presented here is based on the following published methods:

Davis B.J. 1964. Disc electrophoresis. II: Method and application to human serum proteins. *Ann. N.Y. Acad. Sci.* **121:** 404–427.

Hames B.D., ed. 1998. *Gel electrophoresis of proteins: A practical approach,* 3rd edition. Oxford University Press, New York.

Laemmli U.K. 1970. Cleavage of structural proteins during the assembly of the head of bacteriophage T4. *Nature* **227:** 680–685.

Ornstein L. 1964. Disc electrophoresis. I: Background and theory. *Ann. N.Y. Acad. Sci.* **121:** 321–349.

Sharp P.A., Sugden B., and Sambrook J. 1973. Detection of two restriction endonuclease activities in *Haemophilus parainfluenzae* using analytical agarose-ethidium bromide electrophoresis. *Biochemistry* **12:** 3055–3063.

PRELAB PREPARATION

Before performing this Prelab Preparation, please refer to the cautions indicated on the Laboratory Materials list.

1. Prepare aliquots for each experiment:

   5 μg of protein markers (store on ice). Volumes vary depending on supplier.
   40 μl of 2x protein loading dye.

2. Prepare 1x Tris-glycine-SDS buffer for electrophoresis (400–500 ml per gel).
3. Prepare Coomassie Blue protein stain (~100 ml per gel).
4. Adjust water bath to 95°C.

MATERIALS

REAGENTS	SUPPLIES AND EQUIPMENT
Coomassie Blue▼protein stain	Beaker for waste/used tips
HIC cell lysate	Electrophoresis chamber
HIC-purified GFP	Latex gloves
Tris/HCl/SDS polyacrylamide gel (12.5%)▼	Micropipettor (0.5–10 μl) + tips
2x Protein loading dye	Permanent marker
Protein markers	Plastic wrap (optional)
1x Tris/glycine/SDS buffer▼	Power supply
	Test tube rack
	Tubes (1.5-ml)
	White light transilluminator▼ (optional)
	Water bath (95°C)

▼ See Appendix 4 for Cautions list.

METHODS

I. Denature Proteins, Load Gel, and Separate by Electrophoresis

(1 hour, 30 minutes)

Refer to Laboratory 3, DNA Restriction and Electrophoresis, for detailed instructions.

Your instructor may assign you a single sample to prepare for loading into a communal gel.

1. Use a permanent marker to label two 1.5-ml tubes

   CL = cell lysate (from Part A, Step 9)
   GFP = purified GFP (from Part A, Step 16)

2. Transfer 5 μl of cell lysate (CL) and 15 μl of purified GFP (GFP) into the appropriate tubes. Obtain a tube containing protein markers (PM). These volumes correspond to approximately 10 μg of total protein.
3. Add 5 μl of 2x protein loading dye to tube CL. Add 15 μl of 2x protein loading dye to tube GFP.
4. Heat the samples (including markers) for 2 minutes at 95°C to denature the protcins.

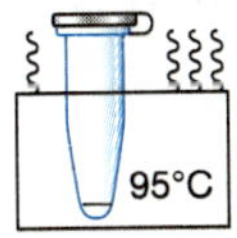

It is often useful to wash the well by repeatedly pipetting running buffer into the wells. This should be done immediately before loading the wells to remove any unpolymerized acrylamide.

5. Load pre-cast 12.5% Tris/HCl/SDS polyacrylamide protein gel into a vertical gel chamber, and add 1x Tris-glycine-SDS buffer to both gel buffer chambers.
6. Load entire contents of each sample tube into a separate well in the gel, as shown in diagram on the following page. *Use a fresh tip for each reaction.*

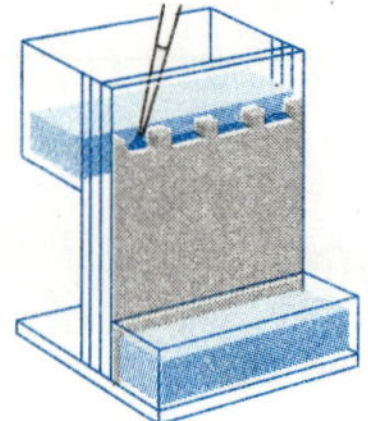

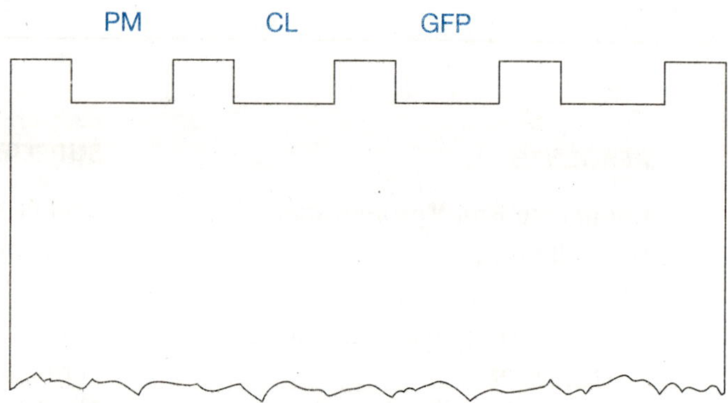

7. Close the tank of the electrophoresis unit and connect the electrical leads to a power supply, anode to anode and cathode to cathode. Electrophorese at 175 volts, until the bromophenol blue band has moved to the bottom of the gel. This should take approximately 1 hour.

8. Turn off power supply, disconnect leads from the inputs, and remove top of electrophoresis box.

9. Carefully remove gel cassette from the electrophoresis chamber. Open the cassette with a spatula, and transfer the gel to a staining tray.

CAUTION
Handle carefully. Polyacrylamide gels are easily torn.

Cover electrophoresis tank and save gel until ready to continue. Gel can be stored in a zip-lock plastic bag and refrigerated overnight for viewing/photographing the next day. However, over longer periods of time, the proteins will diffuse through the gel and the bands will become indistinct or disappear entirely.

Staining may be performed by an instructor in a controlled area when students are not present.

II. Stain Gel with Coomassie Blue and View (Photograph) (3+ hours)

CAUTION
Wear latex gloves when staining, viewing, and photographing gel and during clean up. Confine all staining to a restricted sink area.

1. Flood the gel with Coomassie Blue protein stain, cover the staining tray, and allow the gel to stain from 1 hour to overnight.

2. Wearing gloves, pour off the staining solution, and remove excess stain by rinsing gel in tap water.
3. Destain gel in tap water for 1–2 hours, changing the water every 15 minutes. Protein bands will appear blue. The longer the destaining period, the less intense the background blue stain will appear. Do not destain for too long or the protein bands may lose their intensity.

Gentle rotary shaking hastens destaining, but is not required. A piece of sponge or foam added to the water will absorb stain as it comes off the gel.

4. View the gel on a white background or under a white light transilluminator.
5. Photograph the gel with a Polaroid or digital camera.

RESULTS AND DISCUSSION

1. Describe the different properties and uses of agarose *versus* polyacrylamide gels.
2. Why are the protein samples incubated at 95°C prior to loading onto the polyacrylamide gel?
3. What is the function of the SDS detergent in the gel and running buffer?
4. How does Coomassie Blue stain the proteins in the gel? Why is it important to destain for a sufficient amount of time?
5. View your stained gel on a light box. Compare your gel with the ideal gel shown below, and try to determine which band represents GFP.
   a. Expose the gel to a UV light. Explain why the GFP band does not fluoresce.
   b. Why might you observe some bacterial proteins in your purified GFP lane?
6. Troubleshooting gels. What effect will be observed in the stained bands of protein in a polyacrylamide gel
   a. if the samples are not incubated at 95°C prior to loading?
   b. if the gel is run at too high or too low a voltage?
   c. if too much protein is loaded?
7. Predict the pattern of protein banding observed following SDS-PAGE of the lysate mixed with binding buffer (Part A, Step 10). How would this pattern differ following incubation with HIC beads (Part A, Step 12)? What protein banding pattern would be observed if the wash buffer (Part A, Step 14) was subjected to SDS-PAGE?

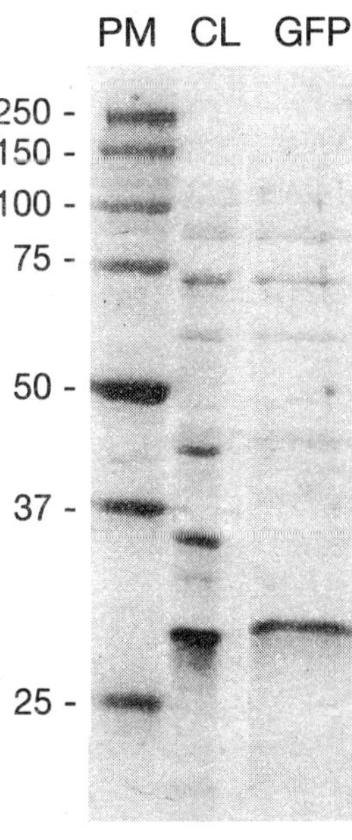

Ideal Gel

FOR FURTHER RESEARCH

Run your purified GFP samples on other types of polyacrylamide gels.

1. Under nondenaturing conditions, native PAGE allows proteins to maintain biological activity, so the GFP band should fluoresce when exposed to UV light. How does the staining pattern differ from SDS-PAGE?
2. Gradient PAGE uses differing gel concentrations along the length of the gel to achieve optimal separation of proteins in a wide range of molecular-weight proteins. How does the staining pattern differ from nongradient PAGE?

Purification and Identification of Plasmid DNA

GROWTH OF *E. COLI* ON AMPICILLIN PLATES demonstrates transformation to an antibiotic-resistant phenotype. In the basic version of Laboratory 5, the observed phenotype was due to uptake of plasmid pAMP, a DNA molecule that is well-characterized.

In experimental situations where numerous recombinant plasmids are generated by joining two or more DNA fragments, the antibiotic resistance marker only functions to indicate which cells have taken up a plasmid bearing the resistance gene. It does not indicate anything about the structure of the new plasmid. Therefore, it is standard procedure to isolate plasmid DNA from transformed cells and to identify the molecular genotype using DNA restriction analysis. In cases where the recombinant molecules are formed by combining well-characterized fragments, restriction analysis is sufficient to confirm the structure of a hybrid plasmid. In other cases, the nucleotide sequence of the insert must be determined. This protocol is divided into two parts: Plasmid Minipreparation of pAMP and Restriction Analysis of Purified pAMP.

- Part A provides a small-scale protocol to purify from transformed *E. coli* enough plasmid DNA for restriction analysis. Cells taken from an ampicillin-resistant colony are grown to stationary phase in suspension culture. The cells from 1 ml of culture are harvested and lysed, and plasmid DNA is separated from the cellular proteins, lipids, and chromosomal DNA. This procedure yields 2–5 μg of relatively crude plasmid DNA, in contrast to large-scale preparations that yield 1 mg or more of pure plasmid DNA from a 1-liter culture.
- Part B provides a protocol using a sample of plasmid DNA isolated in Part A and a control sample of pAMP. These two samples are cut with the restriction enzymes *Bam*HI and *Hin*dIII and coelectrophoresed on an agarose gel, and the restriction patterns are stained and visualized. The purified DNA is shown to have a restriction "fingerprint" identical to that of pAMP. *Bam/Hin*d restriction fragments of the miniprep DNA comigrate with the 784-bp and 3755-bp *Bam/Hin*d fragments of pAMP. This provides genotypic proof that pAMP molecules were successively transformed into *E. coli* in Laboratory 5.

Kits based on this laboratory are available from the Carolina Biological Supply Company.

- Catalog no. 21-1200 (with ethidium bromide stain)
- Catalog no. 21-1205 (with the *Carolina*Blu™ stain)

PART A

Plasmid Minipreparation of pAMP

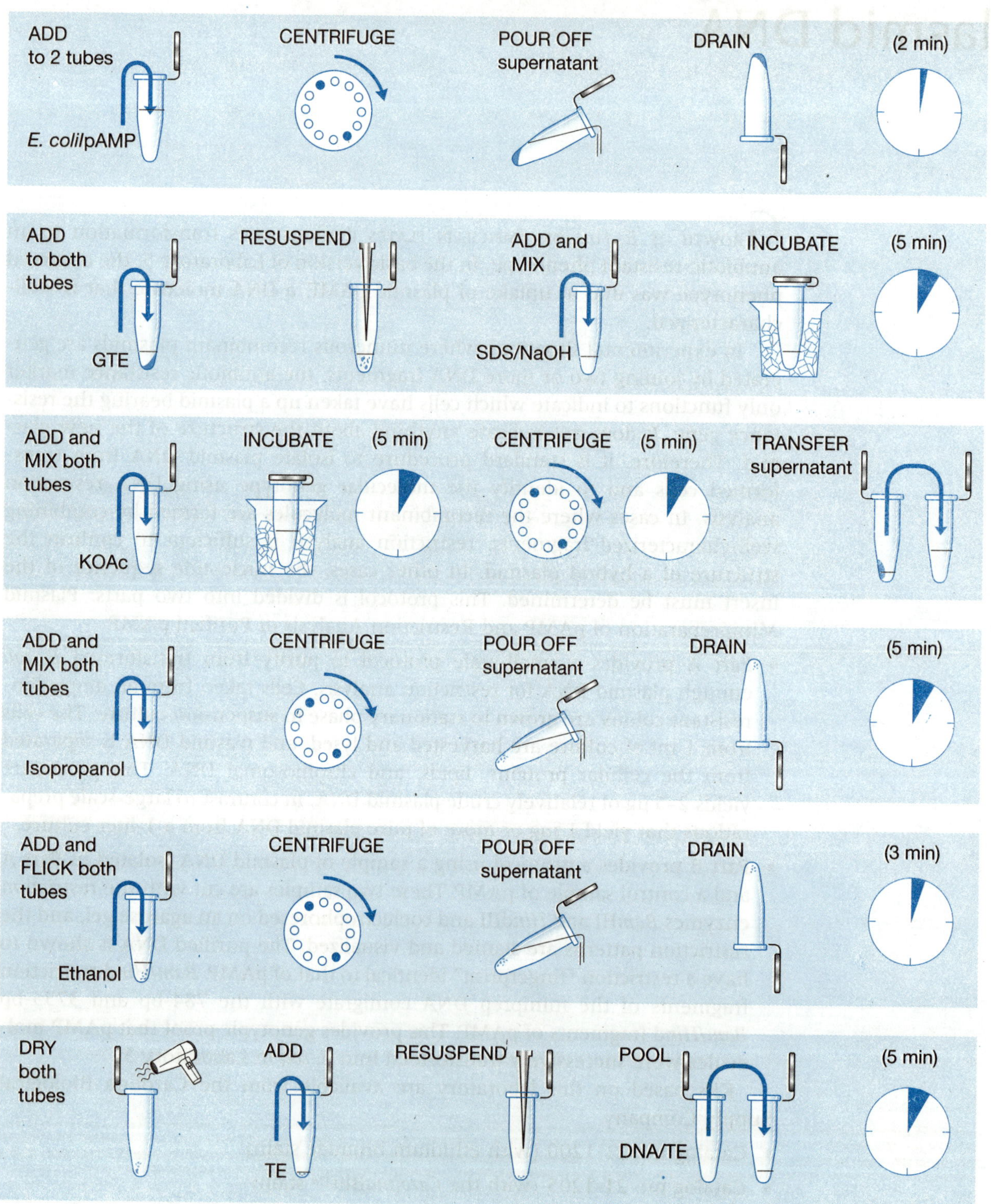

PRELAB NOTES

Optimally, minipreps should be done on cells that have been recently manipulated for transformation. This completes a conceptual stream that firmly cements the genotype-phenotype relationship. Alternatively, use streaked plates of transformed *E. coli* to prepare overnight cultures.

Plasmid Selection

pAMP gives superior yields on minipreps compared to pBR322. A derivative of a pUC expression vector, pAMP is highly amplified—more than 100 copies are present per *E. coli* cell. If substituting a different plasmid for miniprep purposes, select a commercially available member of the pUC family, such as pUC18 or pUC19.

Centrifuge Requirements

A microfuge that generates approximately 12,000 times the force of gravity (12,000*g*) is needed for efficient and rapid purification of plasmid DNA. A slower-spinning clinical or preparatory centrifuge cannot be substituted.

Supplies

Sterile supplies are not required for this protocol. Standard 1-ml pipettes, transfer pipettes, and/or microcapillary pipettes can be used instead of micropipettors. Use good-quality, colorless 1.5-ml tubes, beginning with Step 11. The walls of poor-quality tubes, especially colored tubes, often contain tiny air bubbles that can be mistaken for ethanol droplets in Step 19. We have observed students drying DNA pellets for 15 minutes or more, trying to rid their tubes of these phantom droplets. (Typical drying time is actually several minutes.)

Fine Points of Technique

Be careful not to overmix reagents; excessive manipulation shears chromosomal DNA. The success of this protocol in large part depends on maintaining chromosomal DNA in large pieces that can be differentially separated from intact plasmid DNA. Mechanical shearing increases the amount of short-sequence chromosomal DNA, which is not removed in the purification of plasmid DNA. Make sure that the microfuge will be immediately available for Step 13. If sharing a microfuge, coordinate with other experimenters to begin Steps 12 and 13 together.

For Further Information

The protocol presented here is based on the following published methods:

Birnboim H.C. and Doly J. 1979. A rapid alkaline extraction method for screening recombinant plasmid DNA. *Nucleic Acids Res.* **7:** 1513–1523.

Ish-Horowicz D. and Burke J.F. 1981. Rapid and efficient cosmid cloning. *Nucleic Acids Res.* **9:** 2989–2998.

PRELAB PREPARATION

Before performing this Prelab Preparation, please refer to the cautions indicated on the Laboratory Materials list.

1. This protocol is designed to follow Laboratory 5. Ideally, students should pick colonies from their own transformed plates to begin this experiment. However, the colonies must grow overnight or for at least several hours, so unless the class meets on consecutive days, it may be necessary for the instructor to set up the cultures used in this laboratory. In any case, on the day before the laboratory, prepare an *E. coli* culture according to the protocol in Laboratory 2B, Overnight Suspension Culture. Inoculate the culture with a cell mass scraped from one colony selected from the +LB/amp plate from Laboratory 5, Rapid Colony Transformation of *E. coli* with Plasmid DNA. Maintain antibiotic selection with LB broth plus ampicillin. Alternatively, prepare the culture 2–3 days in advance and store at 4°C or incubate for 24–48 hours at 37°C without shaking. In either case, the cells will settle at the bottom of the culture tube. Shake the tube to resuspend cells before beginning procedure.
2. The SDS/sodium hydroxide solution should be fresh; prepare this solution within a few days of lab. Store solution at room temperature; a soapy precipitate may form at lower temperature. If a precipitate forms, warm the solution by placing the tube in a beaker of hot tap water, and shake gently to dissolve the precipitate.
3. Prepare aliquots for each experiment:

   250 μl of glucose/Tris/EDTA (GTE) solution (store on ice)
   500 μl of SDS/sodium hydroxide (SDS/NaOH) solution
   400 μl of potassium acetate/acetic acid (KOAc) solution (store on ice)
   1000 μl of isopropanol
   500 μl of 95% ethanol
   50 μl of Tris/EDTA (TE) solution

4. Review Part B, Restriction Analysis of Purified pAMP.

MATERIALS

CULTURES AND MEDIA	SUPPLIES AND EQUIPMENT
E. coli/pAMP overnight culture	Beakers for crushed ice and for waste/used tips
Ethanol (95–100%)▼	Bleach▼(10%) or disinfectant
Glucose/Tris▼/EDTA (GTE)	Clean paper towels
Isopropanol▼	Hair dryer
Potassium acetate/acetic acid▼ (KOAc)	Microfuge
SDS/sodium hydroxide (SDS/NaOH)▼	Micropipettor (100–1000 µl and 0.5–10 µl) + tips
Tris▼/EDTA (TE)	Permanent marker
	Test tube rack
	Tubes (1.5-ml)

▼ See Appendix 4 for Cautions list.

METHODS

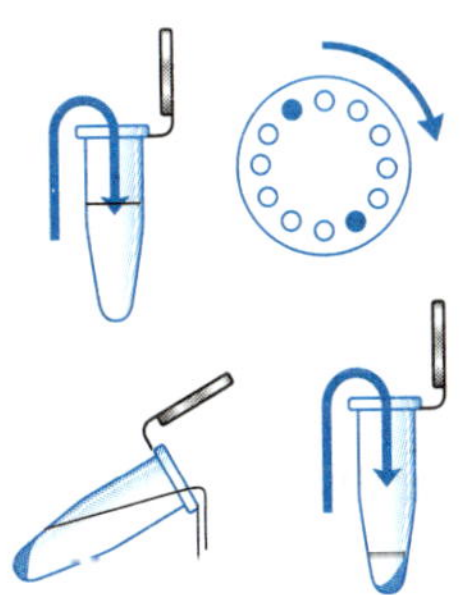

The cell pellet will appear as a small off-white smear on the bottom-side of the tube. Although the cell pellets are readily seen, the DNA pellets in Step 14 are very difficult to observe. Make a habit of aligning the tube with the cap hinges facing outward in the microfuge rotor. Then, pellets should always be located at the tube bottom beneath the hinge.

Accurate pipetting is essential to good plasmid yield. The volumes of reagents are precisely calibrated so that the sodium hydroxide added in Step 6 is neutralized by acetic acid in Step 8.

Prepare Duplicate Minipreps

(50 minutes)

The instructions below are for making duplicate minipreps, which provide balance in the microfuge and insurance if a critical mistake is made.

1. Shake culture tube to resuspend *E. coli* cells.
2. Label two 1.5-ml tubes with your initials. Use a micropipettor to transfer 1000 µl of *E. coli*/pAMP overnight suspension into each tube.
3. Close caps, and place the tubes in a *balanced* configuration in the microfuge rotor. Spin for 1 minute to pellet cells.
4. Pour off supernatant from both tubes into a waste beaker for later disinfection. Alternatively, use a micropipettor to remove supernatant. *Be careful not to disturb the cell pellets.* Invert the tubes, and tap gently on the surface of a clean paper towel to drain thoroughly.
5. Add 100 µl of ice-cold GTE solution to each tube. Resuspend the pellets by pipetting the solution in and out several times. Hold the tubes up to the light to check that the suspension is homogeneous and that no visible clumps of cells remain.
6. Add 200 µl of SDS/NaOH solution to each tube. Close caps, and mix solutions by rapidly inverting tubes five times.
7. Stand tubes on ice for 5 minutes. Suspension will become relatively clear.
8. Add 150 µl of *ice-cold* KOAc solution to each tube. Close caps, and mix solutions by rapidly inverting tubes five times. A white precipitate will immediately appear.

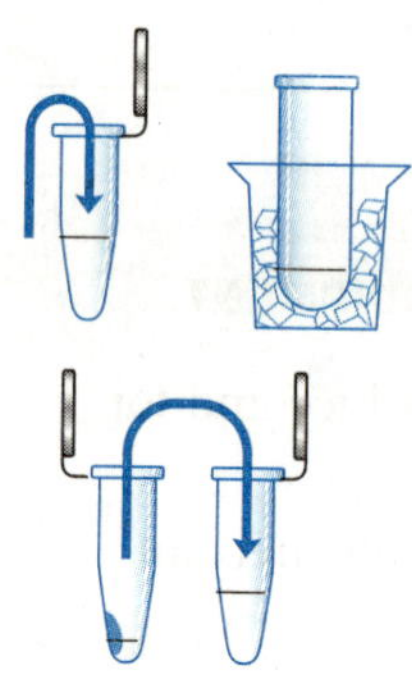

In Step 11, the supernatant is saved and precipitate is discarded. The situation is reversed in Steps 14 and 17, where the precipitate is saved and the supernatant is discarded.

Do Step 12 quickly, and make sure that the microfuge will be immediately available for Step 13.

The pellet may appear as a tiny smear or small particles on the bottom-side of each tube. Do not be concerned if pellet is not visible; pellet size is not a predictor of plasmid yield. A large pellet is composed primarily of RNA, and cellular debris carried over from the original precipitate. A smaller pellet often means a cleaner preparation.

Nucleic acid pellets are not soluble in ethanol and will not resuspend during washing.

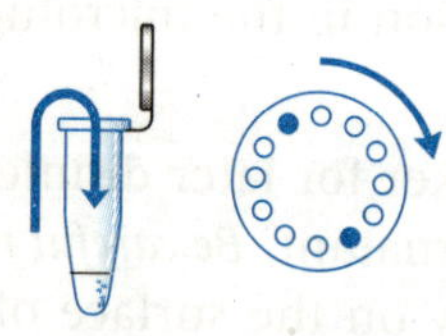

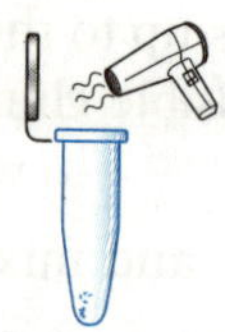

9. Stand tubes on ice for 5 minutes.
10. Place tubes in a *balanced* configuration in the microfuge rotor, and spin them for 5 minutes to pellet the precipitate along the side of the tube.
11. Transfer 400 μl of supernatant from each tube into clean 1.5-ml tubes. If measured correctly, a small amount of supernatant remains behind, which provides a small buffer between the tip and the precipitate. *Avoid pipetting precipitate,* and wipe off any precipitate clinging to the outside of the tip prior to expelling supernatant. Discard old tubes containing precipitate.
12. Add 400 μl of isopropanol to each tube of supernatant. Close caps, and mix vigorously by rapidly inverting tubes five times. *Stand at room temperature for only 2 minutes.* (Isopropanol preferentially precipitates nucleic acids rapidly; however, proteins remaining in solution also begin to precipitate with time.)
13. Place tubes in a *balanced* configuration in the microfuge rotor, and spin for 5 minutes to pellet the nucleic acids. Align tubes in rotor so that cap hinges point outward. The nucleic acid residue, visible or not, will collect on the tube side under the hinge during centrifugation.
14. Pour off supernatant from both tubes. *Be careful not to disturb nucleic acid pellets.* Alternatively, remove the supernatant with a 1000-ml micropipettor. Place tip away from the pellet. If you are concerned that the pellet has been drawn up in the tip, transfer the supernatant to another 1.5-ml tube, recentrifuge, and remove the supernatant again. Invert tubes, and tap gently on the surface of a clean paper towel to drain thoroughly.
15. Add 200 μl of 100% ethanol to each tube, and close caps. Flick tubes several times to wash pellets.

 Store DNA in ethanol at –20°C until ready to continue.

16. Place tubes in a *balanced* configuration in microfuge rotor, and spin for 2–3 minutes.
17. Pour off supernatant from both tubes. *Be careful not to disturb nucleic acid pellets.* Alternatively, remove the supernatant with a 1000-ml micropipettor. Place tip away from the pellet. If you are concerned that the pellet has been drawn up in the tip, transfer the supernatant to another 1.5-ml tube, recentrifuge, and remove the supernatant again. Invert tubes, and tap gently on the surface of a clean paper towel to drain thoroughly.
18. Dry nucleic acid pellets by one of the following methods:
    a. Direct a stream of warm air from a hair dryer across the open ends of the tubes for about 3 minutes. *Be careful not to blow pellets out of tubes.*

    or

    b. Close caps, and pulse tubes in the microfuge to pool remaining ethanol. *Carefully* draw off drops of ethanol using a 1–10-μl micropipettor. Leave cap open and place tube upright in rack, allowing pellets to air-dry for 10 minutes at room temperature.
19. At the end of the drying period, hold each tube up to the light to check that no ethanol droplets remain. If ethanol is still evaporating, an alcohol odor

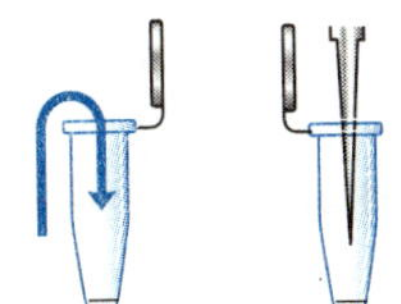

If using a 0.5–10-µl micropipettor, set to 7.5 µl and pipette twice.

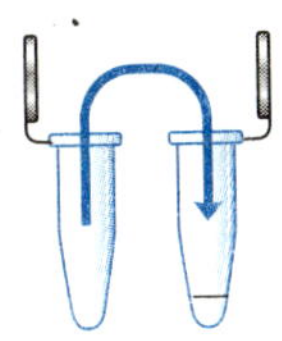

can be detected by sniffing the mouth of the tube. All ethanol must be evaporated before proceeding to Step 20.

20. Add 15 µl of TE to each tube. Resuspend the pellets by smashing with the micropipettor tip and pipetting in and out vigorously. Rinse down the side of tube several times, concentrating on the area where the pellet should have formed during centrifugation (beneath cap hinge). Check that all DNA is dissolved and that no particles remain in the tip or on the side of the tube.
21. Pool DNA/TE solution into one tube.

STOP Freeze DNA/TE solution at –20°C until ready to continue. Thaw before using.

22. Take time for responsible cleanup.
    a. Segregate for proper disposal culture tubes and micropipettor tips that have come in contact with *E. coli*.
    b. Disinfect overnight culture, tips, and supernatant from Step 4 with 10% bleach or disinfectant.
    c. Wipe down lab bench with soapy water, 10% bleach solution, or disinfectant (such as Lysol).
    d. Wash hands before leaving lab.

RESULTS AND DISCUSSION

The minipreparation is a simple and efficient procedure for isolating plasmid DNA. Become familiar with the molecular and biochemical effects of each reagent used in the protocol.

- *Glucose//Tris/EDTA:* The Tris buffers the cells at pH 7.9. EDTA binds divalent cations in the lipid bilayer, thus weakening the cell envelope.
- *SDS/sodium hydroxide:* This alkaline mixture lyses the bacterial cells. The detergent SDS dissolves the lipid components of the cell membrane, as well as cellular proteins. The sodium hydroxide denatures the chromosomal and plasmid DNA into single strands. The intact circles of plasmid DNA remain intertwined.
- *Potassium acetate/acetic acid:* The acetic acid returns the pH to neutral, allowing DNA strands to renature. The large, disrupted chromosomal strands cannot rehybridize perfectly, but instead collapse into a partially hybridized tangle. At the same time, the potassium acetate precipitates the SDS (which is insoluable in potassium) from the cell suspension, along with proteins and lipids with which it has associated. The renaturing chromosomal DNA is trapped in the SDS/lipid/protein precipitate. Only smaller plasmid DNA and RNA molecules escape the precipitate and remain in solution.
- *Isopropanol:* The alcohol rapidly precipitates nucleic acids, but only slowly precipitates proteins. Thus, a quick precipitation preferentially brings down nucleic acids.
- *Ethanol:* A wash with ethanol removes some remaining salts and SDS from the preparation. Ethanol also removes the remaining isopropanol, which has

a higher vapor point than does ethanol. The ethanol-isopropanol evaporates more rapidly in the drying step.

- *Tris/EDTA:* Tris buffers the DNA solution. EDTA protects the DNA from degradation by DNases by binding divalent cations that are necessary cofactors for DNase activity. Buffering DNA is important, as low pH (<6) leads to the loss of purines (adenine and guanine) called depurination. The purines are actually cleaved from their sugars, creating an abasic site. Purine cleavage is a very common occurrence in cells (on the order of 10^5 times per cell per day) and is repaired by specific DNA repair systems. Of course, your DNA is in a tube and there is no DNA repair system present to repair it. Keep in mind that H_2O can have a pH as low as 5.

1. Consider the three major classes of biologically important molecules: proteins, lipids, and nucleic acids. Which steps of the miniprep procedure act on proteins? On lipids? On nucleic acids?
2. What aspect of the plasmid DNA structure allows it to renature efficiently in Step 8?
3. What other kinds of molecules, in addition to plasmid DNA, would you expect to be present in the final miniprep sample? How could you find out?

FOR FURTHER RESEARCH

Determine the approximate mass of plasmid DNA you isolated per milliliter of cells.

1. Set up 20-μl *Hin*dIII restriction reactions using 15 μl of your pAMP preparation and a known mass of λ DNA as a control.
2. Make 1:10, 1:50, and 1:100 dilutions of the digested pAMP and λ DNAs.
3. Separate by electrophoresis equal volumes of each dilution in an agarose gel, and stain with ethidium bromide.

> CAUTION
> Review Responsible Handling of Ethidium Bromide in Laboratory 3. Wear latex gloves when staining, viewing, and photographing gel and during cleanup. Confine all staining to a restricted sink area. For further information, see Appendix 4.

4. Identify a lane of the λ digest where the 4361-bp fragment is *just* visible, and identify a lane of pAMP (4539 bp) that is of equal intensity. These bands should have a nearly equivalent mass of DNA.
5. Determine the mass of λ DNA in the selected fragment, using the formula below. Make sure to account for the dilution factor.

$$\frac{\text{fragment bp (conc. DNA) (vol. DNA)}}{\lambda \text{ bp}}$$

6. Multiply the mass from Step 5 by the dilution factor of the selected pAMP lane.

PART B

Restriction Analysis of Purified pAMP

I. Set Up Restriction Digest

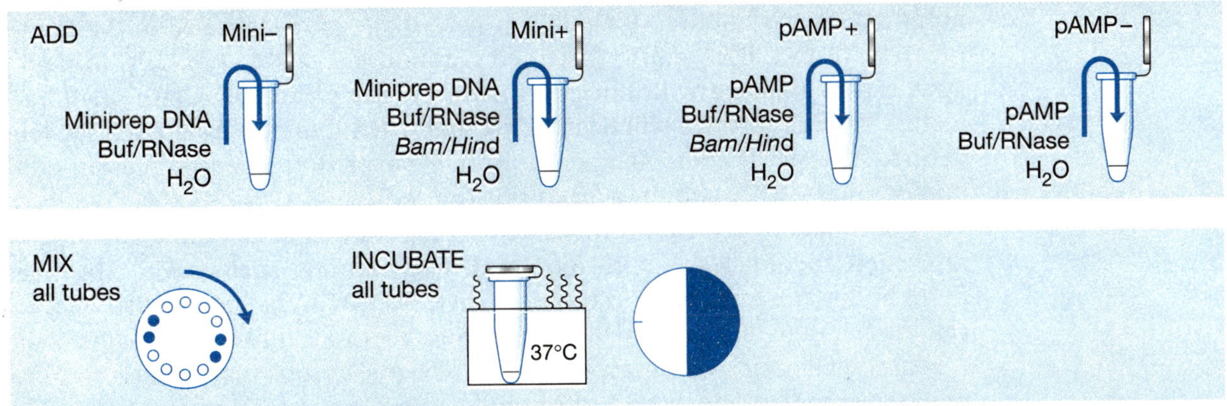

II. Cast 0.8% Agarose Gel

III. Load Gel and Separate by Electrophoresis

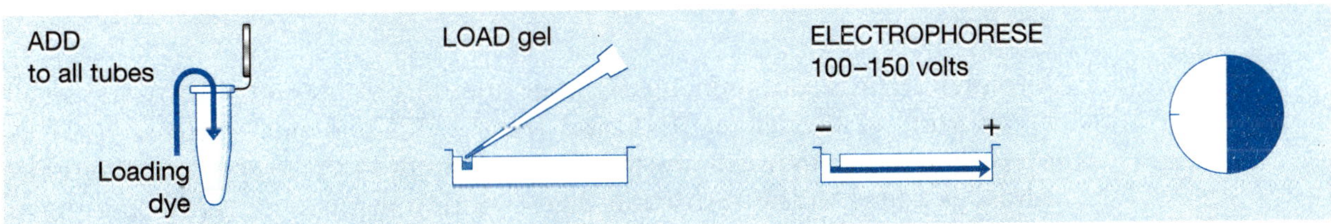

IV. Stain Gel and View (Photograph)

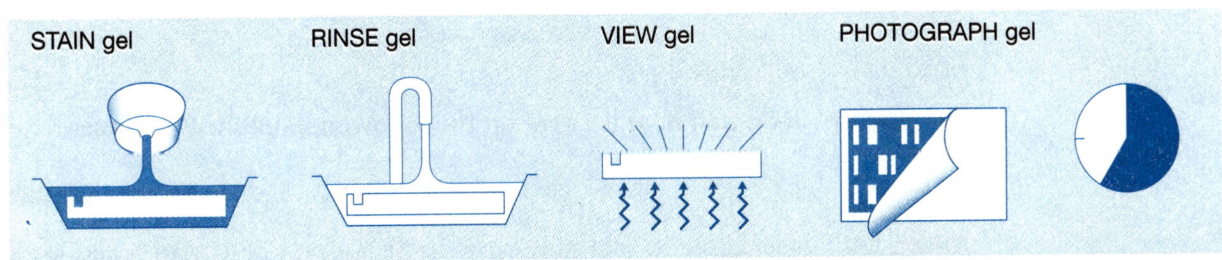

PRELAB NOTES

Review Prelab Notes in Laboratory 3, DNA Restriction Analysis.

Limiting DNase Activity

Unlike highly purified plasmid DNA available from commercial vendors, miniprep DNA is impure. A significant percentage of nucleic acid in the preparation is, in fact, RNA and fragmented chromosomal DNA. Typically, miniprep DNA is contaminated with nucleases (DNases) that cleave DNA into small fragments. Residual DNases will degrade plasmid DNA if minipreps are left for long periods of time at room temperature or even on ice. For this reason, store minipreps at –20°C, and thaw just prior to use.

The situation is further complicated during restriction digestion. DNases and restriction endonucleases both require divalent cations, such as Mg^{++}. Included in TE buffer at a low concentration of 1 mM, Na_2EDTA chelates (binds) divalent cations at a ratio of 2 cations/Na_2EDTA. Thus, a 1 mM solution can chelate about 2 mM of divalent cation. Although some divalent cations may remain free, we are limited to how much Na_2EDTA can be added because higher concentrations would chelate the Mg^{++} necessary for restriction enzyme activity. A balance is thus struck at an EDTA concentration that inhibits most of the contaminating DNases without significantly reducing the activity of the restriction enzymes.

Another balance must be struck. On the one hand, contaminants in the miniprep limit restriction enzyme activity—a 20-minute incubation is not usually sufficient for complete digestion. On the other hand, DNases are activated by Mg^{++} in the restriction buffer and will significantly degrade plasmid DNA if the restriction reaction is incubated too long. Experience has shown that a 30-minute incubation gives optimal results.

RNase

Miniprep DNA is contaminated by large amounts of ribosomal RNA and smaller amounts of messenger RNA and transfer RNA. If not removed from the preparation, this RNA will obscure the DNA bands in the agarose gel. Therefore, RNase is added to the restriction digest; during incubation, the RNase digests RNA into very small fragments (less than 100 nucleotides). These RNA fragments run well ahead of the DNA fragments of interest or are so small that they do not stain.

For Further Information

The protocol presented here is based on the following published methods:

Aaij C. and Borst P. 1972. The gel electrophoresis of DNA. *Biochim. Biophys. Acta* **269:** 192–200.

Helling R.B., Goodman H.M., and Boyer H.W. 1974. Analysis of R-*Eco*RI fragments of DNA from lambdoid bacteriophages and other viruses by agarose-gel electrophoresis. *J. Virol.* **14:** 1235–1244.

Sharp P.A., Sugden B., and Sambrook J. 1973. Detection of two restriction endonuclease activities in *Haemophilus parainfluenzae* using analytical agarose-ethidium bromide electrophoresis. *Biochemistry* **12:** 3055–3063.

PRELAB PREPARATION

Before performing this Prelab Preparation, please refer to the cautions indicated on the Laboratory Materials list.

1. Mix in 1:1 proportion: *Bam*HI + *Hin*dIII (6 μl per experiment). Keep on ice.
2. Prepare aliquots for each experiment:

   12 μl of 0.1 μg/μl pAMP (store on ice)
   12 μl of 5x restriction buffer/RNase (store on ice)
   6 μl of *Bam*HI/*Hin*dIII (store on ice)
   500 μl of distilled water
   500 μl of loading dye

   If another plasmid was substituted for pAMP in the transformation, use that plasmid as a control in the restriction digest.
3. Prepare 0.8% agarose solution (~40–50 ml per experiment). Keep agarose liquid in a hot-water bath (at ~60°C) throughout the experiment. Cover the solution with aluminum foil to retard evaporation.
4. Prepare 1x Tris/Borate/EDTA (TBE) buffer for electrophoresis (400–500 ml per experiment).
5. Prepare ethidium bromide or methylene blue staining solution (100 ml per experiment), or other proprietary stain.
6. Adjust water bath to 37°C.

MATERIALS

REAGENTS

Agarose (0.8%)
*Bam*HI/*Hin*dIII (50:50 mix)
Distilled water
Ethidium bromide▼(1 μg/ml) (or 0.025% methylene blue▼)
Loading dye
Miniprep DNA
pAMP (0.1 μg/μl)
5x Restriction buffer/RNase
1x Tris▼/Borate/EDTA (TBE) buffer

SUPPLIES AND EQUIPMENT

Aluminum foil
Beakers for agarose and for waste/used tips
Camera and film (optional)
Electrophoresis box
Latex gloves
Masking tape
Microfuge (optional)
Micropipettor (0.5–10 μl) + tips
Parafilm or wax paper (optional)
Permanent marker
Plastic wrap (optional)
Power supply
Test tube rack
Transilluminator (optional)▼
Tubes (1.5-ml)
Water baths (37°C and 60°C)

▼ See Appendix 4 for Cautions list.

METHODS

I. Set Up Restriction Digest (40 minutes, including incubation)

Refer to Laboratory 3, DNA Restriction Analysis, for more detailed instructions.

1. Use a permanent marker to label four 1.5-ml tubes, in which restriction reactions will be performed:

   Mini– = miniprep, no enzymes
   Mini+ = miniprep + *Bam*HI/*Hin*dIII
   pAMP+ = pAMP + *Bam*HI/*Hin*dIII
   pAMP– = pAMP, no enzymes

2. Use the matrix below as a checklist while adding reagents to each reaction. Read down each column, adding the same reagent to all appropriate tubes. *Use a fresh tip for each reagent.* Refer to detailed directions that follow.

Tube	Miniprep DNA	pAMP	Buffer/ RNase	*Bam*HI/ *Hin*dIII	H_2O
Mini–	5 µl	—	2 µl	—	3 µl
Mini+	5 µl	—	2 µl	2 µl	1 µl
pAMP+	—	5 µl	2 µl	2 µl	1 µl
pAMP–	—	5 µl	2 µl	—	3 µl

3. Collect reagents, and place in test tube rack on lab bench (*Bam*HI/*Hin*dIII on ice).
4. Add 5 µl of miniprep DNA to tubes labeled Mini– and Mini+.
5. Use a *fresh tip* to add 5 µl of pAMP to tubes labeled pAMP+ and pAMP–.
6. Use a *fresh tip* to add 2 µl of restriction buffer/RNase to a clean spot on each reaction tube.

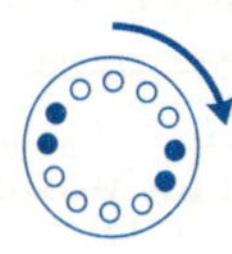

7. Use a *fresh tip* to add 2 µl of *Bam*HI/*Hin*dIII to tubes labeled Mini+ and pAMP+.
8. Use a *fresh tip* to add the proper volumes of distilled water to each tube.

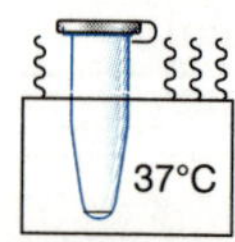

Do not overincubate. During longer incubation, DNases in the miniprep may degrade plasmid DNA.

9. Close tube tops. Pool and mix reagents by pulsing in a microfuge or by sharply tapping the tube bottom on the lab bench.
10. Place reaction tubes in a 37°C water bath, and incubate for 30 minutes only.

> STOP Following incubation, freeze reactions at –20°C until ready to continue. Thaw reactions before continuing to Section III, Step 1.

II. Cast 0.8% Agarose Gel (15 minutes)

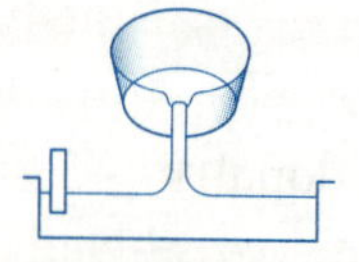

Gel is cast directly in box in some electrophoresis apparatuses.

1. Seal the ends of the gel-casting tray with tape, and insert a well-forming comb. Place gel-casting tray out of the way on the lab bench so that agarose poured in the next step can set undisturbed.
2. Carefully pour enough agarose solution into the casting tray to fill to a depth of about 5 mm. Gel should cover only about one-third the height of comb teeth. Use a pipette tip to move large bubbles or solid debris to the sides or end of tray while the gel is still liquid.

3. Gel will become cloudy as it solidifies (~10 minutes). *Be careful not to move or jar the casting tray while the agarose is solidifying.* Touch corner of agarose *away* from comb to test whether the gel has solidified.

Too much buffer will channel current over top of gel rather than through gel, increasing time required to separate DNA. TBE buffer can be used several times; do not discard. If using buffer remaining in electrophoresis box from a previous experiment, rock chamber back and forth to remix ions that have accumulated at either end.

4. When agarose has set, unseal the ends of the casting tray. Place the tray on the platform of the gel box, so that the comb is at a negative black electrode (cathode).
5. Fill box with TBE buffer to a level that just covers entire surface of gel.
6. Gently remove comb, taking care not to rip wells.
7. Make certain that sample wells left by the comb are completely submerged. If "dimples" appear around the wells, slowly add buffer until they disappear.

 Cover the electrophoresis tank and save the gel until ready to continue. Gel will remain in good condition for at least several days if it is completely submerged in buffer.

Buffer solution helps to lubricate comb. Some gel boxes are designed such that the comb must be removed prior to inserting casting tray into box. In this case, flood casting tray and gel surface with running buffer before removing comb. Combs removed from a dry gel can cause tearing of wells.

III. Load Gel and Separate by Electrophoresis (30–50 minutes)

1. Add loading dye to each reaction. Either

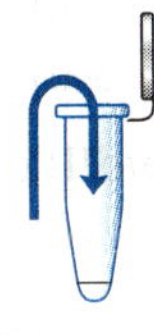

   a. Add 1 μl of loading dye to each reaction tube. Close tube tops, and mix by tapping the tube bottom on the lab bench, pipetting in and out, or pulsing in a microfuge. Make sure that the tubes are placed in a *balanced* configuration in the rotor.

   or

   b. Place four individual droplets of loading dye (1 μl each) on a small square of Parafilm or wax paper. Withdraw contents from reaction tube, and mix with a loading dye droplet by pipetting in and out. Immediately load dye mixture according to Step 2. Repeat successively, *with a clean tip*, for each reaction.

2. Use a micropipettor to load 10 μl of each reaction tube into a separate well in the gel, as shown on the following page. Use a *fresh tip* for each reaction.

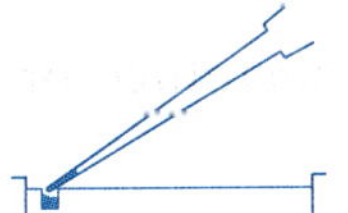

A piece of dark construction paper beneath the gel box will make the wells more visible.

   a. Before loading sample, make sure that there are no bubbles in the wells. If bubbles exist, remove them with a micropipettor tip.
   b. Use two hands to steady the micropipettor over the well.
   c. If there is air in the end of the tip, carefully depress the plunger to push the sample to the end of the tip. (If an air bubble forms a "cap" over the well, DNA/loading dye will flow into the buffer around the edges of the well.)
   d. Dip a micropipettor tip through the surface of the buffer, center it over the well, and gently depress micropipettor plunger to slowly expel the sample. Sucrose in the loading dye weighs down the sample, causing it to sink to the bottom of the well. *Be careful not to punch the tip of the micropipettor through the bottom of the gel.*

3. Close the top of the electrophoresis box, and connect the electrical leads to the power supply, anode to anode (red-red) and cathode to cathode (black-black). Make sure that both electrodes are connected to the same channel of power supply.

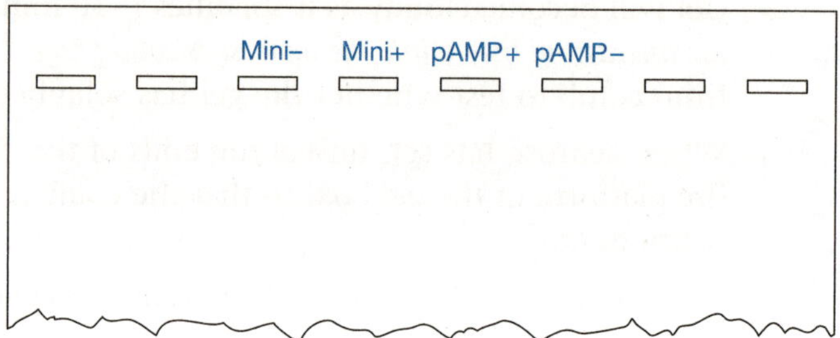

Alternatively, set power supply on lower voltage, and run gel for several hours. When running two gels from the same power supply, current is double that for a single gel at the same voltage.

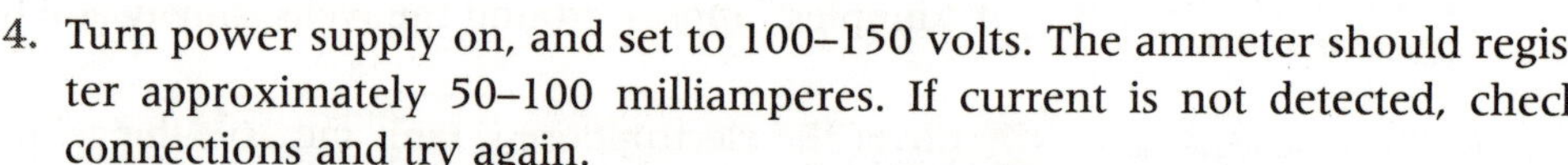

4. Turn power supply on, and set to 100–150 volts. The ammeter should register approximately 50–100 milliamperes. If current is not detected, check connections and try again.

The BamHI/HindIII digest yields two bands containing small fragments of 784 bp and 3755 bp, which are easily resolved during a short electrophoresis run. The 784-bp fragment runs directly behind the purplish band of bromophenol blue (equivalent to ~300 bp), whereas the 3755-bp fragment runs in front of the aqua band of xylene cyanol (equivalent to ~9000 bp).

5. Separate by electrophoresis for 20–40 minutes. Good separation will have occurred when the bromophenol blue band has moved 4–7 cm from the wells. If time allows, carry out electrophoresis until the bromophenol blue band nears the end of the gel. *Stop* electrophoresis before the bromophenol blue band runs off the end of the gel.
6. Turn off power supply, disconnect leads from the inputs, and remove the top of the electrophoresis box.
7. Carefully remove the casting tray from the electrophoresis box, and slide the gel into a disposable weigh boat or other shallow tray. Label staining tray with your name.

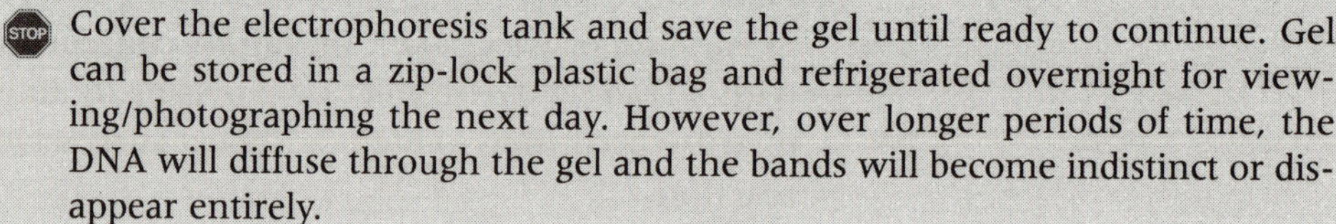

STOP Cover the electrophoresis tank and save the gel until ready to continue. Gel can be stored in a zip-lock plastic bag and refrigerated overnight for viewing/photographing the next day. However, over longer periods of time, the DNA will diffuse through the gel and the bands will become indistinct or disappear entirely.

Staining may be performed by an instructor in a controlled area when students are not present.

8. Stain and view gel using one of the methods described in Sections IVA and IVB.

IVA. Stain Gel with Ethidium Bromide and View (Photograph)

(10–15 minutes)

> CAUTION
> Review Responsible Handling of Ethidium Bromide in Laboratory 3. Wear latex gloves when staining, viewing, and photographing gel and during cleanup. Confine all staining to a restricted sink area. For further information, see Appendix 4.

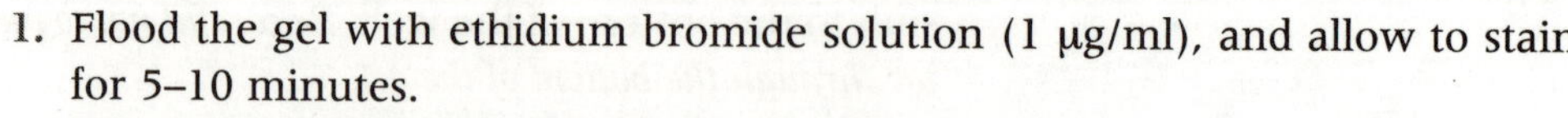

1. Flood the gel with ethidium bromide solution (1 μg/ml), and allow to stain for 5–10 minutes.

Staining time increases markedly for thicker gels.

2. Following staining, use a funnel to decant as much ethidium bromide solution as possible from the staining tray back into the storage container.
3. Rinse gel and tray under running tap water.

Ethidium bromide solution may be reused to stain 15 or more gels. When staining time increases markedly, dispose of ethidium bromide solution as explained in Laboratory 3.

4. If desired, the gel can be destained in tap water or distilled water for 5 minutes or more to help remove background ethidium bromide from the gel.

Staining intensifies dramatically if rinsed gels set overnight at room temperature. Stack staining trays, and cover the top gel with plastic wrap to prevent desiccation.

5. View under UV transilluminator or other UV source.

> CAUTION
> Ultraviolet light can damage eyes. Never look at unshielded UV light source with naked eyes. View only through a filter or safety glasses that absorb harmful wavelengths. For further information, see Appendix 4.

6. Photograph with a Polaroid or digital camera.
7. Take time for responsible cleanup.
   a. Wipe down camera, transilluminator, and staining area.
   b. Wash hands before leaving lab.

IVB. Stain Gel with Methylene Blue and View (Photograph)

(30+ minutes)

1. Wear latex gloves during staining and cleanup.
2. Flood the gel with 0.025% methylene blue, and allow to stain for 20–30 minutes.
3. Following staining, use a funnel to decant as much methylene blue solution as possible from the staining tray back into the storage container.
4. Rinse the gel in running tap water. Let the gel soak for several minutes in several changes of fresh water. DNA bands will become increasingly distinct as gel destains.

Destaining time is decreased by rinsing the gel in warm water, with agitation.

For best results, continue to destain overnight in a *small volume* of water. (Gel may destain too much if left overnight in large volume of water.) Cover staining tray to retard evaporation.

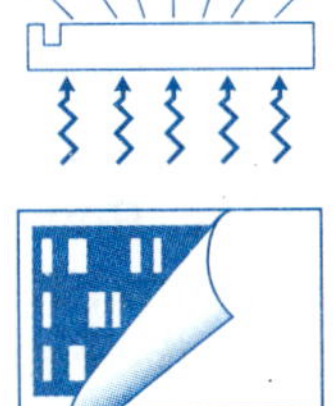

5. View gel over light box; cover the surface of the light box with plastic wrap to prevent staining.
6. Photograph with a Polaroid or digital camera.

RESULTS AND DISCUSSION

Interpreting gels containing plasmid DNA is not always straightforward and is complicated by impurities in the miniprep DNA. Examine the gel and determine which lanes contain cut and uncut control pAMP and which lanes contain miniprep DNA. Even if you have confused the prescribed loading order, the miniprep lanes can be distinguished by the following characteristics:

1. A background "smear" of degraded and partially digested chromosomal DNA, plasmid DNA, and RNA is typically seen running much of the length of the miniprep lanes. The smear is composed of faint bands of virtually every nucleotide length that grade together. A heavy background smear, along with high-molecular-weight DNA at the top of the undigested lane, indicates that the miniprep is contaminated with large amounts of chromosomal DNA.
2. Frequently, undissolved material and high-molecular-weight DNA are seen "trapped" at the front edge of the well. These anomalies are not seen in commercial preparations, where plasmid DNA is separated from degraded nucleic acids by ultracentrifugation in a cesium chloride gradient.
3. A "cloud" of low-molecular-weight RNA is often seen in both the cut and uncut miniprep lanes at a position corresponding to 200 bp or less. Again, variously sized molecules are represented, which are the remnants of larger molecules that have been partially digested by the RNase. However, the majority of RNA is usually digested into fragments that are too small to intercalate the ethidium bromide dye or that migrate off the end of the gel.
4. Only two bands (784 bp and 3755 bp) are expected to be seen in the cut miniprep lane. However, it is common to see one or more faint bands higher up on the gels that comigrate with the uncut plasmid forms described below. Incomplete digestion is usually due to contaminants in the preparation that inhibit restriction enzyme activity or may occur when the miniprep solution contains a very high concentration of plasmid DNA. The plasmid might also be cut at only one site, creating a linear plasmid that will also migrate slower than the 3755-bp band. This is called a partial digest.

It is especially confusing to see several bands in a lane known to contain only uncut plasmid DNA. This occurs because the migration of plasmid DNA in an agarose gel depends on its molecular conformation, as well as its molecular weight (base-pair size). Plasmid DNA exists in one of three major conformations:

- *Form I, supercoiled:* Although a plasmid is usually pictured as an open circle, within the *E. coli* cell (in vivo), the DNA strand is wound around histone-like proteins to make a compact structure. Adding these coils to the coiled DNA helix produces a *super*coiled molecule. The extraction procedure strips proteins from plasmid, causing the molecule to coil about itself. Supercoiling is best demonstrated with a piece of string. Double the string and hold an end in each hand without slack. Now twist the string in one direction. At first, the coils form easily and spread evenly along the length of the string. However, as you add more twists, the string begins to bunch and form knots. If you relax the tension on the string, the string become tangled. This is the equivalent of removing the protein from supercoiled plasmid DNA. Under most gel conditions, the supercoiled plasmid DNA is the fastest-moving form. Its compact molecular shape allows it to move most easily through the agarose matrix. Therefore, the fastest-moving band of uncut plasmid is assumed to be supercoiled.
- *Form II, relaxed or nicked circle:* During DNA replication, the enzyme topoisomerase I introduces a nick into one strand of the DNA helix and rotates the strand to release the torsional strain that holds the molecule in a supercoil.

The relaxed section of the DNA uncoils, allowing access to the replicating enzymes. Introducing nicks into supercoiled plasmid DNA produces the open circular structure with which we are familiar. Physical shearing and enzymatic cleavage during plasmid isolation introduce nicks in the supercoiled plasmid DNA. Thus, the percentage of supercoiled plasmid DNA is an indicator of the care with which the DNA is extracted from the *E. coli* cell. The relaxed circle is the slowest-migrating form of plasmid DNA; its "floppy" molecular shape impedes movement through the agarose matrix.

- *Form III, linear:* Linear DNA is produced when a restriction enzyme cuts the plasmid at a single recognition site or when damage results in strand nicks directly opposite each other on the DNA helix. Under most gel conditions, linear plasmid DNA migrates at a rate intermediate between supercoiled and relaxed circle. The presence of linear DNA in a plasmid preparation is a sign of contamination with nucleases or of sloppy lab procedure (overmixing or mismeasuring SDS/NaOH and KOAc).

MM294 and other strains of *E. coli*, termed *recA*$^+$, have an enzyme system that recombines plasmids to form large concatemers of two or more plasmid units. A general mechanism for shuffling DNA strands, homologous recombination, occurs when identical regions of nucleotides are reciprocally exchanged between two DNA molecules. Homologous recombination occurs frequently between plasmids, which exist as multiple identical copies within the cell.

The RecA protein binds to single-stranded regions of nicked plasmids, promoting crossover and rejoining of homologous sequences. This results in multimeric ("super") plasmids that appear as a series of slow-migrating bands near the top of the gel. Since the concatemers form head-to-tail, they produce restriction fragments identical to those of a monomer (single plasmid) when cut with restriction enzymes. To confuse matters further, multimers can exist in any of the three forms mentioned above. Supercoiled multimers may appear further down on the gel than relaxed or linear plasmids with fewer nucleotides.

1. Examine the photograph of your stained gel (or view on a light box or overhead projector). Compare your gel with the ideal gel. Label the size of fragments in each lane of your gel.
2. Compare the two gel lanes containing miniprep DNA with the two lanes containing control pAMP. Explain possible reasons for variations.
3. A plasmid preparation of pAMP is composed entirely of dimeric molecules (pAMP/pAMP). The two molecules are joined *head-to-head* at a "hot spot" for recombination located 655 bp from the *Hin*dIII site near the origin of replication.
   a. Draw a map of the dimeric plasmid described above.
   b. Draw a map of the dimeric pAMP that actually forms by head-to-tail recombination at the site described above.
   c. Now draw the gel-banding patterns that would result from double digestion of each of these plasmids with *Bam*HI and *Hin*dIII, and label the base-pair size of fragments in each band.
4. Explain why EDTA is an important component of TE buffer in which the miniprep DNA is dissolved.

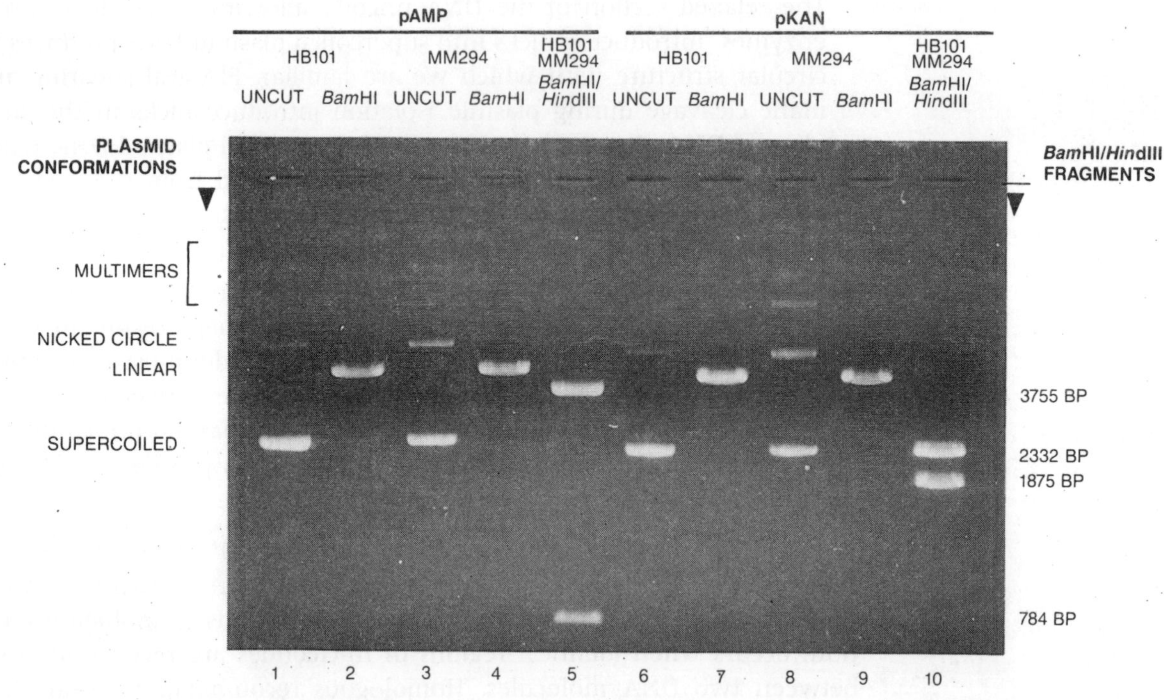

Component of Plasmid DNA Isolated from a *recA*⁻ Strain (HB101) and a *recA*⁺ Strain (MM294)

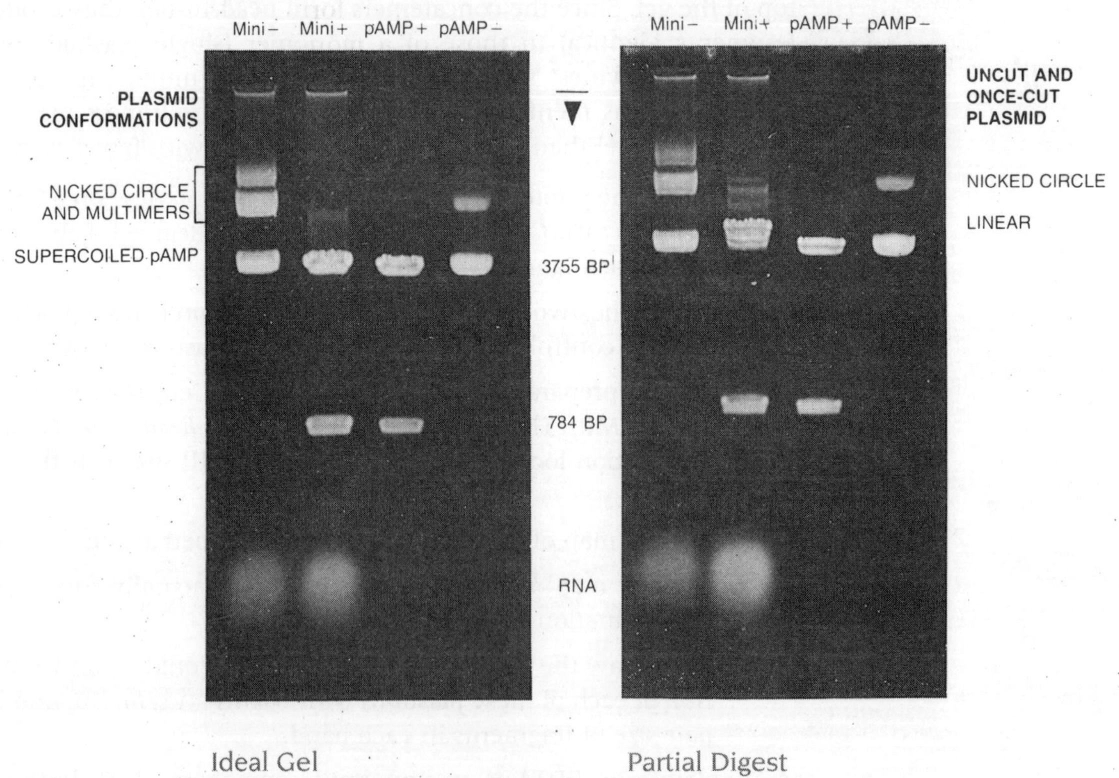

Ideal Gel

Partial Digest

FOR FURTHER RESEARCH

1. Isolate and characterize an unknown plasmid. Make overnight cultures of *E. coli* strains containing any of several commercially available plasmids (such as pAMP, pKAN, pUC19, and pBR322). Digest miniprep and control samples of each plasmid with *Bam*HI/*Hin*dIII, and separate by electrophoresis in an agarose gel.
2. Transform pAMP and/or other plasmids into a *recA*$^+$ strain (MM294) and a *recA*$^-$ strain (HB101). Do minipreps from overnight cultures of each strain, and incubate samples of each with no enzyme, *Hin*dIII, and *Bam*HI+*Hin*dIII. Separate the samples by electrophoresis as far as possible in an agarose gel. Compare the banding patterns of the two strains, especially in the uncut lanes.
3. Research the potential use of homologous recombination in targeted gene therapy.

Recombination of Antibiotic Resistance Genes

LABORATORY 9 BEGINS AN EXPERIMENTAL STREAM designed to construct and analyze a recombinant DNA molecule. The starting reagents are the relaxed plasmids pAMP and pKAN, each of which carries a single antibiotic resistance gene: ampicillin in pAMP and kanamycin in pKAN. The goal is to construct a recombinant plasmid that contains both ampicillin and kanamycin resistance genes. This laboratory is divided into two parts: Restriction Digest of Plasmids pAMP and pKAN and Ligation of pAMP and pKAN Restriction Fragments.

- Part A provides a procedure whereby samples of both plasmids are digested in separate restriction reactions with *Bam*HI and *Hin*dIII. Following incubation at 37°C, samples of digested pAMP and pKAN are analyzed by agarose gel electrophoresis to confirm proper cutting. Each plasmid contains a single recognition site for each enzyme, yielding only two restriction fragments. Cleavage of pAMP yields fragments of 784 bp and 3755 bp, and cleavage of pKAN yields fragments of 1861 bp and 2332 bp.
- Part B provides a technique for ligation of pAMP and pKAN restriction fragments. The restriction digests of pAMP and pKAN are heated to destroy *Bam*HI and *Hin*dIII activity. A sample from each reaction is mixed with DNA ligase plus ATP and incubated at room temperature. Complementary *Bam*HI and *Hin*dIII "sticky ends" hydrogen-bond to align restriction fragments. Ligase catalyzes the formation of phosphodiester bonds that covalently link the DNA fragments to form stable recombinant DNA molecules.

Equipment and materials for this laboratory are available from the Carolina Biological Supply Company (see Appendix 1).

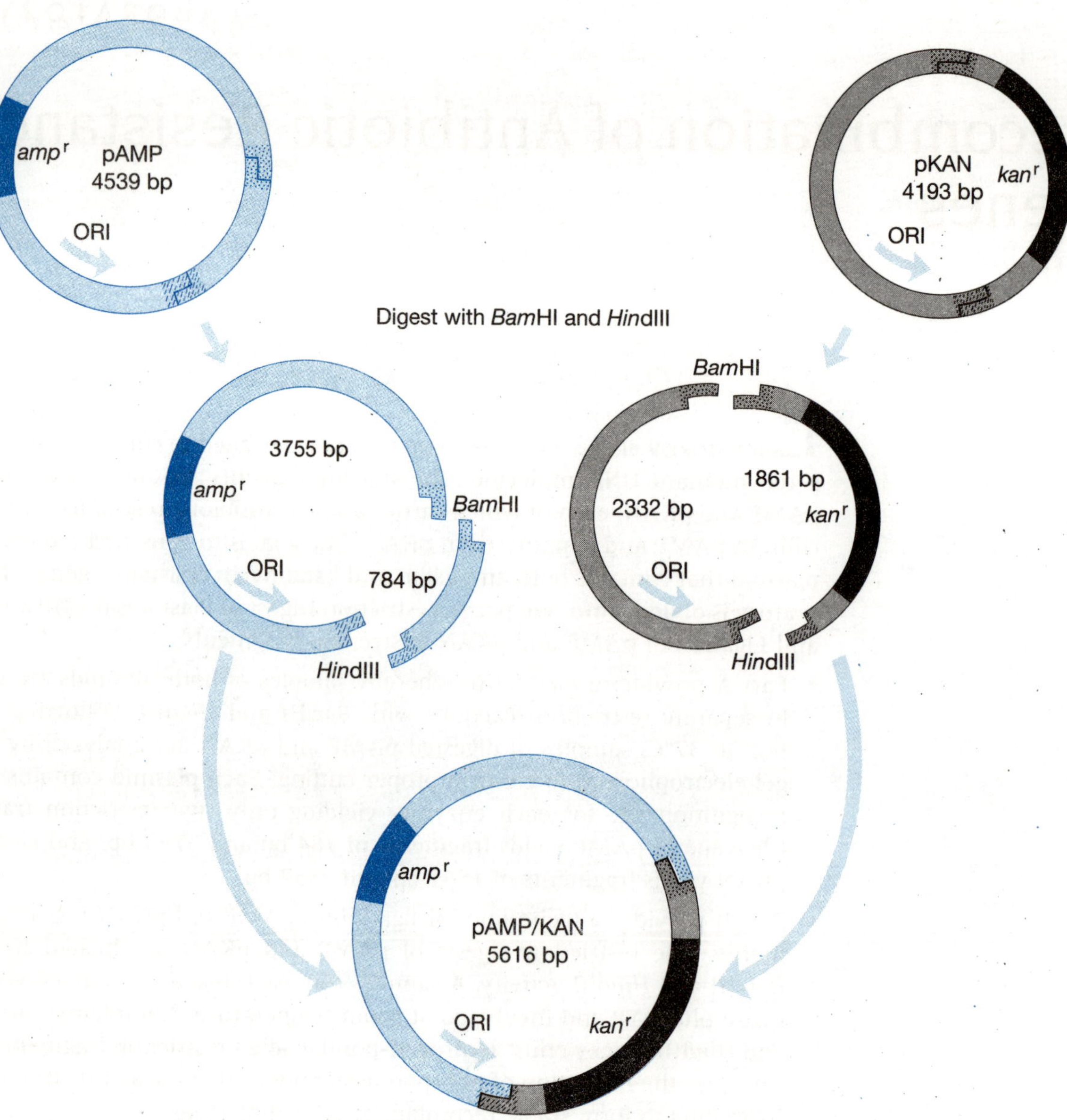

Formation of the "Simple Recombinant" pAMP/KAN

PART A

Restriction Digest of Plasmids pAMP and pKAN

I. Prepare Restriction Digest

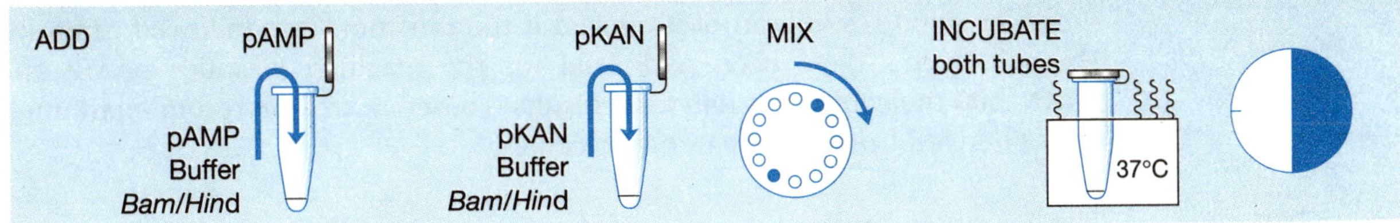

II. Cast 0.8% Agarose Gel

III. Load Gel and Separate by Electrophoresis

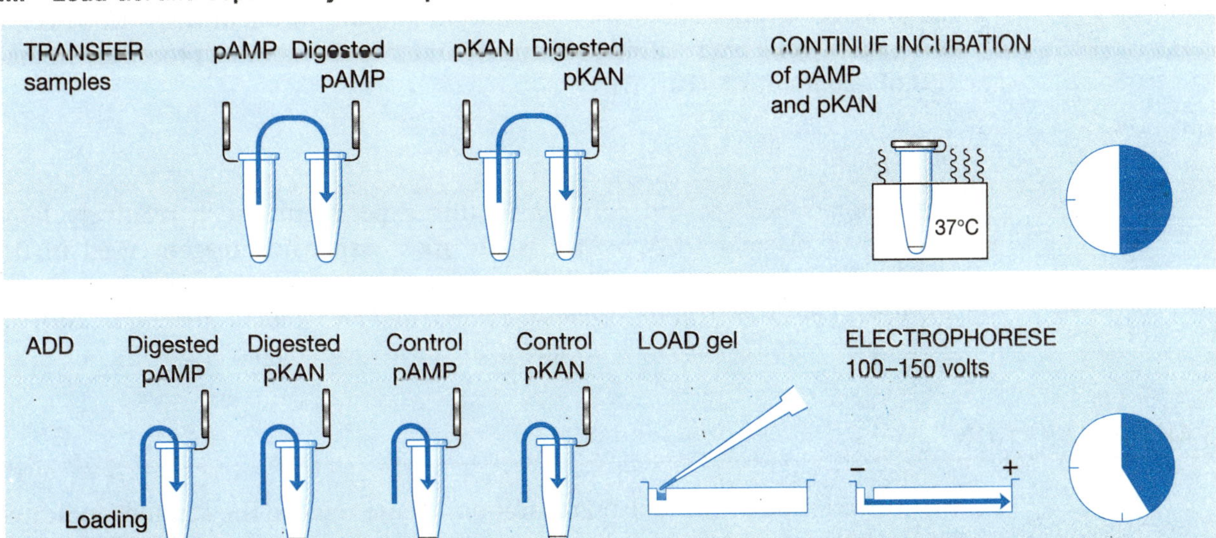

IV. Stain Gel and View (Photograph)

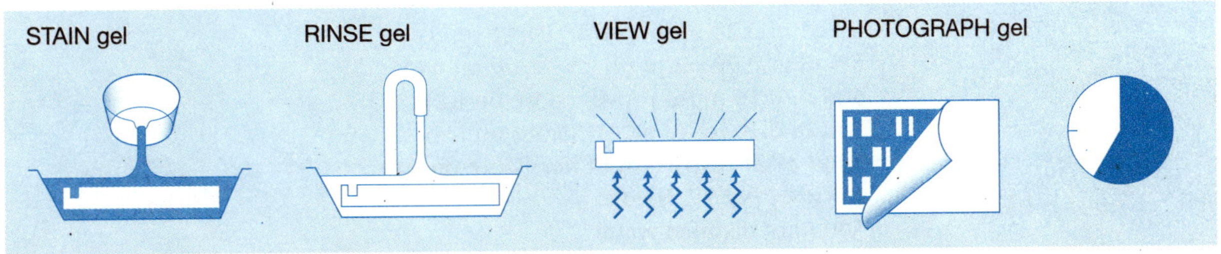

PRELAB NOTES

Review Prelab Notes in Laboratory 3, DNA Restriction Analysis.

Plasmid Substitution

The process of constructing and analyzing recombinant molecules is not trivial. However, good results can be expected if the directions are followed carefully. These protocols have been optimized for the teaching plasmids pAMP and pKAN, and the extensive analysis of results is based *entirely* on recombinant molecules derived from these parent molecules.

The Prudent Control

In Section III, samples of the restriction digests are analyzed by agarose gel electrophoresis, prior to ligation, to confirm complete cutting by the endonucleases. This prudent control is standard experimental procedure. If pressed for time, omit electrophoresis and ligate DNA directly following the restriction digest. However, be sure to pretest the activity of *Bam*HI and *Hin*dIII to determine the incubation time needed for complete digestion.

It is fairly impractical to use methylene blue staining for this step, which demands a rapid and sensitive assay to check for complete digestion of the plasmid DNAs. Methylene blue destaining requires *at least* 30 minutes, and it could fail to detect a small but possibly significant amount of uncut DNA. However, if using methylene blue staining for this lab, refer to the staining procedure in Step IVB of Laboratory 8 (Part B).

Saving DNA

Restriction reactions and controls in this experiment use a relatively large amount of plasmid DNA, which is the most expensive reagent used in the course. *To minimize expense, the protocol directs that the lab be prepared by setting up aliquots of exactly the required volumes of pAMP and pKAN into 1.5-ml tubes. Then the reagents for restriction digestion are added directly to these aliquots of DNA.*

PRELAB PREPARATION

Before performing this Prelab Preparation, please refer to the cautions indicated on the Laboratory Materials list.

1. Mix in 1:1 proportion: *Bam*HI + *Hin*dIII (6 μl per experiment).
2. Prepare aliquots for each experiment:

   5.5 μl of 0.20 μg/μl pAMP (store on ice)
   5.5 μl of 0.20 μg/μl pKAN (store on ice)
   5 μl of 0.10 μg/μl pAMP (store on ice)
   5 μl of 0.10 μg/μl pKAN (store on ice)
   20 μl of 2x restriction buffer (store on ice)
   6.0 μl of *Bam*HI/*Hin*dIII
   500 μl of distilled water
   500 μl of loading dye

3. Prepare 0.8% agarose solution (40–50 ml per experiment). Keep agarose liquid in a hot-water bath (at ~60°C) throughout lab. Cover with aluminum foil to retard evaporation.
4. Prepare 1x Tris/Borate/EDTA (TBE) buffer for electrophoresis (400–500 ml per experiment).
5. Prepare ethidium bromide staining solution (50 ml per experiment).
6. Adjust water bath to 37°C.
7. Review Part B, Ligation of pAMP and pKAN Restriction Fragments.

MATERIALS

REAGENTS	SUPPLIES AND EQUIPMENT
For digest:	Aluminum foil
pAMP (0.20 µg/µl)	Beakers for agarose and for waste/used tips
pKAN (0.20 µg/µl)	Camera and film (optional)
For control:	Electrophoresis box
pAMP (0.1 µg/µl)	Latex gloves
pKAN (0.1 µg/µl)	Masking tape
Agarose (0.8%)	Microfuge (optional)
*Bam*HI/*Hind*III	Micropipettor (0.5–10 µl) + tips
Distilled water	Parafilm or wax paper (optional)
Ethidium bromide▼ (1 µg/µl) (or 0.025% methylene blue▼)	Permanent marker
Loading dye	Plastic wrap (optional)
2x Restriction buffer	Power supply
1x Tris▼/Borate/EDTA (TBE) buffer	Test tube rack
	Transilluminator (optional)▼
	Tubes (1.5-ml)
	Water baths (37°C and 60°C)

▼ See Appendix 4 for Cautions list.

METHODS

I. Set Up Restriction Digest

(40–60 minutes, including incubation through Section III)

Refer to Laboratory 3, DNA Restriction Analysis, for more detailed instructions.

1. Use the matrix below as a checklist while adding reagents to each reaction. Read down each column, adding the same reagent to all appropriate tubes. *Use a fresh tip for each reagent.* Refer to detailed directions that follow.

To avoid confusion, keep 0.1 µg/µl control pAMP and pKAN aside until needed in Section III.

Tube	pAMP 0.2 µg/µl	pKAN 0.2 µg/µl	2x Buffer	*Bam*HI/ *Hind*III
Digested pAMP	5.5 µl	—	7.5 µl	2 µl
Digested pKAN	—	5.5 µl	7.5 µl	2 µl

To minimize waste and expense, aliquots of plasmid DNA can be set up in tubes labeled pAMP and pKAN. Add reagents directly to these tubes.

2. Collect 2x buffer and *Bam*HI/*Hind*III (on ice), and the tubes containing pAMP and pKAN plasmid DNA, and place in a test tube rack on the lab bench.
3. Add 7.5 µl of 2x restriction buffer to the pAMP and pKAN tubes.
4. Use a *fresh tip* to add 2 µl of *Bam*HI/*Hind*III to each tube.
5. Close tube tops. Pool and mix reagents by pulsing in a microfuge or by sharply tapping the tube bottom on the lab bench.
6. Place the reaction tubes in a 37°C water bath, and incubate for a minimum of 30 minutes. Reactions can be incubated for a longer period of time.

37°C

STOP After a full 30-minute incubation (or longer), freeze reactions at –20°C until ready to continue. Thaw reactions before proceeding to Section III, Step 1.

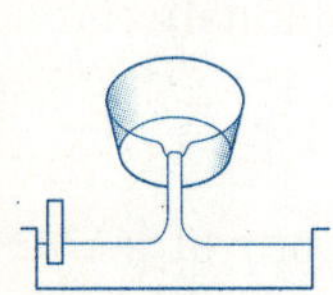

II. Cast 0.8% Agarose Gel (15 minutes)

1. Seal the ends of the gel-casting tray with tape, and insert a well-forming comb. Place the gel-casting tray out of the way on the lab bench so that the agarose poured in the next step can set undisturbed.
2. Carefully pour enough agarose solution into the casting tray to fill to a depth of about 5 mm. Gel should cover only about one-third the height of comb teeth. Use a pipette tip to move large bubbles or solid debris to the sides or end of the tray while gel is still liquid.
3. Gel will become cloudy as it solidifies (~10 minutes). *Be careful not to move or jar the casting tray while the agarose is solidifying.* Touch the corner of the agarose *away* from the comb to test whether the gel has solidified.
4. When the agarose has set, unseal the ends of the casting tray. Place the tray on the platform of the gel box, so that the comb is at negative black electrode (cathode).
5. Fill box with TBE buffer, to a level that just covers the surface of the gel.
6. Gently remove the comb, taking care not to rip the wells.
7. Make sure that sample wells left by the comb are completely submerged. If "dimples" appear around the wells, slowly add buffer until they disappear.

Too much buffer will channel the current over the top rather than through the gel, increasing the time required to separate DNA. TBE buffer can be used several times; do not discard. If using buffer remaining in electrophoresis box from a previous experiment, rock chamber back and forth to remix ions that have accumulated at either end.

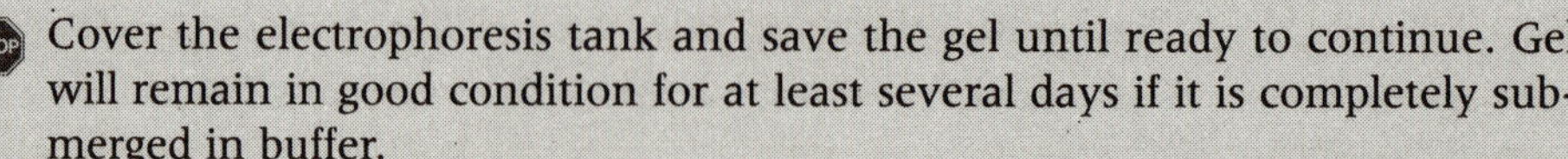

STOP Cover the electrophoresis tank and save the gel until ready to continue. Gel will remain in good condition for at least several days if it is completely submerged in buffer.

III. Load Gel and Separate by Electrophoresis (20–30 minutes)

Only a fraction of the *Bam*HI/*Hin*dIII digests of pAMP and pKAN are separated by electrophoresis to check whether plasmids are completely cut. These restriction samples are separated by electrophoresis along with uncut pAMP and pKAN as controls.

1. Use a permanent marker to label two clean 1.5-ml tubes:

   Digested pAMP
   Digested pKAN

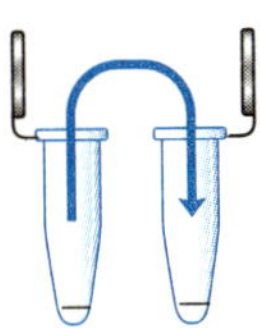

2. Remove original tubes labeled Digested pAMP and Digested pKAN from the 37°C water bath.

   Transfer a 5-μl sample of plasmid from the original Digested pAMP tube into the clean Digested pAMP tube.

   Transfer a 5-μl sample of plasmid from the original Digested pKAN tube into the clean Digested pKAN tube.

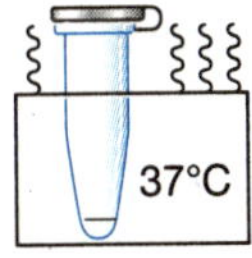

3. *Immediately* return the original Digested pAMP and Digested pKAN tubes to the water bath, and continue incubating at 37°C during electrophoresis.
4. Collect 1.5-ml tubes containing 5 μl each of purified plasmid at 0.1 μg/μl; label tubes:

   Control pAMP
   Control pKAN

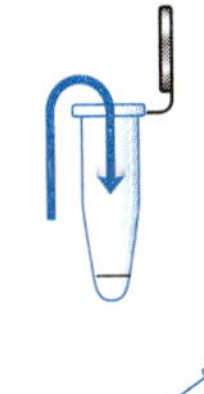

5. Add 1 μl of loading dye to each tube of Digested and Control pAMP and pKAN. Close tube tops, and mix by tapping the tube bottom on the lab bench, pipetting in and out, or pulsing in microfuge.

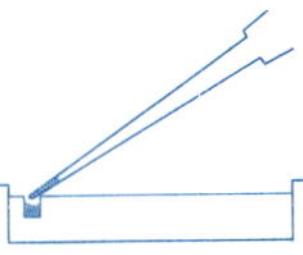

A piece of dark construction paper beneath the gel box will make the wells more visible.

6. Load entire contents of each sample tube into a separate well in the gel, as shown in diagram below. *Use a fresh tip for each sample. Expel any air in the tip before loading, and be careful not to punch the tip of the micropipettor through the bottom of the gel.*

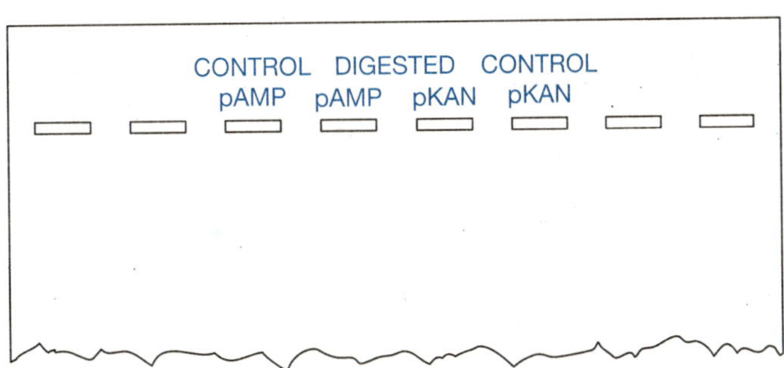

The 784-bp BamHI/HindIII fragment of pAMP migrates just behind the bromophenol blue marker. Stop electrophoresis before the bromophenol blue band runs off the end of the gel or this fragment may be lost.

7. Separate by electrophoresis at 100–150 volts for 15–30 minutes. Adequate separation will have occurred when the bromophenol blue band has moved 2–4 cm from the wells.
8. Turn off power supply, disconnect leads from the inputs, and remove top of electrophoresis box.

Cover electrophoresis tank and save gel until ready to continue. Gel can be stored in a zip-lock plastic bag and refrigerated overnight for viewing/photographing the next day. However, over longer periods of time, the DNA will diffuse through the gel and the bands will become indistinct or disappear entirely.

9. Carefully remove the casting tray from the electrophoresis chamber, and slide the gel into a disposable weigh boat or other shallow tray. Label the staining tray with your name.

Staining may be performed by an instructor in a controlled area when students are not present.

10. Stain and view gel as described in Section IV.

IV. Stain Gel with Ethidium Bromide and View (Photograph)

(10–15 minutes)

CAUTION
Review Responsible Handling of Ethidium Bromide in Laboratory 3. Wear latex gloves when staining, viewing, and photographing gels and during cleanup. Confine all staining to a restricted sink area. For further information, see Appendix 4.

Ethidium bromide solution may be reused to stain 15 or more gels. Dispose of spent staining solution as explained in Laboratory 3.

1. Flood the gel with ethidium bromide solution (1 μg/ml), and allow to stain for 5–10 minutes.
2. Following staining, use a funnel to decant as much ethidium bromide solution as possible from the staining tray back into storage container.
3. Rinse the gel and tray under running tap water.
4. If desired, the gel can be destained in tap water or distilled water for 5 minutes or more to remove background ethidium bromide.
5. View under UV transilluminator or other UV source.

CAUTION
Ultraviolet light can damage eyes. Never look at unshielded UV light source with naked eyes. View only through a filter or safety glasses that absorb harmful wavelengths. For further information, see Appendix 4.

6. Photograph with a Polaroid or digital camera.
7. If both digests look complete, or nearly so (see Results and Discussion), continue on to Part B, Ligation of pAMP and pKAN Restriction Fragments. The reaction will have gone to completion with the additional incubation during electrophoresis.
8. If either or both digests look very incomplete, add another 1 μl of *Bam*HI/*Hin*dIII solution and incubate for an additional 20 minutes. Then continue on to Part B, Ligation of pAMP and pKAN Restriction Fragments.
9. Take time for responsible cleanup.
   a. Wipe down camera, transilluminator, and staining area.
   b. Wash hands before leaving lab.

Freeze *Bam*HI/*Hin*dIII reactions at –20°C until ready to continue. Thaw reactions before proceeding to Part B, Ligation of pAMP and pKAN Restriction Reactions.

RESULTS AND DISCUSSION

Examine the photograph of your stained gel (or view on a light box). Compare your gel with the ideal gel shown below, and check whether both plasmids have been completely digested by *Bam*HI and *Hin*dIII.

1. The Digested pAMP lane should show two distinct fragments: 784 bp and 3755 bp.
2. The Digested pKAN lane should show two distinct fragments: 1861 bp and 2332 bp.
3. Additional bands that comigrate with bands in the uncut Control pAMP and Control pKAN should be faint or absent, indicating that most or all of the pAMP and pKAN plasmid has been completely digested by both enzymes.
4. If both digests look complete, or nearly so, continue on to Part B, Ligation of pAMP and pKAN Restriction Fragments. The reaction will have gone to completion with the additional incubation during electrophoresis.
5. If either or both digests look very incomplete, add another 1 μl of *Bam*HI/*Hin*dIII solution and incubate for an additional 20 minutes before continuing to Part B, Ligation of pAMP and pKAN Restriction Fragments.

> STOP Freeze *Bam*HI/*Hin*dIII reactions at –20°C until ready to continue. Thaw reactions before proceeding to Part B, Ligation of pAMP and pKAN Restriction Fragments.

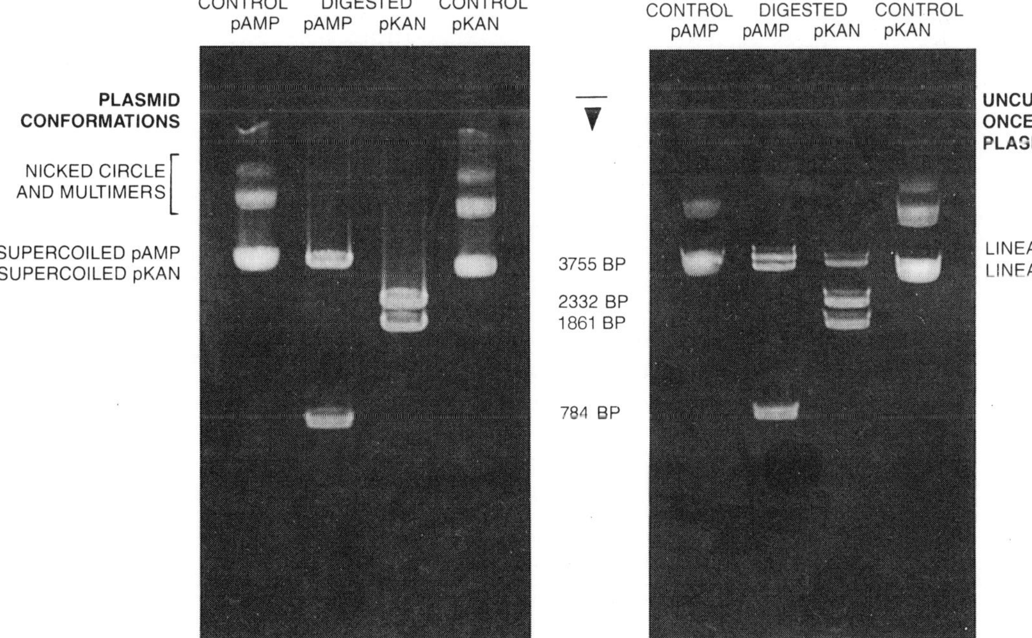

Ideal Gel

Partial Digest

PART B

Ligation of pAMP and pKAN Restriction Fragments

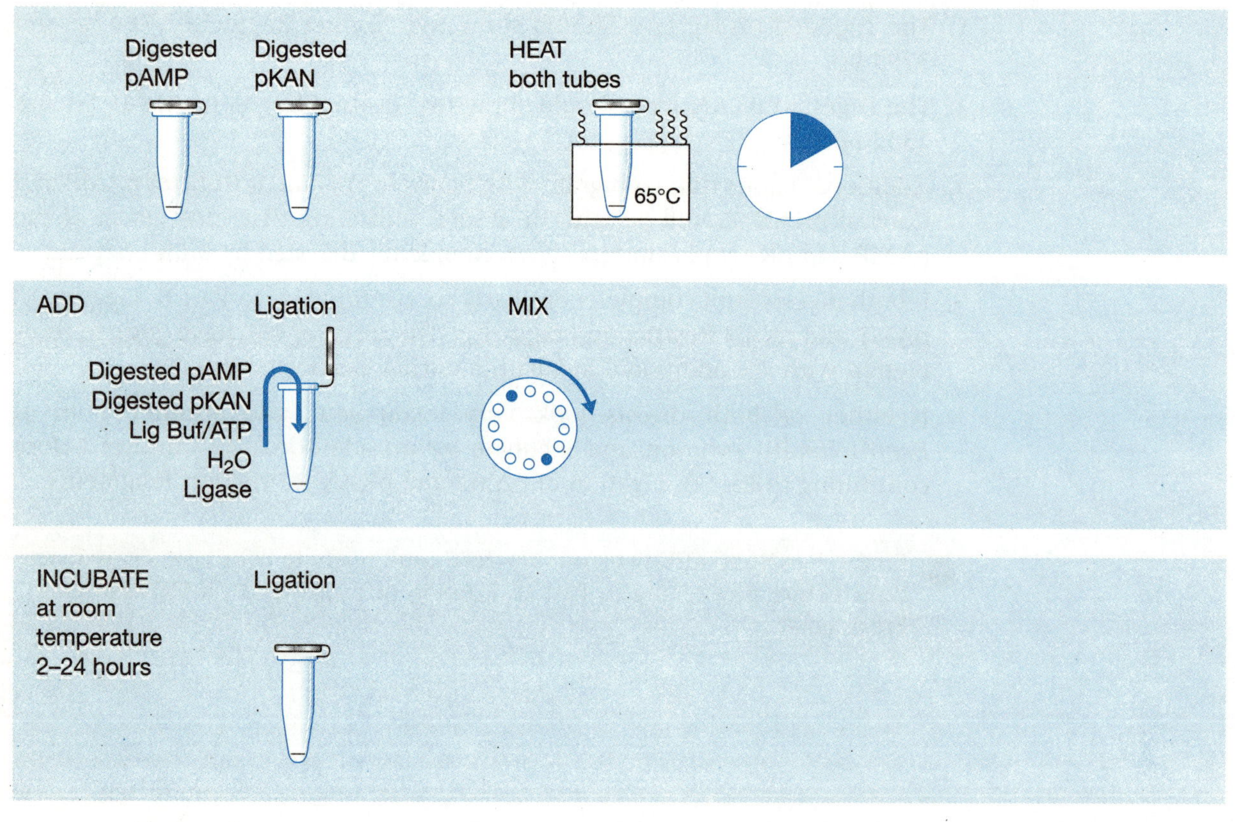

PRELAB NOTES

DNA Ligase

Use only T4 DNA ligase. *E. coli* DNA ligase requires different reaction conditions and cannot be substituted in this experiment. *Cohesive-end units* are used to calibrate ligase activity: One unit of enzyme ligates 50% of *Hin*dIII fragments of λ DNA (6 μg in 20 μl) in 30 minutes at 16°C. This unit is used by New England Biolabs (NEB) and Carolina Biological Supply Company (CBS).

Researchers typically incubate ligation reactions overnight at room temperature. *For brief ligations, down to a minimum of 1 hour, it is essential to choose a high-concentration T4 DNA ligase with at least 100–500 cohesive-end units/μl.*

For Further Information

The protocol presented here is based on the following published method:

Cohen S.N., Chang A.C.Y., Boyer H.W., and Helling R.B. 1973. Construction of biologically functional bacteria plasmids in vitro. *Proc. Natl. Acad. Sci.* **70:** 3240–3244.

PRELAB PREPARATION

1. Obtain fresh 2x ligation buffer/ATP solution. ATP is somewhat unstable in solution, so do not use very old buffer/ATP and take care to keep frozen when not in use.
2. T4 DNA ligase is critical to the experiment and rather expensive. Make one aliquot of ligase sufficient for all experiments, and hold on ice during the laboratory. We suggest that the instructor dispense ligase directly into each experimenter's reaction tube.
3. Prewarm water bath to 65°C.
4. Dispose of 2x restriction buffer from Part A, Restriction Digest of Plasmids pAMP and pKAN, to avoid mistaking it for 2x ligation buffer/ATP.

MATERIALS

REAGENTS	SUPPLIES AND EQUIPMENT
Digested pAMP (from part A)	Beaker for waste/used tips
Digested pKAN (from part A)	Microfuge (optional)
Distilled water	Micropipettor (0.5–10-μl) + tips
2x Ligation buffer/ATP	Test tube rack
T4 DNA ligase	Tube (1.5-ml)
	Water bath (65°C)

METHODS

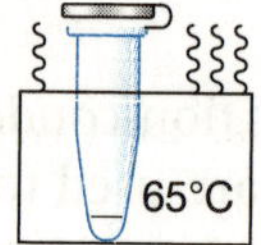

Step 1 is critical. Heat denatures protein, thus inactivating the restriction enzymes.

Ligate pAMP and pKAN (30 minutes, including incubation)

1. Incubate the tubes labeled Digested pAMP and Digested pKAN (from Part A) in a 65°C water bath for 10 minutes.
2. Label a clean 1.5-ml tube LIG (for ligation).
3. Use the matrix below as a checklist while adding reagents to the LIG tube. *Use a fresh tip for each reagent.* Refer to detailed directions that follow.

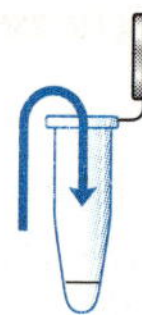

Tube	Digested pAMP	Digested pKAN	2x Ligation Buffer/ATP	Water	Ligase
LIG	3 µl	3 µl	10 µl	3 µl	1 µl

4. Collect reagents (except ligase), and place them in test tube rack on lab bench.
5. Add 3 µl of Digested pAMP.
6. Use a *fresh tip* to add 3 µl of Digested pKAN.
7. Use a *fresh tip* to add 10 µl of 2x ligation buffer/ATP.
8. Use a *fresh tip* to add 3 µl of distilled water.
9. Use a *fresh tip* to add 1 µl of DNA ligase. Carefully check that the droplet of ligase is on the *inside* wall of the tube.
10. Close tube top. Pool and mix reagents by pulsing in a microfuge or by sharply tapping the tube bottom on the lab bench.
11. Incubate the reaction for 2–24 hours at *room temperature.*
12. If time permits, ligation may be confirmed by electrophoresing 5 µl of the ligation reaction, along with *Bam*HI/*Hin*dIII digests of pAMP and pKAN. None of the parent *Bam*HI/*Hin*dIII fragments should be observed in the lane of ligated DNA, which should show multiple bands of high-molecular-weight DNA high up on the gel.

Ligase may be dispensed by your instructor.

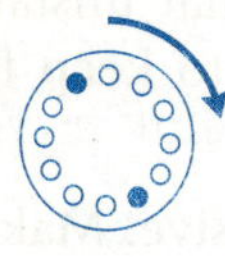

For brief ligations of 2–4 hours, it is essential to use a high-concentration T4 DNA ligase with at least 100–500 cohesive-end units/µl.

Freeze the reaction at –20°C until ready to continue. Thaw the reaction before proceeding to Laboratory 10.

RESULTS AND DISCUSSION

Ligation of the four *Bam*HI/*Hin*dIII restriction fragments of pAMP and pKAN (refer to diagram on page 444) produces many types of hybrid molecules, including plasmids composed of more than two fragments. However, only those constructs possessing an origin of replication will be maintained and expressed. Three different replicating plasmids, with selectable antibiotic resistance, are created by ligating combinations of two *Bam*HI/*Hin*dIII fragments:

- Ligation of the 784-bp fragment to the 3755-bp fragment regenerates pAMP.
- Ligation of the 1861-bp fragment to the 2332-bp fragment regenerates pKAN.
- Ligation of the 1861-bp fragment to the 3755-bp fragment produces the "simple recombinant" plasmid, pAMP/KAN, in which the kanamycin resistance gene has been fused into the pAMP backbone.

The ligation products will be tested in Laboratory 10.

1. Make a scale drawing of the simple recombinant molecule pAMP/pKAN described above. Include fragment sizes, locations of *Bam*HI and *Hin*dIII restriction sites, location of origin(s), and location of antibiotic resistance gene(s).
2. Make scale drawings of other two-fragment recombinant plasmids with the following properties.

   a. Three kinds of plasmids having two origins.

   b. Three kinds of plasmids having no origin.

   Whenever possible ,include fragment sizes, locations of *Bam*HI and *Hin*dIII restriction sites, location of origin(s), and location of antibiotic resistance gene(s).
3. Ligation of the 784-bp fragment, 3755-bp fragment, 1861-bp fragment, and 2332-bp fragment produces a "double plasmid" pAMP/pKAN (or superplasmid). Make a scale drawing of the double plasmid pAMP/pKAN.
4. Make scale drawings of several recombinant plasmids composed of any three of the four *Bam*HI/*Hin*dIII fragments of pAMP and pKAN. Include fragment sizes, locations of *Bam*HI and *Hin*dIII restriction sites, location of origin(s), and location of antibiotic resistance gene(s). What rule governs the construction of plasmids from three kinds of restriction fragments?
5. What kind of antibiotic selection would identify *E. coli* cells that have been transformed with each of the plasmids drawn in Questions 1–4?
6. Explain what is meant by "sticky ends." Why are they so useful in creating recombinant DNA molecules?
7. Why is ATP essential for the ligation reaction?

FOR FURTHER RESEARCH

Clone a DNA fragment using either pUC18 or pBLU as a vector. These vectors contain part or all of a gene coding for β-galactosidase (see discussion in Laboratory 5). The β-galactosidase enzyme acts on the synthetic substrate X-gal to produce a blue product, so cells containing one of these vectors will grow blue on plates containing X-gal. However, the pUC18 and pBLU vectors contain unique restriction sites within the β-galactosidase gene. Cloning a DNA fragment into these sites will disrupt the gene so that it does not make functional β-galactosidase. Cells containing only the disrupted gene will grow white on plates containing X-gal. Thus, pUC18 and pBLU allow you to distinguish the colonies containing plasmids with cloned DNA (white) from the colonies containing plasmids without cloned DNA (blue). Obtain a commercial source of bacteriophage λ DNA or *E. coli* DNA. Digest the DNA with *Bam*HI and *Hin*dIII. Ligate digested DNA to pUC18 or pBLU DNA that is also digested with *Bam*HI and *Hin*dIII. Plate onto X-gal plates and isolate several white colonies. Grow up each colony in a few milliliters of LB and make a miniprep. Carry out a *Bam*HI and *Hin*dIII restriction digest on the miniprep DNA and separate your DNA fragments on an agarose gel to visualize your cloned DNA fragment.

Transformation of *E. coli* with Recombinant DNA

In Part A, Classic Procedure for Preparing Competent Cells, *E. coli* cells are rendered competent to uptake plasmid DNA using a method essentially unchanged since its publication in 1970 by Morton Mandel and Akiko Higa. The procedure begins with vigorous *E. coli* cells grown in suspension culture. Cells are harvested in mid-log phase by centrifugation and incubated at 0°C with two successive changes of calcium chloride solution.

This procedure is more involved than the rapid colony protocol introduced in Laboratory 5. However, the classical procedure typically achieves transformation efficiencies ranging from 5×10^4 to 5×10^6 colonies per microgram of plasmid—a 2–200-fold increase over the colony procedure. The enhanced efficiency is important when transforming ligated DNA (composed primarily of relaxed circular plasmids and linear DNA), which produces 5–100 times fewer transformants than plasmid DNA purified from *E. coli* (containing a high proportion of the supercoiled form).

In Part B, Transformation of *E. coli* with Recombinant DNA, the competent *E. coli* cells are transformed with the ligation products from Laboratory 9, Recombination of Antibiotic Resistance Genes. Ligated plasmid DNA is added to one sample of competent cells, and purified pAMP and pKAN plasmids are added as controls to two other samples. The cell suspensions are incubated with the plasmid DNAs for 20 minutes at 0°C. Following a brief heat shock at 42°C, the cells recover in LB broth for 40–60 minutes at 37°C. Unlike ampicillin selection in Laboratory 5, Rapid Colony Transformation of *E. coli* with Plasmid DNA, the recovery step is essential for the kanamycin selection in this lab. Samples of transformed cells are plated onto three types of LB agar: with ampicillin (LB/amp), with kanamycin (LB/kan), and with both ampicillin and kanamycin (LB/amp+kan).

The ligation reaction produces many kinds of recombinant molecules composed of *Bam*HI/*Hin*dIII fragments, including the religated parental plasmids pAMP and pKAN. The object is to select for transformed cells with dual antibiotic resistance, which must contain a 3755-bp fragment from pAMP containing the ampicillin resistance gene (plus the origin of replication) and a 1861-bp fragment from pKAN containing the kanamycin resistance gene. Bacteria transformed with a single plasmid containing these sequences, or those doubly transformed with both pAMP and pKAN plasmids, form colonies on the LB/amp+kan plate.

Equipment and materials for this laboratory are available from the Carolina Biological Supply Company (see Appendix 1).

PART A

Classic Procedure for Preparing Competent Cells

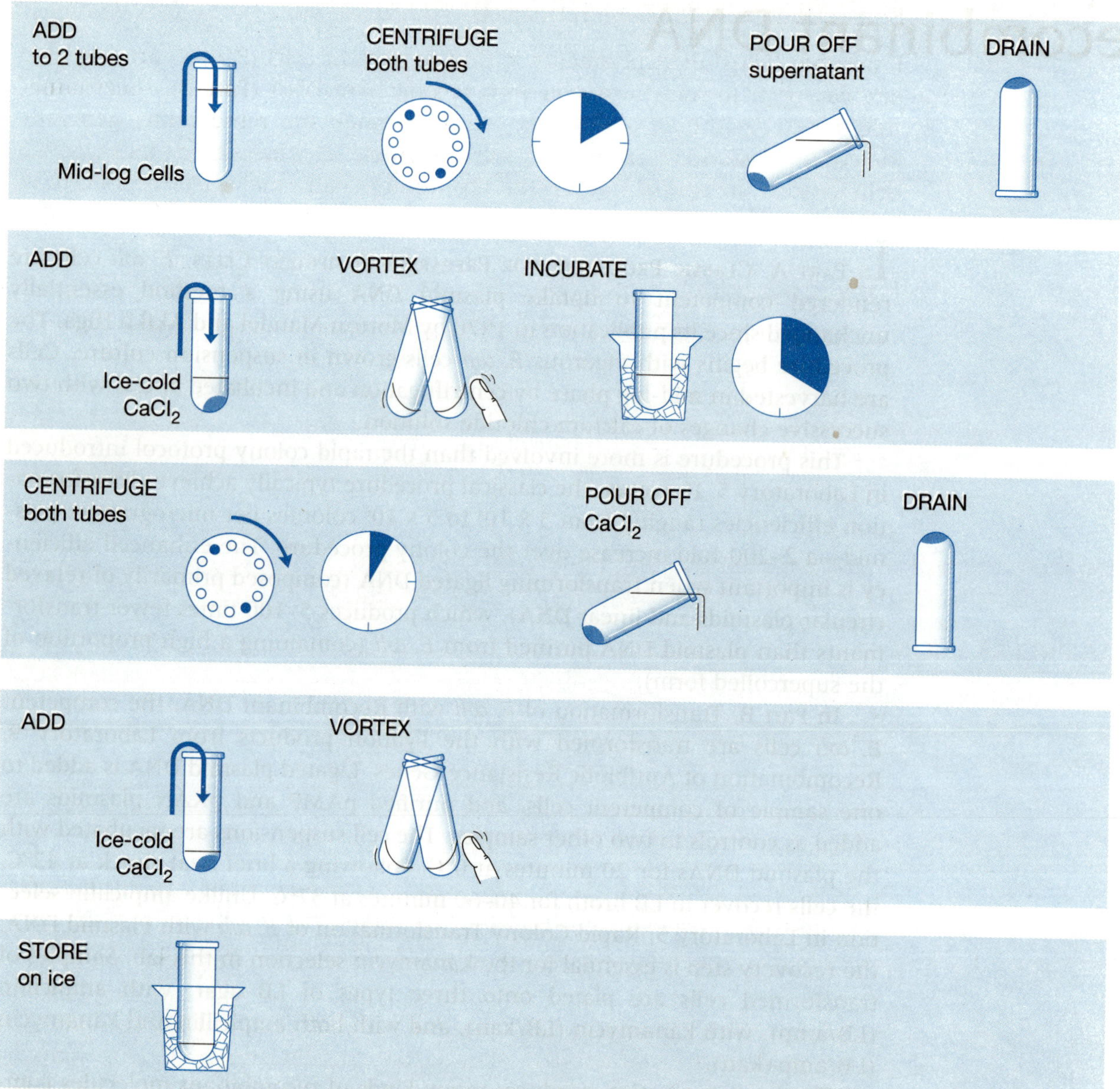

PRELAB NOTES

Review Prelab Notes in Laboratories 1, 2, and 5 regarding sterile techniques, *E. coli* culture, and transformation.

Seasoning Cells for Transformation

If possible, schedule experiments so that competent cells (Part A) are prepared one day prior to transformation with recombinant DNA (Part B). "Seasoning" cells for 12–24 hours at 0°C (an ice bath inside the refrigerator) generally increases transformation efficiency five- to tenfold. This enhanced efficiency will help ensure successful cloning of the recombinant molecules produced in Laboratory 9.

For Further Information

The protocol presented here is based on the following published methods:

Cohen S.N., Chang A.C., and Hsu L. 1972. Nonchromosomal antibiotic resistance in bacteria: Genetic transformation of *Escherichia coli* by R-factor DNA. *Proc. Natl. Acad. Sci.* **69:** 2110–2114.

Dagert M. and Ehrlich S.D. 1979. Prolonged incubation in calcium chloride improves the competence of *Escherichia coli* cells. *Gene* **6:** 23.

Mandel M. and Higa A. 1970. Calcium-dependent bacteriophage DNA infection. *J. Mol. Biol.* **53:** 159–162.

PRELAB PREPARATION

Before performing this Prelab Preparation, please refer to the cautions indicated on the Laboratory Materials list.

1. On the day before this lab, begin an *E. coli* culture from a streaked plate of MM294, according to the protocol in Laboratory 2B, Overnight Suspension Culture.
2. PLAN AHEAD. Make sure that you have a streaked plate of viable *E. coli* cells from which to inoculate overnight. Also, streak your *E. coli* strain on LB/amp and LB/kan plates to ensure that a resistant strain has not been used by mistake.
3. Approximately 2–4 hours before the lab, begin an *E. coli* culture according to the protocol in Laboratory 2C, Mid-log Suspension Culture. Cells are optimal for transformation when the culture reaches an OD_{550} of 0.3–0.40. More simply, cells inoculated into room temperature LB will be ready to transform after 2 hours, 15 minutes. Hold cells in mid-log phase by storing the culture on ice for up to 2 hours prior to beginning calcium chloride treatment. *Each experiment requires 20 ml of mid-log suspension culture.*
4. Sterilize 50 mM calcium chloride ($CaCl_2$) solution and LB broth by autoclaving or filtering through a 0.45-μm or 0.22-μm filter (Nalgene or Corning). Filtered $CaCl_2$ can be stored in a filter collection container or transferred to sterile 50-ml conical tubes.

5. Prepare aliquots for each experiment:

   15 ml of 50 mM $CaCl_2$ in a sterile 50-ml tube (store on ice)
   two 10-ml cultures of mid-log *E. coli* cells in sterile 15-ml culture tubes (store on ice)
   12 µl of 0.005 µg/µl pAMP in a 1.5-ml tube (store on ice)
   12 µl of 0.005 µg/µl pKAN in a 1.5-ml tube (store on ice)

6. Review Part B, Transformation of *E. coli* with Recombinant DNA.

MATERIALS

CULTURE AND REAGENTS	SUPPLIES AND EQUIPMENT
$CaCl_2$ (50 mM)	Beakers for crushed or cracked ice and for waste
Mid-log MM294 cells (two 10-ml cultures)	Bleach (10%)▼ or disinfectant
	Bunsen burner
	Clean paper towels
	Clinical centrifuge (2000–4000 rpm)
	Micropipettor (100–1000 µl) + tips (or 1-ml pipette)
	Pipettes (5-ml or 10-ml)
	Spectrophotometer (optional)
	Sterile pipette aid or bulb
	Test tube rack

▼ See Appendix 4 for Cautions list.

METHODS

Prepare Competent Cells

(40–50 minutes)

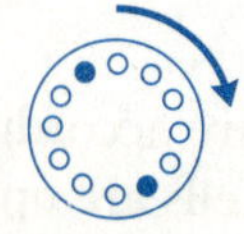

This entire experiment *must be performed under sterile conditions.* Review sterile techniques in Laboratory 1, Measurements, Micropipetting, and Sterile Techniques.

A tight pellet of cells should be easily seen at the bottom of the tube. If pellet does not appear to be consolidated, recentrifuge for an additional 5 minutes.

1. Place sterile tube of $CaCl_2$ solution on ice.
2. Obtain two 15-ml tubes each with 10 ml of mid-log cells, and label with your name.
3. *Securely close caps* and place both tubes of cells in a *balanced* configuration in the rotor of the clinical centrifuge. Centrifuge at 3000 rpm for 10 minutes to pellet cells on the bottom-side of the culture tube.

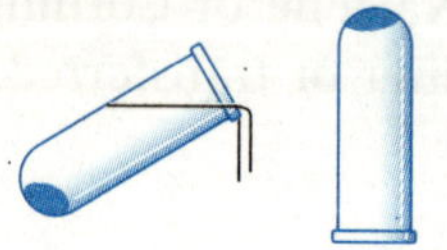

4. Sterilely pour off supernatant from each tube into the waste beaker for later disinfection. *Do not disturb the cell pellet.*

Plan out manipulations for Step 4. Organize lab bench, and work quickly.

Flaming is not necessary since contaminating bacteria do not grow well in $CaCl_2$.

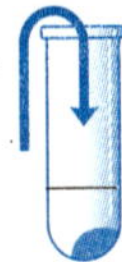

If working as a team, one person handles the pipettes, and the other removes and replaces tube caps.

a. Remove cap from the culture tube, and briefly flame mouth. *Do not place cap on lab bench.*

b. Carefully pour off supernatant. Invert culture tube, and tap gently on the surface of a clean paper towel to drain thoroughly.

c. *Reflame mouth of culture tube, and replace cap.*

5. Use a 5- or 10-ml pipette to sterilely add 5 ml of ice-cold $CaCl_2$ solution to each culture tube:

   a. Remove cap from $CaCl_2$ tube. *Do not place cap on lab bench.*

   b. Withdraw 5 ml of $CaCl_2$ and replace cap.

   c. Remove cap of the culture tube. *Do not place cap on lab bench.*

   d. Expel $CaCl_2$ into culture tube and replace cap.

6. Immediately finger vortex to resuspend pelleted cells in each tube.

   a. Close cap tightly.

   b. Hold upper part of tube securely with thumb and index finger.

   c. With the other hand, vigorously hit the bottom end of the tube with index finger or thumb to create a vortex that lifts the cell pellet off the bottom of the tube. Continue "finger vortexing" until all traces of the cell mass are completely resuspended. This may take a couple of minutes, depending on technique.

   d. Hold the tube up to the light to check that the suspension is homogeneous. No visible clumps of cells should remain.

The cell pellet becomes increasingly difficult to suspend the longer it sits in the $CaCl_2$ solution.

Double check both tubes for complete resuspension of cells, which is probably the most important variable in obtaining good results.

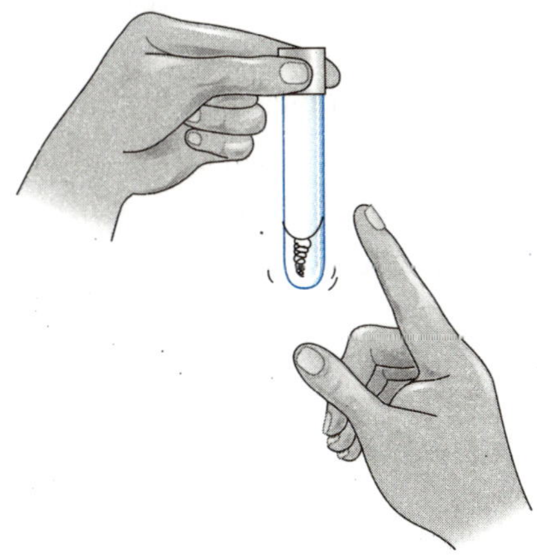

Finger Vortex (Steps 6 and 11)

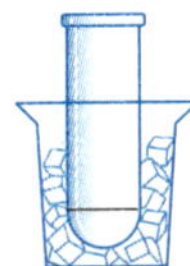

$CaCl_2$ treatment alters adhering properties of E. coli *membranes. The cell pellet is much more dispersed after the second centrifugation.*

7. Return both tubes to ice, and incubate for 20 minutes.

8. Following incubation, respin the cells in a clinical centrifuge for 5 minutes at 2000–4000 rpm. This time the cell pellet will be more spread out on the bottom of the tube due to the $CaCl_2$ treatment.

9. Sterilely pour off $CaCl_2$ from each tube into a waste beaker. *Do not disturb the cell pellet.*

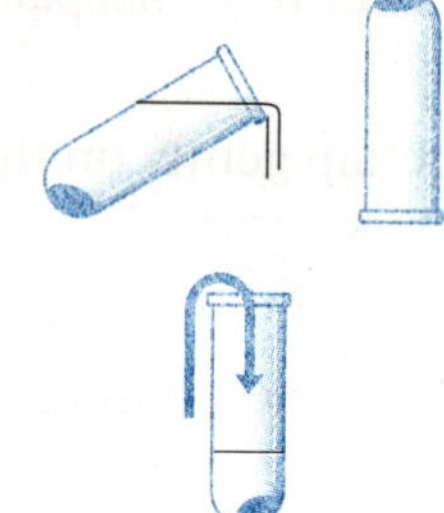

a. Remove the cap from the culture tube. Do not place the cap on the lab bench.

b. Carefully pour off supernatant. Invert the culture tube, and tap gently on the surface of a clean paper towel to drain thoroughly.

c. Replace cap.

10. Use a 100–1000-μl micropipettor (or 1-ml pipette) to sterilely add 1000 μl (1 ml) of fresh, ice-cold $CaCl_2$ to each tube.

    a. Remove cap from $CaCl_2$ tube. *Do not place cap on lab bench.*

    b. Withdraw 1000 μl (1 ml) of $CaCl_2$ and replace cap.

    c. Remove cap of culture tube. *Do not place cap on lab bench.*

    d. Expel $CaCl_2$ into culture tube and replace cap.

11. Close caps tightly, and immediately finger vortex to resuspend pelleted cells in each tube. Hold the tube up to the light to check that the suspension is homogeneous. No visible clumps of cells should remain.

Cell pellet may appear more diffuse than at beginning of procedure and will resuspend more easily. Double check both tubes for complete resuspension of cells.

STOP Store cells in a beaker of ice in the refrigerator (~0°C) until ready for use in Part B. "Seasoning" at 0°C for up to 24 hours increases competency of cells five- to tenfold.

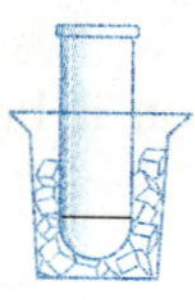

12. If not proceeding immediately to Part B, take time for responsible cleanup:

    a. Segregate for proper disposal culture plates and tubes, pipettes, and micropipettor tips that have come in contact with *E. coli*.

    b. Disinfect mid-log culture, tips, and supernatant from Steps 4 and 9 with 10% bleach or disinfectant.

    c. Wipe down lab bench with soapy water, 10% bleach solution, or disinfectant (such as Lysol).

    d. Wash hands before leaving lab.

PART B

Transformation of *E. coli* with Recombinant DNA

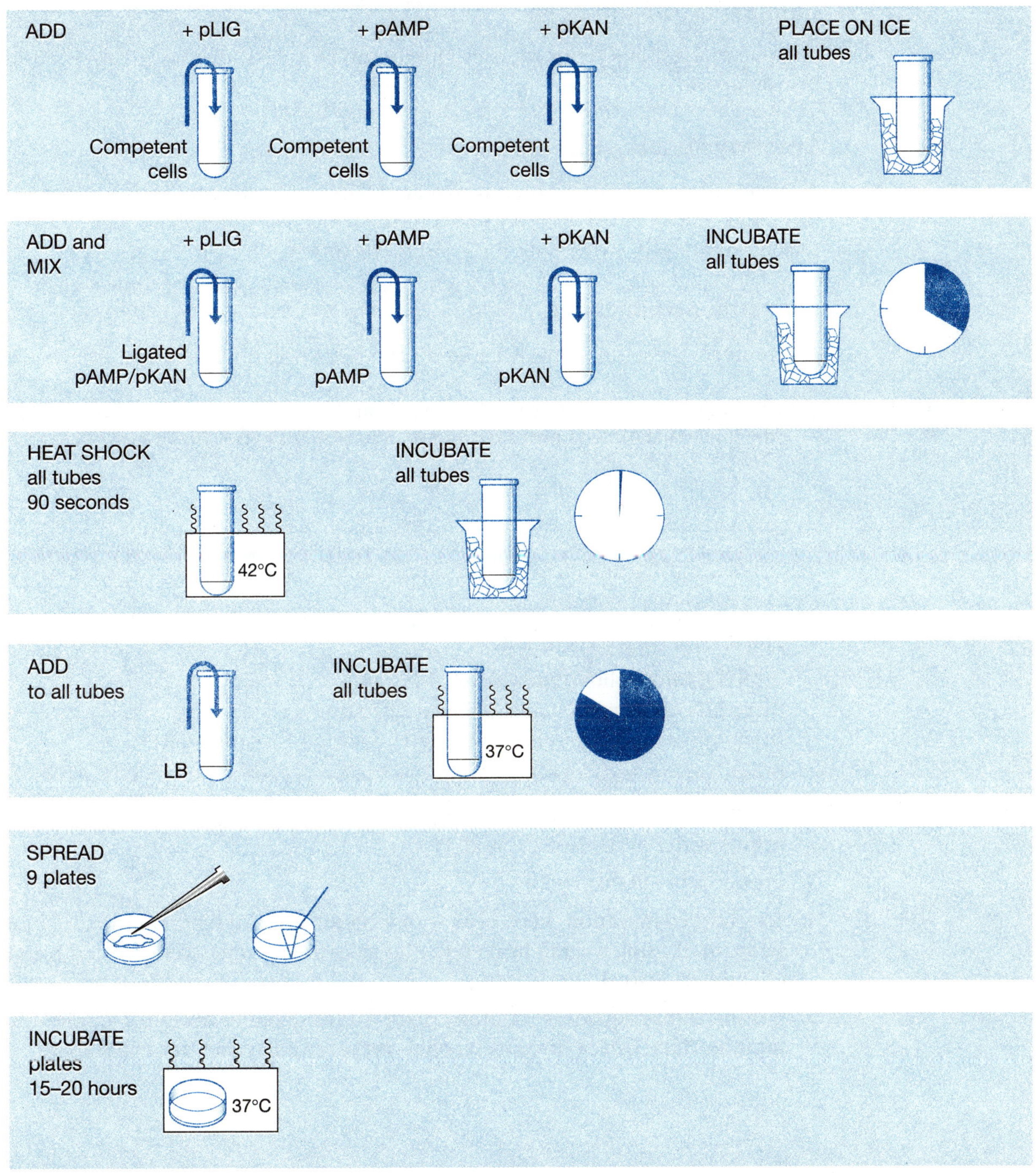

PRELAB NOTES

Review Prelab Notes in Laboratory 5, Rapid Colony Transformation of *E. coli* with Plasmid DNA.

Equipment Substitutions

A standard 1-ml pipette or transfer pipette can be substituted for a 100–1000-μl micropipettor.

Recovery Period

A 40–60-minute postincubation recovery at 37°C, with shaking, is essential prior to plating transformed cells on kanamycin, which acts quickly to kill any cell that is not actively expressing the resistance protein.

For Further Information

The protocol presented here is based on the following published method:

Cohen S.N., Chang A.C.Y., Boyer H.W., and Helling R.B. 1973. Construction of biologically functional bacteria plasmids in vitro. *Proc. Natl. Acad. Sci.* **70:** 3240–3244.

PRELAB PREPARATION

Before performing this Prelab Preparation, please refer to the cautions indicated on the Laboratory Materials list.

1. Prepare for each experiment:

   three LB+ampicillin plates (labeled LB/amp)
   three LB+kanamycin plates (labeled LB/kan)
   three LB+ampicillin+kanamycin plates (labeled LB/amp+kan)

   If only one control transformation, with *either* pAMP *or* pKAN, is done, then one less plate of each type is required.
2. Adjust water baths to 42°C and 37°C.
3. Prewarm incubator to 37°C.
4. To retard evaporation, keep beaker of ethanol covered with Parafilm, plastic wrap, or, if using a small beaker, the lid from a Petri dish. Retrieve and reuse ethanol exclusively for flaming.
5. If using spreading beads, carefully place five to seven beads into a sterile 1.5-ml tube. Tube can be used as scooper. Prepare nine tubes per experiment.

MATERIALS

CULTURE, MEDIA, AND REAGENTS	SUPPLIES AND EQUIPMENT
Competent *E. coli* cells (from Part A)	Beaker of 95% ethanol ▼
LB broth	Beakers for crushed or cracked ice and for waste/used tips
LB/amp plates (3)	"Bio-bag" or heavy-duty trash bag
LB/kan plates (3)	Bleach (10%) ▼ or disinfectant
LB/amp+kan plates (3)	Bunsen burner
Ligation tube (from Laboratory 9)	Cell spreader (or spreading beads)
pAMP (0.005 μg/μl)	Culture tubes (three 15-ml)
pKAN (0.005 μg/μl)	Incubator (37°C)
	Micropipettors (0.5–10-μl and 100–1000-μl) + tips
	Permanent marker
	Shaking water bath (37°C)
	Test tube rack
	Water bath (42°C)

▼ See Appendix 4 for Cautions list.

METHODS

Perform *E. coli* Transformation

(70–90 minutes)

This entire experiment must be performed under sterile conditions. Review sterile techniques in Laboratory 1, Measurements, Micropipetting, and Sterile Techniques.

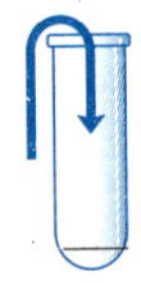

1. Use a permanent marker to label three *sterile* 15-ml culture tubes:

   +pLIG = ligated DNA
   +pAMP = pAMP control
   +pKAN = pKAN control

2. Use a 100–1000-μl micropipettor and *sterile tip* to add 200 μl of competent cells to each tube.

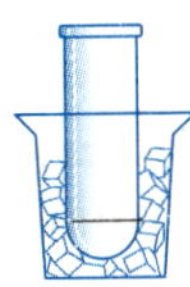

3. Place all three tubes on ice.
4. Use a 1–10-μl micropipettor to add 10 μl of ligated pAMP/KAN solution *directly into cell suspension* in tube labeled +pLIG.

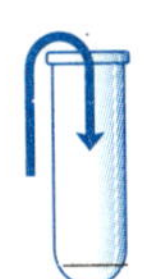

5. Use a *fresh tip* to add 10 μl of 0.005 μg/μl pAMP solution *directly into cell suspension* in tube labeled +pAMP.
6. Use a *fresh tip* to add 10 μl of 0.005 μg/μl pKAN solution *directly into cell suspension* in tube labeled +pKAN.

Store remainder of ligated DNA at 4°C. Separate by electrophoresis with cut pAMP and pKAN controls in Laboratory 12 to observe the products of the ligation reaction.

7. Close caps, and tap tubes with finger to mix. Avoid making bubbles in suspension or splashing suspension up the sides of the tubes.
8. Return all three tubes to ice for 20 minutes.

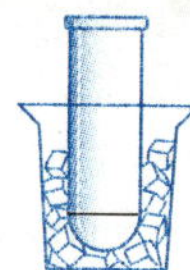

9. While cells are incubating on ice, use a permanent marker to label all nine LB agar plates with your name and the date. Divide plates into three sets of three plates each and mark as follows:

   Set a
   Mark L on one LB/amp, one LB/kan, and one LB/amp+kan plate.

   To save plates, experimenters may omit either Set b or Set c.

   Set b
   Mark A on one LB/amp, one LB/kan, and one LB/amp+kan plate.

   Set c
   Mark K on one LB/amp, one LB/kan, and one LB/amp+kan plate.

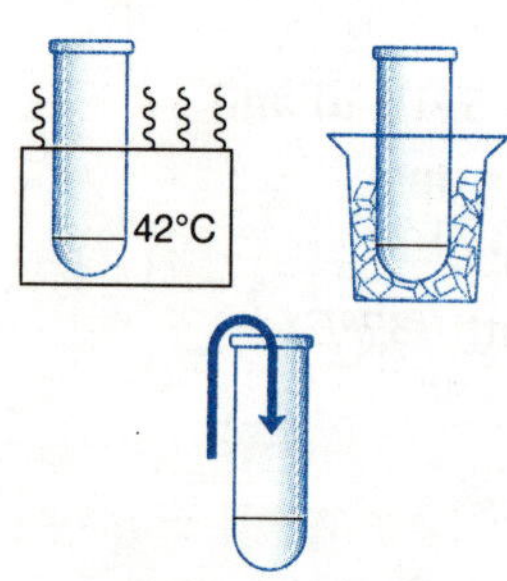

10. Following a 20-minute incubation, heat-shock the cells in all three tubes. *It is critical that cells receive a sharp and distinct shock.*
    a. Carry the ice beaker to the water bath. Remove tubes from ice, and *immediately* immerse in a 42°C water bath for 90 seconds.
    b. Immediately return all three tubes to ice for at least 1 additional minute.
11. Use a 100–1000-µl micropipettor with a sterile tip to add 800 µl of LB broth to each tube. Gently tap tubes with finger to mix.

If a shaking water bath is not available, warm cells for several minutes in a 37°C water bath, and then transfer to a dry shaker inside a 37°C incubator. Alternatively, occasionally swirl tubes by hand in a nonshaking 37°C water bath.

12. Allow cells to recover by incubating all three tubes at 37°C in a shaking water bath (with moderate agitation) for 40–60 minutes. If a shaking water bath is not available, incubate the tubes in a regular 37°C water bath. In this case, gently mix tubes periodically.

STOP Cells may be allowed to recover for up to 2 hours. A recovery period assures the growth of as many kanamycin-resistant recombinants as possible and can help compensate for a poor ligation or cells of low competence.

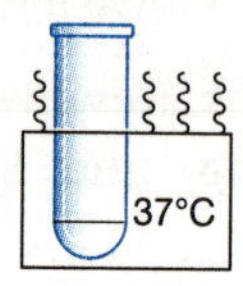

13. Use the matrix below as a checklist as +pLIG, +pAMP, and +pKAN cells are spread on each type of antibiotic plate in Steps 14–20:

	Ligated DNA L	pAMP control A	pKAN control K
LB/amp	100 µl	100 µl	100 µl
LB/kan	100 µl	100 µl	100 µl
LB/amp+kan	100 µl	100 µl	100 µl

If too much liquid is absorbed by agar, cells will not be evenly distributed.

14. Use a micropipettor with a sterile tip to add 100 µl of cell suspension from the tube labeled pLIG onto three plates marked L. *Do not allow suspensions to sit on plates too long before proceeding to Step 15 or 16.* Spread cells using one of the methods described in Steps 15 and 16.
15. Sterilize the cell spreader, and spread cells over the surface of each L plate in succession.

The object is to evenly distribute and separate cells on agar so that each gives rise to a distinct colony of clones. It is essential not to overheat spreader in burner flame and to cool it before touching cell suspensions. A hot spreader will kill E. coli *cells on the plate.*

a. Dip spreader into the ethanol beaker and *briefly* pass it through Bunsen flame to ignite alcohol. Allow alcohol to burn off *away from* the Bunsen flame; spreading rod will become too hot if left in flame.

> CAUTION
> Be extremely careful not to ignite ethanol in the beaker. Do not panic if ethanol is accidentally ignited. Cover the beaker with a Petri lid or other cover to cut off oxygen and rapidly extinguish the fire.

Sterile Spreading Technique (Steps 14 and 15)

b. Lift the lid of the first L plate only enough to allow spreading. *Do not place lid on lab bench.*

c. Cool the spreader by gently rubbing it on surface of the agar *away* from the cell suspension or by touching it to the condensation on the plate lid.

d. Touch the spreader to the cell suspension, and gently drag it back and forth several times across the surface of agar. Rotate plate one-quarter turn, and repeat spreading motion. Try to spread the suspension evenly across the agar surface and be careful not to gouge the agar.

e. Replace plate lid. Return the cell spreader to ethanol *without flaming.*

f. Repeat Steps a through e in succession for the remaining two L plates.

16. Use spreading beads to spread cells over the surface of each L plate in succession.

a. Lift the lid of first L plate enough to allow adding beads. *Do not place lid on lab bench.*

b. Carefully pour five to seven glass spreading beads from a 1.5-ml tube onto the agar surface.

c. Close plate lid and use a swirling motion to move glass beads around the entire surface of the plate. This evenly spreads the cell suspension on the agar suface. Continue swirling until the cell suspension is absorbed into the agar.

d. Repeat Steps a through c in succession for the remaining two L plates.

17. Use a *fresh sterile tip* to add 100 µl of cell suspension from tube labeled +pAMP onto three plates marked A.

18. Repeat Step 15a–f or Step 16a–d to spread cells over the surface of each A plate in succession.

19. Use a *fresh sterile tip* to add 100 µl of cell suspension from tube labeled +pKAN onto three plates marked K.

20. Repeat Step 15a–f or Step 16a–d to sterilize cell spreader and spread cells over the surface of each K plate in succession.

21. If Step 15 was used, reflame spreader one last time before placing it on lab bench.

22. Let plates set for several minutes to allow suspension to become absorbed into agar. If Step 16 was used, invert plates and gently tap plate bottoms, so that the spreading beads fall into plate lids. Carefully pour beads from each lid into a storage container for reuse.

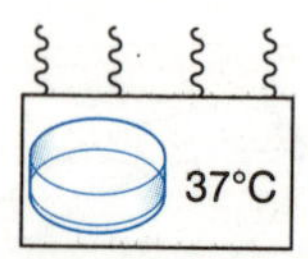

23. Stack plates and tape them in a bundle to keep the experiment together. Place plates upside down in a 37°C incubator, and incubate for 15–20 hours.

Save L LB/amp and L LB/kan plates if planning to do Laboratory 11. Save L LB/amp+kan as a source of colonies to begin overnight suspension cultures if planning to do Laboratory 12.

24. After initial incubation, store plates at 4°C to arrest *E. coli* growth and to slow the growth of any contaminating microbes.

25. Take time for responsible cleanup:

a. Segregate for proper disposal culture plates and tubes, pipettes, and micropipettor tips that have come in contact with *E. coli*.

b. Disinfect overnight cell suspensions, tubes, and tips with 10% bleach or disinfectant.

c. Wipe down lab bench with soapy water, 10% bleach solution, or disinfectant (such as Lysol).

d. Wash hands before leaving lab.

RESULTS AND DISCUSSION

Observe colonies through the bottom of the culture plate, using a permanent marker to mark each colony as it is counted. If the experiment worked well, 5–50 colonies should be observed on the L LB/amp+kan experimental plate, 500–5000 colonies on the A LB/amp control plate, and 200–2000 colonies on the K LB/kan control plate. (If plates are very crowded, draw lines on the bottom of the plate to divide it into equal-sized sections. Count one sector estimated as being representative of the whole plate. After counting, multiply by the number of sectors.) Approximately ten times fewer colonies should be observed on the corresponding L LB/amp plate and L LB/kan plate. An extended recovery period would inflate these numbers. (Question 3 explains how to compute transformation efficiency.) If plates have been overincubated or left at room temperature for several days, "satellite" colonies may be observed on the LB/amp plates. Satellite colonies are never observed on the LB/kan or LB/amp+kan plates.

1. Record your observation of each plate in matrix below. If cell growth is too dense to count individual colonies, record "lawn." Were the results as expected? Explain possible reasons for variations from expected results.

	Ligated DNA L	pAMP control A	pKAN control K
LB/amp			
LB/kan			
LB/amp+kan			

2. Compare and contrast the growth on each of the following pairs of plates. What does each pair of results tell you about transformation and/or antibiotic selection?

   L LB/amp and A LB/amp
   L LB/kan and A LB/kan
   A LB/amp and K LB/kan
   L LB/amp and L LB/kan
   L LB/amp and L LB/amp+kan
   L LB/kan and L LB/amp+kan

3. Calculate transformation efficiencies of A LB/amp and K LB/kan positive controls. Remember that transformation efficiency is expressed as the number of antibiotic resistant colonies per microgram of intact plasmid DNA. The object is to determine the mass of pAMP or pKAN that was spread on each plate and was therefore responsible for the transformants observed.

   a. Determine the total mass (in micrograms) of pAMP used in Step 5 and of pKAN used in Step 6: concentration x volume = mass.

   b. Determine the fraction of cell suspension spread onto the A LB/amp plate (Step 17) and K LB/kan plate (Step 19): volume suspension spread/*total* volume suspension (Steps 2 and 11) = fraction spread.

   c. Determine the mass of plasmid pAMP and pKAN in the cell suspension spread onto the A LB/amp plate and K LB/kan plate: total mass plasmid (*a*) x fraction spread (*b*) = mass plasmid spread.

   d. Determine the number of colonies per microgram of pAMP and pKAN. Express answer in scientific notation: colonies observed/mass plasmid spread (*c*) = transformation efficiency.

4. Calculate transformation efficiencies of the L LB/amp, L LB/kan, and L LB/amp+kan plates.

   a. Calculate the mass of pAMP and pKAN used in the restriction reactions of Laboratory 9 (see matrix in Part A, Step 1). Then calculate the concentration of plasmid in each restriction reaction.

   b. Calculate the mass of pAMP and pKAN used in the ligation reaction (Laboratory 9, Part B, Step 3). Then calculate the *total* concentration of plasmid in the ligation mixture.

   c. Use this concentration in calculations following Steps a–d of Question 3 above.

5. Compare the transformation efficiencies that you calculated for the A LB/amp plate in this laboratory and the +pAMP plate in Laboratory 5. By what factor is the classical procedure more or less efficient than colony transformation? What differences in the protocols contribute to the increase in efficiency?

6. Compare the transformation efficiencies that you calculated for control pAMP and pKAN *versus* the ligated pAMP and pKAN. How can you account for the differences in efficiency? Take into account the formal definition of transformation efficiency.

FOR FURTHER RESEARCH

Interpretable experimental results can only be achieved when the classic transformation protocol can be repeated with reproducible results. Only attempt the experiments below when you are able to routinely achieve 500–2000 colonies on the A LB/amp plate.

1. Design and execute an experiment to compare the transformation efficiencies of linear versus circular plasmid DNAs. Keep molecular weight constant.

2. Design and execute a series of experiments to test the relative importance of each of the four major steps of most transformation protocols: (1) preincubation, (2) incubation, (3) heat shock, and (4) recovery. Which steps are absolutely necessary?

3. Design and execute a series of experiments to compare the transforming effectiveness of $CaCl_2$ *versus* salts of other monovalent (+), divalent (++), and trivalent (+++) cations.

   a. Make up 50 mM solutions of each salt.

   b. Check the pH of each solution, and buffer to pH 7 when necessary.

   c. Is $CaCl_2$ unique in its ability to facilitate transformation?

   d. Is there any consistent difference in the transforming ability of monovalent *versus* divalent *versus* trivalent cations?

4. Test the effect of pH differences on transformation. First, prepare a series of 50 mM $CaCl_2$/20 mM Tris transformation solutions at pH 5.0, 5.5, 6.0, 6.5, 7.0, 7.5, 8.0, 8.5, and 9.0. Next grow a 100-ml culture of *E. coli* to mid-log phase (OD_{550} 0.4–0.6). Remove nine 10-ml aliquots and prepare competent cells with the above series of transformation solutions. Then, transform each set of competent cells using a plasmid DNA, such as pAMP. For consistency, select from only one dilution tube of plasmid DNA for all of your transformations. Follow identical protocols for each transformation so that pH is the only variable. Plate transformations on selective media and incubate overnight. The next day, count and determine which pH transformation solution has the highest efficiency.

5. Carry out a similar experiment to determine the effect of adding dithiothreitol (DTT) to the transformation at various concentrations. Use 50 mM

$CaCl_2$/20 mM Tris transformation solution at pH 7 (or whichever pH you have determined works best). Try DTT at 0 mM, 0.5 mM, 1 mM, 2 mM, 5 mM, and 10 mM.

6. Design a series of experiments to determine saturating conditions for transformation reactions.

   a. Transform *E. coli* using the following DNA concentrations:

      0.00001 µg/µl
      0.00005 µg/µl
      0.0001 µg/µl
      0.0005 µg/µl
      0.001 µg/µl
      0.005 µg/µl
      0.01 µg/µl
      0.05 µg/µl
      0.1 µg/µl

   b. Plot a graph of DNA mass *versus* colonies per plate.

   c. Plot a graph of DNA mass *versus* transformation efficiency.

   d. At what mass does the reaction appear to become saturated?

   e. Repeat the experiment with concentrations clustered on either side of the presumed saturation point to produce a fine saturation curve.

7. Repeat Experiment 6 above, but transform with a 1:1 mixture of pAMP and pKAN at each concentration. Plate transformants on LB/amp, LB/kan, and LB/amp+kan plates. *Be sure to include a 40–60-minute recovery, with shaking.*

   a. Calculate the percentage of double transformations at each mass.

$$\frac{\text{colonies amp+kan plate}}{\text{colonies amp plate + colonies kan plate}}$$

   b. Plot a graph of DNA mass *versus* colonies per plate.

   c. Plot a graph of DNA mass *versus* percentage of double transformations. Under saturating conditions, what percentage of bacteria are doubly transformed?

8. Plot a recovery curve for *E. coli* transformed with pKAN. Allow cells to recover for 0–120 minutes at 20-minute intervals.

   a. Plot a graph of recovery time *versus* colonies per plate.

   b. At what time point is antibiotic expression maximized?

   c. Can you discern a point at which the cells began to replicate?

9. Attempt to isolate pAMP/KAN recombinants using the colony transformation protocol in Laboratory 5. What trick would increase the likelihood of retrieving ampicillin/kanamycin-resistant colonies?

$CaCl_2$/20 mM Tris transformation solution at pH 7 (or whichever pH you have determined works best). Try DTT at 0 mM, 0.5 mM, 1 mM, 2 mM, 5 mM, and 10 mM.

6. Design a series of experiments to determine saturating conditions for transformation reactions.

   a. Transform *E. coli* using the following DNA concentrations:

      0.00001 μg/μl
      0.00005 μg/μl
      0.0001 μg/μl
      0.0005 μg/μl
      0.001 μg/μl
      0.005 μg/μl
      0.01 μg/μl
      0.05 μg/μl
      0.1 μg/μl

   b. Plot a graph of DNA mass *versus* colonies per plate.

   c. Plot a graph of DNA mass *versus* transformation efficiency.

   d. At what mass does the reaction appear to become saturated?

   e. Repeat the experiment with concentrations clustered on either side of the presumed saturation point to produce a fine saturation curve.

7. Repeat Experiment 6 above, but transform with a 1:1 mixture of pAMP and pKAN at each concentration. Plate transformants on LB/amp, LB/kan, and LB/amp+kan plates. Be sure to include a 40–60-minute recovery with shaking.

   a. Calculate the percentage of double transformations at each mass:

$$\frac{\text{colonies amp+kan plate}}{\text{colonies amp plate + colonies kan plate}}$$

   b. Plot a graph of DNA mass *versus* colonies per plate.

   c. Plot a graph of DNA mass *versus* percentage of double transformants. Under saturating conditions, what percentage of bacteria are double transformed?

8. Plot a recovery curve for *E. coli* transformed with pKAN. Allow cells to recover for 0–120 minutes at 20-minute intervals.

   a. Plot a graph of recovery time *versus* colonies per plate.

   b. At what time point is antibiotic expression maximized?

   c. Can you discern a point at which the cells began to replicate?

9. Attempt to isolate pAMP/KAN recombinants using the colony transformation protocol in Laboratory 5. What trick would increase the likelihood of retrieving ampicillin/kanamycin-resistant colonies?

LABORATORY 11

Replica Plating to Identify Mixed *E. coli* Populations

LIGATION OF *BAM*HI/*HIND*III FRAGMENTS of pAMP and pKAN in Laboratory 9 created a number of recombinant plasmids containing either an ampicillin resistance gene (*amp*^{r}), a kanamycin resistance gene (*kan*^{r}), or both genes together (*amp*^{r}/*kan*^{r}). Samples of this ligated DNA (L) were then transformed into competent *E. coli* cells, which were plated onto LB/amp, LB/kan, and LB/amp+kan plates in Laboratory 10. The results clearly indicate that colonies growing on the L LB/amp+kan plate (having an *amp*^{r}/*kan*^{r} phenotype) contain both resistance genes (the *amp*^{r}/*kan*^{r} genotype).

It is not possible, however, to be certain of the *amp*^{r}/*kan*^{r} genotypes of bacteria growing on the L LB/amp or L LB/kan plates. Although a colony growing on the L LB/amp plate possesses an *amp*^{r} gene, it is not possible to say whether it also possesses a *kan*^{r} gene. Conversely, although a colony growing on the L LB/kan must possess a *kan*^{r} gene, it is not possible to know whether it also possesses an *amp*^{r} gene. A conclusion can be made about the presence of an antibiotic resistance gene only when the organism has been challenged with that antibiotic.

In this laboratory, replica plating provides a rapid means to distinguish between single- and dual-resistant colonies growing on the L LB/amp and L LB/kan plates. Cells from 12 colonies on the L LB/amp plate and from 12 colonies on the L LB/kan plate are transferred onto one fresh LB/amp plate and one fresh LB/kan plate to which numbered grids have been attached. An L colony is scraped with a sterile toothpick (or inoculating loop), and a sample of cells is streaked successively into the same numbered squares of the fresh LB/amp and LB/kan plates. Following overnight incubation at 37°C, colonies that grow in the same squares of both the LB/amp and LB/kan plates have the *amp*^{r}/*kan*^{r} genotype.

Equipment and materials for this laboratory are available from the Carolina Biological Supply Company (see Appendix 1).

INOCULATE same number square on both plates from each colony

squares 1–12

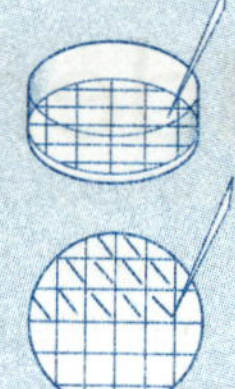

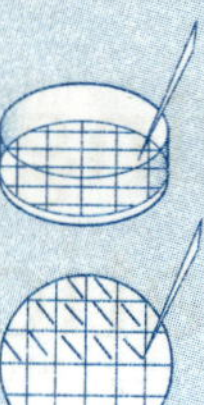

INOCULATE same number square on both plates from each colony

squares 13–24

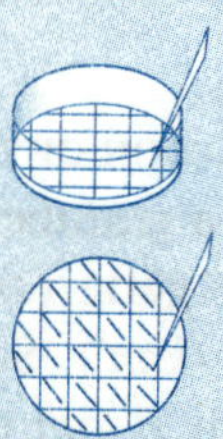

INCUBATE
15–20 hours

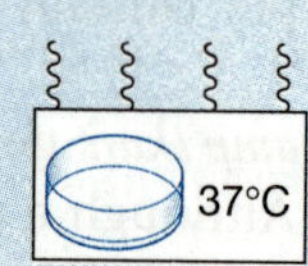

PRELAB NOTES

Review Prelab Notes in Laboratory 2A, Isolation of Individual Colonies.

Replica plating provides a rapid means to screen L LB/amp and L LB/kan plates for dual-resistant colonies that potentially contain pAMP/KAN recombinant plasmids. If colonies were not obtained on the L LB/amp+kan plate, replica plating provides another chance to identify dual-resistant colonies from which to isolate recombinant plasmids for Laboratory 12, Purification and Identification of Recombinant DNA.

For Further Information

The protocol presented here is based on the following published method:

Lederberg J. and Lederberg E.M. 1952. Replica plating and indirect selection of bacterial mutants. *J. Bacteriol.* **63:** 399.

PRELAB PREPARATION

Before performing this Prelab Preparation, please refer to the cautions indicated on the Laboratory Materials list.

1. Prepare for each experiment:

   one LB + ampicillin plate (LB/amp)
   one LB + kanamycin plate (LB/kan)

2. Sterilize 30 toothpicks per experiment. Put toothpicks in 50-ml beaker, cover with aluminum foil, and autoclave for 15 minutes at 121°C. (Although much less rapid, a flamed and cooled inoculating loop can be used to transfer colonies.)

3. Make two copies of the replica-plating grid (below) per experiment.

		1	2		
	3	4	5	6	
7	8	9	10	11	12
13	14	15	16	17	18
	19	20	21	22	
		23	24		

4. Prewarm incubator to 37°C.

MATERIALS

CULTURES AND MEDIA	SUPPLIES AND EQUIPMENT
L LB/amp plate w/colonies (from Laboratory 10)	Beaker for waste
L LB/kan plate w/colonies (from Laboratory 10)	"Biobag" or heavy-duty trash bag
LB/amp plate (1)	Bleach▼(10%) or disinfectant
LB/kan plate (1)	Incubator (37°C)
	Permanent marker
	Replica-plating grids (2)
	Sterile toothpicks (or inoculating loop + Bunsen burner)

▼ See Appendix 4 for Cautions list.

METHODS

Prepare Replica Plates

(10 minutes)

A 24-square grid may be drawn on the bottom of the plate with a permanent marker.

Lift plate lids only enough to select colony and streak. Do not place lids on the lab bench.

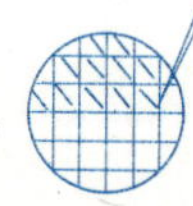

1. Attach a replica-plating grid to the *bottom* of an LB/amp plate and to the *bottom* of an LB/kan plate. Use a permanent marker to label each plate with your name and the date.
2. Replica plate a sample of cells from one colony on the L LB/amp plate onto the fresh LB/amp and LB/kan plates.
   a. Use a sterile toothpick (or inoculating loop) to scrape up a cell mass from a well-defined colony on the L LB/amp plate.
   b. Immediately drag the *same* toothpick (or loop) gently across the agar surface to make a short diagonal (/) streak *within Square 1* of the LB/amp plate.
   c. Immediately use the *same* toothpick (or loop) to make a diagonal (/) streak *within Square 1* of the LB/kan plate.
   d. Discard the toothpick in a waste beaker (or reflame and cool inoculating loop).
3. Repeat Step 2a–d *using fresh toothpicks* (or *flamed* and *cooled* inoculating loop) to streak cells from 11 *different* L LB/amp colonies onto Squares 2–12 of both LB/amp and LB/kan plates.

If you have fewer than 12 colonies on either plate, obtain a plate from another experimenter.

4. Repeat Step 2a–d *with fresh toothpicks* (or a *flamed* and *cooled* inoculating loop) to streak cells from 12 *different* L LB/kan colonies onto Squares 13–24 of both LB/amp and LB/kan plates.
5. Place plates upside down in a 37°C incubator, and incubate for 15–20 hours.
6. After initial incubation, store plates at 4°C to arrest *E. coli* growth and to slow the growth of any contaminating microbes.

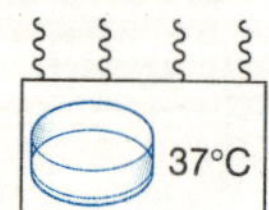

7. Take time for responsible cleanup:

Save replica plates as a source of cells from which to isolate plasmid DNA in Laboratory 12 if your L LB/amp+kan plate has fewer than two colonies.

a. Segregate for proper disposal bacterial cultures *and* used toothpicks.

b. Disinfect plates and toothpicks with 10% bleach or disinfectant.

c. Wipe down lab bench with soapy water, 10% bleach solution, or disinfectant (such as Lysol).

d. Wash hands before leaving lab.

RESULTS AND DISCUSSION

In general, the results of replica plating indicate the success of the ligation in Laboratory 9 and parallel the results observed on the L LB/amp+kan plate from Laboratory 10. Thus, if there were a large number of colonies on the LB/amp+kan plate, it is likely that there will be a high percentage of dual-resistant colonies that grow on both the LB/amp and LB/kan replica plates. Experience indicates that 30–70% of transformants selected with only ampicillin *or* kanamycin actually have dual resistance. Roughly equal numbers of dual-resistant colonies are identified from the L LB/amp and L LB/kan plates.

1. Observe the LB/amp and LB/kan plates. Use the matrix below to record as + the squares in which new bacterial growth has expanded the width of the initial streak. Record as – the squares in which no new growth has expanded the initial streak. Remember that nonresistant cells may survive, separated from the antibiotic on top of a heavy initial streak; however, no *new growth* will be observed.

Colony source	Replica plates		*Colony source*	Replica plates	
L LB/amp	LB/amp	LB/kan	*L LB/kan*	LB/amp	LB/kan
1			13		
2			14		
3			15		
4			16		
5			17		
6			18		
7			19		
8			20		
9			21		
10			22		
11			23		
12			24		

2. On the basis of your observations:

   a. Calculate the percentage of dual-resistant colonies taken from the L LB/amp plate (Squares 1–12).

   b. Calculate the percentage of dual-resistant colonies taken from the L LB/kan plate (Squares 13–24).

   c. Give an explanation for the similarity or difference in the percentages of dual-resistance colonies taken from the two source plates.

3. Draw restriction maps for different plasmid molecules that could be responsible for the dual resistance phenotype.

FOR FURTHER RESEARCH

1. We have discussed the phenomenon of satellite colonies on LB/amp plates—small nonresistant colonies that grow in a halo around large resistant colonies. To prove this, replica plate satellite colonies from an LB/amp plate onto fresh LB and LB/amp plates.
2. The Amp^r protein, β-lactamase, is not actively secreted into the medium, but it is believed to "leak" through the cell wall of *E. coli*. Satellite colonies do not form on kanamycin plates because the antibiotic kills all nonresistant cells outright. The following experiment tests whether resistance protein escapes from ampicillin- and kanamycin-resistant cells.
   a. Grow separate overnight cultures of an *amp*^{r} colony from the A LB/amp plate and a *kan*^{r} colony from the K LB/kan plate (from Laboratory 10, or use other *amp*^{r} and *kan*^{r} strains). Inoculate 5 ml of *plain* LB broth, according to the protocol in Laboratory 2B, Overnight Suspension Culture.
   b. Pass each overnight culture through a 0.22-μm or 0.45-μm filter, and collect the filtrate in a clean, sterile 15-ml tube. Filtering removes all *E. coli* cells.
   c. Use a permanent marker to mark one LB/amp plate and one LB/kan plate. Draw a line on the plate bottom to divide each plate into two equal parts; mark one half +.
   d. Sterilely spread 100 μl of the A filtrate onto the + *half only* of the LB/amp plate. Sterilely spread 100 μl of the K filtrate onto the + *half only* of the LB/kan plate. Allow filtrates to soak into plates for 10–15 minutes.
   e. Sterilely streak wild-type (nontransformed) *E. coli* cells on each filtrate-treated plate, taking care to streak back and forth across the dividing line.
   f. Incubate plates for 15–20 hours at 37°C. Compare growth on the treated sides *versus* untreated sides of each plate.
3. The Amp^r protein is believed to leak primarily from stationary-stage cells. The following experiment tests the hypothesis that leakage of β-lactamase is growth-phase-dependent.
   a. Grow an overnight culture of an ampicillin-resistant colony from the A LB/amp plate in Laboratory 10 (or other *amp*^{r} strain). Inoculate 1 ml of *plain* LB broth, according to the protocol in Laboratory 2B, Overnight Suspension Culture.
   b. Use an overnight culture to inoculate 100 ml of fresh LB broth, and grow according to the protocol in Laboratory 2C, Mid-log Suspension Culture.
   c. Sterilely withdraw 10-ml aliquots from the culture after 1, 2, and 4 hours, holding aliquots on ice.
   d. Take the OD_{550} of each aliquot.

NOTE The objective is to test resistance protein "leakage" as a function of culture age, *not as a function of cell number.* Because cell number increases over time, it must be equalized by diluting the 2-hour and 4-hour samples with sterile LB to the *E. coli* concentration of the 1-hour sample. Since the OD_{550} values are proportional to the cell number, they can be used to compute the dilution factor.

e. Pass each of the three samples through a 0.22-µm or 0.45-µm filter to remove the bacteria.

f. Prepare a 10-fold and 100-fold dilution for each filtrate, using sterile LB broth.

g. Use a permanent marker to draw a line dividing each of six LB/amp plates into equal parts.

h. Sterilely spread 100 µl of undiluted filtrate from the 1-hour sample over half of the first plate. Label the plate with time point and dilution factor. Allow filtrate to soak into the plate for 10–15 minutes.

i. Spread 100 µl of undiluted 2-hour and 4-hour filtrates over separate halves of the second plate, as described above.

j. Repeat spreading procedure for the 10-fold and 100-fold dilutions, as described in Steps h and i.

k. After filtrates have soaked into the plates, sterilely streak wild-type (non-transformed) *E. coli* cells on each half of each plate.

l. Incubate the plates for 15–20 hours at 37°C. Compare growth for each time point across each dilution.

Purification and Identification of Recombinant DNA

GROWTH OF *E. COLI* COLONIES ON THE L LB/AMP+KAN PLATE in Laboratory 10 confirms that they have been transformed to a dual resistance (Amp^r/Kan^r) phenotype. This resistance is expressed by one or more replicating plasmids, which were assembled in Laboratory 9 by ligating four *Bam*HI/*Hin*dIII restriction fragments of the parental plasmids pAMP and pKAN:

- a 784-bp pAMP fragment
- a 3755-bp pAMP fragment containing an origin of replication and an *amp*^{r} gene
- a 1861-bp pKAN fragment containing a *kan*^{r} gene
- a 2332-bp pKAN fragment containing an origin of replication

The goal of Laboratory 12 is to determine the genotype responsible for dual resistance; that is, the number and probable arrangement of any two or more pAMP and pKAN fragments. This laboratory is divided into two parts: Plasmid Minipreparation of pAMP/pKAN Recombinants and Restriction Analysis of Purified Recombinant DNA.

- Part A provides a procedure to isolate plasmid DNA from overnight cultures of two different colonies from an L LB/amp+kan plate (Laboratory 10) or from replica plates (Laboratory 11).
- Part B provides a procedure to incubate samples of the plasmids isolated in Part A and a control sample of pAMP+pKAN with *Bam*HI and *Hin*dIII. The three digested samples and samples of uncut minipreps are coelectrophoresed in an agarose gel, along with uncut pAMP and λ/*Hin*dIII size markers. The comigration of *Bam*/*Hin*d fragments in the lanes of miniprep DNA and pAMP/pKAN controls, along with an evaluation of the relative sizes of uncut supercoiled DNAs, gives evidence of the structure, size, and number of plasmids present in each of the transformed strains.

Equipment and materials for this laboratory are available from the Carolina Biological Supply Company (see Appendix 1).

PART A

Plasmid Minipreparation of pAMP/pKAN Recombinants

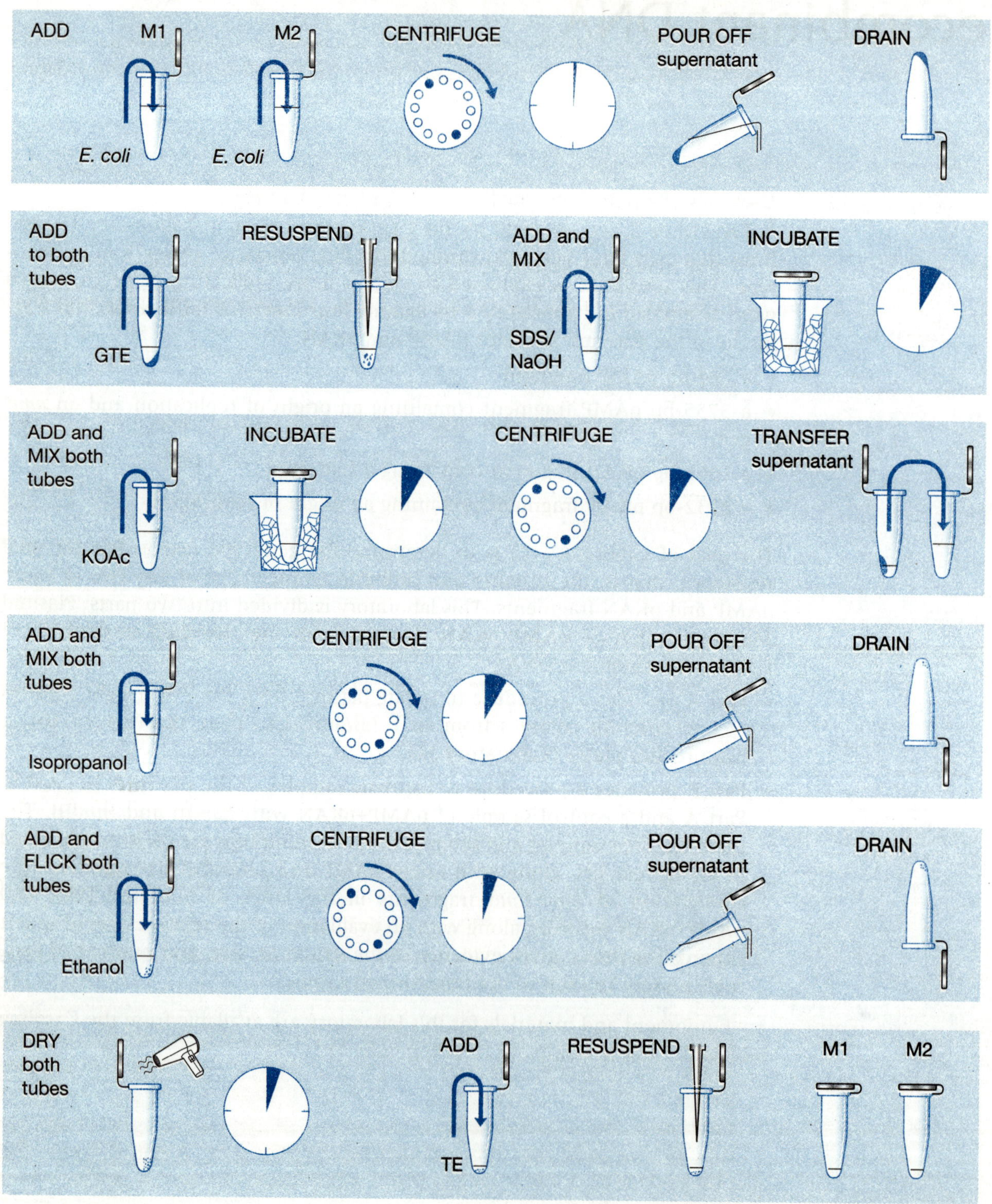

PRELAB NOTES

Review Prelab Notes in Laboratory 8A, Plasmid Minipreparation of pAMP.

Antibiotic Selection

Although plain LB broth may be used, it is safest to maintain antibiotic selection when culturing overnight cultures of the *E. coli* transformed with recombinant plasmid. Single selection in LB/amp is sufficient.

Alternate Sources of Dual-resistant Colonies

If colonies were not obtained on the L LB/amp+kan plate:

1. Use dual-resistant colonies identified by replica plating in Laboratory 11. These may be picked from either the LB/amp or LB/kan replica plate.
2. If there is not enough time to replica plate, prepare overnight cultures from five or more different colonies. The following selection strategies allow only dual-resistant colonies to grow; there should be at least two cloudy cultures in the morning.
   a. Inoculate LB/amp with colonies from the L LB/kan plate.
   b. Inoculate LB/kan with colonies from the L LB/amp plate.
   c. Inoculate LB/amp+kan with colonies from either the L LB/amp plate or L LB/kan plate.

For Further Information

The protocol presented here is based on the following published methods:

Birnboim H.C. and Doly J. 1979. A rapid alkaline extraction method for screening recombinant plasmid DNA. *Nucleic Acids Res.* **7:** 1513–1523.

Ish-Horowicz D. and Burke J.F. 1981. Rapid and efficient cosmid cloning. *Nucleic Acids Res.* **9:** 2989–2988.

PRELAB PREPARATION

Before performing this Prelab Preparation, please refer to the cautions indicated on the Laboratory Materials list.

1. The day before the laboratory, prepare two *E. coli* cultures according to the protocol in Laboratory 2B, Overnight Suspension Culture. Inoculate each overnight culture with a cell mass scraped from a different colony on the L LB/amp+kan plate from Laboratory 10. Maintain antibiotic selection with LB broth plus ampicillin. Alternatively, prepare the culture 2–3 days in advance and store at 4°C or incubate at 37°C without shaking for 24–48 hours. In either case, the cells will settle at the bottom of the culture tube. Shake the tube to resuspend cells before beginning the procedure. If colonies were not obtained on the L LB/amp+kan plate, see Alternate Sources of Dual-resistant Colonies in the Prelab Notes.

2. Circle colonies on the plate used to inoculate each overnight culture. Label one colony and its overnight M1 (miniprep 1). Label the other colony and its overnight M2 (miniprep 2). This will allow you to refer back to the original colony.
3. Prepare SDS/sodium hydroxide solution within a few days of the lab. Store at room temperature; a soapy precipitate may form at lower temperature. Warm solution by placing the tube in a beaker of hot tap water, and shake gently to dissolve the precipitate.
4. Prepare aliquots for each experiment:

   250 μl of glucose/Tris/EDTA (GTE) solution (store on ice)
   500 μl of SDS/sodium hydroxide (SDS/NaOH) solution
   400 μl of potassium acetate/acetic acid (KOAc) solution (store on ice)
   1000 μl of isopropanol
   500 μl of 100% ethanol
   500 μl of Tris/EDTA (TE) solution

5. Review Part B, Restriction Analysis of Purified DNA Recombinant DNA.

MATERIALS

REAGENTS	SUPPLIES AND EQUIPMENT
E. coli pAMP/pKAN overnight cultures (2)	Beakers for crushed ice and for waste/used tips
Ethanol▼ (95% and 100%)	Bleach▼(10%) or disinfectant
Glucose/Tris▼/EDTA (GTE)	Clean paper towels
Isopropanol▼	Hair dryer
Potassium acetate/acetic acid▼ (KOAc)	Microfuge
SDS/sodium hydroxide (SDS/NaOH)▼	Micropipettors (0.5–10 μl and 100–1000 μl) + tips
Tris▼/EDTA (TE)	Permanent marker
	Test tube rack
	Tubes (1.5-ml)

▼ See Appendix 4 for Cautions list.

METHODS

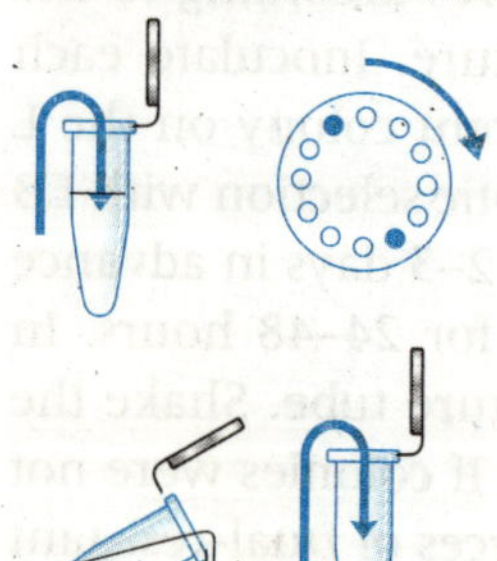

Perform Plasmid Miniprep (50 minutes)

1. Shake culture tubes to resuspend *E. coli* cells.
2. Label two 1.5-ml tubes with your initials. Label one tube M1, and label the other tube M2. Use a micropipettor to transfer 1000 μl of overnight suspension M1 and 1000 μl of overnight suspension M2 into appropriate tubes.
3. Close caps, and place the tubes in a *balanced* configuration in the microfuge rotor. Spin for 1 minute to pellet cells.

The cell pellet will appear as small off-white smear on bottom-side of tube. Although cell pellets are readily seen, DNA pellets in Step 14 are very difficult to observe. Make a habit of aligning tube with cap hinges facing outward in the microfuge rotor. Then, pellets should always be located at tube bottom beneath hinge.

Accurate pipetting is essential to good plasmid yield. Volumes of reagents are precisely calibrated so that sodium hydroxide added in Step 6 is neutralized by acetic acid in Step 8.

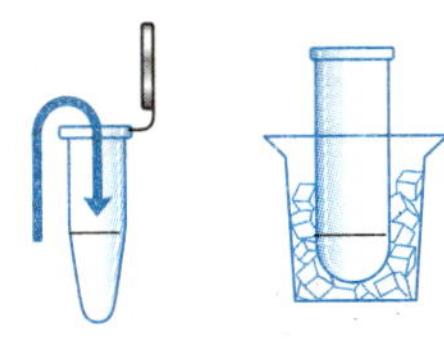

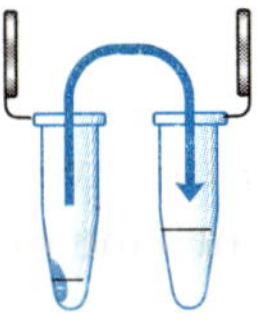

In Step 11, the supernatant is saved and precipitate discarded. The situation is reversed in Steps 14 and 17, where the precipitate is saved and supernatant discarded.

Do Step 12 quickly, and make sure that the microfuge will be immediately available for Step 13.

The pellet may appear as a tiny smear or small particles on the bottom-side of each tube. Do not be concerned if the pellet is not visible; pellet size is not a predictor of plasmid yield. A large pellet is composed primarily of RNA and cellular debris carried over from the original precipitate. A smaller pellet often means a cleaner preparation.

Nucleic acid pellets are not soluble in ethanol and will not resuspend during washing.

4. Pour off the supernatant from both tubes into a waste beaker for later disinfection. Alternatively, use a micropipettor to remove supernatant. *Do not disturb the cell pellets.* Invert the tubes, and tap gently on the surface of clean paper towel to drain thoroughly.
5. Add 100 μl of ice-cold GTE solution to each tube. Resuspend the pellets by pipetting the solution in and out several times. Hold the tubes up to the light to check that the suspension is homogeneous and no visible clumps of cells remain.
6. Add 200 μl of SDS/NaOH solution to each tube. Close caps, and mix the solutions by rapidly inverting the tubes five times.
7. Stand the tubes on ice for 5 minutes. The suspension will become relatively clear.
8. Add 150 μl of *ice-cold* KOAc solution to each tube. Close caps, and mix the solutions by rapidly inverting the tubes five times. A white precipitate will immediately appear.
9. Stand the tubes on ice for 5 minutes.
10. Place the tubes in a *balanced* configuration in the microfuge rotor. Spin the tubes for 5 minutes to pellet the precipitate along the side of the tube.
11. Transfer 400 μl of supernatant from M1 into a clean 1.5-ml tube labeled M1. Transfer 400 μl of supernatant from M2 into a clean 1.5-ml tube labeled M2. *Avoid pipetting precipitate,* and wipe off any precipitate clinging to the outside of tip prior to expelling supernatant. Discard old tubes containing precipitate.
12. Add 400 μl of isopropanol to each tube of supernatant. Close caps, and mix the supernatant vigorously by rapidly inverting the tubes five times. *Stand the tubes at room temperature for 2 minutes only.* (Isopropanol preferentially precipitates nucleic acids rapidly; however, proteins remaining in solution also begin to precipitate with time.)
13. Place the tubes in a *balanced* configuration in the microfuge rotor. Spin the tubes for 5 minutes to pellet the nucleic acids. Align tubes in the rotor so that the cap hingcs point outward. The nucleic acid residue, visible or not, will collect under the hinge during centrifugation.
14. Pour off supernatant from both tubes. *Do not disturb the nucleic acid pellets.* Alternatively, remove the supernatant with a 1000-ml micropipettor. Place the tip away from the pellet. If the pellet has been drawn up in the tip, transfer the supernatant to another 1.5-ml tube, recentrifuge, and remove the supernatant again. Invert tubes, and tap gently on the surface of a clean paper towel to drain thoroughly.
15. Add 200 μl of 100% ethanol to each tube, and close caps. Flick tubes several times to wash pellets.

STOP Store DNA/ethanol solution at –20°C until ready to continue.

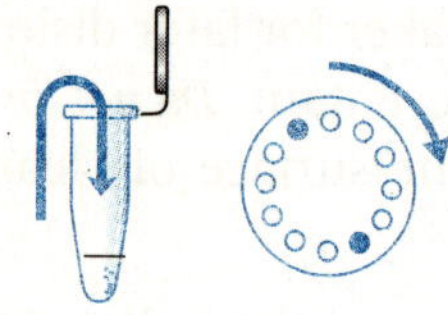

16. Place the tubes in a *balanced* configuration in the microfuge rotor, and spin them for 2–3 minutes.

17. Pour off supernatant from both tubes. *Do not disturb the nucleic acid pellets.* Aternatively, remove the supernatant with a 1000-ml micropipettor. Place the tip away from the pellet. If the pellet has been drawn up in the tip, transfer the supernatant to another 1.5-ml tube, recentrifuge, and remove the supernatant again. Invert tubes, and tap gently on the surface of a clean paper towel to drain thoroughly.

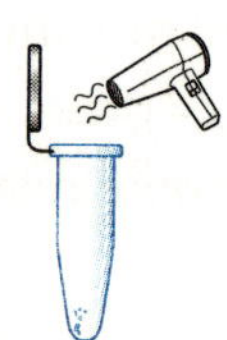

18. Dry nucleic acid pellets by one of following methods:

    a. Direct a stream of warm air from a hair dryer across the open ends of the tube for about 3 minutes. *Do not blow the pellets out of the tubes.*

    or

    b. Close caps, and pulse the tubes in a microfuge to pool remaining ethanol. *Carefully* draw off drops of ethanol using a 1–10-μl micropipettor. Leave cap open, and place the tube upright in the rack, allowing the pellets to air-dry for 10 minutes at room temperature.

19. All ethanol must be evaporated before proceeding to Step 20. Hold each tube up to the light to check that no ethanol droplets remain. If ethanol is still evaporating, an alcohol odor can be detected by sniffing the mouth of the tube.

If using a 0.5–10-μl micropipettor, set to 7.5 μl and pipette twice.

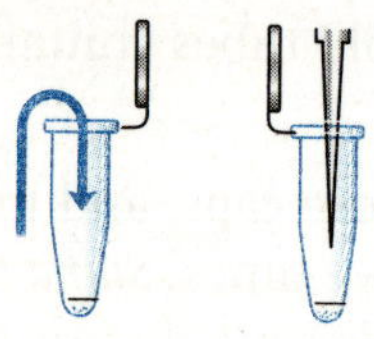

20. Add 15 μl of TE to each tube. Resuspend the pellets by smashing with the pipette tip and pipetting in and out vigorously. Rinse down the side of tube several times, concentrating on the area where the pellet should have formed during centrifugation (beneath the cap hinge). Check that all DNA is dissolved and that no particles remain in the tip or on the side of the tube.

21. Keep the two DNA/TE solutions *separate. DO NOT* pool into one tube.

Freeze DNA/TE solution at –20°C until ready to continue. Thaw before using.

22. Take time for responsible cleanup.

    a. Segregate for proper disposal culture tubes and micropipettor tips that have come in contact with *E. coli*.

    b. Disinfect overnight culture, tips, and supernatant from Step 4 with 10% bleach or disinfectant.

    c. Wipe down lab bench with soapy water, 10% bleach solution, or disinfectant (such as Lysol).

    d. Wash hands before leaving lab.

PART B

Restriction Analysis of Purified Recombinant DNA

I. Set Up Restriction Digest

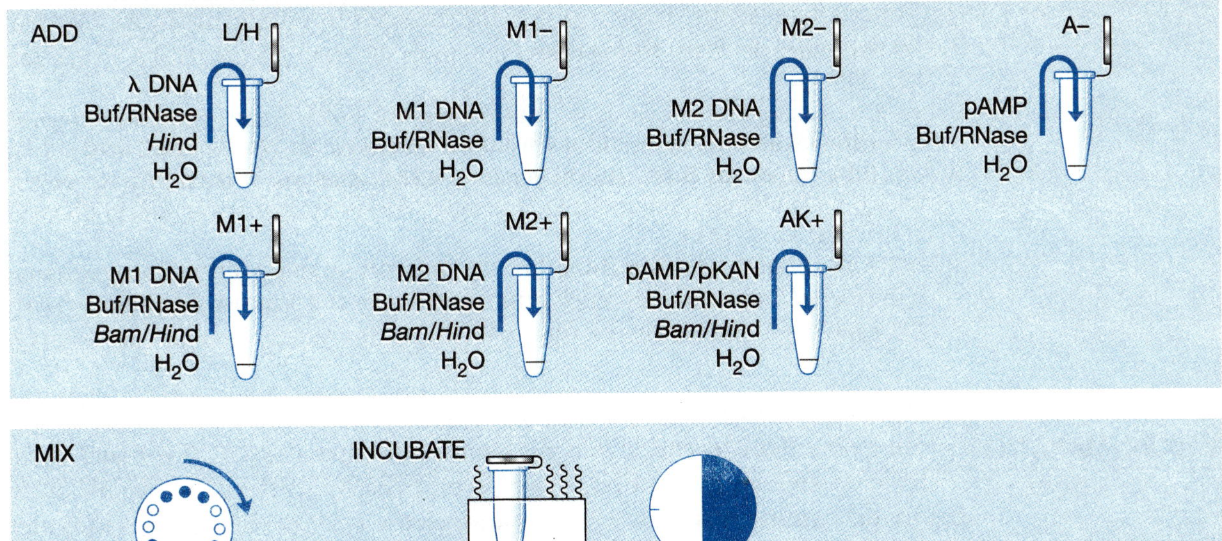

II. Cast 0.8% Agarose Gel

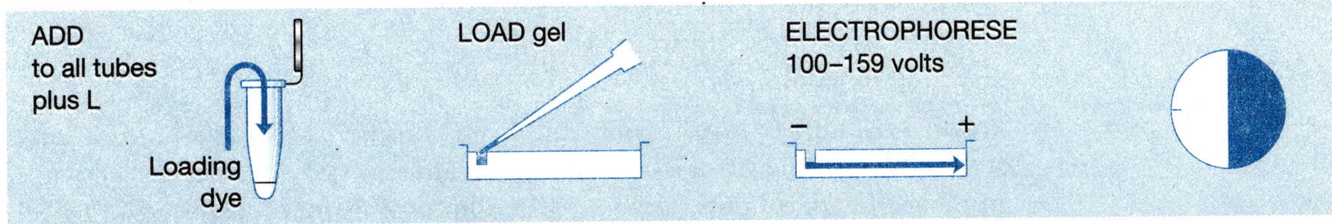

III. Load Gel and Separate by Electrophoresis

IV. Stain Gel and View (Photograph)

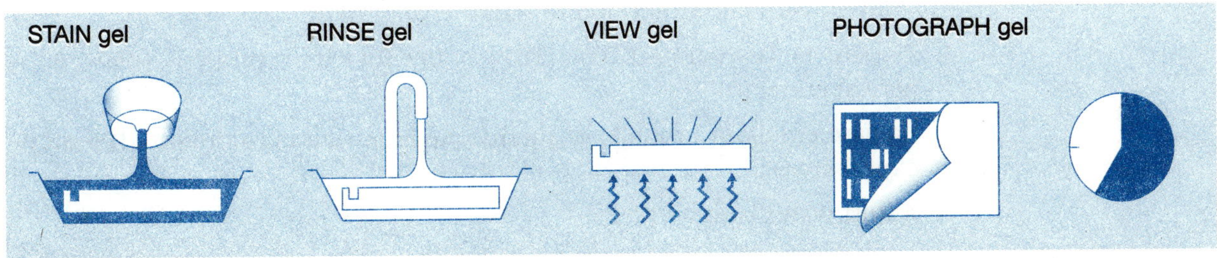

PRELAB NOTES

Review Prelab Notes in Laboratory 8B, Restriction Analysis of Purified pAMP.

For Further Information

The protocol presented here is based on the following published methods:

Aaij C. and Borst P. 1972. The gel electrophoresis of DNA. *Biochim. Biophys. Acta* **269:** 192–200.

Cohen S.N., Chang A.C., Boyer H.W., and Helling R.B. 1973. Construction of biologically functional bacteria plasmids in vitro. *Proc. Natl. Acad. Sci.* **70:** 3240–3244.

Helling R.B., Goodman H.M., and Boyer H.W. 1974. Analysis of R-*Eco*RI fragments of DNA from lambdoid bacteriophages and other viruses by agarose-gel electrophoresis. *J. Virol.* **14:** 1235–1244.

Sharp P.A., Sugden B., and Sambrook J. 1973. Detection of two restriction endonuclease activities in *Haemophilus parainfluenzae* using analytical agarose-ethidium bromide electrophoresis. *Biochemistry* **12:** 3055–3063.

PRELAB PREPARATION

Before performing this Prelab Preparation, please refer to the cautions indicated on the Laboratory Materials list.

1. Mix in 1:1 proportion:

   *Bam*HI + *Hin*dIII (8 µl per experiment)

   pAMP + pKAN (6 µl per experiment)

2. Prepare aliquots for each experiment:

   6 µl of 0.1 µg/µl pAMP/pKAN (store on ice)
   6 µl of 0.1 µg/µl pAMP (store on ice)
   6 µl of 0.1 µg/µl λ DNA (store on ice)
   16 µl of 5x restriction buffer/RNase (store on ice)
   8 µl of *Bam*HI/*Hin*dIII (store on ice)
   2 µl of *Hin*dIII (store on ice)
   500 µl of deionized/distilled water
   500 µl of loading dye

   Precut λ/*Hin*dIII is readily available from commercial suppliers, or a large-scale λ/*Hin*dIII digest can be done in advance to provide size markers for a number of experiments. Omit *Hin*dIII aliquot if using predigested λ DNA.

3. Prepare 0.8% agarose solution (~40–50 ml per experiment). Keep agarose liquid in a hot-water bath (at ~60°C) throughout the experiment. Cover the solution with aluminum foil to retard evaporation.
4. Prepare 1x Tris/borate/EDTA (TBE) buffer for electrophoresis (400–500 ml per experiment).
5. Prepare ethidium bromide or methylene blue staining solution (100 ml per experiment).
6. Adjust water bath to 37°C.

MATERIALS

REAGENTS

Agarose (0.8%)
*Bam*HI/*Hin*dIII
Distilled water
Ethidium bromide▼(1 μg/ml) (or 0.025% methylene blue▼)
*Hin*dIII
λ DNA (0.1 μg/μl)
Loading dye
Miniprep DNA/TE (M1, M2)
pAMP (0.1 μg/μl)
pAMP/pKAN (0.1 μg/μl)
5x Restriction buffer/RNase
1x Tris▼/Borate/EDTA (TBE) buffer

SUPPLIES AND EQUIPMENT

Aluminum foil
Beakers for agarose and for waste/used tips
Camera (optional)
Electrophoresis box
Latex gloves
Masking tape
Microfuge (optional)
Micropipettor (0.5–10 μl) + tips
Parafilm or wax paper (optional)
Permanent marker
Plastic wrap (optional)
Power supply
Test tube rack
Transilluminator (optional)▼
Tubes (1.5-ml)
Water baths (37°C and 60°C)

▼ See Appendix 4 for Cautions list.

METHODS

I. Set Up Restriction Digest
(40 minutes, including incubation)

Refer to Laboratory 3, DNA Restriction Analysis, for more detailed instructions.

Skip L/H tube if using predigested λ/HindIII markers.

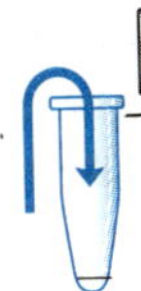

Return unused miniprep DNA (M1 and M2) to freezer at –20°C for possible use in further experiments suggested at the end of this laboratory.

1. Use a permanent marker to label seven 1.5-ml tubes, in which restriction reactions will be performed:

   L/H = λ DNA, *Hin*dIII
   M1– = miniprep 1, no enzyme
   M2– = miniprep 2, no enzyme
   A– = pAMP, no enzyme
   M1+ = miniprep 1, *Bam*HI/*Hin*dIII
   M2+ = miniprep 2, *Bam*HI/*Hin*dIII
   AK+ = pAMP/pKAN, *Bam*HI/*Hin*dIII

2. Use the matrix below as a checklist while adding reagents to each reaction. Read down each column, adding the same reagent to all appropriate tubes. *Use a fresh tip for each reagent.* Refer to detailed directions that follow.

Tube	λ DNA	M1	M2	pAMP	pAMP/ pKAN	Buffer/ RNase	*Hin*dIII	*Bam*HI/ *Hin*dIII	H_2O
L/H	5 µl	–	–	–	–	2 µl	1 µl	–	2 µl
M1–	–	5 µl	–	–	–	2 µl	–	–	3 µl
M2–	–	–	5 µl	–	–	2 µl	–	–	3 µl
A–	–	–	–	5 µl	–	2 µl	–	–	3 µl
M1+	–	5 µl	–	–	–	2 µl	–	2 µl	1 µl
M2+	–	–	5 µl	–	–	2 µl	–	2 µl	1 µl
AK+	–	–	–	–	5 µl	2 µl	–	2 µl	1 µl

3. Collect the reagents, and then place them in a test tube rack on the lab bench (*Bam*HI/*Hin*dIII on ice).
4. Add 5 µl of λ DNA to the tube labeled L/H.
5. Use a *fresh tip* to add 5 µl of M1 DNA to the tubes labeled M1– and M1+.
6. Use a *fresh tip* to add 5 µl of M2 DNA to the tubes labeled M2– and M2+.
7. Use a *fresh tip* to add 5 µl of pAMP to the tube labeled A–.
8. Use a *fresh tip* to add 5 µl of pAMP/pKAN to the tube labeled AK+.
9. Use a *fresh tip* to add 2 µl of restriction buffer/RNase to a clean spot on each reaction tube.
10. Use a *fresh tip* to add 1 µl of of *Hin*dIII to the tube labeled L/H.
11. Use a *fresh tip* to add 2 µl of *Bam*HI/*Hin*dIII to the tubes labeled M1+, M2+, and AK+.
12. Use a *fresh tip* to add proper volumes of distilled water to each tube.
13. Close tube tops. Pool and mix reagents by pulsing in a microfuge or by sharply tapping the tube bottom on the lab bench.
14. Place reaction tubes in a 37°C water bath, and incubate for 30 minutes only.

Do not overincubate. During longer incubation, DNases in miniprep may degrade plasmid DNA.

Following incubation, freeze reactions at –20°C until ready to continue. Thaw reactions before continuing to Section III, Step 1.

II. Cast 0.8% Agarose Gel

(15 minutes)

1. Seal the ends of the gel-casting tray with tape, and insert a well-forming comb. Place gel-casting tray out of the way on the lab bench so that agarose poured in the next step can set undisturbed.
2. Carefully pour enough agarose solution into the casting tray to fill to a depth of about 5 mm. Gel should cover only about one-third the height of comb teeth. Use a pipette tip to move large bubbles or solid debris to the slides or end of the tray, while the gel is still liquid.
3. Gel will become cloudy as it solidies (~10 minutes). *Be careful not to move or jar casting tray while agarose is solidifying.* Touch corner of agarose *away* from the comb to test whether gel has solidified.
4. When agarose has set, unseal the ends of the casting tray. Place the tray on the platform of the gel box, so that the comb is at negative black electrode (cathode).

Too much buffer will channel the current over the top rather than through the gel, increasing the time required to separate DNA. TBE buffer can be used several times; do not discard. If using buffer remaining in electrophoresis box from a previous experiment, rock chamber back and forth to remix ions that have accumulated at either end.

5. Fill box with TBE buffer, to a level that just covers the surface of the gel.
6. Gently remove the comb, taking care not to rip the wells.
7. Make sure that sample wells left by the comb are completely submerged. If "dimples" appear around the wells, slowly add buffer until they disappear.

Cover the electrophoresis tank and save the gel until ready to continue. Gel will remain in good condition for at least several days if it is completely submerged in buffer.

III. Load Gel and Separate by Electrophoresis (30–50 minutes)

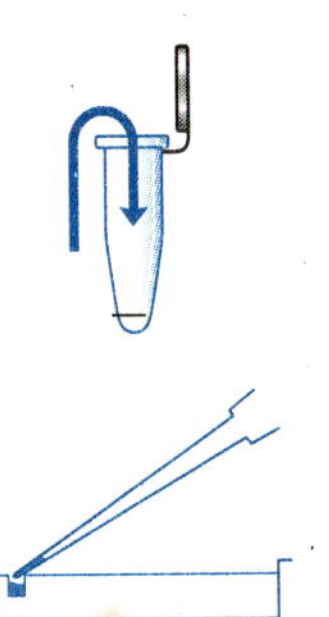

1. Add 1 µl of loading dye to each reaction tube. Close tube tops, and mix by tapping the tube bottom on the lab bench, pipetting in and out, or pulsing in a microfuge.
2. Load 10 µl of each reaction tube into a separate well in the gel, as shown in the diagram below. Use a *fresh tip* for each reaction. *Expel any air in the tip before loading, and be careful not to punch the tip of the micropipettor through the bottom of the gel.*

A piece of dark construction paper beneath the gel will make the wells more visible.

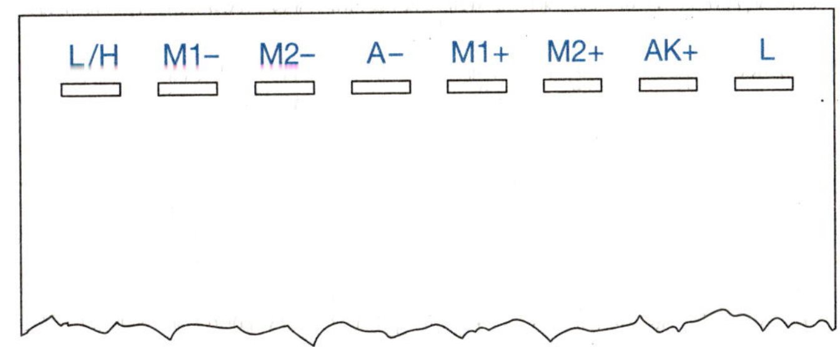

3. Add loading dye to ligated DNA saved from Laboratory 9, Recombination of Antibiotic Resistance Genes. Load entire contents of this L tube (5–10 µl) into well 8.

4. Electrophorese at 100–150 volts for 20–40 minutes. Good separation will have occurred when the bromophenol blue band has moved 4–6 cm from the wells.
   - If time allows, electrophorese until the bromophenol blue band nears the end of the gel. This will allow maximum separation of uncut DNA, which is important in differentiating a large "superplasmid" from a double transformation of two smaller plasmids.
   - Stop electrophoresis before the bromophenol blue band runs off the end of the gel or the 784-bp *Bam*HI/*Hin*dIII fragment of pAMP, which migrates just behind the bromophenol blue marker, may be lost.
5. Turn off the power supply, disconnect the leads from the inputs, and remove the top of the electrophoresis box.

6. Carefully remove the casting tray from electrophoresis box, and slide the gel into a disposable weigh boat or other shallow tray. Label the staining tray with your name.

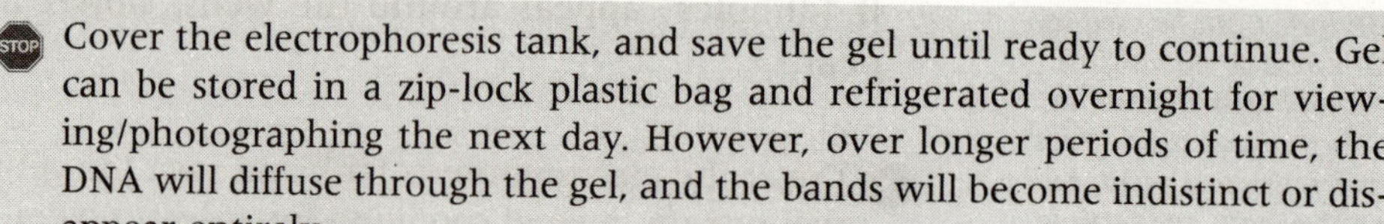

Cover the electrophoresis tank, and save the gel until ready to continue. Gel can be stored in a zip-lock plastic bag and refrigerated overnight for viewing/photographing the next day. However, over longer periods of time, the DNA will diffuse through the gel, and the bands will become indistinct or disappear entirely.

Staining may be performed by an instructor in a controlled area when students are not present.

7. Stain and view the gel using one of the methods described in Sections IVA and IVB.

IVA. Stain Gel with Ethidium Bromide and View (Photograph)

(10–15 minutes)

> CAUTION
> Review Responsible Handling of Ethidium Bromide in Laboratory 3. Wear latex gloves when staining, viewing, and photographing gel and during cleanup. Confine all staining to a restricted sink area. For further information, see Appendix 4.

Ethidium bromide solution may be reused to stain 15 or more gels. Dispose of spent staining solution as explained in Laboratory 3.

1. Flood gels with ethidium bromide solution (1 μg/ml), and allow to stain for 5–10 minutes.
2. Following staining, use a funnel to decant as much of the ethidium bromide solution as possible from the staining tray back into the storage container.
3. Rinse the gel and tray under running tap water.
4. If desired, destain the gel in tap water or distilled water for 5 minutes or more to help remove the background ethidium bromide.

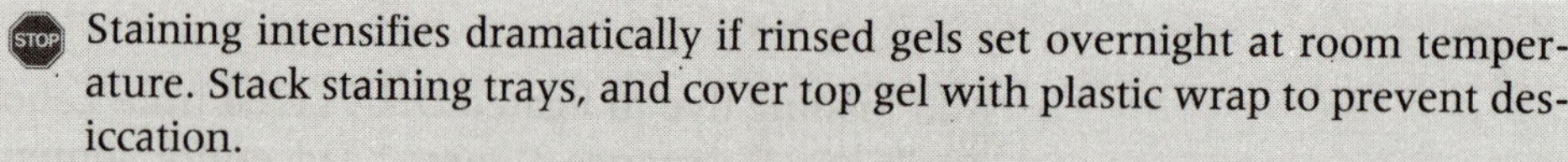

Staining intensifies dramatically if rinsed gels set overnight at room temperature. Stack staining trays, and cover top gel with plastic wrap to prevent desiccation.

5. View under a UV transilluminator or other UV source.

> CAUTION
> Ultraviolet light can damage eyes. Never look at unshielded UV light source with naked eyes. View only through a filter or safety glasses that absorb harmful wavelengths. For further information, see Appendix 4.

6. Photograph the gel with a Polaroid and digital camera.
7. Take time for responsible cleanup.
   a. Wipe down camera, transilluminator, and staining area.
   b. Wash hands before leaving lab.

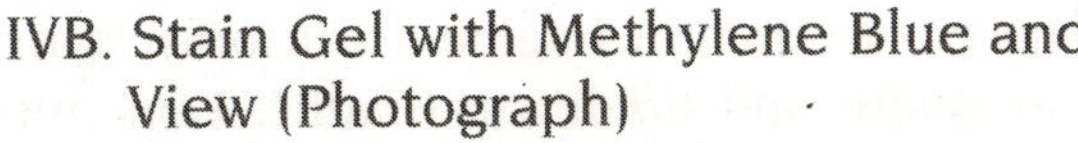

IVB. Stain Gel with Methylene Blue and View (Photograph)

(30+ minutes)

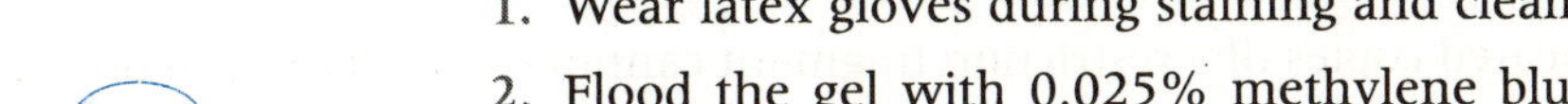

1. Wear latex gloves during staining and cleanup.

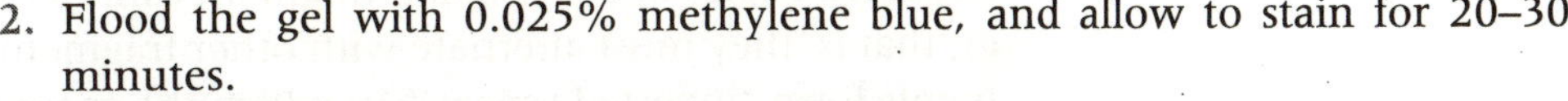

2. Flood the gel with 0.025% methylene blue, and allow to stain for 20–30 minutes.
3. Following staining, use a funnel to decant as much of the methylene blue solution as possible from the staining tray back into the storage containiner.

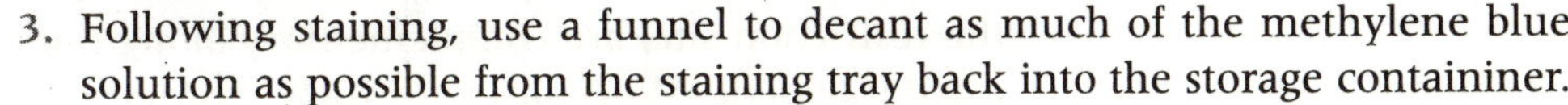

Destaining time is decreased by rinsing the gel in warm water, with agitation.

4. Rinse the gel in running tap water. Let the gel soak for several minutes in several changes of fresh water. DNA bands will become increasingly distinct as gel destains.

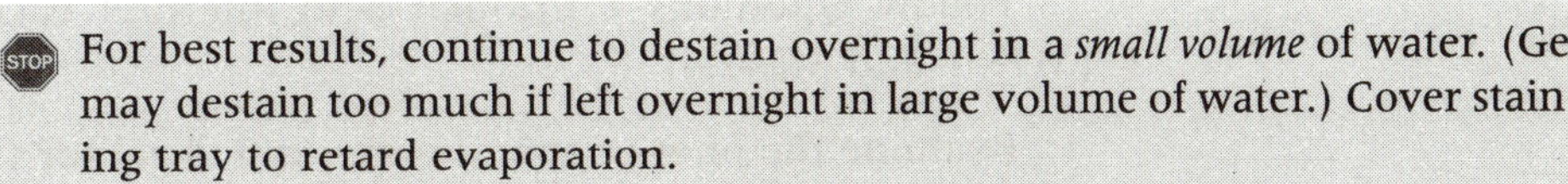

STOP For best results, continue to destain overnight in a *small volume* of water. (Gel may destain too much if left overnight in large volume of water.) Cover staining tray to retard evaporation.

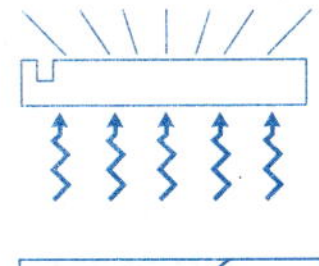

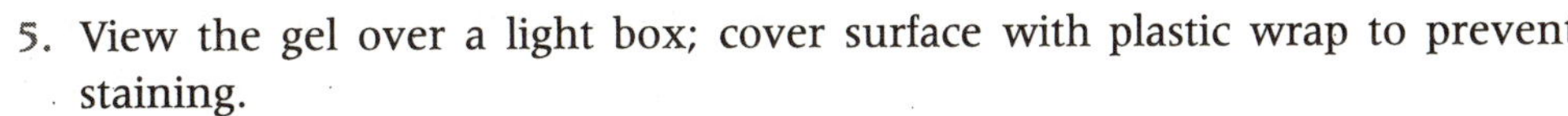

5. View the gel over a light box; cover surface with plastic wrap to prevent staining.

6. Photograph the gel with a Polaroid or digital camera.

RESULTS AND DISCUSSION

Observe your gel and determine which lanes contain control pAMP/pKAN and which lanes contain minipreps M1 and M2. Even if you have confused the prescribed loading order, the miniprep lanes can be distinguished by the following characteristics:

- a background "smear" of degraded and partially digested chromosomal DNA, plasmid DNA, and RNA
- undissolved material and high-molecular-weight DNA "trapped" at the front edge of the well
- a "cloud" of low-molecular-weight RNA at a position corresponding to 100–200 bp
- presence of high-molecular-weight bands of uncut plasmid in lanes of digested miniprep DNA

Refer to the Results and Discussion section of Laboratory 8B, Restriction Analysis of Purified pAMP, for more details about interpreting miniprep gels and plasmid conformations.

Remember these three facts when considering possible constructions of the ligated plasmids in M1 and M2.

1. Every replicating plasmid must have an origin of replication. Recombinant plasmids with more than one origin also replicate normally; however, only one origin is active.

2. Each adjacent restriction fragment can only ligate at a like restriction site: *Bam*HI to *Bam*HI and *Hin*dIII to *Hin*dIII. Thus, intact plasmid must be constructed of an *even* number of fragments (2, 4, 6, 8, etc.).
3. Repeated copies of a restriction fragment cannot exist adjacent to one another; that is, they must alternate with other fragments. Adjacent duplicate fragments form "inverted repeats" in which the sequences, one on either side of the restriction site, are complementary along the entire length of the duplicated fragment. Molecules with such inverted repeats cannot replicate properly. As the plasmid opens up to allow access to DNA polymerase, the single-strand regions on either side of the restriction site base pair to one another to form a large "hairpin loop," which fouls replication.

Follow Questions 1 through 8 to interpret each pair of miniprep results (M1+/– and M2+/–).

1. Examine the photograph of your stained gel (or view on a light box or overhead projector). Compare your gel with the ideal gel on the following page. Label the size of the fragments in each lane of your gel.
2. Label the fragment sizes of the four bands in the AK+ lane (cut control pAMP and pKAN) from top of gel to bottom: 3755 bp, 2332 bp, 1861 bp, and 784 bp. Every miniprep must contain the 3755-bp fragment containing the *amp*^{r} gene and the 1861-bp fragment containing the *kan*^{r} gene. Locate these bands by comparing the M+ lane (cut miniprep) with the AK+ lane (cut control).
3. Now look for evidence of any other bands in the M+ lane. Compare the M+ lane with the AK+ lane. The 2332-bp fragment and/or the 784-bp fragment may be present. If neither of these two additional bands is present, the molecule is termed a "simple recombinant."
4. If a third band of 784 bp is present, the molecule may be
   a. a "superplasmid" in which one of the three fragments is repeated
   or
   b. a double transformation of a simple recombinant *and* a religated pAMP.
5. If a third band of 2332 bp is present, the molecule may be
   a. a superplasmid in which one of the three fragments is repeated
   or
   b. a double transformation of the simple recombinant *and* religated pKAN
   or
   c. a double transformation of the simple recombinant *and* ligated 3755-bp + 2332-bp fragments
   or
   d. a double transformation of religated pKAN *and* ligated 3755-bp + 2332-bp fragments.
6. If all four bands are present, the molecule may be
   a. a superplasmid containing all four fragments
   or

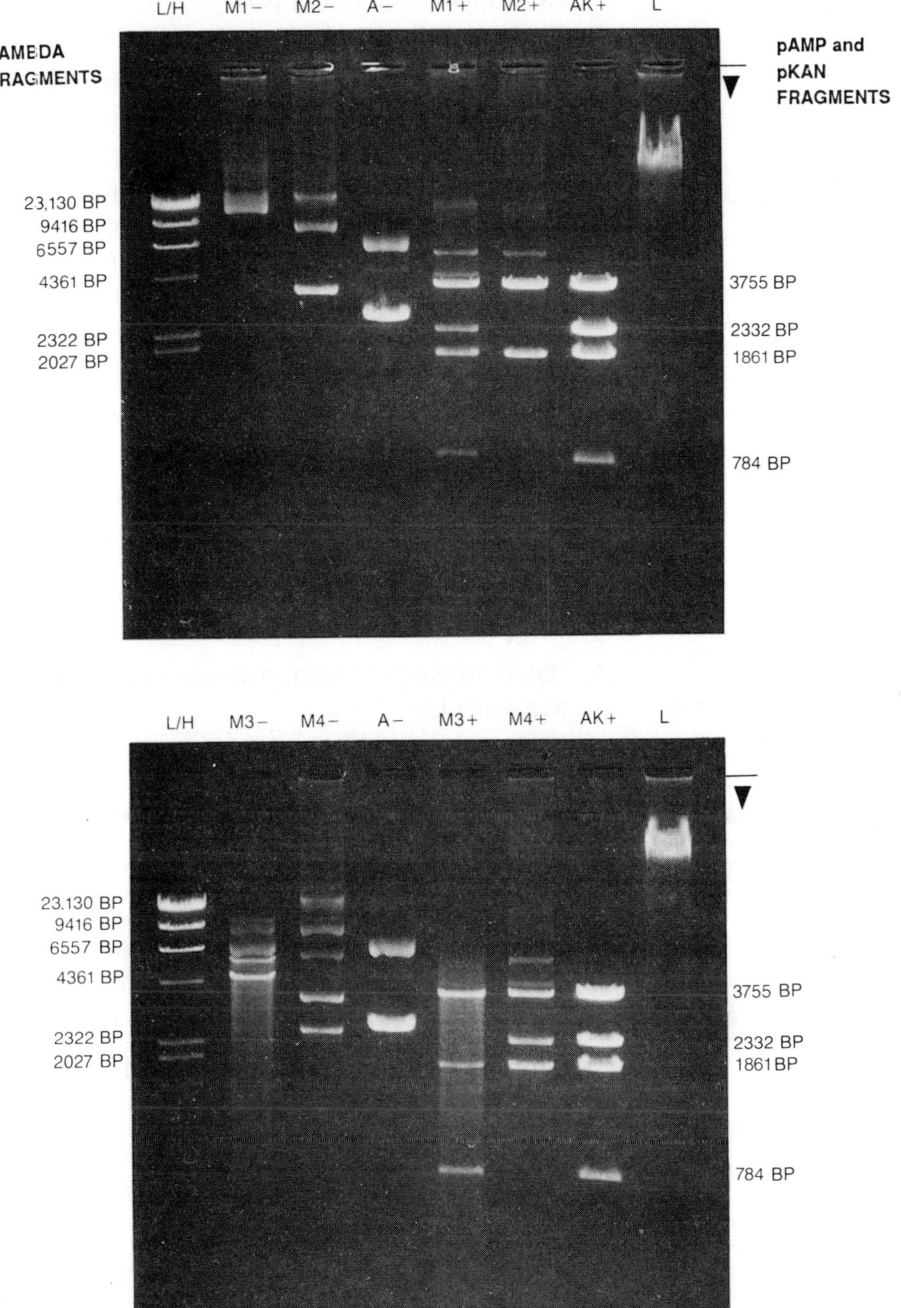

Restriction Analysis of Four pAMP/KAN Recombinants (M1, M2, M3, M4)

b. a double transformation of religated pAMP *and* religated pKAN

or

c. a double transformation of a simple recombinant *and* ligated 2332-bp + 784-bp fragments.

7. To gauge the size of the miniprep plasmid, compare the M– lane (uncut miniprep) with the A– lane (uncut pAMP) and the L/H lane (λ markers).

Remember that uncut plasmid can assume several conformations, but that the fastest-moving form is supercoiled.

a. Locate the band that has migrated furthest in the A– lane; this is the supercoiled form of pAMP.

b. Now examine the band(s) furthest down the M– lane. If this band and the pAMP band have comigrated similar distances, your miniprep is likely a double transformation. The possible molecules present in a double transformation range in size from 3116 bp (Step 6c above) to 6087 bp (Step 5c,d above), and thus may appear noticeably lower or higher on the gel than supercoiled pAMP.

c. If the fastest moving band of the uncut miniprep is very high on the gel, your molecule is likely a superplasmid. Compute the possible sizes of superplasmids composed of three or four fragments.

8. When bacteria are transformed with two different plasmids having related origins of replication, one of the two plasmids can be preferentially replicated within the host cell. Over generations, one of the two plasmids is eventually lost. Thus, in double transformations with four different fragments, one pair of fragments should be fainter than the other pair.

9. Based on your evaluation above, make scale restriction maps of your M1 and M2 plasmids.

FOR FURTHER RESEARCH

Further research may reveal with certainty the structure of perplexing recombinant plasmids. To obtain additional plasmid for further experimentation, carry out double minipreps from your master colonies (M1 and M2) in Laboratory 10 or from colonies of interest from replica plates in Experiment 1 below.

1. Perform a series of experiments to distinguish between a superplasmid and a double transformation in Questions 4, 5, and 6 in Results and Discussion.

   a. Make a 1:10 dilution of your miniprep DNA.

   b. Use the dilute miniprep DNA to transform competent *E. coli* cells, and plate onto LB/amp and LB/kan.

   c. Replica plate colonies from each master plate onto fresh LB/amp and LB/kan plates.

   d. Examine the proportion of dual-resistant colonies.

   - If all the restriction fragments are contained in a single superplasmid, all transformants will have dual resistance.
   - If three or four restriction fragments are distributed among separate plasmids, the transformants will have mixed antibiotic resistance. Matching the observed pattern of anitibiotic resistance with alternate two-gene recombinants can often reveal the structure of the two plasmids involved.

Digesting miniprep DNA with the restriction enzyme *Xho*I can elucidate some of the structures of superplasmids and plasmids in double transformations. This enzyme has a single recognition site within the 1861-bp pKAN fragment and *no sites* within any of the other three *Bam*HI/*Hin*dIII fragments. Carry out electrophoresis of *Xho*I digests of miniprep DNA with samples of uncut pAMP and λ/*Hin*dIII size markers.

2. If your miniprep DNA shows three fragments, including the 784-bp fragment (Question 4 in Results and Discussion), then you may have either a superplasmid with one repeated fragment or a double transformation.
   a. The results of a *Xho*I digest of a superplasmid will differ according to which fragment is repeated:
      - If the 784-bp fragment is repeated, a linear 7184-bp plasmid is produced.
      - If the 3755-bp fragment is repeated, a linear 10,155-bp plasmid is produced.
      - If the 1861-bp fragment containing the *Xho*I site is repeated, two fragments of 2645 bp and 5616 bp are produced.
   b. A double transformation of the simple recombinant and religated pAMP produces a linear 5616-bp plasmid *plus* an *uncut* pAMP plasmid.
3. If your miniprep DNA shows three bands, including the 2332-bp fragment (Question 5 in Results and Discussion), then you may have a superplasmid with one repeated fragment or a double transformation.
   a. The results of a *Xho*I digest of a superplasmid will differ according to which fragment is repeated:
      - If the 2332-bp fragment is repeated, a linear 10,280-bp plasmid is produced.
      - If the 3755-bp fragment is repeated, a linear 11,703-bp plasmid is produced.
      - If the 1861-bp fragment containing the *Xho*I site is repeated, two fragments of 4193 bp and 5616 bp are produced.
   b. A double transformation of the simple recombinant and religated pKAN produces two fragments of 5616 bp and 4193 bp.
   c. A double transformation of the simple recombinant and ligated 3755-bp + 2332-bp fragments produces a linear 5616-bp plasmid *plus* an *uncut* 6087-bp plasmid.
   d. A double transformation of religated pKAN and ligated 3755-bp + 2332-bp fragments produces a linear 4193-bp plasmid *plus* an *uncut* 6087-bp plasmid.
4. If your miniprep DNA contains all four fragments (Question 6 in Results and Discussion), the *Xho*I digest will discriminate between a superplasmid and double transformations.
   a. A superplasmid produces a linear 8732-bp plasmid.

b. A double transformation of religated pAMP and religated pKAN produces a linear 4193-bp pKAN plasmid *plus* an *uncut* pAMP plasmid.

c. A double transformation of the simple recombinant and ligated 2332-bp + 784-bp fragments produces a linear 5616-bp plasmid *plus* an *uncut* 3102-bp plasmid.

5. Make a restriction map of the simple recombinant plasmid using *Bam*HI, *Hin*dIII, and *Pvu*I. Prior experiments showed that the *Bam*HI and *Hin*dIII sites are separated by 1861 bp. *Pvu*I cuts the recombinant plasmid at two positions.

   a. Carry out double minipreps to obtain additional plasmid from a master colony known to contain the simple recombinant.

   b. Digest aliquots of the miniprep DNA with

      *Pvu*I
      *Pvu*I+*Bam*HI
      *Pvu*I+*Hin*dIII
      *Bam*HI+*Hin*dIII
      *Bam*HI+*Hin*dIII+*Pvu*I

   c. Separate the digested samples by electrophoresis on a 1.2% agarose gel, stain, and photograph.

   d. The expected number of fragments and their sizes are shown in the diagram below.

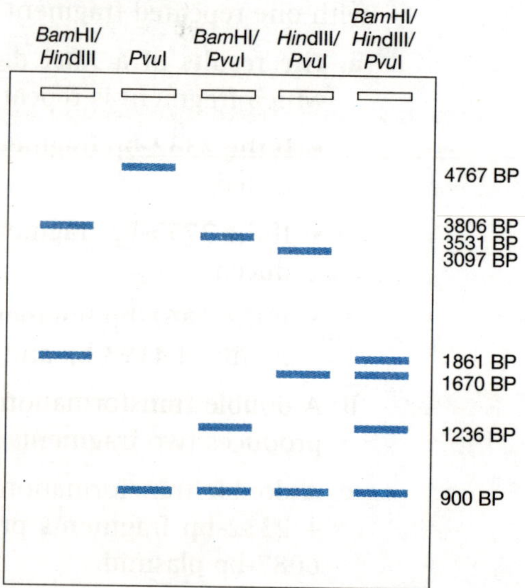

6. Using the data from Experiment 5 above and applying a little logic, the relative positions of the restriction sites can be positioned around a circle to produce a restriction map of the simple recombinant plasmid.

   a. The *Bam*HI/*Hin*dIII digest reveals that the *Bam*HI and *Hin*dIII sites are separated by 1861 bp.

   b. The *Pvu*I digest reveals that the two *Pvu*I sites are separated by 900 bp.

c. The *Bam*HI/*Hin*dIII/*Pvu*I digest shows both 1861-bp and 900-bp fragments. This means that the 1670-bp and 1236-bp fragments separate the 900-bp *Pvu*I fragment from the 1861-bp *Bam*HI/*Hin*dIII fragment.

d. The *Pvu*I/*Bam*HI digest shows a 3531-bp fragment that must be composed of the 1861-bp fragment plus the 1670-bp fragment.

e. The *Pvu*I/*Hin*dIII digest shows a 3097-bp fragment that must be composed of the 1861-bp fragment plus the 1236-bp fragment.

f. Results from Steps d and e indicate that the 1236-bp fragment is adjacent to the *Bam*HI site and the 1670-bp fragment is adjacent to the *Hin*dIII site.

g. Complete the restriction map showing all restriction sites and the distances between them.

Answers to Discussion Questions

LABORATORY 1: MEASUREMENTS, MICROPIPETTING, AND STERILE TECHNIQUES

1. **Why must tubes be balanced in a microfuge rotor?**

   An unbalanced microfuge can damage or ruin the motor.

2. **Use the rotor diagrams below to show how to balance 3–11 tubes (for the 12-place rotor) or 3–15 tubes (for the 16-place rotor).**

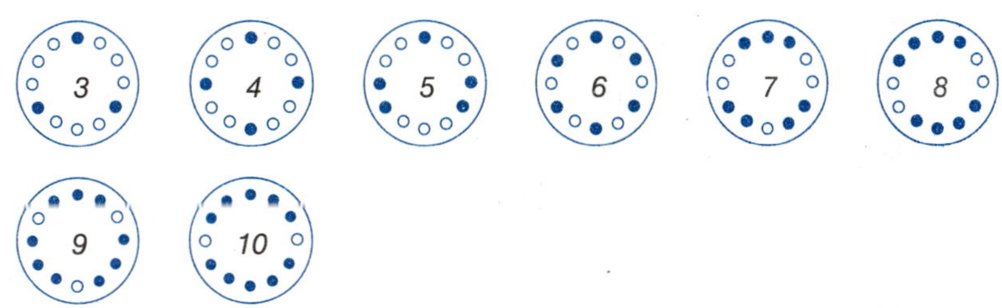

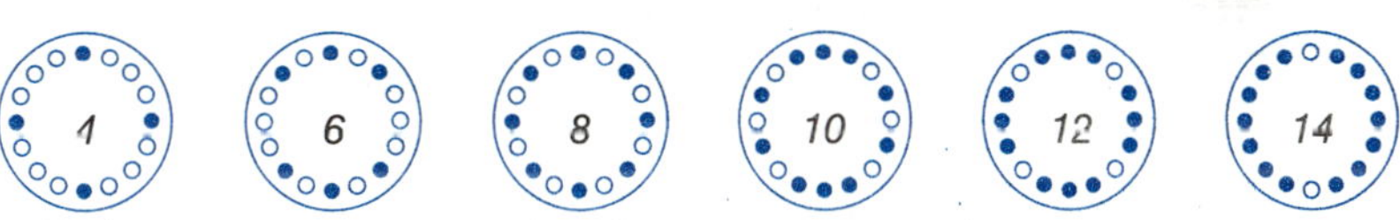

(Reprinted with permission of Pearson Education Inc.; ©1996 by The Benjamin/Cummings Publishing Company, Inc.)

   Which number of tubes cannot be balanced in the 12- or 16-place rotors?

   One tube or 11 tubes cannot be balanced in the 12-place rotor, whereas any odd number of tubes cannot be balanced in the 16-place rotor.

3. **What common error in handling a micropipettor can account for pipetting too much reagent into a tube? What errors account for underpipetting?**

   The most common cause of overpipetting is initially pushing past position one to position two, and withdrawing too much fluid. Underpipetting may be caused by releasing thumb pressure too quickly and drawing air into the micropipettor or by failing to completely dislodge the last drop from the micropipettor tip.

4. **When is it necessary to use sterile technique?**

   If you are working with bacteria that will still be living following your manipulations, then sterile technique should be observed.

5. **What does flaming accomplish?**

The chief purpose of flaming is not to heat-kill microorganisms on the mouth of the container, but to create outward-directed convection currents that help prevent microorganisms from falling in.

6. **Convert the following into microliters (μl): 0.130 ml, 0.002 ml, 0.025 ml, 1.034 ml**

   130 μl, 2 μl, 25 μl, 1034 μl

   Convert the following into milliliters (ml): 0.036 liter, 345 μl, 0.803 liter, 1345 μl

   36 ml, 0.345 ml, 803 ml, 1.345 ml

LABORATORY 2A: ISOLATION OF INDIVIDUAL COLONIES

1. **Were results as expected? Explain possible causes for variations from expected results.**

   If growth is observed where it is not expected (wild-type cells on LB/amp plate) or if growth is not observed where it is expected (transformed cells on LB/amp plate), then verify that the proper type of cells were spread.

2. **In Step 7:**

   a. **What is the reason for the zigzag streaking pattern?**

   The long zigzag streaking pattern ensures that the cells are adequately diluted and spread out on the plate surface.

   b. **Why is the inoculating loop resterilized between each new streak?**

   It is necessary to kill bacteria remaining on the inoculating loop following a streak. This ensures a progressive dilution of cells with each new streak.

   c. **Why should a new streak intersect the end of the previous one only at a single point?**

   To ensure that only a small number of cells are being spread in the new streak.

3. **Describe the appearance of a single *E. coli* colony. Why can it be considered genetically homogeneous?**

   An *E. coli* colony is round, with a slightly irregular border, and off-white or brownish-yellow in color. If the cells are spread properly, then each colony is assumed to be derived by mitotic division from a single cell. This means that each cell in the colony is a clone and therefore genetically identical to the progenitor cell and every other cell in the colony.

4. **Upcoming laboratories use cultures of *E. coli* cells derived from a single colony or from several discrete parental colonies isolated as described in this experiment. Why is it important to use this type of culture in genetic experiments?**

   By using cells that are genetically identical, it can be assumed that observed changes in phenotypes are due to experimental manipulation, rather than to genetic diversity in the starting culture.

5. ***E. coli* strains containing the plasmid pAMP are resistant to ampicillin. Describe how this plasmid functions to bring about resistance.**

The plasmid pAMP contains a gene that encodes the resistance protein, β-lactamase. This enzyme catalyzes cleavage of ampicillin's β-lactam ring, which renders the drug incapable of interfering with cell wall synthesis.

6. **A major medical problem is the ever-increasing number of bacterial strains that are resistant to specific antibiotics. Antibiotic resistance is carried on circular DNA molecules, called plasmids, which are generated separately from the cell's chromosome. With this in mind, suggest a mechanism through which new antibiotic-resistant strains of bacteria arise.**

   Nonresistant bacteria can take up (be transformed by) plasmids from resistant bacteria, even if the bacteria are different strains. The newly transformed bacteria then become resistant as well.

   Incompatibility of replication origins and promoters between species may sometimes hinder this process, but bacteria are incredibly adaptive. If a bacterium can take up the information for resistance, then it can probably rearrange it into an appropriate plasmid by recombination.

LABORATORY 2B: OVERNIGHT SUSPENSION CULTURE

1. **Why is 37°C the optimum temperature for *E. coli* growth?**

   E. coli is a constituent of the normal bacterial fauna that inhabits the human colon and has therefore adapted itself to best grow at the temperature found there.

2. **Give two reasons why it is ideal to provide continuous shaking for a suspension culture.**

   Shaking a suspension culture provides aeration and facilitates exchange of nutrients and flushing away of waste products.

3. **What growth phase is reached by a suspension of *E. coli* following overnight shaking at 37°C?**

   Following an overnight incubation, a suspension of *E. coli* will be stationary phase.

4. **Approximately how many *E. coli* cells are in a 5-ml suspension culture at stationary phase?**

   There are about 5 billion *E. coli* cells in a 5-ml suspension culture at stationary phase.

LABORATORY 2C: MID-LOG SUSPENSION CULTURE

1. **What variables influence the length of time for an *E. coli* culture to reach mid-log phase?**

   The time required for an *E. coli* culture to reach mid-log phase is influenced by the size of the initial incoculum, the volume of the flask, the shaking rate, and the temperature of the nutrient medium.

2. **What are the disadvantages of beginning a mid-log culture from a colony scraped off a plate, as opposed to an inoculum of overnight culture?**

   Cells scraped off a plate have a larger proportion of nonviable or dead cells. In addition, the scraped cells must be completely dispersed throughout the medium before beginning the mid-log culture.

LABORATORY 3: DNA RESTRICTION ANALYSIS

1. **Why is water added to tube labeled "–" in Part I, Step 7?**

   This tube is the unrestricted control and shows the appearance of uncut DNA. Water is added to the tube at a volume equivalent to that of the restriction enzyme, thereby ensuring that the salt concentration remains equal to that in the restricted samples.

2. **What is the function of compromise restriction buffer?**

   Restriction buffer maintains the pH in a range suitable for enzyme activity, as well as supplying salt cofactors required for catalysis. Since different restriction enzymes require varying salt conditions and pH, a single compromise buffer can be used that strikes a balance between conditions preferred by the various restriction enzymes.

3. **What are the two functions of loading dye?**

   Loading dye contains sucrose (or glycerol) that increases sample density, making it sink to the bottom of the sample well. The dyes bromophenol blue and xylene cyanol do not bind to DNA but are visible during electrophoresis. The migration of the visible dyes helps one to estimate the position of the invisible DNA fragments.

4. **How does ethidium bromide stain DNA? How does this relate to the need to minimize exposure to humans?**

   Ethidium bromide intercalates into the stacked regions of the double helix in a sequence-independent fashion. Since the presence of ethidium bromide bound to the DNA can interfere with accurate DNA replication, ethidium bromide is a mutagen and suspected carcinogen in humans.

5. **Troubleshooting electrophoresis. What will occur**

   a. **if the gel box is filled with water instead of TBE buffer?**

      Electrophoresis will proceed very slowly. When running at constant voltage, the higher resistance of water will reduce the current by a factor of 3 or 4.

   b. **if water is used to prepare the gel instead of TBE buffer?**

      Electrophoresis will proceed more or less normally. However, the DNA bands will have a pronounced smeared appearance.

   c. **if the electrodes are reversed?**

      DNA fragments will electrophorese in the opposite direction. In a short time, they migrate out of the gel and into the TBE buffer.

6. **Examine the photograph of your stained gel (or view on a light box or overhead projector). Compare your gel with ideal gel shown, and try to account for fragments of λ DNA in each lane. How can you account for differences in separation and band intensity between your gel and the ideal gel?**

   Differences in band separation are largely a function of the voltage, time of electrophoresis, and amount of buffer covering the gel. Increased voltage and/or running time will increase band separation, whereas increased buffer over the gel will reduce separation, as more of the electricity will travel through the buffer instead of through the gel. Obviously, band intensity will increase as more DNA is loaded into the gel. Within a gel lane, smaller fragments will appear less intense since they bind proportionately less ethidium bromide (or other stain).

7. **Troubleshooting gels. What effect will be observed in the stained bands of DNA in an agarose gel**

   a. **if the casting tray is moved or jarred while agarose is solidifying in Part II, Step 3?**

      The gel may solidify unevenly, producing bands with a diffuse or smeared appearance.

   b. **if the gel is run at very high voltage?**

      Bands will have a smeared appearance due to heating effects associated with high voltage.

   c. **if a large air bubble or clump is allowed to set in agarose?**

      All DNA fragments that pass through the bubble or clump will take on a "V" shape, with the bubble/clump at the vertex. A very large bubble may split the DNA band, causing a gap to appear.

   d. **if too much DNA is loaded in a lane?**

      Bands will appear streaked and smeared, with the effect most noticeable for larger fragments.

8. a–i No answers.

   j. **For which fragment sizes was your graph most accurate? For which fragment sizes was it least accurate? What does this tell you about the resolving ability of agarose gel electrophoresis?**

      The graph should produce most accurate results for fragments between 1,000 and 10,000 bp. The least accurate results will be obtained with fragments far outside this size range. These results indicate that for a given agarose concentration, there is a resolving range where migration distance of linear DNA fragments is inversely proportional to base-pair size.

9. **What could be done to resolve the doublet fragments?**

   Carrying out electrophoresis for a longer time may be of some benefit. However, the best solution may be to re-run the samples in a gel whose agarose concentration places the doublet within its resolving range.

10. **Determine a range of sensitivity of DNA detection by ethidium bromide by comparing the mass of DNA in the bands of the largest and smallest detectable fragments on the gel.**

    The largest DNA fragment seen in the ideal gel is, of course, the uncut λ DNA sample. The mass of DNA in this band can be calculated as:

    $$0.1\ \mu g/\mu l \times 4\ \mu l = 0.4\ \mu g$$

    The smallest DNA fragment seen in the ideal gel is the 2,027-bp fragment of the *Hin*dIII digest of λ DNA. The mass of DNA in this fragment is calculated as:

    $$\frac{2{,}027\ \text{bp}\ (0.1\ \mu g/\mu l)(4\ \mu l)}{48{,}502\ \text{bp}} = 0.042(0.4\ \mu g) = 0.017\ \mu g$$

11. a–d No answers.

    e. **What *Bam*HI and *Eco*RI fragments are created in the circular molecules? Why (or why not) can you locate each of these fragments on your gel or the ideal gel?**

The circular λ molecule creates a 12,275-bp fragment in the *Bam*HI digest and a 24,756-bp fragment in the *Eco*RI digest. The 12,275-bp fragment (*Bam*HI digest) can be seen in a 0.8% gel migrating just ahead of the largest (16,841 bp) fragment, but the 24,756-bp fragment (*Eco*RI digest) will not resolve from the 21,226-bp fragment.

LABORATORY 4: EFFECTS OF DNA METHYLATION ON RESTRICTION

1. **How can you account for differences in separation and band intensity?**

   Differences in band separation are largely a function of the voltage, time of electrophoresis, and amount of buffer covering the gel. Increased voltage and/or running time will increase band separation, whereas increased buffer over the gel will reduce separation since more of the electricity will travel through the buffer instead of through the gel. Obviously, band intensity will increase as more DNA is loaded into the gel. Within a gel lane, smaller fragments will appear less intense since they bind less ethidium bromide (or other stain).

2. **What does the M+H+ control tell you about *Eco*RI methylation?**

   The *Eco*RI methylase is sequence-specific and will not protect DNA against the activity of *Hin*dIII.

3. **What does the M+E− control tell you about methylation?**

   Methylation does not affect DNA migration through agarose gels.

4. **What biological function do methylases perform in bacteria? What adaptive value do they have for a bacterium?**

   Methylases serve to protect the bacteria's own DNA from the activity of the restriction enzymes they produce. This allows bacteria to specifically destroy the DNA from an invading bacteriophage while preserving its own DNA.

5. a. **Draw a diagram of this cloning experiment.**

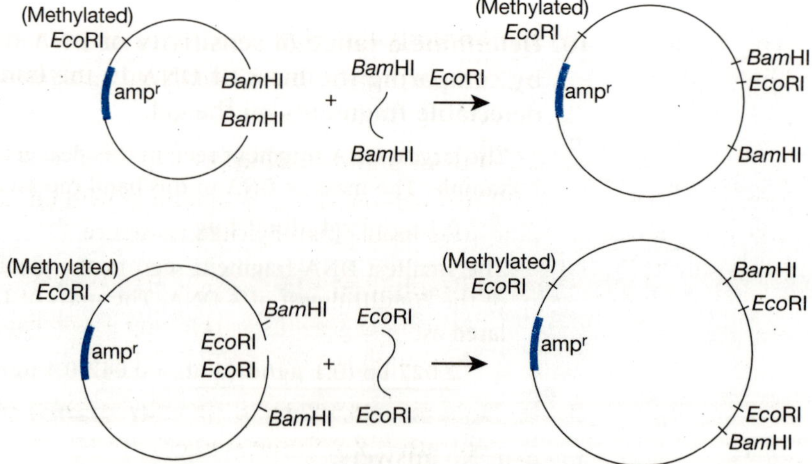

b. **Explain how *Eco*RI methylase could be used to solve this experimental problem.**

First, *Eco*RI methylase is used to methylate (and protect) the *Eco*RI site on pAMP. Next, the *Bam*HI fragment containing an unmethylated *Eco*RI site is cloned into pAMP. Finally, *Eco*RI is used to cut at the unprotected *Eco*RI site within the *Bam*HI insert, so that the *Eco*RI insert can be ligated.

6. **Which nucleotide(s) is (are) methylated by *Eco*RI methylase? Draw the structure of the newly methylated base.**

The site of methylation is the second A of the recognition sequence GAATTC.

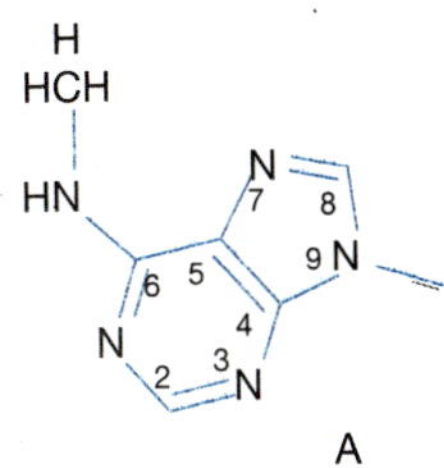

LABORATORY 5: RAPID COLONY TRANSFORMATION OF *E. COLI* WITH PLASMID DNA

1. **Explain possible reasons for variations from expected results.**

If growth is observed where it is not expected (negative control on LB/amp plate) or if growth is not observed where it is expected (transformed cells on LB/amp plate), then verify that the proper type of cells were spread on the appropriate plates. If growth appears on all plates, verify that the ampicillin in the plates is still active.

2. **Compare and contrast the growth on each of the following pairs of plates. What does each pair of results tell you about the experiment?**

a. **+LB and –LB**

Growth on these plates confirms that the cells used in the experiment are viable.

b. **–LB/amp and –LB**

No growth on the –LB/amp plate confirms that the ampicillin in the plates is active.

c. **+LB/amp and –LB/amp**

Growth on +LB/amp and no growth on –LB/amp confirm that pAMP is responsible for ampicillin resistance.

d. **+LB/amp and +LB**

Much less growth on +LB/amp as compared to +LB confirms that transformation is a rare event.

3. a. **Determine total mass (in micrograms) of plasmid used in Step 9: concentration x volume = mass.**

0.005 µg/µl x 10 µl = 0.05 µg

b. **Determine the fraction of the cell suspension spread onto +LB/amp plate (Step 19): volume suspension spread/*total* volume suspension (Steps 2 and 14) = fraction spread.**

   100 μl/500 μl = 0.2

c. **Determine the mass of plasmid in the cell suspension spread onto +LB/amp plate: total mass plasmid (*a*) x fraction spread (*b*) = mass plasmid spread.**

   0.05 μg x 0.2 = 0.01 μg

d. **Determine number of colonies per microgram of plasmid. Express answer in scientifc notation: colonies observed/mass plasmid spread (*c*) = transformation efficiency.**

   500 colonies (substitute your result)/0.01 μg =
   5×10^4 transformants/μg

4. **What factors might influence transformation efficiency?**

   Factors that can influence transformation efficiency include the health of the *E. coli* cells, how well the cells are resuspended in $CaCl_2$, the temperature and duration of the heat shock; amount of plasmid DNA used, spreading technique, and length of recovery period.

5. **Your favorite gene (*YFG*) is cloned into pAMP, and 0.2 μg of pAMP/YFG is used to transform *E. coli* according to the protocol described in this laboratory. Using the information below, calculate the number of molecules of pAMP/YFG that are present in a culture 200 minutes after transformation.**

   a. Since the transformation efficiency is 1×10^6 colonies per μg plasmid and 0.2 μg plasmid was used in the experiment, there must be $0.2 \times 10^6 = 2 \times 10^5$ transformants in the culture.

   b. Since the average copy number of the plasmid is 100, there must be 100 $(2 \times 10^5) = 2 \times 10^7$ copies of YFG in the culture.

   c. The culture entered log phase after 60 minutes and doubled every 20 minutes until it entered stationary phase at 200 minutes. This represents seven doublings (2^7) or a factor of 128. Therefore, at 200 minutes, the culture contains 128 $(2 \times 10^7) = 2.56 \times 10^9$ copies of YFG.

6. **The transformation protocol above is used with 10 μl of intact plasmid DNA at different concentrations. The following numbers of colonies are obtained when 100 μl of transformed cells are plated on selective medium:**

0.00001 μg/μl	**4 colonies**
0.00005 μg/μl	**12 colonies**
0.0001 μg/μl	**32 colonies**
0.0005 μg/μl	**125 colonies**
0.001 μg/μl	**442 colonies**
0.005 μg/μl	**542 colonies**
0.01 μg/μl	**507 colonies**
0.05 μg/μl	**475 colonies**
0.1 μg/μl	**516 colonies**

a. **Calculate transformation efficiencies at each concentration.**

Concentration	Transformation Efficiency
0.00001 μg/μl	2×10^5
0.00005 μg/μl	1.2×10^5
0.0001 μg/μl	1.6×10^5
0.0005 μg/μl	1.25×10^5
0.001 μg/μl	2.21×10^5
0.005 μg/μl	5.42×10^4
0.01 μg/μl	2.54×10^4
0.05 μg/μl	4.75×10^3
0.1 μg/μl	2.58×10^3

b. **Plot a graph of DNA mass versus colonies.**

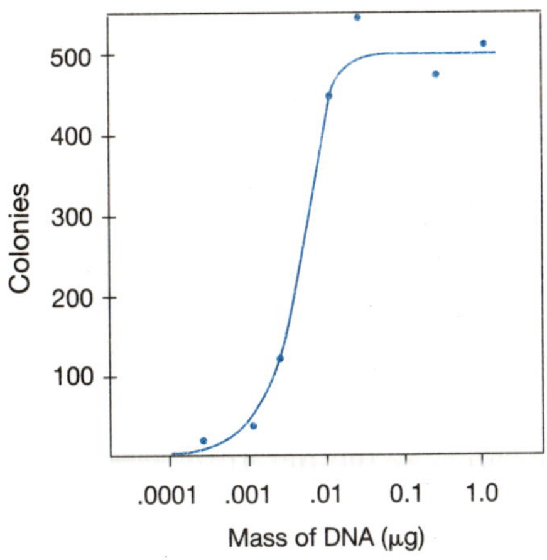

(Reprinted with permission of Pearson Education Inc.; ©1996 by The Benjamin/Cummings Publishing Company, Inc.)

c. **Plot a graph of DNA mass versus transformation efficiency.**

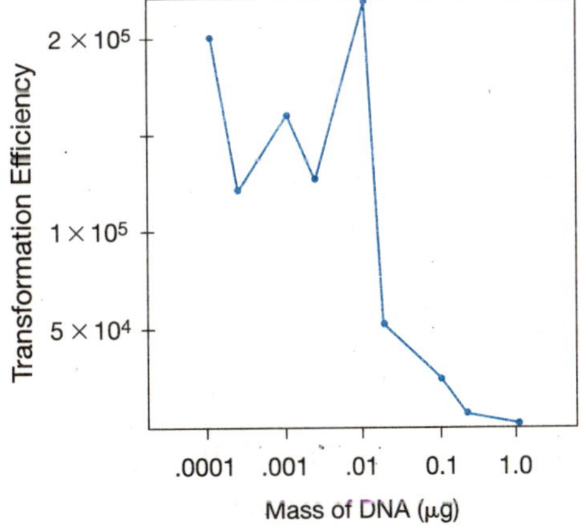

(Reprinted with permission of Pearson Education Inc.; ©1996 by The Benjamin/Cummings Publishing Company, Inc.)

d. **What is the relationship between mass of DNA transformed and transformation efficiency?**

Transformation efficiency remains relatively constant until the cells become saturated with respect to DNA. Thereafter, transformation efficiency declines with increasing mass of DNA.

e. **At what point does the transformation reaction appear to be saturated?**

The saturation point appears to be at a concentration of about 0.001 μg/μl.

f. **What is the true transformation efficiency?**

The true transformation efficiency is that calculated at the saturation point.

7. **For cells transformed with pBLU or pGREEN, what color would you expect nontransformed satellite colonies to be?**

Nontransformed satellite colonies should be colorless or clear.

LABORATORY 6: ASSAY FOR AN ANTIBIOTIC RESISTANCE ENZYME

1. **Why is the phenol red solution made up in 10 mM Tris-HCl (pH 7.9)?**

Tris-HCl is a buffer used to maintain pH. If the phenol red is made up in water, the pH of the water can affect the color of the phenol red. By using Tris-HCl (pH 7.9), we set the reaction to begin at a basic pH. We use a low concentration because this will still allow us to detect a change in pH.

2. **Write out the equation for the enzyme reaction that occurs in this experiment.**

The reaction is: Penicillin ∅ penicilloic acid + H^+

3. **How would you predict the following factors to affect the rate of the reaction:**

a. **Decreasing the temperature.**

Decreasing the temperature will slow down the reaction.

b. **Using a higher concentration penicillin-G.**

The reaction will go faster as the rate is dependent on substrate concentration.

c. **Heating the β-lactamase to 70°C for 10 minutes.**

The enzyme will be inactivated and thus the reaction can no longer proceed.

d. **Adding more β-lactamase.**

The reaction will speed up because the reaction rate is dependent on enzyme concentration.

e. **Adding more phenol red.**

No effect on the rate of the reaction, only the color intensity.

f. **Using 50 mM Tris (pH 7.9) to make up the phenol red.**

Increases the time it takes for the yellow color to appear, as the increased concentration of Tris-HCl will buffer the pH change.

g. **Using 10 mM Tris (pH 9.0) to make up the phenol red.**

When starting at this higher pH, it will take longer for the pH to go below 7 and for the color to change.

4. **Plot the spectrophotometer data for "pAMP sup" β-lactamase reaction. Plot time in minutes on the *x* axis. Plot absorbance on the *y* axis.**

Plots should be obtained with a linear diagonal line that eventually flattens.

5. **Predict what the plots would look like if you took readings at 560 nm (the peak absorbance of phenol red at basic pH).**

   560 nm is the peak absorbance of phenol red at basic pH. A plot of these data should be the inverse of the reading at 430 nm—it starts at a high absorbance, and the values decrease with time.

6. **The rate of the reaction is defined as the amount of substrate formed over time. In this experiment, you indirectly measured substrate formation as an increase in [H^+], which you measured as the increase in absorbance at 430 nm. Look at your plot and describe the point at which the rate of the reaction is greatest. Write the rate of the reaction as change in absorbance per min.**

   The rate is greatest where the line has the greatest slope. The rate is the difference in absorbance at two time points that are 1 minute apart.

LABORATORY 7A: PURIFICATION OF GFP BY HIC

1. **What class of molecules does the lysis buffer interact with to release GFP from *E. coli* cells?**

   The detergent SDS interacts with the fatty (hydrophobic) portion of the phospholipids in the cell membrane and disrupts the lipid bilayer. This breaks apart the cell membrane and releases GFP and other cellular proteins.

2. **What aspect of GFP structure allows it to interact so strongly with the HIC beads?**

   The hydrophobic chromophore of the GFP molecule is exposed following incubation in high salt and binds tightly to the hydrophobic resin.

3. **How does the TE buffer release the GFP molecules from the HIC beads in Step 15?**

   TE is a low-salt buffer that allows GFP molecules to return to their normal structure. The hydrophobic chromophore is no longer exposed and interaction with the hydrophobic resin is lost.

4. **HIC chromatography does not yield 100% pure GFP. What other types of cellular proteins would most likely be found in the GFP preparation?**

   Bacterial proteins with hydrophobic domains that are exposed following incubation in high salt will also bind to HIC resin and will purify along with GFP.

LABORATORY 7B: PAGE ANALYSIS OF PURIFIED GFP

1. **Describe the different properties and uses of agarose *versus* polyacrylamide gels.**

   Polyacrylamide and agarose gels separate molecules based on size. Smaller molecules migrate more quickly through both types of gels. Almost all analytical electrophoresis of proteins is carried out in polyacrylamide gels. Most commonly, the anionic detergent SDS is used in combination with heat to dissociate and denature proteins before they are loaded on the gel.

   Polyacrylamide gels allow the separation of DNA fragments less than 500 nucleotides long. However, the pores in polyacrylamide gels are too small to permit very large DNA molecules to pass through. To separate large DNA molecules by size, more porous gels made from dilute solutions of agarose are used.

2. **Why are the protein samples incubated at 95°C prior to loading onto the polyacrylamide gel?**

   Samples are heated to 95°C to completely dissociate proteins prior to gel electrophoresis.

3. **What is the function of the SDS detergent in the gel and running buffer?**

   SDS is a negatively charged detergent that binds to hydrophobic regions of protein molecules, causing them to unfold into extended polypeptide chains. When heated, SDS-coated proteins are released from their association with other proteins or lipid molecules and are soluble. SDS binds to proteins in proportion to their molecular size, and the net negative charge of the bound detergent overwhelms the intrinsic charge of the protein. During SDS-PAGE, the SDS-polypeptide complexes migrate through polyacrylamide gels according to the size of the polypeptide. Smaller proteins migrate faster. SDS is present in the gel and running buffer to maintain proteins in this negatively charged state and to prevent protein aggregation.

4. **How does Coomassie Blue stain the proteins in the gel? Why is it important to destain for a sufficient amount of time?**

   Coomassie blue stains the entire gel matrix. The dye is removed from the gel matrix during the destaining procedure. However, the dye remains bound to proteins due to an ionic interaction between the dye and basic amino acids. Sufficient destaining is important to remove dye from the gel matrix so that stained proteins can be visualized.

5. **View your stained gel on a light box. Compare your gel with the ideal gel shown, and try to determine which band represents GFP.**

   a. **Expose the gel to a UV light. Explain why the GFP band does not fluoresce.**

      Treatment with SDS and heat denatures proteins. Fluorescence of GFP is dependent on its native conformation.

   b. **Why might you observe some bacterial proteins in your purified GFP lane?**

      Bacterial proteins with exposed hydrophobic domains copurify with GFP because they bind to HIC resin.

6. **Troubleshooting gels. What effect will be observed in the stained bands of protein in a polyacrylamide gel**

   a. **if the samples are not incubated at 95°C prior to loading?**

      If samples are not incubated to 95°C, protein aggregates may be detected. The band that represents GFP may migrate more slowly and appear to be of a larger molecular mass.

   b. **if the gel is run at too high or too low a voltage?**

      When electrophoresis is carried out at too high a voltage, protein bands may appear smeared. If overheating of the electrophoresis unit occurs, distortion of the protein bands may occur. If electrophoresis is carried out for a prolonged period of time, proteins may run off the end of the gel and visualization may be impossible. If the applied voltage is too low, protein bands may appear diffuse.

c. **if too much protein is loaded?**

Protein bands appear smeared and widened when too much protein is loaded. Lanes adjacent to overloaded lanes may appear "squeezed" or compressed. Additionally, protein may remain stuck at the interface between the stacking and separating gels.

7. **Predict the pattern of protein banding observed following SDS-PAGE of the lysate mixed with binding buffer (Part A, Step 10). How would this pattern differ following incubation with HIC beads (Part A, Step 12)? What protein banding pattern would be observed if the wash buffer (Part A, Step 14) was subjected to SDS-PAGE?**

   Following SDS-PAGE of lysate mixed with binding buffer, the banding pattern would appear similar to that observed following SDS-PAGE of the bacterial lysate. However, protein bands will appear wider. After incubation with HIC beads, the intensity of staining of the bands corresponding to GFP and other proteins with exposed hydrophobic domains will be decreased. Wash buffer will contain bacterial proteins that only weakly associate with HIC beads and are removed by mid salt wash.

LABORATORY 8A: PLASMID MINIPREPARATION OF pAMP

1. **Consider the three major classes of biologically important molecules: proteins, lipids, and nucleic acids. Which steps of the miniprep procedure act on proteins? On lipids? On nucleic acids?**

   SDS-sodium hydroxide helps to solubilize proteins and lipids, which are precipitated upon addition of potassium acetate. Additional proteins are removed into the isopropanol. Both isopropanol and ethanol precipitate nucleic acid.

2. **What aspect of plasmid DNA structure allows it to renature efficiently in Step 8?**

   The two DNA strands are interlinked.

3. **What other kinds of molecules, in addition to plasmid DNA, would you expect to be present in the final miniprep sample? How could you find out?**

   RNA molecules should also be present. They can be visualized on ethidium-bromide-stained agarose gels and removed following incubation with RNase.

LABORATORY 8B: RESTRICTION ANALYSIS OF PURIFIED pAMP

2. **Compare the two gel lanes containing miniprep DNA with the two lanes containing control pAMP. Explain possible reasons for variations.**

   At the bottom of each miniprep gel lane, a small cloud of RNA is visible, which should not be seen in the control lanes. Likewise, a heterologous background of stained DNA may be visible near the top of each miniprep lane. This represents residual chromosomal DNA that was not removed during the plasmid purification and will not be seen in the control lanes.

3. **A plasmid preparation of pAMP is composed entirely of dimeric molecules (pAMP/pAMP). The two molecules are joined *head-to-head* at a "hot spot" for recombination located 655 bp from the *Hin*dIII site near the origin of replication.**

   a. **Draw a map of the dimeric plasmid described above.**

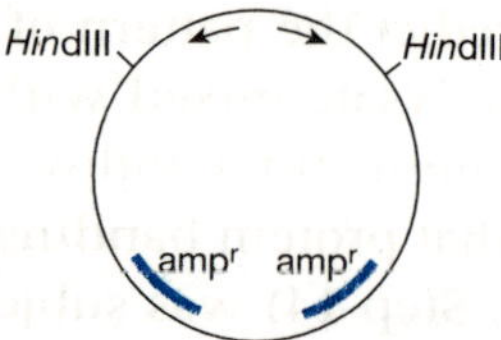

(Reprinted with permission of Pearson Education Inc.; ©1996 by The Benjamin/Cummings Publishing Company, Inc.)

   b. **Draw a map of the dimeric pAMP that actually forms by head-to-tail recombination at the site described above.**

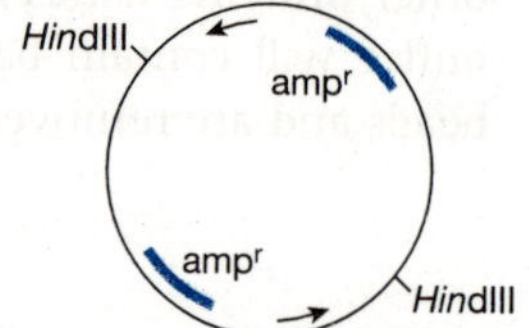

(Reprinted with permission of Pearson Education Inc.; ©1996 by The Benjamin/Cummings Publishing Company, Inc.)

   c. **Now draw the gel-banding patterns that would result from double digestion of each of these plasmids with *Bam*HI and *Hin*dIII, and label the base-pair size of fragments in each band.**

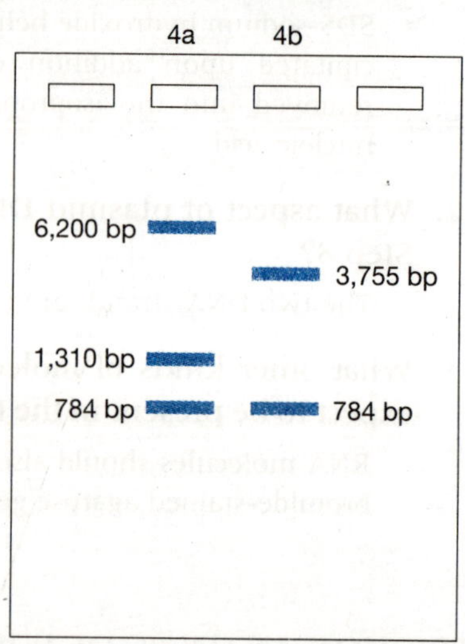

(Reprinted with permission of Pearson Education Inc.; ©1996 by The Benjamin/Cummings Publishing Company, Inc.)

4. **Explain why EDTA is an important component of TE buffer in which the miniprep DNA is dissolved.**

   EDTA binds divalent cations, in particular Mg^{++}, which is a cofactor for exonucleases released from the cells during the plasmid purification. Removal of Mg^{++} helps preserve the stability of the DNA.

LABORATORY 9B: LIGATION OF pAMP AND pKAN RESTRICTION FRAGMENTS

1. **Make a scale drawing of the simple recombinant molecule pAMP/KAN described above. Include fragment sizes, locations of *Bam*HI and *Hin*dIII restriction sites, location of origins(s), and location of antibiotic resistance gene(s).**

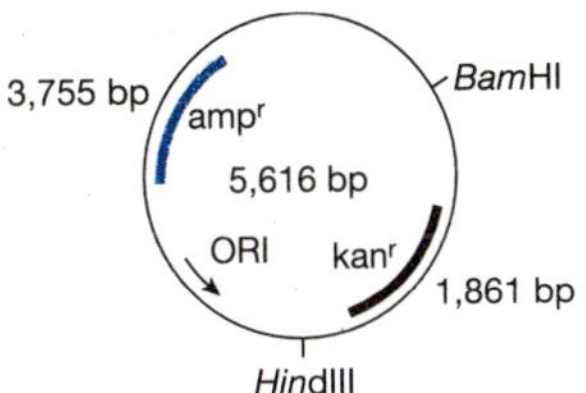

(Reprinted with permission of Pearson Education Inc.; ©1996 by The Benjamin/Cummings Publishing Company, Inc.)

2. **Make scale drawings of other two-fragment recombinant plasmids having the following properties.**

   a. **Three kinds of plasmids having two origins.**

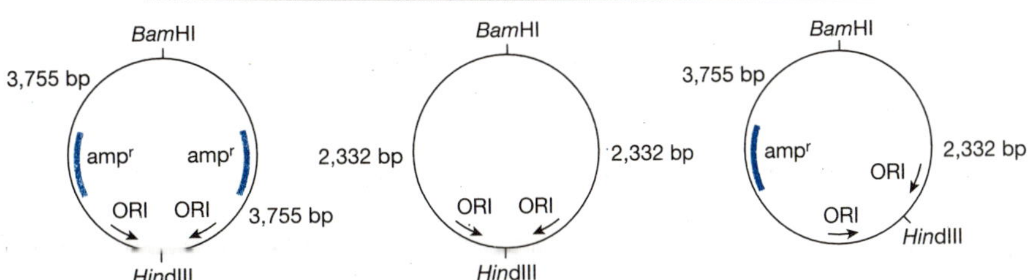

(Reprinted with permission of Pearson Education Inc.; ©1996 by The Benjamin/Cummings Publishing Company, Inc.)

   b. **Three kinds of plasmids having no origin.**

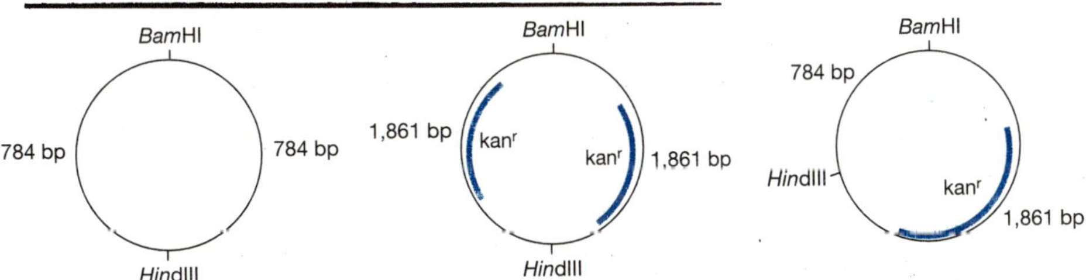

(Reprinted with permission of Pearson Education Inc.; ©1996 by The Benjamin/Cummings Publishing Company, Inc.)

3. **Ligation of the 784-bp fragment, 3755-bp fragment, 1861-bp fragment, and 2332-bp fragment produces a "double plasmid" pAMP/pKAN (or super-plasmid). Make a scale drawing of the double plasmid pAMP/ pKAN.**

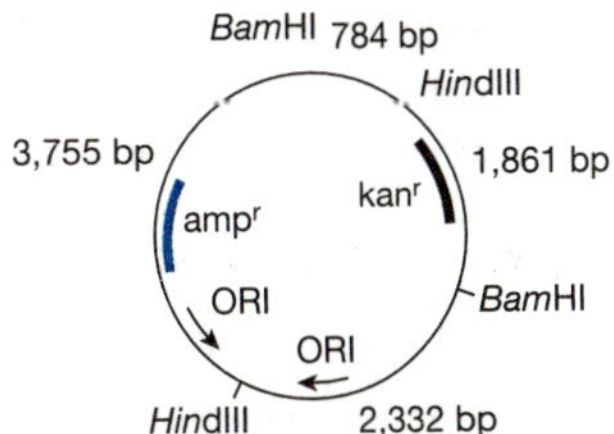

(Reprinted with permission of Pearson Education Inc.; ©1996 by The Benjamin/Cummings Publishing Company, Inc.)

4. **Make scale drawings of several recombinant plasmids composed of any three of the four *Bam*HI/*Hin*dIII fragments of pAMP and pKAN. Include fragment sizes, locations of *Bam*HI and *Hin*dIII restriction sites, location of origin(s), and location of antibiotic resistance gene(s). What rule governs the construction of plasmids from three kinds of restriction fragments?**

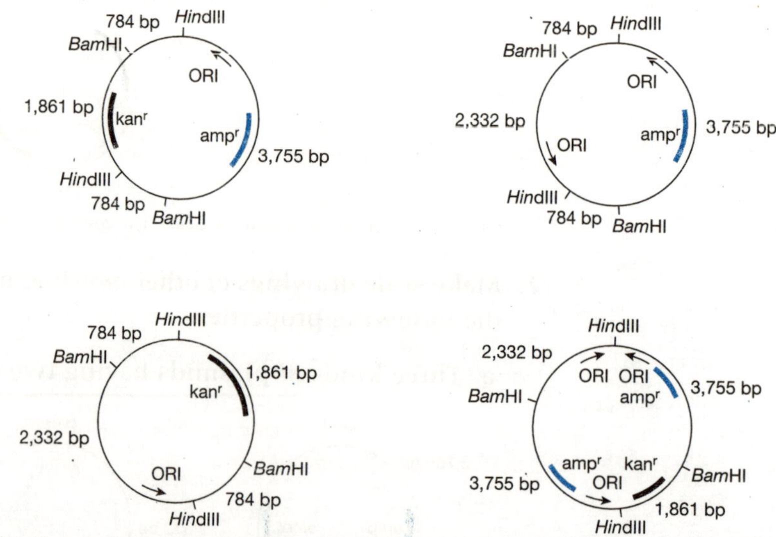

Plasmids composed of fragments containing one *Bam*HI and one *Hin*dIII end must contain an even number of fragments. Therefore, in recombinant plasmids containing three types of restriction fragments, one of them must be present twice. However, in order for plasmids to replicate, copies of the same fragment should not be in head-to-head orientation.

5. **What kind of antibiotic selection would identify *E. coli* cells that have been transformed with each of the plasmids drawn in Questions 1–4?**

Examine each plasmid and ascertain which antibiotic resistance genes (*amp*ʳ, *kan*ʳ, or both) are present. Those plasmids containing a single antibiotic resistance gene should be selected with that antibiotic, whereas plasmids containing both resistance genes should be selected using ampicillin and kanamycin.

6. **Explain what is meant by "sticky ends." Why are they so useful in creating recombinant DNA molecules?**

Sticky ends are protruding single-stranded nucleotide sequences produced when a restriction enzyme cleaves off center in its recognition sequence. Digestion of various DNA molecules with the same restriction enzyme produces compatible sticky ends that can base-pair to facilitate formation of a recombinant DNA molecule.

7. **Why is ATP essential for the ligation reaction?**

ATP is a cofactor for T4 DNA ligase. It serves as an energy source for the covalent linkage that occurs between the ligated DNA fragments.

LABORATORY 10B: TRANSFORMATION OF *E. COLI* WITH RECOMBINANT DNA

1. **Were the results as expected? Explain possible reasons for variations from expected results.**

   Transformation with the pAMP control should produce hundreds of colonies on the LB/amp plate and no growth on the LB/kan and LB/amp+kan plates. Conversely, transformation with the pKAN control should produce hundreds of colonies on the LB/kan plate and no growth on the LB/amp and LB/amp+kan plates. Transformation with the ligated DNA should show growth on all three types of plates, with about tenfold fewer colonies than on the controls.

   If a few colonies are found on plates where they are unexpected, then inadequate sterilization of the spreading rod or poor sterile technique are possible explanations. If these colonies are always found in one permutation (e.g., pAMP transformation growing on LB/kan plate), then the pAMP DNA may be contaminated with a small quantity of pKAN. If large numbers of colonies are seen on plates where no growth is expected, then consider whether the antibiotic is still active in the plates. A more trivial explanation is that cells were spread on the wrong plate.

   If no growth is observed on a plate where you expect to see colonies, then one possibility is that the cells were spread with a spreader that was too hot. If the transformations with ligated DNA show no growth, but the controls exhibit the expected number of colonies, then the ligation reaction may not have gone to completion. If samples of the ligated DNA are still available, separate them by electrophoresis to gauge the extent of the ligation. If all plates show much less growth than expected, then the bacteria may not have been rendered sufficiently "competent" to accept plasmid DNA.

2. **Compare and contrast the growth on each of the following pairs of plates. What does each pair of results tell you about transformation and/or antibiotic selection?**

   L LB/amp and A LB/amp

   There should be about ten times less growth on the L LB/amp plate as compared to the A LB/amp plate. This is because the total mass of ligated DNA is not suitable for transformation. For example, DNA fragments lacking an origin of replication may ligate to each other.

   L LB/kan and A LB/kan

   Growth should be observed on the L LB/kan plate, but not on the A LB/kan plate. This comparison indicates that the kanamycin is active and able to inhibit growth of bacteria lacking the appropriate resistance gene.

   A LB/amp and K LB/kan

   There should be somewhat less growth on the K LB/kan plate as compared to the A LB/amp plate. This indicates that selection by kanamycin is more severe than that by ampicillin. A long recovery period may permit the growth on the two plates to appear more equal.

L LB/amp and L LB/kan

Here again, less growth should be seen on the L LB/kan plate as compared to the L LB/amp plate, reflecting the tougher selection by kanamycin.

L LB/amp and L LB/amp+kan

Less growth should be seen on the L LB/amp+kan plate as compared to the L LB/amp plate, since selection by two antibiotics is more restrictive than that by a single antibiotic.

L LB/kan and L LB/amp+kan

Although you should expect less growth on the L LB/amp+kan plate, since double selection is tougher than single selection, you may observe that the two plates have about equal numbers of colonies. This indicates that the amount of growth is largely determined by the action of the tougher antibiotic, kanamycin.

3. a. **Determine total mass (in micrograms) of pAMP used in Step 5 and of pKAN used in Step 6: concentration x volume = mass.**

   0.005 µg/µl x 10 µl = 0.05 µg

   b. **Determine the fraction of the cell suspension spread onto the A LB/amp plate (Step 17) and K LB/kan plate (Step 19): volume suspension spread/*total* volume suspension (Steps 2 and 11) = fraction spread**

   100 µl/1000 µl = 0.1

   c. **Determine the mass of plasmid pAMP and pKAN in cell suspension spread onto the A LB/amp plate and K LB/kan plate: total mass plasmid (*a*) x fraction spread (*b*) = mass plasmid spread.**

   0.05 µg x 0.1 = 0.005 µg

   d. **Determine the number of colonies per microgram of pAMP and pKAN. Express answer in scientific notation: colonies observed/mass plasmid spread (*c*) = transformation efficiency.**

   2000 colonies (substitute your result)/0.005 µg =
   4×10^5 transformants/µg

4. a. **Calculate the mass of pAMP and pKAN used in the restriction reactions of Laboratory 9 (see matrix in Part A, Step 1). Then calculate the concentration of plasmid in each restriction reaction.**

   0.2 µg/µl x 5.5 µl = 1.1 µg

   concentration = 1.1 µg/15 µl = 0.073 µg/µl

   b. **Calculate the mass of pAMP and pKAN used in the ligation reaction (Laboratory 9, Part B, Step 3). Then calculate the *total* concentration of plasmid in the ligation mixture.**

   mass (each plasmid) = 0.073 µg/µl x 3 µl = 0.22 µg

   total plasmid concentration = 0.44 µg/20 µl = 0.022 µg/µl

c. **Use this concentration in calculations following Steps a–d of Question 3 above.**

mass of plasmid = 0.022 μg/μl x 10 μl = 0.22 μg

fraction cell suspension spread = 100 μl/1000 μl = 0.1

mass of plasmid spread = 0.22 μg x 0.1 = 0.022 μg

transformation efficiency = 50 colonies (substitute your results)/ 0.022 μg = 2.27 x 10^3 transformants/μg

5. **Compare the transformation efficiencies that you calculated for the A LB/amp plate in this laboratory and the +pAMP plate in Laboratory 5. By what factor is the classical procedure more or less efficient than colony transformation? What differences in the protocols contribute to the increase in efficiency?**

The classical procedure is typically found to be 5–10 times more efficient than the colony procedure. The increased efficiency of the classical procedure is largely due to using cells at the mid-log growth phase as opposed to cells growing on a plate. If the classical procedure employs an extended recovery period as compared to the colony procedure, then the increase in efficiency may appear even larger. (Transformation efficiencies should not be calculated under conditions where cell doubling occurs during the recovery period.)

6. **Compare the transformation efficiencies that you calculated for control pAMP and pKAN *versus* the ligated pAMP and pKAN. How can you account for the differences in efficiency? Take into account the formal def inition of transformation efficiency.**

Very often, transformation efficiencies calculated from experiments using ligated DNA are ten times less efficient than the corresponding experiment using control (unligated plasmid). This decrease in efficiency results from lowering the mass of transformable DNA in the ligated sample. Following ligation, many molecular species are produced that will not give rise to transformants, either because they do not contain the appropriate resistance gene or because they lack an origin of replication. Furthermore, some fragments may fail to ligate to another, or fail to produce a circular plasmid.

LABORATORY 11: REPLICA PLATING TO IDENTIFY MIXED *E. COLI* POPULATIONS

2. **On the basis of your observations:**

a. **Calculate the percentage of dual-resistant colonies taken from the L LB/amp plate (Squares 1–12).**

Since the cells placed in Squares 1–12 were originally growing in the presence of ampicillin, all 12 should exhibit growth on the LB/amp plate. Check to see how many of Squares 1–12 show growth on the LB/kan plate and divide this number by 12 to get a percentage.

b. Calculate the percentage of dual-resistant colonies taken from the LB/kan plate (Squares 13–24).

Since the cells placed in Squares 13–24 were originally growing in the presence of kanamycin, all 12 should exhibit growth on the LB/kan plate. Check to see how many of Squares 13–24 show growth on the LB/amp plate and divide by 12 to get a percentage.

c. Give an explanation for the similarity or difference in the percentage of dual-resistant colonies taken from the two source plates.

Published reports suggest that under saturating conditions, with respect to DNA, there is about a 50% probability that a given cell will take up two plasmid molecules. Since the pAMP, pKAN, and recombinant plasmid are nearly the same size, they are about equally likely to be taken up by a bacterial cell. Therefore, in this experiment, one might expect approximately half of the colonies from each source plate to be doubly resistant.

3. Draw restriction maps for different plasmid molecules that could be responsible for the dual resistance phenotype.

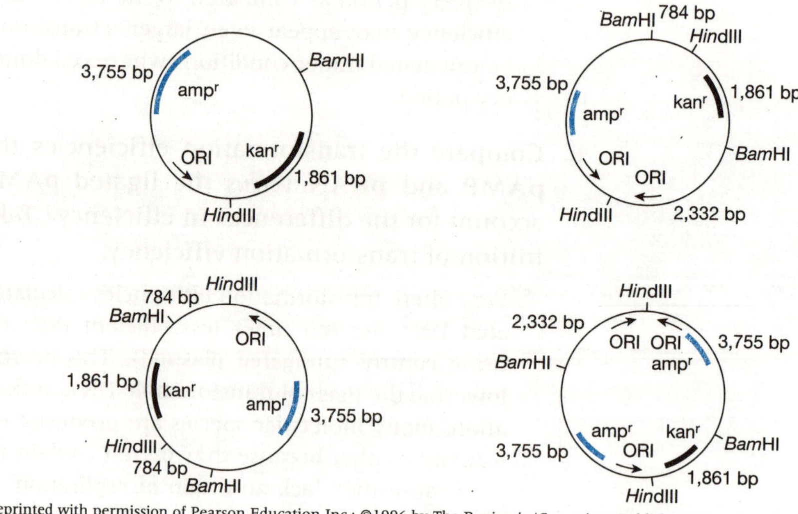

Equipment, Supplies, and Reagents

I. EQUIPMENT

Materials needed	Laboratory number												CBS catalog number*
	1	2	3	4	5	6	7	8	9	10	11	12	
Cell spreader					X					X			70-3412
Clinical centrifuge[a]						X				X			21-4075
Electrophoresis box[a]			X	X				X	X			X	21-3668
Electrophoresis power supply[a]			X	X			X	X	X			X	21-3673
Glass spreading beads					X					X			21-5821
Incubator[a]		X			X	X[b]		X[b]		X	X	X[b]	21-5868
Inoculating loop[a]		X			X	X[b]	X[b]	X[b]		X[b]	X	X[b]	70-3060
Microfuge[a]	X		X	X			X	X	X			X	21-4050
Micropipettor (adjustable volume)[a]													
0.5–10 μl	X		X	X	X			X	X	X		X	21-4651
20–200 μl			X[b]	X[b]	X[b]			X[b]	X[b]	X[b]		X[b]	73-6877
100–1000 μl	X				X	X	X	X		X		X	21-4659
Water bath			X	X	X		X	X	X	X		X	21-6248
Pipette pump[a] (10 ml)	X	X				X[b]		X[b]		X		X[b]	21-4684
Pipette aid[a] (10 ml)	X	X				X[b]		X[b]		X		X[b]	73-6877
Shaking water bath		X					X[b]	X[b]		X		X[b]	21-6256
Spectrophotometer		X				X				X[b]			65-3300
Test tube rack[a]	X	X	X	X	X	X	X	X	X	X		X	21-5570
Ultraviolet lamp							X						15-4683
Ultraviolet transilluminator/camera			X	X				X	X			X	21-3678
Vertical electrophoresis chamber							X						21-3671
White light transilluminator/camera			X	X			X	X	X			X	21-3680

*Carolina Biological Supply (CABISCO) Company Catalog.

[a]Options available. For more information, see Carolina Biological Supply Company Catalog.

[b]Used in laboratory preparation.

II. SUPPLIES

Materials needed	Laboratory number												CBS catalog number*
	1	2	3	4	5	6	7	8	9	10	11	12	
Beaker[a]	X		X	X	X	X	X	X	X	X	X	X	72-1475
Bunsen burner[a]	X	X			X					X	X		70-6702 21-6115
Camera film (Polaroid)			X	X			X	X	X			X	21-3679
Conical tube (50-ml) sterile, screwcap	X	X				X[b]	X[b]	X[b]		X[b]		X[b]	21-5095
Culture tube (15-ml) sterile, 2-position cap[a]	X				X	X				X			21-5090
Gloves, latex			X	X				X	X			X	70-6380 70-6381 70-6382
Micropipettor tip[a]													
0.5–10 μl	X		X	X	X			X	X	X		X	21-5070
1–200 μl			X[b]	X[b]	X[b]			X[b]	X[b]	X[b]		X[b]	21-5050
101–1000 μl	X				X	X[b]	X	X		X		X	21-5060
Parafilm								X	X			X	21-5600
Petri dish, sterile (100 x 15 mm)		X			X					X	X		74-1250
Pipette, sterile[a]													
1 ml transfer					X					X			21-4550
10 ml (serological)	X	X				X[b]		X[b]		X		X[b]	21-4576
Precast acrylamide gel (12%) (8 x 10 cm)							X						21-1375
Replica plating grid										X			70-3442
Sterile filter (.45 μm)[a]					X					X			73-4701 or 19-9595A
Tube (1.5 ml)[a]	X		X	X	X[b]	X	X	X	X	X		X	21-5232

*Carolina Biological Supply (CABISCO) Company Catalog.

[a]Options available. For more information, see Carolina Biological Supply Company Catalog.

[b]Used in laboratory preparations.

III. MEDIA (LB)

Materials needed	Laboratory number 1	2	3	4	5	6	7	8	9	10	11	12	CBS catalog number
A. Ready-to-pour, sterile (20 plates)													
LB agar		X			X								21-6620
LB agar + ampicillin		X			X					X	X		21-6621
LB agar + kanamycin										X	X		21-6622
LB agar + amp + kan										X			21-6623
LB agar + amp + X-gal					X								21-6624
B. Ready-to-use, prepoured (10 plates)													
LB agar		X			X								21-6600
LB agar + ampicillin		X			X					X	X		21-6601
LB agar + kanamycin										X	X		21-6602
LB agar + amp + kan										X			21-6603
LB agar + amp + X-gal					X								21-6604
C. Media solution, sterile (5–50-ml aliquots)													
LB broth		X			X	X[a]				X			21-6650
LB broth + ampicillin		X				X[a]	X[a]	X[a]				X[a]	21-6652

[a]Used in laboratory preparations.

IV. MEDIA COMPONENTS

Materials needed	Laboratory number 1	2	3	4	5	6	7	8	9	10	11	12	CBS catalog number
LB agar base		X			X					X	X		21-6700
LB base		X			X	X[a]	X[a]	X		X	X	X	21-6710
Tryptone		X			X	X[a]	X[a]	X		X	X	X	21-6741
Yeast extract		X			X	X[a]	X[a]	X		X	X	X	21-6745
Sodium hydroxide 1 M		X			X	X[a]	X[a]	X		X	X	X	88-9571
Ampicillin													
solution, 10 mg/ml		X			X	X[a]	X[a]	X		X	X	X	21-6860
powder		X			X	X[a]	X[a]	X		X	X	X	21-6880
Ampicillin + X-gal					X								21 6874
Kanamycin													
solution, 10 mg/ml										X	X		21-6861
powder										X	X		21-6881

[a]Used in laboratory preparations.

V. BIOLOGICALS AND ENZYMES

Materials needed	Laboratory number												CBS catalog number
	1	2	3	4	5	6	7	8	9	10	11	12	
A. DNA													
λ DNA			X	X								X	21-1408
Plasmid pAMP					X			X	X	X		X	21-1430
Plasmid pKAN									X	X		X	21-1440
Plasmid pBLU					X								21-1420
Plasmid pGREEN					X								21-1449
B. Bacterial Strains													
MM294		X			X	X				X			21-1530
MM294/pAMP		X				X		X					21-1540
C. Enzymes													
*Bam*HI			X					X	X			X	21-1660
*Eco*RI			X	X									21-1670
*Eco*RI methylase (with 10x with restriction buffer/SAM)				X									21-1671
*Hin*dIII			X	X				X	X			X	21-1690
T4 DNA ligase									X				21-1740
RNase A, pancreatic (5 mg/ml)								X				X	21-1745
Protein markers						X							21-1510

VI. REAGENTS

Materials needed	Laboratory number												CBS catalog number
	1	2	3	4	5	6	7	8	9	10	11	12	
Agarose, LE (low EEO)			X	X				X	X			X	21-7080
Calcium chloride solution sterile (50 mM)					X					X			21-1320
Coomassie destain							X						21-9785
Coomassie stain							X						21-9784
Ethanol (95%)[a]					X			X		X		X	86-1281
Ethidium bromide solution (5 μg/ml)[b]			X	X				X	X			X	21-7422
Glucose-Tris-EDTA (GTE) solution								X				X	21-7710
Isopropanol								X				X	88-4890
Ligation buffer/ATP (2x)									X				included with T4 DNA ligase
DNA loading dye			X	X				X	X			X	21-8200
Phenol Red solution (0.04%)													
120 ml						X							87-9873
500 ml						X							87-9875
Protein loading dye (4x)							X						21-8660
Restriction buffer (10x compromise)			X	X				X	X			X	21-8770
Sodium dodecyl sulfate (SDS) 10% solution							X	X				X	21-8822
Tris/Borate/EDTA (TBE) buffer (20x mix)			X	X				X	X			X	21-9027
Tris-EDTA (TE) buffer			X				X	X				X	21-9026
Tris-glycine-SDS							X						21-9030

[a]Be careful not to ignite the ethanol in the beaker.

[b]Review Responsible Handling of Ethidium Bromide in Laboratory 3. Wear latex gloves when staining, viewing, and photographing gels and during cleanup. Confine all staining to a restricted sink area.

VII. REAGENT COMPONENTS

Materials needed	Laboratory number												CBS catalog number
	1	2	3	4	5	6	7	8	9	10	11	12	
Acetic acid, glacial								x				x	84-1290
Ammonium sulfate							x						84-4360
Boric acid			x	x				x	x			x	84-8440
Bromophenol blue			x	x				x	x			x	84-9080
Calcium chloride, dihydrate					x					x			85-1800
Ethylenediaminetetraacetate (EDTA)-disodium salt, dihydrate			x	x			x	x	x			x	86-1790
Glucose, anhydrous								x				x	85-7450
Magnesium chloride, 6-hydrate			x	x				x	x			x	87-3248
Potassium acetate								x				x	88-2540
Sodium chloride		x	x	x	x	x	x	x	x	x	x	x	88-8880
Sodium dodecyl sulfate (SDS)							x	x				x	21-8820
Sodium hydroxide pellets								x				x	88-9470
Sucrose			x	x				x	x			x	89-2870
Tris base			x	x		x		x	x			x	89-6970

VIII. READY-TO-USE KITS

Phenotypic Bacterial ID Kit

21-1136 Student Kit
21-1136C Perishable Refill

Colony Transformation Kit

21-1140 Teacher Demonstration Kit
21-1141 Three-station Student Kit
21-1142 Six-station Student Kit
21-2730V Instructional Video

Green Gene Colony Transformation Kit

21-1080 Teacher Demonstration Kit
21-1081 Three-station Student Kit
21-1082 Six-station Student Kit
21-1088 Glow-in-the-Dark Transformation Kit

pBLU Colony Transformation Kit

21-1145X Teacher Demonstration Kit
21-1145Y Three-station Student Kit
21-1146 Six-station Student Kit

Restriction Enzyme Cleavage of DNA Kit

21-1148Y Three-station Student Kit
21-1149 Six-station Student Kit
21-2732V Instructional Video
21-1149B Kit with Video
21-1149A DNA only

DNA Restriction Analysis Kit

21-1103 Kit with Ethidium Bromide
21-1106 Kit with *Carolina*BLU
21-2732V Instructional Video

EZ Gene Splicer DNA Recombination and Transformation Kit

21-1160 Teacher Demonstration Kit
21-1162 Student Kit

Assay for Antibiotic Resistance Enzyme

21-1137 Teacher Demonstration Kit
21-1139 Six-station Student Kit

Purification and Identification of Recombinant GFP

Module 1: Purification
21-1070 Teacher Demonstration Kit
21-1072 Six-station Student Kit
Module 2: Identification
21-1071 Teacher Demonstration Kit
21-1073 Six-station Student Kit

SUPPLIERS

With the exception of those supplies listed in the text with their addresses, all suppliers mentioned in this manual can be found in the BioSupplyNet Source Book and on the Web Site at http://www.biosupplynet.com.

- Complete the Free Source Book Request Form found at the Web Site at:
  http://www.biosupplynet.com
- E-mail a request to:
  info@biosupplynet.com
- Fax a request to 1-919-659-2199.

APPENDIX 2

Recipes for Media, Reagents, and Stock Solutions

THE SUCCESS OF THE LABORATORIES IN THIS BOOK depends on the use of uncontaminated reagents. Follow the recipes with care and pay scrupulous attention to cleanliness. Use a clean spatula for each ingredient or carefully pour each ingredient from its bottle.

The recipes are organized in eight sections. Stock solutions that are used in more than one laboratory are listed once, according to their *first use* in the laboratories.

> CAUTION
>
> See Appendix 4 for appropriate handling of hazardous materials marked with ▼.

I. BACTERIAL CULTURE

4 N Sodium Hydroxide (NaOH)
10 mg/ml Ampicillin
10 mg/ml Kanamycin
Luria-Bertani (LB) Broth
LB Broth + Antibiotic
LB Agar Plates
LB Agar + Antibiotic
Stab Cultures

II. DNA RESTRICTION

1 M Tris (pH 8.0)
5 M Sodium Chloride (NaCl)
1 M Magnesium Chloride ($MgCl_2$)
1 M Dithiothreitol (DTT)
10x Compromise Restriction Buffer
2x Restriction Buffer
30 mM *S*-Adenosyl Methionine (SAM)
2x Restriction Buffer/SAM
0.05% Glacial Acetic Acid
5 mg/ml RNase A (Pancreatic RNase)
5x Restriction Buffer/RNase

III. GEL ELECTROPHORESIS

10x Tris/Borate/EDTA (TBE) Electrophoresis Buffer
1x Tris/Borate/EDTA (TBE) Electrophoresis Buffer

0.8% Agarose
Loading Dye
1 μg/ml Ethidium Bromide Staining Solution
1% Methylene Blue
0.025% Methylene Blue Staining Solution

IV. BACTERIAL TRANSFORMATION

1 M Calcium Chloride ($CaCl_2$)
50 mM Calcium Chloride ($CaCl_2$)

V. ENZYME ASSAY

0.04% Phenol Red
4 mg/ml Penicillin-G

VI. PROTEIN PURIFICATION

Binding Buffer
Equilibration Buffer
Wash Buffer
Tris/EDTA (TE) Buffer (Dilution Buffer)
Elution Buffer
Lysis Buffer
Methyl Hydrophobic Interaction Chromatography (HIC) Resin

VII. PLASMID MINIPREPARATION

0.5 M Ethylenediaminetetraacetic Acid (EDTA)
Tris/EDTA (TE) Buffer
Glucose/Tris/EDTA (GTE)
5 M Potassium Acetate (KOAc)
Potassium Acetate/Acetic Acid
10% Sodium Dodecyl Sulfate (SDS)
1% SDS/0.2 N NaOH

VIII. DNA LIGATION

1 M Tris (pH 7.6)
10x Ligation Buffer
0.1 M Adenosine Triphosphate (ATP)
2x Ligation Buffer + ATP

Notes on Buffers

1. Typically, solid reagents are dissolved in a volume of deionized or distilled water equivalent to 70–80% of the finished volume of buffer. This leaves room for the addition of acid or base to adjust the pH. Finally, water is added to bring the solution up to the final volume.
2. The final concentration of each liquid reagent is given in the right-hand column of the reagent list.
3. Buffers are used as 2x, 5x, or 10x solutions. Buffers are diluted when mixed with other reagents to produce a working concentration of 1x.

I. BACTERIAL CULTURE

4 N Sodium Hydroxide (NaOH)

Makes 100 ml.
Store at room temperature (indefinitely).

1. Slowly add 16 g of NaOH▼ pellets (m.w. = 40.00) to 80 ml of deionized or distilled water, with stirring. The solution will get very hot.
2. When NaOH pellets are completely dissolved, add water to a final volume of 100 ml.

10 mg/ml Ampicillin

Makes 100 ml.
Store at –20°C (1 year) or 4°C (3 months).

1. Add 1 g of ampicillin▼ (sodium salt, m.w. = 371.40) to 100 ml of deionized or distilled water in a clean 250-ml flask. (The sodium salt dissolves readily; however, the free acid form is difficult to dissolve.)
2. Stir to dissolve.
3. Prewash a 0.45- or 0.22-µm sterile filter (Nalgene or Corning) by drawing through 50–100 ml of deionized or distilled water. Pass ampicillin solution through the washed filter.
4. Dispense 10-ml aliquots in sterile 15-ml tubes (Falcon 2059 or equivalent), and freeze at –20°C.

10 mg/ml Kanamycin

Makes 100 ml.
Store at –20°C (1 year) or 4°C (3 months).

1. Add 1.0 g of kanamycin sulfate▼ (m.w. = 582.60) to 100 ml of deionized or distilled water in a clean 250-ml flask.
2. Stir to dissolve.
3. Prewash a 0.45- or 0.22-µm sterile filter (Nalgene or Corning) by drawing through 50–100 ml of deionized or distilled water. Pass kanamycin solution through the washed filter.
4. Dispense 10-ml aliquots in sterile 15-ml tubes (Falcon 2059 or equivalent), and freeze at –20°C.

Luria-Bertani (LB) Broth

Makes 1 liter.
Store at room temperature (indefinitely).

1. Weigh out

   10 g of tryptone
   5 g of yeast extract
   10 g of NaCl (m.w. = 58.44)

   Alternatively, use 25 g of premix containing all of these ingredients.

2. Add all ingredients to a clean 2-liter flask that has been rinsed with deionized or distilled water.
3. Add 1 liter of deionized or distilled water to flask.
4. Add 0.5 ml of 4 N NaOH▼.
5. Stir to dissolve the dry ingredients, preferably using a magnetic stir bar.
6. *If preparing for mid-log cultures:* Split LB broth into two 500-ml aliquots in 2-liter flasks. Plug top with cotton or foam, and cover with aluminum foil. (Alternatively, cover with aluminum foil only.) Autoclave for 15–20 minutes at 121°C.

   or

   If preparing for general use in transformations: Dispense 100-ml aliquots into sterile 150–250-ml bottles using one of the following methods:

   a. Loosely put on the caps. Autoclave for 15–20 minutes at 121°C. (To help guard against breakage, autoclave the bottles in a shallow pan with a small amount of water.)

   or

   b. Prewash a 0.45- or 0.22-μm sterile filter (Nalgene or Corning) by drawing through 50–100 ml of deionized or distilled water. Pass LB broth through the filter, and dispense aliquots into sterile bottles.

NOTE LB broth can be considered sterile as long as the solution remains clear. Cloudiness is a sign of contamination by microbes. Always swirl solution to check for bacterial or fungal cells that may have settled at the bottom of the flask or bottle.

LB Broth + Antibiotic

Makes 100 ml.
Store at 4°C (3 months).

1. Sterilely add 1 ml of 10 mg/ml antibiotic to 100 ml of *cool* LB broth.
2. Swirl to mix.

LB Agar Plates

Makes 35–40 plates.
Store at 4°C (3 months) or room temperature (3 months).

1. Weigh out

   10 g of tryptone
   5 g of yeast extract
   10 g of NaCl (m.w. = 58.44)
   15 g of agar

   Alternatively, use 40 g of premix containing all of these ingredients.
2. Add all ingredients to a clean 2-liter flask that has been rinsed with deionized or distilled water.
3. Add 1 liter of deionized or distilled water.
4. Add 0.5 ml of 4 N NaOH▼.

5. Stir to dissolve dry ingredients, preferably using a magnetic stir bar. Any undissolved material will dissolve during autoclaving.
6. Cover flask mouth with aluminum foil, and autoclave solution for 15 minutes at 121°C.
7. During autoclaving, the agar may settle to the bottom of flask. Swirl to mix agar evenly.
8. Allow solution to cool just until flask can be held in bare hands (55–60°C). (If solution cools too long and the agar begins to solidify, remelt by briefly autoclaving for 5 minutes or less or heating it in a microwave oven for a few minutes.)
9. While agar is cooling, mark *culture plate bottoms* with the date and description of the media (e.g., LB). If using presterilized polystyrene plates, carefully cut the end of plastic sleeves, and save the sleeves for storing the poured plates. Spread plates out on lab bench.
10. When agar flask is cool enough to hold, lift lid of culture plate only enough to pour solution. *Do not place lid on lab bench.* Quickly pour in agar to just cover plate bottom (~25–30 ml). Tilt the plate to spread the agar, and immediately replace lid.
11. Continue pouring agar into plates. Occasionally flame mouth of flask to maintain sterility.
12. To remove bubbles in the surface of the poured agar, touch plate surface with the flame from the Bunsen burner while agar is still liquid.
13. Allow agar to solidify undisturbed.
14. If possible, incubate plates *lidside down* for several hours at 37°C (overnight if convenient). This dries the agar, limiting condensation when plates are stored under refrigeration. It also allows the ready detection of any contaminated plates.
15. Stack plates in their original sleeves for storage.

LB Agar + Antibiotic

Makes 30–45 plates.
Store at 4°C (3 months).

1. Follow recipe above for LB agar plates through Step 9.
2. When agar flask is cool enough to hold, sterilely add 10 ml of 10 mg/ml antibiotic. Ampicillin and kanamycin are destroyed by heat; therefore, *it is essential to cool agar before adding antibiotic.* (For LB/amp + kan plates, add 10 ml *each* of 10 mg/ml ampicillin▼ and kanamycin▼.)
3. Swirl flask to mix antibiotic.
4. Resume recipe above with Step 10.

NOTE In a pinch, antibiotic-containing plates can be made quickly by evenly spreading 200 μl of 10 mg/ml antibiotic on the surface of an LB agar plate. Allow the antibiotic to absorb into agar for 10–20 minutes before using. Outdated antibiotic plates can be refurbished in this manner.

Stab Cultures

Makes 30–40 stab cultures.
Store in dark at room temperature (1 year).

To prepare vials:

1. Weigh out

   1.0 g of tryptone
   0.5 g of yeast extract
   1.0 g of NaCl (m.w. = 58.44)
   0.7 g of agar

NOTE This recipe is the same as the LB agar recipe above, but the lower percentage of agar makes the stab culture easier to use. Standard LB agar or 4 g of premix can also be used.

2. Add all ingredients to a clean 250-ml flask that has been rinsed with deionized or distilled water.
3. Add 100 ml of deionized or distilled water.
4. Add 50 μl of 4 N NaOH▼.
5. Stir while heating to dissolve dry ingredients; preferably, use a magnetic stir bar and a hot plate.
6. Pour dissolved solution into 4-ml vials (15 x 45 mm) until two-thirds filled, and loosely replace caps.
7. Autoclave vials for 15 minutes at 121°C. (Alternatively, pour sterilized agar into presterilized vials.)
8. Allow agar to solidify undisturbed.
9. Prior to storage, tighten caps completely.
10. Seal caps with Parafilm for longer shelf life.

To inoculate:

1. Sterilely scrape up cell mass from a single colony of desired genotype.
2. Stab inoculating loop several times into agar.
3. Loosely replace cap, and incubate stab culture overnight at 37°C.
4. Following incubation, tighten cap, and store in the dark at room temperature. Wrap the cap with Parafilm for long-term storage.

II. DNA RESTRICTION

1 M Tris (pH 8.0)

Makes 100 ml.
Store at room temperature (indefinitely).

1. Dissolve 12.1 g of Tris▼ base (m.w. = 121.10) in 70 ml of deionized or distilled water.

2. Adjust the pH by slowly adding ~5.0 ml of concentrated hydrochloric acid▼ (HCl); monitor with a pH meter. (If a pH meter is not available, adding 5.0 ml of concentrated HCl will yield a solution of ~pH 8.0.)
3. Add deionized or distilled water to make a total volume of 100 ml of solution.

NOTES

- A yellow-colored solution indicates poor-quality Tris. Discard it, and obtain a Tris solution from a different source.
- Many types of electrodes do not accurately measure the pH of Tris solutions; check with manufacturer to obtain a suitable one.
- The pH of Tris solutions is temperature-dependent; measure pH at room temperature.
- pH can also be measured using strips of pH paper.

5 M Sodium Chloride (NaCl)

Makes 100 ml.
Store at room temperature (indefinitely).

1. Dissolve 29.2 g of NaCl (m.w. = 58.44) in 70 ml of deionized or distilled water.
2. Add deionized or distilled water to make a total volume of 100 ml of solution.

1 M Magnesium Chloride ($MgCl_2$)

Makes 100 ml.
Store at room temperature (indefinitely).

1. Dissolve 20.3 g of $MgCl_2$▼ (6-hydrate, m.w. = 203.30) in 80 ml of deionized or distilled water.
2. Add deionized or distilled water to make a total volume of 100 ml of solution.

1 M Dithiothreitol (DTT)

Makes 10 ml.
Store at –20°C (indefinitely).

1. Dissolve 1.5 g of dithiothreitol▼ (DTT) (m.w. = 154.25) in 8 ml of deionized or distilled water.
2. Add deionized or distilled water to make a total volume of 10 ml of solution.
3. Dispense into 1-ml aliquots in 1.5-ml tubes.

NOTE Do not autoclave DTT or solutions containing it.

10x Compromise Restriction Buffer

Makes 1 ml.
Store at –20°C (indefinitely).

Mix ingredients in a 1.5-ml tube.

100 µl of 1 M Tris▼ (pH 8.0)	(100 mM)
100 µl of 1 M $MgCl_2$▼	(100 mM)
200 µl of 5 M NaCl	(1 M)
7 µl of 14.3 M β-mercaptoethanol▼	(100 mM)
593 µl of deionized water	

or

100 µl of 1 M Tris▼ (pH 8.0)	(100 mM)
100 µl of 1 M $MgCl_2$▼	(100 mM)
200 µl of 5 M NaCl	(1 M)
10 µl of 1 M DTT▼	(10 mM)
590 µl of deionized water	

NOTES

- β-Mercaptoethanol, also called 2-mercaptoethanol, is a 14.3 M liquid at room temperature.
- Compromise buffer provides salt conditions that allow relatively high activity of a number of different restriction enzymes, including all those used in these laboratories. Consider using the specific buffer provided by the manufacturer when using expensive enzymes in critical experiments.
- DTT used in the second recipe may precipitate from the solution with repeated freezing and thawing. Vortex vigorously to redissolve.

2x Restriction Buffer

Makes 1 ml.
Store at –20°C (indefinitely).

Mix 200 µl of 10x restriction buffer with 800 µl of deionized or distilled water.

30 mM *S*-Adenosyl Methionine (SAM)

Makes 1 ml.
Store at –20°C (1 year).

1. Obtain 1 M sulfuric acid▼ (H_2SO_4) or prepare by *carefully* mixing 1 part concentrated acid (18 M) into 17 parts of deionized or distilled water.
2. Prepare a 5 mM solution by adding 5 µl of 1 M H_2SO_4 to 995 µl of deionized or distilled water.
3. Add 900 µl of 5 mM H_2SO_4 solution to 100 µl of ethanol▼ (100%).
4. Dissolve 15.8 mg of SAM▼ (iodide salt, grade I, m.w. = 526.30) in 1 ml of 5 mM H_2SO_4/10% ethanol solution.
5. Dispense 100-µl aliquots in 1.5-ml tubes. Use each solution once, and discard.

NOTES

- SAM is very unstable, and activity diminishes rapidly with freezing and thawing. Store SAM solution on ice at all times. SAM solutions have half-lives on the order of 30 minutes at room temperature.
- 30 mM SAM is often supplied with *Eco*RI methylase.

2x Restriction Buffer/SAM

Makes 1 ml.
Hold on ice or at –20°C, and use within several hours of mixing.

1. Mix in a 1.5-ml tube:

   10 μl of 30 mM SAM▼

   200 μl of 10x restriction buffer

   790 μl of deionized water

2. Use within several hours, and discard.

NOTE Store restriction buffer/SAM on ice at all times during manipulations. SAM solutions have half-lives on the order of 30 minutes at room temperature.

0.05% Glacial Acetic Acid

Makes 50 ml.
Store at room temperature (indefinitely).

1. Add 25 μl of the glacial acetic acid▼to 50 ml of deionized or distilled water.
2. The solution should be at approximately pH 4.0.

5 mg/ml RNase A (Pancreatic RNase)

Makes 20 ml.
Store at –20°C (indefinitely).

1. Dissolve 100 mg of RNase A in 20 ml of 0.05% glacial acetic acid▼, and transfer to a 50-ml conical tube.
2. Place the tube in a boiling-water bath for 15 minutes.
3. Cool the solution, and neutralize by adding 120 μl of 1 M Tris▼(pH 8.0).
4. Dispense 1-ml aliquots in 1.5-ml tubes.

NOTES

- Use only RNase A from bovine pancreas.
- Dissolving RNase in the acetic acid prevents subsequent precipitation of the RNase. The solution can be prepared by simply dissolving RNase in deionized or distilled water; however, the RNase will occasionally precipitate from the solution and activity will be lost.

5x Restriction Buffer/RNase

Makes 1 ml.
Store at –20°C (several months).

Mix in a 1.5-ml tube:

500 μl of 10x restriction buffer
100 μl of 5 mg/ml RNase
400 μl of water

II. GEL ELECTROPHORESIS

10x Tris/Borate/EDTA (TBE) Electrophoresis Buffer

Makes 1 liter.
Store at room temperature (indefinitely).

1. Add the following dry ingredients to 700 ml of deionized or distilled water in a 2-liter flask.

   1 g of NaOH▼ (m.w. = 40.00)
   108 g of Tris▼ base (m.w. = 121.10)
   55 g of boric acid▼ (m.w. = 61.83)
   7.4 g of EDTA (disodium salt, m.w. = 372.24)

2. Stir to dissolve, preferably using a magnetic stir bar.
3. Add deionized water to bring total solution to 1 liter.

NOTE If stored 10x TBE comes out of solution, place the flask in a water bath (37°C to 42°C) and stir occasionally until all solid matter goes back into solution.

1x Tris/Borate/EDTA (TBE) Electrophoresis Buffer

Makes 10 liters.
Store at room temperature (indefinitely).

1. Into a spigotted carboy, add 9 liters of deionized or distilled water to 1 liter of 10x TBE electrophoresis buffer.
2. Stir to mix.

0.8% Agarose

Makes 200 ml.
Use fresh or store solidified agarose at room temperature (several weeks).

1. Add 1.6 g of agarose (low EEO electrophoresis grade) to 200 ml of 1x TBE electrophoresis buffer in a 600-ml beaker or Erlenmeyer flask.
2. Stir to suspend agarose.
3. Cover beaker with aluminum foil, and heat in a boiling-water bath (double boiler) or on a hot plate until all agarose is dissolved (~10 minutes).

   or

   Heat *uncovered* in a microwave oven at high setting until all agarose is dissolved (3–5 minutes per beaker).

NOTE Agarose will go into solution as the liquid begins to boil. Do not allow the solution to boil for more than a few seconds as this will alter the final concentration of the agarose.

4. Swirl solution and check bottom of beaker to make sure that all agarose has dissolved. (Just prior to complete dissolution, particles of agarose appear as translucent grains.) Reheat for several minutes if necessary.

NOTES

- 1.6-g samples of agarose powder can be preweighed and stored in capped test tubes until ready for use.
- Solidified agarose can be stored at room temperature and then remelted over a boiling-water bath (15–20 minutes) or in a microwave oven (5–7 minutes per beaker) prior to use. Always loosen the cap when remelting the agarose in a bottle.

5. Cover with aluminum foil, and hold in a hot-water bath (at ~60°C) until ready for use. Remove any "skin" of solidified agarose from surface prior to pouring.

Loading Dye

Makes 100 ml.
Store at room temperature (indefinitely).

1. Dissolve the following ingredients in 60 ml of deionized or distilled water.

   0.25 g of bromophenol blue▼(m.w. = 669.96)
   0.25 g of xylene cyanol▼(m.w. = 538.60)
   50.00 g of sucrose (m.w. = 342.30)
   1.00 ml of 1 M Tris▼(pH 8.0)

2. Add deionized or distilled water to make a total volume of 100 ml of solution.

NOTE Glycerol may be used instead of sucrose. Dissolve the xylene cyanol, bromophenol blue, and Tris in 40 ml of deionized or distilled water and stir in 50 ml of glycerol to make a total volume of 100 ml of solution.

1 µg/ml Ethidium Bromide Staining Solution

Makes 500 ml.
Store in dark at room temperature (indefinitely).

CAUTION
Ethidium bromide is a mutagen by the Ames microsome assay and a suspected carcinogen. Wear latex gloves when preparing and using ethidium bromide solutions. Review Responsible Handling of Ethidium Bromide in Laboratory 3.

1. Add 100 µl of 5 mg/ml ethidium bromide▼to 500 ml of deionized or distilled water.
2. Store in unbreakable bottles (preferably opaque). Label bottle CAUTION: Ethidium Bromide. Mutagen and cancer-suspect agent. Wear latex gloves when handling.

NOTE Ethidium bromide is light-sensitive; store in dark container or wrap container in aluminum foil.

1% Methylene Blue

Makes 50 ml.
Store at room temperature (indefinitely).

Dissolve 0.5 g of methylene blue▼ in 50 ml of deionized or distilled water.

0.025% Methylene Blue Staining Solution

Makes 400 ml.
Store at room temperature (indefinitely).

Add 10 ml of 1% methylene blue▼ to 390 ml of deionized or distilled water.

IV. BACTERIAL TRANSFORMATION

1 M Calcium Chloride ($CaCl_2$)

Makes 100 ml.
Store at room temperature (indefinitely).

1. Dissolve 11.1 g of the anhydrous $CaCl_2$ (m.w. = 110.99) or 14.7 g of the dihydrate (m.w. = 146.99) in 80 ml of deionized or distilled water.
2. Add deionized or distilled water to make a total volume of 100 ml of solution.

50 mM Calcium Chloride ($CaCl_2$)

Makes 1000 ml.
Store at 4°C or room temperature (indefinitely).

1. Mix 50 ml of 1 M $CaCl_2$ with 950 ml of deionized water.
2. Prerinse a 0.45- or 0.22-µm sterile filter by drawing through 50–100 ml of deionized or distilled water.
3. Pass $CaCl_2$ solution through prerinsed filter.
4. Dispense aliquots into presterilized 50-ml conical tubes or autoclaved 150–250-ml bottles.

NOTES

- Alternatively, dispense 100-ml aliquots into 150–250-ml bottles; autoclave 15 minutes at 121°C.
- With storage at 4°C, solution is precooled and ready for making competent cells.

V. ENZYME ASSAY

0.04% Phenol Red

Makes 50 ml.
Store at room temperature indefinitely.

Add together the following:

2 ml of 1% phenol red▼
0.5 ml of 1 M Tris▼(pH 7.9)
47.5 ml of deionized or distilled water

4 mg/ml Penicillin-G

Makes 50 ml.
Keep on ice for same day use or store at –20°C (1 month).

1. Add 200 mg of penicillin-G▼ (procaine salt) to 50 ml of deionized or distilled water.
2. Stir until dissolved.
3. After stirring for several minutes, the penicillin-G should be dissolved but the solution may remain cloudy. This is normal and can be used this way.

VI. PROTEIN PURIFICATION

Binding Buffer

Makes 100 ml.
Store at room temperature (indefinitely).

Weigh out 52.8 g of ammonium sulfate▼.
Dissolve in TE (pH 8) for a final volume of 100 ml.

Equilibration Buffer

Makes 100 ml.
Store at room temperature (indefinitely).

Weigh out 26.4 g of ammonium sulfate▼.
Dissolve in TE (pH 8) for a final volume of 100 ml.

Wash Buffer

Makes 100 ml.
Store at room temperature (indefinitely).

Weigh out 17.2 g of ammonium sulfate▼.
Dissolve in TE (pH 8) for a final volume of 100 ml.

Tris/EDTA (TE) Buffer

Makes 100 ml.
Store at room temperature (indefinitely).

Mix:

1 ml of 1 M Tris▼ (pH 7.9)	(10 mM)
200 µl of 0.5 M EDTA	(1 mM)
99 ml of distilled water	

Elution Buffer

Elution buffer is TE (pH 8).
Refer to TE recipe.

Lysis Buffer

Makes 50 ml.
Store at room temperature (indefinitely).

Mix:

10 ml of 1 M Tris▼(pH 8.0)
2.5 ml of 5 M NaCl
2.5 ml of 0.5 M EDTA
2.5 ml of 10% SDS▼
32.5 ml of deionized water

Methyl Hydrophobic Interaction Chromatography (HIC) Resin

Makes 50 ml.
Store at room temperature (indefinitely).

Mix:

25 ml of HIC resin
25 ml of Equilibration Buffer

It is important to mix the HIC resin before pipetting. Distribute 1.5-ml aliquots into 1.5-ml tubes. Microfuge and leave supernatant until ready to use. Remove supernatant shortly before applying cell lysate (with Binding Buffer) to the hydrophobic beads.

VII. PLASMID MINIPREPARATION

0.5 M Ethylenediaminetetraacetic Acid (EDTA, pH 8.0)

Makes 100 ml.
Store at room temperature (indefinitely).

1. Add 18.6 g of EDTA (disodium salt, m.w. = 372.24) to 80 ml of deionized or distilled water.
2. Adjust pH by slowly adding ~2.2 g of sodium hydroxide▼ pellets (m.w. = 40.00). (If a pH meter is not available, adding 2.2 g of NaOH pellets will make a solution of ~pH 8.0.)
3. Mix vigorously with a magnetic stirrer or by hand. EDTA will only dissolve when the pH has reached 8.0 or higher.
4. Add deionized or distilled water to make a total volume of 100 ml of solution.

NOTE Use only the disodium salt of EDTA.

Tris/EDTA (TE) Buffer

Makes 100 ml.
Store at room temperature (indefinitely).

Mix:

1 ml of 1 M Tris▼(pH 8.0) (10 mM)
200 μl of 0.5 M EDTA (1 mM)
99 ml of deionized water

Glucose/Tris/EDTA (GTE)

Makes 100 ml.
Store at 4°C or room temperature (indefinitely).

Mix:

0.9 g of glucose (m.w. = 180.16)	(50 mM)
2.5 ml of 1 M Tris▼ (pH 8.0)	(25 mM)
2 ml of 0.5 M EDTA	(10 mM)
94.5 ml of deionized water	

NOTE With storage at 4°C, the solution is precooled and ready for minipreparations.

5 M Potassium Acetate (KOAc)

Makes 200 ml.
Store at room temperature (indefinitely).

1. Add 98.1 g of potassium acetate (m.w. = 98.14) to 160 ml of deionized water.
2. Add deionized or distilled water to make a total volume of 200 ml of solution.

Potassium Acetate/Acetic Acid

Makes 100 ml.
Store at 4°C or room temperature (indefinitely).

Add 60 ml of 5 M potassium acetate and 11.5 ml of glacial acetic acid▼ to 28.5 ml of deionized or distilled water.

NOTES

- The sharp odor of the acetic acid distinguishes the finished KOAc/acetic acid solution from the KOAc stock. The two are easily confused.
- With storage at 4°C, the solution is precooled and ready for minipreparations.

10% Sodium Dodecyl Sulfate (SDS)

Makes 100 ml.
Store at room temperature (indefinitely).

1. Dissolve 10 g of electrophoresis-grade SDS▼ (m.w. = 288.37) in 80 ml of deionized water.
2. Add deionized or distilled water to make a total volume of 100 ml of solution.

NOTES

- Avoid inhaling SDS powder; wear a mask that covers both nose and mouth.
- SDS is the same as sodium lauryl sulfate.
- Always use fresh SDS/NaOH solution.
- A precipitate may form at colder temperatures. Warm solution in a water bath, and shake gently to dissolve precipitate.

1% SDS/0.2 N NaOH

Makes 10 ml.
Store at room temperature (several days).

Mix 1 ml of 10% SDS▼ and 0.5 ml of 4 N NaOH▼ into 8.5 ml of distilled water.

NOTES

- Always use fresh SDS/NaOH solution.
- A precipitate may form at colder temperatures. Warm solution in a water bath, and shake gently to dissolve precipitate.

VIII. DNA LIGATION

1 M Tris (pH 7.6)

Makes 100 ml.
Store at room temperature (indefinitely).

1. Dissolve 12.1 g of Tris▼ base (m.w. = 121.10) in 70 ml of deionized or distilled water.
2. Adjust the pH by slowly adding ~6.3 ml of concentrated hydrochloric acid (HCl)▼; monitor with a pH meter. (If a pH meter is not available, adding 6.3 ml of concentrated HCl will yield a solution of ~pH 7.6.)
3. Add deionized or distilled water to make a total volume of 100 ml of solution.

NOTES

- A yellow-colored solution indicates poor-quality Tris. Discard it, and obtain a Tris solution from a different source.
- Many types of electrodes do not accurately measure the pH of Tris solutions; check with manufacturer to obtain a suitable one.
- The pH of Tris solutions is temperature-dependent; measure pH at room temperature.

10X Ligation Buffer

Makes 1000 µl.
Store at –20°C (indefinitely).

Mix the following ingredients in a 1.5-ml tube:

600 µl of 1 M Tris▼ (pH 7.6)	(600 mM)
100 µl of 1 M $MgCl_2$▼	(100 mM)
70 µl of 1 M DTT▼	(70 mM)
230 µl of deionized water	

NOTE DTT may precipitate from the solution with repeated freezing and thawing. Vortex vigorously to redissolve.

0.1 M Adenosine Triphosphate (ATP)

Makes 5 ml.
Store at –20°C (1 year).

1. Dissolve 0.3 g of ATP (disodium salt, m.w. = 605.19) in 5 ml of deionized or distilled water.
2. Dispense 500-µl aliquots into 1.5-ml tubes.

NOTE ATP loses activity with repeated freezing and thawing. Discard aliquots that have been thawed several times.

2x Ligation Buffer + ATP

Makes 500 µl.
Store at –20°C (1 month).

1. Mix 100 µl of 10x ligation buffer and 10 µl of 0.1 M ATP in a 1.5-ml tube.
2. Add 390 µl of deionized or distilled water.

NOTE ATP loses activity in dilute solution; use a fresh solution.

APPENDIX 3

Restriction Map Data for pAMP, pKAN, pBLU, pGREEN, and Bacteriophage λ

I. Restriction Maps of pAMP, pKAN, pBLU, and pGREEN

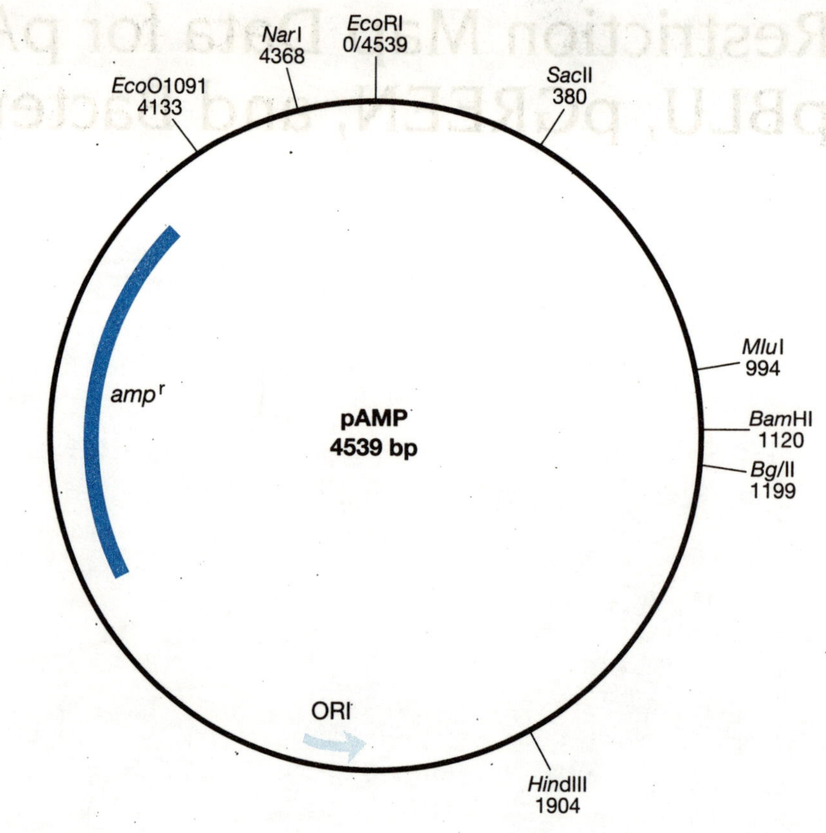

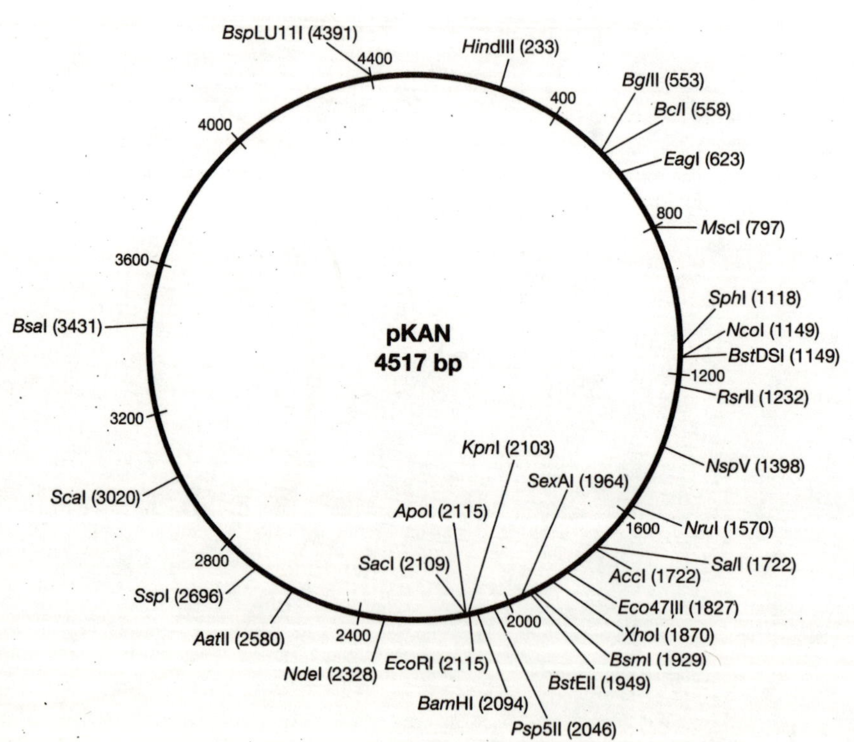

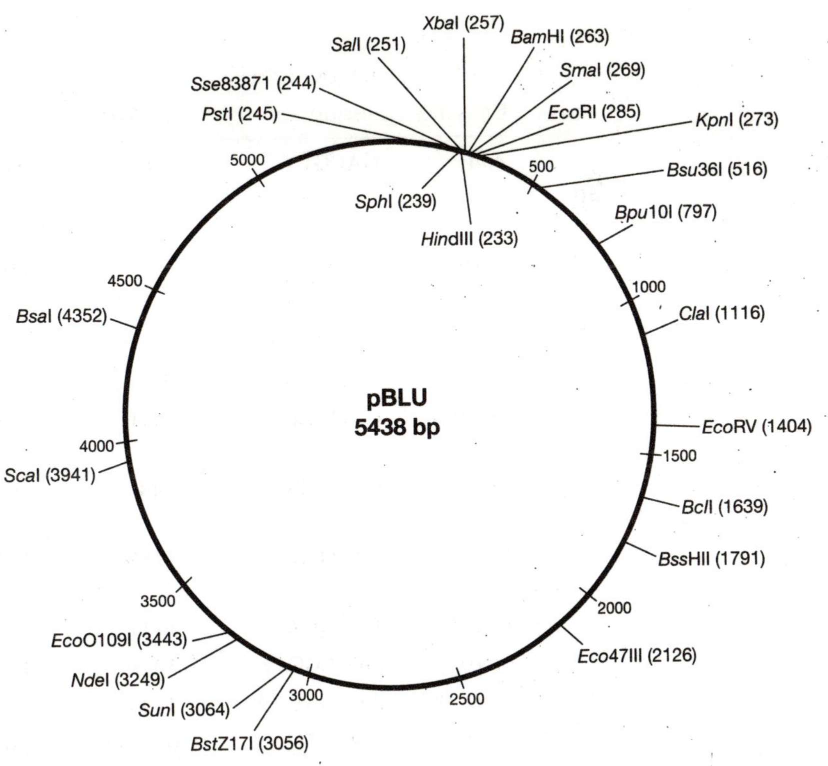
Sal I (251)
Xba I (257)
BamH I (263)
Sse83871 (244)
Sma I (269)
Pst I (245)
EcoR I (285)
Kpn I (273)
5000
500
Bsu36 I (516)
Sph I (239)
Hind III (233)
Bpu10 I (797)
4500
1000
Bsa I (4352)
Cla I (1116)
pBLU
5438 bp
EcoR V (1404)
4000
1500
Sca I (3941)
Bcl I (1639)
BssH II (1791)
3500
2000
EcoO109 I (3443)
Eco47 III (2126)
Nde I (3249)
3000
2500
Sun I (3064)
BstZ17 I (3056)

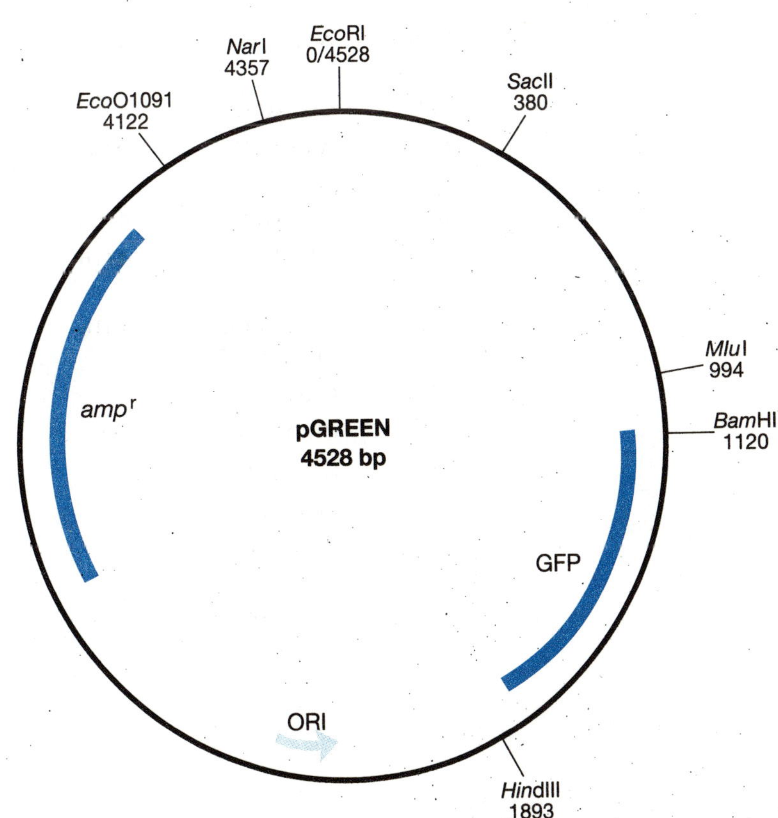
EcoR I
0/4528
Nar I
4357
Sac II
380
EcoO1091
4122
Mlu I
994
amp^r
pGREEN
4528 bp
BamH I
1120
GFP
ORI
Hind III
1893

II. Restriction Enzymes That Cut Once in pAMP or pKAN and Corresponding Sites in Bacteriophage λ

Enzyme	Recognition sequence	Location		
		pAMP	pKAN	bacteriophage λ
*Aat*II	GACGTˇC	4078	4138	5105; 9394; 11,243; 14,974; 29,036; 40,806; 41,113; 42,247; 45,563; 45,592
*Afl*III	AˇCACGT AˇCGTGT	–	2651	none
*Alw*NI	CAGNNNˇCTG	2676	3062	none
*Bal*I	TGGˇCCA	34	1727	18 sites
*Bam*HI	GˇGATCC	1120	418	5505; 22,346; 27,972; 34,499; 41,732
*Ban*II	GPuGCPyˇC	344	402	581; 10,086; 19,763; 21,570; 24,772; 25,877; 39,453
*Bgl*II	AˇGATCT	1199	1973	415; 22,425; 35,711; 38,103; 38,754; 38,814
*Bsp*MII	TˇCCGGA	1117	–	24 sites
*Cfr*10I	PuˇCCGGPy	3238	3624	none
*Eco*O109I	PuGˇGNCCPy	4133	–	2815; 28,797; 48,473
*Eco*RI	GˇAATTC	0/4539	396	21,226; 26,104; 31,747; 39,168; 44,972
*Hae*II	PuGCGCˇPy	–	1902	none
*Hin*dIII	AˇAGCTT	1904	2293	23,130; 25,157; 27,479; 36,895; 37,459; 37,584; 44,141
*Kpn*I	GGTACˇC	–	408	17,053; 18,556
*Mlu*I	AˇCGCGT	994	–	458; 5548; 15,372; 17,791; 19,996; 20,952; 22,220
*Nar*I	GGˇCGCC	4368	235	45,679
*Nde*I	CAˇTATG	4316	183	27,630; 29,883; 33,679; 36,112; 36,668; 38,357; 40,131
*Rsr*II	CGˇGACCG CGˇGTCCG	1016	–	3800; 6041; 13,983; 19,288; 22,242
*Sac*I	GAGCTˇC	–	402	24,772; 25,877
*Sac*II	CCGCˇGG	380	–	20,320; 20,530; 21,606; 40,386
*Sal*I	GˇTCGAC	–	804	32,745; 33,244
*Sma*I	CCCˇGGG	–	412	19,397; 31,617; 39,888
*Sph*I	GCATGˇC	–	1405	2212; 12,002; 23,942; 24,371; 27,374; 39,418
*Xho*I	CˇTCGAG	–	662	33,498

III. Construction of pAMP

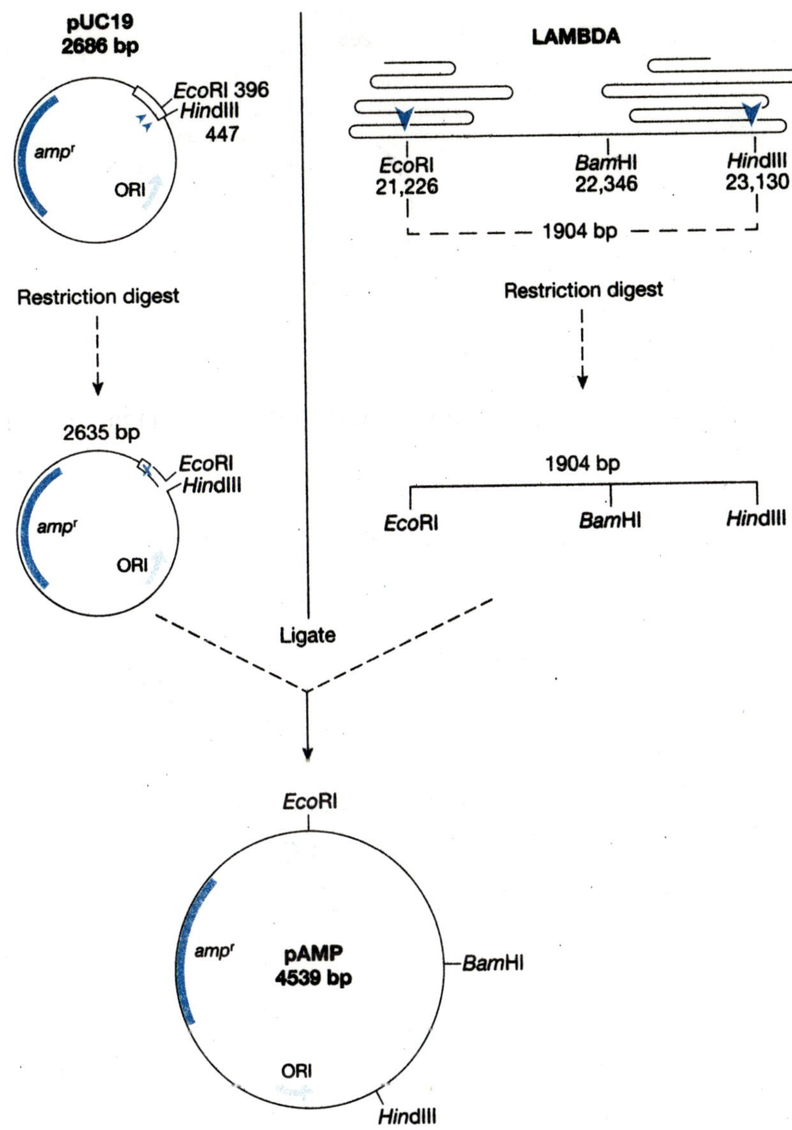

IV. Construction of pKAN

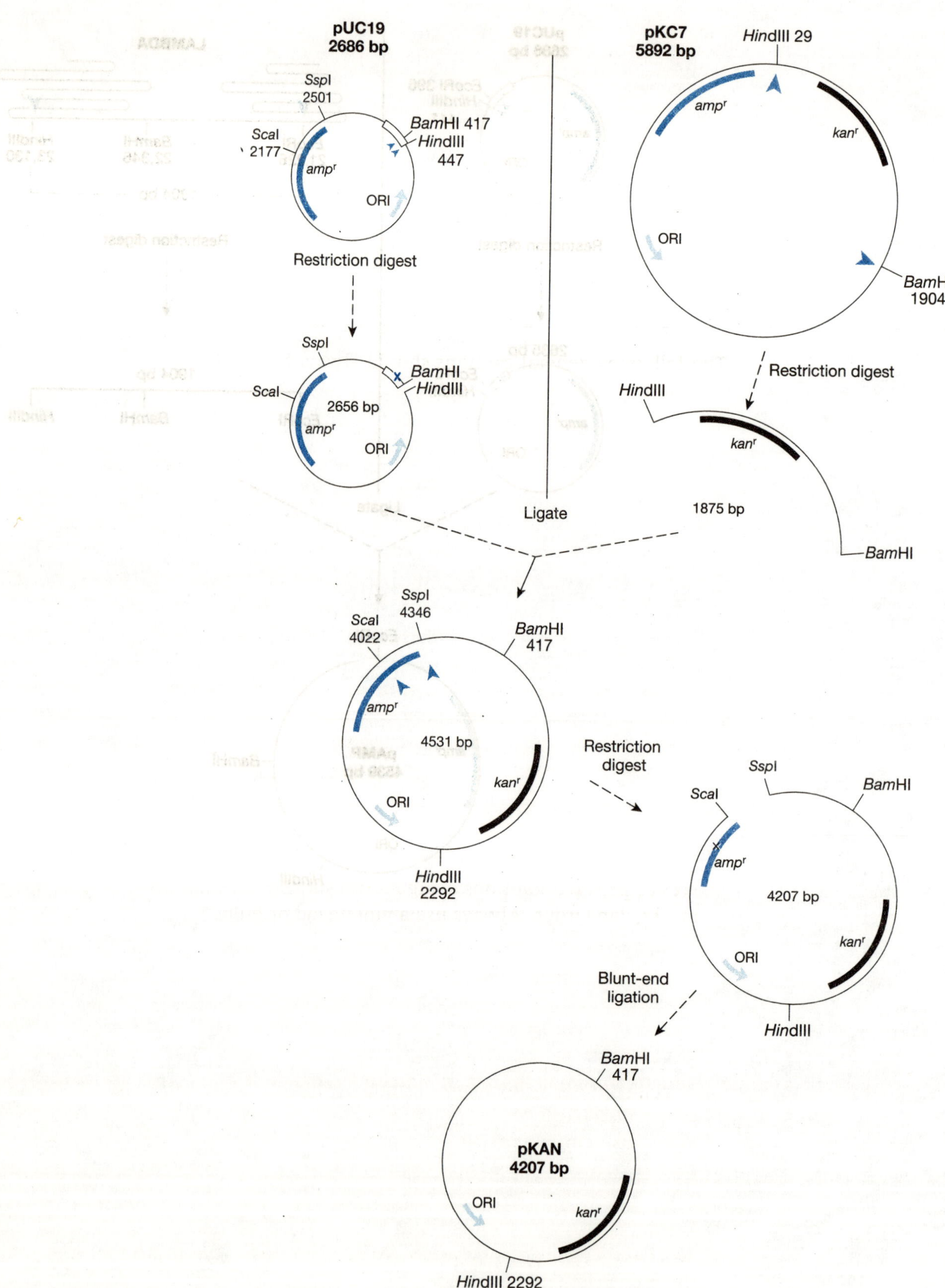

APPENDIX 4

Cautions

GENERAL CAUTIONS

The following general cautions should always be observed.

- **Become completely familiar with the properties of substances** used before beginning the procedure.
- **The absence of a warning** does not necessarily mean that the material is safe, since information may not always be complete or available.
- **If exposed to toxic substances,** contact your local safety office immediately for instructions.
- **Use proper disposal procedures** for all chemical, biological, and radioactive waste.
- **For specific guidelines on appropriate gloves,** consult your local safety office.
- **Handle concentrated acids and bases with great care.** Wear goggles and appropriate gloves. Wear a face shield when handling large quantities.

  Do not mix strong acids with organic solvents as they may react. Sulfuric acid and nitric acid especially may react highly exothermically and cause fires and explosions.

  Do not mix strong bases with halogenated solvent as they may form reactive carbenes which can lead to explosions.
- **Never pipette** solutions using mouth suction. This method is not sterile and can be dangerous. Always use a pipette aid or bulb.
- **Keep halogenated and nonhalogenated solvents separately** (e.g., mixing chloroform and acetone can cause unexpected reactions in the presence of bases). Halogenated solvents are organic solvents such as chloroform, dichloromethane, trichlorotrifluoroethane, and dichloroethane. Some nonhalogenated solvents are pentane, heptane, ethanol, methanol, benzene, toluene, *N,N*-dimethylformamide (DMF), dimethyl sulfoxide (DMSO), and acetonitrile.
- **Laser radiation**, visible or invisible, can cause severe damage to the eyes and skin. Take proper precautions to prevent exposure to direct and reflected beams. Always follow manufacturer's safety guidelines and consult your local safety office. See caution below for more detailed information.
- **Flash lamps**, due to their light intensity, can be harmful to the eyes. They also may explode on occasion. Wear appropriate eye protection and follow the manufacturer's guidelines.

- **Photographic fixatives and developers** also contain chemicals that can be harmful. Handle them with care and follow manufacturer's directions.
- **Power supplies and electrophoresis equipment** pose serious fire hazard and electrical shock hazards if not used properly.
- **Microwave ovens and autoclaves** in the lab require certain precautions. Accidents have occurred involving their use (e.g., melting agar or bacto-agar stored in bottles or to sterilize). If the screw top is not completely removed and there is not enough space for the steam to vent, the bottles can explode and cause severe injury when the containers are removed from the microwave or autoclave. Always completely remove bottle caps before microwaving or autoclaving. An alternative method for routine agarose gels that do not require sterile agar is to weigh out the agar and place the solution in a flask.
- **Use extreme caution when handling cutting devices** such as microtome blades, scalpels, razor blades, or needles. Microtome blades are extremely sharp! Use care when sectioning. If unfamiliar with their use, have someone demonstrate proper procedures. For proper disposal, use the "sharps" disposal container in your lab. Discard used needles *unshielded*, with the syringe still attached. This prevents injuries (and possible infections; see Biological Safety) while manipulating used needles since many accidents occur while trying to replace the needle shield. Injuries may also be caused by broken Pasteur pipettes, coverslips, or slides.

GENERAL PROPERTIES OF COMMON CHEMICALS

The hazardous materials list can be summarized in the following categories:

- Inorganic acids, such as hydrochloric, sulfuric, nitric, or phosphoric, are colorless liquids with stinging vapors. Avoid spills on skin or clothing. Dilute spills with large amounts of water. The concentrated forms of these acids can destroy paper, textiles, and skin as well as cause serious injury to the eyes.
- Inorganic bases such as sodium hydroxide are white solids which dissolve in water and under heat development. Concentrated solutions will slowly dissolve skin and even fingernails.
- Salts of heavy metals are usually colored powdered solids which dissolve in water. Many of them are potent enzyme inhibitors and therefore toxic to humans and to the environment (e.g., fish and algae).
- Most organic solvents are flammable volatile liquids. Avoid breathing the vapors which can cause nausea or dizziness. Also avoid skin contact.
- Other organic compounds, including organosulphur compounds such as mercaptoethanol or organic amines, can have very unpleasant odors. Others are highly reactive and should be handled with appropriate care.
- If improperly handled, dyes and their solutions can stain not only your sample, but also your skin and clothing. Some of them are also mutagenic (e.g., ethidium bromide), carcinogenic, and toxic.

- All names ending with "ase" (e.g., catalase, β-glucuronidase, or ligase) refer to enzymes. There are also other enzymes with nonsystematic names like pepsin. Many of them are provided by manufacturers in preparations containing buffering substances, etc. Be aware of the individual properties of materials contained in these substances.
- Toxic compounds are often used to manipulate cells. They can be dangerous and should be handled appropriately.
- Be aware that several of the compounds listed have not been thoroughly studied with respect to their toxicological properties. Handle each chemical with the appropriate respect. Although the toxic effects of a compound can be quantified (e.g., LD_{50} values), this is not possible for carcinogens or mutagens where one single exposure can have an effect. Also realize that dangers related to a given compound may also depend on its physical state (fine powder vs. large crystals/diethylether vs. glycerol/dry ice vs. carbon dioxide under pressure in a gas bomb). Anticipate under which circumstances during an experiment exposure is most likely to occur and how best to protect yourself and your environment.

HAZARDOUS MATERIALS

Acetic acid (concentrated) must be handled with great care. It may be harmful by inhalation, ingestion, or skin absorption. Wear appropriate gloves and goggles and use in a chemical fume hood.

***S*-Adenosyl methionine (SAM)** is toxic and may be harmful by inhalation, ingestion, or skin absorption. Wear appropriate gloves and safety glasses and use in a chemical fume hood. Do not breathe the dust.

Ammonium sulfate, $(NH_4)_2SO_4$, may be harmful by inhalation, ingestion, or skin absorption. Wear appropriate gloves and safety glasses.

Ampicillin may be harmful by inhalation, ingestion, or skin absorption. Wear appropriate gloves and safety glasses and use in a chemical fume hood.

Bacterial strains (shipping of): The Department of Health, Education, and Welfare (HEW) has classified various bacteria into different categories with regard to shipping requirements (see Sanderson and Zeigler, *Methods Enzymol. 204:* 248–264 [1991]). Nonpathogenic strains of *E. coli* (such as K12) and *B. subtilis* are in Class 1 and are considered to present no or minimal hazard under normal shipping conditions. However, *Salmonella, Haemophilus,* and certain strains of *Streptomyces* and *Pseudomonas* are in Class 2. Class 2 bacteria are "Agents of ordinary potential hazard: agents which produce disease of varying degrees of severity...but which are contained by ordinary laboratory techniques." For detailed regulations regarding the packaging and shipping of Class 2 strains, see Sanderson and Ziegler (*Methods Enzymol. 204:* 248–264 [1991] or the instruction brochure by Alexander and Brandon (*Packaging and Shipping of Biological Materials at ATCC* [1986]) available from the American Type Culture Collection (ATCC), Rockville, Maryland.

BCIG, *see* **5-Bromo-4-chloro-3-indolyl-β-D-galactopyranoside**

Bleach (Sodium hypochlorite), NaOCl, is poisonous, can be explosive, and may react with organic solvents. It may be fatal by inhalation and is also harmful by ingestion and destructive to the skin. Wear appropriate gloves and safety glasses and use in a chemical fume hood to minimize exposure and odor.

BME, *see* **β-Mercaptoethanol (2-Mercaptoethanol), $HOCH_2CH_2SH$**

Boric acid, H_3BO_3, may be harmful by inhalation, ingestion, or skin absorption. Wear appropriate gloves and goggles.

5-Bromo-4-chloro-3-indolyl-β-D-galactopyranoside (BCIG), *see* **X-gal.**

Bromophenol blue may be harmful by inhalation, ingestion, or skin absorption. Wear appropriate gloves and safety glasses and use in a chemical fume hood.

CH_3CH_2OH, *see* **Ethanol**

Coomassie brilliant blue may be harmful by inhalation, ingestion, or skin absorption. Wear appropriate gloves and safety glasses.

***N,N*-Dimethylformamide (DMF), $HCON(CH_3)_2$,** is irritating to the eyes, skin, and mucous membranes. It can exert its toxic effects through inhalation, ingestion, or skin absorption. Chronic inhalation can cause liver and kidney damage. Wear appropriate gloves and safety glasses and use in a chemical fume hood.

Dithiothreitol (DTT) is a strong reducing agent that emits a foul odor. It may be harmful by inhalation, ingestion, or skin absorption. When working with the solid form or highly concentrated stocks, wear appropriate gloves and safety glasses and use in a chemical fume hood.

DMF, *see* ***N,N*-Dimethylformamide**

DTT, *see* **Dithiothreitol**

Ethanol (EtOH), CH_3CH_2OH, may be harmful by inhalation, ingestion, or skin absorption. Wear appropriate gloves and safety glasses.

Ethidium bromide is a powerful mutagen and is toxic. Consult the local institutional safety officer for specific handling and disposal procedures. Avoid breathing the dust. Wear appropriate gloves when working with solutions that contain this dye.

EtOH, *see* **Ethanol**

Glassware, pressurized, must be used with extreme caution. Autoclave and cool sealed bottles in metal containers, pressurize bottles behind Plexiglas shields, and encase 20-liter bottles in wire mesh. Handle glassware under vacuum, such as desiccators, vacuum traps, drying equipment, or a reactor for working under argon atmosphere, with appropriate caution. Always wear safety glasses.

Glycine may be harmful by inhalation, ingestion, or skin absorption. Wear gloves and safety glasses. Avoid breathing the dust.

H_3BO_3, *see* **Boric acid**

HCl, ***see*** **Hydrochloric acid**

$HOCH_2CH_2SH$, ***see*** **β-Mercaptoethanol**

H_2SO_4, ***see*** **Sulfuric acid**

Hydrochloric acid, HCl, is volatile and may be fatal if inhaled, ingested, or absorbed through the skin. It is extremely destructive to mucous membranes, upper respiratory tract, eyes, and skin. Wear appropriate gloves and safety glasses and use with great care in a chemical fume hood. Wear goggles when handling large quantities.

Isopropanol is irritating and may be harmful by inhalation, ingestion, or skin absorption. Wear appropriate gloves and safety glasses. Do not breathe the vapor. Keep away from heat, sparks, and open flame.

Kanamycin may be harmful by inhalation, ingestion, or skin absorption. Wear appropriate gloves and safety glasses. Use only in a well-ventilated area.

Lysozyme is caustic to mucous membranes. Wear appropriate gloves and safety glasses.

Magnesium chloride, $MgCl_2$, may be harmful by inhalation, ingestion, or skin absorption. Wear appropriate gloves and safety glasses and use in a chemical fume hood.

β-Mercaptoethanol (2-Mercaptoethanol), $HOCH_2CH_2SH$, may be fatal if inhaled or absorbed through the skin and is harmful if ingested. High concentrations are extremely destructive to the mucous membranes, upper respiratory tract, skin, and eyes. β-Mercaptoethanol has a very foul odor. Wear appropriate gloves and safety glasses and always use in a chemical fume hood.

Methylene blue is irritating to the eyes and skin. It may be harmful by inhalation, ingestion, or skin absorption. Wear appropriate gloves and safety glasses.

$MgCl_2$, ***see*** **Magnesium chloride**

NaOCl, ***see*** **Bleach**

NaOH, ***see*** **Sodium hydroxide**

$(NH_4)_2SO_4$, ***see*** **Ammonium sulfate**

Penicillin G (Procaine Salt) may cause allergic respiratory and skin reactions and may be harmful by inhalation, ingestion, or skin absorption. Wear appropriate gloves. Do not breathe the dust.

Phenol red may be harmful by inhalation, ingestion, or skin absorption. Wear appropriate gloves and safety glasses and use in a chemical fume hood.

Radioactive substances: When planning an experiment that involves the use of radioactivity, include the physicochemical properties of the isotope (half-life, emission type and energy), the chemical form of the radioactivity, its radioactive concentration (specific activity), total amount, and its chemical concentration. Order and use only as much as really needed. Always wear appropriate gloves, lab coat, and safety goggles when handling radioactive material. **X-rays** and

gamma rays are electromagnetic waves of very short wavelengths either generated by technical devices or emitted by radioactive materials. They may be emitted isotropically from the source or may be focused into a beam. Their potential dangers depend on the time period of exposure, the intensity experienced, and the wavelengths used. Be aware that appropriate shielding is usually of lead or other similar material. The thickness of the shielding is determined by the energy(s) of the X-rays or gamma rays. Consult the local safety office for further guidance in the appropriate use and disposal of radioactive materials. Always monitor thoroughly after using radioisotopes. A convenient calculator to perform routine radioactivity calculations can be found at:

http://www. graphpad.com/quickcalcs/radcalcform.cfm

SAM, *see S*-Adenosyl methionine

Sodium hydroxide, NaOH, and solutions containing NaOH are highly toxic and caustic and should be handled with great care. Wear appropriate gloves and a face mask. All other concentrated bases should be handled in a similar manner.

Sulfuric acid, H_2SO_4, is highly toxic and extremely destructive to tissue of the mucous membranes and upper respiratory tract, eyes, and skin. It causes burns, and contact with other materials (e.g., paper) may cause fire. Wear appropriate gloves, safety glasses, and lab coat and use in a chemical fume hood.

Transilluminator, *see* UV light

Tris may be harmful by inhalation, ingestion, or skin absorption. Wear appropriate gloves and safety glasses.

UV light and/or **UV-radiation** is dangerous and can damage the retina of the eyes. Never look at an unshielded UV light source with naked eyes. Examples of UV light sources that are common in the laboratory include hand-held lamps and transilluminators. View only through a filter or safety glasses that absorb harmful wavelengths. UV radiation is also mutagenic and carcinogenic. To minimize exposure, make sure that the UV light source is adequately shielded. Wear protective appropriate gloves when holding materials under the UV light source.

X-gal is toxic to the eyes and skin and may be harmful by inhalation, ingestion, or skin absorption. Wear appropriate gloves and safety goggles. Note that stock solutions of X-gal are prepared in DMF, an organic solvent. For details, see ***N,N*-dimethylformamide (DMF)**. See also **5-Bromo-4-chloro-3-indolyl-β-D-galactopyranoside (BCIG).**

Xylene is flammable and may be narcotic at high concentrations. It may be harmful by inhalation, ingestion, or skin absorption. Wear appropriate gloves and safety glasses and use only in a chemical fume hood. Keep away from heat, sparks, and open flame.

Xylene cyanol, *see* Xylene

Name Index

Italic page numbers indicate photos also shown.

Subject Index

Page references in *italics* denote figures and tables.